T0205568

# Lecture Notes in Physics

## Volume 983

# The Lecture Notes in Physics

The series Lecture Notes in Physics (LNP), founded in 1969, reports new developments in physics research and teaching-quickly and informally, but with a high quality and the explicit aim to summarize and communicate current knowledge in an accessible way. Books published in this series are conceived as bridging material between advanced graduate textbooks and the forefront of research and to serve three purposes:

- to be a compact and modern up-to-date source of reference on a well-defined topic;
- to serve as an accessible introduction to the field to postgraduate students and nonspecialist researchers from related areas;
- to be a source of advanced teaching material for specialized seminars, courses and schools.

Both monographs and multi-author volumes will be considered for publication. Edited volumes should however consist of a very limited number of contributions only. Proceedings will not be considered for LNP.

Volumes published in LNP are disseminated both in print and in electronic formats, the electronic archive being available at springerlink.com. The series content is indexed, abstracted and referenced by many abstracting and information services, bibliographic networks, subscription agencies, library networks, and consortia.

Proposals should be sent to a member of the Editorial Board, or directly to the responsible editor at Springer:

Dr Lisa Scalone
Springer Nature
Physics
Tiergartenstrasse 17
69121 Heidelberg, Germany
lisa.scalone@springernature.com

More information about this series at http://www.springer.com/series/5304

Nicolas Michel • Marek Płoszajczak

# Gamow Shell Model

The Unified Theory of Nuclear Structure and Reactions

 Springer

Nicolas Michel
Nuclear Physics Research Center
Institute of Modern Physics
Chinese Academy of Sciences
Lanzhou, China

Marek Płoszajczak
Grand Accélérateur National d'Ions Lourds
(GANIL), CEA/DSM - CNRS/IN2P3
Caen, France

ISSN 0075-8450          ISSN 1616-6361   (electronic)
Lecture Notes in Physics
ISBN 978-3-030-69355-8     ISBN 978-3-030-69356-5   (eBook)
https://doi.org/10.1007/978-3-030-69356-5

This Springer imprint is published by the registered company Springer Nature Switzerland AG.
The registered company address is: Gewerbestrasse 11, 6330 Cham, Switzerland

# Preface

*Every great and deep difficulty bears in itself its own solution.*
*It forces us to change our thinking in order to find it.*

[Niels Bohr (1919)]

Atomic nucleus is a self-organized, many-body quantum system emerging from complex interactions between quarks and gluons. At low energies, the nucleus is seen as an aggregation of neutrons and protons held together by the short-range nuclear force. The complexity of interactions, different regimes of binding energy, and the large number of degrees of freedom make the computation of nuclear properties from first principles exceedingly difficult. Nuclear properties are strongly affected by coupling to the many-body continuum of scattering and decay channels. In this context, resonance phenomena play a prominent role. A simultaneous understanding of the structural and reaction aspects of nuclear many-body problem in a unified framework is at the core of understanding of the short-lived nuclear states.

Atomic nucleus is the excellent example of an open quantum system. The complexity of discrete many-body states embedded in the continuum, and a strong variation of both the effective interaction among nucleons and the nuclear spectra with the excitation energy, makes the unitarity of the theoretical description of atomic nucleus to the key theoretical issue.

This textbook is the first ever book on the open quantum system formulation of the configuration interaction approach, the Gamow shell model, which provides a unitary description of dynamics of the many-body system in different regimes of binding energy. The book is intended for graduate students and experienced researchers. The aim is to fill the gap between standard textbooks on quantum many-body physics and nuclear theory, and the specialized articles on open quantum systems in the context of nuclear physics, atomic physics, or nanoscience.

The course is accompanied by analytical exercises of varying difficulties. Several numerical exercises make use of available computer programs, and there are prepared input files as examples of different possible applications. The exercises provided throughout the body text are linked to computations that can be found at https://github.com/GSMUTNSR. At the end of each chapter, the solutions to the exercises are discussed. In this way, this course or its parts can be used as basis for advanced courses in quantum mechanics of open many-body systems, or the theory of weakly bound or unbound nuclear and atomic systems.

The difficulty of exercises is in the scale from one star (simple exercise) to three stars (difficult exercise of technical nature). All problems given in exercises are being solved or commented. We hope that these exercises will provide a better understanding of the foundations of the Gamow shell model and explain various methods to calculate observable quantities related to resonances. Some difficult exercises have been included to introduce reader to additional details of the theory which are omitted from the main part of the textbook. Note however that in principle the material presented in this book should be accessible without specialized background other than the advanced courses on quantum mechanics and quantum many-body theory, and a basic course on nuclear physics. Also the mathematical parts do not require more than the graduate level on the functional analysis, linear algebra, and the theory of differential equations. It should be noted that the important part of this textbook can be read and understand with the background in mathematics and quantum mechanics at the undergraduate level.

Even if the precise examples are rather specific to gain homogeneity and complete the basic course, the approach and results developed in this book are by no means attached to a field of nuclear physics. Applications exist and are possible in atomic physics, molecular physics, nanoscience, to cite just a few the most evident areas.

It is impossible to acknowledge all influences on our present understanding and ideas that we have profited from our collaborators and other theorists who contributed with discussion, opposition, and alternative points of view. We thank in particular Witek Nazarewicz for the long-term exciting and intense collaboration which lead to the formulation of Gamow shell model and its further developments.

This book would never be completed without the significant contribution of our long-time collaborators at GANIL, Oak Ridge National Laboratory and Michigan State University. Here we want to acknowledge in particular Kevin Fossez, Yannen Jaganathen, Alexis Mercenne, George Papadimitriou, Jimmy Rotureau, and Simin Wang.

We would like to thank warmly our friends and collaborators: Bruce R. Barrett, Karim Bennaceur, Jacek Dobaczewski, Jorge Dukelsky, W. Ray Garrett, Morten Hjorth-Jensen, Kiyoshi Kato, Jacek Okołowicz, Ingrid Rotter, Tamas Vertse, Furong Xu, and Wei Zuo for many discussions and friendly exchanges.

We also gratefully acknowledge the support and encouragement we have received at GANIL, Caen and Institute of Modern Physics, Lanzhou throughout the preparation of this book.

Caen,  Lanzhou                                            Nicolas Michel    Marek Płoszajczak
December 2020

# Contents

# About the Authors

**Nicolas Michel** is a Chief Scientist in Nuclear Physics at the Institute of Modern Physics in Lanzhou, China. He obtained his MSc at the University of Strasbourg (France) and PhD at the University of Caen Normandy (France). He worked in several nuclear research centers, including the GANIL in Caen, Michigan State University, Oak Ridge National Laboratory, Kyoto University and CEA, Center of Saclay. He was awarded a Standard Fellowship for Foreign Researchers by the Japanese Society for the Promotion of Science. He is the author of 1 monograph and over 65 scientific papers. His research focuses on the study of resonant nuclei and molecules using the continuum shell model formalism. He also developed parallel codes to efficiently calculate many-body weakly bound and unbound nuclear states.

**Marek Płoszajczak** is a Director of Research at the French Atomic Energy Commission, currently working at the GANIL SPIRAL2 accelerator center, Caen (France). He obtained his MSc at the Jagiellonian University in Kraków (Poland) and PhD at the Rhenish Friedrich-Wilhelm University Bonn (Germany). He has held several research positions, including at the Niels Bohr Institute in Copenhagen, SUNY Stony Brook, CRN Strasbourg, and GSI Darmstadt. In 1990, he became Professor of Theoretical Physics at the Institute of Nuclear Physics in Kraków (Poland). He is a Fellow of the American Physical Society. He was a co-Director of the French-U.S. Theory Institute for Physics with Exotic Nuclei at GANIL, Caen, and co-organizer of the Training in Advanced Low Energy Nuclear Theory initiative. He authored/edited 3 books, is the author of 7 monographs, and over 200 scientific papers. His research interests focuses on open quantum system

theory of atomic nucleus. He worked on fragmenta-
tion theory and order-parameter fluctuations in finite
systems, quantum theory of molecular dynamics for
fermions, quantization of large amplitude non-linear
theories by time-periodic solutions, and microscopic
theory of rapidly rotating nuclei.

# Introduction: From Bound States to the Continuum

The paramount complexity of nuclear many-body problem and its great importance for an understanding of various systems ranging from the subnucleon to astronomical scales makes it to the intellectual challenge of first importance [1]. Complexity of this problem is the reason why the theory of atomic nucleus does not exist and one deals with various models, describing selected features of the nuclear many-body problem. As a result, our understanding of the nuclear properties and structure remains incomplete and in certain aspects incoherent. Let us take an example of the low-energy excitations in light atomic nuclei, where the states are described using either the nuclear shell model or the cluster model. Organization of nucleons according to these two models is quite different. Shell model is based on a concept of single-particle mean field and associated to it the shell structure. On the other hand, cluster model builds the wave functions from the correlated substructures, such as the $\alpha$-particles. Can such two different pictures of nuclear structure coexist, and can they be reconciled?

Resonances are one of the most striking phenomena in Nature. They are genuine intrinsic properties of quantum systems, associated with their natural frequencies, and describing preferential decays of unbound states. Experimental manifestation of the resonances is either the sharp peak in the cross section or the exponential decay of the probability to find the unstable particle. The sharp peaks in the cross section are characterized by their energy and width, whereas the decay of a particle is characterized by its energy and lifetime. Resonances should find a satisfactory formulation in quantum mechanics. However, the standard quantum mechanics formulated in Hilbert space does not allow the description of state vectors with exponential growth and exponential decay, such as resonance states, and they are simply discarded as unphysical. In Hilbert space, the usual procedure to deal with resonance states is either to extract the trace of resonances from the real-energy continuum level density or describe the resonances by joining the bound state solution in the interior region with an asymptotic solution. This does not yield a

© Springer International Publishing AG 2021
N. Michel, M. Płoszajczak, *Gamow Shell Model*, Lecture Notes in Physics 983,
https://doi.org/10.1007/978-3-030-69356-5_1

sufficient legitimacy to resonance phenomena such as the unstable atomic nuclei or decaying particles.

The aforementioned difficulties with Hilbert space formulation of quantum mechanics have been resolved by extending the Hilbert space to a rigged Hilbert space [2–9] within which the resonance wave functions find a natural place. Resonance wave functions are given by Gamow states, also called Siegert states [10], which are the eigenvectors of the Hamiltonian with a complex eigenvalue, and their time evolution follows the exponential decay law. The Gamow states can describe both sharp peaks in the cross section and decay of resonances or unstable particles, unifying these two facets of the resonance phenomenon.

Open quantum systems, whose properties are affected by the environment of scattering states, are intensely studied in nuclear physics, atomic and molecular physics, nanoscience, quantum optics, etc. In spite of their specific features, they have generic properties, which are common to all weakly bound or unbound systems. Specific experimental conditions of their studies make them complementary. For example, nuclei and atoms can be prepared experimentally in a well-defined state; however, one can hardly tune their individual properties. On the other hand, in artificial open quantum system, such as quantum dots or atomic clusters, to achieve identical experimental conditions is virtually impossible, but these systems can be easily tunable by varying an external parameter.

What is being identified as a quantum environment of the system depends on the considered physics problem. For example, the quantum environments in quantum cosmology [11], quantum biology [12], or quantum information science [13] differ one from another and from the environment which is relevant in nuclear physics [14–16]. Consequently, many different theoretical and numerical approaches have been proposed for the description of those different open quantum systems. The standard approach is based on approximations for the exact master equation. Tracking an exact evolution of the combined system-plus-environment is usually impossible and also unwanted, as it involves the large amount of redundant information which can be traced neither experimentally nor theoretically. Hence, it is natural to develop reduced descriptions where the dynamics of the system is considered explicitly, whereas the dynamics of the environment is described implicitly. From this postulate follows the attempt to describe the system evolution in terms of the reduced density obtained by taking partial trace over the exact density of a combined system-plus-environment; hence, the evolution of combined system-plus-environment is unitary. Depending on the nature of coupling between system and environment, one obtains either Markovian or non-Markovian equations of motion which describe the evolution of the open quantum system.

The main interest in studies using reduced density matrices is the energy transfer to environment (the quantum dissipation) and the loss of coherence of considered state(s) (the quantum decoherence). These are not quantities of interest in the nuclear physics which deals with the well-defined quantum states and where precise experiments provide detailed information about structure of these states and their decays. Hence, the main emphasis in nuclear case is on the unitarity at the opening of quantum system, i.e., at the transition from well-bound states (the closed quantum systems) to the weakly bound or unbound states (the open quantum systems) while approaching the limit of nuclear stability with respect to the particle emission. For this purpose, the approaches based on the equation of motions for the reduced density of the combined system-plus-environment are not suitable and the configuration-interaction approaches are preferred.

The shell structure and single-particle motion is a cornerstone of nuclear structure [17]. The interacting shell model consists of the single-particle potential, characterized by a strong spin-orbit term [18, 19] and supplemented by a two-body residual interaction [20, 21]. This successful model in describing low-energy excitations in well-bound nuclei describes nucleus as a closed quantum system with nucleons occupying bound levels isolated from the scattering states and decay channels. But is this picture physically correct? Indeed, low-lying states of well-bound atomic nuclei from the valley of beta-stability can be considered as the closed quantum systems. However, in the vicinity of driplines and close to the lowest particle-emission threshold in well-bound nuclei, the continuum coupling becomes gradually more important, changing the nature of weakly bound states. Finally, in the particle-unbound states, couplings to reaction channels and a continuum of scattering states have a direct impact on their properties. In this regime, nuclear states in neighboring nuclei form a network of interconnected states via the continuum, with the clusters of correlated states in different domains of excitation energy, angular momentum, and nucleon number (see Fig. 1.1). Depending on the network activity caused by the coupling of states to reaction thresholds in different nuclei, different phases of connectivity may exist in this network. Moreover, the divide between the discrete resonant states and the nonresonant scattering continuum leads to the artificial separation of nuclear structure from the nuclear reactions, and hinders a deeper understanding of nuclear properties. Indeed, many structural properties of the nucleus are determined indirectly and heavily depend on the nuclear reaction theory, and vice versa, and this cries out for a unified theoretical framework.

Correlations among nucleons become crucial when the particle separation energy is either small or negative. In this situation, the nucleon-nucleon correlations (e.g., the pairing field) and the single-particle field become equally important. Consequently, the configuration mixing, involving also continuum states, can no longer be treated as a small perturbation [22]. On the contrary, it is an essential ingredient for the existence and stability of these exotic systems. This has far going consequences not only for the proprieties of weakly bound and resonance many-body states, but also for the shell structure and the survival of magic numbers of nucleons. Away from the beta-stability line toward the nucleon driplines, experimental data suggest that standard magic numbers gradually disappear and new magic numbers appear [23]. In the region of superheavy nuclei, the spin-orbit energy gaps diminish as a result of the fast growth of the density of states and, consequently, the strong shell effects at magic nucleon numbers weaken. Nuclear properties in this limit are extremely sensitive both to the nucleon-nucleon interaction and the continuum coupling [24].

Unbound states have significant impact on spectroscopic properties of nuclei, especially those close to the particle driplines. The continuum of scattering states,

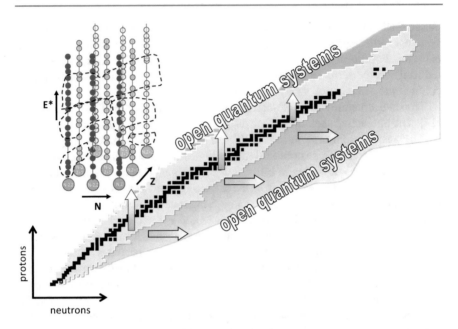

**Fig. 1.1** (Color online) Illustration of transition to an open quantum system regime of nuclear states at higher excitation energies above the lowest particle emission threshold, and in the vicinity of nucleon driplines. In the open quantum system regime, many-body states in neighboring nuclei form the domains of correlated states interconnected via the coupling to decay channels

the decay thresholds, and the double-poles (exceptional points) of the scattering matrix, which are essential ingredients of the configuration mixing, are all neglected in the traditional shell model. What can be said about the structure of many-body states in the narrow range of energies around the reaction threshold? Are those properties independent of any particular realization of the Hamiltonian? Is there a connection between the branch point singularity at the particle emission threshold and the appearance of cluster states?

Incompleteness of the shell model description of atomic nucleus has been noticed very early. For instance, Wigner [25] explained the universal properties of reaction cross sections at the particle-emission threshold(s) by changing boundary conditions. Similarly, the change in boundary conditions at the nuclear surface due to Coulomb wave function distortion in the external region explained relative displacement of states in the mirror nuclei [26–28]. Fano noticed [29] that the ordinary perturbation theory is inadequate for the description of resonances because the continuum states of different configurations coincide in energy exactly. It was therefore clear that one needs a conceptual revolution to resolve various inconsistencies of the shell model picture of the atomic nucleus. In this context, application of time-dependent Schrödinger equation for the description of resonances is not an alternative due to the huge difference of time scales involved in nuclear resonances, which ranges from $\sim 10^{-22}$ s for an average passage time of nucleon in a nucleus to

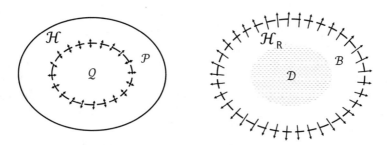

**Fig. 1.2** A schematic illustration of the two different formulations of the nuclear open quantum system. The left panel shows building blocks of the continuum shell model in the Hilbert space. A large Hilbert space $\mathcal{H}$ is divided into orthogonal subspaces $Q$ and $P$ which denote, respectively, subspace of the *system* consisting of localized wave functions and embedding subspace of the *environment* containing scattering states and decay channels. The open quantum system in $Q$ is modified by couplings to the external environment in $P$. The system-plus-environment remains the closed system in Hilbert space. The right panel depicts the structure of the Gamow shell model in the rigged Hilbert space $\mathcal{H}_R$. The shaded area shows the subspace $\mathcal{D}$ of discrete resonant states. The unshaded area depicts the background part $\mathcal{B}$ consisting of complex-energy scattering states. Gamow shell model eigenvectors are expanded in the complete basis of Slater determinants with nucleons in bound, resonance and scattering single-particle states of the Berggren ensemble. In this formulation, one obtains the isolated, open quantum system in the rigged Hilbert space, described by Hermitian Hamiltonian. Both the continuum shell model in Hilbert space and the Gamow shell model in rigged Hilbert space respect the unitarity when changing from the bound levels to the unbound ones

$\sim 10^{19}$ years for a lifetime of an isotope of bismuth $^{209}$Bi, i.e., the variation of 48 orders of magnitude! For very narrow resonances, a direct time-propagation of the time-dependent Schrödinger equation is impossible. On the other end of the time scale, for very broad resonances, even the notion of a nuclear state may lose its meaning.

The first promising efforts were to reconcile the shell model with the theory of nuclear reactions within a framework of the Hilbert space, by replacing the paradigm of a closed quantum system as in standard shell model by the paradigm of a system interacting with the environment (see the left panel of Fig. 1.2) of scattering states and decay channels. Using projection operator technique, Feshbach expressed the collision matrix of the optical model in terms of matrix elements of the nuclear Hamiltonian [30, 31]. This has, on one hand, given a strong push to the shell model approach to nuclear reactions [29, 32–35] and, on the other hand, led to various formulations of the continuum shell model in Hilbert space [16, 36–40]. A recent version of the continuum shell model, the shell model embedded in the continuum [14, 15, 41–43], provides a unified description of structure and reactions with up to two nucleons in the scattering continuum using the shell model Hamiltonian. In this approach, one divides the Fock space of an $A-$particle system into two subspaces: the subspace $Q$ of the system, which consists of square-integrable functions of the standard shell model, and the subspace $P$ of the environment, embedding the system and consisting of the scattering states and the decay channels. Description of internal dynamics (in $Q$) is given by the energy-dependent effective Hamiltonian which includes couplings to the environment (in $P$). Below the lowest reaction threshold, the effective Hamiltonian is Hermitian, whereas above the first threshold, the non-Hermitian part describes the irreversible decay from the system to the environment. Coupling to the continuum generates effective many-body interactions in $Q$, even if in the combined system-plus-environment it is the two-body interaction. One should stress that the combined system-plus-environment remains the closed quantum system which is described by the Hermitian Hamiltonian.

In this formulation of continuum shell model, a direct consequence of the opening of quantum system is the replacement of Hermitian Hamiltonian of the closed quantum system by the non-Hermitian, complex-symmetric effective Hamiltonian. The proper framework for the continuum shell model in Hilbert space is provided by non-Hermitian quantum mechanics which became an important alternative to the standard Hermitian quantum mechanics [16, 44]. Eigenvalues and eigenfunctions of this Hamiltonian may be identified with physical quantities if and only if the subspaces of the Hilbert space describing the system and the environment are defined adequately. The subspace of the environment contains scattering wave functions with the asymptotic conditions of the scattering matrix, whereas the subspace of the system is defined by shell model wave functions constructed from bound single-particle states having large amplitude inside the nucleus. Adding them up, one immediately notice a conceptual difficulty which concerns single-particle resonances which may have large amplitude inside the nucleus and, at the same time, their well-defined asymptotic behavior make them a part of the environment. Hence, the environment should be redefined by excluding from it a large amplitude part of the single-particle resonances, which are localized in the nucleus. The so redefined environment contains only the scattering states and the tails of the single-particle resonances, whereas the system includes the bound states and the localized parts of resonance wave functions inside the nucleus. In this way, one obtains the resonance-free single-particle basis which can be used for the construction of a complete many-body basis of the open quantum system.

For appropriately defined subspaces of the system and the environment, the continuum shell model in Hilbert space provides the unitary description of weakly bound and unbound states. In practice, however, the unitarity can be hardly guaranteed in view of both the large number of reaction channels, which involve also complicate cluster channels, and the intricate conditions, which physical particle continua should abide. For that reason, most applications were restricted to the one- or two-nucleon scattering continuum, severely limiting the number of considered physical cases.

The origin of difficulties encountered in Hilbert space formulation of continuum shell model is related to how resonances are considered in Hilbert space. In this respect, the mathematical formulation of rigged Hilbert space [2] opens new perspectives [45–48]. The breakthrough in this field was a result of a series of unrelated developments in mathematics and theoretical physics over more than 40 years. The theory of rigged Hilbert space, also called a Gel'fand triplet, has been developed in the works of Gel'fand and Vilenkin [2] (see also e.g., Refs. [4, 8]) and was motivated by the necessity to accommodate Dirac's formalism of bras and kets [5, 6]. Soon it was realized that the rigged Hilbert space is also a natural setting for Gamow states [10, 49], and therefore provides a rigorous mathematical framework for extending the domain of quantum mechanics into the time-asymmetric processes like decays or captures. The resonance amplitude associated with the Gamow states is proportional to the complex $\delta$-function and such amplitude can be approximated in the near resonance region by the Breit-Wigner amplitude [8, 50]:

$$\mathscr{A}(E_n \to E) \propto i\sqrt{2\pi}\delta(E - E_n) \sim \frac{1}{2\pi}\frac{1}{E - E_n} \tag{1.1}$$

An important change with respect to the standard Hilbert space formulation is that one can accommodate a more general completeness relation, called the Berggren completeness relation [3, 51–54], where the contribution of real-energy scattering states is substituted by the resonance contribution and the background contribution of complex-energy scattering states. In this way, the resonance spectrum is treated in the same way as bound state spectrum, what has far going consequences, both conceptual and practical ones. For example, when by changing parameters of the Hamiltonian, the complex resonance energy comes close to the real-energy of a bound state, then the eigenfunction of a Gamow state becomes a bound state eigenfunction. Hence, the only difference between narrow resonances and bound states is purely quantitative, namely resonances have nonzero width whereas the bound states have no width.

The configuration-interaction approach based on Gamow states, the so-called Gamow shell model [1, 45–48, 55, 56], is a natural generalization of the standard shell model in which the harmonic oscillator basis is replaced by the Berggren basis which includes bound states, resonances, and (complex-energy) scattering states. In this way, one obtains formulation of the shell model respecting unitarity in all regimes of binding energy, which is built on a skeleton of the $S$-matrix and the many-body completeness relation where bound, resonance, and scattering states enter on equal footing. For well-bound states, the Gamow shell model wave functions become virtually identical with the corresponding shell model wave functions. In this sense, the Gamow shell model fulfills the goals of shell model to provide a comprehensive description of both the configuration interaction and the shell structure, while removing various inconsistencies and limitations of the standard shell model. Similarly to its precursor, and contrary to the continuum shell model in Hilbert space, Gamow shell model describes nucleus as an isolated system. However, in contrast to the standard shell model, the Gamow shell model, which is formulated in the open quantum system formalism, preserves the unitarity, the fundamental principle of quantum mechanics, in the description of eigenvalues of bound and unbound states. It is also important to mention that in contrast to the continuum shell model in Hilbert space, the Hamiltonian in the Gamow shell model is Hermitian, as in the standard shell model. No interaction with an external quantum environment is necessary to describe the system decay, as the Gamow shell model is formulated as an isolated open quantum system. Moreover, the quantum decoherence of wave function(s) comes out naturally as a result of the interference between localized part of the system and its structureless background.

Application of the Gamow shell model for a description of reaction cross sections requires its reformulation [57–59]. Indeed, any eigenfunction of the Gamow shell model is a superposition of Slater determinants , and as such the Gamow shell model is the tool par excellence for structure studies. However, in this representation, reaction channels, in particular entrance and exit channel functions, cannot be accurately defined. To describe nuclear reactions, one has to express the Gamow shell model in the representation of coupled channels which, formally, is equivalent to the Slater determinant representation. An advantage of such a double representation is that one may always compare the Gamow shell model eigenvalues calculated in both representations and in this way check that the important reaction channels are included.

In the coupled-channels representation of Gamow shell model, the nuclear structure and nuclear reactions become unified because the same Hamiltonian and the same many-body approach describes both the discrete part of the energy spectrum and the reaction cross sections at low excitation energies [57–60]. This provides the unique opportunity to remove uncertainties in the interpretation of results of indirect experiments where, for example, the measured cross sections are interpreted by a model to obtain another quantity of interest. Such indirect measurements include the reactions of transfer, breakup, or knockout, the Coulomb excitations, and many others. An example of this kind concerns the interpretation of the ratio of experimental $\sigma_{exp}$ and theoretical $\sigma_{th}$ inclusive one-nucleon removal cross sections for a large number of projectiles which shows a strong dependence on the asymmetry of the neutron and proton separation energies [61–63].

Different formulations of the Gamow shell model open a possibility for a vast number of applications in nuclear physics and related fields. In a core-plus-valence particle approximation [64], one may study spectra and reactions in different mass regions using effective interactions. The no-core formulation of Gamow shell model [56, 65] opens a possibility to test the realistic two- and three-nucleon forces on examples of the resonances in light nuclei. Using interchangeably the representations of Slater determinant or coupled-channels representations of the Gamow shell model wave functions, formulated in Jacobi coordinates or in cluster orbital shell model variables, allows to unify many apparently unrelated aspects of the nuclear structure and reactions in the well-bound states, weakly bound states, or the unbound states. Those different rigorous formulations of the Gamow shell model can be a convenient starting point to derive multiple simpler models to study consequences of the flux conservation (unitarity) at around reaction thresholds. On the practical level, the Gamow shell model can provide foundation of the modern nuclear structure and reaction theory, helping to understand the wealth of data on nuclear levels, moments, collective excitations, various decays, and different low-energy cross sections [17].

Let us now review different chapters in some details.

**Chapter 2** Here the essential mathematical formalism for the discussion of discrete states and continuum is presented along with the examples of various useful potentials. The emphasis is put on the analytical properties of the wave functions, their asymptotic behavior, and analyticity of the complex momentum wave functions. These properties are used in a subsequent chapter to prove the one-body completeness relations in general case. Much attention is devoted to a discussion of the Coulomb potential and Coulomb wave functions. This basic material is then referred to in different contexts all over this textbook.

**Chapter 3** This chapter is devoted to the derivation of the one-body completeness relation in different special cases, and in the general case of the Coulomb wave function and the finite angular momentum. In particular, the Berggren completeness relation for Gamow states will be derived and discussed for real and complex local potentials, as well as for the nonlocal potentials. Important practical question concerns the domains of applicability of the Berggren completeness relation. This discussion, which begins in this chapter, will be further continued in Chap. 5, where the many-body completeness relation is formulated.

Different aspects of Gamow states, necessary for proving the completeness relation or perform practical calculations with Gamow functions, such as the normalization and orthogonality of Gamow states, and the calculation matrix elements of one- and two-body potentials are discussed as well. Together with the analytical discussion of completeness relations, we will also analyze different aspects of the numerical implementation of the Berggren relation, in a standard case which includes resonances, and also in the special case of antibound neutron state. The last part of this chapter includes the presentation of basic properties of complex-symmetric matrices which replace Hermitian matrices in the general Gamow shell model problem.

**Chapter 4** Various physically important two-particle systems are discussed in this chapter. These include unbound dineutron, diproton, and weakly bound deuteron, which will be described in Berggren basis using a realistic chiral interaction. In this context, it is instructive to see a qualitative difference between singlet ($S = 0$) and triplet ($S = 1$) two-nucleon channels. The second class of two-particle systems is the dipolar and quadrupolar anions which are described using the extension of the particle-plus-rotor model in Berggren ensemble. For certain ranges of dipole and/or quadrupole moments, the anions form extremely extended, weakly bound systems. The rich spectrum of resonances with a characteristic band structure is foreseen in these anions. The resonance spectra depend strongly on the asymptotic properties of the pseudo-potentials in each studied case. The particle-plus-rotor model is also used to describe the spectra of favored and unfavored resonance bands in a one-neutron halo nucleus of $^{11}$Be.

**Chapter 5** After formulating in the Berggren basis the two-particle problem and the particle-rotor model, a general multi-particle configuration mixing approach of the Gamow shell model in the Slater determinants representation is discussed in detail. Its mathematical setting in the rigged Hilbert space, the many-body Berggren relation, and the interpretational issues related to complex observables are presented. Gamow shell model can be formulated either in the core-plus-valence-particles model space using the translationally invariant scheme of cluster orbital shell model or, as will be discussed in Chap. 8, having all particles active. In this chapter, most part of the discussion deals with the first option.

The optimization of single-particle basis and an appropriate truncation strategy of many-body Berggren basis in an extended model space, including the single-particle shells of discretized continuum, are the topics of great importance for any practical application of the Gamow shell model. The diagonalization of large Gamow shell model matrices is discussed using the density matrix renormalization approach and the Jacobi-Davidson method. Related to that is the issue of calculation of the two-particle matrix elements and their optimal storage in the large-scale calculations.

A separate important problem is the quantification of theoretical errors on quantities calculated in the Gamow shell model. The statistical evaluation of the quality of the Gamow shell model predictions includes calculation of the penalty function and Bayesian inference of parameters.

**Chapter 6** How the matrix elements of the effective interaction in Gamow shell model depend on angular momentum and continuum coupling is an issue of great importance for understanding of the complex spectra in atomic nuclei. Do they follow in weakly bound states or resonances a similar dependence as found in a standard shell model for bound states? This important topic is addressed by analyzing the dependence of renormalized two-body matrix elements on the semiclassical angle between angular momenta of the coupled nucleons, and by calculating the two-particle spectra in the shells of different angular momenta. Moreover, selected examples of Gamow shell model spectra in light nuclei are discussed and compared with the experimental information.

The Borromean halo nuclei and the emitters of two-protons or two-neutrons are intriguing examples of near-threshold phenomena. One can grasp the structure of these near-threshold states either in the Gamow three-body model in Jacobi relative coordinates, or in the Gamow shell model in the cluster orbital shell model coordinates. Comparison of these approaches on selected observables provides the convergence test for these two Berggren bases.

Residual pairing interaction and the mean field are essential ingredients of our understanding of the shell structure and the properties of heavy nuclei. In the vicinity of continuum, the role of pairing becomes essential. Exact solution of the pairing Hamiltonian in a schematic model exists for bound single-particle levels [66, 67]. The generalization of this solution in Berggren basis may help to describe pairing correlations in systems with a large number of fermions, such as the ultra-small superconducting grains or nuclei with large number of valence particles, which cannot be discussed in the Gamow shell model applications.

**Chapter 7** Nuclear halo states with one or more nucleons in the classically inaccessible region are one of the most spectacular manifestations of the nuclear open quantum system. Formation of halo depends on whether external nucleon is a proton or neutron, and on its angular momentum. Structure of the halo depends also on the deformation of a core and on the nucleon-nucleon correlations which in some situations may prevent formation of an extended halo. Another salient features are the properties of the asymptotic normalization coefficient and one-nucleon overlap functions which are essential ingredients in understanding of the direct reactions and the astrophysical factor in radiative capture reactions.

Near-threshold behavior of the wave functions has an influence on the reaction cross sections and on the spectroscopic factors. In fact, as will be discussed in this chapter, they are closely related. The relation between them is evident in the open quantum system formulation of the shell model which assures a unitary description of bound state and resonance wave functions in the vicinity of the particle emission threshold.

The near-threshold effects depend on the nature of the nearby reaction channel. In particular, it crucially depends on whether it is charged or neutral particle emission channel. This has a direct consequence on the properties of mirror nuclei, their spectra, and the isospin symmetry breaking in mirror states.

**Chapter 8** Elimination of the core in the Gamow shell model opens a possibility for the ab initio description of the resonances in light nuclei and multi-neutron composites. Important issue in this case is the treatment of the problem of center-of-mass excitations. The no-core Gamow shell model with realistic chiral interactions is benchmarked first against results of other many-body methods in bound nuclei $^3$H, $^3$He, $^4$He, and then applied to study the unbound $^5$He nucleus and the multi-neutron composites $^3$n and $^4$n. Using the renormalized interactions with a $\hat{Q}$-box method, the Gamow shell model in a core-plus-valence-particle approximation is then also applied to study oxygen and fluorine isotopes.

**Chapter 9** The Gamow shell model in Slater determinant representation does not allow to describe entrance and final reaction channels. It is therefore a theory adapted for studies of discrete bound states and resonances. To describe reaction cross sections, one has to formulate the Gamow shell model in coupled-channel representation. The resulting coupled-channel Hamiltonian can then be used not only to describe reaction cross sections but, when diagonalized, can provide discrete eigenvalues, i.e., can be used for studies of spectra and transition probabilities in bound and unbound systems. In this representation, Gamow shell model becomes the unified theory of nuclear structure and reactions, whereby using the same Hamiltonian and the same coupled-channel equations, one describes all different aspects of nuclear phenomena. Up to now, this unifying formalism has been successfully applied to the simple direct reactions in light nuclei at low energies, but this formalism is completely general and can be used to describe various cross sections, spectra, transition probabilities, and decays also in heavy systems.

Study of the clustering $^3$H, $^3$He, $^4$He and various multi-neutron clusterization in the near-threshold configurations and the low-energy reactions with multi-nucleon projectiles, like the radiative capture of $\alpha$ particle, are the exciting challenges for future studies.

Applications of the Gamow shell model in both the Slater determinant and coupled-channel representations which are discussed, among others, for mirror radiative capture reactions $^7$Be(p,$\gamma$), $^7$Li(n,$\gamma$) and elastic scattering of deuteron on $^4$He, are allowing to grasp essential features of this double approach.

All chapters of this textbook give a fairly compact description of the generalization of the phenomenologically successful shell model in rigged Hilbert space, to obtain the shell model for open quantum systems. Despite various limitations of this textbook, we believe that the Gamow shell model deserves further detailed studies for several reasons.

Firstly, it overcomes the unwanted separation between the nuclear structure and the nuclear reaction models, what often plagued an understanding of nuclear phenomena.

Secondly, as the Gamow shell model is an extension of a well-known nuclear shell model, it could become the workhorse of comprehensive studies of nuclear structure, allowing to systematize the wealth of data on spectra, transition probabilities, nuclear moments, particle decay, etc.

Thirdly, the Gamow shell model, like its predecessor the traditional shell model, could become the foundation of many microscopic theories of nuclear structure. The construction of a realistic interaction for finite nuclei could also be put on a solid ground once unitarity is respected in calculating the eigenfunctions and eigenvalues.

Finally, as the coupled-channel Gamow shell model is the reaction theory rooted in the configuration-mixing approach, it could provide a starting point for simpler approaches, like those using optical potentials, or help to assess importance of nonlocal effects in nuclear reactions.

## References

1. N. Michel, W. Nazarewicz, M. Płoszajczak, T. Vertse, J. Phys. G. Nucl. Part. Phys. **36**, 013101 (2009)
2. I.M. Gel'fand, N.Y. Vilenkin, *Generalized Functions*, vol. 4 (Academic Press, New York, 1961)
3. K. Maurin, *Generalized Eigenfunction Expansions and Unitary Representations of Topological Groups* (Polish Scientific Publishers, Warsaw, 1968)
4. A. Bohm, *Lecture Notes in Physics*, vol. 78 (Springer Verlag, New York, 1978)
5. G. Ludwig, *Foundations of Quantum Mechanics* , vol. I and II (Springer, New York, 1983)
6. G. Ludwig, *An Axiomatic Basis of Quantum Mechanics*, vol. I and II (Springer, New York, 1983)
7. A. Bohm, M. Gadella, *Lecture Notes in Physics*, vol. 348 (Springer, Berlin, 1989)
8. R. de la Madrid, Eur. J. Phys. **26**, 287 (2005)
9. R. de la Madrid, J. Math. Phys. **53**, 102113 (2012)
10. A. Siegert, Phys. Rev. **56**, 750 (1939)
11. J. Halliwell, in *Proceedings of the 7th Winter School for Theoretical Physics on Quantum Cosmology and Baby Universes, Israel, December 27–January 4, 1990*, ed. by T. P., R. Coleman, J.B. Hartle, S. Weinberg (World Scientific, Singapore, 1991)

12. J. Brookes, Proc. Roy. Soc. A **473**, 20160822 (2017)
13. C. Bennett, P. Shor, IEEE Trans. Inf. Theory **44**, 2724 (1998)
14. J. Okołowicz, M. Płoszajczak, W. Nazarewicz, Prog. Theor. Phys. Supp. **196**, 230 (2012)
15. J. Okołowicz, W. Nazarewicz, M. Płoszajczak, Fortschr. Phys. **61**, 66 (2013)
16. J. Okołowicz, M. Płoszajczak, I. Rotter, Phys. Rep. **374**, 271 (2003)
17. A.N. Bohr, B.R. Mottelson, *Nuclear Structure* (World Scientific, Singapore, 1998)
18. M.G. Mayer, Phys. Rev. **75**, 1969 (1949)
19. O. Haxel, J.H.D. Jensen, H.E. Suess, Phys. Rev. **75**, 1766 (1949)
20. A.M. Lane, Proc. Phys. Soc. A **68**, 189 (1955)
21. D. Kurath, Phys. Rev. **101**, 216 (1956)
22. J. Dobaczewski, N. Michel, W. Nazarewicz, M. Płoszajczak, J. Rotureau, Prog. Part. Nucl. Phys. **59**, 432 (2007)
23. O. Sorlin, M.G. Porquet, Prog. Part. Nucl. Phys. **61**, 602 (2008)
24. P. Jerabek, B. Schuetrumpf, P. Schwerdtfeger, W. Nazarewicz, Phys. Rev. Lett. **120**, 053001 (2018)
25. E.P. Wigner, Phys. Rev. **73**, 1002 (1948)
26. J.B. Ehrman, Phys. Rev. **81**, 412 (1951)
27. R.G. Thomas, Phys. Rev. **81**, 148 (1951)
28. R.G. Thomas, Phys. Rev. **88**, 1109 (1952)
29. U. Fano, Phys. Rev. **124**, 1866 (1961)
30. H. Feshbach, Ann. Phys. (NY) **5**, 357 (1958)
31. H. Feshbach, Ann. Phys. (NY) **19**, 287 (1962)
32. W. Brenig, Nucl. Phys. **13**(3), 333 (1959)
33. L.S. Rodberg, Phys. Rev. **124**, 210 (1961)
34. W.M. MacDonald, Nucl. Phys. **56**, 636 (1964)
35. W.M. MacDonald, Nucl. Phys. **54**, 393 (1964)
36. C. Mahaux, H. Weidenmüller, *Shell-Model Approach to Nuclear Reactions* (North-Holland, Amsterdam, 1969)
37. H.W. Barz, I. Rotter, J. Höhn, Nucl. Phys. A **275**, 111 (1977)
38. R. Philpott, Fizika **9**, 109 (1977)
39. I. Rotter, H.W. Barz, J. Höhn, Nucl. Phys. A **297**, 237 (1978)
40. A. Volya, V. Zelevinsky, Phys. Rev. C **74**, 064314 (2006)
41. K. Bennaceur, F. Nowacki, J. Okołowicz, M. Płoszajczak, Nucl. Phys. A **651**, 289 (1999)
42. K. Bennaceur, F. Nowacki, J. Okołowicz, M. Płoszajczak, Nucl. Phys. A **671**, 203 (2000)
43. J. Rotureau, J. Okołowicz, M. Płoszajczak, Nucl. Phys. A **767**, 13 (2006)
44. N. Moiseyev, *Non-Hermitian Quantum Mechanics* (Cambridge University, New York, 2011)
45. N. Michel, W. Nazarewicz, M. Płoszajczak, K. Bennaceur, Phys. Rev. Lett. **89**, 042502 (2002)
46. R.M. Id Betan, R.J. Liotta, N. Sandulescu, T. Vertse, Phys. Rev. Lett. **89**, 042501 (2002)
47. N. Michel, W. Nazarewicz, M. Płoszajczak, J. Okołowicz, Phys. Rev. C **67**, 054311 (2003)
48. N. Michel, W. Nazarewicz, M. Płoszajczak, Phys. Rev. C **70**, 064313 (2004)
49. G. Gamow, Z. Phys. **51**, 204 (1928)
50. R. de la Madrid, Nucl. Phys. A **885**, 3 (2007)
51. T. Berggren, Nucl. Phys. A **109**, 265 (1968)
52. T. Berggren, Phys. Lett. B **73**, 389 (1978)
53. T. Berggren, P. Lind, Phys. Rev. C **47**, 768 (1993)
54. T. Berggren, Phys. Lett. B **373**, 1 (1996)
55. R.M. Id Betan, R.J. Liotta, N. Sandulescu, T. Vertse, Phys. Rev. C **67**, 014322 (2003)
56. G. Papadimitriou, J. Rotureau, N. Michel, M. Płoszajczak, B.R. Barrett, Phys. Rev. C **88**, 044318 (2013)
57. Y. Jaganathen, N. Michel, M. Płoszajczak, Phys. Rev. C **89**, 034624 (2014)
58. K. Fossez, N. Michel, M. Płoszajczak, Y. Jaganathen, R.M. Id Betan, Phys. Rev. C **91**, 034609 (2015)
59. A. Mercenne, N. Michel, M. Płoszajczak, Phys. Rev. C **99**, 044606 (2019)

60. G.X. Dong, N. Michel, K. Fossez, M. Płoszajczak, Y. Jaganathen, R.M. Id Betan, J. Phys. G. Nucl. Part. Phys. **44**, 045201 (2017)
61. A. Gade, D. Bazin, B.A. Brown, C.M. Campbell, J.A. Church, D.C. Dinca, J. Enders, T. Glasmacher, P.G. Hansen, Z. Hu, K.W. Kemper, W.F. Mueller, H. Olliver, B.C. Perry, L.A. Riley, B.T. Roeder, B.M. Sherrill, J.R. Terry, J.A. Tostevin, K.L. Yurkewicz, Phys. Rev. Lett. **93**, 042501 (2004)
62. A. Gade, P. Adrich, D. Bazin, M.D. Bowen, B.A. Brown, C.M. Campbell, J.M. Cook, T. Glasmacher, P.G. Hansen, K. Hosier, S. McDaniel, D. McGlinchery, A. Obertelli, K. Siwek, L.A. Riley, J.A. Tostevin, D. Weisshaar, Phys. Rev. C **77**, 044306 (2008)
63. J.H. Kelley, S.M. Austin, R.A. Kryger, D.J. Morrissey, N.A. Orr, B.M. Sherrill, M. Thoennessen, J.S. Winfield, J.A. Winger, B.M. Young, Phys. Rev. Lett. **74**, 30 (1995)
64. K. Fossez, W. Nazarewicz, Y. Jaganathen, N. Michel, M. Płoszajczak, Phys. Rev. C **93**, 011305 (2016)
65. K. Fossez, J. Rotureau, N. Michel, M. Płoszajczak, Phys. Rev. Lett. **119**, 032501 (2017)
66. R.W. Richardson, Phys. Lett. **3**, 277 (1963)
67. R.W. Richardson, N. Sherman, Nucl. Phys. **52**, 221 (1964)

# The Discrete Spectrum and the Continuum

**2**

One-body Hamiltonians play an important role in the study of many-body quantum systems. Indeed, the multiple inter-nucleon correlations between the $A$ nucleons inside the nucleus can be approximated by $A$ one-body Hamiltonians, whose potentials, acting on a single nucleon, represent the averaged interactions induced by the remaining $A - 1$ nucleons. As a zeroth-order approximation, one can also deem these potentials as spherical, as the bulk properties of nuclei can be accounted for with spherical potentials.

The three-dimensional Schrödinger equation associated to spherical one-body potentials can be separated in three one-dimensional equations: two depending on angular coordinates, whose eigenstates are expressed in terms of spherical harmonics, and one depending on the radial coordinate, which is the so-called radial Schrödinger equation. As the latter equation is an integro-differential equation depending on the radial coordinate $r$ only, it is possible to determine the properties of its eigenstates using the mathematical apparatus provided by the theory of analytical functions [1]. In particular, one can demonstrate that the spectrum of one-body Hamiltonians whose potential vanishes for $r \to +\infty$ contains a discrete part, made of bound states of negative energy, and a continuous part of scattering states, which bear real positive energy. Similarly, one can show that one-body Hamiltonians whose potential goes to infinity for $r \to +\infty$, such as the harmonic oscillator potential, possess an infinity of bound states.

Radial Schrödinger equation can be solved in closed form in a few particular cases of the spherically symmetric potentials. This is the case for the well-known harmonic oscillator potential, of fundamental importance in quantum physics. The point-particle Coulomb Hamiltonian is also analytically solvable, as the solutions of its one-body radial Schrödinger equation can be expressed in terms of confluent hypergeometric functions. Eigenfunctions of this Hamiltonian, the so-called Coulomb wave functions, are the asymptotic solutions of the radial Schrödinger equation for charged particles. Finally, the Pöschl-Teller-Ginocchio potential, which is a spherical potential mimicking the nuclear interior is analytically solvable for all

© Springer International Publishing AG 2021
N. Michel, M. Płoszajczak, *Gamow Shell Model*, Lecture Notes in Physics 983,
https://doi.org/10.1007/978-3-030-69356-5_2

energies. As it possesses bound, resonance, and scattering eigenstates, contrary to the harmonic oscillator potential, it is of great importance for the study of continuum degrees of freedom in nuclei.

In this chapter, one will also deal with analytic properties of the eigenfunctions of the radial Schrödinger equation with respect to their linear momentum, denoted $k$, in the complex momentum plane. Indeed, analyticity of eigenfunctions in $k$ is necessary to demonstrate that bound and scattering states form a complete set of states (see Sect. 3). One will also introduce resonance states which are eigenstates of complex energy embedded in the continuum of scattering states. As they physically correspond to long-lived states, they are of fundamental importance for the study of weakly bound and unbound nuclei.

## 2.1   Definition of One-Body States

For one-body Hamiltonians bearing spherical symmetry, the solution of the Schrödinger equation separates into two parts, a spherical harmonics and a radial wave function [2]:

$$\phi(k, r) = \left( \frac{u(k, r)}{r} \right) Y_{\ell m}(\theta, \varphi) . \tag{2.1}$$

The properties of the spherical harmonics $Y_m^\ell(\theta, \varphi)$ are well known and described for example in Ref. [2]. The radial wave function $u(k, r)$ obeys the Schrödinger equation:

$$u''(k, r) = \left( \frac{\ell(\ell + 1)}{r^2} + v_l(r) - k^2 \right) u(k, r) , \tag{2.2}$$

where $v_l(r)$ is the potential, $\ell$ is the (positive) orbital angular momentum, and $k$ is the linear momentum which can be complex. Potentials will be considered to be real on the real axis unless stated otherwise. The Hamiltonian of Eq. (2.2) can be formally written as an operator:

$$h = -\frac{d^2}{dr^2} + \frac{\ell(\ell + 1)}{r^2} + \hat{v}_l . \tag{2.3}$$

Equation (2.2) bear the limiting cases for $r \to 0$ and $r \to +\infty$:

$$u''(k, r) = \left[ \frac{\ell_0(\ell_0 + 1)}{r^2} + \frac{v_{c0}}{r} + v_0 - k^2 \right] u(k, r) , \quad r \to 0 \tag{2.4}$$

$$u''(k, r) = \left[ \frac{\ell(\ell + 1)}{r^2} + \frac{v_c}{r} - k^2 \right] u(k, r) , \quad r \to +\infty \tag{2.5}$$

where $v_c$ is the strength of the Coulomb potential for $r \rightarrow +\infty$, which vanishes for neutral particles but is proportional to the charge of the potential $v_l(r)$ for charged particles, and where $v_{c_0}$ and $v_0$ are real. One demands that potentials have a Coulomb asymptote for both $r \rightarrow 0$ and $r \rightarrow +\infty$. These two conditions typically occur in practical applications.

$\ell_0$ and $v_{c_0}$ in (2.4) can be different from $\ell$ and $v_c$, as for example in the Pöschl-Teller-Ginocchio potential (see Sect. 2.4), but one must have $\ell_0 \geq 0$. The resonant and scattering $u(k, r)$ states form the spectrum of $h$, in which only bound states are eigenstates of $h$, as they are integrable. However, as is commonly done in quantum physics, the function $u(k, r)$ of Eq. (2.2) and its energy proportional to $k^2$ will always be referred to as eigenstate and eigenvalue of $h$ in Eq. (2.3), respectively.

For simplicity, one will consider in the following that Eqs. (2.4) and (2.5) are obtained for $0 < r < r_0$ and $r > R$, respectively, with $0 < r_0 < R$. Indeed, $v_l(r) - v_c/r$ typically vanishes exponentially for $r \rightarrow +\infty$, so that it can be neglected after a finite radius $R$. Moreover, $v_l(r)$ can only have a divergence of Coulomb type, that is, of the form $1/r$, when $r \rightarrow 0$. In fact, the reduction of Eq. (2.2) to Eqs. (2.4) and (2.5) for $0 < r < r_0$ and $r > R$, respectively, covers all practical purposes. As all calculations are done up to a given numerical precision, taking sufficiently small $r_0$ and sufficiently large $R$ provides with the same result as that arising by taking the values $r_0 = 0$ and $R = +\infty$. For the same reason, one can also demand $v_l(r)$ to be twice differentiable $\forall r \geq 0$ (see Eq. (2.2)). Indeed, even though $v_l(r)$ is often only piecewise continuous on the real axis, as for a square well potential, it can always be approximated up to an arbitrarily small precision by a twice differentiable function.

Hamiltonians bearing an effective mass when $r_0 < r \leq R$ are implicitly accounted for in Eq. (2.2) as they can always be rewritten via a point-canonical transformation to verify a Schrödinger equation without effective mass [3]. Evidently, the new Schrödinger equation induced by the point-canonical transformation must also reduce to Eqs. (2.4) and (2.5) when $0 < r < r_0$ and $r > R$, respectively.

$\phi(k, r)$ in Eq. (2.1) must be finite for all $r$. As a consequence, the boundary conditions for $u(k, r)$ read:

$$u(k, r) = C_0 \, F_{\ell_0, \eta_0}(k_0 r) \, , \quad 0 \leq r \leq r_0 \tag{2.6}$$

$$u(k, r) = C^+ H^+_{\ell, \eta}(kr) + C^- H^-_{\ell, \eta}(kr) \, , \quad r \geq R \tag{2.7}$$

where $F_{\ell_0, \eta_0}(k_0 r)$ and $H^\pm_{\ell, \eta}(kr)$ are Coulomb wave functions (see Sect. 2.3). In Eqs. (2.6) and (2.7), $k_0 = k\sqrt{1 - v_0/k^2}$, $\eta$ and $\eta_0$ are the Sommerfeld parameters equal to $v_c/(2k)$ and $v_{c_0}/(2k_0)$, respectively. $C_0$, $C^+$, and $C^-$ are constants to be determined from the continuity, derivability, and normalization of $u(k, r)$. Equations (2.2) and (2.6) univocally define $u(k, r)$ up to a normalization constant.

One demonstrates the existence and unicity of $u(k, r)$ with the Picard method by rewriting Eqs. (2.2) and (2.6) as an integral equation:

$$u(k, r) = C_0 \, F_{\ell_0, \eta_0}(k_0 r_0) + C_0 \, k_0 \, F'_{\ell_0, \eta_0}(k_0 r_0) \, (r - r_0) \tag{2.8}$$

$$+ \int_{r_0}^{r} (r - r') \mathscr{L}(k, r') u(k, r') \, dr' \,,$$

with $r \geq r_0$. In this equation, $\mathscr{L}(k, r)$ is an operator representing the action of potentials and linear momentum in Eq. (2.2):

$$\mathscr{L}(k, r) = \frac{\ell(\ell + 1)}{r^2} + v_l(r) - k^2 \,. \tag{2.9}$$

The demonstration is effected in Exercise I. It is, however, rather technical, so that readers can omit it if they find it too difficult and directly assume that $u(k, r)$ exists and is unique.

---

**Exercise  I ★ ★ ★**

In this exercise, one will show that Eq. (2.2) has a unique solution for every linear momentum $k$.

**A.** Define $u^{(n+1)}(k, r)$ as a function of $u^{(n)}(k, r)$ from Eq. (2.9), so as to obtain a recurrence relation between the two functions. In order to fix normalization constants, one will pose $C_0 = k_0^{-\ell_0 - 1} C_{\ell_0}(\eta_0)^{-1}$ in Eqs. (2.6) and (2.9).
Deduce from the integral equation verified by $u^{(n)}(k, r)$ that $u(k, r)$ exists and that $u^{(n)}(k, r) \rightarrow u(k, r)$ uniformly in every finite domain of the complex $k$-plane.

**B.** Show that the solution of Eq. (2.2) built from Eq. (2.9) is unique.
For this, let us consider the $\Delta u(k, r)$ function, difference of two solutions of Eq. (2.9) whose functions and derivatives are equal in one point.
Using the same recurrence relation scheme as in **A**, show that $\Delta u(k, r) = 0$ and conclude that $u(k, r)$ is unique.

---

## 2.2    The Harmonic Oscillator Potential

The harmonic oscillator Hamiltonian and wave functions pervade all quantum mechanics. It is the simplest nontrivial Hamiltonian to bear analytical solutions and related to many physical processes. Similarly to its classical analog, harmonic vibration around equilibrium is the most basic motion, and can describe a large variety of physical systems at low energy. Moreover, the harmonic oscillator Hamiltonian is analytical for all its eigenstates, which is very rare in quantum mechanics.

The harmonic oscillator potential in one dimension is a simple quadratic function of $x$:

$$V_{HO}(x) = \frac{1}{2}m\omega^2 x^2 , \tag{2.10}$$

where $\omega$ is the angular frequency of the harmonic oscillator.

The radial Schrödinger equation with a harmonic oscillator potential reads:

$$\left(\frac{\hbar^2}{2m}\right) u_{HO}^{n_x}{}''(x) = \left(\frac{1}{2}m\omega^2 x^2 - e_{n_x}\right) u_{HO}^{n_x}(x) . \tag{2.11}$$

$n_x$ in this equation is the radial quantum number of the harmonic oscillator state, and $e_{n_x}$ is its energy. Eigenfunction of the harmonic oscillator potential reads:

$$u_{HO}^{n_x}(x) = \left(\frac{\pi^{-1/4}}{\sqrt{2^{n_x} n_x! b}}\right) e^{-x^2/(2b^2)} H_{n_x}(x/b) , \tag{2.12}$$

where $b$ is the harmonic oscillator length:

$$b = \sqrt{\frac{\hbar}{m\omega}} , \tag{2.13}$$

and $H_n(x)$ is a Hermite polynomial. The energy $e_{n_x}$ of the considered harmonic oscillator state reads:

$$e_{n_x} = \left(n_x + \frac{1}{2}\right)\hbar\omega . \tag{2.14}$$

The spherical harmonic oscillator potential in three dimensions is the sum of the one-dimensional harmonic oscillator potentials for the three space directions:

$$V_{HO}(r) = V_{HO}(x) + V_{HO}(y) + V_{HO}(z) = \frac{1}{2}m\omega^2 r^2 . \tag{2.15}$$

The radial Schrödinger equation with a harmonic oscillator potential then reads:

$$\left(\frac{\hbar^2}{2m}\right) u_{HO}^{n}{}''(r) = \left(\left(\frac{\hbar^2}{2m}\right)\left(\frac{\ell(\ell+1)}{r^2}\right) + \frac{1}{2}m\omega^2 r^2 - e_{n\ell}\right) u_{HO}^{n}(r) , \tag{2.16}$$

where $n$ is the radial quantum number of the harmonic oscillator state, and $e_{n\ell}$ is its energy. It is one of the only potentials to bear the analytic solutions for all orbital angular momenta $\ell$, which are simple functions of exponentials and polynomials:

$$u_{HO}^{n}(r) = N_{n\ell}(r/b)^{\ell+1} e^{-r^2/(2b^2)} L_n^{(\ell+1/2)}(r^2/b^2) . \tag{2.17}$$

In this expression $L_n^{(\alpha)}(x)$ is the generalized Laguerre polynomial and $N_{n\ell}$ is a normalization constant:

$$N_{n\ell} = \sqrt{\frac{2}{b}\left(\frac{\Gamma(n+1)}{\Gamma(n+\ell+3/2)}\right)}, \qquad (2.18)$$

where $\Gamma(z)$ is the Euler Gamma function. The energy $e_{n\ell}$ of the considered harmonic oscillator state reads:

$$e_{n\ell} = \left(2n + \ell + \frac{3}{2}\right)\hbar\omega. \qquad (2.19)$$

Comparison of the analytical formulas of energy and wave functions in the one- and three-dimensional cases will be effected in Exercise II.

---

**Exercise  II ⋆**

One will compare the harmonic oscillator wave functions and energies in the one- and three-dimensional cases and point out analogies between them.

For this, show that the formulas of Eqs. (2.12), (2.14), and (2.17)–(2.19) provide with normalized solutions of Eqs. (2.11) and (2.16).

Explain from Eq. (2.15) why Eqs. (2.12), (2.14), and (2.17)–(2.19) are very similar in the one- and three-dimensional cases.

---

As the harmonic oscillator potential goes to $+\infty$ for increasing $x$ and $r$ (see Eqs. (2.10) and (2.15)), it possesses only bound states. Its set of eigenstates is then discrete. Therefore, it is straightforward to show their completeness properties from variational arguments [1]. However, due to the rapid decrease of harmonic oscillator wave functions on the real axis (see Eqs. (2.12) and (2.17)), the harmonic oscillator states in practice are efficient to expand only well-bound states.

The basis of harmonic oscillator states is fundamental to the standard shell model, which has been very effective to describe well-bound states of stable nuclei. However, weakly bound states and resonances cannot be expanded in this basis as the squared modulus of their wave functions is either slowly decreasing or increasing along the real axis. For that purpose, the use of the Berggren basis, possessing bound states, resonances, and scattering states, is much better suited. Nevertheless, harmonic oscillator states are very important for weakly bound and resonance nuclei as well, not for the expansion their wave functions, but to expand two-body matrix elements of nuclear interaction. Indeed, on the one hand, the nuclear interaction is localized, so that it can be efficiently expanded with harmonic oscillator states. On the other hand, two-body harmonic oscillator states whose coordinates are defined with laboratory coordinates, can always be written as a finite sum of products of two-body harmonic oscillator states in the relative or center-of-mass coordinates of the two-particle system. This is the objective of the Talmi-Brody-Moshinsky transformation [4, 5]. Indeed, as nuclear interactions depend only on relative degrees of freedom, their matrix elements which are cumbersome to calculate in laboratory coordinates, are straightforward to obtain with relative coordinates.

The Talmi-Brody-Moshinsky transformation allows to calculate efficiently the two-body matrix elements, for which a direct calculation with Berggren basis states would be prohibitive (see Sect. 5.6). One will write this transformation for the case of two spinless particles of identical mass in independent harmonic oscillator potentials. Indeed, the introduction of the spin degree of freedom only generates additional recoupling terms using the Wigner–Eckart theorem.

The Talmi-Moshinsky-Brody transformation reads:

$$|n_1\ell_1\rangle \, |n_2\ell_2\rangle = \sum_{n\ell NL} \langle n\ell NL|n_1\ell_1 n_2\ell_2\rangle \, |n\ell\rangle \, |NL\rangle \ , \tag{2.20}$$

where $|n_1\ell_1\rangle$ and $|n_2\ell_2\rangle$ are two harmonic oscillator states of coordinates $\mathbf{r}_1$ and $\mathbf{r}_2$, respectively, and $|n\ell\rangle$ and $|NL\rangle$ are two harmonic oscillator states of coordinates defined from their relative and center-of-mass coordinates, which respectively read:

$$\mathbf{r}_{\mathrm{rel}} = \mathbf{r}_1 - \mathbf{r}_2 \tag{2.21}$$

$$\mathbf{R}_{\mathrm{CM}} = \frac{\mathbf{r}_1 + \mathbf{r}_2}{2} \ . \tag{2.22}$$

Associated linear momenta come forward:

$$\mathbf{p}_{\mathrm{rel}} = \frac{\mathbf{p}_1 - \mathbf{p}_2}{2} \tag{2.23}$$

$$\mathbf{P}_{\mathrm{CM}} = \mathbf{p}_1 + \mathbf{p}_2 \ . \tag{2.24}$$

Coefficients $\langle n\ell NL|n_1\ell_1 n_2\ell_2\rangle$ in (2.20) are called the Talmi-Brody-Moshinsky coefficients, and can be expressed analytically [4, 5]. One must also have:

$$2n + \ell + 2N + L = 2n_1 + \ell_1 + 2n_2 + \ell_2 \ . \tag{2.25}$$

Consequently, Eq. (2.20) has a finite number of terms, which is furthermore reduced for small both radial quantum numbers and orbital angular momenta, as is the case in practical applications (see Exercise III).

---

**Exercise  III ⋆**

One will demonstrate that the Talmi-Brody-Moshinsky transformation of Eq. (2.20) exists.

**A.** Show that the harmonic oscillator Hamiltonian for two independent particles of coordinates $\mathbf{r}_1$ and $\mathbf{r}_2$.

$$\frac{\mathbf{p}_1^2}{2m} + \frac{\mathbf{p}_2^2}{2m} + \frac{1}{2}m\omega^2\mathbf{r}_1^2 + \frac{1}{2}m\omega^2\mathbf{r}_2^2 \tag{2.26}$$

can be expressed with the coordinates of Eqs. (2.21) and (2.22).

> Deduce that the eigenstates of Eq. (2.26) can be written in terms of $\mathbf{r}_1$, $\mathbf{r}_2$ or $\mathbf{r}_{rel}$, $\mathbf{R}_{CM}$ coordinates.
> **B.** Deduce from Eq. (2.26) that an eigenstate of the Hamiltonian written in laboratory coordinates is a finite linear combination of those written in relative/center-of-mass coordinates. One will use Eq. (2.25) for that matter. From the latter result, prove the existence of the Talmi-Brody-Moshinsky transformation and the energy condition of Eq. (2.25).

Consequently, the use of harmonic oscillator states in the Gamow shell model is not anecdotal in the Gamow Shell Model but, on the contrary, has been essential to its development.

## 2.3   Coulomb Potential and Coulomb Wave Functions

Coulomb wave functions are among the most fundamental tools of quantum physics. They describe the behavior of a charged particle in a Coulomb field, and are therefore present in almost all domains of quantum physics. Their dimensionless Schrödinger equation reads:

$$W''(z) = \left( \frac{\ell(\ell+1)}{z^2} + \frac{2\eta}{z} - 1 \right) W(z) , \qquad (2.27)$$

where $W(z)$ is a Coulomb wave function, $\ell$ is the orbital angular momentum, and $\eta$ the Sommerfeld parameter:

$$\eta = \frac{e\,Z\,m}{\hbar^2\,k} , \qquad (2.28)$$

where $e$ is the elementary charge, $m$ is its mass, $k$ is its linear momentum and $Z$ is the charge of the external Coulomb potential. All parameters are considered as complex in this section (see Refs. [6,7] for details about the numerical evaluation of Coulomb wave functions).

In the following, one will discuss how to define and calculate Coulomb wave functions with complex parameters. For this, it is convenient to consider that $\Re(\ell) \geq -1/2$. There is no loss of generality in this restriction because Eq. (2.27) is invariant with the change $\ell \to -\ell - 1$.

The Coulomb wave functions are initially defined with confluent hypergeometric functions [8]. The regular Coulomb wave function takes the form:

$$F_{\ell,\eta}(z) = C_\ell(\eta)\, z^{\ell+1}\, e^{i\omega z}\, {}_1F_1\left(1 + \ell + i\omega\eta; 2\ell + 2; -2i\omega z\right) \qquad (2.29)$$

$$C_\ell(\eta) = 2^\ell \exp\left[ \frac{-\pi\eta + [\ln(\Gamma(1+\ell+i\eta)) + \ln(\Gamma(1+\ell-i\eta))]}{2} \right. \qquad (2.30)$$

$$\left. - \ln(\Gamma(2\ell+2))\right] ,$$

where $\omega = \pm 1$ and $C_\ell(\eta)$ is the Gamow factor. One defines outgoing ($\omega = 1$) and incoming ($\omega = -1$) Coulomb wave functions similarly:

$$H_{\ell,\eta}^\omega(z) = e^{i\omega[z - \eta \ln(2z) - \ell\frac{\pi}{2} + \sigma_\ell(\eta)]} \tag{2.31}$$

$$\times {}_2F_0\left(1 + \ell + i\omega\eta, -\ell + i\omega\eta; ; -\frac{i}{2\omega z}\right)$$

$$\sigma_\ell(\eta) = \frac{\ln(\Gamma(1 + \ell + i\eta)) - \ln(\Gamma(1 + \ell - i\eta))}{2i}, \tag{2.32}$$

where $\sigma_\ell(\eta)$ is the Coulomb phase shift [8]. The cut on the negative real axis of the function $\ln(\Gamma(z))$ occurring in $C_\ell(\eta)$ (see Eq. (2.31)) is defined as in Refs. [9,10]. It guarantees consistent values even when the negative real axis branch cut of complex variables $1 + \ell + i\eta$ and $1 + \ell - i\eta$ is crossed.

The regular Coulomb wave function $F_{\ell,\eta}(z)$ and the irregular Coulomb wave function $G_{\ell,\eta}(z)$ are linear combinations of $H_{\ell,\eta}^+(z)$ and $H_{\ell,\eta}^-(z)$ [8]:

$$F_{\ell,\eta}(z) = \frac{H_{\ell,\eta}^+(z) - H_{\ell,\eta}^-(z)}{2i} \tag{2.33}$$

$$G_{\ell,\eta}(z) = \frac{H_{\ell,\eta}^+(z) + H_{\ell,\eta}^-(z)}{2}. \tag{2.34}$$

The following formulas allow to calculate the Coulomb wave functions defined in Eqs. (2.29), (2.32), and (2.34) when $\ell$, $\eta$, and $z$ are complex. $H_{\ell,\eta}^\omega(z)$ can be expanded in asymptotic series if $\Re(z) > 0$ or $\Re(z) < 0$ and $\omega\Im(z) > 0$ [6,11], so that, for a fixed integer $N$ and $|z|$ sufficiently large, $H_{\ell,\eta}^\omega(z)$ reads:

$$H_{\ell,\eta}^\omega(z) \simeq e^{i\omega[z - \eta \ln(2z) - \ell\frac{\pi}{2} + \sigma_\ell(\eta)]} \sum_{n=0}^{N-1} a_n z^{-n}, \tag{2.35}$$

where

$$a_0 = 1$$

$$a_{n+1} = \frac{n(n + 1 + 2i\omega\eta) + i\eta(i\eta + \omega) - \ell(\ell + 1)}{2i\omega(n + 1)} a_n \qquad \forall n \geq 0. \tag{2.36}$$

Even though Eq. (2.35) is not correct if the conditions $\Re(z) > 0$, or $\Re(z) < 0$ and $\omega\Im(z) > 0$ are not fulfilled, nevertheless it gives a linear combination of $H_{\ell,\eta}^+(z)$ and $H_{\ell,\eta}^-(z)$, which is useful to calculate $H_{\ell,\eta}^\omega(z)$ in these regions of the complex plane (see Sect. 2.3.2).

The regular solution $F_{\ell,\eta}(z)$ is conveniently represented by a power series [8]:

$$F_{\ell,\eta}(z) = C_\ell(\eta) \sum_{n=0}^{+\infty} b_n \, z^{n+\ell+1} \qquad (2.37)$$

$$b_0 = 1$$

$$b_1 = \frac{\eta}{\ell+1}$$

$$b_n = \frac{2\eta \, b_{n-1} - b_{n-2}}{n(n+2\ell+1)} \qquad \forall n \geq 2 \, . \qquad (2.38)$$

For small $z$, the latter expression is preferable to the expression of $F_{\ell,\eta}(z)$ in Eq. (2.33). Indeed, the expression of $F_{\ell,\eta}(z)$ in Eq. (2.33) consists of the difference of two very large numbers providing with a very small value for $F_{\ell,\eta}(z)$ and hence cannot be used in practice. Conversely, Eq. (2.37) converges very quickly for $|z| < 1/2$. One should mention that Eq. (2.37) becomes unstable if $|z|$ increases, even though its radius of convergence is infinite theoretically.

This phenomenon regularly occurs in the power series expansion of entire functions oscillating at infinity, as the general term of the series there becomes very large, so that important numerical cancellations occur before the series starts to converge. Consequently, Eqs. (2.33) and (2.37) are complementary even though they provide with the same value numerically.

$H_{\ell,\eta}^\omega(z)$ can be calculated using the following formula if $2\ell$ is not an integer [6, 11]:

$$H_{\ell,\eta}^\omega(z) = \frac{F_{\ell,\eta}(z) \, e^{i\omega\chi} - F_{-\ell-1,\eta}(z)}{\sin \chi} \qquad (2.39)$$

$$\chi = \sigma_\ell(\eta) - \sigma_{-\ell-1}(\eta) - (\ell+1/2)\pi \, . \qquad (2.40)$$

Note that Eq. (2.40) is not always stable numerically. In the case of the numerical inaccuracy, to determine $\chi$ one uses another formula:

$$\sin \chi = -(2\ell+1) \, C_\ell(\eta) \, C_{-\ell-1}(\eta) \, , \qquad (2.41)$$

which can be demonstrated using the Wronskian of $F_{\ell,\eta}(z)$ and $F_{-\ell-1,\eta}(z)$, as well as Eq. (2.37) for $z \to 0$.

## 2.3.1   Continued Fractions

While Coulomb wave functions in the complex plane vary exponentially as a function of $\eta$ and $z$, it is not the case for their logarithmic derivatives. Indeed, the logarithmic derivatives of Coulomb wave functions typically have a rational dependence on $\eta$ and $z$, so that it is more stable numerically to calculate logarithmic derivatives of Coulomb wave functions instead of calculating Coulomb wave functions directly. Due to the analytic properties of the confluent hypergeometric functions present in Coulomb wave functions (see Eqs. (2.29) and (2.32)), it is possible to express the logarithmic derivatives of Coulomb wave functions with continued fractions [6, 11]:

$$f^{\omega}(z) = \frac{\ell+1}{z} + i\omega + \frac{1}{z}\left[\frac{-2i\omega az}{b+2i\omega z+}\frac{-2i\omega(a+1)z}{b+1+2i\omega z+\cdots}\right] \quad (2.42)$$

$$h^{\omega}(z) = i\omega\left(1-\frac{\eta}{z}\right) + \frac{i\omega}{z}\left[\frac{ac}{2(z-\eta+i\omega)+}\frac{(a+1)(c+1)}{2(z-\eta+2i\omega)+\cdots}\right], \quad (2.43)$$

where the standard notations $a = 1 + \ell + i\omega\eta$, $b = 2\ell + 2$, and $c = -\ell + i\omega\eta$ are used [6, 11].

The value of $f^{\omega}(z)$ (also denoted as $f(z)$) is derived from Eq. (2.29) and thus does not depend on $\omega$. Continued fractions can be evaluated numerically using the Lentz method [6, 11]. The convergence domain of $f^{\omega}(z)$ is the whole complex plane besides zeros of $F_{\ell,\eta}(z)$, while the convergence domain of $h^{\omega}(z)$ is that of $_2F_0$, so that it is the whole complex plane minus the half-axis $[0 : -i\omega\infty)$, where $_2F_0$ has a branch cut discontinuity.

The asymptotic series of Eq. (2.35), very precise to calculate $H_{\ell,\eta}^{\omega}(z)$ when $|z|$ is large, is unavailable for small and moderate $z$ values. Moreover, Eq. (2.37) can be used in practice to calculate $F_{\ell,\eta}(z)$ only if $|z|$ is small, as important numerical cancellations occur even for moderate $z$ values. In fact, Eqs. (2.42) and (2.43) are fundamental for a precise calculation of the Coulomb wave functions defined in Eqs. (2.29), (2.32), and (2.34) [6, 7]. Indeed, $f^{\omega}(z)$ and $h^{\omega}(z)$ (see Eqs. (2.42) and (2.43)), which correspond to the logarithmic derivatives of Coulomb functions $F_{\ell,\eta}(z)$ and $H_{\ell,\eta}^{\omega}(z)$, are typically very stable numerically.

The value of $F_{\ell,\eta}(z)$ is given by Eq. (2.37) for small $z$ and by direct integration for moderate $z$. However, the numerical integration of Eq. (2.27) is stable only if the modulus of the integrated Coulomb wave function increases or remains close to a constant. Increase or decrease of $|F_{\ell,\eta}(z)|$ along the integration path is determined by its second-order Taylor expansion at $z = z_0 + h$, where $z_0$ is the point from which direct integration starts, and $h$ is the integration step. The knowledge of $f^{\omega}(z)$ (see Eq. (2.42)) allows to use direct integration without loss of numerical precision when $|F_{\ell,\eta}(z)|$ increases along the integration path. One may note that integrating Eq. (2.27) backwards, that is, from $z$ to $z_0$, is stable because $|F_{\ell,\eta}(z)|$ increases in modulus in this direction. Thus, one numerically integrates Eq. (2.27) from $z$ to $z_0$ using the initial conditions: $C\,F_{\ell,\eta}(z) = 1$ and $C\,F'_{\ell,\eta}(z) = f^{\omega}(z)$, where $f^{\omega}(z)$ is given by Eq. (2.42) and $C = 1/F_{\ell,\eta}(z)$ must be determined afterwards. The backward integration process provides with the values $C\,F_{\ell,\eta}(z_0)$ and $C\,F'_{\ell,\eta}(z_0)$. As $F_{\ell,\eta}(z_0)$ and $F'_{\ell,\eta}(z_0)$ are already known, the constant $C$ is immediate to obtain. So it follows directly that the values of $F_{\ell,\eta}(z)$ and $F'_{\ell,\eta}(z)$ are equal to $C^{-1}$ and $C^{-1}\,f^{\omega}(z)$, respectively.

Using the values of $F_{\ell,\eta}(z)$, $F'_{\ell,\eta}(z)$ obtained by direct integration, that of $h^\omega(z)$ arising from Eq. (2.43), and the Wronskian relation between $F_{\ell,\eta}(z)$ and $H^\omega_{\ell,\eta}(z)$, one can determine $H^\omega_{\ell,\eta}(z)$ and its derivative. Conversely, in the rare cases where $f^\omega(z)$ (see Eq. (2.42)) is numerically imprecise, $H^\omega_{\ell,\eta}(z)$ is obtained similarly by direct integration using $h^\omega(z)$ (see Eq. (2.43)), while $F_{\ell,\eta}(z)$ is obtained from the Wronskian relation between $F_{\ell,\eta}(z)$, $H^\omega_{\ell,\eta}(z)$, and $h^{\pm\omega}(z)$. As the continued fraction providing $h^{\pm\omega}(z)$ is typically more robust numerically than that of $f^\omega(z)$ (see Eqs. (2.42) and (2.43)), this procedure leads to a numerically stable algorithm in practice. As a consequence, along with the use of direct integration, the continued fractions (2.42), (2.43) allow to precisely implement the regular and irregular Coulomb wave functions and their derivatives in the difficult case of large Sommerfeld parameter.

### 2.3.2  Analytic Continuation of Coulomb Wave Functions

The analytic continuation of Coulomb wave functions in the complex plane is straightforward for $\Re(z) \geq 0$, as all previously used equations are analytical therein. However, this is no longer the case for $\Re(z) < 0$, as the presence of different cuts implies that the latter expressions of Coulomb wave functions do not provide the same values. Hence, analytic continuation for $\Re(z) < 0$ has to be considered separately.

One will firstly consider the analytic continuation of the regular function $F_{\ell,\eta}(z)$ for $\Re(z) < 0$. It is straightforward to implement because $F_{\ell,\eta}(z)$ and $F_{\ell,-\eta}(-z)$ are mutually proportional (see Eqs. (2.31) and (2.37)):

$$F_{\ell,\eta}(z) = -e^{-\pi(\eta-i\ell)} F_{\ell,-\eta}(-z) \text{ if } \arg(z) > 0$$

$$F_{\ell,\eta}(z) = -e^{-\pi(\eta+i\ell)} F_{\ell,-\eta}(-z) \text{ if } \arg(z) \leq 0 . \tag{2.44}$$

Hence, $F_{\ell,\eta}(z)$ can always be obtained from $F_{\ell,-\eta}(-z)$, so to determine $F_{\ell,\eta}(z)$ in the whole complex plane, it is sufficient to restrict calculations so that $\Re(z) \geq 0$.

The analytic continuation of $H^\omega_{\ell,\eta}(z)$, on the contrary, is not so simple. Indeed, if $\Re(z) < 0$, the direct evaluation of Eqs. (2.35) and (2.43) provides correct function values only if $\omega\Im(z) > 0$ [6, 11]. Let us denote as $H^{\omega\,(\Sigma_d)}_{\ell,\eta}(z)$ and $H^{\omega\,(h_d)}_{\ell,\eta}(z)$ the functions arising from the direct implementation of Eqs. (2.35) and (2.43), respectively, when $\Re(z) < 0$ and $\omega\Im(z) < 0$. As $H^{\omega\,(\Sigma_d)}_{\ell,\eta}(z)$ and $H^{\omega\,(h_d)}_{\ell,\eta}(z)$ solve Eq. (2.27), but have different branch cuts in the complex plane, they are linear combinations of $H^+_{\ell,\eta}(z)$ and $H^-_{\ell,\eta}(z)$ when $\Re(z) < 0$ and $\omega\Im(z) < 0$. As a matter of fact, the coefficients entering the latter linear combinations are elementary functions of $\ell$ and $\eta$ and are related to circuital properties of Coulomb wave functions [12].

Let us calculate the coefficients of the linear combinations of $H^+_{\ell,\eta}(z)$ and $H^-_{\ell,\eta}(z)$. For this, it is convenient to have $|z| \to +\infty$. Eqs. (2.32) and (2.35) imply in this situation that:

$$H_{\ell,\eta}^{\omega}(z) = H_{\ell,\eta}^{\omega\,(\Sigma_d)}(z) + c_{\omega}(\ell,\eta)\,H_{\ell,\eta}^{-\omega\,(\Sigma_d)}(z)\,,\,\Re(z) < 0\,,\,\omega\Im(z) < 0\,, \qquad (2.45)$$

where $c_{\omega}(\ell,\eta)$ is a function of $\ell$ and $\eta$ only and the equality $H_{\ell,\eta}^{-\omega}(z) = H_{\ell,\eta}^{-\omega\,(\Sigma_d)}(z)$ in the considered part of the complex plane has been used. The following equalities arise from Eqs. (2.33) and (2.45):

$$\mp 2i\,F_{\ell,\eta}(x_{\pm}) = H_{\ell,\eta}^{\mp\,(\Sigma_d)}(x_{\pm}) + (c_{\mp}(\ell,\eta) - 1)\,H_{\ell,\eta}^{\pm\,(\Sigma_d)}(x_{\pm}) \qquad (2.46)$$

where $\epsilon > 0$, $x < 0$, and $x_{\pm} = x \pm i\epsilon$. $F_{\ell,\eta}(z)$ and $H_{\ell,\eta}^{\omega\,(\Sigma_d)}(z)$ bear simple branch cut discontinuities (see Eqs. (2.35) and (2.37)), so that from Eq. (2.46) one has:

$$2i\,e^{-2i\pi\ell}\,F_{\ell,\eta}(x_+) = e^{-2\pi\eta}\,H_{\ell,\eta}^{+\,(\Sigma_d)}(x_+) \qquad (2.47)$$

$$+e^{2\pi\eta}\,(c_+(\ell,\eta) - 1)H_{\ell,\eta}^{-\,(\Sigma_d)}(x_+) + \mathscr{O}(\epsilon)\,.$$

Hence, using Eqs. (2.46) and (2.48), one obtains :

$$c_{\omega}(\ell,\eta) = 1 - e^{2i\pi(i\eta - \ell\omega)}\,. \qquad (2.48)$$

The formula providing the analytic continuation of $H_{\ell,\eta}^{\omega}(z)$ for $\Re(z) < 0$ and $\omega\Im(z) < 0$ is then deduced from Eqs. (2.46), (2.48), and (2.48):

$$H_{\ell,\eta}^{\omega}(z) = H_{\ell,\eta}^{\omega\,(\Sigma_d)}(z) + \left[1 - e^{2i\pi(i\eta - \ell\omega)}\right] H_{\ell,\eta}^{-\omega\,(\Sigma_d)}(z)\,. \qquad (2.49)$$

The analytic continuation of $H_{\ell,\eta}^{\omega\,(h_d)}(z)$ arises from the different branch cuts of Eqs. (2.35) and (2.43) when $\Re(z) < 0$ and $\omega\Im(z) < 0$ :

$$H_{\ell,\eta}^{\omega\,(h_d)}(z) = e^{2i\pi(\ell\omega - i\eta)}\,H_{\ell,\eta}^{\omega\,(\Sigma_d)}(z)\,. \qquad (2.50)$$

Using Eqs. (2.33), (2.49), and (2.50), one derives the formulas analogous to Eq. (2.49) when continued fractions are utilized:

$$H_{\ell,\eta}^{\omega}(z) = H_{\ell,\eta}^{\omega\,(h_d)}(z) \qquad (2.51)$$

$$-2i\omega\left[e^{2i\pi(\ell\omega - i\eta)} - 1\right] F_{\ell,\eta}(z)\,,\,\Re(z) < 0\,,\,\omega\Im(z) < 0$$

$$H_{\ell,\eta}^{\omega}(z) = H_{\ell,\eta}^{-\omega\,(h_d)}(z) \qquad (2.52)$$

$$+2i\omega\,e^{-2i\pi(\ell\omega + i\eta)}\,F_{\ell,\eta}(z)\,,\,\Re(z) < 0\,,\,\omega\Im(z) > 0\,.$$

### 2.3.3  Asymptotic Forms of Coulomb Wave Functions for Small and Large Arguments

Coulomb wave functions appear as a function of $z = kr$ in radial wave functions (see Eq. (2.7)). Therefore, it is convenient to have asymptotic forms of Coulomb wave functions for $k \to 0$ and $r > 0$ fixed.

If $v_c \geq 0$ and $k > 0$, then equivalents of $F_{\ell\eta}(kr)$ and $G_{\ell\eta}(kr)$ write [8]:

$$F_{\ell\eta}(kr) \sim C_\ell(0) \, (kr)^{\ell+1} \,, \quad (v_c = 0) \tag{2.53}$$

$$G_{\ell\eta}(kr) \sim (2\ell+1)^{-1} \, C_\ell(0)^{-1} \, (kr)^{-\ell} \,, \quad (v_c = 0) \tag{2.54}$$

$$F_{\ell\eta}(kr) \sim \Gamma(2\ell+2) \, C_\ell(\eta) \, (2\eta)^{-\ell-1} \, \pi^{-1/2} \, (v_c r)^{\frac{1}{4}} \, (2\sqrt{v_c r}) \, i_{2\ell+\frac{1}{2}} \tag{2.55}$$

$$\times (2\sqrt{v_c r}) \,, \quad (v_c > 0)$$

$$G_{\ell\eta}(kr) \sim (2\eta)^\ell \, \Gamma(2\ell+2)^{-1} \, C_\ell(\eta)^{-1} \, \pi^{1/2} \, (v_c r)^{\frac{1}{4}} \, (2\sqrt{v_c r}) \, k_{2\ell+\frac{1}{2}} \tag{2.56}$$

$$\times (2\sqrt{v_c r}) \,, \quad (v_c > 0)$$

where $i_{2\ell+\frac{1}{2}}(x)$ and $k_{2\ell+\frac{1}{2}}(x)$ are the modified spherical Bessel functions of the first and second kind, respectively. One can also verify by direct insertion of Eqs. (2.53)–(2.57) in Eq. (2.5) that their $r$-dependent parts are solutions of Eq. (2.5) for $k = 0$.

$H^\omega_{\ell,\eta}(kr)$ bears the same asymptotic form as $G_{\ell,\eta}(kr)$ when $k \to 0$ because $H^\omega_{\ell,\eta}(kr) = G_{\ell\eta}(kr) + i\omega F_{\ell\eta}(kr)$ and $|F_{\ell\eta}(kr)| \ll |G_{\ell,\eta}(kr)|$ therein. Note that $C_\ell(\eta)$, appearing in Eqs. (2.56) and (2.57), has a simple equivalent when $\eta \to +\infty$. Indeed, the $\Gamma(1 + \ell \pm i\eta)$ functions in Eq. (2.31) can be evaluated with the Stirling formula. One then obtains:

$$C_\ell(\eta) \sim \frac{(2\eta)^\ell \, \exp(-\pi\eta) \, \sqrt{2\pi\eta}}{\Gamma(2\ell+2)} \,. \tag{2.57}$$

If $v_c < 0$, the solutions of Eq. (2.5) with $k = 0$ also write concisely:

$$u(k = 0, r) = C_j \, r^{1/4} \left(2\sqrt{|v_c|r}\right) j_{2\ell+1/2}\left(2\sqrt{|v_c|r}\right) \tag{2.58}$$

$$+ C_y \, r^{1/4} \left(2\sqrt{|v_c|r}\right) y_{2\ell+1/2}\left(2\sqrt{|v_c|r}\right) \,,$$

where $j_{2\ell+\frac{1}{2}}(x)$ and $y_{2\ell+\frac{1}{2}}(x)$ are the spherical Bessel functions of the first and second kind, respectively, and $C_j$ and $C_y$ are integration constants. It is straightforward to show that Eq. (2.59) bears a simple asymptotic form for large arguments $r$:

$$u(k = 0, r) = C \, r^{1/4} \sin\left(2\sqrt{|v_c|r} + \delta\right) + \mathcal{O}(r^{-1/4}) \,, \tag{2.59}$$

where $C$ and $\delta$ are the normalization constant and the phase shift at zero energy, respectively.

The asymptotic forms of Coulomb wave functions for $r$ fixed and $k \to 0$ in the complex plane can be expressed from their analytical formulas, function of confluent hypergeometric functions (see Eqs. (2.29), (2.32), and (2.34) and Ref. [8]). Contrary to real $k$ values, these equivalents are complex functions of $k$ and $\ell$ and possess branch cuts in the complex plane. However, one can derive useful equalities using the analytical properties of Coulomb wave functions:

$$F_{\ell\eta}(kr) = k^{\ell+1} \, C_\ell(\eta) \, f_B(\ell, k, r) \,, \tag{2.60}$$

$$H^+_{\ell\eta}(kr) = k^{-\ell} \, C_\ell(\eta)^{-1} \, h^+_B(\ell, k, r) \,, \tag{2.61}$$

where $f_B(\ell, k, r)$ and $h^+_B(\ell, k, r)$ are bounded when $k \to 0$ if $v_c \geq 0$ and $\Im(k) \geq 0$ in Eq. (2.61). It is straightforward to obtain Eq. (2.60) using the power series expansion of $F_{\ell\eta}(kr)$ (see Eq. (2.37)). If $2\ell$ is not an integer, Eq. (2.61) can be then derived using Eqs. (2.39) and (2.60). Conversely, when $2\ell$ is an integer, it is preferable to use the analytical expression of the $_2F_0$ function based on power series [8, 13] in Eq. (2.32) for that matter. Note that Eqs. (2.60) and (2.61) are also correct if $v_c < 0$ provided that one avoids the close vicinity of the poles associated to bound states in the $k$-plane [13].

Asymptotic forms of Coulomb wave functions and their derivatives for $|kr| \to +\infty$ are obtained from Eqs. (2.33)–(2.35):

$$F_{\ell,\eta}(kr) = \sin\left(kr - \eta \ln(2kr) - \ell\frac{\pi}{2} + \sigma_\ell(\eta)\right) + \mathcal{O}(\exp(|\Im(k)|r)\,(kr)^{-1}) \tag{2.62}$$

$$G_{\ell,\eta}(kr) = \cos\left(kr - \eta \ln(2kr) - \ell\frac{\pi}{2} + \sigma_\ell(\eta)\right) + \mathcal{O}(\exp(|\Im(k)|r)\,(kr)^{-1}) \tag{2.63}$$

$$H^\omega_{\ell,\eta}(kr) = e^{i\omega[kr - \eta \ln(2kr) - \ell\frac{\pi}{2} + \sigma_\ell(\eta)]}(1 + \mathcal{O}((kr)^{-1})) \tag{2.64}$$

$$F_{\ell,\eta}'(kr) = \cos\left(kr - \eta \ln(2kr) - \ell\frac{\pi}{2} + \sigma_\ell(\eta)\right) + \mathcal{O}(\exp(|\Im(k)|r)\,(kr)^{-1}) \tag{2.65}$$

$$G_{\ell,\eta}'(kr) = -\sin\left(kr - \eta \ln(2kr) - \ell\frac{\pi}{2} + \sigma_\ell(\eta)\right) + \mathcal{O}(\exp(|\Im(k)|r)\,(kr)^{-1}) \tag{2.66}$$

$$H^\omega_{\ell,\eta}{}'(kr) = i\omega \, e^{i\omega[kr - \eta \ln(2kr) - \ell\frac{\pi}{2} + \sigma_\ell(\eta)]}(1 + \mathcal{O}((kr)^{-1})) \,, \tag{2.67}$$

where one must have $\Re(kr) > 0$ or $\Re(kr) < 0$ and $\omega\Im(kr) > 0$. Eqs. (2.62)–(2.67) along with Eqs. (2.33), (2.34), and (2.49) allow to derive similar asymptotic forms in the other quadrants of the complex plane.

Let us consider that $\ell \geq 0$, $v_c \geq 0$ and $k > 0$. One also demands that $\ell$ and $v_c$ are not equal to zero at the same time. In this case, an important region of the real $r$-axis is the vicinity of the Coulomb turning point, that is, the smallest value $r_t(k) > 0$ for which $F''_{\ell,\eta}(kr_t(k)) = 0$ and $G''_{\ell,\eta}(kr_t(k)) = 0$ in Eq. (2.27). $r_t(k)$ is easily obtained from Eq. (2.27):

$$kr_t(k) = \eta + \sqrt{\eta^2 + \ell(\ell + 1)} . \tag{2.68}$$

$r_t(k)$ separates the real $r$-axis in two zones: the non-oscillatory region and the oscillatory region. Indeed, $F_{\ell,\eta}(kr)$ increases and $G_{\ell,\eta}(kr)$ decreases when $0 < r \leq r_t(k)$. This is clearly the case when $r \ll r_t(k)$ (see Eqs. (2.53)–(2.57)). $F_{\ell,\eta}(kr)$ and $G_{\ell,\eta}(kr)$ both oscillate when $r \geq r_t(k)$, and their behavior for $kr \to +\infty$ is provided by Eqs. (2.62), and (2.63).

One will derive approximate forms of Coulomb wave functions when the radius $r$ is close to the turning point $r_t(k)$. This domain of the real $r$-axis is handled by linearizing Eq. (2.27) around $r_t(k)$, which provides with approximate solutions proportional to Airy functions [8]. The procedure is standard, as it allows to connect different WKB approximations, valid for small and large $r$ values, respectively [2]. For $r \sim r_t(k)$, one has [8]:

$$F_{\ell,\eta}(kr) = \sqrt{\pi}\, a^{-1} \left( \mathrm{Ai}\,(a(kr_t(k) - kr)) + \mathcal{O}((1 - r/r_t(k))^2) \right) \tag{2.69}$$

$$G_{\ell,\eta}(kr) = \sqrt{\pi}\, a^{-1} \left( \mathrm{Bi}\,(a(kr_t(k) - kr)) + \mathcal{O}((1 - r/r_t(k))^2) \right) , \tag{2.70}$$

where $\mathrm{Ai}(z)$ and $\mathrm{Bi}(z)$ are the regular and irregular Airy functions, respectively, and $a$ reads:

$$a = \left( k^{-1} r_t(k)^{-1} + k^{-3} r_t(k)^{-3} \ell(\ell + 1) \right)^{1/3} . \tag{2.71}$$

An important consequence of Eqs. (2.69) and (2.70) is that $F_{\ell,\eta}(kr)$ and $G_{\ell,\eta}(kr)$ for $r \sim r_t(k)$ and $k \to 0$ are both $\mathcal{O}(\eta^{1/6})$ when $v_c > 0$. Moreover, as one enters the oscillatory region when $r > r_t(k)$, the amplitude of $F_{\ell,\eta}(kr)$ and $G_{\ell,\eta}(kr)$ cannot be larger than $\mathcal{O}(\eta^{1/6})$ therein, as $\ell(\ell + 1)/r^2 + v_c/r - k^2$ becomes more and more negative when $r$ increases. In fact, the amplitude of Coulomb wave functions eventually becomes of the order of $\mathcal{O}(1)$ when $kr \to +\infty$ (see Eqs. (2.62) and (2.63)). Numerical illustrations of the different behavior of Coulomb wave functions in the complex plane are done in Exercise IV.

---

**Exercise IV $\star$**

    We will numerically calculate Coulomb wave functions in a few examples in order to exhibit their rapid variations and cut discontinuity in the complex plane.

    Run the code calculating Coulomb wave functions for several values of $\ell$ and $\eta$ and plot results. Typical values are $0 \leq \ell \leq 10 - 20$, $0 < \Re[\eta] < 100$, and $0 < |\Im[\eta]| < 5$.

Notice that the Coulomb wave functions can vary by several orders of magnitude in a small region of the complex plane. Check their precision in the considered region of the complex plane with the same code.

Notice the discontinuity of Coulomb wave functions at their cut on the negative real axis. Explain the appearance of discontinuities using the analytic properties of Coulomb wave functions.

## 2.4 Pöschl-Teller-Ginocchio Potential

The solvable Hamiltonians are worth considering, as they provide physical insight that is difficult to access with numerical calculations only. Analytical solutions of the Schrödinger equation have been derived for several nonrelativistic Hamiltonians in Refs. [14–20]. Such solutions exist also for the Klein-Gordon [21] and Dirac equations [22–24]. Solvable Hamiltonians corresponds mainly to one-dimensional potentials. The harmonic oscillator Hamiltonian is the only one-body Hamiltonian which is solvable in three-dimensions for all partial waves. However, if one removes artificially the centrifugal and Coulomb barriers of the potential at infinity, it is possible to devise several Hamiltonians which are fully solvable in the three-dimensional case as well. In the nonrelativistic case, the hypergeometric potentials of Natanzon type are particularly important due to their six defining parameters, allowing these potentials to model a wide range of physical situations [19, 25–27].

The most interesting potential of this kind is the Pöschl-Teller-Ginocchio potential [28–30]. This potential resembles typical nuclear potential and its bound or unbound eigenstates are solvable analytically for all energies [31–34]. Hence, contrary to the harmonic oscillator potential, the Pöschl-Teller-Ginocchio potential yields bound, resonance, and scattering states, so that it can provide with important information about the physics of the continuum.

Therefore, before dealing with the general case of spherical potential introduced in Sect. 2.1, one will study the one-body eigenstates generated by a Pöschl-Teller-Ginocchio potential. As all wave functions and resonant eigenenergies are analytical, the different nature of resonant and scattering wave functions is explicitly from in the formulas providing with radial solutions of the Schrödinger equation. As resonant energies are simple functions of the parameters entering the Pöschl-Teller-Ginocchio potential, one can directly see how bound states can be generated in the presence of a sufficiently binding potential. Added to that, resonance eigenstates can exist only for a certain class of Pöschl-Teller-Ginocchio potentials, which is embedded in a simple inequality involving the parameters of the Pöschl-Teller-Ginocchio potential. Hence, the complex conditions leading to resonance formation can be modeled in a simple analytical manner within the Pöschl-Teller-Ginocchio potential. Hence, as many properties related to bound and continuum states can be analytically expressed using a Pöschl-Teller-Ginocchio potential in Eq. (2.2), its study provides an interesting introduction to the physics of continuum at one-body level before considering the general spherical potential, which will be the topic of the following sections.

The Pöschl-Teller-Ginocchio potential $V_{PTG}$ depends on four parameters, which is standard to denote as $\Lambda$, $s$, $\nu$, and $a$. $\Lambda$ determines the overall shape of the potential. For small $\Lambda$, the Pöschl-Teller-Ginocchio potential is very diffuse. $\Lambda = 1$ corresponds to a standard Pöschl-Teller potential. If $\Lambda$ is large, a flat bottom is generated in the Pöschl-Teller-Ginocchio potential. $s$ is a scaling parameter as $V_{PTG}(r) \propto s^2$. $\nu$ fixes the depth of the potential and $a$ is the strength of the effective mass, with $0 \leq a \leq 1$. The effective mass at $r = 0$ is given by $(1 - a)m_0$.

These four parameters allow the Pöschl-Teller-Ginocchio potential to cover many physical situations. While $\Lambda \leq 1$ provides with a potential mimicking that of a crystal acting on an electron, potentials with $\Lambda \sim 5$ are very close to a Woods-Saxon potential and hence are more suitable for the study of nuclei. The parameters $s$ and $\nu$ play the same role as the Woods-Saxon potential depth, so that all the situations that can be described with a Woods-Saxon potential can also be modeled by a Pöschl-Teller-Ginocchio potential. However, the $\Lambda$, $s$, and $\nu$ parameters are not as intuitive as the diffuseness, radius, and depth of the Woods-Saxon potential, so that the physical properties of the Pöschl-Teller-Ginocchio potential are rather implicit. Consequently, in practical calculations, the $\Lambda$, $s$, and $\nu$ parameters of the Pöschl-Teller-Ginocchio potential are typically fitted from a given Woods-Saxon potential.

The Schrödinger equation with Pöschl-Teller-Ginocchio potential [29] reads:

$$\left[ \frac{\hbar^2}{2m_0} \left( -\frac{d}{dr} \frac{1}{\mu(r)} \frac{d}{dr} + \frac{\ell(\ell+1)}{r^2\mu(r)} \right) + V_{PTG}(r) \right] u(r) = e\, u(r), \quad (2.72)$$

where $m_0$ is the mass of the particle, $\mu(r)$ is the dimensionless effective mass, $\ell$ is the orbital angular momentum, and $V_{PTG}(r)$ is the Pöschl-Teller-Ginocchio potential. Eq. (2.72) closely resembles the Schrödinger equation with a Woods-Saxon potential used along with an effective mass. Consequently, it is expected that the Pöschl-Teller-Ginocchio potential may provide useful insight into realistic nuclear wave functions for both bound and scattering states.

$\mu(r)$ and $V_{PTG}(r)$ in (2.72) are expressed through the variable $y$ which depends on $r$ and is defined by way of an implicit equation [28, 29]:

$$\Lambda^2 s\, r = \mathrm{arctanh}(y) + \sqrt{\Lambda^2 - 1}\ \arctan(\sqrt{\Lambda^2 - 1}\ y), \Lambda > 1 \quad (2.73)$$

$$= \mathrm{arctanh}(y) - \sqrt{1 - \Lambda^2}\ \mathrm{arctanh}(\sqrt{1 - \Lambda^2}\ y), \Lambda \leq 1. \quad (2.74)$$

The parameter $y(r)$ defined in Eqs. (2.73) and (2.74) can be expressed in closed form only for the Pöschl-Teller potential, where $\Lambda = 1$, so that $y(r) = \tanh(sr)$. In other situations, $y(r)$ must be calculated numerically. However, the numerical calculation of $y(r)$ is not straightforward and demands the use of its asymptotes at small and large $r$. Consequently, one will firstly describe the properties of the

$y$ variable. The expressions of $\mu(r)$ and of the Pöschl-Teller-Ginocchio potential $V_{PTG}(r)$ will be provided afterwards. The methods used to determine $y(r)$ are complicated, nevertheless, so that readers can omit to read the next section if they are not interested in this calculation.

## 2.4.1 Calculation of the Function y(r)

One has to solve two problems in order to calculate $y(r)$ with Eqs. (2.73) and (2.74). The first one arises from the rapid variations of $\mathrm{arctanh}(y)$ when $y \sim 1$, occurring for large $r$. The second problem is induced by the fact that $y \to 1$ when $r \to +\infty$, because a direct computation of $1 - y^2$ then becomes incorrect. In fact, a precise value of $1 - y^2$ is necessary to calculate precisely $\mu(r)$ and $V_{PTG}(r)$ (see Eqs. (2.88) and (2.92)).

One will concentrate on the numerical implementation of $y(r)$. Indeed, while $r$ is immediate to calculate when $y$ is fixed (see Eqs. (2.73) and (2.74)), there is no closed formula for the function $y(r)$, which must be calculated numerically. Even though no formula exists for $y(r)$ in the general case, the minimal and maximal values of $y(r)$, belonging to the $[0:1)$ interval, can be analytically determined (see Exercise V).

Asymptotic expansions of $y(r)$ when $y \sim 0$ or $y \sim 1$ can be derived (see Eqs. (2.85)–(2.87) in Exercise V), and $y'(r)$ has an analytical expression (see Eqs. (2.73) and (2.74)). Consequently, it is possible to use the Newton method in order to calculate $y(r)$ numerically. In this procedure, one uses the following starting point, denoted as $y_{st}$:

$$y_{st} = y_d \quad \text{if} \quad y_d > 0.5 \ , \quad \Lambda > 1 \tag{2.75}$$

$$= y_e \quad \text{if} \quad y_e > 0.5 \ , \quad \Lambda \leq 1 \tag{2.76}$$

$$= \min(sr, 0.99) \quad \text{otherwise} . \tag{2.77}$$

It can happen in rare cases that the Newton method fails to converge, that is, one obtains $y < 0$ or $y > 1$ during iterations. In this case, $y(r)$ is recalculated using bisection in the interval $[y_d : y_e]$.

The latter procedure is not sufficiently precise if $y \sim 1$. Hence, if the Newton method provides with a $y$ value verifying $|y - 1| < 10^{-5}$, $y$ must be refined with another method. For this, one calculates instead $x = \mathrm{arctanh}(y)$ using an iterative fixed-point algorithm:

$$x_n = \Lambda^2 sr - \sqrt{\Lambda^2 - 1} \ \arctan(\sqrt{\Lambda^2 - 1} y_n) \ , \quad \Lambda > 1 \tag{2.78}$$

$$= \Lambda^2 sr + \sqrt{1 - \Lambda^2} \ \mathrm{arctanh}(\sqrt{1 - \Lambda^2} y_n) \ , \quad \Lambda \leq 1 \tag{2.79}$$

$$y_{n+1} = \tanh(x_n) \ , \quad n \geq 0 \ , \tag{2.80}$$

and $y_0 = 1$ as a starting point. As $y \sim 1$, $x_n \to x$ and $y_n \to y$ rapidly for $n \to +\infty$. Added to that, $1 - y^2$ can be precisely evaluated with this algorithm, because $1 - y^2 = \dfrac{4e^{-2x}}{(1+e^{-2x})^2}$, which is clearly devoid of numerical inaccuracy when $y \sim 1$.

---

**Exercise  V ★ ★ ★**

Using Eqs. (2.73) and (2.74), show that $y$ belongs to the interval $[y_d : y_e]$, where $y_d$ and $y_e$ read:

$$y_d = \max(\tanh(\Lambda^2 sr - \sqrt{\Lambda^2 - 1}\ \arctan(\sqrt{\Lambda^2 - 1})), 0)\,, \quad \Lambda > 1 \quad (2.81)$$

$$= \tanh(\Lambda^2 sr)\,, \quad \Lambda \le 1. \tag{2.82}$$

$$y_e = \tanh(\Lambda^2 sr)\,, \quad \Lambda > 1 \tag{2.83}$$

$$= \tanh(\Lambda^2 sr + \sqrt{1 - \Lambda^2}\ \mathrm{arctanh}(\sqrt{1 - \Lambda^2}))\,, \quad \Lambda \le 1. \tag{2.84}$$

For this, one will replace $\arctan(\sqrt{\Lambda^2 - 1}\ y)$ and $\mathrm{arctanh}(\sqrt{1 - \Lambda^2}\ y)$ in Eqs. (2.73) and (2.74) by values independent of $y$. Eqs. (2.73) and (2.74) then become inequalities which can be solved analytically in $y$.

---

**Exercise  VI ★ ★ ★**

Show that $y(r)$ verifies simple expansions for $y \sim 0$ and $y \sim 1$:

$$y(r) = sr + \mathcal{O}(y^3)\,, \quad r \to 0 \tag{2.85}$$

$$= y_d + \mathcal{O}\left((1 - y)^2\right)\,, \quad r \to +\infty\,, \quad \Lambda > 1 \tag{2.86}$$

$$= y_e + \mathcal{O}\left((1 - y)^2\right)\,, \quad r \to +\infty\,, \quad \Lambda \le 1. \tag{2.87}$$

---

### 2.4.2   Different Terms of the Pöschl-Teller-Ginocchio Potential

It is convenient to write $V_{PTG}(r)$ from the sum of its $V_\mu(r)$, $V_\ell(r)$ and $V_c(r)$ potential parts, where $V_\mu(r)$ is proportional to $a$, $V_\ell(r)$ is the $\ell$-dependent part, and $V_c(r)$ is the principal central part. Along with the effective mass $\mu(r)$, the latter potential parts and hence $V_{PTG}(r)$ read [29]:

$$\mu(r) = 1 - a(1 - y^2) \tag{2.88}$$

$$V_\mu(r) = \left[1 - a + \left[a(4 - 3\Lambda^2) - 3(2 - \Lambda^2)\right]y^2 - (\Lambda^2 - 1)(5(1 - a) + 2ay^2)\ y^4\right]$$
$$\times \frac{a}{\mu(r)^2}(1 - y^2)\left[1 + (\Lambda^2 - 1)y^2\right] \tag{2.89}$$

$$V_\ell(r) = \ell(\ell+1)\left[\frac{(1-y^2)(1+(\Lambda^2-1)y^2)}{y^2} - \frac{1}{s^2r^2}\right], \quad r > 0 \tag{2.90}$$

$$V_c(r) = (1-y^2)\left[-\Lambda^2 v(v+1) - \frac{\Lambda^2-1}{4}\left(2 - (7-\Lambda^2)y^2 - 5(\Lambda^2-1)y^4\right)\right] \tag{2.91}$$

$$V_{\text{PTG}}(r) = \frac{\hbar^2 s^2}{2m_0\mu(r)}\left(V_\mu(r) + V_\ell(r) + V_c(r)\right) \tag{2.92}$$

The value of $V_\ell(r)$ provided by Eq. (2.90) becomes numerically unstable when $r \to 0$. It is possible to suppress this numerical inaccuracy by devising a power series representation of $V_\ell(r)$ (see Eqs. (2.73), (2.74), and (2.90)):

$$V_\ell(r) = \ell(\ell+1)\left[\frac{y}{\Lambda^2 sr}\left(1 + \frac{y}{sr}\right)P_\Lambda(y) + \Lambda^2 - 1 - (1 + (\Lambda^2-1)y^2)\right], \quad r > 0 \tag{2.93}$$

$$P_\Lambda(y) = \sum_{n=0}^{+\infty} \frac{1 - (1-\Lambda^2)^{n+2}}{2n+3} y^{2n} \tag{2.94}$$

$$V_\ell(0) = \ell(\ell+1)\left(\frac{\Lambda^2-2}{3}\right) \tag{2.95}$$

If $r = 0$, one uses directly Eq. (2.95) which is providing with $V_\ell(0)$. If $r > 0$ and if the parameters entering the arctangent and hyperbolic arctangent functions in Eqs. (2.73) and (2.74) are smaller than 0.01, the power series $P_\Lambda(y)$ of Eq. (2.94) converges quickly, so that Eq. (2.93) is the method of choice to calculate $V_\ell(r)$ in this situation. In all other cases, one can use the initial expression of $V_\ell(r)$ of Eq. (2.90), as it is sufficiently precise in practice. $V_\ell(r)$ can then be very precisely calculated $\forall r \geq 0$.

---

**Exercise  VII ⋆**

One will exhibit the main limitation of the Pöschl-Teller-Ginocchio potential, namely that its centrifugal potential has unphysical properties.

Show that the centrifugal potential of $V_{\text{PTG}}(r)$ is only approximate, in the sense that $V_{\text{PTG}}(r) + \ell(\ell+1)/r^2$ behaves like $\ell(\ell+1)/r^2$ for $r \to 0$ but vanishes exponentially for $r \to +\infty$.

Deduce that all bound and resonance states are independent of $\ell$ in the asymptotic region.

Show that all bound and resonance states then behave as if they belonged to the neutron $s$-partial wave in the asymptotic region.

---

The fast decay of the centrifugal part of the Pöschl-Teller-Ginocchio potential is common to all hypergeometric potentials, as it is related to Eqs. (2.73) and (2.74).

Indeed, in order to transform the initial Schrödinger equation of Eq. (2.72) into a hypergeometric equation, all parts of the Pöschl-Teller-Ginocchio potential in Eq. (2.92) must be rational functions of $y$. This prevents the existence of both a centrifugal potential in $1/r^2$ and a Coulomb potential in $1/r$ in the Pöschl-Teller-Ginocchio potential (see Exercise VII).

This is the fundamental deficiency of the Pöschl-Teller-Ginocchio potential, which otherwise would be as general as the Woods-Saxon potential. In fact, the only known analytical wave functions, which are able to accommodate an exact centrifugal potential, are the harmonic oscillator wave functions, the modified Bessel functions, and the Coulomb wave functions. One will nevertheless see at the end of this chapter that adding a proper centrifugal+Coulomb part to the Pöschl-Teller-Ginocchio potential leads to a quasi-analytical potential whose bound, resonance, and scattering states bear physical asymptotes.

### 2.4.3   Wave Functions of the Pöschl-Teller-Ginocchio Potential

The general expression of Pöschl-Teller-Ginocchio wave functions for resonances and scattering states, respectively, reads [29]:

$$\phi(r) = \mathscr{N} \; A_\mu(r) B^+(r) F_0(r) \tag{2.96}$$

$$= \mathscr{N} \; A_\mu(r) \left( \mathscr{C}^+ B^+(r) F^+(r) + \mathscr{C}^- B^-(r) F^-(r) \right), \tag{2.97}$$

where $\mathscr{N}$ is a normalization constant. The functions in Eq. (2.97) read:

$$B^\pm(r) = f^{\pm \frac{\bar{\beta}}{2}} \tag{2.98}$$

$$A_\mu(r) = \sqrt{\mu(r)} \; (g + \Lambda^2 f)^{1/4} \; g^{\frac{\ell+1}{2}} \tag{2.99}$$

$$F_0(r) = {}_2F_1 \left( v^-, v^+, \ell + \frac{3}{2}; g \right) \tag{2.100}$$

$$F^+(r) = {}_2F_1 \left( v^-, v^+, 1 + \bar{\beta}; f \right) \tag{2.101}$$

$$F^-(r) = {}_2F_1 \left( \mu^-, \mu^+, 1 - \bar{\beta}; f \right), \tag{2.102}$$

where the used parameters and functions are defined in the following way:

$$\bar{\beta} = -\frac{ik}{\Lambda^2 s}, \; v^\pm = \frac{1}{2} \left( \ell + \frac{3}{2} + \bar{\beta} \pm \bar{v} \right),$$

$$\mu^\pm = \frac{1}{2} \left( \ell + \frac{3}{2} - \bar{\beta} \pm \bar{v} \right) \tag{2.103}$$

$$\bar{v} = \sqrt{\left(v + \frac{1}{2}\right)^2 - \bar{\beta}^2 \left(\Lambda^2(1-a) - 1\right)} \qquad (2.104)$$

$$\mathscr{C}^+ = \frac{\Gamma(\ell + \frac{3}{2})\Gamma(-\bar{\beta})}{\Gamma(\mu^+)\Gamma(\mu^-)}, \qquad \mathscr{C}^- = \frac{\Gamma(\ell + \frac{3}{2})\Gamma(\bar{\beta})}{\Gamma(\nu^+)\Gamma(\nu^-)} \qquad (2.105)$$

$$f = \frac{1 - y^2}{1 - y^2 + \Lambda^2 y^2}, \qquad g = \frac{\Lambda^2 y^2}{1 - y^2 + \Lambda^2 y^2} \qquad (2.106)$$

The $r$-dependence in these equations arises implicitly from the $f$ and $g$ functions, depending on the $y(r)$ parameter of Eqs. (2.73) and (2.74). Both Eqs. (2.96) and (2.97) define the same $\phi(r)$ wave function. However, it is necessary to use both these equations in practice, because formulas (2.96), and (2.97) are numerically stable only for small and large $r$, respectively. Moreover, Eq. (2.97) allows to explicitly identify the outgoing ("+" terms) and incoming parts ("−" terms) of the $\phi(r)$ wave function.

Scattering states are normalized with the Dirac delta normalization. The $\mathscr{N}$ normalization then reads:

$$\mathscr{N} = \left(\frac{1}{\Gamma(\ell + \frac{3}{2})}\right)$$

$$\times \sqrt{\frac{2\Lambda^2 s\bar{\beta}\,(\ell + \frac{3}{2} + \bar{\beta} + 2n)\,\Gamma(\ell + \frac{3}{2} + \bar{\beta} + n)\,\Gamma(\ell + \frac{3}{2} + n)}{(\ell + \frac{3}{2} + \bar{\beta}\Lambda^2(1-a) + 2n)\,\Gamma(n+1)\,\Gamma(\bar{\beta} + n + 1)}} \qquad \text{(poles)}$$

$$\qquad (2.107)$$

$$= \left(\frac{1}{\Gamma(\ell + \frac{3}{2})}\right)\sqrt{\frac{\Gamma(\nu^+)\Gamma(\nu^-)\Gamma(\mu^+)\Gamma(\mu^-)}{2\pi\,\Gamma(\bar{\beta})\Gamma(-\bar{\beta})}} \qquad \text{(scattering states)}.$$

$$\qquad (2.108)$$

The expressions resulting from Eqs. (2.96) and (2.97) can be obtained by successive changes of variables in the initial Schrödinger equation (2.72) so that it becomes a hypergeometric equation. Its solutions are Jacobi polynomials in the bound case and hypergeometric functions otherwise. The normalization constant of Eq. (2.107) arises from an integral of Jacobi polynomials multiplied by elementary power function, which is analytical [28, 29], while that of Eq. (2.108) is straightforward to obtain with the Dirac delta normalization. The demonstration of Eqs. (2.96) and (2.97) is done in Exercise VIII. Derivations therein are, however, tedious, so that they can be omitted in a first reading. Equations (2.96) and (2.97) can also be derived with path integrals [30].

**Exercise VIII ★ ★ ★**

One will show that the analytical formulas of Eqs. (2.96) and (2.97) are indeed solving Eq. (2.72).

Show that the wave function defined in Eq. (2.96) is a solution of Eq. (2.72) by a direct insertion of Eq. (2.96) in Eq. (2.72).

Deduce Eq. (2.97) from Eq. (2.96) using the following standard property of the hypergeometric function:

$$
{}_2F_1(a, b, c; z) = \frac{\Gamma(c)\Gamma(c - a - b)}{\Gamma(c - a)\Gamma(c - b)} \, {}_2F_1(a, b, a + b - c + 1; 1 - z)
$$

$$
+ (1 - z)^{c - a - b} \frac{\Gamma(c)\Gamma(a + b - c)}{\Gamma(a)\Gamma(b)} \tag{2.109}
$$

$$
\times {}_2F_1(c - a, c - b, c - a - b + 1; 1 - z)
$$

$\phi(r)$ has analytic asymptotic expressions for $r \to 0$ or $r \to +\infty$, which arise from Eqs. (2.85)–(2.87) and Eqs. (2.96) and (2.97):

$$
\phi(r) \sim \mathcal{N} \sqrt{\Lambda(1 - a)} \, (\Lambda s)^{\ell+1} \, r^{\ell+1} = C_0 \, r^{\ell+1} \, , r \to 0 \, , \tag{2.110}
$$

$$
\phi(r) \sim \mathcal{N} \, \mathscr{C}^+ e^{ik(r - r_1)} + \mathcal{N} \, \mathscr{C}^- e^{-ik(r - r_1)}
$$

$$
= C^+ e^{ikr} + C^- e^{-ikr} \, , r \to +\infty \tag{2.111}
$$

with:

$$
\Lambda^2 s \, r_1 = \sqrt{\Lambda^2 - 1} \, \arctan(\sqrt{\Lambda^2 - 1}) - \ln\frac{\Lambda}{2} \, , \quad \Lambda > 1 \tag{2.112}
$$

$$
= -\sqrt{1 - \Lambda^2} \, \mathrm{arctanh}(\sqrt{1 - \Lambda^2}) - \ln\frac{\Lambda}{2} \, , \quad \Lambda \leq 1 \tag{2.113}
$$

### 2.4.3.1 Energies of Bound, Antibound, and Resonance States

The Pöschl-Teller-Ginocchio potential possesses bound, antibound, virtual, and resonance states (see definition of unbound states in Sect. 2.6.6), whose energies are expressible in a closed form [29]. The linear momentum $k$ and energy $e$ of each state:

$$
k = is\left(\frac{-\left(N + \frac{1}{2}\right) \pm \sqrt{\Delta}}{1 - a}\right), \tag{2.114}
$$

$$
e = \left(\frac{\hbar^2 s^2}{2m_0(1 - a)^2}\right)\left(-\left(N + \frac{1}{2}\right)^2 - \Delta \pm (2N + 1)\sqrt{\Delta}\right) \tag{2.115}
$$

$$
\Delta = \Lambda^2\left(v + \frac{1}{2}\right)^2(1 - a) - \left[(1 - a)\Lambda^2 - 1\right]\left[N + \frac{1}{2}\right]^2 \tag{2.116}
$$

depend on the value of the integer $N = 2n + \ell + 1$, where $n$ is its radial quantum number (see Exercise IX. Note that this exercise is rather difficult and can be omitted during first reading).

---

**Exercise IX ★ ★ ★**

Demonstrate Eqs. (2.114), (2.115), and (2.116).

---

The nature of the state depends on the following boundary values:

$$B_\Lambda = \Lambda \sqrt{\frac{1-a}{\Lambda^2(1-a)-2}} \left(\nu + \frac{1}{2}\right) - \frac{1}{2}, \quad \Lambda > \sqrt{\frac{2}{1-a}} \tag{2.117}$$

$$= +\infty, \quad \Lambda \leq \sqrt{\frac{2}{1-a}} \tag{2.118}$$

$$B_\nu = [\nu], \quad \nu \in \mathbb{R}, \nu \notin \mathbb{N} \tag{2.119}$$

$$B_\nu = \nu - 1, \quad \nu \in \mathbb{N} \tag{2.120}$$

Below, we will show that bound states occur when $N \leq B_\Lambda$ and $N \leq B_\nu$. Antibound states and complex states of negative real energy occur for $N \leq B_\Lambda$ and $N > B_\nu$ ($N > \nu$ if $\nu \in \mathbb{N}$). Resonances of positive energy and width are found for $N > B_\Lambda$. The $\pm$ sign in Eqs. (2.114) and (2.115) is "+" for bound states and antibound states of physical importance, that is, whose linear momentum $k$ is close to the real $k$-axis, and states of negative real energy and width. It is "−" for resonance, other antibound states, and virtual states.

One will first consider the case $\Delta < 0$ in Eq. (2.116). Using Eq. (2.117) and assuming that $\Lambda^2(1 - a) > 2$, the real and imaginary parts of Eq. (2.115) read:

$$\Re(e) = \left(\frac{\hbar^2 s^2}{2m_0(1-a)^2}\right)\left(\Lambda^2(1-a) - 2\right)(B_\Lambda + N + 1)(N - B_\Lambda) \tag{2.121}$$

$$\Im(e) = \pm \left(\frac{\hbar^2 s^2}{2m_0(1-a)^2}\right)(2N+1)\sqrt{-\Delta}. \tag{2.122}$$

Hence, $\Re(e) > 0$ occurs only if $\Lambda^2(1 - a) > 2$ and $N > B_\Lambda$, and $\Im(e) < 0$ can happen only in the class of states denoted by the "−" sign in Eqs. (2.114) and (2.115). Thus one obtains resonances of positive energy and width if $N > B_\Lambda$ and $\Lambda^2(1 - a) > 2$. The class of states denoted by "+" sign contains capturing resonances, which are a complex conjugate of the physical decaying resonances. All cases different from physical resonances clearly bear $N \leq B_\Lambda$.

The only possibility for $k$ to lie on the positive imaginary axis is to have $\Delta > (N + 1/2)^2$ while using the "+" sign, as can be seen from Eq. (2.114). This directly translates in the $N < \nu$ equation, where $N = \nu$ is suppressed for $\nu \in \mathbb{N}$ as it provides the value $k = 0$ with which the Pöschl-Teller-Ginocchio wave function vanishes identically (see Eq. (2.107)). $N > \nu$ and $N \le B_\Lambda$ thus provide with all other types of states.

Equation (2.114) implies that $k$ lies on the negative imaginary axis if $0 \le \Delta < (N + 1/2)^2$ and if $k$ is closest to the real axis by using the "+" sign. Hence, an antibound state occurs therein. There is another antibound state for "−" sign, but it is usually far from the real axis as $\sqrt{\Delta}$ cannot partially cancel with $N + 1/2$.

The last remaining case of complex states of negative energy and positive width occurs for $N > \nu$, $N \le B_\Lambda$, and $\Delta < 0$ either using the "+" or "−" sign in Eqs. (2.121) and (2.122). Even though these states do not represent unbound states, they can have a physical importance and they can influence cross sections if their linear momentum is sufficiently close to the real $k$-axis (see Sect. 4.2).

One may notice that all states bearing $2n + \ell + 1 = N$ are degenerate, as for the harmonic oscillator potential. This is due to the SU(3) symmetry obeyed by the Pöschl-Teller-Ginocchio potential. Moreover, one can see in Eqs. (2.117), (2.114), and (2.115) that $\Lambda = \sqrt{2/(1-a)}$ is a critical value for resonances. Indeed, on the one hand, no resonance can exist if $\Lambda \le \sqrt{2/(1-a)}$ as $B_\Lambda = +\infty$ in this case. On the other hand, resonance states exist when $\Lambda > \sqrt{2/(1-a)}$. This is consistent with the fact that the Pöschl-Teller-Ginocchio potential can properly describe a nucleus if $\Lambda$ is sufficiently large. This also points out to the importance of flat bottom of a potential for a purpose of generating the resonances.

### 2.4.4 Modified Pöschl-Teller-Ginocchio Potential

The Pöschl-Teller-Ginocchio potential cannot provide physical values of widths due to the exponential decrease of its centrifugal part. In the following, one will discuss a modification of the Pöschl-Teller-Ginocchio potential which yields a proper centrifugal+Coulomb asymptote. Let us consider the potential:

$$V_{PTG-mod}(r) = V_{PTG}(r) + E_b , \quad 0 < r < r_{as} \tag{2.123}$$

$$V_{PTG-mod}(r) = \frac{v_c}{r} , \quad r > r_{as} \tag{2.124}$$

where the barrier energy $E_b$ is fixed and the radius defining the asymptoting region $r_{as}$ is chosen so that $V_{PTG-mod}(r)$ is continuous in $r = r_{as}$. In order to determine the parameters entering $V_{PTG-mod}(r)$, one considers a Woods-Saxon potential with the centrifugal part included,

$$\frac{\ell(\ell + 1)}{r^2} + \frac{v_c}{r} + V_{WS}(r)$$

and one finds its maximal value in the asymptotic zone, which is the barrier energy $E_b$. Parameters of the Pöschl-Teller-Ginocchio potential $V_{PTG}(r)$ of Eq. (2.123) (see Eq. (2.88)–(2.92)) are defined so that the Pöschl-Teller-Ginocchio potential closely resembles the Woods-Saxon potential $V_{WS}(r)$.

The wave function $u(k, r)$ is provided by Eq. (2.96) for $0 < r < r_{as}$, while one must use Eq. (2.7) to obtain $u(k, r)$ when $r > r_{as}$. As $V_{PTG-mod}(r)$ is continuous for all radii, Eq. (2.2) implies that $u(k, r)$, $u'(k, r)$, and $u''(k, r)$ are all continuous.

As $V_{PTG-mod}(r)$ is not differentiable in $r = r_{as}$, $u''(k, r)$ is not differentiable therein as well, so that the third and higher derivatives of $u(k, r)$ are all discontinuous in $r = r_{as}$. The Jost functions read:

$$J^{\pm}(k) = u'_{PTG}(r_{as})H^{\pm}_{\ell,\eta}(kr_{as}) - u_{PTG}(r_{as})kH^{\pm}_{\ell,\eta}{}'(kr_{as}) . \qquad (2.125)$$

Its second-order Taylor expansion is:

$$J^{\pm}(k + \delta k) \simeq J^{\pm}(k) + \delta k\, J^{\pm\,'}(k) + \frac{\delta k^2}{2} J^{\pm\,''}(k) , \qquad (2.126)$$

so that the equation $J^{\pm}(k + \delta k) = 0$ leads to the second-order value of $\delta k$:

$$\delta k = -\frac{J^{\pm\,'}(k)}{J^{\pm\,''}(k)} \left( 1 - \sqrt{1 - 2\frac{J^{\pm}(k)J^{\pm\,''}(k)}{J^{\pm\,'}(k)^2}} \right) . \qquad (2.127)$$

The choice of $k$ associated to $E = E_{PTG} + E_b$ corresponds to the eigenstate of $V_{PTG}(r) + E_b$, which is the closest analytical potential to $V_{PTG-mod}(r)$. If one considers a narrow resonance, one has $E < E_b$ so that the zeroth-order approximation of the exact eigenstate is bound, hence without a width. But, as $E > 0$, one has $k > 0$ and thus $H^{+}_{\ell,\eta}(kr_{as})$ is complex. Consequently, $\delta k$ is complex in Eq. (2.127) and provides a nonzero approximation for the width.

The second-order approximation of energies works well except very close or very far from the particle-emission threshold. Indeed, in this latter case the width becomes very large and the width arising from the second-order expansion can no longer be correct. Even though the second-order approximation of energies close to zero energy is usually good, the width might become imprecise as it is very small. Here, it is better to calculate energies numerically but to evaluate widths with the current formula (2.196). The eigenenergies and phase shifts of the modified Pöschl-Teller-Ginocchio potential are studied in numerical examples in Exercise X.

---

**Exercise  X ★**

   In this exercise, we will concentrate on the ability of the modified Pöschl-Teller-Ginocchio potential to provide with physical values of eigenenergies and phase shifts.

**A.** Calculate analytically the phase shifts of the scattering states generated by the $V_{PTG}(r)$ (see Eq. (2.92)).

**B.** Run the one-particle code of radial wave functions to calculate phase shifts numerically for the exact $V_{PTG}(r)$ and modified $V_{PTG-mod}(r)$ Pöschl-Teller-

Ginocchio potentials (see Eqs. (2.92), (2.123), and (2.124). Compare them to those obtained with a Woods-Saxon potential using the Woods-Saxon code.

**C.** Run the one-particle code of radial wave functions to calculate energies and widths of resonances with both the second-order approximation for $V_{PTG-mod}(r)$ and with a direct integration of Eq. (2.2). From numerical calculations, determine the typical domain of validity of the second-order approximation.

**D.** Locate the phase shift changes in the obtained figure and relate them to energies and widths of calculated resonances.

## 2.5    Basic Properties of Bound States

From a mathematical and physical point of view, bound one-body states are important, as they are related to the properties of localized quantum systems. Moreover, these states must always be included in the set of states entering the completeness relations generated by a one-body Hamiltonian (2.3). Due to the special role played by bound states in the eigenspectrum of one-body Hamiltonians, we will state the general properties of bound states and discuss the most important of them. We will also show that bound states form a discrete set of states and that they are orthogonal to each other. These properties are fundamental, as they are directly related to the fact that the experimental spectrum of bound levels of the quantum system is discrete.

Another feature associated to bound eigenstates is that there is a finite number of them if their generating potential is short-range or repulsive at large distances. Conversely, one has an infinity of bound eigenstates accumulating at zero energy if their generating potential is of infinite range and attractive, as is the case of the hydrogen spectrum. One will also demonstrate this property due to its mathematical and physical importance. Finally, one will state the virial theorem associated to the bound eigenstates of the Hamiltonian of Eq. (2.3). One will see that this theorem allows to explain why the hydrogen atom is stable, on the one hand, and, on the other hand, why halos and resonance states of energy close to zero can develop in atomic nuclei.

Due to the $k \leftrightarrow -k$ symmetry of Eq. (2.2), it is sufficient to consider the upper half of the complex plane. Thus, one can deduce from Eq. (2.7) that $u(k, r)$ is a bound state only if $C^- = 0$, as $H^{\pm}_{\ell, \eta}(kr) \sim \exp(\pm(ikr - \eta \ln(2kr)))$ up to a function of $kr$ behaving as a rational function for $k \neq 0$ and $r \to +\infty$ (see Sect. 2.3). Hence, both $\int |u(k, r)|^2 dr$ and integrals of $u(k, r)$ and of its derivatives are finite as well. As $k^2$ is real for bound states (see Exercise XI), one can always write $k = i\kappa$, with $\kappa \geq 0$. This notation will be commonly used in the following. For real $k^2$, Eq. (2.2) involves only real numbers. Hence, $u(k, r)$ can be always chosen so that it is a real-valued function.

> **Exercise XI** ⋆
>     Show that for the Hermitian Hamiltonian (2.5), $k^2$ has to be real if $u(k, r)$ is a bound state.

It is sufficient to study the asymptotic behavior of wave functions at infinity to be able to know whether a one-body system can sustain a zero-energy bound state or not. In the neutron case, the solutions of Eq. (2.5) for $k = 0$ read (see Eqs. (2.53) and (2.54)):

$$u(k, r) = Ar^{\ell+1} + Br^{-\ell}, \quad r \geq R. \tag{2.128}$$

If $\ell = 0$, then Eq. (2.128) never vanishes for $r \to +\infty$, so that no bound state of zero energy can exist therein. Conversely, if $A = 0$ in (2.128) when $\ell > 0$, then $u(k, r)$ is the square-integrable wave function of a bound state.

For proton states, solutions of Eq. (2.5) when $k = 0$ are provided by Eqs. (2.56) and (2.57)):

$$u(k, r) = A\, r^{1/4} \, (2\sqrt{v_c r})\, i_{2\ell+1/2}(2\sqrt{v_c r})$$
$$+ B\, r^{1/4} \, (2\sqrt{v_c r})\, k_{2\ell+1/2}(2\sqrt{v_c r}), \quad r \geq R. \tag{2.129}$$

In the above equation, $(2\sqrt{v_c r})\, k_{2\ell+1/2}(2\sqrt{v_c r}) \sim \exp(-2\sqrt{v_c r})$ up to an unimportant constant when $r \to +\infty \; \forall \ell \geq 0$. Therefore, $u(k, r)$ is a bound state if $A = 0$. Charged particles can then bear zero-energy bound states in all partial waves.

## 2.5.1 Number of Bound States of a One-Body Hamiltonian

One will show now that if $v_c \geq 0$ $u(0, r)$ is bound, then the number of bound states is finite and equals to the number of zeros of the $u(0, r)$, including the zero at infinity. $u(0, r)$ is either a zero-energy bound state (see Eqs. (2.128) and (2.129)) or $|u(0, r)| \to +\infty$ for $r \to +\infty$, due to the properties of Coulomb wave functions for $k \to 0$. Thus, $u(0, r)$ cannot oscillate along the real $r$-axis in the asymptotic region, so that its number of zeros is necessarily finite.

To show that the number of bound states is finite, one needs to consider the partial overlaps of two different states $u(k_a, r)$ and $u(k_b, r)$, not necessarily bound, but with $k_a^2 < k_b^2 \leq 0$. Note that no complex conjugate of wave functions will enter radial overlaps because $u(k_a, r)$ and $u(k_b, r)$ wave functions can be chosen to bear real values. A straightforward manipulation of Eq. (2.2) provides with the following equation:

$$\int_{r_1}^{r_2} \left( u''(k_b, r)u(k_a, r) - u''(k_a, r)u(k_b, r) \right) \, dr$$
$$= (k_a^2 - k_b^2) \int_{r_1}^{r_2} u(k_a, r)u(k_b, r) \, dr$$

$$\Rightarrow W(u(k_b, r_2), u(k_a, r_2)) - W(u(k_b, r_1), u(k_a, r_1))$$

$$= -(\kappa_a^2 - \kappa_b^2) \int_{r_1}^{r_2} u(k_a, r) u(k_b, r) \, dr \tag{2.130}$$

involving $u(k_a, r)$ and $u(k_b, r)$ for the two radii $0 \le r_1 < r_2$. In this equation, $W(u(k_b, r), u(k_a, r))$ is the Wronskian of $u(k_a, r)$ and $u(k_b, r)$:

$$W(u(k_b, r), u(k_a, r)) = u'(k_b, r) u(k_a, r) - u'(k_a, r) u(k_b, r) \, . \tag{2.131}$$

Taking $r_1 = 0$ and $r_2 = +\infty$ for two different (real) bound states $u(k_a, r)$ and $u(k_b, r)$, one obtains their standard orthogonality relation:

$$\int_0^{+\infty} u(k_a, r) u(k_b, r) \, dr = 0 \, , \tag{2.132}$$

where one has used the fact that $u(k_a, 0) = 0$ and $u(k_a, r) \to 0$ for $r \to +\infty$, and the same for $k_b$.

One can deduce from Eq. (2.132) that the set of bound states is discrete. Indeed, if one has arbitrarily close bound states $u(k_a, r)$ and $u(k_b, r)$, with $k_a$ fixed, their overlap can be made arbitrarily close to $\int |u(k_a, r)|^2 \, dr > 0$ by continuity of $u(k_b, r)$ with $k_b$ (see Exercise XII), what is contradictory to their orthogonality. We will show in this case that the zeros of the $u(k, r)$ functions interlace, which means that between two zeros of $u(k_a, r)$, $u(k_b, r)$ there is at least one zero.

Let $r_1$ and $r_2$ be two consecutive zeros of $u(k_a, r)$, which is assumed to be strictly positive in an interval $]r_1 : r_2[$. Hence, as the zeros of $u(k_a, r)$ are simple, which arises from the fact that $u(k_a, r)$ is a solution of a second-order differential equation, the continuity arguments imply that $u'(k_a, r_1) > 0$ and $u'(k_a, r_2) < 0$. One will assume that also $u(k_b, r) > 0$ in this region. Consequently, Eq. (2.130) becomes:

$$-u'(k_a, r_2) u(k_b, r_2) + u'(k_a, r_1) u(k_b, r_1) = -(\kappa_a^2 - \kappa_b^2) \int_{r_1}^{r_2} u(k_a, r) u(k_b, r) \, dr \, . \tag{2.133}$$

It is immediate to see that the left-hand side of Eq. (2.133) is strictly positive, whereas its right-hand side is strictly negative. Consequently, $u(k_b, r)$ bears one zero in $]r_1 : r_2[$. Moreover, Eq. (2.2) clearly provides with $u''(k, r) \ne 0 \, \forall r > 0$ with $\kappa$ sufficiently large, so that the nodes of $u(k, r)$ eventually disappear for energies becoming more and more negative.

A direct consequence of the interlacing property is that a zero of $u(k, r)$ at $r_1 > 0$ will move forward on the real axis without converging to a finite value when $\kappa$ increases. By continuity of $C^-$ with respect to $k$ (see Exercise XII), one obtains a bound state when $r_1$ is pushed to infinity, as then $C^- = 0$. This process can be repeated for all radii $(r > 0)$ where $u(0, r) = 0$, so that the number of zeros of the

$u(0, r)$ is equal to the number of bound states. Hence, the number of bound states is finite and equal to the number of zeros of the $u(0, r)$.

If $v_c < 0$, the Hamiltonian models a hydrogenoid system, and one can show that it has an infinite number of bound states. By continuity of $u(k, r)$ in $k = 0$ for $r$ fixed (see Exercise XII) and from Eq. (2.59), there exists $\kappa > 0$ so that $u(k, r)$ has $n$ nodes for $n > 0$. Hence, one can then generate a bound state $u_{n-1}(r)$ of $n - 1$ nodes by pushing the last node of an unbound $u(k, r)$ state to $r \to +\infty$. This proves that one has an infinity of bound states, which accumulate in $k = 0$.

## 2.5.2   Virial Theorem

Average kinetic and potential energies in bound eigenstates of a Hamiltonian are related in a simple way which is provided by the virial theorem. The virial theorem originates from classical mechanics, where one can show that the temporal averages of kinetic and potential energies of a point particle verify:

$$\frac{1}{2}m \langle v^2 \rangle = \langle \mathbf{r} \cdot \mathbf{F} \rangle \ . \tag{2.134}$$

In this equation, $m$ is the mass of the particle, $\mathbf{r}$ is its position, $\mathbf{v}$ its velocity, and $\mathbf{F}$ is the force acting on the particle. Interestingly, the extension of the virial theorem to quantum mechanics with local potentials provides with an equation formally identical to Eq. (2.134).

To demonstrate the virial theorem in the quantum case [35, 36], let us consider a bound state, denoted as $|\phi\rangle$, of a spherical one-body Hamiltonian $h$ (see Eq. (2.3)). Note, however, that the virial theorem also holds for a general nonspherical potential. To simplify notation, one will write in this section: $v_l(r) = v(r)$. As $|\phi\rangle$ is an eigenstate of $h$, then every matrix element of the form $\langle \phi|[O, h]|\phi\rangle$, with $O$ being an arbitrary operator, equals to zero. Therefore, one obtains:

$$\langle \phi|[\mathbf{r} \cdot \mathbf{p}, h]|\phi\rangle = 0 \ . \tag{2.135}$$

Separating kinetic and potential parts of $h$ in Eq. (2.135), one obtains:

$$\langle \phi| \left[ \mathbf{r} \cdot \mathbf{p}, \frac{\mathbf{p}^2}{2m} \right] |\phi\rangle + \langle \phi|[\mathbf{r} \cdot \mathbf{p}, v]|\phi\rangle = 0$$

$$\Rightarrow -2 \langle \phi| \frac{\mathbf{p}^2}{2m}|\phi\rangle + \langle \phi|r \frac{\partial v}{\partial r}(r)|\phi\rangle = 0$$

$$\Rightarrow 2 \langle t \rangle = \langle \mathbf{r} \cdot \nabla v \rangle \ , \tag{2.136}$$

where $t$ denotes the kinetic part. The calculation of commutators in (2.136) is straightforward and averages are integrals over $\mathbf{r}$. It can be easily seen that Eqs. (2.134) and (2.136) are identical, if one formally replaces time averages by space averages, and if one considers that Eq. (2.134) is used with conservative forces which always derive from a potential.

Equation (2.136) is physically important as it shows that quantum particle in an attractive Coulomb potential, for example, an electron in an atom, cannot fall into the center of atom despite the infinitely attractive character of the potential in $r = 0$. Indeed, the integral involving $v(r)$ in Eq. (2.136) would become infinite if the electron would collapse to $r = 0$. Consequently, one would obtain an infinite kinetic energy from Eq. (2.136), which is forbidden by energy conservation.

It is convenient to rewrite Eq. (2.136) as a function of $e$ and $v(r)$ only:

$$2 \int_0^{+\infty} u(r)^2 v(r) \, dr + \int_0^{+\infty} u(r)^2 r \frac{\partial v}{\partial r}(r) \, dr = 2e \, . \qquad (2.137)$$

Equation (2.137) is obtained by adding $2v(r)$ to the left- and right-hand sides of Eq. (2.136) and integrating over $r$. Moreover, one used Eq. (2.2) and the fact that $|\phi\rangle$ is normalized. As $u(k, r)^2$ is always positive for bound states, therefore Eq. (2.136) allows to explicitly exhibit the role of the signs of $v(r)$ and $\partial v/\partial r$ on the value of $e$.

Equation (2.137) is particularly insightful if $e \to 0$ for the case of nuclear potentials which are attractive at short distances and negligible or repulsive in the asymptotic region. Indeed, Eq. (2.137) shows that for bound states with energy close to zero, the first integral on the left-hand side of Eq. (2.137), which is negative, is compensated by the second integral on the left-hand side of this equation, which is positive. This means, in particular, that nuclear halo states can exist because of the relatively constant negative value of the potential inside of the nucleus, on the one hand, and, on the other hand, because of the rapid increase of the nuclear potential at the surface.

## 2.6   Analytical Properties of the Wave Functions

In the description of nuclear reactions, scattering states of real-energy play a prominent role as they represent a projectile being scattered on a target. Even though cross sections are defined in terms of real-energy states, they are also influenced by the structure of one-body states in the complex energy plane. Indeed, reaction cross sections are large at real energies in the vicinity of complex-energy narrow resonance states. At these real energies, the composite system formed by the target and the projectile is in fact long-lived, so that the collision probability is much enhanced therein. Consequently, the study of scattering states in the complex-momentum plane is of physical interest and will be done in this section.

In the following, one will consider the analytic dependence of scattering states with respect to their linear momentum $k$, as well as their asymptotic behavior for $|k| \to +\infty$, which is simply expressible in terms of Coulomb wave functions. The demonstration of the completeness relation of the eigenstates of Eq. (2.3), where both bound and scattering states enter, heavily rely on the analytical properties of the scattering wave functions in the whole complex $k$-plane. For this, one will also determine the properties of wave functions in a potential screened of its Coulomb part at finite distance. The scattering wave functions generated by an infinite-range potential are singular in the complex $k$-plane, so that the use of a screened potential simplifies their theoretical study in the complex $k$-plane.

Mathematical tools of fundamental importance for scattering theory are the Jost function and the scattering matrix ($S$-matrix). They are defined from the incoming and outgoing components of scattering states, so that they are directly related to the probability to form a composite system in a nuclear reaction. The dependence of nuclear cross sections on nuclear structure is in fact present only in the $S$-matrix. The $S$-matrix enters the dispersion relations [37], which are of fundamental importance as they allow to connect real and imaginary parts of $S$-matrix and, hence, calculate the complex $S$-matrix from the knowledge of its real part. The $S$-matrix dispersion relation formulas will then be also derived in this section.

As said earlier, narrow resonance states play a special role in the complex $k$-plane. Moreover, the numerical calculation of resonances is more difficult than the calculation of bound states. To calculate bound states, one can either diagonalize Eq. (2.3) in a basis of bound states, for example, in a basis of harmonic oscillator states, or use a bisection method, as bound states belong to a finite segment of the negative real energy axis (see Sect. 2.5). On the other hand, the resonance energy is complex, which precludes the use of the bisection method. Moreover, resonances diverge on the real axis, so that they cannot be expanded with the basis of bound states. Thus, the numerical method to calculate resonances, based on the use of the Jost function, will be stated in this section.

An additional problem arises when the width of a narrow resonance is much smaller than its energy, hence when one deals with very long-lived resonance. In this case, it is impossible to precisely calculate both the real and imaginary parts of the resonance energy simultaneously. In order to solve this problem, one recurs to the flux equation which provides the width of a narrow resonance from the real part of its energy and the wave function. This equation and its ability to describe widths which are several orders of magnitude smaller than associated energies will be demonstrated in this section.

## 2.6.1   Decomposition in Incoming and Outgoing Wave Functions

Physically, $u^+(k, r)$ and $u^-(k, r)$ represent the state of a particle which can only leave the potential zone, or enter it, respectively. $u^+(k, r)$ and $u^-(k, r)$ are defined as solutions of Eq. (2.2), but with the following boundary condition for $r \geq R$ (see Eq. (2.7)):

$$u^\pm(k, r) = H^\pm_{\ell, \eta}(kr) \, . \tag{2.138}$$

In principle, Eq. (2.138) is not defined in general if $k < 0$, as $k$ lies on the branch cut of irregular Coulomb wave functions. Nevertheless, as $H^\pm_{\ell, \eta}(kr)$ possesses a limit starting from the upper complex plane, one will assume that this limiting process is implicitly done when $k < 0$.

The existence and unicity of $u^{\pm}(k, r)$ $\forall k$ can be demonstrated by the same method as used for $u(k, r)$ in Exercise I. For this, one has to consider the integral equation verified by $u^{\pm}(k, r)$, which reads:

$$u^{\pm}(k, r) = H^{\pm}_{\ell,\eta}(kR) + k \, H^{\pm'}_{\ell,\eta}(kR) \, (r - R) + \int_R^r (r - r')\mathscr{L}(k, r')u^{\pm}(k, r') \, dr' \,, \tag{2.139}$$

where $0 < r \leq R$. As this equation is formally identical to the integral equation verified by $u(k, r)$, up to the replacements of $r_0$ by $R$ and $C_0 \, F_{\ell_0,\eta_0}(k_0 r_0)$ by $H^{\pm}_{\ell,\eta}(kR)$ (see Eq. (2.9)), the techniques used in Exercise I with $u(k, r)$ can be applied mutatis mutandis to $u^{\pm}(k, r)$. Using Eqs. (2.2), (2.7), (2.138), and (2.139), one obtains $\forall r > 0$:

$$u(k, r) = C^+ u^+(k, r) + C^- u^-(k, r) \,. \tag{2.140}$$

One will now consider $u^{\pm}(k, r)$ when $0 < r \leq r_0$. Indeed, $u^{\pm}(k, r)$ cannot verify Eq. (2.6) $\forall k$, otherwise it would be proportional to $u(k, r)$, which is impossible as this would imply the presence of a continuum of bound states (see Sect. 2.5.1). Consequently, $u^{\pm}(k, r)$ is in general an irregular solution of Eq. (2.2) in $r = 0$, so that $u^{\pm}(k, 0) \neq 0$ for $\ell_0 = 0$ and $u^{\pm}(k, r) \propto r^{-\ell_0}$ for $\ell_0 > 0$ and $r \to 0$ (see Eq. (2.4)).

## 2.6.2  Asymptotic Behavior of Complex-Momentum Wave Functions

In order to demonstrate the completeness relation of complex-momentum wave functions $u(k, r)$ as well as calculate dispersion relations related to the $S$-matrix (see below), it will be necessary to know the asymptotic behavior of $u(k, r)$ for $k \to 0$ and $|k| \to +\infty$. Note that all the derivations and exercises in this section are cumbersome. Hence, the reader can admit all results if they do not want to delve into complicated derivations and directly go to Sect. 2.6.3.

One will firstly consider the case $k \to 0$. It is clear that one can choose $|k|$ sufficiently small so that $R$ is smaller than its turning point, that is, the smallest radius $r_t(k) > 0$ for which $u''(k, r_t(k)) = 0$ (see Sect. 2.3.3). Hence, $F_{\ell,\eta}(kr)$ and $G_{\ell,\eta}(kr)$ cannot vanish therein. Consequently, Eq. (2.5) implies that for $r \geq R$:

$$u(k, r) = \mathscr{N}_k(F_{\ell,\eta}(kr) + A_k G_{\ell,\eta}(kr)) \,, \tag{2.141}$$

where $\mathscr{N}_k$ and $A_k$ are functions of $k$.

If one has no bound state for $k = 0$, one cannot have $u(k, r) \propto G_{\ell,\eta}(kr)$ when $k \to 0$ for $r > R$. Indeed, the limit of the $G_{\ell,\eta}(kr)$ function when $k \to 0$

is an integrable function of $r$. Let us now assume that there is a bound state at $k = 0$. The zero-energy bound state is proportional to $G_{\ell,\eta}(kr)$ when $r > R$ (see Sect. 2.3.3). As $u(k, r)$ is orthogonal to the zero-energy bound state $\forall k > 0$, the relation $u(k, r) \propto G_{\ell,\eta}(kr)$ when $k \to 0$ for $r > R$ cannot hold. Consequently, $A_k G_{\ell,\eta}(kR) = \mathcal{O}(F_{\ell,\eta}(kR))$ when $k \to 0$, so that one has for $r \geq R$:

$$u(k, r) = \mathcal{N}_k \, F_{\ell,\eta}(kr) \, f(k, r) \,, \tag{2.142}$$

where $f(k, r)$ is bounded for $k \to 0$, but can diverge for $r \to 0$.

In order to devise the behavior of $f(k, r)$ for $0 < r < R$, let us rewrite $u(k, r)$ as:

$$u(k, r) = u(k, R) \left( \frac{F_{\ell,\eta}(kr)}{F_{\ell,\eta}(kR)} \right) \left( \frac{f(k, r)}{f(k, R)} \right) . \tag{2.143}$$

$f(k, R)$ is bounded and the ratio of Coulomb wave functions has a finite limit in Eq. (2.143) for $k \to 0$ (see Eq. (2.37)). $u(k, r)$ is also continuous and finite for $k \to 0$ (see Exercise XII). Consequently, $f(k, r)$ for $k \to 0$ and $0 < r < R$ is bounded as well. This implies that Eq. (2.142) is also valid for $0 < r < R$.

For a fixed $r > 0$ and $k \to +\infty$, one will consider the uniform WKB approximation [38], that is, based on the use of a special function removing turning point singularity and hence valid $\forall r > 0$. $k$ is supposed to be real for the moment. For this, let us write Eq. (2.2) similarly to Eq. (2.4), with the $\ell_0$ and $v_{c0}$ values appearing explicitly:

$$u''(k, r) = \left( \frac{\ell_0(\ell_0 + 1)}{r^2} + \frac{v_{c0}}{r} + v_0(r) - k^2 \right) u(k, r) \,, \tag{2.144}$$

where $v_0(r)$ is finite $\forall r \geq 0$ and $v_0(r) = v_0$ for $0 \leq r \leq r_0$ (see Sect. 2.1). One can then choose $k$ so that $|v_0(r)|$ is arbitrarily small against $k^2$ $\forall r \geq 0$ so that one can introduce the following functions:

$$u_{app}^{\pm}(k, r) = k_0^{1/2} \, \lambda^{-1/2}(r) \, H_{\ell_0,\eta_0}^{\pm}(\Lambda(r)) \,, \quad r > 0 \tag{2.145}$$

where:

$$\lambda(r) = k \sqrt{1 - \frac{v_0(r)}{k^2}} \tag{2.146}$$

$$\Lambda(r) = \int_0^r \lambda(r') \, dr' \,. \tag{2.147}$$

Eq. (2.145) provides with an approximate solution of Eq. (2.2), what can be checked by inserting Eq. (2.145) in Eq. (2.144). This arises due to the standard properties of WKB approximation and from the fact that the functions entering Eq. (2.145) are bounded on the real $k$-axis.

The error term is conveniently handled by a Neumann series generated by Eq. (2.145) [39]. For this, one writes $u^\pm(k, r) = u^\pm_{\text{app}}(k, r)(1 + \epsilon^\pm(k, r))$. The Neumann series representing $\epsilon(k, r)$ is generated iteratively using intermediate $\epsilon^\pm_n(k, r)$ functions:

$$\epsilon^\pm_{n+1}(k, r) = \int_{r_d}^r u^\pm_{\text{app}}(k, r')^{-2} \int_R^{r'} u^\pm_{\text{app}}(k, r'')^2 F(k, r'', \epsilon^\pm_n) \, dr'' \, dr', \quad n \geq 0,$$

(2.148)

where $r_d = r_0$ for the calculation of $u(k, r)$ and $r_d = R$ for that of $u^\pm(k, r)$. Different values of $r_d$ are used in Eq. (2.148) to efficiently handle the boundary conditions verified by $u(k, r)$ and $u^\pm(k, r)$ (see Eqs. (2.6) and (2.7)). Moreover, $\epsilon^\pm_0(k, r) = 0$ and $F(k, r'', \epsilon^\pm_n)$ is the rest term of Eq. (2.144):

$$F(k, r, \epsilon^\pm_n) = \left(1 + \epsilon^\pm_n(k, r)\right)\left(\frac{1}{2}\lambda''(r)\lambda(r)^{-1} - \frac{3}{4}\lambda'(r)^2\lambda(r)^{-2}\right)$$

$$+ \left(1 + \epsilon^\pm_n(k, r)\right)\left(\frac{\ell_0(\ell_0 + 1)}{r^2}\left(1 - \frac{r^2\lambda^2(r)}{\Lambda(r)^2}\right)\right)$$

$$+ \frac{v_{c_0}}{r}\left(1 - \frac{2\eta_0 r \lambda^2(r)}{\Lambda(r)v_{c_0}}\right)\right).$$

(2.149)

One can then prove from Eqs. (2.148) and (2.149) that the Neumann series defining $\epsilon^\pm(k, r)$ converges if $k$ is sufficiently large, and that $\epsilon^\pm(k, r) = \mathcal{O}(k^{-2})$. For $k \to +\infty$, one obtains:

$$u^\pm_{\text{app}}(k, r) = \exp\left(\pm i\,(kr - \ell_0\pi/2)\right)$$

$$\times \left(1 \mp i \frac{r\mathcal{V}_0(r) + rv_{c_0}(\ln(2kr) - \Psi(\ell_0 + 1)) - \ell_0(\ell_0 + 1)}{2kr}\right)$$

$$\times \left(1 + \mathcal{O}\left(\ln^2(k)\,k^{-2}\right)\right)$$

(2.150)

$$u^{\pm\prime}_{\text{app}}(k, r) = \left(\pm ik\, u^\pm_{\text{app}}(k, r) + \exp\left(\pm i\,(kr - \ell_0\pi/2)\right)\right.$$

$$\times \left(\mp i \frac{r^2 v_0(r) + rv_{c_0} + \ell_0(\ell_0 + 1)}{2kr^2}\right)\right)\left(1 + \mathcal{O}\left(\ln^2(k)\,k^{-2}\right)\right),$$

(2.151)

where $\Psi(x)$ is the digamma function, and $\mathcal{V}_0(r)$ is:

$$\mathcal{V}_0(r) = \int_0^r v_0(r')\, dr'.$$

(2.152)

All asymptotic expansions of interest will consist of rational functions involving Eqs. (2.150) and (2.151). Consequently, one just has to check that the dominant term of the denominators occurring in the asymptotic expansions derived using Eqs. (2.150) and (2.151) is nonzero.

One can write $u(k, r)$ and $u^\pm(k, r)$ as a linear combination of the functions $u_{app}^\pm(k, r)(1 + \epsilon^\pm(k, r))$. Indeed, it is straightforward to check that the Wronskian of $u_{app}^\pm(k, r)(1 + \epsilon^\pm(k, r))$ functions is nonzero for $k$ sufficiently large. One will use the notation $u^\pm(k, r) = u_{app}^\pm(k, r)(1 + \epsilon_u^\pm(k, r))$ when calculating the $u(k, r)$ function, while the initial notation $u_{app}^\pm(k, r)(1 + \epsilon^\pm(k, r))$ remains unchanged when considering $u^\pm(k, r)$.

Boundary conditions are provided by Eqs. (2.6) and (2.138) in $r = r_0$ and $r = R$ for $u(k, r)$ and $u^\pm(k, r)$, respectively. One then obtains:

$$u^\pm(k, r) = A^\pm(k) \, u_{app}^\pm(k, r)(1 + \epsilon^\pm(k, r)) + B^\pm(k) \, u_{app}^\mp(k, r)(1 + \epsilon^\mp(k, r))$$

(2.153)

$$u(k, r) = \frac{C_0}{2i} \left( u_{app}^+(k, r)(1 + \epsilon_u^+(k, r)) - u_{app}^-(k, r)(1 + \epsilon_u^-(k, r)) \right) ,$$

(2.154)

where:

$$A^\pm(k) = \left( \frac{H_{\ell,\eta}^\pm(kR)}{u_{app}^\pm(k, R)} \right) \left( \frac{h_{\ell,\eta}^\pm(kR) - h_{\ell_0,\eta_0}^\mp(\Lambda(R))}{h_{\ell_0,\eta_0}^\pm(\Lambda(R)) - h_{\ell_0,\eta_0}^\mp(\Lambda(R))} \right) \left( 1 + \mathscr{O}\left(k^{-2}\right) \right)$$

(2.155)

$$B^\pm(k) = \left( \frac{H_{\ell,\eta}^\pm(kR)}{u_{app}^\mp(k, R)} \right) \left( \frac{h_{\ell_0,\eta_0}^\pm(\Lambda(R)) - h_{\ell,\eta}^\pm(kR)}{h_{\ell_0,\eta_0}^\pm(\Lambda(R)) - h_{\ell_0,\eta_0}^\mp(\Lambda(R))} \right) \left( 1 + \mathscr{O}\left(k^{-2}\right) \right) ,$$

(2.156)

and $h_{\ell,\eta}^\pm(z)$ is the logarithm derivative of $H_{\ell,\eta}^\pm(z)$. It can be shown using Eqs. (2.145), (2.150), and (2.151) that $h_{\ell_0,\eta_0}^\pm(\Lambda(r))$ obeys the following asymptotic expansion:

$$h_{\ell_0,\eta_0}^\pm(\Lambda(r)) = \left( \frac{u_{app}^{\pm\,'}(k, r)}{k\, u_{app}^\pm(k, r)} \right) \left( 1 + \mathscr{O}\left(k^{-2}\right) \right) .$$

(2.157)

One has $u_{app}^\pm(k, R) \sim \exp(\pm ikr \mp i\ell_0\pi/2)$ and $h_{\ell_0,\eta_0}^\pm(\Lambda(r)) \sim \pm i$ when $k \to +\infty$ (see Eqs. (2.150), (2.151), and (2.157)), so that $A^\pm(k)$, $B^\pm(k)$ and $h_{\ell_0,\eta_0}^\pm(\Lambda(r))$ in Eqs. (2.155)–(2.157) are well defined.

By matching Eqs. (2.140) and (2.154) in $r = R$, one may derive asymptotic equations for the $C^{\pm}$ constants (see Eq. (2.7)) when $k \to +\infty$:

$$2iC^{\pm} = -C_0 \left( \frac{u_{\text{app}}^{\pm}(k, R)}{H_{\ell,\eta}^{\pm}(kR)} \right) \left( \frac{h_{\ell,\eta}^{\mp}(kR) - h_{\ell_0,\eta_0}^{\pm}(\Lambda(R))}{h_{\ell,\eta}^{+}(kR) - h_{\ell,\eta}^{-}(kR)} \right) \left( 1 + \mathcal{O}\left(k^{-2}\right) \right)$$

$$+ C_0 \left( \frac{u_{\text{app}}^{\mp}(k, R)}{H_{\ell,\eta}^{\pm}(kR)} \right) \left( \frac{h_{\ell,\eta}^{\mp}(kR) - h_{\ell_0,\eta_0}^{\mp}(\Lambda(R))}{h_{\ell,\eta}^{+}(kR) - h_{\ell,\eta}^{-}(kR)} \right) \left( 1 + \mathcal{O}\left(k^{-2}\right) \right) ,$$

$$\text{(2.158)}$$

where the asymptotic expansions are clearly well defined.

---

**Exercise XII ★ ★ ★**

One will show that $u(k, r)$ cannot grow faster than a given exponential function for large values of $|k|$ in the complex plane.

**A.** Show that $u(k, r)$ can be defined from the following integral equation:

$$u(k, r) = C_0 \, F_{\ell_0, \eta_0}(k_0 r) + \int_{r_0}^{r} g(r, r') \, \Delta V(r') \, u(k, r') \, dr' , \qquad \text{(2.159)}$$

where the Green's function of Eq. (2.2) is used [40]. In this expression, $g(r, r')$ and $\Delta V(r)$ are defined by the following equations:

$$g(r, r') = \frac{F_{\ell_0, \eta_0}(k_0 r) G_{\ell_0, \eta_0}(k_0 r') - F_{\ell_0, \eta_0}(k_0 r') G_{\ell_0, \eta_0}(k_0 r)}{k_0} \qquad \text{(2.160)}$$

$$\Delta V(r) = v_0(r) - v_0 . \qquad \text{(2.161)}$$

**B.** Show that, for $r_0 \le r' \le r \le R$:

$$\left| g(r, r') \, \Delta V(r') \right| \le C \, e^{|\Im(k)|(r-r')} \, |k|^{-1} , \qquad \text{(2.162)}$$

where $|k|$ is sufficiently large and $C > 0$ is independent of $k$, $r$, and $r'$.

Using the same method and notations as in Exercise I, show from the previous equations that:

$$\left| u^{(n+1)}(k, r) - u^{(n)}(k, r) \right| \le 2 \, |C_0| \, e^{|\Im(k)|r} \left( \frac{C^n \, r^n \, |k|^{-n}}{n!} \right). \qquad \text{(2.163)}$$

Deduce that $u(k, r)$ cannot diverge faster than $|C_0| \, \exp(|\Im(k)|r)$ when $|k| \to +\infty$ for $0 < r \le R$.

**C.** Devise a similar result for $u^{\pm}(k, r)$ when $|k| \to +\infty$ in the zones of the complex plane where it diverges.

We will now derive the equivalents of Eq. (2.153) in the upper complex $k$-plane. One will have to integrate wave functions in the whole upper complex $k$-plane (see Sect. 3.2.1). As $u_{app}^+(k, r)$ is bounded if $\Im(k) > 0$, $u_{app}^+(k, r)(1 + \epsilon^+(k, r))$ is a converging Neumann series solution of Eq. (2.144). However, the iterative process of Eq. (2.148) fails to produce a solution of Eq. (2.144) when applied to $u_{app}^-(k, r)$, as the latter function diverges in the upper complex $k$-plane when $\Im(k) \to +\infty$. In order to counteract this divergence, one will consider the following function:

$$f_{app}(k, r) = u_{app}^+(k, r) + e^{2ikR} u_{app}^-(k, r) . \tag{2.164}$$

One can see that the term depending on $u_{app}^-(k, r)$ is now smaller or comparable to $u_{app}^+(k, r)$ in Eq. (2.164) when $\Im(k) \to +\infty$, because both terms behave as $\exp(ikr)$ and $\exp(ik(2R-r))$, respectively. Moreover, Eq. (2.150) implies that $f_{app}(k, r) \neq 0$ if $|k|$ is sufficiently large. Consequently, as $f_{app}(k, r)$ is bounded when $\Im(k) \to +\infty$, its associated Neumann series, denoted as $f(k, r) = f_{app}(k, r)(1 + \epsilon_f(k, r))$, converges and is then a solution of Eq. (2.144). As $f(k, r)$ is linearly independent of $u^+(k, r)$ for $|k|$ sufficiently large, $u(k, r)$ and $u^+(k, r)$ can be written as a linear combination of the Neumann series $u_{app}^+(k, r)(1 + \epsilon^+(k, r))$ and $f_{app}(k, r)(1 + \epsilon_f(k, r))$.

Let us now derive the equivalent of Eq. (2.154) in the upper complex $k$-plane. The rest function $\epsilon_f(k, r)$ vanishes when $r = r_0$. Using Eq. (2.164), one obtains that $u(k, r)$ can diverge as $\exp(\Im(k)(2R - r)) |k|^{-\ell_0 - 3}$ when $\Im(k) \to +\infty$ and $r_0 \leq r \leq R$. This is due to the appearance of an additional term in Eq. (2.154), equal to $C_0/(2i) e^{-2ikR} u_{app}^+(r) (\epsilon_u^+(k, r) - \epsilon_f(k, r))$. However, $u(k, r)$ cannot increase faster than $|C_0| \exp(\Im(k) r)$ when $\Im(k) \to +\infty$ (see Exercise XII). Consequently, one has $\epsilon_u^+(k, r) - \epsilon_f(k, r) = \mathcal{O}(\exp(2ik(R-r)) k^{-2})$ when $\Im(k) \to +\infty$ and $r_0 \leq r \leq R$ (see Eqs. (2.148)–(2.151)). Therefore, Eq. (2.154) is still valid in the upper complex $k$-plane, except that $\epsilon_u^\pm(k, r)$ therein is now a complicated function of the initial $\epsilon_u^+(k, r)$ and $\epsilon_f(k, r)$ rest terms. Clearly, one still has $\epsilon_u^\pm(k, r) = \mathcal{O}(k^{-2})$ when $|k| \to +\infty$ for this newly defined $\epsilon_u^\pm(k, r)$ function.

Because Eq. (2.158) was derived from Eq. (2.154), it is also verified when $\Im(k) \to +\infty$. One can then devise the asymptotic behavior of the constant $C^\pm(k)$ (see Eqs. (2.155), (2.156), and (2.158)) when $|k| \to +\infty$ in the upper complex $k$-plane. One will denote as $C_1^\pm(k)$ and $C_2^\pm(k)$ the two consecutive terms defining $C^\pm(k)$ in Eq. (2.158). One obtains from Eq. (2.158) that:

$$2i C_1^\pm(k) = \pm C_0 \exp(\pm i(\ell - \ell_0)(\pi/2)) \left(1 + \mathcal{O}(\ln(k) k^{-1})\right) \tag{2.165}$$

$$C_2^\pm(k) = \mathcal{O}(C_0 \exp(\mp 2ikR) \ln(k) k^{-1}) . \tag{2.166}$$

We will now deal with $u^{\pm}(k, r)$ in the upper complex plane, so that $\epsilon_f(k, R) = 0$. Using Eq. (2.164), one has for $u^{\pm}(k, r)$:

$$u^+(k, r) = \left(A^+(k)(1 + \epsilon^+(k, r)) + B^+(k) e^{-2ikR}(\epsilon_f(k, r) - \epsilon^+(k, r))\right) u_{app}^+(k, r)$$

$$+ B^+(k) u_{app}^-(k, r)(1 + \epsilon_f(k, r)) \tag{2.167}$$

$$u^-(k, r) = \left(B^-(k)(1 + \epsilon^-(k, r)) + A^-(k) e^{-2ikR}(\epsilon_f(k, r) - \epsilon^-(k, r))\right) u_{app}^+(k, r)$$

$$+ A^-(k) u_{app}^-(k, r)(1 + \epsilon_f(k, r)) . \tag{2.168}$$

The coefficients of $u_{app}^-(r)$, that is, $B^+(k)$ and $A^-(k)$, are the same as in Eq. (2.153) for both functions $u^+(k, r)$ and $u^-(k, r)$. Inspection of Eqs. (2.155) and (2.156) shows that

$$A^{\pm}(k) \to \exp(\pm i(\ell_0 - \ell)(\pi/2))$$

and

$$B^{\pm}(k) = \mathcal{O}(\exp(\pm 2ikR) \ln(k) k^{-1}) .$$

Hence, the presence of a term depending on $B^+(k)$ does not change the asymptotic behavior of the coefficient of $u_{app}^+(k, r)$ in Eq. (2.167) when $|k| \to +\infty$, with $k$ belonging to the upper part of the complex plane. The asymptotic expansion of $u^{\pm}(k, r)$ in Eq. (2.167) then bears the same properties as that of Eq. (2.153), which is derived in the real-$k$ case, that is, when $k \to +\infty$. On the contrary, $u^-(k, r)$ exhibits in general a strong divergence when $\Im(k) \to +\infty$, due to the presence of an overall factor $e^{-2ikR}$ in the coefficient of $u_{app}^+(k, r)$ (see Eq. (2.168)). Indeed, when $\Im(k) \to +\infty$, $u^-(k, r)$ can diverge faster than $u_{app}^-(k, r)$, behaving as $\exp(-ikr)$ (see Eq. (2.150) and Exercise XII).

### 2.6.3  Analyticity of Complex Momentum Wave Functions

After considering the properties of one-body wave functions in the bound state region, one can now exhibit their analytic properties in the scattering region. Indeed, the dependence of Eq. (2.2) on $k$ implies that $u(k, r)$ is analytic with respect to $k$ [40]. This property is fundamental as it is tantamount to showing that the set of $u(k, r)$ functions is complete, that is, that it can be used to expand any integrable function. The analyticity of $u(k, r)$ will be demonstrated in Exercise XIII.

---

**Exercise XIII** ★★

One will demonstrate that $u(k, r)$ is analytical in the complex $k$-plane and that its domain of analyticity depends on the values of $\ell$ and $\eta$.

**A.** Using the recurrence relation for $u^{(n)}(k, r)$, which is defined in Exercise I, and setting $C_0 = k_0^{-\ell_0-1} C_{\ell_0}(\eta_0)^{-1}$ in Eq. (2.6) as in Exercise I, show that the complex momentum wave function $u(k, r)$ (see Eq. (2.2)) is an entire function of $k$.

**B.** Using Eqs. (2.138) and (2.139), formulate the analytic properties of the complex momentum wave function $u^\pm(k, r)$. Show that the domain of analyticity of $u^\pm(k, r)$ is at least that of $H_{\ell,\eta}^\pm(kr)$. Precise the cases where $u^\pm(k, r)$ is entire or analytic everywhere in the complex plane except in $k = 0$.

---

### 2.6.4 Jost Functions

Fundamental objects for the study of $u(k, r)$ functions in the complex plane are the Jost functions [41]. They are defined as the Wronskians between $u(k, r)$ and $u^\pm(k, r)$ functions (see Eq. (2.131)) and is related to the bound and scattering states of Eq. (2.2):

$$J^\pm(k) = u(k, r)u^{\pm\prime}(k, r) - u'(k, r)u^\pm(k, r) . \tag{2.169}$$

$J^\pm(k)$ is independent of $r$, as can be demonstrated directly using Eq. (2.2). Therefore, one can calculate $J^\pm(k)$ by taking $r \to +\infty$ in Eq. (2.169). $u(k, r)$, $u^\pm(k, r)$ and their derivatives indeed bear a simple asymptotic form at large distance (see Eq. (2.7) and Sect. 2.3), so that $J^\pm(k)$ is immediate to calculate:

$$J^\pm(k) = \pm 2ikC^\mp . \tag{2.170}$$

It is clear that $J^+(k) = 0$ for bound and resonance states only, as then $u(k, r) = C^+ u^+(k, r)$ in the asymptotic region. The domain of analyticity of $J^\pm(k)$ is that of $H_{\ell,\eta}^\pm(kR)$ if one poses $C_0 = k_0^{-\ell_0-1} C_{\ell_0}(\eta_0)^{-1}$ in Eq. (2.6) (see Exercises I and VIII).

It is of interest to know the derivative of $J^+(k)$ if $k$ corresponds to a bound state, which will be denoted as $k_n$. For this, let us differentiate Eq. (2.130) with respect to $k_b$, for $r_1 = 0$ and $r_2 = r$ or $r_1 = r$ and $r_2 > r_1$ [42]:

$$\dot{W}(u(k_b, r_2), u(k_a, r_2)) - \dot{W}(u(k_b, r_1), u(k_a, r_1))$$

$$= -2k_b \int_{r_1}^{r_2} u(k_a, r')u(k_b, r') \, dr' + (k_a^2 - k_b^2)\frac{d}{dk_b} \int_{r_1}^{r_2} u(k_a, r')u(k_b, r') \, dr' .$$

$$\tag{2.171}$$

By replacing $u(k_a, r)$ by $u^+(k_a, r)$ if $r_1 = 0$ and $r_2 = r$, and $u(k_b, r)$ by $u^+(k_b, r)$ if $r_1 = r$ and $r_2 > r_1$, and putting $k_a = k_b = k_n$ in Eq. (2.171), one obtains:

$$u^{+'}(k_n, r)\dot{u}(k_n, r) - \dot{u}'(k_n, r)u^+(k_n, r) = 2k_n \int_0^r u(k_n, r')u^+(k_n, r') \, dr' \tag{2.172}$$

$$\dot{u}^{+'}(k_n, r)u(k_n, r) - u'(k_n, r)\dot{u}^+(k_n, r) = 2k_n \int_r^{+\infty} u(k_n, r')u^+(k_n, r') \, dr' , \tag{2.173}$$

where the limit $r_2 \to +\infty$ has been effected, and the dot above $u$ indicates differentiation with respect to $k_b$. One has also used the fact that $u(k_n, 0) = 0$ and $u(k_n, r') \to 0$ for $r' \to +\infty$, arising from the bound character of $u(k_n, r)$. One then obtains from Eqs. (2.172) and (2.173) the following equality:

$$\left[ \frac{dJ^+}{dk}(k) \right]_{k=k_n} = 2k_n \int_0^{+\infty} u(k_n, r)u^+(k_n, r) \, dr , \tag{2.174}$$

which also proves that the zeros of $J^+(k)$ are simple if $k_n \neq 0$ as $u(k_n, r) \propto u^+(k_n, r)$ therein.

One will show in Sect. 3.3 that Eq. (2.174) can be generalized to resonance states as well. Consequently, the determination of bound and resonance states amounts to finding the zeros of an analytic complex function, which is done numerically with the quickly converging Newton method.

## 2.6.5   Wave Functions in a Screened Potential

The infinite range of the Coulomb potential induces an essential singularity in the $k$-plane for $k = 0$. Consequently, it is customary to consider a screened potential, vanishing after a finite large radius $R^{(s)}$, to suppress the singularity induced by the Coulomb potential at large distances. Unfortunately, the introduction of a screened potential is not appropriate in all circumstances. For example, it is not sufficient to screen the potential in Eq. (2.3) to demonstrate the completeness relation for proton states (see Sect. 3.2). However, a screened potential is a useful intermediary function between finite-range potentials and infinite-range potentials containing a Coulomb potential. The centrifugal potential poses no theoretical problem as it is integrable on the real $r$-axis.

To demonstrate the completeness relation involving proton states, one will start from that generated by a screened potential, whose radius $R^{(s)}$ will go to infinity afterwards to recover the infinite range of the Coulomb potential. For simplicity, even though it is not necessary, one will also suppress the centrifugal potential when $r > R^{(s)}$, so that Eq. (2.5) becomes that of a neutron $s$ state ($\ell = 0$) for $r > R^{(s)}$.

The asymptotes of $u^{\pm}(k, r)$ and $u^{(s)\pm}(k, r)$ can be related one to another using Eq. (2.169) and properties of $H_{\ell,\eta}^{\pm}(kr)$ (see Sect. 2.3):

$$u^{(s)\pm}(k, r) = u^{\pm}(k, r) \exp[\mp i(-\eta \ln(2kr) - \ell\pi/2 + \sigma_{\ell}(\eta))] + \mathcal{O}(r^{-1}),$$
$$(2.175)$$

as $u^{(s)\pm}(k, r) = \exp(\pm ikr)$, $u^{\pm}(k, r) = H_{\ell,\eta}^{\pm}(kr)$ (see Eq. (2.138)), and $r \geq R^{(s)}$. Consequently, by calculating the Jost functions $J^+(k)^{(s)}$ and $J^+(k)$ at $r = R^{(s)}$ from Eq. (2.169), and using Eq. (2.175) and the fact that $u(k, r)$ and $u^{(s)}(k, r)$ are proportional to each other, imply that $J^+(k)^{(s)}$ and $J^+(k)$ are mutually proportional up to a small term equal to $\mathcal{O}(R^{(s)-1})$. Equation (2.7) implies as well that:

$$C^{+(s)}e^{ikR^{(s)}} + C^{-(s)}e^{-ikR^{(s)}} = A(C^+ H_{\ell,\eta}^+(kR^{(s)}) + C^- H_{\ell,\eta}^-(kR^{(s)})), \quad (2.176)$$

where $A$ is the coefficient of proportionality between $u(k, r)$ and $u^{(s)}(k, r)$. Consequently, as $R^{(s)} \to +\infty$, Eq. (2.176) becomes:

$$C^{+(s)}e^{ikR^{(s)}} \pm C^{-(s)}e^{-ikR^{(s)}} = A\, C^+ \exp\left(i\Sigma(\ell, k, R^{(s)})\right)\left(1 + \mathcal{O}\left(R^{(s)-1}\right)\right)$$
$$\pm A\, C^- \exp\left(-i\Sigma(\ell, k, R^{(s)})\right)\left(1 + \mathcal{O}\left(R^{(s)-1}\right)\right),$$

where $\Sigma(\ell, k, R^{(s)}) = kR^{(s)} - \eta \ln(2kR^{(s)}) - \ell\pi/2 + \sigma_{\ell}(\eta)$ and where one has used the continuity of $u'(k, r)$ and $u^{(s)'}(k, r)$ in $r = R^{(s)}$.

Scattering wave functions $u(k, r)$ and $u^{(s)}(k, r)$ have to be normalized so that

$$2\pi C^+ C^- = 2\pi C^{+(s)} C^{-(s)} = 1.$$

As will be shown in Sect. 3.1, the latter condition is equivalent to the Dirac delta normalization. Equation (2.177) then implies that:

$$A = 1 + \mathcal{O}\left(R^{(s)-1}\right). \tag{2.177}$$

Therefore, if wave functions are normalized with a Dirac delta, then $u^{(s)}(k, r) \to u(k, r)$ when $R^{(s)} \to +\infty$. An immediate consequence is that $C_0{}^{(s)} \to C_0$ when $R^{(s)} \to +\infty$ (see Eq. (2.6)).

One will now reconsider the limit $k \to 0$ using a screened potential, that is, with Eq. (2.2) reducing to $u''(k, r) + k^2 u(k, r) = 0$ for $r > R^{(s)}$. It is only necessary to consider $v_c \geq 0$, as the singularity induced by attractive Coulomb potentials therein is not as strong as in the repulsive case (see Sect. 3.2.3). For this, let us introduce the $u_c^{(s)}(k, r)$ functions, which are normalized regular Coulomb wave functions screened at $R^{(s)}$. From Eq. (2.142) one then obtains:

$$u^{(s)}(k,r) = \mathcal{N}_k^{(s)}(F_{\ell,\eta}(kr) + A_k^{(s)}G_{\ell,\eta}(kr)), \quad R \le r \le R^{(s)} \quad (2.178)$$

$$u^{(s)}(k,r) = \mathcal{N}_k^{(s)} F_{\ell,\eta}(kr) f^{(s)}(k,r), \quad 0 < r \le R^{(s)} \quad (2.179)$$

$$u_c^{(s)}(k,r) = \mathcal{N}_k^{(s)(c)} F_{\ell,\eta}(kr), \quad 0 \le r \le R^{(s)} \quad (2.180)$$

where $A_k^{(s)}G_{\ell,\eta}(kR) = \mathcal{O}(F_{\ell,\eta}(kR))$, $f^{(s)}(k,r)$ is bounded for $k \to 0$ (see Eqs. (2.141) and (2.142)), and the normalization constants $\mathcal{N}_k^{(s)}$ and $\mathcal{N}_k^{(s)(c)}$ can be chosen to be positive.

$A_k^{(s)}$ in Eq. (2.178) is determined by matching $u^{(s)}(k,r)$ in $r = R$, and is independent of $R^{(s)}$. The behavior of $A_k^{(s)}$ when $k \to 0$ can be determined from Eqs. (2.53)–(2.57)). One obtains that $A_k^{(s)} = \mathcal{O}(k^{2\ell+1})$ ($v_c = 0$) or $A_k^{(s)} = \mathcal{O}(\exp(-2\pi\eta))$ ($v_c > 0$) when $k \to 0$. Hence, $A_k^{(s)}$ can be made arbitrarily small for $k \in ]0 : k_s]$, with $k_s > 0$ sufficiently small.

When $r \ll r_t(k)$ (see Eq. (2.68)), $F_{\ell,\eta}(kR)$ increases as $r^{\ell+1}$ ($v_c = 0$) or $r^{3/4} i_{2\ell+\frac{1}{2}}(2\sqrt{v_c r})$ ($v_c > 0$), while $G_{\ell,\eta}(kR)$ decreases as $r^{-\ell}$ ($v_c = 0$) or $r^{3/4} k_{2\ell+\frac{1}{2}}(2\sqrt{v_c r})$ ($v_c > 0$) (see Eqs. (2.53)–(2.57)). These estimates clearly hold $\forall k \in ]0 : k_s]$ (see Eqs. (2.53)–(2.57)). $F_{\ell,\eta}(kr)$ and $G_{\ell,\eta}(kr)$ are $\mathcal{O}(1)$ ($v_c = 0$) or $\mathcal{O}(\eta^{1/6})$ ($v_c > 0$) when $r \sim r_t(k)$ (see Eqs. (2.69) and (2.70)), while their amplitude becomes $\mathcal{O}(1)$ without increasing in all cases when $r \to +\infty$ (see Sect. 2.3.3). Consequently, $F_{\ell,\eta}(kR^{(s)}) + A_k^{(s)}G_{\ell,\eta}(kR^{(s)})$ can be arbitrarily close to $F_{\ell,\eta}(kR^{(s)})$ $\forall k \in ]0 : k_s]$ for $R^{(s)}$ sufficiently large. $\mathcal{N}_k^{(s)}$ and $\mathcal{N}_k^{(s)(c)}$ in Eqs. (2.178) and (2.180) can then be made arbitrarily close $\forall k \in ]0 : k_s]$.

One will now consider $r$ fixed. Clearly, $k_s$ can always be chosen small enough so that $r < r_t(k_s)$ (see Eq. (2.68)). As a consequence, by taking $k_s$ sufficiently small, using the fact that $\mathcal{N}_k^{(s)}$ and $\mathcal{N}_k^{(s)(c)}$ can be arbitrarily close, and noticing that $f^{(s)}(k,r)$ is bounded (see Eq. (2.179)) when $k \to 0$, implies that:

$$|u^{(s)}(k,r)| \le M(r) u_c^{(s)}(k,r), \quad (2.181)$$

where $k \in ]0 : k_s]$, and $M(r) > 0$ is independent of $k$. Moreover, the positivity of $F_{\ell,\eta}(kr)$ functions in the non-oscillatory region has been used to derive the above relation.

## 2.6.6   The Scattering Matrix

The $S$-matrix is very important from a theoretical point of view as it is directly related to the calculation of reaction cross sections. In the one-body case, $S$-matrix is defined from Jost functions and hence $C^{\pm}$ constants:

$$S_{\ell}(k) = \frac{J^{-}(k)}{J^{+}(k)} = -\frac{C^{+}}{C^{-}}.$$                                    (2.182)

The fact that the potential $v_l(r)$ is real in Eq. (2.2) implies that $C^{+} = C^{-*}$ if $k > 0$ (see Eq. (2.7)). One cannot have $C^{\pm} = 0$ for $k > 0$, as otherwise $u(k, r) = 0$ $\forall r \geq 0$. Thus, $|S_{\ell}(k)| = 1$ on the real $k$-axis.

This relation is important physically as it implies the flux conservation in a reaction process. As a consequence, using Eq. (2.169), it is possible to define $S_{\ell}(k)$ on the real $k$-axis as:

$$S_{\ell}(k) = \exp(2i\delta_{\ell}(k)),$$                                                          (2.183)

where $\delta_{\ell}(k)$ is called the phase shift of the scattering state $u(k, r)$. In particular, it varies very quickly close to a narrow resonance, so that one can identify long-lived states in the positive energy region. The bound states and narrow resonances of linear momentum $k$ are associated to the poles of the $S$-matrix, because $S(k)$ is infinite therein, as can be seen from Eq. (2.182). The residues of the poles of the $S$-matrix are calculated in Exercise XIV.

---

**Exercise  XIV** ⋆

   Calculate the residues of $S_{\ell}(k)$ at its nonzero poles $k_n$ using Eq. (2.174).

---

Due to Eq. (2.169), the $S$-matrix is analytical over the complex plane, except for its poles and a cut along the real negative $k$-axis in the case of a Coulomb potential. In fact, the analyticity of $S$-matrix, that is, the fact the $S$-matrix is a function of $k$ only, is a prerequisite for it to be able to describe physical observables. Indeed, as the state having $k^*$ as linear momentum is the time reversed state of a resonance bearing $k$ as linear momentum, a function of both $k$ and $k^*$ would be influenced by both past and future, thus violating the causality principle.

The nature of the poles of the $S$-matrix is of importance, as they correspond to bound states or resonances. One has already demonstrated in the previous section that bound states can exist only if $k$ is purely imaginary (see Exercise XI). Consequently, the only poles that the $S$-matrix can have in the upper complex plane correspond to real-energy bound states and lie on the positive imaginary-$k$ axis. There is no such requirement in the lower half-plane so that complex poles can exist therein, and they indeed correspond to unbound states.

In the first place, one will consider $\Re(k) \geq 0$. The $S$-matrix poles of physical importance are those close to the real $k$-axis as they represent the resonance states bearing a long lifetime. In the interval $-\pi/4 < \arg(k) < 0$, real and imaginary parts of $k$ and $k^2$ have the same sign, so in the expression: $k^2 = E - i\frac{\Gamma}{2}$, both $E$ and $\Gamma$ are positive. The physical significance of $E$ and $\Gamma$ becomes explicit when put in the time-dependent factor of the full wave function:

$$\exp\left(it\left(E - i\frac{\Gamma}{2}\right)\right) = \exp\left(iEt\right)\exp\left(-\frac{\Gamma}{2}t\right).\tag{2.184}$$

The $E$-dependent exponential term of Eq. (2.184) behaves as in real-energy states, so that $E$ is interpreted as the energy of the considered state. Conversely, the $\Gamma$-dependent term of Eq. (2.184) induces a decay factor of $\exp\left(-\Gamma t\right)$ in observables which is typical of particle decay. Consequently, the states of complex energy whose associated linear momentum is close to the real $k$-axis can be interpreted as decaying states, of half-life proportional to $1/\Gamma$.

As all states belonging to the sector $-\pi/4 < \arg(k) < 0$ of the complex plane have $E > 0$ and $\Gamma > 0$, therefore that part of the complex plane is deemed as the physical sheet. States whose $k$ value is situated below the sector defined by $-\pi/4 < \arg(k) < 0$ cannot be interpreted as unbound, because $E \leq 0$ and $\Gamma > 0$. This part of the complex plane is then called the unphysical sheet. However, states lying on the negative imaginary $k$-axis can be close to the real axis, so that they can be expected to have an influence on the reaction cross section. They are called antibound, as their wave function increase exponentially on the real $r$-axis, in direct opposition to the exponential decrease of wave functions for bound states positive imaginary $k$-axis. If an antibound state is close to the real axis, this means that the considered Hamiltonian is very close to have an additional bound state. Hence, scattering states at low energy have large amplitudes inside the nucleus. Consequently, reaction cross sections in the low energy region are typically large if nuclear potentials bear antibound states.

As the Hamiltonian is Hermitian, $-k^*$ is also a pole of the $S$-matrix, which is readily seen from Eq. (2.2) by applying complex conjugation therein. The one-body state associated to the complex linear momentum $-k^*$ corresponds physically to a time-conjugate resonance state, because complex conjugation is equivalent to time reversal. Consequently, for each resonance state of complex momentum $k$, representing a particle leaving the nucleus, it exists a capturing state of complex momentum $-k^*$, which describes a particle captured from infinity by the nucleus.

Different types of the $S$-matrix poles are depicted in Fig. 2.1 on the example of a $1s_{1/2}$ one-body state. The bound and antibound states in Fig. 2.1 are neutron $1s_{1/2}$ states, whereas a proton $1s_{1/2}$ state has been considered for the resonance. One can see that the behavior of the bound state wave function and of the real part of the resonance wave function inside the nucleus is similar and close to that of a harmonic oscillator state. The fundamental difference is the unbound character of the resonance state, as it clearly increases in modulus in the asymptotic region. While small inside the nucleus, the imaginary part of the resonance state increases outside the nucleus to become of the same order of magnitude as that of its real part. A similar analysis can be done for the antibound state, except that it is purely imaginary and increases monotonously along the real axis.

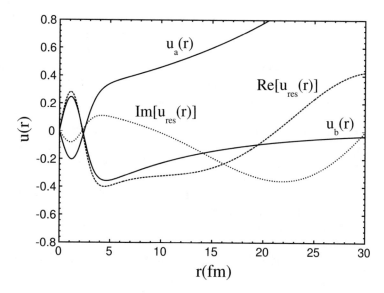

**Fig. 2.1** Illustration of bound, resonance, and antibound one-body wave functions of a Woods-Saxon potential. Neutron $1s_{1/2}$ wave functions are shown with solid lines for the bound and antibound states. The bound (antibound) state wave function is purely real (imaginary). Proton $1s_{1/2}$ resonance wave function is depicted with dashed lines. The proton wave function is complex, so that both its real and imaginary parts are shown

In fact, the one-body states shown in Fig. 2.1 can be naturally divided into:

- Bound states—lying on the imaginary positive $k$-axis (or negative real energy axis).

- Narrow decaying states—lying close to and below the real $k$-axis (or below the positive real-energy axis). Those states can be interpreted as physical resonances of the system.

- Other unbound states—such as broad resonance, antibound and virtual (i.e., bearing a negative real energy and a positive width) states. Due to their large width and/or negative energy, they cannot be associated to physical states. Unless one aims at explicitly studying their properties, these states are typically not considered in practical applications.

These definitions can be extended to many-body states in the complex energy (or momentum) plane. Numerical examples of bound, resonance or antibound on-body states are studied in Exercise XV.

Another important observable arising from the $S$-matrix is the scattering length. It is defined from phase shift at vanishing linear momentum via the so-called effective range expansion. It reads in the neutron case:

$$k \cot(\delta_\ell(k)) = -\frac{1}{a} + \frac{1}{2}r_0^2 k^2 \,, \tag{2.185}$$

where $a$ is the scattering length and $r_0$ is the effective range, as it is related to the distance up to which the nuclear interaction can act. This definition arises from the equation $u(k, r) = 0$, solved for $k \to 0$ and moderate $r$, using the asymptotic wave function of Eq. (2.7). Following the results of Sect. 2.5.1, there will be a zero of $u(k, r)$ for small $k > 0$ at large distance if there is a bound state of energy close to zero. Consequently, the scattering length is large and positive therein. Conversely, a negative scattering length implies that there is no bound state at low energy. This typically indicates the presence of a low energy antibound state for neutron $s$ states, as no resonance state can exist therein. As the scattering length can be measured experimentally, it can give insights to the position of its bound, resonance, and antibound states close to the particle emission threshold.

Equation (2.185) can be extended to the case of charged particles. However, due to its infinite range, additional terms have to be added to suppress Coulomb divergences at zero energy [43]:

$$\left( \frac{2\pi \eta}{\exp(2\pi \eta) - 1} \right) k \cotan(\delta_\ell(k)) + 2k\eta(-\ln(\eta) + \Re[\Psi(i\eta)] + \gamma) = -\frac{1}{a} + \frac{1}{2}r_0^2 k^2 ,$$

$$(2.186)$$

where $\eta$ is the Sommerfeld parameter of the charged particle, $\Psi(x)$ is the digamma function, and $\gamma$ is the Euler constant.

---

**Exercise  XV** ⋆

One will consider numerical examples of $u(k, r)$ when the one-body state is bound, resonance, or antibound and point out the dependence of width on orbital angular momentum $\ell$ and number of protons in the resonance case.

**A.** Run the one-particle code of radial wave functions to generate bound, resonance, antibound, and scattering states of a Woods-Saxon potential by varying its depth. Concentrate on states which are well-bound, close to zero with negative energy (bound and antibound states only) or positive energy (resonances), and on resonances with energies far from zero. One recalls that antibound states can only be obtained with $\ell = 0$ neutron states around zero energy, whereas resonance states of low energy can be obtained with $\ell > 0$ or in the case of charged particles for arbitrary partial wave. Indeed, a resonance state of low energy can develop only in the presence of a Coulomb or centrifugal barrier.

Plot obtained radial wave functions and comment on their asymptotic behavior.

**B.** Run the same code to generate plots of energies and linear momenta by changing the depth of the central part of Woods-Saxon potential, so that the wave function changes smoothly from bound to unbound. Plot obtained energies and linear momenta for a few angular momenta $\ell$ and a few different numbers of protons defining the Coulomb part of the Woods-Saxon potential and comment on obtained results.

Check that wave functions become resonance states above particle-emission threshold and explain the value of width as a function of $\ell$ and number of protons.

An important relation borne by the $S$-matrix is the dispersion relation [37] which provides the relation between $\Re(S_\ell(k))$ and $\Im(S_\ell(k))$. In particular, the dispersion relation allows to determine univocally phase shift in Eq. (2.183), because its experimental value is known only up to a $k$-dependent phase. It would be convenient to have a dispersion relation for both cases of charged and uncharged particles. However, as one has to integrate $S_\ell(k)$ in $k = 0$, therefore having a Coulomb potential in the one-body Hamiltonian is precluded because $S_\ell(k)$ possesses an essential singularity in $k = 0$ in this situation. Consequently, one will consider that the potential vanishes for $r > R$, that is, one considers only the neutron case.

One starts from the Cauchy integral expression of $(S_\ell(k) - 1)e^{ikR^{(s)}}$, with $2R < R^{(s)}$, using a contour $\mathscr{C}_K$ formed by the $[-K : K]$ segment and closed by a circle of radius $K$ in the upper half plane, with $K$ sufficiently large so that all the bound poles of $S_\ell(k)$ are taken into account:

$$
\begin{aligned}
(S_\ell(k) - 1)e^{ikR^{(s)}} = & \frac{1}{i\pi}\mathscr{P}\int_{-K}^{K}\frac{(S_\ell(k') - 1)e^{ik'R^{(s)}}}{k' - k}\,dk' \\
& + \frac{1}{i\pi}\int_{\mathscr{C}_K}\frac{(S_\ell(k') - 1)e^{ik'R^{(s)}}}{k' - k}\,dk' \\
& - 2\sum_{n}\mathrm{Res}\left[\frac{(S_\ell(k') - 1)e^{ik'R^{(s)}}}{k' - k}\right]_{k'=k_n}.
\end{aligned}
\tag{2.187}
$$

In this expression, $k$ is positive ($k > 0$), $\mathscr{P}$ stands for the principal part of the integral, and residues correspond only to the poles of $S_\ell(k)$. One will consider the asymptotic behavior of $S_\ell(k)$ for $|k| \to +\infty$ (see Eq. (2.182) and Sect. 2.6.2):

$$
S_\ell(k) = \mathscr{O}(e^{2|\Im(k)|R}).
\tag{2.188}
$$

The integrand on the right-hand side of Eq. (2.187) behaves as $\mathscr{O}(e^{-|\Im(k)|(R^{(s)}-2R)})$, so that the component of the Cauchy integral on the contour $\mathscr{C}_K$ vanishes for $K \to +\infty$, and one obtains:

$$
\begin{aligned}
(S_\ell(k) - 1)e^{ikR^{(s)}} = & \frac{1}{i\pi}\mathscr{P}\int_{-\infty}^{+\infty}\frac{(S_\ell(k') - 1)e^{ik'R^{(s)}}}{k' - k}\,dk' \\
& - 2\sum_{n}\mathrm{Res}\left[\frac{(S_\ell(k') - 1)e^{ik'R^{(s)}}}{k' - k}\right]_{k'=k_n}.
\end{aligned}
\tag{2.189}
$$

The crossing relation of $S_\ell(k)$ arises from Eqs. (2.7) and (2.182):

$$S_\ell(-k^*) = S_\ell(k)^* . \tag{2.190}$$

Using (2.190) in Eq. (2.189) one obtains:

$$
\begin{aligned}
(S_\ell(k) - 1)e^{ikR^{(s)}} &= \frac{1}{i\pi} \mathscr{P} \int_0^{+\infty} \left( \frac{(S_\ell(k') - 1)e^{ik'R^{(s)}}}{k' - k} + \frac{(S_\ell(k')^* - 1)e^{-ik'R^{(s)}}}{-k' - k} \right) dk' \\
&\quad - 2 \sum_n \text{Res} \left[ \frac{(S_\ell(k') - 1)e^{ik'R^{(s)}}}{k' - k} \right]_{k'=k_n} .
\end{aligned}
\tag{2.191}
$$

Equation (2.191) can then be written as a function of real and imaginary parts of $S_\ell(k)$, which one will also express as an energy integral:

$$
\begin{aligned}
\Re((S_\ell(E) - 1)e^{ikR^{(s)}}) &= \frac{1}{\pi} \mathscr{P} \int_0^{+\infty} \Im \left( \frac{(S_\ell(E') - 1)e^{ik'R^{(s)}}}{E' - E} \right) dE' \\
&\quad -2 \sum_n \Re \left( \text{Res} \left[ \frac{(S_\ell(k') - 1)e^{ik'R^{(s)}}}{k' - k} \right]_{k'=k_n} \right) .
\end{aligned}
\tag{2.192}
$$

Let us consider the convergence properties of the integral of Eq. (2.192). One has $|S_\ell(E)| = 1$ on the real $E$-axis, so that one only has to consider the convergence of the integral of Eq. (2.192) for $E' \to +\infty$. According to Eqs. (2.165) and (2.166):

$$S_\ell(E') - 1 = \mathcal{O}(\ln(E')\, E'^{-1/2}) , \tag{2.193}$$

so that integrand in Eq. (2.189) is equal $\mathcal{O}(\ln(E')\, E'^{-3/2})$ for $E' \to \pm\infty$ and is absolutely converging. However, the latter convergence might be too slow, as experimentally accessible energies are only a few hundreds of MeV. Hence, it would be convenient to accelerate the convergence of the integral in Eq. (2.192). For this, following the method used in Ref. [44] in the context of the dispersion relations of charged particles, one will consider Eq. (2.192) at two different energies $E$ and $E_0$, with the $E_0$-dependent formula subtracted from the $E$-dependent equation:

$$
\begin{aligned}
\Re((S_\ell(E) - 1)&e^{ikR^{(s)}} - (S_\ell(E_0) - 1)e^{ik_0R^{(s)}}) \\
&= (E - E_0)\frac{1}{\pi} \mathscr{P} \int_0^{+\infty} \Im \left( \frac{(S_\ell(E') - 1)e^{ik'R^{(s)}}}{(E' - E)(E' - E_0)} \right) dE' \\
&\quad -2(k - k_0) \sum_n \Re \left( \text{Res} \left[ \frac{(S_\ell(k') - 1)e^{ik'R^{(s)}}}{(k' - k)(k' - k_0)} \right]_{k'=k_n} \right) .
\end{aligned}
\tag{2.194}
$$

$k_0$ in this equation is the linear momentum associated to $E_0$. The integrand in Eq. (2.194) is now $\mathcal{O}(\ln(E') \, E'^{-5/2})$, so that the gain in precision is about one to two orders of magnitude. As $\Re(S_\ell(k))$ in Eq. (2.194) is a function of an integral involving $\Im(S_\ell(k))$, hence the knowledge of $|S_\ell(k)|$ implies $S_\ell(k)$.

The $S$-matrix is often considered for complex potentials as well. The addition of an imaginary part to the real potential of Eq. (2.2) breaks the unitarity of the $S$-matrix, so that the particle flux is no longer conserved. This idea is behind the optical model potential, where its absorptive, negative imaginary part accounts for the flux lost to the compound nucleus component in the scattering process. Hence, the complex optical potential arises from an attempted simplification of the description of the reaction [45–47], and can be considered as an effective way to take into account all of the physical processes which one does not want to consider explicitly.

In the exact formulation using a Hermitian Hamiltonian, a huge number of reaction channels should be taken into account to generate particle absorption into the compound nucleus configurations. This strategy in the description of direct reactions is not only impractical but also impossible to realize in any practical application. The phenomenological replacement of those reaction channels by the (negative) imaginary part of optical potential can be justified by the fact that the individual compound nucleus configurations cannot be studied in practice. Therefore, their effects can be represented by simple functions, which take the form of an imaginary part of the complex optical potential in practice.

### 2.6.7 Radial Equation and Shooting Method

The one-body bound, antibound, and resonances are calculated numerically by integrating Eq. (2.2) with the shooting method which is based on finding the zeros of the Jost function $J^+(k)$ of Eq. (2.169). For this, one has to provide a starting value $k_{\text{start}}$ for the linear momentum of the one-body state to calculate, which is then refined using the Newton method on $k$:

1. One considers $k_{\text{start}}$, close to the exact $k$ value of the resonant state. It is typically given by a diagonalization of the potential with a harmonic oscillator basis for well-bound states, and by a bisection method in the real $E$-axis or complex $k$-plane for loosely bound, antibound, and resonance states.
2. Equation (2.2) is integrated from zero to a matching radius $R_m$ and Eq. (2.6) is used for the boundary condition in $r = 0$. This provides with $u(R_m)$.
3. Equation (2.2) is integrated from $R$ to the matching radius $R_m$ and Eq. (2.7) is used for the boundary condition in $r = R$, along with $C^+ = 1$ and $C^- = 0$. This provides with $u^+(R_m)$.
4. $J^+(k)$ is then calculated from $u(R_m)$, $u^+(R_m)$ and their derivatives. $dJ^+(k)/dk$ is calculated from the standard difference formula $(J^+(k_p) - J^+(k))/(k_p - k)$, where $k_p$ is the linear momentum issued from the previous iteration. The linear

momentum $k$ is then updated from the standard Newton method formula $k \rightarrow k - J(k)/(dJ^+(k)/dk)$.

Points 1–4 are then iterated until convergence.

### 2.6.8    Calculation of the Width of a Metastable State

When a narrow resonance has a width much smaller than typically 1 keV, it is possible to calculate its width from the current formula. This property has an immediate physical interest, as it relates the width of the resonance to the particle flux, the latter being directly provided by the current formula. In particular, the current formula is of interest when the real and imaginary parts of energy differ by many orders of magnitude. Very narrow widths occur, for example, in proton emitters [48, 49], whose width is of the order of $10^{-22}$ MeV, as compared to their excitation energy, of the order of 1 MeV. Typical examples of proton emitters are $^{131}$Eu and $^{141}$Ho, which are unbound by 0.947 MeV and 1.190 MeV with respect to proton-emission threshold, and bear half-lives of 17.8 ms and 4.1 ms, respectively [48, 50]. In fact, particle emission widths in this situation are so small that they cannot be calculated by the direct integration of the radial wave function of Eq. (2.2) (see Sect. 2.6.7). In the following, we will see how to obtain the particle-emission width value from the current formula.

Let us consider a radius $r \geq R$, where $R$ is a large radius after which the nuclear potential is negligible and only Coulomb and centrifugal parts remain. The continuity equation for $u(k, r)$ implies that

$$u^*(k, r)u'(k, r) - u'^*(k, r)u(k, r) = (k^{*2} - k^2) \int_0^r |u(k, r')|^2 \, dr' . \qquad (2.195)$$

This equation is obtained by multiplying Eq. (2.2) by $u^*(k, r)$, subtracting the obtained equation from its complex conjugate, and integrating over $r$. Noticing that $k^{*2} - k^2$ is proportional to the width $\Gamma$ of $u(k, r)$, and using the standard mirror relation for Coulomb wave functions:

$$H_{\ell,\eta}^+(z)^* = H_{\ell,\eta^*}^-(z^*) ,$$

arising from the fact that both functions obey the same differential equation and behave as $\exp(-iz^* + i\eta^* \ln(2z^*))$ for $\Im(z) \rightarrow +\infty$, one obtains:

$$\Gamma = \left( \frac{\hbar^2}{m} \right) |C^+|^2 \left( \frac{k H_{\ell,\eta^*}^-(k^*r) H_{\ell,\eta}^+(kr)' - k^* H_{\ell,\eta^*}^-(k^*r)' H_{\ell,\eta}^+(kr)}{2i \int_0^r |u(k, r')|^2 \, dr'} \right) .$$

$$(2.196)$$

To get rid of the explicit $r$-dependence in Eq. (2.196), further approximations are necessary [48,49]. Neglecting $\Im(k)$ in the Coulomb wave functions of the numerator implies that their Wronskian becomes equal to $2i\Re(k)$. Moreover, as $u(k, r)$ is a metastable state, it decreases exponentially along the real $r$-axis in the asymptotic region. Thus, the integral in the denominator is almost equal to one when $r$ is chosen in the asymptotic region. Under these assumptions, valid for narrow resonances, Eq. (2.196) simplifies to:

$$\Gamma = \left(\frac{\hbar^2}{m}\right)|C^+|^2\Re(k).\qquad(2.197)$$

Numerical applications of Eq. (2.197) are effected in Exercise XVI.

Equation (2.197) can be further simplified so that it can be expressed as the product of a simple function of $k$ multiplied by a constant function of Hamiltonian parameters only. As proton and neutron states have very different asymptotes for $k \to 0$ (see Sect. 2.3), they give rise to different asymptotic formulas for the width therein. One will rewrite Eq. (2.7) in $r = R$ for $k \to 0$ using Eqs. (2.54) and (2.57), and omitting $k$-independent factors:

$$u(k, R) = C^+ H^+_{\ell,\eta}(kR) \propto C^+ C_\ell(\eta)^{-1} k^{-\ell}.\qquad(2.198)$$

In this expression, $\eta$ is the Sommerfeld parameter, equal to $(m/\hbar^2)Ze^2/k$ and $C_\ell(\eta)$ is the Gamow factor defined in Eq. (2.31).

If one considers neutrons, $\eta = 0$ and $C_\ell(\eta)^{-1} = (2\ell + 1)!!$. Thus, Eq. (2.198) implies that $C^+ \sim A_0 k^\ell$, with $A_0$ a constant depending on Hamiltonian parameters only. Indeed, $u(k, R)$ has a finite limit in $k = 0$, as one has a zero energy bound state due to the presence of a barrier, so that Eq. (2.197) has a well-defined limit for $k \to 0$. Equation (2.197) thus becomes for neutrons:

$$\Gamma \sim A_n \Re(k)^{2\ell+1},\qquad(2.199)$$

where $A_n$ is a constant depending on the neutron Hamiltonian only (see Exercise XVII for numerical applications).

---

**Exercise XVI** ⋆

One will calculate numerically proton resonance states in order to show that Eq. (2.197) is valid.

**A.** Run the one-particle code of radial wave functions to generate proton resonance states very close to zero energy, so that their width is very small ($\Gamma < 0.1$ keV). For this, one can fit a potential width for a loosely bound state, for example, $-0.1$ keV, and then slowly vary the potential depth so that the calculated state becomes a resonance.

**B.** Notice that both numerical integration and current formula give the same results when $\Gamma > 10^{-6}$ keV. Check that only the current formula can provide with a nonzero width when $\Gamma < 10^{-6}$ keV.

In the proton case, $u(k, R)$ becomes bound at zero energy as well, so that one obtains similarly to the neutron case:

$$C^+ \sim A_0 \, C_\ell(\eta) \, k^\ell \sim A_1 \, \exp(-\pi \eta) \, \eta^{1/2} , \tag{2.200}$$

where $A_0$ and $A_1$ are constants depending on Hamiltonian parameters only, and where Eq. (2.57) has been used. Thus Eq. (2.197) for protons becomes:

$$\Gamma \sim A_p \, \exp(-2\pi \, \Re(\eta)) , \tag{2.201}$$

where $A_p$ is a constant depending on the proton Hamiltonian only (see Exercise XVII for numerical applications). Note that approximate expressions for the energy and width of resonance states have been developed in the context of the two-potential method [51], where only bound and real-energy scattering solutions of Eq. (2.2) are needed for their evaluation.

---

**Exercise  XVII ⋆**

One will demonstrate in a numerical example that the relations between energy and width in the resonance case of Eqs. (2.199) and   (2.201) hold in practice.

**A.** Run the one-particle code of radial wave functions to generate plots of complex linear momenta, energies, and widths close to the particle emission threshold, for both proton and neutron case, and different orbital angular momenta $\ell$.

**B.** Determine the overall factor $A$ in Eqs. (2.199) and   (2.201), so that the asymptotic behavior of width in the vicinity of particle emission threshold is reproduced.

**C.** Check that the expressions in Eqs. (2.199), and (2.201) provide with the correct behavior of $\Gamma$ in the vicinity of particle emission threshold.

---

## Solutions to Exercises[1]

**Exercise I. A.** The Picard method is an iterative process determining the solution of Eq. (2.2) from the repeated action of Eq. (2.9):

$$u^{(n+1)}(k,r) = C_0 \, F_{\ell_0,\eta_0}(k_0 r_0) + C_0 \, k_0 \, F'_{\ell_0,\eta_0}(k_0 r_0) \, (r - r_0)$$

$$+ \int_{r_0}^{r} (r - r') \mathcal{L}(k, r') u^{(n)}(k, r') \, dr' , \qquad (2.202)$$

where $n \geq 0$ and $u^{(0)}(k, r) = 0$. As one imposes $C_0 = k_0^{-\ell_0 - 1} C_{\ell_0}(\eta_0)^{-1}$, Eq. (2.37) implies that $C_0 \, F_{\ell_0,\eta_0}(k_0 r)$ is an entire power series in $k^2$. Therefore, $u^{(n)}(k, r)$ is well defined in the whole complex $k$-plane.

One will show that $u^{(n)}(k, r) \to u(k, r)$ for $n \to +\infty$. An immediate recurrence in Eq. (2.202) implies that $\forall n \geq 1$:

$$|u^{(n+1)}(k, r) - u^{(n)}(k, r)| = \left| \int_{r_0}^{r} (r - r^{(1)}) \mathcal{L}(k, r^{(1)}) \; \ldots \right.$$

$$\left. \times \int_{r_0}^{r^{(n-1)}} (r - r^{(n)}) \mathcal{L}(k, r^{(n)}) [u^{(1)}(k, r^{(n)}) - u^{(0)}(k, r^{(n)})] \, dr^{(n)} \; \ldots \; dr^{(1)} \right|$$

$$\leq r^n ||\mathcal{L}||^n ||u^{(1)} - u^{(0)}||_\infty \int_{r_0}^{r} \ldots \int_{r_0}^{r^{(n-1)}} dr^{(n)} \; \ldots \; dr^{(1)}$$

$$\leq \frac{r^{2n} ||\mathcal{L}||^n}{n!} ||u^{(1)} - u^{(0)}||_\infty , \qquad (2.203)$$

where the norm $||\mathcal{L}||$ is defined as:

$$||\mathcal{L}|| = \frac{\ell(\ell + 1)}{r_0^2} + ||v_l||_\infty + |k|^2 , \qquad (2.204)$$

and all supremum norms are defined for $r_0 \leq r \leq R$ (same for r'). Equation (2.203) implies that:

$$\sum_n |u^{(n+1)}(k, r) - u^{(n)}(k, r)| \leq \exp(r^2 ||\mathcal{L}||) \, ||u^{(1)} - u^{(0)}||_\infty . \qquad (2.205)$$

---

[1]The input files, codes and code user manual associated to computer-based exercises can be found at https://github.com/GSMUTNSR.

Consequently, $u^{(n)}(k, r) \to u(k, r)$ for $n \to +\infty$ and the existence of $u(k, r)$ is demonstrated. It is straightforward to show that one has $\forall N > n$:

$$|u^{(n)}(k, r) - u(k, r)| \le |u^{(N)}(k, r) - u(k, r)| + \sum_{m=n}^{N-1} |u^{(m+1)}(k, r) - u^{(m)}(k, r)| .$$

(2.206)

Hence, one obtains for $N \to +\infty$:

$$|u^{(n)}(k, r) - u(k, r)| \le \sum_{m=n}^{+\infty} \left( \frac{r^{2m} ||\mathcal{L}||^m}{m!} \right) ||u^{(1)} - u^{(0)}||_\infty ,$$

(2.207)

where the point-wise convergence of $u^{(N)}(k, r)$ to $u(k, r)$ for $N \to +\infty$ and Eq. (2.203) have been used. Therefore, $\sup_{k \in \mathcal{D}} |u^{(n)}(k, r) - u(k, r)| \to 0$ in every finite domain $\mathcal{D}$ of the complex $k$-plane, proving uniform convergence therein.

**B.** The unicity of $u(k, r)$ will be demonstrated using the recurrence relation scheme of Eq. (2.9). One will consider two solutions of Eq. (2.9), whose functions and derivatives are equal in $r = r_0$, and whose difference is equal to a function $\Delta u(k, r) \ne 0$. From Eq. (2.9), $\Delta u(k, r)$ obeys the following integral equation:

$$\Delta u(k, r) = \int_{r_0}^{r} (r - r') \mathcal{L}(k, r') \Delta u(k, r') \, dr' ,$$

(2.208)

as $\Delta u(k, r_0) = \Delta u'(k, r_0) = 0$. Similarly, using Eq. (2.203) one obtains:

$$|\Delta u(k, r)| \le \frac{r^{2n} ||\mathcal{L}||^n}{n!} ||\Delta u||_\infty ,$$

as $\Delta u(k, r)$ is also a solution of Eq. (2.9). One obtains immediately $\Delta u(k, r) = 0$ with $n \to +\infty$, so that assuming $\Delta u(k, r) \ne 0$ leads to a contradiction. This implies that $u(k, r)$ is unique.

**Exercise II.** The methods to calculate the harmonic oscillator eigenstates and eigenenergies in the one- and three-dimensional cases are analogous so that they will be described altogether. The inclusion of Eq. (2.12) in Eqs. (2.11) and (2.17) in Eq. (2.16) provides the Hermite and Laguerre equations:

$$\frac{d^2 H_{n_x}}{dx^2}(x) - 2x \frac{d H_{n_x}}{dx}(x) + 2n_x H_{n_x}(x) = 0$$

$$x \frac{d^2 L_n^{(\ell+1/2)}}{dx^2}(x) + \left( \ell + \frac{3}{2} - x \right) \frac{d L_n^{(\ell+1/2)}}{dx}(x) + n L_n^{(\ell+1/2)}(x) = 0 ,$$

which are obtained using the energies of the harmonic oscillator states of Eqs. (2.14) and (2.19). The constant factor in Eqs. (2.12), and (2.18) arise from the orthonormalization of Hermite and Laguerre polynomials:

$$\int_{-\infty}^{+\infty} e^{-x^2} H_{n_x}(x) \, H_{m_x}(x) \, dx = \sqrt{\pi} \, 2^{n_x} \, n_x! \, \delta_{n_x \, m_x}$$

$$\int_0^{+\infty} x^{\ell+1/2} e^{-x} L_n^{(\ell+1/2)}(x) \, L_m^{(\ell+1/2)}(x) \, dx = \left( \frac{\Gamma(n+\ell+3/2)}{\Gamma(n+1)} \right) \delta_{nm} .$$

As $V_{HO}(r) = V_{HO}(x) + V_{HO}(y) + V_{HO}(z)$, the three-dimensional harmonic oscillator eigenstate $u_{HO}^n(r)/r \, Y_m^\ell(\theta, \varphi)$ (see Eq. (2.1)), of energy $e_{n\ell}$, has to be a linear combination of the states $u_{HO}^{n_x}(x) \, u_{HO}^{n_y}(y) \, u_{HO}^{n_z}(z)$, for which $e_{n_x} + e_{n_y} + e_{n_z} = e_{n\ell}$.

Clearly, several triplets $(n_x, n_y, n_z)$ occur in general. Firstly, this explains the appearance of a Gaussian term in Eq. (2.17) from those of Eq. (2.12), as $e^{-r^2/(2b^2)} = e^{-x^2/(2b^2)} e^{-y^2/(2b^2)} e^{-z^2/(2b^2)}$. Furthermore, $L_n^{(\ell+1/2)}(r^2/b^2)$ is a multivariate polynomial in $x^2$, $y^2$, and $z^2$ of degree $n$, and $r^\ell \, Y_m^\ell(\theta, \varphi)$, a solution of the Laplace equation, is a multivariate polynomial in $x$, $y$, and $z$ of degree $\ell$. Consequently, the function $r^\ell \, Y_m^\ell(\theta, \varphi) \, L_n^{(\ell+1/2)}(r^2/b^2)$ is a multivariate polynomial in $x$, $y$, and $z$ of degree $2n + \ell$. Hence, it can be expanded with the set of multivariate polynomials $H_{n_x}(x/b) \, H_{n_y}(y/b) \, H_{n_z}(z/b)$ whose triplets $(n_x, n_y, n_z)$ verify $n_x + n_y + n_z \leq 2n + \ell$, and hence $n_x + n_y + n_z = 2n + \ell$, as one must have $e_{n_x} + e_{n_y} + e_{n_z} = e_{n\ell}$. Consequently, the similarity between Eqs. (2.12), (2.14) and (2.17)–(2.19) explicitly appears. Indeed, the three-dimensional Gaussian term is simply the product of one-dimensional Gaussian terms, while the rest of wave functions in Eqs. (2.12) and (2.17) can always be written as multivariate polynomials in $x$, $y$ and $z$.

**Exercise III.**

**A.** Using Eqs. (2.21)–(2.24), one immediately obtains that Eq. (2.26) equals:

$$\frac{\mathbf{p}_{rel}^2}{2m_{rel}} + \frac{\mathbf{P}_{CM}^2}{2M} + \frac{1}{2} m_{rel} \omega^2 \mathbf{r}_{rel}^2 + \frac{1}{2} M \omega^2 \mathbf{R}_{CM}^2 \tag{2.209}$$

where $m_{rel} = m/2$ and $M = 2m$. One can see that Eq. (2.209) has the same structure as Eq. (2.26), that is, that Eq. (2.209) is made of two independent relative and center of mass harmonic oscillator Hamiltonians. Consequently, the eigenstates of Eq. (2.26) can be written as the tensor products of harmonic oscillator states $|n_1 \ell_1\rangle \otimes |n_2 \ell_2\rangle$ or $|n_{rel} \ell_{rel}\rangle \otimes |N_{CM} L_{CM}\rangle$.

**B.** Two eigenstates of a Hamiltonian with two different energies are orthogonal, so that Eq. (2.25) holds. The number of eigenstates of Eq. (2.26) with energy $E$ is finite, as it is the combination of all one-body harmonic oscillator $|n_1 \ell_1\rangle$ and $|n_2 \ell_2\rangle$ states of energy $e_1$ and $e_2$ for which $e_1 + e_2 = E$. According to **A**, the situation is similar if one considers the combination of all one-body harmonic oscillator $|n_{rel}\ell_{rel}\rangle$ and $|N_{CM}L_{CM}\rangle$ states, which are eigenstates of the relative and center of mass parts of Eq. (2.209), respectively. The respective eigenenergies of $|n_{rel}\ell_{rel}\rangle$ and $|N_{CM}L_{CM}\rangle$ indeed verify $e_{rel} + E_{CM} = E$.

Consequently, for a given energy $E$, an eigenstate of the Hamiltonian of Eq. (2.26) defined with laboratory coordinates is then a finite linear combination of those defined with relative/center-of-mass coordinates of the same energy using Eq. (2.25). This proves the existence of Eq. (2.20), as the sum therein is necessarily finite.

**Exercise IV.** Coulomb wave functions are in general not analytic on the negative real axis and there is a cut therein. $F_{\ell,\eta}(z)$ is clearly not analytic for $\ell$ non-integer, while it is entire when $\ell$ is an integer (see Eq. (2.37)). Conversely, $H^{\omega}_{\ell,\eta}$ always has a cut on the negative real axis for $\eta \neq 0$ (see Sect. 2.3.2). Thus, Coulomb wave functions present a discontinuity when crossing the negative real axis. Evidently, if Coulomb wave functions vary by several orders of magnitude for slowly varying $z$, the magnitude of the discontinuity on the negative real axis can be expected to be large as well.

**Exercise V.** Determine inequalities from Eqs. (2.73) and (2.74) replacing $y$ by 1 in their $\Lambda$ dependent term and conclude using the fact that $y \in [0 : 1]$.

**Exercise VI.** To derive Eq. (2.85), use the Taylor expansions of arctan and arctanh in Eqs. (2.73) and (2.74) for $r \to 0$.

As $y \to 1$ when $r \to +\infty$, Eqs. (2.86) and (2.87) are directly obtained from Eqs. (2.73), (2.74), (2.81), and (2.84).

**Exercise VII.** The fact that $V_{PTG}(r) + \ell(\ell + 1)/r^2 \sim \ell(\ell + 1)/r^2$ for $r \to 0$ and $V_{PTG}(r) \to 0$ tend to zero exponentially for $r \to +\infty$ is a direct implication of the results of Exercises V and VI. As $V_{PTG}(r) + \ell(\ell + 1)/r^2$ is negligibly small in the asymptotic region, the effective orbital angular momentum in the asymptotic region is equal to zero, so that all bound and resonance states behave like neutron $s$-states.

**Exercise VIII.** As there is an effective mass in Eq. (2.72), it is customary to consider the reduced wave function $u_0(r) = u(r)/\sqrt{\mu(r)}$. Inserting this ansatz in Eq. (2.72) provides with a simpler equation:

$$-u_0''(r) + \frac{\ell(\ell + 1)}{r^2}u_0(r) + s^2(V_\ell(r) + V_c(r))u_0(r) = \left(\frac{2m_0}{\hbar^2}\right)e\,\mu(r)\,u_0(r) ,$$

$$(2.210)$$

where one has used Eq. (2.92) and where the potential $V_\mu(r)$ is no longer present. One then has to check that the wave function of Eq. (2.96) verifies Eq. (2.210).

As Eq. (2.96) explicitly depends on $g$, it is preferable to differentiate with respect to $g$. It is convenient to use the following concise expressions:

$$y'(r) = s(1 - y^2)(1 + \Lambda^2 y^2 - y^2) \tag{2.211}$$

$$\frac{d^2}{dr^2} = \frac{d^2 g}{dr^2} \frac{d}{dg} + \left(\frac{dg}{dr}\right)^2 \frac{d^2}{dg^2} \tag{2.212}$$

$$\frac{dg}{dr} = 2\Lambda^2 s\, y\, f \tag{2.213}$$

$$\frac{d^2 g}{dr^2} = 2\Lambda^2 s^2 \left((1 - y^2)^2 - 2\Lambda^2\, y^2\, f\right), \tag{2.214}$$

where Eqs. (2.73), (2.74), (2.88), and (2.106) have been used. Note that the differentiation of $f$ with respect to $g$ is immediate as $f + g = 1$.

Equation (2.210) is then obtained after a tedious but straightforward calculation, in which one has to use the differential equation verified by the hypergeometric function:

$$z(1 - z)\, {}_2F_1''(a, b, c; z) + (c - (a + b + 1)z)\, {}_2F_1'(a, b, c; z)$$
$$- ab\, {}_2F_1(a, b, c; z) = 0. \tag{2.215}$$

Equation (2.97) is straightforward to derive from (2.96) by using a property (2.110) of the hypergeometric function.

**Exercise IX.** Equations (2.114) and (2.115) are valid for the resonant eigenstates of the Pöschl-Teller-Ginocchio Hamiltonian. Hence, one must have $C^- = 0$ in Eq. (2.111). Equations (2.103)–(2.105) provide with $C^- = 0$ if $1/\Gamma(\nu^+) = 0$ or $1/\Gamma(\nu^-) = 0$, so that $\nu^\pm = -n$, $n \in \mathbb{N}$. Using Eqs. (2.103) and (2.104), one obtains a quadratic equation in $\bar{\beta}$:

$$\Lambda^2(1 - a)\bar{\beta}^2 + 2\left(N + \frac{1}{2}\right)\bar{\beta} + \left(N + \frac{1}{2}\right)^2 - \left(\nu + \frac{1}{2}\right)^2 = 0, \tag{2.216}$$

where $N = 2n + \ell + 1$. Solving Eq. (2.216) and using Eq. (2.103) provides with Eqs. (2.114)–(2.116).

**Exercise X.**

**A.** Phase shifts are immediate from Eqs. (2.111) and (2.183):

$$\delta = \frac{1}{2i} \ln\left(-\frac{C^+}{C^-}\right) = \frac{1}{2i} \ln\left(-e^{-2ikr_1} \frac{\mathscr{C}^+}{\mathscr{C}^-}\right) = \frac{1}{2i} \ln\left(-\frac{\mathscr{C}^+}{\mathscr{C}^-}\right) - kr_1, \tag{2.217}$$

with $\mathscr{C}^+$, $\mathscr{C}^-$, and $r_1$ provided by Eqs. (2.105) and (2.113).

**B.** Phase shifts vary smoothly using $V_{PTG}(r)$, due to the absence of centrifugal or Coulomb barrier. This implies that no narrow resonance can exist, as narrow resonances always generate important variations of phase shifts near the resonance energy. Conversely, the numerical examples considered present a narrow resonance in calculations using $V_{PTG-mod}(r)$ and Woods-Saxon potentials, which is visible on phase shifts by their rapid change therein.

**C.** The second-order approximation is, in principle, valid when $|\delta k| \ll |k|$ in Eq. (2.127), that is, when one considers narrow resonances. Indeed, numerical calculations show that widths issued from the second-order approximation whose value is slightly smaller or close to $100\,\text{keV}$ are almost exact, whereas their precision deteriorates after $1\,\text{MeV}$. However, very small widths are not well reproduced in absolute value, as the width provided by the second-order approximation is typically too large. Indeed, even if the width arising from the second-order approximation is small, it varies by several orders of magnitude when $\Gamma \ll E$. The current formula of Sect. 2.6.8 is, in fact, needed in this case, as it is very precise when widths are much smaller than $1\,\text{keV}$.

**D.** As the Woods-Saxon and modified Pöschl-Teller-Ginocchio potentials are not exactly equal, their phase shifts and resonance energies slightly differ. The quick change of phase shifts is directly related to the small value of the width of the narrow resonance.

**Exercise XI.** To show this property, let us multiply the Schrödinger equation (2.2) by $u^*(k,r)$ and integrate from 0 to $+\infty$:

$$\int_0^{+\infty} u''(k,r)u^*(k,r)\,dr = \int_0^{+\infty} \left(\frac{\ell(\ell+1)}{r^2} + v_l(r) - k^2\right) u(k,r)u^*(k,r)\,dr$$

$$\Rightarrow k^2 \int_0^{+\infty} |u(k,r)|^2\,dr = \int_0^{+\infty} |u'(k,r)|^2\,dr$$

$$+ \int_0^{+\infty} \left(\frac{\ell(\ell+1)}{r^2} + v_l(r)\right) |u(k,r)|^2\,dr\ .\quad (2.218)$$

In order to obtain Eq. (2.218), one effected an integration by parts:

$$\int_0^{+\infty} u''(k,r)u^*(k,r)\,dr = \left[u'(k,r)u^*(k,r)\right]_{r=0}^{r=+\infty} - \int_0^{+\infty} |u'(k,r)|^2\,dr\ .$$

$$(2.219)$$

The first term on the right-hand-side vanishes because $u(k,0) = 0$ and $u(k,r) \to 0$ for $r \to +\infty$. It is then clear from Eq. (2.218) that $k^2$ has to be real if $u(k,r)$ is a bound state.

**Exercise XII.**

A. Let us show that Eq. (2.159) is equivalent to Eq. (2.2) with the boundary conditions of Eq. (2.6). If one differentiates twice Eq. (2.159), one obtains Eq. (2.2). Moreover, the boundary conditions of Eq. (2.6) are shown to be fulfilled by Eq. (2.159) by considering $u(k, r)$ and $u'(k, r)$ in $r = r_0$. Therefore, due to the unicity of solutions of Eq. (2.2) for fixed boundary conditions (see Exercise I), the solutions of Eq. (2.159), and of Eq. (2.2) using the boundary conditions of Eq. (2.6), are the same.

B. The inequality majorizing $\left| g(r, r') \, \Delta V(r') \right|$ is obtained using the asymptotic expansions of $F_{\ell_0, \eta_0}(k_0 r)$ and $G_{\ell_0, \eta_0}(k_0 r)$ of Eqs. (2.62) and (2.63) for $|k| \to +\infty$.

One will reuse the method and notations of Exercise I to show the requested inequality. Let us firstly consider the $n = 0$ case. One has $u^{(0)}(k, r) = 0$ and $u^{(1)}(k, r) = C_0 \, F_{\ell_0, \eta_0}(k_0 r)$, so that $|u^{(1)}(k, r) - u^{(0)}(k, r)| \leq 2 \, |C_0| \, e^{|\Im(k)| r}$ when $|k|$ is sufficiently large (see Eq. (2.62)). The inequality involving $|u^{(n+1)}(k, r) - u^{(n)}(k, r)|$ is then obtained similarly to that of I. Following the method described in Exercise I, one has then:

$$|u(k, r)| \leq 2 \, \exp\left( Cr|k|^{-1} \right) \, |C_0| \, e^{|\Im(k)| r} . \tag{2.220}$$

The first exponential term is equal to $1 + \mathcal{O}(k^{-1})$ when $|k| \to +\infty$. Consequently, $u(k, r)$ cannot diverge faster than $|C_0| \, e^{|\Im(k)| r}$ when $|k| \to +\infty$ and $r_0 \leq r \leq R$. Therefore, this is the case if $0 < r \leq R$ because $u(k, r) = C_0 \, F_{\ell_0, \eta_0}(k_0 r)$ when $0 < r \leq r_0$.

C. Let us devise the analogous integral equation of Eq. (2.159) defining $u^{\pm}(k, r)$:

$$u^{\pm}(k, r) = H_{\ell, \eta}^{\pm}(kr) + \int_R^r \left( \frac{H_{\ell, \eta}^{+}(kr) H_{\ell, \eta}^{-}(kr') - H_{\ell, \eta}^{+}(kr') H_{\ell, \eta}^{-}(kr)}{2ik} \right)$$
$$\times \left( v_l(r') - \frac{v_c}{r'} \right) u(k, r') \, dr' . \tag{2.221}$$

$u^{\pm}(k, r)$ diverges when $\mp \Im(k) \to +\infty$, so that one considers only these zones of the complex $k$-plane.

One can see now that $r' \geq r$ in Eq. (2.221), so that the exponential term of Eq. (2.162) becomes $e^{|\Im(k)|(r'-r)}$. Thus, the exponential term to insert in the Picard method must be of the form $e^{|\Im(k)|(r_1-r)}$, with $r_1$ fixed. One can check that the inequality valid therein in the $n = 0$ case if $r_0 \leq r \leq R$ in the Picard algorithm is:

$$\left| u^{\pm}(k, r)^{(1)} - u^{\pm}(k, r)^{(0)} \right| \leq 2 \, e^{|\Im(k)|(2R-r)} .$$

Thus, applying the Picard algorithm to Eq. (2.221) would provide a $e^{|\Im(k)|(2R-r)}$ factor instead of $e^{|\Im(k)|r}$ as in Eq. (2.220). Consequently, $u^{\pm}(k, r)$ functions can be expected to diverge as $e^{|\Im(k)|(2R-r)}$ when $\mp\Im(k) \to +\infty$.

## Exercise XIII.

**A.** If $0 \le r \le r_0$, then $u(k, r) = C_0 \, F_{\ell_0, \eta_0}(k_0 r)$ (see Eq. (2.6)). As one imposes $C_0 = k^{-\ell_0-1} C_{\ell_0}(\eta_0)^{-1}$, therefore $C_0 \, F_{\ell_0, \eta_0}(k_0 r)$ is an entire power series in $k^2$ (see Exercise I). Consequently, $u(k, r)$ is an entire function of $k$ if $0 \le r \le r_0$.

If $r > r_0$, the analyticity of $u^{(n)}(k, r)$ is firstly proved by induction using Eq. (2.202). For this, one notices that $u^{(n+1)}(k, r)$ is the sum of $C_0 \, F_{\ell_0, \eta_0}(k_0 r_0)$, $C_0 \, k_0 \, F'_{\ell_0, \eta_0}(k_0 r_0) \, (r - r_0)$ and of the $r$-integral of the analytic function $(r - r')\mathscr{L}(k, r')u^{(n)}(k, r')$. The domain of analyticity of $u^{(n)}(k, r)$ is clearly identical to that of $C_0 \, F_{\ell_0, \eta_0}(k_0 r_0)$, that is, that $u^{(n)}(k, r)$ is an entire function. $u^{(n)}(k, r)$ converges uniformly to $u(k, r)$ on any finite sector of the complex plane (see Exercise I). Thus, for $n \to +\infty$ by integrating $u(k, r)$ along a closed complex contour $\mathscr{C}$, one obtains :

$$\left| \oint_{\mathscr{C}} u(k, r) \, dk \right| = \left| \oint_{\mathscr{C}} \left( u(k, r) - u^{(n)}(k, r) \right) dk \right|$$

$$\le L(\mathscr{C}) \sup_{k \in \mathscr{C}} |u^{(n)}(k, r) - u(k, r)| \to 0 , \qquad (2.222)$$

where $L(\mathscr{C})$ is the length of the $\mathscr{C}$ contour and the analyticity of $u^{(n)}(k, r)$ has been used. $u(k, r)$ is then an entire function of $k$ if $C_0 = k_0^{-\ell_0-1} C_{\ell_0}(\eta_0)^{-1}$.

**B.** One now considers the analytic properties of $u^{\pm}(k, r)$ function as a function of $k$. It is straightforward to verify that one can follow the same method as for $u(k, r)$ in **A**, up to the replacements of $r_0$ by $R$ and $C_0 \, F_{\ell_0, \eta_0}(k_0 r)$ by $H^{\pm}_{\ell, \eta}(kr)$. However, the fundamental difference here is that $H^{\pm}_{\ell, \eta}(kr)$ is not analytic in general, as it bears a cut on the negative $k$-axis unless $\ell$ is an integer and $\eta = 0$. Consequently, the domain of analyticity of $u^{\pm}(k, r)$ follows that of $H^{\pm}_{\ell, \eta}(kr)$, that is, $u^{\pm}(k, r)$ bears in general a cut on the negative $k$-axis, while it is analytic in the rest of the complex $k$-plane.

$u^{\pm}(k, r)$ is an entire function of $k$ if $\ell = 0$ and $\eta = 0$, as then $u^{\pm}(k, r) = \exp(\pm ikr)$. $u^{\pm}(k, r)$ is an analytic function of $k$ everywhere in the complex plane except in $k = 0$ if $\eta = 0$ and $\ell \in \mathbb{N}^*$. In this latter case, $u^{\pm}(k, r)$ is indeed a spherical Bessel function of the third kind.

**Exercise XIV.** One has shown in this chapter that the nonzero poles $k_n$ of the $S$-matrix generated by a real potential are simple. Thus, residues of the $S$-matrix arise from Eq. (2.182) by differentiating its denominator only and evaluating it at $k = k_n$:

$$
\begin{aligned}
\text{Res}\,[S_\ell(k)]_{k=k_n} &= (-2ik_nC^+)\left(\left[\frac{dJ^+}{dk}(k)\right]_{k=k_n}\right)^{-1} \\
&= (-2ik_nC^+)\left(2k_n\int_0^{+\infty} u(k_n,r)u^+(k_n,r)\,dr\right)^{-1} \\
&= \left(i\int_0^{+\infty} u^+(k_n,r)^2\,dr\right)^{-1}.
\end{aligned}
\tag{2.223}
$$

## Exercise XV.

**A.** Bound wave functions are real and decrease exponentially at infinity.
Inside of a nucleus, wave function of a narrow resonance is similar to bound state wave function, as the real part of the resonance wave function resembles the bound state having the same number of nodes in the nuclear interior and its imaginary part is very small therein. Real and imaginary parts become of the same order of magnitude in the asymptotic region, as the outgoing Hankel and Coulomb wave functions are therein equal to $\exp(ikr)$ multiplied by a smoothly varying function. They increase exponentially with oscillations on the real axis, but this increase can be mild if the width is very small.
Wave functions of broad resonances always have large real and imaginary parts, and increase quickly in modulus on the real axis. There is no resemblance with bound states.
Wave functions of low energy antibound states are similar to bound state wave functions for very small radii, but quickly reach their asymptote, which is exponential without oscillations. The antibound wave functions are also purely imaginary.

**B.** One can see from the obtained plots that widths depend strongly on orbital angular momentum $\ell$ and total charge Z. Indeed, the increase of width with energy is less and less visible when $\ell$ and Z are augmented, with the effect more pronounced for the charge than for the angular momentum $\ell$. The centrifugal barrier, in $\ell(\ell + 1)/r^2$, is less confining than the Coulomb barrier, in $Z/r$. In the absence of barrier in the asymptotic region, that is, for neutron $s$ states, one cannot generate resonance states when the potential becomes more and more shallow. In fact, when decreasing the potential depth one goes from the bound to antibound region.

## Exercise XVI.

**A.** It is necessary to slowly change potential depth because width varies quickly with energy.

**B.** Calculations show that energy and width must be of the same order of magnitude for direct integration to be able to provide both of them accurately.

Both direct integration and current formula can be used when $\Gamma > 10^{-6}$ keV, as the current formula is very precise for widths smaller than typically 1 keV and $\Gamma$ is still sufficiently large for direct integration to be numerically stable.

Values provided by the current formula when $\Gamma < 10^{-6}$ keV are virtually exact, as the real part of the wave function can always be calculated precisely with direct integration and the current formula has already reached its domain of validity when $\Gamma \sim 10^{-6}$ keV. Conversely, the imaginary part of the energy quickly becomes too imprecise to provide with $\Gamma$ when $\Gamma \ll 10^{-6}$.

**Exercise XVII.**

**A.** The different behavior of width as a function of energy is clearly visible and changes according to the proton or neutron case, and to the considered partial wave.

Width decreases quickly with decreasing energy for large orbital angular momentum in the proton or neutron case, and for nuclei bearing an important proton charge in the proton case.

**B.** The polynomial relation between energy and width for neutrons and exponential relation depending on $\eta$ for protons is obtained by fitting plots of widths as a function of $\Re(k)$.

**C.** The considered expression is valid as long as the width provided by the current formula is sufficiently small, that is, typically smaller than 10 keV.

# References

1. P.M. Morse, H. Feshbach, *Methods of Theoretical Physics* (Mc Graw-Hill, New York, 1953)
2. A. Messiah, *Quantum Mechanics*, vol. 1 and 2 (North Holland, Amsterdam, 1961)
3. A.D. Alhaidari, Phys. Rev. A **66**, 042116 (2002)
4. I. Talmi, Helv. Phys. Acta **25**, 185 (1952)
5. M. Moshinsky, Nucl. Phys. **13**, 104 (1959)
6. I.J. Thompson, A.R. Barnett, Comput. Phys. Comm. **36**, 363 (1985)
7. N. Michel, Comput. Phys. Comm. **176**, 232 (2007)
8. M. Abramowitz, *Handbook of Mathematical Functions, National Bureau of Standards, Applied Mathematics*, vol. 55, ed. by M. Abramowitz, I.A. Stegun (National Bureau of Standards, Gaithersburg, 1972)
9. J. Humblet, L. Rosenfeld, Nucl. Phys. **26**, 529 (1961)
10. K.S. Kölbig, Comput. Phys. Comm. **4**, 221 (1972)
11. I.J. Thompson, A.R. Barnett, J. Comput. Phys. **64**, 490 (1986)
12. A. Dzieciol, S. Yngve, P.O. Fröman, J. Math. Phys. **40**, 6145 (1999)
13. N. Mukunda, Am. J. Phys. **49**, 910 (1978)
14. C. Eckart, Phys. Rev. **35**, 1303 (1930)
15. N. Rosen, P.M. Morse, Phys. Rev. **42**, 210 (1932)
16. L. Hulthén, Ark. Mat. Astron. Fys. **28**, 5 (1942)
17. M.F. Manning, N. Rosen, Phys. Rev. **44**, 953 (1933)
18. S.H. Dong, J. Garcia-Ravelo, Phys. Scr. **75**, 307 (2007)
19. G.A. Natanzon, Vestn. Leningr. Univ. Fiz. **10**, 22 (1971)
20. G.A. Natanzon, Theor. Math. Phys. **38**, 146 (1979)

21. C. Rojas, V.M. Villalba, Phys. Rev. A **71**, 052101 (2005)
22. G. Jian-You, F.X. Zheng, X. Fu-Xin, Phys. Rev. A **66**, 062105 (2002)
23. G. Jian-You, J. Meng, X. Fu-Xin, Chin. Phys. Lett. **20**, 602 (2003)
24. Y. Sucu, N. Ünal, J. Math. Phys. **48**, 052503 (2007)
25. F. Cooper, J.N. Ginocchio, A. Khare, Phys. Rev. D **36**, 2458 (1987)
26. P. Cardero, S. Salamo, J. Math. Phys. **35**, 3301 (1994)
27. L. Chetouani, L. Guechi, A. Lecheheb, T.F. Hammann, J. Math. Phys. **34**, 1257 (1993)
28. J.N. Ginocchio, Ann. Phys. **152**, 203 (1984)
29. J.N. Ginocchio, Ann. Phys. **159**, 467 (1985)
30. L. Chetouani, L. Guechi, A. Lecheheb, T.F. Hammann, Czech. J. Phys. **45**, 699 (1995)
31. K. Bennaceur, J. Dobaczewski, M. Płoszajczak, Phys. Rev. C **60**, 034308 (1999)
32. M.V. Stoitsov, S. Dimitrova, S. Pittel, P. Van Isacker, Phys. Lett. B **415**, 1 (1997)
33. M.V. Stoitsov, S. Pittel, J. Phys. G: Nucl. Part. Phys. **24**, 1461 (1998)
34. M.V. Stoitsov, N. Michel, K. Matsuyanagi, Phys. Rev. C **77**, 054301 (2008)
35. N. Moiseyev, Phys. Rev. A **24**, 2824 (1981)
36. Y. Li, F.L. Zhang, J.L. Chen, J. Phys. A: Math. Theor. **44**, 365306 (2010)
37. A.I. Baz', Y.B. Zel'dovich, A.M. Perelomov, *Scattering, Reactions and Decay in Nonrelativistic Quantum Mechanics* (Israel Program for Scientific Translations, Jerusalem, 1969)
38. M.V. Berry, K.E. Mount, Rep. Prog. Phys. **35**, 315 (1972)
39. N. Michel, J. Math. Phys. **49**, 022109 (2008)
40. R. Newton, J. Math. Phys. **1**, 319 (1960)
41. R. Jost, Helv. Phys. Acta **20**, 256 (1947)
42. R. Newton, *Scattering Theory of Waves and Particles*, 2nd edn. (Dover Publications, New York, 2013)
43. H.A. Bethe, Phys. Rev. **76**, 38 (1949)
44. R.D. Viollier, G.R. Plattner, D. Trautmann, K. Alder, Nucl. Phys. A **206**, 498 (1973)
45. P. Fröbrich, R. Lipperheide, *Theory of Nuclear Reactions* (Oxford Science Publications, Clarendon Press, Oxford, 1996)
46. G.R. Satchler, *Direct Nuclear Reactions* (Clarendon Press, Oxford, 1983)
47. N.K. Glendenning, *Direct Nuclear Reactions* (Academic Press, New York, 1983)
48. B. Barmore, A.T. Kruppa, W. Nazarewicz, T. Vertse, Phys. Rev. C **62**, 054315 (2000)
49. A.T. Kruppa, W. Nazarewicz, Phys. Rev. C **69**, 054311 (2004)
50. http://www.nndc.bnl.gov/ensdf
51. S.A. Gurvitz, P.B. Semmes, W. Nazarewicz, T. Vertse, Phys. Rev. A **69**, 042705 (2004)

# Berggren Basis and Completeness Relations

**3**

The main interest of the one-body states introduced in Chap. 2 is that they form a complete set of states. The complete character of the solutions of differential equations defined in a finite radial interval is straightforward to demonstrate within the regular Sturm-Liouville theory [1, 2]. The regular Sturm-Liouville problem with nonlocal potentials is evidently more complex and was treated in Refs. [3, 4]. Extension to the singular case of infinite intervals in the frame of differential equations only has been considered in Refs. [2, 5]. Demonstrating the general completeness relation provided by the spectral decomposition of a self-adjoint operator, that is, of its set of bound and scattering eigenstates, however, demands the use of Lebesgue measure and the Riesz representation theorem [6].

The Hamiltonian $h$ of Eq. (2.3) is self-adjoint and defined in a dense subset of the Hilbert space of square-integrable functions. Thus, it can be written as an integral involving its spectral decomposition, that is, the bound and scattering $u(k, r)$ states of Eq. (2.2), embedded in a Stieltjes-Lebesgue measure, denoted as $d(u(k, r))$ [6]. The existence and unicity of $u(k, r)$ functions are a direct consequence of the existence of the Stieltjes-Lebesgue measure [6]. One has:

$$h = \int_k k^2 \, d(u(k, r)) \, .$$

The Stieltjes-Lebesgue measure $d(u(k, r))$ also gives rise to the resolution of identity:

$$\int_k d(u(k, r)) = \hat{\mathbf{1}} \, .$$

The resolution of identity is, in fact, equivalent to the completeness of $u(k, r)$ eigenstates. However, the nature of the spectrum of the considered operator is implicit therein, so that one cannot assess the convergence properties associated with the expansion in $u(k, r)$ eigenstates for a given physical situation. Consequently, it

© Springer International Publishing AG 2021
N. Michel, M. Płoszajczak, *Gamow Shell Model*, Lecture Notes in Physics 983,
https://doi.org/10.1007/978-3-030-69356-5_3

is preferable for our purpose to directly demonstrate the one-body completeness relation from their analytical properties. One can then show explicitly that the one-body completeness relation possesses the same properties as the Fourier transform. Conversely, demonstrating the one-body completeness relation along these lines can be performed only when potentials have a Coulomb asymptote, as the asymptotic behavior of scattering states must follow Eq. (2.7). The latter condition is sufficient for applications as only Coulomb and centrifugal potentials remain in the asymptotic region, the nuclear part of the Hamiltonian being of short range.

This one-body completeness relation involves real-energy states only. However, the Gamow shell model is built from the completeness relation of complex energy eigenstates (2.3), the Berggren completeness relation [7], where resonance and scattering states enter. The Berggren completeness relation will be demonstrated in this chapter from the Newton completeness relation of real-energy one-body states. Generalizations involving the use of complex-valued potentials or nonlocal potentials will be provided in this chapter as well.

The Berggren completeness relation is a practical tool, aimed to be used in numerical calculations. Thus, after demonstrating the Berggren completeness relation, one will demonstrate that it can be efficiently implemented to integrate the many-body Schrödinger equation and, furthermore discuss numerical examples in details. Hamiltonians represented in the Berggren basis take the form of complex-symmetric matrices. The numerical methods needed for their diagonalization are different from the case of Hermitian matrices and are not standard. One will show, however, that it is possible to diagonalize numerically the complex-symmetric Hamiltonian matrices almost as efficiently as the real-symmetric or Hermitian matrices.

Let us emphasize that the following demonstrations of completeness are technical and might be too cumbersome for the reader. In this latter case, we advise the reader to go directly to Sect. 3.3, where the Berggren completeness relation will be devised using the previously demonstrated Newton completeness relation.

## 3.1    Normalization of Gamow Functions

The standard method to normalize scattering wave functions is to use Dirac delta normalization [8]. In the case of neutral particle, Dirac delta normalization is immediate to perform as scattering states behave asymptotically as $\exp(\pm ikr)$ for $r \rightarrow +\infty$. On the contrary, the appearance of a logarithm in the asymptote $\exp(\pm i(kr - \eta \ln(2kr)))$ of the charged particle wave function (see Eq. (2.35)) demands care.

One can show from Eq. (2.130) that bound states and scattering states are orthogonal to each other. However, one has not proved yet that scattering states are orthonormal. As they are not integrable (see Eq. (2.7)), the orthonormalization of scattering states has to be understood in a weak sense, that is, with a Dirac delta distribution:

$$\int_0^{+\infty} u(k_a, r) u(k_b, r) \, dr = \delta(k_a - k_b) , \tag{3.1}$$

where $k_a > 0$ and $k_b > 0$ are linear momenta. Dirac delta normalization will be shown as equivalent to the normalization condition $2\pi\, C^+ C^- = 1$, which is then the same equality both in the charged particle case and the neutral particle case. For this, it will be demonstrated that the partial overlap $I_{ab}(R_\delta)$ in $[0 : R_\delta]$ between $u(k_a, r)$ and $u(k_b, r)$ weakly converges to a Dirac delta distribution when $R_\delta \to +\infty$. As $R_\delta$ enters integrated functions via $\ln(R_\delta)$, a generalization of the Riemann-Lebesgue lemma including this dependence will be stated for clarity. This lemma will be used to prove that all terms not leading to a Dirac delta vanish for $R_\delta \to +\infty$.

Let us consider a differentiable function $f_{R_\delta}(k)$ defined for $k \in [k_{\min} : k_{\max}]$, verifying:

$$|f_{R_\delta}(k)| = \mathcal{O}\left(\ln^n(R_\delta)\right) \forall k$$

$$\int_{k_{\min}}^{k_{\max}} |f'_{R_\delta}(k)| \, dk = \mathcal{O}\left(\ln^n(R_\delta)\right) , \tag{3.2}$$

where $n \in \mathbb{N}$. Integration by parts provides:

$$\int_{k_{\min}}^{k_{\max}} f_{R_\delta}(k)\, e^{ikR_\delta} \, dk = \frac{1}{iR_\delta} \left( \left[ f_{R_\delta}(k)\, e^{ikR_\delta} \right]_{k_{\min}}^{k_{\max}} - \int_{k_{\min}}^{k_{\max}} f'_{R_\delta}(k)\, e^{ikR_\delta} \, dk \right) . \tag{3.3}$$

Thus majorizing Eq. (3.3), one finds:

$$\left| \int_{k_{\min}}^{k_{\max}} f_{R_\delta}(k)\, e^{ikR_\delta} \, dk \right| \leq \frac{1}{R_\delta} \left( |f_{R_\delta}(k_{\min})| + |f_{R_\delta}(k_{\max})| + \int_{k_{\min}}^{k_{\max}} |f'_{R_\delta}(k)| \, dk \right)$$

$$= \mathcal{O}\left( \frac{\ln^n(R_\delta)}{R_\delta} \right) \to 0 .$$

The partial overlap $I_{ab}(R_\delta)$ between $u(k_a, r)$ and $u(k_b, r)$ in $[0 : R_\delta]$ arises directly from Eq. (2.130):

$$I_{ab}(R_\delta) = \int_0^{R_\delta} u(k_a, r) u(k_b, r) \, dr = \frac{u'(k_b, R_\delta) u(k_a, R_\delta) - u'(k_a, R_\delta) u(k_b, R_\delta)}{k_a^2 - k_b^2} . \tag{3.4}$$

In order to demonstrate that $u(k_a, r)$ and $u(k_b, r)$ are orthogonal to each other when $k_a \neq k_b$, one will show that $I_{ab}(R_\delta)$ converges weakly to a Dirac delta when $R_\delta \to +\infty$. Note that if $ka = k_b$ then a priori Eq. (3.4) is undefined, as both its numerator and denominator are equal to zero therein. In fact, Eq. (3.4) for $k_a = k_b$ can be evaluated with l'Hôpital rule. In the following, one will then always assume that the l'Hôpital rule is implicitly applied in this case.

The asymptotic behavior of $u(k, R_\delta)$ and $u'(k, R_\delta)$ when $R_\delta \to +\infty$ is obtained from Eqs. (2.7), (2.62), (2.63), (2.65), and (2.66):

$$u(k, R_\delta) = C_k \sin\left(k R_\delta - \eta_k \ln(2k R_\delta) + \delta_k^{(tot)}\right) + \mathcal{O}(R_\delta^{-1}) \tag{3.5}$$

$$u'(k, R_\delta) = k \, C_k \cos\left(k R_\delta - \eta_k \ln(2k R_\delta) + \delta_k^{(tot)}\right) + \mathcal{O}(R_\delta^{-1}), \tag{3.6}$$

where $C_k^2 = 4C_k^+ C_k^-$ and $\delta_k^{(tot)} = -\ell\frac{\pi}{2} + \sigma_\ell(\eta_k) + \delta_k$, with $\delta_k$ the phase shift associated to $u_k(r)$. Inserting (3.5) and (3.6) in Eq. (3.4), one obtains (see Exercise I):

$$I_{ab}(R_\delta) = C_{k_a} C_{k_b} \frac{\sin\left(\Delta_k R_\delta + \beta_{ab} \, \Delta_k \, \ln(R_\delta)\right)}{2\Delta_k}$$

$$+ C_{k_a} C_{k_b} \sin\left(\Delta_k R_\delta + \beta_{ab} \, \Delta_k \, \ln(R_\delta)\right) \left(\frac{\cos\left(f_-(k_a, k_b)\right) - 1}{2\Delta_k}\right)$$

$$+ C_{k_a} C_{k_b} \cos\left(\Delta_k R_\delta + \beta_{ab} \, \Delta_k \, \ln(R_\delta)\right) \left(\frac{\sin\left(f_-(k_a, k_b)\right)}{2\Delta_k}\right)$$

$$- C_{k_a} C_{k_b} \frac{\sin\left((k_a + k_b) R_\delta - (\eta_{k_a} + \eta_{k_b}) \ln(R_\delta) + f_+(k_a, k_b)\right)}{2(k_a + k_b)}$$

$$+ \mathcal{O}(R_\delta^{-1}), \tag{3.7}$$

where

$$\Delta_k = k_b - k_a \tag{3.8}$$

$$\beta_{ab} = \frac{v_c}{2k_a k_b} \tag{3.9}$$

$$f_\pm(k_a, k_b) = -(\eta_{k_b} \ln(2k_b) \pm \eta_{k_a} \ln(2k_a)) + \delta_{k_b}^{(tot)} \pm \delta_{k_a}^{(tot)}. \tag{3.10}$$

Note that the derivatives of all values with respect to $k_a$, necessary to apply the l'Hôpital rule, are finite, because $k_a > 0$ and $u(k_a, r)$ is analytic with respect to $k_a$ (see Exercise XIII of Sect. 2.6.3).

---

**Exercise I ⋆⋆⋆**

Demonstrate Eq. (3.7) by using Eqs. (3.5) and (3.6) in Eq. (3.4).

The four terms of Eq. (3.7) will be denoted respectively as $J_{ab}^{(i)}(\Delta_k, R_\delta)$, $i \in \{1, 2, 3, 4\}$. Only $J_{ab}^{(1)}(\Delta_k, R_\delta)$ will provide a Dirac delta when $R_\delta \to +\infty$. All other terms will be shown to vanish in a weak sense when $R_\delta \to +\infty$.

In order to show that $I_{ab}(R_\delta)$ converges to a Dirac delta, one will use a smooth test function $F(\Delta_k)$ of compact support: $\Delta_k \in [\Delta_{k_{\min}} : \Delta_{k_{\max}}]$, where $-k_a < \Delta_{k_{\min}} < 0$ and $\Delta_{k_{\max}} > 0$. These conditions are consistent with the requirements $k_a > 0$ (fixed) and $k_b = k_a + \Delta_k > 0$ (see Eq. (3.8)). One now integrates $F(\Delta_k)$ with $I_{ab}(R_\delta)$:

$$I_F(R_\delta) = \int_{\Delta_{k_{\min}}}^{\Delta_{k_{\max}}} F(\Delta_k) I_{ab}(R_\delta) \, d\Delta_k = \sum_{i=1}^{4} I_F^{(i)}(R_\delta) , \tag{3.11}$$

where

$$I_F^{(i)}(R_\delta) = \int_{\Delta_{k_{\min}}}^{\Delta_{k_{\max}}} F(\Delta_k) J^{(i)}(\Delta_k, R_\delta) \, d\Delta_k , \tag{3.12}$$

with $i \in [1 : 4]$. Integrals $I_F^{(i)}(R_\delta)$ with $i \geq 2$ can be written as the real or imaginary part of $\int_{\Delta_{k_{\min}}}^{\Delta_{k_{\max}}} f_{R_\delta}^{(i)}(\Delta_k) \, e^{i \Delta_k R_\delta} \, d\Delta_k$, where:

$$f_{R_\delta}^{(2)}(\Delta_k) = C_{k_a} C_{k_b} e^{i\beta_{ab} \Delta_k \ln(R_\delta)} \left( \frac{\cos\left( f_-(k_a, k_b) \right) - 1}{2\Delta_k} \right)$$

$$f_{R_\delta}^{(3)}(\Delta_k) = C_{k_a} C_{k_b} e^{i\beta_{ab} \Delta_k \ln(R_\delta)} \left( \frac{\sin\left( f_-(k_a, k_b) \right)}{2\Delta_k} \right) \tag{3.13}$$

$$f_{R_\delta}^{(4)}(\Delta_k) = C_{k_a} C_{k_b} e^{2ik_a R_\delta - i(\eta_{k_a} + \eta_{k_b}) \ln(R_\delta)} \left( \frac{e^{if_+(k_a, k_b)}}{2(k_a + k_b)} \right) .$$

One can verify that $f_{R_\delta}^{(i)}(\Delta_k)$ with $i \geq 2$ always verifies Eq. (3.2). Therefore, one can apply the generalized Riemann-Lebesgue lemma to $I_F^{(i)}(R_\delta)$ when $i \geq 2$, so that $I_F^{(i)}(R_\delta) \to 0$ when $R_\delta \to +\infty$.

Let us show that $I_F^{(1)}(R_\delta)$ in Eq. (3.14) vanishes when $R_\delta \to +\infty$. For that purpose, one will expand the sine function in $I_F^{(1)}(R_\delta)$ in products of sine and cosine functions:

$$I_F^{(1)}(R_\delta) = \frac{C_{k_a} C_{k_b}}{2} \int_{\Delta_{k_{\min}}}^{\Delta_{k_{\max}}} F(\Delta_k) \left( \frac{\sin\left( \Delta_k R_\delta \right)}{\Delta_k} \right) \cos\left( \beta_{ab} \Delta_k \ln(R_\delta) \right) d\Delta_k$$

$$+ \frac{C_{k_a} C_{k_b}}{2} \int_{\Delta_{k_{\min}}}^{\Delta_{k_{\max}}} F(\Delta_k) \cos\left( \Delta_k R_\delta \right) \left( \frac{\sin\left( \beta_{ab} \Delta_k \ln(R_\delta) \right)}{\Delta_k} \right) d\Delta_k .$$

$$\tag{3.14}$$

The second integral in (3.14) vanishes when $R_\delta \to +\infty$ as $f_{R_\delta}(\Delta_k) = F(\Delta_k) \sin(\beta_{ab} \, \Delta_k \, \ln(R_\delta))/\Delta_k$ verifies Eq. (3.2).

One will now determine the limit of the first integral of $I_F^{(1)}(R_\delta)$ in Eq. (3.14) when $R_\delta \to +\infty$. For this, one introduces the function $G_{R_\delta}(\Delta_k)$ via:

$$G_{R_\delta}(\Delta_k) = \frac{F(\Delta_k) \, \cos(\beta_{ab} \, \Delta_k \, \ln(R_\delta)) - F(0)}{\Delta_k}.$$  (3.15)

Then, $I_F^{(1)}(R_\delta)$ becomes:

$$I_F^{(1)}(R_\delta)$$

$$= \frac{C_{k_a} C_{k_b}}{2} \left( F(0) \int_{\Delta_{k_{\min}} R_\delta}^{\Delta_{k_{\max}} R_\delta} \frac{\sin(x)}{x} \, dx + \int_{\Delta_{k_{\min}}}^{\Delta_{k_{\max}}} G_{R_\delta}(\Delta_k) \, \sin\left(\Delta_k R_\delta\right) d\Delta_k \right),$$  (3.16)

where the change of variable $x = \Delta_{k_{\max}} R_\delta$ has been effected.

Clearly, the first integral of Eq. (3.16) has $\pi$ as a limit when $R_\delta \to +\infty$. As one can apply the generalized Riemann-Lebesgue lemma to $G_{R_\delta}(\Delta_k)$ (see Eqs. (3.2) and (3.15)) the second integral of Eq. (3.16) vanishes when $R_\delta \to +\infty$. Thus, one can evaluate the weak limit of $I_{ab}(R_\delta)$ in Eq. (3.7) for $R_\delta \to +\infty$:

$$I_{ab}(R_\delta) \to \frac{\pi}{2} C_{k_a}^2 \delta(k_a - k_b),$$  (3.17)

and from Eq. (3.4) one obtains:

$$\int_0^{+\infty} u(k_a, r) u(k_b, r) \, dr = \frac{\pi}{2} C_{k_a}^2 \delta(k_a - k_b) = 2\pi \, C_{k_a}^+ C_{k_a}^- \delta(k_a - k_b).$$  (3.18)

The Dirac delta normalization of $u(k, r)$ thus arises from the following equality:

$$\int_0^{+\infty} u(k_a, r) u(k_b, r) \, dr = \delta(k_a - k_b) \Leftrightarrow 2\pi \, C^+ C^- = 1 \quad \forall u_k,$$  (3.19)

where Eq. (2.7) has been used for its derivation.

Another formula equivalent to Eq. (3.19) can be written using the expansion $u(k, r) = C_F \, F_{\ell,\eta}(kr) + C_G \, G_{\ell,\eta}(kr)$ for $r \geq R$:

$$\int_0^{+\infty} u(k_a, r) u(k_b, r) \, dr = \delta(k_a - k_b) \Leftrightarrow C_F^2 + C_G^2 = \frac{2}{\pi} \quad \forall u_k.$$  (3.20)

which is immediate from Eqs. (2.33) and (2.34).

The asymptotic expansion of $u(k, r)$ for $|k| \to +\infty$ (see Sect. 2.6.2) is indispensable to demonstrate the completeness of $u(k, r)$ functions (see Sect. 3.2).

Therefore, one will discuss now the properties of $u(k, r)$ functions normalized to a Dirac delta in this range of $|k|$-values. Let us first consider the case $\Re(k) \to +\infty$ and $\Im(k) = \mathcal{O}(1)$. For this, one will write the expansion $u(k, R) = C_F\, F_{\ell,\eta}(kR) + C_G\, G_{\ell,\eta}(kR)$ when $\Re(k) \to +\infty$ using Eq. (2.150):

$$u(k, r) = C_F\, (\sin(kR - \ell\pi/2) - A(k, R) \cos(kR - \ell\pi/2))$$

$$+ C_G\, (\cos(kR - \ell\pi/2) + A(k, R) \sin(kR - \ell\pi/2)) + \mathcal{O}\left(\ln^2(k)\, k^{-2}\right)$$

$$= (C_F + C_G\, A(k, R))\, \sin(kR - \ell\pi/2)$$

$$+ (C_G - C_F\, A(k, R))\, \cos(kR - \ell\pi/2) + \mathcal{O}\left(\ln^2(k)\, k^{-2}\right), \qquad (3.21)$$

where $A(k, R) = \mathcal{O}\left(\ln(k)\, k^{-1}\right)$.

In order to relate Eq. (3.21) to the asymptotic relations derived in Sect. 2.6.2, one will now write Eq. (2.154) in $r = R$ as a linear combination of the sine and cosine functions appearing in Eq. (3.21):

$$u(k, r) = C_0\, (\cos(\phi) + A_0(k, R) \sin(\phi))\, \sin(kR - \ell\pi/2)$$

$$+ C_0\, (\sin(\phi) - A_0(k, R) \cos(\phi))\, \cos(kR - \ell\pi/2)$$

$$+ \mathcal{O}\left(C_0\, \ln^2(k)\, k^{-2}\right), \qquad (3.22)$$

where $\phi = (\ell - \ell_0)\, (\pi/2)$, and $A_0(k, R) = \mathcal{O}\left(\ln(k)\, k^{-1}\right)$. To derive (3.22), one has used Eq. (2.150). Identifying the coefficients of the sine and cosine functions in Eqs. (3.21) and (3.22), one has:

$$C_F + C_G\, A(k, R) = C_0\, (\cos(\phi) + A_0(k, R) \sin(\phi)) + \mathcal{O}\left(C_0\, \ln^2(k)\, k^{-2}\right)$$

$$(3.23)$$

$$C_G - C_F\, A(k, R) = C_0\, (\sin(\phi) - A_0(k, R) \cos(\phi)) + \mathcal{O}\left(C_0\, \ln^2(k)\, k^{-2}\right).$$

$$(3.24)$$

By squaring and summing both sides of Eqs. (3.23) and (3.24), as well as using Dirac delta normalization, one obtains the asymptotic expansion of $C_0$ for $\Re(k) \to +\infty$:

$$C_0 = \sqrt{\frac{2}{\pi}} + \mathcal{O}\left(\ln^2(k)\, k^{-2}\right). \qquad (3.25)$$

What is left over is the case of $\Im(k) \to +\infty$ and an arbitrary $\Re(k)$. For $\Im(k)$ sufficiently large, $|\sin(kR - \ell\pi/2)| \geq 1$ and $|\cos(kR - \ell\pi/2)| \geq 1$, so that these functions can be multiplied by $1 + \mathscr{O}\left(\ln^2(k)\, k^{-2}\right)$ in all equations (see Sect. 2.6.2):

$$u(k, r) = (C_F + C_G\, A(k, R))\, \sin(kR - \ell\pi/2)\left(1 + \mathscr{O}\left(\ln^2(k)\, k^{-2}\right)\right)$$
$$+ (C_G - C_F\, A(k, R))\, \cos(kR - \ell\pi/2)\left(1 + \mathscr{O}\left(\ln^2(k)\, k^{-2}\right)\right).$$
(3.26)

Consequently, one obtains similarly to the real case:

$$C_F + C_G\, A(k, R) = C_0\, (\cos(\phi) + A_0(k, R)\sin(\phi))\left(1 + \mathscr{O}\left(\ln^2(k)\, k^{-2}\right)\right)$$
(3.27)

$$C_G - C_F\, A(k, R) = C_0\, (\sin(\phi) - A_0(k, R)\cos(\phi))\left(1 + \mathscr{O}\left(\ln^2(k)\, k^{-2}\right)\right).$$
(3.28)

Equation (3.25) follows immediately from Eqs. (3.27) and (3.28) when $\Im(k) \to +\infty$ and $\Re(k)$ is arbitrary. Consequently, the asymptotic expansion of Eq. (3.25) is valid for complex $k$ verifying $|k| \to +\infty$.

## 3.2    One-Body Completeness Relation

The first demonstration of the one-body completeness relation using the analytical properties of one-body states has been done by R. Newton [9]. The Newton completeness relation is effected using the Cauchy theorem in the complex plane, by closing a segment on the real axis by half a circle in the upper part of the $k$-complex plane and using the asymptotic form of $u(k, r)$ functions when $k \to +\infty$. While very effective and intuitive, this demonstration can only be used for the neutron case, as the $S$-matrix (see Eq. (2.182)) must be analytical in $k$ in the upper half plane including the real $k$-axis. The potentials with a Coulomb asymptotic are thus prohibited.

In order to remove this restriction, the one-body completeness relation has been demonstrated starting from a box completeness relation [10], where the one-body states $u(k, r)$ are all bound and form a discrete set of states, so that their completeness property is standard [11]. However, the limit when the radius of the box goes to infinity (the continuum limit) demands a careful monitoring of the $u(k, r)$ states, which become infinitely dense. Moreover, the completeness of Coulomb wave functions is needed therein to properly treat $u(k, r)$ scattering states in the vicinity of $k = 0$.

Hence, one will follow another route, which borrows ideas from both approaches. Firstly, the Newton completeness relation will be shown for the neutron $\ell = 0$ case. Secondly, the completeness of the Coulomb wave functions will be demonstrated. For this, one will firstly screen the centrifugal and Coulomb parts of the considered potential, so that this case can be treated within the neutron $\ell = 0$ case. The general case of potentials possessing a centrifugal and/or a Coulomb part in the asymptotic region will then be recovered by taking the limit $R_s \rightarrow +\infty$, with $R_s$ the screening radius after which the potential vanishes. The completeness of the Coulomb wave functions will play a prominent role in the treatment of the essential singularity in the complex momentum plane at $k = 0$, arising when $R_s \rightarrow +\infty$.

### 3.2.1  One-Body Completeness Relation for $\ell = 0$ Neutrons

To demonstrate the Newton completeness relation in the neutron $\ell = 0$ case, one considers the following integral [9, 12]:

$$I(K) = \int_0^K \frac{u(k, r)u(k, r')}{2\pi \, C^+ C^-} \, dk \,, \tag{3.29}$$

where $0 < r' \le r$ and $K > 0$. $u(k, r)$ is a solution of Eq. (2.2) with $C_0 = k^{-\ell_0 - 1} C_{\ell_0}(\eta_0)^{-1}$ in Eq. (2.6) (see Exercise XIII in Sect. 2.6.3). Therefore, $u(k, r) \sim r^{\ell_0 + 1}$ for $r \rightarrow 0$ and $u(k, r)$ is analytic in the upper half of the complex $k$-plane except for the poles of the $S$-matrix (see Sects. 2.6.3, 2.6.4, and 2.6.6). The additional factor $2\pi \, C^+ C^-$ in Eq. (3.29) is related to the Dirac delta normalization of scattering states (see Sect. 3.1).

Equation (3.29) will be calculated with complex integration [9, 12]. One will show now that this equation has no singularity, that is, that the integrand of $I(K)$ is finite $\forall k \in [0 : K]$. Equation (2.7) implies that $u(k, r) = C \sin(kr + \delta)$ if $k > 0$, where $C$ and $\delta$ are amplitude and phase shift in the asymptotic region, respectively. Thus, $u(k, R)^2 = C^2 \sin^2(kR + \delta)$ and $u'(k, R)^2 = C^2 k^2 \cos^2(kR + \delta)$ in $r = R$. One readily obtains: $C^2 = u(k, R)^2 + u'(k, R)^2 / k^2$. As a consequence, it is impossible to have $C \rightarrow 0$ in $k \ge 0$, as $u(k, R)$ and $u'(k, R)$ cannot be simultaneously equal to zero, which would imply that $u(k, r) = 0 \,\forall r \ge 0$. Hence, as $|C^\pm| = |C|/2$, the integrand of $I(K)$ in (3.29) is finite $\forall k \in [0 : K]$.

As (2.2) is invariant with respect to the change $k \rightarrow -k$, one has:

$$u(k, r) = u(-k, r) \tag{3.30}$$

$$u^\pm(k, r) = u^\mp(-k, r) \tag{3.31}$$

$$C^\pm(k) = C^\mp(-k) \,, \tag{3.32}$$

where Eqs. (2.2), (2.6), (2.7), and (2.138) have been used. The $k$-dependence of $C^+$ and $C^-$ constants has been explicitly written in Eq. (3.32) for readability. Note that $u^\pm(k, r)$ is analytic $\forall k$, because $H_{\ell, \eta}^\pm = \exp(\pm ikr)$ in Eq. (2.138) (see Sect. 2.6.3).

The symmetry relations of Eqs. (3.30), (3.31), and (3.32), relating wave functions and integration constants at linear momenta $k$ and $-k$, allow to rewrite Eq. (3.29) as an integral on $[-K : K]$:

$$I(K) = \int_{-K}^{K} \frac{u^+(k,r)u(k,r')}{2\pi\, C^-(k)}\, dk\,.\tag{3.33}$$

---

**Exercise  II** ⋆

Calculate the residues of Eq. (3.33) using Eqs. (2.182) and (2.223).

---

One can then calculate Eq. (3.33) with the Cauchy theorem by closing the segment $[-K : K]$ by a half-circle $\mathscr{C}_K$ of radius $K$ in the upper part of the complex $k$-plane. The only possible poles in Eq. (3.33) are situated in the upper half plane and arise from the bound states $u(k,r)$, whose residues are straightforward to calculate (see Sect. 2.5 and Exercise II). One has shown in Sect. 2.5.1 that the number of bound states is finite, so that the number of poles in Eq. (3.33) is finite as well. Note that one cannot have a zero-energy bound state with neutron $s$-states (see Sect. 2.5).

---

**Exercise  III** ⋆⋆

In this exercise, we show that $k$-integrals depending on $K$ have simple limits when $K \to +\infty$, which will be used to demonstrate the Berggren completeness relation of neutron states.

Calculate the limit $K \to +\infty$ of the following two integrals:

$$-\frac{1}{2\pi} \int_{\mathscr{C}_K} e^{ikx}\, dk\,, x \in \mathbb{R}$$

$$\int_0^{+\infty} f(x) \int_{\mathscr{C}_K} e^{ikx}\, \frac{a(k)}{k}\, dk\, dx\,,$$

where $f(x)$ is a smooth test function, and $|a(k)| \le C \ln(K)\ \forall k \in \mathscr{C}_K$, with $C$ independent of $K$.

Deduce the weak limit of the integral $\int_{\mathscr{C}_K} e^{ikx}\, \frac{a(k)}{k}\, dk$, when $K \to +\infty$ and $x \ge 0$.

---

**Exercise  IV** ⋆⋆⋆

One will demonstrate the Berggren completeness relation for the case of neutron $s$-states.

**A.** In order to demonstrate the Berggren completeness relation, one will derive the asymptotic expansion of Eq. (3.35) when $K \to +\infty$. One will firstly consider that $0 < r' \le r \le R$ in Eq. (3.35). Using the results of Sect. 2.6.2, write Eq. (3.35) when $K \to +\infty$ as the sum of four integrals involving $C_0$, $A^\pm(k)$, $B^\pm(k)$, and $u_{\mathrm{app}}^\pm(k,r)$. The functions $\epsilon^\pm(k,r)$ and $\epsilon_u^\pm(k,r)$ (see Sect. 2.6.2) can be replaced by $1 + \mathcal{O}(k^{-2})$ when $K \to +\infty$. Explain why one can do this replacement.

**B.** Deduce that the obtained four integrals can be put in a simple form using the asymptotic expansions of $A^{\pm}(k)$, $B^{\pm}(k)$ and $u^{\pm}_{\text{app}}(k, r)$ for $|k| \to +\infty$. For that, the explicit calculation of the functions entering integrals is not necessary, one will only explicitly state their $k$-dependence. Using the integral limits devised in Exercise III, one can then demonstrate the Berggren completeness relation when $0 < r, r' \le R$, that is, that $I(\mathscr{C}_K) \to \delta(r - r')$ (see Eq. (3.35)) for $K \to +\infty$.

**C.** Show that the case $0 < r' \le R$ and $r \ge R$ can be treated as in problems **A** and **B**, so that $I(\mathscr{C}_K) \to \delta(r - r')$ as well therein.

**D.** By using a similar method as in **A** and **B**, demonstrate that $I(\mathscr{C}_K) \to \delta(r - r')$ if $r \ge r' \ge R$.

It is convenient to introduce the normalized bound state $u_n(r)$ of linear momentum $k_n$ from the value of the residues of Eq. (3.33). As a consequence, applying the Cauchy theorem to Eq. (3.33) provides:

$$I(K) = -\sum_n u_n(r)u_n(r') + I(\mathscr{C}_K) , \tag{3.34}$$

where

$$I(\mathscr{C}_K) = -\int_{\mathscr{C}_K} \frac{u^+(k, r)u(k, r')}{2\pi \, C^-(k)} \, dk . \tag{3.35}$$

The demonstration of the completeness relation of the Berggren one-body states will be obtained by showing that $I(\mathscr{C}_K) \to \delta(r - r')$ in Eq. (3.35). The limit $K \to +\infty$ of integrals involving $u^{\pm}(k, r) = \exp(\pm ikr)$ will then be necessary for that matter (see Exercise III).

One can now demonstrate the Berggren completeness relation for the case of neutron $s$-states ($\ell = 0$) (see Exercise IV). For this, one derives the asymptotic limit of Eq. (3.35) when $K \to +\infty$ using the asymptotic expansions of scattering wave functions presented in Sect. 2.6.2. Using the results of Sect. 2.6.2, Eq. (3.35) is written as the sum of several integrals involving elementary functions, which will either vanish or weakly converge to a Dirac delta distribution when $K \to +\infty$ (see Exercise IV). As demonstrated in Exercise IV, one obtains the limit $I(\mathscr{C}_K) \to \delta(r - r')$ (see Eq. (3.35)) when $K \to +\infty$. Using normalized scattering functions $u(k, r)$, implying $2\pi \, C^+ C^- = 1$, the Newton completeness relation then follows from Eq. (3.35) and the limit $I(\mathscr{C}_K) \to \delta(r - r')$:

$$\sum_n u_n(r)u_n(r') + \int_0^{+\infty} u(k, r)u(k, r') \, dk = \delta(r - r') . \tag{3.36}$$

As the Dirac delta distribution $\delta(r - r')$ of Eq. (3.42) arises from the limit of $\sin(K(r - r'))/K$ for $K \to +\infty$ (see Exercise IV), the functions that can be expanded with Eq. (3.36) are those that possess a Fourier transform.

### 3.2.2   Completeness Relation of Coulomb Wave Functions

The demonstration of the completeness of Coulomb wave functions can be done using the analytical properties of confluent hypergeometric functions. Let us consider potentials for which $\ell \geq 0$ and $v_c \in \mathbb{R}$ (see Eq. (2.5)). One will follow the method originally presented by Mukunda [13] in an attractive Coulomb potential and adapted to the repulsive Coulomb potential in Ref. [10]. For clarity, one will also replace confluent hypergeometric functions by Coulomb wave functions.

Let us define an integral similar to (3.29):

$$
I_c(k_s, K) = \frac{2}{\pi} \int_{k_s}^{K} F_{\ell,\eta}(kr) F_{\ell,\eta}(kr') \, dk ,
\tag{3.37}
$$

where $K > k_s > 0$ and where the integration constant $C_0$ of Eq. (2.6) is equal to $\sqrt{2/\pi}$. The introduction of $k_s$ is necessary as $k = 0$ is an essential singularity for Coulomb wave functions [9].

In order to write Eq. (3.37) as an integral in $[-K : K]$, one has to write $F_{\ell,\eta}(kr')$ as a function of $H_{\ell,\eta}^{\pm}(kr')$ [13] (see Eq. (2.33)). Consequently, one has to pay attention to the cut of Coulomb wave functions on the negative real axis. For this, one considers that $k$ belongs to the lower half of the complex plane with strictly negative imaginary part. $H_{\ell,\eta}^{-}(kr)$ and $H_{\ell,-\eta}^{+}(-kr)$ are minimal for $r \to +\infty$ (see Eq. (2.32)). As they are solutions of the same differential equation, they are mutually proportional, so that one can determine their coefficient of proportionality from their asymptote for $r \to +\infty$ (see Eq. (2.64)):

$$
H_{\ell,-\eta}^{+}(-kr) = H_{\ell,\eta}^{-}(kr) \exp(-i\pi \ell) \exp(-\pi \eta) .
\tag{3.38}
$$

A similar equation arises from the power series defining $F_{\ell,\eta}(kr)$ (see Eq. (2.37)):

$$
F_{\ell,-\eta}(-kr) = -F_{\ell,\eta}(kr) \exp(i\pi \ell) \exp(\pi \eta) .
\tag{3.39}
$$

Hence, Eq. (3.37) reads:

$$
I_c(k_s, K) = \frac{1}{i\pi} \int_{-K}^{-k_s} F_{\ell,\eta}(kr) H_{\ell,\eta}^{+}(kr') \, dk + \frac{1}{i\pi} \int_{k_s}^{K} F_{\ell,\eta}(kr) H_{\ell,\eta}^{+}(kr') \, dk ,
\tag{3.40}
$$

where $k$ is considered as $|k| \exp(i\pi)$ if $k < 0$ and where Eqs. (3.38) and (3.39) have been used. As a consequence, $F_{\ell,\eta}(kr) H_{\ell,\eta}^{+}(kr')$ can be considered as an analytic function of $k$ if $\Im(k) \geq 0$ and $k \neq 0$, up to the eventual presence of the poles [13]. The Cauchy theorem can then be applied if the segments $[-K : -k_s]$ and $[k_s : K]$ are complemented by a small half-circle $\mathscr{C}_{k_s}$ of radius $k_s$ and by a large semicircle $\mathscr{C}_K$ of radius $K$ in order to form a closed contour in the upper half-plane.

Let us consider the eventual presence of non-analyticities of $F_{\ell,\eta}(kr)H^+_{\ell,\eta}(kr')$ inside the considered closed contour. One can see from Eq. (2.37) that this can occur only for $F_{\ell,\eta}(kr)$ due to $C_\ell(\eta)$, because the remaining power series in Eq. (2.37) is analytic in the upper complex $k$-plane. Equation (2.32) implies that $H^+_{\ell,\eta}(kr)$ can be nonanalytic only through $\sigma_\ell(\eta)$, as $_2F_0$ is analytic in the upper part of the complex $k$-plane as well [13]. Hence, for $F_{\ell,\eta}(kr)H^+_{\ell,\eta}(kr')$ to exhibit a non-analyticity, the Gamma functions in Eqs. (2.31) and (2.32) must be infinite.

$\Gamma(2\ell+2)$ is always finite for $\ell \geq 0$. One can see from Eqs. (2.31), (2.32), (2.35), and (2.37) that the factor $\Gamma(1+\ell-i\eta)$ cancels out in the product $F_{\ell,\eta}(kr)H^+_{\ell,\eta}(kr')$ and that the $\Gamma(1+\ell+i\eta)$ function appears therein. Therefore, non-analyticities may occur only from $\Gamma(1+\ell+i\eta)$, that is from its poles, occurring if $1+\ell+i\eta$ is a negative integer. This is impossible if $\ell \geq 0$ and $v_c \geq 0$. Conversely, the $\Gamma(1+\ell+i\eta)$ function has an infinite number of poles if $v_c < 0$, which correspond to the infinite number of bound states of Coulomb attractive potential [14].

This series of bound states converges absolutely as bound state wave functions behave as $\mathcal{O}(n^{-3/2})$ [14] when $n \to +\infty$. Therefore, it poses no problem when $k_s \to 0$. Equation (3.40) thus reads:

$$I_c(k_s, K) = \frac{1}{i\pi} \int_{\mathscr{C}_{k_s}} F_{\ell,\eta}(kr)H^+_{\ell,\eta}(kr')\, dk - \frac{1}{i\pi} \int_{\mathscr{C}_K} F_{\ell,\eta}(kr)H^+_{\ell,\eta}(kr')\, dk \ . \tag{3.41}$$

The main difference with Eq. (3.34) is the integral along the $\mathscr{C}_{k_s}$ contour. For $k \to 0$, Eqs. (2.60) and (2.61) imply that $F_{\ell,\eta}(kr)H^+_{\ell,\eta}(kr') \to 0$, so that the integral along the $\mathscr{C}_{k_s}$ contour vanishes when $k_s \to 0$. The limit $K \to +\infty$ of the integral of Eq. (3.41) is stated in Exercise IV where the case of Coulomb wave functions is considered (for this, only consider points **A** and **B** in Exercise IV of Sect. 3.2.1 by using Coulomb wave functions instead of neutron $s$-states). The Coulomb completeness relations thus follow:

$$\frac{2}{\pi} \int_0^{+\infty} F_{\ell,\eta}(kr)F_{\ell,\eta}(kr')\, dk = \delta(r-r'), \quad v_c \geq 0 \tag{3.42}$$

$$\sum_{n=0}^{+\infty} u_n(r)u_n(r') + \frac{2}{\pi} \int_0^{+\infty} F_{\ell,\eta}(kr)F_{\ell,\eta}(kr')\, dk = \delta(r-r'), \quad v_c < 0 \tag{3.43}$$

where $u_n(r)$ is a bound state bearing $n$ nodes. The functions that can be expanded with Eqs. (3.42) and (3.43) are those possessing a Fourier transform, following the same argument as in Sect. 3.2.1.

### 3.2.3   Completeness Relation for the General Case

From the Newton completeness relation (3.36) for neutron $\ell = 0$ case and the Coulomb completeness relation of Eq. (3.42), it is possible to demonstrate the completeness relation of the $u(k, r)$ wave functions generated by the Schrödinger equation of Eq. (2.2) in the case of repulsive Coulomb potential $v_c \geq 0$ and $\ell \geq 0$. The completeness relation for an attractive Coulomb potential ($v_c < 0$) will also be discussed afterwards in this section.

Before one considers the demonstration itself, one will explain why one could not proceed as in the two previous demonstrations. The problem lies solely in the $C^{\pm}$ constants. Indeed, in order to use the Cauchy theorem, one has to calculate an integral whose interval of integration is $[-K : K]$ (see Eqs. (3.33) and (3.40)). A necessary condition for this method to hold is that the integrated function has to bear a $k \rightarrow -k$ symmetry (see Eqs. (3.33) and (3.37)). The $k \rightarrow -k$ symmetry arises from Eq. (3.32) in the neutron one-body completeness relation for $\ell = 0$, and from the analytical properties of Coulomb wave functions with a point-particle Coulomb potential (see Sect. 3.2.2). However, there is no longer any symmetry between $C^{\pm}(k)$ and $C^{\mp}(-k)$ in the general case. Indeed, $u(k, r)$ is a complicate linear combination of $H_{\ell,\eta}^{\pm}(kr)$ functions, whose coefficients are nonanalytic functions of $k$ in $k = 0$. Consequently, Eq. (3.29) is not symmetric in $k$, even though the $u(k, r)$ states are by construction.

Thus, instead of using the Cauchy theorem with an integral defined in $[-K : K]$, one will only consider an integral of $u(k, r)$ functions defined in $[0 : K]$ for which $v_c \geq 0$. One will consider screened $u^{(s)}(k, r)$ functions at a radius $R^{(s)}$ (the subscript $(s)$ will always refer to the screened case) in the following functional of $u(k, r)$ and $u^{(s)}(k, r)$ functions:

$$I_s(r, r', R^{(s)}) = \sum_n u_n(r)u_n(r') - \sum_n u_n^{(s)}(r)u_n^{(s)}(r')$$

$$+ \int_0^{+\infty} \left[ u(k, r)u(k, r') - u^{(s)}(k, r)u^{(s)}(k, r') \right] dk , \quad (3.44)$$

where $r > 0$ and $r' > 0$ are fixed radii and $R^{(s)}$ is chosen larger than $R$. As $r$ and $r'$ are fixed, the radial coordinate that can take arbitrary positive values will always be denoted as $r''$ to avoid confusion. The Schrödinger equation (2.2) is then the same for $u(k, r'')$ and $u^{(s)}(k, r'')$ if $0 \leq r'' \leq R^{(s)}$, which implies that $u(k, r'') \propto u^{(s)}(k, r'')$ if $0 \leq r'' \leq R^{(s)}$. One also supposes that $u(k, r'')$ and $u^{(s)}(k, r'')$ are normalized so that $2\pi \, C^+C^- = 2\pi \, C^{+(s)}C^{-(s)} = 1$. One will show in the following that $I_s(r, r', R^{(s)})$ vanishes, which is clearly equivalent to the completeness of $u(k, r'')$ states as $I_s(R^{(s)})$ amounts to the difference between two completeness relations in Eq. (3.44).

Let us consider the limit of the terms involving bound states in Eq. (3.44) when $R^{(s)} \rightarrow +\infty$, denoted as $\delta_b(R^{(s)})$. For this, let us show that $u_n^{(s)}(r'') \rightarrow u_n(r'')$ for $R^{(s)} \rightarrow +\infty$. Firstly, one notices that the number of bound states $u_n(r'')$ and

$u_n^{(s)}(r'')$ is finite and is the same when $R^{(s)} \to +\infty$. This arises because the nodes of $u(k = 0, r'')$ are in finite number, $u^{(s)}(k = 0, r'') \propto u(k = 0, r'')$ if $0 \le r'' \le R^{(s)}$ and $u^{(s)}(k = 0, r'') \propto u'(k = 0, R^{(s)})(r'' - R^{(s)}) + u(k = 0, R^{(s)})$ if $r'' \ge R^{(s)}$ (see Eq. (2.5)). Indeed, if $u(k = 0, r'')$ is not a bound state, $|u(k = 0, r'')| \to +\infty$, so that, for $R^{(s)}$ sufficiently large, $|u^{(s)}(k = 0, r'')| \to +\infty$ for $r'' \to +\infty$ and all zeros of $u(k = 0, r'')$ and $u^{(s)}(k = 0, r'')$ can be situated in $[0 : R^{(s)}]$. Alternatively, if $u(k = 0, r'')$ is a bound state, for $R^{(s)}$ sufficiently large, $u^{(s)}(k = 0, r'')$ possesses a single zero for $r'' > R^{(s)}$, so that the number of zeros of $u(k = 0, r'')$ (including the one at infinity) and $u^{(s)}(k = 0, r'')$ are the same. Therefore, $u_n^{(s)}(r'')$ and $u_n(r'')$ are normalized bound states of $n$ nodes which solve Eq. (2.2), respectively, for $0 \le r'' \le R^{(s)}$ and for all radii. Consequently, $u_n^{(s)}(r'') \to u_n(r'')$ when $R^{(s)} \to +\infty$ by unicity of $u_n(r'')$ (see Exercise I of Chap. 2). Therefore, $\delta_b(R^{(s)})$ vanishes when $R^{(s)} \to +\infty$.

One will now turn to the scattering part of Eq. (3.44). As $u(k, r'')$ and $u^{(s)}(k, r'')$ are well defined for $k > 0$ (see Sect. 2.6.6), one just has to check if the integral of Eq. (3.44) is well defined for $k \to 0$ and $k \to +\infty$. Due to the essential singularity occurring in Coulomb wave functions at $k = 0$, one introduces $k_s > 0$ and separates the integral of Eq. (3.44) in two parts: one from 0 to $k_s$ and the other from $k_s$ to $+\infty$:

$$I_s(r, r', R^{(s)}) = \int_0^{k_s} u(k, r)u(k, r') \, dk - \int_0^{k_s} u^{(s)}(k, r)u^{(s)}(k, r') \, dk \qquad (3.45)$$

$$+ \int_{k_s}^{+\infty} \left[ u(k, r)u(k, r') - u^{(s)}(k, r)u^{(s)}(k, r') \right] \, dk + \delta_b(R^{(s)}) \, .$$

Two limiting procedures have to be done in Eq. (3.46), which involve $k_s$ and $R^{(s)}$. Firstly, let us take $R^{(s)} \to +\infty$. The limit $k_s \to 0$ will be effected afterwards. The treatment of integrals of Eq. (3.46) is standard (see Exercise IV), except for the integral in $k$ going from 0 to $k_s$ involving $u^{(s)}(k, r'')$. Owing to the singular character of $H_{\ell,\eta}^{\pm}(kr)$, the equalities $A = 1 + \mathcal{O}(R^{(s)^{-1}})$ and $C_0 = C_0^{(s)} + \mathcal{O}(R^{(s)^{-1}})$ of Sect. 2.6.5 do not hold uniformly for $k \in ]0 : k_s]$. In fact, the turning point of $u(k, r'')$ goes to infinity when $k \to 0$, so that its oscillatory region goes to infinity as well, whereas the oscillatory region of $u^{(s)}(k, r'')$ starts at least at $r'' = R^{(s)}$. Thus, for a sufficiently large $R^{(s)}$, there will always be a sufficiently small $k$ for which $u(k, r'')$ and $u^{(s)}(k, r'')$ will be significantly different for $r'' > R^{(s)}$, so that their normalization constant will differ as well. Then having $R^{(s)} \to +\infty$ inside the second integral of Eq. (3.46) for $k \in ]0 : k_s]$ cannot be handled with standard mathematical methods.

---

**Exercise  V ★★★**

Show that the first and third integrals of Eq. (3.46) vanish for $R^{(s)} \to +\infty$ and $k_s \to 0$, with the limits taken in that order.

A direct calculation of this limit is in principle possible by calculating asymptotic expansions of $u^{(s)}(k, r'')$ $\forall k \in ]0 : k_s]$, using the results of Sects. 2.3.3 and 2.6.2. However, this method is very cumbersome, as one would have to (1) devise three different WKB approximations for $u^{(s)}(k, r'')$, respectively, valid for $r'' \ll r_t(k)$, $r'' \sim r_t(k)$, and $r'' \gg r_t(k)$ (see Eq. (2.68)), (2) generate a Neumann series solutions from these ansatz (see Sect. 2.6.2), and (3) connect them at the limits of their respective domains of validity.

In fact, it is more convenient to proceed indirectly, by using the previously demonstrated completeness of Coulomb wave functions in Sect. 3.2.2. For this, one considers the standard and screened Coulomb wave functions $u_c(k, r)$ and $u_c^{(s)}(k, r'')$ functions defined in Sect. 2.6.5. The part of the integral of Eq. (3.46) with $k \in ]0 : k_s]$ can be majorized by $u_c^{(s)}(k, r'')$ functions (see Eq. (2.181)):

$$\left| \int_0^{k_s} u^{(s)}(k, r) u^{(s)}(k, r') \, dk \right| \leq M(r) M(r') \int_0^{k_s} u_c^{(s)}(k, r) u_c^{(s)}(k, r') \, dk .$$

$$(3.46)$$

This suggests to consider $u(k, r'') = u_c(k, r'')$ and $u^{(s)}(k, r'') = u_c^{(s)}(k, r'')$ in Eq. (3.44). One will then consider $I_s^{(c)}(r, r', R^{(s)})$, which is the analog of $I_s(r, r', R^{(s)})$ function defined in Eq. (3.44):

$$I_s^{(c)}(r, r', R^{(s)}) = \int_0^{k_s} u_c(k, r) u_c(k, r') \, dk - \int_0^{k_s} u_c^{(s)}(k, r) u_c^{(s)}(k, r') \, dk$$

$$+ \int_{k_s}^{+\infty} \left[ u_c(k, r) u_c(k, r') - u_c^{(s)}(k, r) u_c^{(s)}(k, r') \right] dk . \quad (3.47)$$

$I_s^{(c)}(r, r', R^{(s)})$ is continuous in $r, r'$. Indeed, the $u_c(k, r'')$ and $u_c^{(s)}(k, r'')$ are bounded and the integral of Eq. (3.47) converges uniformly with respect to $r, r'$ (see Exercise V), so that:

$$\int_0^{+\infty} I_s^{(c)}(r, r', R^{(s)})^2 \, f(r') \, dr'$$

$$= \int_0^{+\infty} I_s^{(c)}(r, r', R^{(s)}) \, f(r') \int_0^{+\infty} \left[ u_c(k, r) u_c(k, r') - u_c^{(s)}(k, r) u_c^{(s)}(k, r') \right] dk \, dr'$$

$$= \int_0^{+\infty} u_c(k, r) \int_0^{+\infty} I_s^{(c)}(r, r', R^{(s)}) \, f(r') \, u_c(k, r') \, dr' \, dk$$

$$- \int_0^{+\infty} u_c^{(s)}(k, r) \int_0^{+\infty} I_s^{(c)}(r, r', R^{(s)}) \, f(r') \, u_c^{(s)}(k, r') \, dr' \, dk$$

$$= I_s^{(c)}(r, r, R^{(s)}) \, f(r) - I_s^{(c)}(r, r, R^{(s)}) \, f(r) = 0 , \quad (3.48)$$

where $f(r)$ in the above equation is an integrable function and where one has used the fact that the $u_c(k, r'')$ functions form a complete set of states (same for the $u_c^{(s)}(k, r'')$ functions). Thus, as Eq. (3.48) is valid for all functions $f(r)$, the integral $I_s^{(c)}(r, r', R^{(s)})$ vanishes for all screening radii $R^{(s)}$ larger than $R$. Consequently, as one has already shown that the integral of Eq. (3.46) with $k > k_s$ vanishes for $R^{(s)} \to +\infty$, one obtains:

$$\lim_{k_s \to 0} \lim_{R^{(s)} \to +\infty} \int_0^{k_s} u_c^{(s)}(k, r) u_c^{(s)}(k, r') \, dk = 0 \,. \tag{3.49}$$

Eq. (3.46) thus implies the same property for the $u^{(s)}(k, r'')$ functions. As a result, $I_s(r, r', R^{(s)}) \to 0$ when $R^{(s)} \to +\infty$. The completeness relation in the general Coulomb plus centrifugal potential case can then be proved, letting $R^{(s)} \to +\infty$ (see Exercise VI):

$$\sum_n u_n(r) u_n(r') + \int_0^{+\infty} u(k, r) u(k, r') \, dk = \delta(r - r') \,. \tag{3.50}$$

Eq. (3.50) is one-dimensional, that is, it is valid for one-body states with fixed $\ell$ and $j$ quantum numbers. The completeness relation in three dimensions is then the sum of the completeness relations of Eq. (3.50) involving all possible partial waves.

---

**Exercise  VI ⋆**

Show that Eq. (3.50) is fulfilled when $I_s(r, r', R^{(s)}) \to 0$ for $R^{(s)} \to +\infty$.

---

As the normalization of $u(k, r'')$ scattering states is the same for both charged and neutral particle cases, that is, $2\pi \, C^+ C^- = 1$, this general completeness relation writes similarly to Eq. (3.36). Its domain of validity is also that of the Fourier transform (see Sect. 3.2.1).

The consideration of an attractive Coulomb potential, for which $v_c < 0$ is, in fact, conceptually simpler. Indeed, as can be demonstrated in the frame of quantum defect theory [14], the infinity of $u_n(r)$ bound states verify $u_n(r) = \mathcal{O}(n^{-3/2})$ for $n \to +\infty$, so that their series is absolutely convergent. Moreover, due to the attractive character of the Coulomb potential therein, no Coulomb barrier can develop, so that $u(k, r) = \mathcal{O}(1)$ and $u^{(s)}(k, r) = \mathcal{O}(1)$ for $k \to 0^+$, which implies that the integral of $u(k, r)$ and $u^{(s)}(k, r)$ functions in Eq. (3.44) is uniformly convergent in $k = 0$ when $R^{(s)} \to +\infty$. The continuum part of Eq. (3.44) for $k \to +\infty$ can be treated as in the repulsive case. Thus, there is no problem to demonstrate with the presented methods the completeness relation in the general case of attractive Coulomb plus centrifugal potential.

## 3.3    Normalization and Orthogonality of Gamow States and One-Body Matrix Elements

The norm $N_i$ of a resonance state:

$$N_i^2 = \int_0^{+\infty} u_i^2(r) dr ,    \tag{3.51}$$

and the radial matrix elements calculated in the Berggren basis:

$$O_{if} = \int_0^{+\infty} u_f(r)\, \hat{\mathcal{O}}(r)\, u_i(r)\, dr    \tag{3.52}$$

are diverging, but this difficulty can be avoided by means of a regularization procedure [7, 15–19]. Zel'dovich proposed to multiply the integrand of a radial matrix element by a Gaussian convergence factor [15]:

$$\text{Reg} \int_0^{+\infty} u_f(r)\, \hat{\mathcal{O}}(r)\, u_i(r)\, dr = \lim_{\epsilon \to 0} \int_0^{+\infty} e^{-\epsilon r^2} u_f(r)\, \hat{\mathcal{O}}(r)\, u_i(r)\, dr ,    \tag{3.53}$$

where Reg indicates that a regularization of the initially diverging integral is effected with the Gaussian term equal to $\exp(-\epsilon r^2)$. In this expression, $u_f(r)$ and $u_i(r)$ stand for single-particle states, and $\hat{\mathcal{O}}(r)$ is a radial part of a one-body operator. Note that there is no complex conjugation in Eqs. (3.51)–(3.53). This arises because one uses analytic continuation to define radial matrix elements involving resonances from the radial matrix elements involving real bound states. Indeed, when $u_i(r)$, $u_f(r)$ are bound, one can make them real, so that $u_i^*(r) = u_i(r)$ and $u_f^*(r) = u_f(r)$. In fact, the absence of complex conjugation in Eq. (3.53) is related to the time-dependent character of resonance states and will be further detailed in Sect. 5.

The method of Zel'dovich, even though important on formal grounds, cannot be used in numerical applications due to the difficulty in approaching the limit in (3.53) for diverging integrals.

An equivalent and more practical procedure was proposed by Gyarmati and Vertse [18]. For that, one considers the functional $F(V_o)$:

$$F(V_o) = \frac{O_{if}}{N_i N_f} ,    \tag{3.54}$$

where $V_o$ is the depth of the potential generating single-particle wave functions $u_i(r)$ and $u_f(r)$, and:

$$O_{if} = \int_0^{+\infty} u_f(r)\, \hat{\mathcal{O}}(r)\, u_i(r)\, dr ,$$

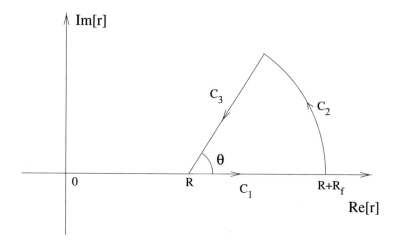

**Fig. 3.1** The path in the complex coordinate space corresponding to the complex rotation by angle $\theta$. $R$ is the point from which the exterior complex rotation starts. $R$ is large as compared to the nuclear radius. Hence it is assumed that the nuclear potential is negligible for $r > R$ (from Ref. [20])

and where $N_i^2$ and $N_f^2$ are, respectively, the square-roots of norms of wave functions $u_i(r)$ and $u_f(r)$. One demands that $V_o < V_{\text{lim}}$, with $V_{\text{lim}}$ is the depth of the potential for which one of these functions is bound and the other one is at zero energy. The $F(V_o)$ functional is then well defined because the integral converges in the domain $(-\infty; V_{\text{lim}})$. It represents the radial matrix element of Eq. (3.52) between two not necessarily normalized wave functions.

In Ref. [21], the analytical continuation of $F(V_o)$ is made using the Padé approximants. In the present work, we shall refer to the technique of the complex rotation [18] which allows the calculation of $F(V_o)$ with $V_o > V_{\text{lim}}$. To see that, let us call $f(r)$ one of three integrands $u_f(r)\mathcal{O}(r)u_i(r)$, $u_f^2(r)$, or $u_i^2(r)$ and let us take $V_o < V_{\text{lim}}$. Since $f(z)$ is analytical in the upper complex plane (see Fig. 3.1), then, following the Cauchy theorem, one has:

$$\int_{C_1} f(z)\,dz + \int_{C_2} f(z)\,dz + \int_{C_3} f(z)\,dz = 0 . \tag{3.55}$$

Since $f(z)$ decreases exponentially for $\text{Re}[z] > 0$, the integral $\int_{C_2} f(z)\,dz \to 0$ if $R_f \to +\infty$. For the same reason, the integrals $\int_{C_1} f(z)\,dz$ and $\int_{C_3} f(z)\,dz$ converge if $R_f \to +\infty$. Consequently, for $R_f \to +\infty$ one obtains:

$$\int_R^{+\infty} f(r)\,dr = \int_0^{+\infty} f(R + x \cdot e^{i\theta})e^{i\theta}\,dx . \tag{3.56}$$

Hence, on the interval $(-\infty; V_{\lim})$, one can define $F(V_o)$ by Eq. (3.54), with the norm (3.51) given by:

$$N_i = \sqrt{\int_0^R u_i^2(r)\, dr + \int_0^{+\infty} u_i^2(R + x \cdot e^{i\theta})\, e^{i\theta}\, dx} \qquad (3.57)$$

and with the matrix element (3.52) of the form:

$$O_{if} = \int_0^R u_f(r)\, \hat{O}(r)\, u_i(r)\, dr$$

$$+ \int_0^{+\infty} \left[ u_f(R + x \cdot e^{i\theta})\, \hat{O}(R + x \cdot e^{i\theta}) u_i(R + x \cdot e^{i\theta}) e^{i\theta}\, dx \right] . \quad (3.58)$$

---

**Exercise  VII** ★★

One will present the fundamental features of the complex scaling method from a theoretical point of view.

**A.** In the Gamow shell model, analytical calculations of integrals are not feasible, and all integrals involving complex scaling must be computed numerically. Consequently, in order to prove the efficiency of complex scaling on formal grounds, one will concentrate on a simple case where all integrals can be calculated analytically. For this, the integral of $\exp(ikr)$ will be calculated analytically and numerically using complex rotation on the interval $[R : +\infty[$. Explain why this situation qualitatively corresponds to those encountered in practical cases.

**B.** Prove that:

$$\text{Reg}\int_R^{+\infty} \exp(ikz)dz = \int_0^{+\infty} \exp(ik(R + xe^{i\theta}))e^{i\theta}dx , \qquad (3.59)$$

where Reg means that the diverging integral is regularized with complex scaling and where $\theta$ is the rotation angle, chosen so that the integral converges.

**C.** Show that one must have $\theta \in ]\theta_k : \theta_k + \pi[$ with $k = |k| \exp(-i\theta_k)$, for the integral of **B** to converge. Note that $\theta$ is defined modulo $2\pi$. Show that:

$$\text{Reg}\int_R^{+\infty} \exp(ikz)dz = -\frac{\exp(ikR)}{ik} \qquad (3.60)$$

and explain the $\theta$-independence of the integral using the arguments of analytic continuation.

If $u_i(r)$ and $u_f(r)$ are wave functions of bound or decaying states, one can write $k_i = |k_i|e^{-i\alpha_i}$ and $k_f = |k_f|e^{-i\alpha_f}$. As $u_i \sim a_i(z)e^{ik_iz}$ and $u_f(z) \sim a_f(z)e^{ik_fz}$ when $\Re(z) \to +\infty$, with $a_i$ and $a_f$ the algebraic increasing functions, the integrals defining $F(V_o)$ converge if one takes:

$$\theta > \alpha_f + \alpha_i .\tag{3.61}$$

In addition, the expression for $F(V_o)$ is analytical because $F(V_o)$ is a function of converging integrals of analytical functions. Square root in Eq. (3.57) causes no problems because $N_i^2$ and $N_f^2$ never cross the negative real axis when $V_o$ varies. Consequently, following the theorem of analytic continuation, Eq. (3.54) defines also $F(V_o)$ for $V_o > V_{\lim}$. In this way, one may calculate the radial matrix elements of resonance states which a priori are not normalizable.

Complex scaling in Eq. (3.58) can be readily applied to Eq. (2.174) when $u(k_n, r)$ is unbound, so that Eq. (2.174) can be immediately generalized to resonance $u(k_n, r)$ states. Similarly, one can show that $u(k_n, r)$ bound and resonance states are orthogonal to each other, by using the method explained in Sect. (2.5.1) and by integrating the complex-rotated functions. The fundamental features of complex scaling at theoretical level are depicted in Exercise VII, while the numerical accuracy of complex scaling in practical applications is demonstrated in Exercise VIII.

---

**Exercise VIII** ⋆⋆

One will now describe how to use the method of complex scaling numerically and show its usefulness in practice.

**A.** In the Gamow shell model, the radial matrix elements are calculated with complex scaling numerically. Numerical implementation involves the use of the Gauss-Legendre quadrature to evaluate integrals.

One will now present how complex scaling is numerically implemented. For this, one will calculate numerically the integral of **B** in Exercise VII. Show that the latter integral is equal to:

$$\int_0^{+\infty} \exp(ik(R + xe^{i\theta}))e^{i\theta}dx = 4e^{i\theta} \int_0^{R^{-1/4}} \exp(ik(R + x^{-4}e^{i\theta}))x^{-5}dx .\tag{3.62}$$

Explain why the integral on the right-hand side can be discretized with Gauss-Legendre quadrature and why this is impossible on the left-hand side. Show that the change of variable in $x \to x^{-4}$ provides with a quickly converging integral even for very small $k$. Notice that in practice, one has $\sin(\theta - \theta_k) > 0.1$, $|k| > 10^{-5}$ fm$^{-1}$ and the first Gaussian point on $[0 : R^{-1/4}]$ is about 0.01 (see Exercise VII for the definition of $\theta_k$).

B. Run the Gauss-Legendre integration code calculating complex scaled integrals for several values of $R$ and $k$. Typical values are $15 < R < 30$ fm and $0.01 < |k| < 0.1$ fm$^{-1}$ for bound and resonance states.

   Check that complex scaling is numerically precise only for $\theta$ values around the middle of an interval $]\theta_k : \theta_k + \pi[$.

C. A fundamental application of complex scaling is to calculate the norm of resonance states (see Eq. (3.57)). Explain why this norm is not equal to one in general and how it is used to normalize resonance states.

D. One will now numerically calculate the norm of a one-body neutron resonance eigenstate generated by a Woods-Saxon potential.

   For this, one will choose a Woods-Saxon potential depth so that the considered eigenstate varies from narrow resonance to broad resonance.

   Run the one-particle code of radial wave functions in these conditions in order to obtain the norm of one-body resonance states. Explain the dependence of the norm of the calculated resonance eigenstate with the depth of the used Woods-Saxon potential.

## 3.4   Matrix Elements Involving Scattering States

The calculation of matrix elements involving the scattering states is based on the (3.57) and (3.58). The one-body matrix element can be written as:

$$F(k_f) = \int_0^R u_f(r)V(r)u_i(r)\,dr + A_f A_i F_{++}(k_f)$$
$$+ A_f B_i F_{+-}(k_f) + B_f A_i F_{-+}(k_f) + B_f B_i F_{--}(k_f), \quad (3.63)$$

where

- $u_f(r) = A_f u_f^+(r) + B_f u_f^-(r)$
- $u_i(r) = A_i u_i^+(r) + B_i u_i^-(r)$, where $k_i$ in $F(k_f)$ is fixed and, in general, $u_i(r)$ can be either bound, resonance, or scattering state
- $F_{s_f s_i}(k_f) = \int_0^{+\infty} u_f^{s_f}(R+x)\hat{\mathcal{O}}(R+x)u_i^{s_i}(R+x)dx$ with $s_f, s_i \in (+, -)$ .

This separation is necessary because the presence of incoming and outgoing waves in the same integral does not allow one to find a unique path in the complex plane along which the integrand decreases exponentially. Consequently, for each $F_{s_f s_i}(k_f)$ one has to consider the domain of the complex plane where it converges, and then one performs an analytical continuation with the appropriate angle $\theta_{s_f s_i}$.

Certain integrals cannot be regularized in the above sense. Those include $F_{+-}(k_f)$ and $F_{-+}(k_f)$ with $u_i(r) = u_f(r)$. For $\hat{\mathcal{O}}(r) = 1$, the integrand tends toward a constant value at $+\infty$, independently of the values of $\theta_{+-}$ and $\theta_{-+}$. This can be immediately seen for neutrons, because with $z = R + x \cdot e^{i\theta}$ and with $|z| \rightarrow +\infty$, the product $u^+(z) \cdot u^-(z) \rightarrow \text{const} \times e^{ikz} \times e^{-ikz} = \text{const}$, and the corresponding integral diverges. In this case, however, it is easy to see that the integral is in fact a $\delta$-distribution, and it can be calculated by using a discrete representation of the Dirac $\delta$-function,

$$\delta(k - k') \rightarrow \frac{\delta_{k,k'}}{\Delta k}, \tag{3.64}$$

with $\Delta k$ being the discretization step in $k$.

## 3.5 Completeness Relation Involving Single-Particle Gamow States

As one aims at expanding many-body resonance wave functions, one has to devise the completeness relation of complex-energy one-body eigenstates. The first step was to devise a metric with which unbound states can be normalized and matrix elements involving resonance states can be numerically calculated. This has been effected in Sects. 3.3 and 3.4. In the following step, the real-energy completeness relation of Eq. (3.50) will be generalized to complex-energy states, so that the resonances can enter the completeness relation of one-body eigenstates.

A convenient method for that purpose is to use complex integration, by deforming the integration contour of Eq. (3.50) into the complex $k$−plane, as shown in Fig. 3.2. Following the residuum theorem, one obtains:

$$-\int_0^{+\infty} u_k(r)u_k(r')\,dk + \int_{L^+} u_k(r)u_k(r')\,dk = 2i\pi \sum_{k_n} \text{Res}\left(u_{k_n}(r)u_{k_n}(r')\right)_{k=k_n},$$
$$\tag{3.65}$$

where $k_n$ are the poles of $u_k(r)u_k(r')$ lying between the real axis and the complex contour. The Berggren completeness relation follows immediately:

$$\sum_n u_n(r)u_n(r') + \int_{L^+} u_k(r)u_k(r')\,dk = \delta(r - r'). \tag{3.66}$$

In the above equation, $u_n(r)$ are the bound states and resonances present between the real $k$-axis and the $L^+$ contour of complex-energy scattering states defined in the complex $k$-plane.

Figure 3.2 illustrates the ingredients entering Eq. (3.66). The resonant states, that is, the poles of the $S$ matrix, are represented by the dots. They are divided into the bound, decaying resonances, capturing resonances, and antibound states (see, e.g., Refs. [7, 22, 23]). The relation (3.66) involves the bound and resonance states and the contour $L^+$ lying in the fourth quadrant of the complex-$k$ plane. One may notice

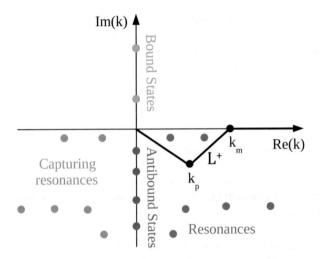

**Fig. 3.2** (Color online) Representation of the complex $k$-plane, showing the positions of bound states (green), resonances (blue), capturing resonances (orange), and antibound states (red). $L^+$ is the contour representing the nonresonant continuum, consisting of the two complex segments $[0 : k_p]$, $[k_p : k_m]$, and of the half-line $[k_m : +\infty)$ of the real $k$-axis. The Berggren completeness relation involves the bound states, resonance states lying between $L^+$ and real $k$–axis, and the scattering states on $L^+$. The contour $L^+$ must encompass all the poles in the discrete sum in Eq. (3.66), which are contained in the domain between $L^+$ and the real $k$-axis

that the resonances in Eq. (3.51) are normalized using the squared wave function and not the modulus of the squared wave function.

### 3.5.1 Domain of Applicability of the Berggren Completeness Relation

Equation (3.66) allows to expand states with complex $k$ in the upper half plane and inside the zone between the real $k$-axis and the $L^+$ contour of Eq. (3.66). This property is not trivial and will be hereby demonstrated.

One considers an analytical function $f(r)$ which will be expanded with the Berggren basis:

$$f(r) \sim e^{ik_f r} f_0(r), \tag{3.67}$$

where $k_f$ belongs to the aforementioned zone, $f_0(r)$ has an asymptotical rational behavior at infinity, and $f_0(r) \sim r^{\ell_0+1}$ for $r \to 0$ (see Eq. (2.4)). Evidently, $f(r)$ is integrable with complex scaling.

One will firstly consider a well-bound function $f(\alpha, r) = f(r)e^{-\alpha r}$, with $\alpha > 0$, lying in the upper part of the complex $k$-plane, so that the overlaps between $f(\alpha, r)$ and the one-body states in Eq. (3.66) are all finite without having to use complex scaling. As $f(\alpha, r)$ is bound, it can be expanded with Eq. (3.50). Moreover, as the products $f(\alpha, r')u_k(r')$ decay exponentially, the integrals in $k$ converge uniformly with respect to $r'$. One can then multiply Eq. (3.66) by $f(\alpha, r')$, integrate over $r'$ and interchange integrals in $r'$ and $k$. Equations (3.50) and (3.65) thus provide with the expansion of $f(\alpha, r)$:

$$f(\alpha, r) = \sum_n c_n(\alpha)\, u_n(r) + \int_{L^+} c_k(\alpha)\, u_k(r)\, dk \,, \tag{3.68}$$

where

$$c_n(\alpha) = \int_0^\infty f(\alpha, r')\, u_n(r')\, dr' \,, \tag{3.69}$$

and

$$c_k(\alpha) = \int_0^\infty f(\alpha, r')\, u_k(r')\, dr' \,. \tag{3.70}$$

Following the same method as in Sect. 3.3, one can replace the real-axis integrals of Eqs. (3.69) and (3.70) by complex-scaled integrals of rotation angle $\theta$ so that the latter integrals converge $\forall \alpha \geq 0$. This is possible because $k_f + i\alpha$ never crosses the $L^+$ contour when $\alpha \to 0$. Indeed, in the opposite case, one would obtain integrals for which no complex scaling can regularize integrals, similarly to the Dirac delta normalization encountered in Sect. 3.3. Consequently, $c_n(\alpha)$ and $c_k(\alpha)$ in Eqs. (3.69) and (3.70) are analytical functions of $\alpha$, with $c_k(\alpha) = \mathcal{O}(k^{-2})$ along the $L^+$ contour due to the analyticity and boundary conditions verified by $f(\alpha, r)$. As a consequence, the Berggren expansion of $f(\alpha, r)$ is an analytical function of $\alpha$ as well. Hence, Eq. (3.68) is valid $\forall \alpha \geq 0$ and $f(r)$ belongs to the domain of applicability of the Berggren completeness relation:

$$f(r) = \sum_n c_n\, u_n(r) + \int_{L^+} c_k\, u_k(r)\, dk \,. \tag{3.71}$$

$c_n$ and $c_k$ coefficients in Eq. (3.71) represent the components of the $f(r)$ function expanded in the Berggren basis, calculated using the complex scaling integration method of Eq. (3.63):

$$c_n = \int f(z')\, u_n(z')\, dz' \,, \tag{3.72}$$

and

$$c_k = \int f(z')\, u_k(z')\, dz' \,. \tag{3.73}$$

Conversely, let us consider a function $f(r)$ of the form (3.67) where $k_f$ belongs to the lower half plane outside the zone between the real $k$-axis and the $L^+$ contour of Eq. (3.66). Below, one will show that it is not possible to expand this function using Eq. (3.66). A simple example of such functions $f(r)$ are the resonances of the Berggren basis lying below the $L^+$ contour. Indeed, they are orthogonal to all bound states, resonances, and scattering states of the Berggren basis (see Eq. (3.66)), so that the expansion of $f(r)$ generated by Eq. (3.66) vanishes identically.

For the general case, let us assume that Eqs. (3.71)–(3.73)) are valid. If the $L^+$ is moved toward the real axis, Eq. (3.73) can still be used because the $L^+$ contour never crosses $k_f$ while being deformed back to the real axis. The residuum theorem must be considered similarly as in Eq. (3.71), so that the resonance poles of Eq. (3.71) must be removed from its sum every time the $L^+$ contour reaches them. As a result, Eq. (3.71) can be reduced to the Newton completeness relation, so that all basis functions are either bound states or scattering states of real energy. Consequently:

$$|f(r)| \leq M \left( \sum_n |c_n| + \int_{L^+} |c_k| \, dk \right) , \qquad (3.74)$$

where $M$ is a constant for which $|u_n(r)| \leq M$ and $|u_k(r)| \leq M$ and the $k$-integral converges because $c_k = \mathcal{O}(k^{-2})$ due to the analyticity of $f(r)$.

Equation (3.74) cannot be correct because $|f(r)| \to +\infty$ for $r \to +\infty$. Consequently, a function bearing an asymptote in $e^{ikr}$, with $k$ in the lower half plane outside the zone between the real $k$-axis and the complex contour, does not belong to the domain of applicability of the Berggren completeness relation when the $L^+$ contour is fixed. Evidently, that function can still be expanded with a Berggren completeness relation whose $L^+$ contour encompasses its complex momentum $k$.

### 3.5.2   Berggren Completeness Relation for Complex Potentials

Up to now one has considered the Berggren completeness relation for eigenstates of real potentials. However, bound states, resonances, and scattering states generated by complex potentials also form a complete set of states, with the same properties as in the case of real potentials. A demonstration of this property can be found using a formula similar to Eq. (3.44):

$$I(r, r', \lambda) = \sum_n u_n(r)u_n(r') - \sum_n u_n^{(R)}(r)u_n^{(R)}(r')$$

$$+ \int_{L^+} \left[ u(k, r)u(k, r') - u^{(R)}(k, r)u^{(R)}(k, r') \right] dk , \qquad (3.75)$$

where the one-body states entering the Berggren completeness generated by a real potential $V_R(r)$ are denoted as $u^{(R)}(k, r)$, and the bound states, resonances and scattering states generated by a complex potential of real and imaginary parts equal to $V_R(r)$ and $V_I(i)$, respectively, are denoted as $u(k, r)$.

One will write the potential generating the basis as a function of the complex variable $\lambda$:

$$V(\lambda, r) = V_R(r) + \lambda V_I(r) . \tag{3.76}$$

Obviously, the potential in Eq. (3.76) is real for real values $\lambda$. Equation (3.75) then vanishes identically in this case, as it is the difference of two completeness relations generated by two real potentials. However, as $V_R(r)$ and $V_I(i)$ are the real and imaginary parts of the complex potential of interest, one has to have $\lambda = i$ in Eq. (3.76). One then has to recur to analytical continuation in $\lambda$ to recover the complex-potential case.

Equation (2.2) implies that $u(k, r)$ functions are defined for complex values of $\lambda$. However, the resonant states $u_n(r)$ of Eq. (3.75) can no longer be normalized to one if the norm provided by complex scaling in Eq. (3.58) is equal to zero or diverges. Moreover, scattering states cannot be normalized with a Dirac delta if $C^\pm = 0$ in Eq. (3.18). Consequently, one will assume for the moment that all wave functions remain normalizable when $\lambda$ varies in the complex $k$-plane. The special conditions leading to diverging $u_n(r)$ and $u(k, r)$ functions in Eq. (3.75) will be dealt with in the following.

$I(r, r', \lambda)$ in Eq. (3.75), which is defined by an absolutely converging integral, is then an analytical function of $\lambda$, hence vanishing identically. The Berggren completeness relation is then obtained for $\lambda = i$ by an analytical continuation with respect to $\lambda$. Consequently, the bound, resonances, and scattering states generated by a complex potential form a complete set as long as its bound and resonance states can be normalized with complex scaling and the integral of scattering states is well defined.

A situation where analytic continuation must be used with caution is when one has spectral singularities [24], that is, resonances on the real axis, which thus generate nonanalytic points on the real-energy contour. This situation can be quickly fixed by deforming the contour in the complex plane. Indeed, spectral singularities can be normalized with complex scaling, as they bear the same properties as the resonance states generated by real potentials.

Let us consider Eq. (3.76) with real potentials, that is, with $\lambda$ real. One demands therein that Eq. (3.76) is defined with resonance $u_n(r)$ states and a complex contour of $u(k, r)$ scattering states encompassing these resonances. When $\lambda$ goes from the real axis to the demanded value of $\lambda = i$, resonance $u_n(r)$ states will become spectral singularities. Moreover, as one always integrates along an $L^+$ complex contour, the integral of Eq. (3.75) is always well defined. Hence, analytic continuation can be used when $\lambda$ goes from the real axis to $\lambda = i$. Therefore, $I(r, r', \lambda) = 0$ for $\lambda = i$, and the Berggren completeness relation also holds in the presence of spectral singularities.

Another situation where analytic continuation cannot be used arises when one obtains exceptional points [25–30]. Exceptional points arise when two resonances coalesce in the complex $k$-plane. In this case, the Berggren set of states is no longer complete. Nevertheless, one can recover a complete set by adding to the initial set of resonant $u_n(r)$ and scattering $u(k, r)$ states the derivative with respect to $k$ of the wave function associated to exceptional states [31]. In order to show the latter property, let us consider two resonance states $u_n(r)$ and $u_m(r)$, where $m \neq n$, which coalesce and form an exceptional point for a certain value of $\lambda$. Wave functions associated to exceptional points have a vanishing Berggren norm (see Eq. (3.58)). Consequently, one will consider in the following that $u_n(r)$ and $u_m(r)$ are no longer normalized using Eq. (3.58). This obviously does not change the completeness properties of that newly defined Berggren set of states. As $u_n(r) \propto u_m(r)$ when $k_m = k_n$, the expansion coefficients of $u_n(r)$ and $u_m(r)$ of a given function can diverge when $k_m - k_n \to 0$. All other components are clearly well defined in that limit. Hence, one replaces $u_m(r)$ by the finite difference $(u_n(r) - u_m(r))/(k_n - k_m)$ in Eq. (3.66), with $k_n$ and $k_m$ the linear momenta associated to $u_n(r)$ and $u_m(r)$, respectively. As $k_n \to k_m$, the latter finite differences become equal to $\dot{u}_n(r)$, with the dot denoting derivative with respect to $k$.

One will show that $u_n(r)$ and $\dot{u}_n(r)$ are linearly independent. For this, one notices that $\dot{u}_n(r)$ is solution of a differential equation, which is derived by differentiating Eq. (2.2) with respect to $k$:

$$\dot{u}_n''(r) = \left( \frac{\ell(\ell + 1)}{r^2} + v_l(r) - k_n^2 \right) \dot{u}_n(r) - 2k_n u_n(r) . \tag{3.77}$$

Let us suppose that $\dot{u}_n(r) = C u_n(r)$, where $C$ is a complex number. By multiplying Eq. (2.2) by $C$ and subtracting this equation from Eq. (3.77), one obtains that $k_n u_n(r) = 0$, so that $u_n(r) = 0$ for $k_n \neq 0$, which is impossible. One cannot have $k_n = 0$ either, as $u_n(r)$ would be bound in this case (see Sect. 2.5). Therefore, $u_n(r)$ and $\dot{u}_n(r)$ are linearly independent. This demonstration is straightforward to generalize if a derivative of higher order with respect to $k$ must be utilized.

The only overlap that might not vanish when $k_m - k_n \to 0$ is that involving $u_n(r)$ and $\dot{u}_n(r)$. There is a fixed constant $\alpha$ so that $\dot{u}_n(r) + \alpha u_n(r)$ is orthogonal to all states in Eq. (3.66), up to an error term which can be made arbitrarily small when $k_m - k_n \to 0$. Consequently, one can replace $u_m(r)$ by $\dot{u}_n(r) + \alpha u_n(r)$ in Eq. (3.66) when $k_m \sim k_n$, so that all the newly defined states present in Eq. (3.66) form a complete set of states and are linearly independent when $k_m \sim k_n$. This proves that the introduction of the derivative of exceptional states with respect to $k$ is sufficient to restore completeness of the Berggren basis in the presence of exceptional points.

This procedure can be clearly generalized if one has more than one exceptional point, or with three states coalescing to form a threefold exceptional state, etc., in which case higher derivatives of $u_n(r)$ with respect to $k$ must be included.

### 3.5.3 Overcompleteness Relations Involving Single-Particle Gamow States

Other expansions as Eq. (3.66) have been devised several decades ago in order to expand radial functions of unbound states. A remarkable basis expansion involving unbound states is that arising from Mittag-Leffler theory [32]:

$$\frac{1}{2} \sum_n u_n(r)u_n(r') = \delta(r - r') , \qquad (3.78)$$

where $u_n(r)$ includes all poles of the $S$-matrix (see Sect. 2.6.6), so that $u_n(r)$ is either a bound, antibound, resonance, or capturing state. The domain of validity of Eq. (3.78) is rather large, as Eq. (3.78) can be used to expand all radial functions vanishing beyond a given radius $R_c$ smaller than the range of the potential generating $u_n(r)$ basis functions. Apparently, Eq. (3.78) seems to be of more practical interest than Eq. (3.66) due to the absence of contour integral. In reality, Eq. (3.78) is difficult to apply in numerical calculations due to its overcomplete character, that is, the fact that basis states are not linearly independent, which is embodied by the $1/2$ factor in Eq. (3.78).

Non-orthogonality of basis states is usually difficult to consider in numerical applications, as one then has to solve generalized eigensystems. Moreover, the Mittag-Leffler expansion of Eq. (3.78) makes use of antibound states, which generate numerical instabilities when used in a basis expansion (see Sect. 3.7.2) for the study of Berggren basis expansions including antibound states). Added to that, it is difficult to study extended nuclear wave functions such as halo and narrow resonant states, with the Mittag-Leffler expansion, as all functions must be equal to zero for $r > R_c$. A large value for $R_c$ would then be required in practice, which make calculations even more unstable, as antibound states increase exponentially on the real $r$-axis (see Fig. 2.1). Consequently, despite its attractive features, notably its discrete character, the Mittag-Leffler expansion is mainly of theoretical interest, as the generalized eigensystem to solve generated by Eq. (3.78) would be highly unstable numerically.

An overcompleteness relation involving complex-energy states and closely related to the Mittag-Leffler expansion of Eq. (3.78) has been obtained by Romo [33]:

$$\frac{1}{2} \sum_n u_n(r)u_n(r') + \int_U u_k(r)u_k(r')\, dk = \delta(r - r') , \qquad (3.79)$$

where $U$ is the complex contour built from the infinite intervals $(-\infty : K]$, $[K : +\infty)$, completed by a half-circle of radius $K$ in the lower complex $k$-plane, denoted as $U$, and the $1/2$ factor arises from the overcomplete character of Eq. (3.79). Thus, the discrete part in Eq. (3.79) contains all bound states and poles of the $S$-matrix between the $U$ contour and the real $k$-axis. Other overcompleteness relations can be generated by deforming the real $k$-axis in many different ways, which will result in different discrete parts as that of Eq. (3.79) (see Ref. [34] for the derivation of overcompleteness relations similar to Eq. (3.79). In particular, Eq. (3.79) reduces to Eq. (3.78) when expanded functions vanish after a radius $R_c$ [34].

One can see that the number of antibound states in Eq. (3.79) can be reduced by using a sufficiently small $K$ value compared to that of Eq. (3.78). However, the continuum part of Eq. (3.79) does not vanish in general for not truncated radial functions, even for large $K$ values [34]. Thus, similarly to Eqs. (3.78) and (3.79) is mainly interesting from a theoretical point of view. Indeed, even though expanded radial function can be extended along the real $r$-axis, the overcomplete character of Eq. (3.79) and the fact that its continuum part is sizable renders its applicability very difficult. In fact, the orthogonal character of the basis states as well as the possibility to include only narrow or mildly broad resonance states in a discrete sum of Eq. (3.66) made the Berggren basis both physically relevant and numerically stable in Gamow shell model applications.

## 3.6    Newton and Berggren Completeness Relations Generated by Nonlocal Potentials

Local potentials are the most commonly used potentials in nuclear physics, with, in particular, the widely used Woods-Saxon potential [35]. Potentials used in atomic and molecular physics are typically also local, and are deemed as pseudo-potentials (see Sect. 4.3.1.1 for applications to dipolar and quadrupolar anions). However, nonlocal potentials often arise in quantum physics as well. This is typically due to the Pauli principle, whose effects are nonlocal. Hence, the Hartree-Fock potential generated by a finite range interaction, of nuclear or Coulomb type, is nonlocal [36]. The Pauli principle also plays an important role in nuclear collisions, so that nonlocal potentials are frequently used in the reaction theory [37].

However, as one will see in this section, the nonlocal potentials greatly augment the mathematical difficulties related to the use of one-body states. A nonlocal Schrödinger equation suitable for physical applications is obtained from Eq. (2.2) by adding a nonlocal potential to its local part:

$$v_l(r)\, u(k, r) \;\rightarrow\; v_l(r)\, u(k, r) + \int_{r_0}^{R} v_{\mathrm{nl}}(r, r')\, u(k, r')\, dr' \,, \tag{3.80}$$

where $v_{\mathrm{nl}}(r, r')$ is a smooth function symmetric in $r$ and $r'$, behaving as $r^{\ell_0+1}$ when $r \rightarrow 0$ (same for $r'$).

Bound states of the nonlocal potential are real, which can be shown using a method similar to that of Exercise VI in Sect. 2.5. As the non-locality of the Hartree-Fock potential is induced by a finite range interaction, which plays no role far from the nucleus or molecule, it is physically justified to have a vanishing nonlocal potential for $r, r' \geq R$. One can then show that the boundary conditions verified by $u(k, r)$ are still valid in the nonlocal case (see Sect. 2.1). Moreover, as Eq. (2.2) is local for $r > R$, one can define the Jost function associated to $u(k, r)$ using $r \geq R$ as in Sect. 2.6.4. The definition of Jost functions in the general nonlocal case is indeed not simple, as Wronskians are not independent of $r$ therein. One can also note that the demonstration followed in Sect. 2.6.8 to obtain the expressions for a narrow width in Eqs. (2.197), (2.199), and (2.201) can be utilized as well in the nonlocal case.

The existence and unicity of $u(k, r)$ are not guaranteed in the nonlocal case. Indeed, Eqs. (2.9) and (2.139) for nonlocal potentials become Fredholm equations of the second kind, so that the mathematical properties of $u(k, r)$ rely on Fredholm theory [38]. In particular, a $u(k, r)$ function verifying the boundary condition of Eq. (2.6) does not exist if Eq. (2.9) possesses a nonvanishing solution whose function and derivative vanish in $r = r_0$ [38]. Similarly, a $u^{\pm}(k, r)$ function satisfying the boundary condition of Eq. (2.7) in $r = R$ cannot be constructed if Eq. (2.9) provides a nonzero solution whose function and derivative vanish in $r = R$ [38]. These situations arise when the Fredholm operator acting on $u(k, r)$ functions is non-invertible for certain energy values [38].

Moreover, the number of bound states generated by nonlocal potentials is not necessarily equal to the number of zeros of $u(k = 0, r)$ in the case of local potentials (see Sect. 2.5). In fact, using convexity arguments on $u(k, r)$ for $r > R$ in the bound energy region, one can show that the number of bound states is equal or smaller than the number of zeros of $u(k = 0, r)$ [10]. Nevertheless, the nonlocal integral operators entering Eqs. (2.9) and (2.139) are diagonally dominant in practice, and hence invertible for every $k$ value in problems of physical interest. Therefore, Eqs. (2.9) and 2.139) in practical applications always bear a unique solution, and the relation between bound states and the number of zeros of $u(k = 0, r)$ is fulfilled, as in the local case. As the nonlocal integral operators in Eqs. (2.9) and (2.139) are invertible in practice and analytic in $k$, hence $u(k, r)$ bears the same analytic properties as in the local case.

In fact, the main problem encountered with nonlocal Schrödinger equation is related to the Newton completeness relation of Sect. 3.2.1. This arises because, in general, the asymptotic solutions for $|k| \to +\infty$, devised in Sect. 2.6.2, no longer hold in the nonlocal potential case. Real values of $k$ pose no problem in this context. One may show with integration by parts that the nonlocal term of Eq. (3.80) is $\mathcal{O}(k^{-2})$ on the real $k$-axis using an analogous method as that presented in Ref. [10]. Nevertheless, the asymptotic expansion of $u^{\pm}(k, r)$ and $u(k, r)$ (see Sect. 2.6.2) strongly diverges when $\Im(k) \to +\infty$, by the way of additional terms diverging exponentially with $\Im(k)$, induced by the nonlocal kernel (see Sect. 2.6.2).

Consequently, several integrals entering the demonstration of Newton completeness relation diverge in general when $K \to +\infty$, so that this procedure fails in the non-local case (see Sect. 3.2.1 and Exercise IV). Therefore, the Newton completeness relation involving non-local potentials must be demonstrated either with an integral based on the Stieltjes-Lebesgue measure (see Sect. 3.2), or by using a set of discrete eigenstates defined in a box, whose radius goes to infinity to recover the continuum [10].

One can check that the methods used in the sections following Sect. 3.2.2, with the exception of Sect. 3.5.3, can be applied without modification to the nonlocal case, because one always has $\Im(k) \neq 0$ in a finite zone of the complex plane. As a consequence, the completeness properties of the eigenstates generated by local and nonlocal potentials are identical and both can be used to generate Berggren basis states for Gamow shell model applications. This is particularly interesting when using a Hartree-Fock potential to generate the Berggren basis, because this potential recaptures a significant part of the strength provided by the nuclear interaction.

## 3.7  Numerical Implementation of the Berggren Completeness Relation

In practical applications of the Berggren completeness relation, one has to discretize the integral in Eq. (3.66) [39, 40]:

$$\int_{L^+} u(k,r)u(k,r')\,dk \simeq \sum_{i=1}^{N_d} u_i(r)u_i(r') , \qquad (3.81)$$

where $u_i(r) = \sqrt{\Delta_{k_i}}\,u(k_i, r)$ and $\Delta_{k_i}$ is the discretization step. It follows from Eq. (3.81) that the $u_i(r)$ are orthonormalized so that the discretized Berggren relation (3.66) takes the form:

$$\sum_n u_n(r)u_n(r') + \sum_{i=1}^{N_d} u_i(r)u_i(r') \simeq \delta(r - r') . \qquad (3.82)$$

This relation is formally identical to the standard completeness relation in a discrete basis. However, as the formalism of Gamow states is non-Hermitian, the Hamiltonian matrix $H$ is complex symmetric.

Up to this point, the choice of the contour in Eq. (3.2) has been arbitrary. In practice, however, one wants to minimize the number of discretization points $N_d$ along $L^+$. This can be achieved if phase shifts of the scattering functions on the contour change smoothly from point to point. This condition can be met if the contour does not lie in the vicinity of a pole, especially the narrow resonance (see Sect. 2.6.6). The completeness relations derived above hold in every $(\ell, j)$ partial wave. Consequently, in practical calculations, one has to take different contours for different partial waves. As discussed below, the choice of the contour depends on the distribution of resonance states in the complex $k$-plane.

In order to obtain the zeroth-order approximation of Gamow shell model eigenstates (see Sect. 5), it is customary to neglect the $L^+$ contour in Eq. (3.82). Indeed, the contribution of scattering states in the Berggren basis expansion is typically much smaller than that arising from the bound and resonance states. To suppress the $L^+$ contour in Eq. (3.82) is indeed very convenient numerically, because the number of bound and resonance states is one to two orders of magnitude smaller than the number of discretized basis scattering states. The use of a truncated Berggren basis, where one removes all scattering states, is called the pole approximation, as one leaves only $S$-matrix poles in the Berggren basis (3.82).

### 3.7.1  Completeness Relations Involving Proton States

In this section, we shall discuss numerical examples of the Berggren completeness relation in the one-proton case, as it was theoretically considered in Sect. 3.5. Contrary to the previous sections, the efficiency of the Berggren completeness relation from a computational point of view will be demonstrated. Indeed, while Eq. (3.82) directly leads to a matrix of finite dimension to be diagonalized, one still has to demonstrate that it is of practical use, that is, that the method described in Sect. 3.7 leads to a fast and stable numerical algorithm.

For this, one will consider the diagonalization of a Woods-Saxon potential with the Berggren basis generated by another Woods-Saxon potential. Indeed, one can then directly check the precision of diagonalized eigenstates by comparing them to the quasi-exact wave functions obtained from a direct integration of Eq. (2.2). Dealing with the resonances is difficult in the proton case due to the singular character of the scattering functions in the complex $k$-plane (see Sect. 2.3).

The single-particle basis will be generated by the spherical Woods-Saxon + Coulomb potential:

$$V(r) = -V_0 f(r) - 4V_{SO} (\mathbf{l} \cdot \mathbf{s}) \left(\frac{1}{r}\right) \left(\frac{df(r)}{dr}\right) + V_{Coul}(r), \qquad (3.83)$$

where

$$f(r) = \left[1 + \exp\left(\frac{r - R_0}{d}\right)\right]^{-1}. \qquad (3.84)$$

The Coulomb potential $V_{Coul}$ is assumed to be given by a uniformly charged sphere of radius $R_0$ and charge $Q=+20e$. The depth of the central part is varied to simulate different situations.

We shall begin by expanding the weakly bound $2p_{3/2}$ state, $u_{WS}(r)$, in the basis $u_{WS^B}(k, r)$ which is generated by the Woods-Saxon potential of a different depth:

$$u_{WS}(r) = \sum_i c_{k_i} u_{WS^B}(k_i, r) + \int_{L^+} c(k) u_{WS^B}(k, r) dk. \qquad (3.85)$$

In the above equation, the first term represents contributions from the resonant states while the second term is the background contribution from nonresonant continuum states. Since the basis is properly normalized, the expansion amplitudes meet the

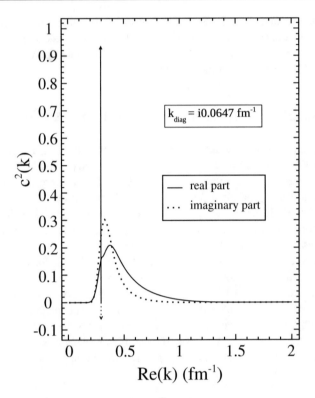

**Fig. 3.3** Distribution of the squared amplitudes $c^2(k)$ of the $2p_{3/2}$ proton state of a Woods-Saxon potential with a depth $V_0 = 75\,\text{MeV}$, in the single-particle basis generated by a Woods-Saxon potential with a depth $V_0^{(B)} = 70\,\text{MeV}$. In this example, the Woods-Saxon potential has radius $R_0=5.3$ fm, diffuseness $d=0.65$ fm, and spin-orbit strength $V_{SO} = 5\,\text{MeV}$. The Coulomb potential is assumed to be that of a uniformly charged sphere. The amplitudes of both real (solid line) and imaginary (dotted line) components of the wave function are plotted as a function of $\Re[k]$. The height of the arrow gives the squared amplitude of the $2p_{3/2}$ state contained in the basis, which is resonant so that its linear momentum verifies $\Re(k) > 0$. The value of the linear momentum of the diagonalized $2p_{3/2}$ state is shown in an insert. It is imaginary as this state is bound (adapted from Ref. [20])

condition:

$$\sum_i c_{k_i}^2 + \int_{L^+} c^2(k)\, dk = 1 \,. \tag{3.86}$$

In the example, shown in Fig. 3.3, the basis is that of the Woods-Saxon potential with a depth of $V_0^{(B)}$=70 MeV, and the expanded state corresponds to a Woods-Saxon potential with $V_0$=75 MeV. This is an interesting case since one expresses a bound state (real) wave function in the basis which contains only complex wave functions. In the considered example, $0p_{3/2}$ and $1p_{3/2}$ orbitals that are bound by

~50 MeV and ~20 MeV, respectively, play a minor role in the expansion of a weakly bound $2p_{3/2}$ state. Completeness of the expansion is guaranteed by the scattering states along the contour $L^+$ which corresponds to three straight segments in the complex $k$−plane, joining the points: $k_0 = 0.0 - i0.0$, $k_1 = 0.3 - i0.1$, $k_2 = 1.0 - i0.0$, and $k_3 = 2.0 - i0.0$ (all in fm$^{-1}$). The contour is discretized with $n$=60 points using a Gauss-Legendre quadrature:

$$u_{WS}(r) \simeq \sum_{n \in b,d} c_n\, u_n(r) + \sum_{i=1}^{n} c_{k_i}\, u(k_i, r) . \tag{3.87}$$

The exact energy of the expanded state is $E$=–0.0923 MeV. The diagonalization provides with almost the same energy, as the precision of diagonalization is about $10^{-4}$ keV for both energy and width.

One can see that the contribution of $2p_{3/2}$ basis resonance dominates in the expansion. The scattering states become important when their energies approach the $2p_{3/2}$ resonance in the basis, indicating the stronger variation of phase shift therein. There are no components of states with energies close to the energy of the expanded state $E = -0.0923$ MeV (see Fig. 3.3). Indeed, the only bound states of the Berggren basis are the well bound $0p_{3/2}$ and $1p_{3/2}$ basis states. Consequently, the expanded weakly bound $2p_{3/2}$ state is built almost exclusively from unbound states. Hence, one did not plot in Fig. 3.3 bound state components nor indicate linear momenta of bound states. In fact, the bound state asymptote of the expanded state arises from subtle cancellations between the unbound $2p_{3/2}$ resonance and complex scattering states of the Berggren basis.

Even though energy of the expanded $2p_{3/2}$ weakly bound state is real, its eigenvector in the used Berggren basis is complex (see Fig. 3.3). Hence, the value of its imaginary part which theoretically equals to zero, is a good measure of the numerical precision of calculated expanded states. Indeed, the eigenvalues of the Hamiltonian matrix related to bound states have a small imaginary part due the finite discretization of the contour of scattering states. For other numerical tests, see Ref. [40] for a study of the single-particle level density and Ref. [41] for a study of the Berggren expansion in the pole approximation, as well as Exercises IX and X.

In the above discussion, the Hamiltonian containing a Woods-Saxon and Coulomb potentials was diagonalized in the basis generated by a similar Hamiltonian, that is, in both Hamiltonians the Coulomb potential was identical. Consequently, besides the energies of basis states forming the diagonal, the Hamiltonian to diagonalize was the difference of two finite-range Woods-Saxon potentials. Matrix elements of the Hamiltonian then converge quickly on the real axis due to the fast exponential decrease of the nuclear part of the Woods-Saxon potential. Thus, no complex scaling is necessary to calculate the matrix elements.

The situation changes if the Coulomb potentials in the basis Hamiltonian and in the diagonalized Hamiltonian are different. Indeed, the difference of two non-identical Coulomb potentials is infinite-range, so that complex scaling is necessary to calculate the matrix elements of the difference of Coulomb potentials. Due to the slow fall-off of the Coulomb potential, in $1/r$, the matrix elements involving two scattering states $u(k, r)$ and $u(k', r)$ with $k \sim k'$ diverge as $\ln(k - k')$. In the absence of contour discretization, this divergence poses no theoretical problem as it is integrable in the complex $k$-plane. However, if one discretizes the contour, one obtains infinite matrix elements on the diagonal of the Hamiltonian matrix, because one has exactly $k = k'$ therein. Consequently, these infinite matrix elements have to be regularized and a precise scheme to calculate the matrix elements of a Coulomb potential with the Berggren basis has to be devised.

Let us generate the single-particle basis states using a spherical Woods-Saxon potential and a Coulomb potential proportional to an error function:

$$V_{\text{Coul}}(Z, r) = \frac{C_{\text{Coul}} \; Z \; \text{erf}(3\sqrt{\pi} r/(4R_0))}{r} , \tag{3.88}$$

defined with the Coulomb constant $C_{\text{Coul}}$ and the charge acting on the proton $Z$. $R_0$ in Eq. (3.88) is the radius of the used Woods-Saxon potential. The radial function entering the Coulomb potential $V_{\text{Coul}}(Z, r)$ is standard and arises from the use of a Gaussian charge density for the closed core [42–44]. The coefficient $3\sqrt{\pi}/(4R_0)$ is chosen in order for $V_{\text{Coul}}(Z, r = 0)$ to have the same value as that of the Coulomb potential for a uniformly charged sphere of radius $R_0$ [45]. For simplicity, the Hamiltonian which is used to generate the basis and the Hamiltonian to diagonalize will have the same finite-range part, so that they differ only through their charge $Z_c$.

---

**Exercise  IX $\star$**

One will show the main features of the eigenvector components issued from the diagonalization of a one-body Hamiltonian with the Berggren basis in a numerical example.

Run the one-particle diagonalization code for the case described above, using the potential depths: $(V_0^{(B)}, V_0) = (70, 65), (75, 70), (75, 80)$ (all in MeV). Plot the $c(k)$ components in Berggren space as a function of $k$, similarly to Fig. 3.3.

Explain differences with the example shown in the text, namely the importance of continuum coupling according to the values of $V_0^{(B)}$ and $V_0$.

---

**Exercise  X $\star$**

In this numerical example, one will point out the differences of the diagonalization of a one-body Hamiltonian when using a Berggren basis generated either by a Woods-Saxon potential or by a Pöschl-Teller-Ginocchio potential.

**A.** Run the one-particle diagonalization code in which a basis generated by a Pöschl-Teller-Ginocchio potential diagonalizes a Woods-Saxon potential. Explain why one can only consider the expansion of bound states therein.

---

**B.** The central depth of the Woods-Saxon potential is now decreased by 10%. Run the code to calculate the new eigenstate of the Woods-Saxon potential. The calculated state must have its energy close to 0 MeV.

**C.** Explain why one cannot generate unbound states by diagonalizing the Hamiltonian using a real-energy basis generated by a Pöschl-Teller-Ginocchio potential.

---

One will denote the charge of the basis generating potential as $Z^{(b)}$, that of the diagonalized potential as $Z^{(d)}$, and their difference will be denominated as $\Delta Z = Z^{(d)} - Z^{(b)}$. The eigenstate $u_{\text{eig}}(r)$ of the diagonalized Hamiltonian will be expanded using Eq. (3.66):

$$u_{\text{eig}}(r) = \sum_i c_{k_i} u(k_i, r) + \int_{L^+} c_k u(k, r)\, dk , \qquad (3.89)$$

where the coefficients $c_{k_i}$ and $c_k$ will be determined by diagonalization.

Due to the unbound character of basis states, the matrix elements of Eq. (3.88) have to be calculated using a complex rotation of $r$ [18, 46]:

$$\text{Reg} \int_0^{+\infty} u_a(r)\, V_{\text{Coul}}(\Delta Z, r)\, u_b(r)\, dr$$

$$= \int_0^R u_a(r)\, V_{\text{Coul}}(\Delta Z, r)\, u_b(r)\, dr$$

$$+ \sum_{\substack{\omega_a = \pm \\ \omega_b = \pm}} e^{i\theta} \int_0^{+\infty} u_a^{\omega_a}(z(x)) \left( \frac{C_{\text{Coul}}\, \Delta Z}{R + x e^{i\theta}} \right) u_b^{\omega_b}(z(x))\, dx , \qquad (3.90)$$

where Reg indicates regularization using the complex scaling (see Eq. (3.58)), $u_a(r)$ and $u_b(r)$ are two Berggren basis states, and $R$ is a radius above which the complex rotation is applied. In the following, the integrals of Eq. (3.90) going from 0 to $+\infty$ will be denoted as the improper integrals.

Equation (3.90) cannot be used for $u_a(r) = u_b(r) = u(k, r)$, with $u(k, r)$ a scattering wave function, as no such rotation angle $\theta$ exist so that the improper integrals with $\omega_a \omega_b = -1$ become converged. On the contrary, for $\omega_a \omega_b = 1$ the improper integrals are always well defined with complex rotation of $r$. This is straightforward to demonstrate using the asymptotic expressions of $u^{\pm}(z)$ for $|z| \to +\infty$ (see Sect. 2.3.3). Consequently, the diagonal matrix elements involving scattering states are infinite, that is, they cannot be regularized with complex rotation of $r$. Modification of diagonal matrix elements in Eq. (3.90) is then necessary. One will now describe the three methods to handle the infinite range character of the Coulomb potential, that is, the cut method, the subtraction method, and the off-diagonal method.

The cut method is the crudest one, as it simply removes all improper integrals in Eq. (3.90), leaving only the finite integral on $[0 : R]$. This removes all singularities in the Hamiltonian kernel of Eq. (3.90), but at the price of introducing a cut-dependence on $R$ which has to be assessed.

The subtraction method is the standard method to treat integrable singularities in Fredholm kernels [47]. The main idea is to separate the kernel integral into two parts, one whose integrand is regular and the other singular but analytically integrable. For this, let us firstly write the Fredholm equations verified by the $c_{k_i}$ and $c_k$ constants of Eq. (3.89):

$$\sum_{i'} c_{k_{i'}} \text{Reg} \int_0^{+\infty} u(k_{i'}, r) \, V_{\text{Coul}}(\Delta Z, r) \, u(k_i, r) \, dr$$

$$+ \int_{L^+} c_{k'} \text{Reg} \int_0^{+\infty} u(k', r) \, V_{\text{Coul}}(\Delta Z, r) \, u(k_i, r) \, dr \, dk' = (E - e_i) c_{k_i}$$

$$\tag{3.91}$$

$$\sum_{i'} c_{k_{i'}} \text{Reg} \int_0^{+\infty} u(k_{i'}, r) \, V_{\text{Coul}}(\Delta Z, r) \, u(k, r) \, dr$$

$$+ \int_{L^+} c_{k'} \text{Reg} \int_0^{+\infty} u(k', r) \, V_{\text{Coul}}(\Delta Z, r) \, u(k, r) \, dr \, dk' = (E - e_k) c_k \, ,$$

$$\tag{3.92}$$

where $u(k_i, r)$ is a bound or resonance basis state of energy $e_i$, $u(k, r)$ is a scattering basis state of energy $e_k$, and $E$ is the energy of $u_{\text{eig}}(r)$ in Eq. (3.89) to be determined. All matrix elements of Eq. (3.91) converge so that they are directly calculated with complex scaling using Eq. (3.58). Conversely, the matrix elements of Eq. (3.92) can diverge, so that one will apply the subtraction method therein. One will consider only the integral part of Eq. (3.92), as it is the only one which leads to divergences:

$$\int_{L^+} c_{k'} \text{Reg} \int_0^{+\infty} u(k', r) \, V_{\text{Coul}}(\Delta Z, r) \, u(k, r) \, dr \, dk'$$

$$= \int_{L^+} \left[ \text{Reg} \int_0^{+\infty} \left( c_{k'} \, u_{k'}(r) \, V_{\text{Coul}}(\Delta Z, r) \, u_k(r) \right. \right.$$

$$\left. \left. - c_k \, s_{k'}(r) \left( \frac{C_{\text{Coul}} \, \Delta Z}{r} \right) s_k(r) \right) dr \right] dk'$$

$$+ c_k \int_{L^+} \text{Reg} \int_0^{+\infty} s(k', r) \, V_{\text{Coul}}(\Delta Z, r) \, s(k, r) \, dr \, dk' \, , \tag{3.93}$$

where

$$s(k, r) = \sqrt{\frac{2}{\pi}} \sin(kr) , \tag{3.94}$$

so that one has just added and subtracted the $\ell = 0$ Fourier-Bessel transform of the Coulomb potential $(C_{\text{Coul}} \Delta Z)/r$. The converging character of Eq. (3.93) will be demonstrated in Exercise XI.

---

**Exercise XI ⋆⋆**

   One will demonstrate that the subtraction method allows to integrate out the singularities occurring due to the infinite range of the Coulomb potential.

**A.** Show that the first radial integral of Eq. (3.93) can always be calculated with the complex rotation.
**B.** Calculate the second integral of Eq. (3.93) analytically.
**C.** Based on the results obtained in **A** and **B**, conclude that Eq. (3.93) can always be calculated precisely, noticing that singularities occur only in the second integral of Eq. (3.93).

---

In the off-diagonal method, diagonal infinite matrix elements are replaced by off-diagonal matrix elements situated in the vicinity of the diagonal of the matrix. This is motivated by the analytical approximation of the integral of $\ln |k - k'|$ between $k' = 0$ and $k' = k_{\max}$ acquired from the trapezoidal rule (see Exercise XII). The discretization scheme is defined in the following way. First one calculates all off-diagonal matrix elements, where complex scaling can always be applied. Diagonal matrix elements involving resonances are also straightforward to calculate with complex scaling, as divergences can only occur when integrating two scattering states of close $k$ values. Afterwards one replaces the initially diverging diagonal matrix elements, that is, those involving scattering states, by off-diagonal matrix elements lying close to the diagonal:

$$\text{Reg} \int_0^{+\infty} u(k, r) V_{\text{Coul}}(\Delta Z, r) u(k, r) dr$$

$$\rightarrow \text{Reg} \int_0^{+\infty} u(k^+, r) V_{\text{Coul}}(\Delta Z, r) u(k^-, r) dr, \tag{3.95}$$

where

$$k^\pm = k \pm \frac{\Delta k}{4\pi} , \tag{3.96}$$

as suggested by the Exercise XII. However, $\Delta k$ in (3.96) remains to be fixed.

In Exercise XII, the trapezoidal rule is used for the discretization of the $k$-contour. While the trapezoidal rule gives a theoretical ground for the off-diagonal method, it cannot be applied in practice because it would lead to poor numerical accuracy. In fact, as seen in Sect. 3.3, the Gauss-Legendre quadrature leads to the most precise results. For this, it is most precise to take $\Delta k$ equal to the Gauss-Legendre weight of the considered linear momentum $k$ of the Berggren basis contour discretized with Gauss-Legendre quadrature. Note that the off-diagonal method used with Gauss-Legendre quadrature is empirical and that there is no mathematical demonstration for this recipe.

Let us consider now the Berggren basis expansion of three proton single-particle wave functions, namely $1s_{1/2}$, $0d_{5/2}$, and $0d_{3/2}$. The parameters defining the Woods-Saxon potential $V_{WS}(r)$ and the Coulomb potential $V_{Coul}(Z, r)$ (see Eqs. (3.83) and (3.88)) and Berggren basis contour can be found in Ref. [48]. The basis-generating Hamiltonian and the Hamiltonian whose matrix is diagonalized differ only through the value of $Z$ in Eq. (3.88), in which $Z^{(b)} = 10$ and $Z^{(d)} = 8$. When wielding the cut method, the radius $R$ of Eq. (3.90) is $R = 75$ fm for $s_{1/2}$ and $d_{5/2}$ partial waves, and $R = 35$ fm for the $d_{3/2}$ partial wave. These values yield the best precision for the cut method. The Berggren basis contour (see Eq. (3.66)) consists of three segments in the complex $k$-plane, delimited by the four points $k_{min} = 0$ fm$^{-1}$, $k = 0.25-0.1i$ fm$^{-1}$ ($s_{1/2}$ and $d_{5/2}$ partial waves) or $k = 0.4-0.39i$ fm$^{-1}$ ($d_{3/2}$ partial wave), $k = 1$ fm$^{-1}$, and $k_{max} = 4$ fm$^{-1}$.

---

**Exercise  XII** ⋆⋆
     One will demonstrate that the condition (3.96) is exact if the second integral of Eq. (3.93) is discretized with the trapezoidal rule.

**A.** Calculate the integral of $\ln |k - k'|$ for $k' \in [0 : k_{max}]$ with the trapezoidal rule, where $k > 0$ is fixed. For this, one will separate the $[0 : k_{max}]$ interval into two intervals, equal to $[0 : k)$ and $(k : k_{max}]$. One will also pose $N = k/\Delta k$ and $M = (k_{max} - k)/\Delta k$ as the respective number of discretized points of the intervals $[0 : k)$ and $(k : k_{max}]$, with $\Delta k > 0$. The infinite end point contributions of the trapezoidal rule occurring at $k' = k$ in $[0 : k)$ and $(k : k_{max}]$ will not be considered.

**B.** Show that the integral of $\ln |k - k'|$ for $k' \in [0 : k_{max}]$ can be made as precise as in the regular case of the trapezoidal rule, i.e. up to an error of $\mathcal{O}(\Delta k^2)$, by adding $\Delta k \ln[\Delta k/(2\pi)]$ to the integral. Show that the integral of $\ln |k - k'|$ for $k' \in [0 : k_{max}]$ is the same as the exact integral up to $\mathcal{O}(\Delta k^2)$ if one replaces $k$ by $k - \Delta k/(4\pi)$ and $k'$ by $k' + \Delta k/(4\pi)$ if $k = k'$.

**C.** Show that the replacement of $k$ by $k - \Delta k/(4\pi)$ and $k'$ by $k' + \Delta k/(4\pi)$ if $k = k'$ can be also performed to precisely calculate the integrals of functions of the form $f(k, k') + \ln |k - k'|$, where $f(k, k')$ is regular and symmetric in $k, k'$. Conclude that this method can be applied not only to the point-particle

> Coulomb potential, proportional to $1/r$, which was considered in **A** and **B**, but also to the arbitrary Coulomb potential, which becomes proportional to $1/r$ in the asymptotic region only.

In Eq. (3.90), one uses in the subtraction and off-diagonal methods the values $R$ = 15 fm and $\theta = -3\pi/4$, $-\pi/4$, $\pi/4$ or $3\pi/4$. Improper integrals of Eq. (3.90) are indeed guaranteed to converge employing one of these angular values. The Berggren basis contour used is very similar to that of the cut method, except that $k_{\min}$ is chosen so that $F_{\ell\eta}(k_{\min}R) \sim 10^{-5}$. The latter condition arises from the very small values of proton scattering wave functions with $k \sim 0$. Indeed, proton scattering wave functions are very close to regular Coulomb wave functions $\sqrt{2/\pi}\, F_{\ell\eta}(kr)$ when $k \to 0$ (see Eqs. (2.142) and (2.143)). However, complex rotation demands the use of $H_{\ell\eta}^{\pm}(kr)$ irregular Coulomb wave functions (see Eq. (3.90)). This leads to important numerical cancellations, because $|H_{\ell\eta}^{\pm}(kr)|$ becomes very large when $k \to 0$, whereas $F_{\ell\eta}(kr) \to 0$ when $k \to 0$. Consequently, one cannot evaluate numerically Eq. (3.90) using complex scaling with $u(k, r)$ functions whose $k$ value is small. On the other hand, the Coulomb matrix element (see Eq. (3.90)) becomes negligible when $k \to 0$. Due to the extreme smallness of their amplitudes, $u(k, r)$ functions whose $k$ value is close to zero play virtually no role in the completeness of Berggren basis. Hence, proton scattering states become important only when $k_{\min}$ become sufficiently large, and the condition above has been shown to mitigate the numerical instability occurring for $k \sim 0$ while yielding precise results.

The precision of the radial wave functions issued from the Hamiltonian diagonalization is evaluated by calculating the root mean square deviation of their real and imaginary parts with respect to the exact functions, issued from the direct integration:

$$\mathrm{rms}(\Re[u]) = \sqrt{\frac{\displaystyle\sum_{i=1}^{N} (\Re[u(r_i)] - \Re[u_e(r_i)])^2}{N}} \tag{3.97}$$

$$\mathrm{rms}(\Im[u]) = \sqrt{\frac{\displaystyle\sum_{i=1}^{N} (\Im[u(r_i)] - \Im[u_e(r_i)])^2}{N}}. \tag{3.98}$$

In these expressions, $N$ is equal to 512, $r_i = i \cdot (R/N)$ for $1 \le i \le N$ is a set of uniformly distributed radii in an interval $[0 : R]$, $u(r)$ is the diagonalized wave function, and $u_e(r)$ is the exact wave function.

The root-mean-square deviations (3.97), (3.98) for proton $1s_{1/2}$, $0d_{5/2}$, and $0d_{3/2}$ wave functions are illustrated in Fig. 3.4 for all studied cases, that is, the cut method, the subtraction method and the off-diagonal method. One can see that the cut method produces very poor results, as it does not even reach the precision acquired with

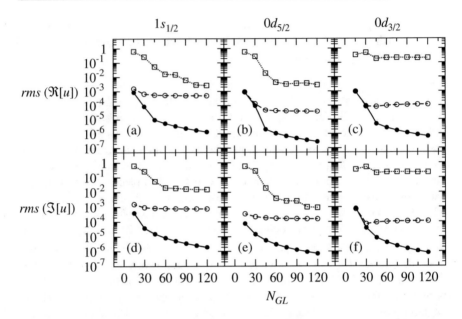

**Fig. 3.4** Root-mean-square deviation between the real rms($\Re[u]$) and imaginary rms($\Im[u]$) parts of the diagonalized wave functions $u(r)$ and the corresponding exact wave functions obtained by a direct integration. Results for proton $1s_{1/2}$, $0d_{5/2}$, and $0d_{3/2}$ wave functions are plotted. rms($\Re[u]$) and rms($\Im[u]$) are plotted as a function of the number of discretized scattering states $N_{GL}$ used in the Gauss-Legendre integration. Dotted lines with empty squares refer to the cut method. Dashed lines with empty circles show results of the subtraction method, and solid lines with filled circles correspond to the off-diagonal method (adapted from Ref. [48])

the off-diagonal method when the smallest value of the number of Gauss-Legendre scattering states $N_{GL}$ is employed. Moreover, for the $0d_{3/2}$ proton state, although a rather good description of energy and width occurs, the reproduction of the wave function is mediocre. Added to that, the choice of the cut radius could only be effected by comparison with exact results, whereas it has been checked that the two other methods are very robust when the $L^+$ contour parameters are changed.

The subtraction method, while not being completely inaccurate, saturates very quickly to a wrong value for energies, widths and wave functions when $N_{GL}$ increases.

The best reproduction of the considered proton states clearly arises with the off-diagonal method. When $N_{GL}$ augments, the exponential convergence occurs for both real and imaginary parts of the wave function. This is an intriguing observation, as the subtraction scheme is based on an exact calculation of the integral exhibiting singularities, leaving a well-defined function to be integrated numerically, whereas the off-diagonal method, which could be expected at best to be comparable to the subtraction method, surprisingly surpasses the latter by several orders of magnitude (see Fig. 3.4)).

This curious feature of the off-diagonal method can be explained by noticing that the integrand of the first integral on the right-hand side of Eq. (3.93), which is formed by radial integrals converging with the complex rotation of $r$ (see Eq. (3.90)), while everywhere finite, is not analytic at $k' \sim k$. Indeed, it is equivalent, up to an unimportant constant, to $(k - k') \ln(k - k')$, which does not even possess a finite derivative with respect to $k'$ at $k' = k$. Consequently, a Gauss-Legendre discretization of this integral will be far less precise than that of analytic functions, which can be usually well approximated by polynomials. On the contrary, the use of $u_i^+(r)$ and $u_i^-(r)$ to calculate the diagonal matrix element of $V_{Coul}(\Delta Z, r)$, effectively replaces the matrix elements involving $u(k, r)$ and $u(k', r)$ with $k' \sim k$ by an analytic function of $k$. Indeed, the moduli of matrix elements involving $u(k, r)$ and $u(k', r)$ with $k' \sim k$ are very large, because matrix elements have a singularity in $k = k'$ due to the infinite range of the Coulomb potential. Conversely, the analytic function of $k$ replacing these matrix elements has smooth variations when $k' \sim k$. Therefore, the Gauss-Legendre quadrature yields fast convergence to the numerical value of the integral of Eq. (3.95). The efficiency of the off-diagonal method in practical applications is considered in more details in Exercise XIII.

---

**Exercise XIII ★**

Based on numerical examples, one will show that the off-diagonal method is efficient for practical applications.

Recalculate previously shown results with the one-particle diagonalization code.

Change the number of protons of the Coulomb potential of the diagonalized Woods-Saxon potential by a few units. Notice that the Berggren basis expansion of radial wave functions is precise in this case.

Explain why the numerical precision of the Berggren basis expansion of radial wave functions in the presence of a Coulomb potential is independent of the charge defining the Coulomb potential.

---

## 3.7.2 Completeness Relations Involving Antibound $s_{1/2}$ States

An important question in the Gamow shell model concerns the inclusion of antibound states in the Berggren ensemble [49]. Antibound states have real and negative energy eigenvalues that are located in the second Riemann sheet of the complex energy plane (the corresponding momentum lies on the negative imaginary $k$-axis) [9, 50–52]. Contrary to bound states, the radial wave functions of antibound states increase exponentially at large distances. As often discussed in the literature, it is difficult to give a clear physical interpretation to antibound states. Strictly speaking, as the second energy sheet is unphysical and inaccessible for direct experiments, the antibound pole of the scattering matrix is not actually a state but rather a feature of the system. Antibound states with small energy greatly increase a localization of real-energy scattering states just above the threshold [53, 54].

Related to this is the influence of antibound states on the behavior of the scattering cross section at low energies. Classic examples include the low-energy $^1S_0$ neutron-neutron scattering phase shift, characterized by a large and negative scattering length [51], the scattering of slow electrons on molecules [55–57], and the *eep*-Coulomb system [58].

Coming back to nuclear structure, it was argued that as a result of the inversion of $0p_{1/2}$ and $1s_{1/2}$ shells [59], the neutron-unbound $^{10}$Li nucleus sustains a low-lying $1s_{1/2}$ antibound state very close to the one-neutron (1n) emission threshold [60]. Although many theoretical calculations predict the $0p_{1/2} - 1s_{1/2}$ shell inversion, the phenomenon still remains a matter of debate [61, 62]. Experimentally, several groups have reported evidence of the $\ell = 0$ strength at the 1n-threshold in $^{10}$Li [63–68]; however, no evidence of a weakly bound $1s_{1/2}$ state has been found. Since, experimentally, the n+$^9$Li has a large and negative scattering length, this may indicate the presence of an antibound $1s_{1/2}$ state in $^{10}$Li close to the 1n-threshold, though the presence of a low-lying $0p_{1/2}$ state cannot be ruled out [69].

The Gamow shell model calculations that explicitly consider the antibound $1s_{1/2}$ single-particle state were performed for the ground state of $^{11}$Li [70, 71]. It was argued that the presence of an antibound state was important for the formation of a neutron halo. It was also noted that the bound state wave function of $^{11}$Li could be expanded in terms of the real-energy, nonresonant $\ell=0$ continuum, that is, without explicit inclusion of the antibound state. Moreover, a destructive interference between the $1s_{1/2}$ antibound single-particle state and the associated complex-energy, nonresonant $s_{1/2}$ background was noticed.

The question of how important is the $\ell=0$ antibound state for the description of a neutron halo in $^{11}$Li, can be answered by analyzing results for several complementary Berggren basis expansions. The standard Berggren completeness relation consists of a discrete sum over bound and resonance states, and an integral over nonresonant scattering states from the contour $L_b^+$ (see Fig. 3.5 where a discrete sum runs over all bound (b) and decaying resonance (d) states lying above the complex contour $L_b^+$). The continuous part takes into account the nonresonant scattering states lying on the contour. In the particular case of $\ell=0$ neutron partial wave, there are no $s_{1/2}$ resonances due to the absence of both Coulomb and centrifugal barriers. Consequently, a real-$k$ contour would have been sufficient to describe the $s_{1/2}$ neutron channel. However, to investigate the convergence of imaginary part components of the expanded wave functions, one takes the complex-$k$ contour $L_b^+$ close to the real-$k$ axis.

One will perform now studies of the Berggren expansion in a more general case when an $\ell=0$ antibound single-particle state is included in the basis. In this case, the Berggren completeness relation takes the form [23, 70]:

$$\sum_{n\in(a,b,d)} u_n(r)u_n(r') + \int_{L_a^+} u(k,r)u(k,r')\,dk = \delta(r-r'),  \qquad (3.99)$$

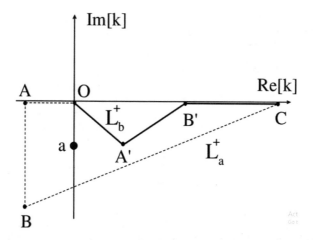

**Fig. 3.5** Contours in the complex $k$-plane used in the Berggren completeness relations for the $s_{1/2}$ partial wave. The $L_a^+$ contour (OABC) is only used in the $s_{1/2}$ channel; it allows wielding the antibound state $1s_{1/2}$ (marked as 'a'). The $L_b^+$ contour (OA'B'C) is employed for $s_{1/2}$ channel and permits expansions of bound and resonance states only (adapted from Ref. [49])

where the sum also includes antibound ($a$) states lying above the complex contour $L_a^+$ (see Fig. 3.5). It is to be noted that, as discussed in Ref. [23], the contour $L_a^+$ is obtained by deforming continuously the contour $L_b^+$ placed in the fourth quadrant of the complex $k$-plane so that it encompasses the antibound states of interest.

To test the Berggren completeness relations as applied for the antibound states, the Woods-Saxon Hamiltonian with the reduced mass of a neutron with respect to the $^9$Li core is used to generate the single-particle states entering Eq. (3.99). The depth of the Woods-Saxon potential is adjusted to yield the $1s_{1/2}$ eigenstate respectively antibound, loosely bound, and well bound. The corresponding Woods-Saxon potentials are denoted as $WS_1$, $WS_2$, and $WS_3$ in the following. The $1s_{1/2}$ eigenstate of a given Woods-Saxon potential is expanded in the basis generated by another Woods-Saxon potential ($WS^{(0)}$) of different depth:

$$u_{WS}(r) = \sum_n c_n u_n(r) + \int_{L^+} c(k) u(k, r) \, dk , \qquad (3.100)$$

where $n$ is running over the bound and antibound basis states in Eq. (3.99), and $L^+$ is $L_a^+$ or $L_b^+$, respectively. All combinations of the potentials studied are listed in Table 3.1. The contours $L_a^+$ and $L_b^+$ used in this section are defined by vertices (all in fm$^{-1}$): $[O = (0.0, 0.0); A = (-0.01, 0); B = (-0.01, -i0.02); C = (3.5, 0.0)]$ and $[O = (0.0, 0.0); A' = (0.1, -i0.01); B' = (1.5, 0.0); C = (3.5, 0.0)]$, respectively.

**Table 3.1** Set of different Woods-Saxon potentials and $1s_{1/2}$ states used in numerical tests of the Berggren completeness relations (3.99)

| Case | $WS^{(0)}$ | $V_0$ [MeV] | Contour | $1s_{1/2}$ ($WS^{(0)}$) | $\varepsilon_{1s_{1/2}}$ [MeV] |
|------|-----------|-------------|---------|-------------------------|--------------------------------|
| (i)   | $WS_1$ | 50.5 | $L_a^+$ | Antibound   | $-0.002955$ |
| (ii)  | $WS_1$ | 52.5 | $L_b^+$ | No pole     | $-0.0329$   |
| (iii) | $WS_3$ | 60.5 | $L_b^+$ | Well bound  | $-1.0372$   |

The Berggren ensemble generated by a potential $WS^{(0)}$ (second column) consists of the $0s_{1/2}$ bound single-particle state, the contour in the nonresonant continuum (third column), and—possibly—the $1s_{1/2}$ single-particle state (fourth column). "No pole" denotes a situation where the virtual $1s_{1/2}$ state is not included in the basis. In all cases, the expansion has been carried out for the loosely bound $1s_{1/2}$ single-particle state of $WS_2$ (adapted from Ref. [49])

To assess the quality of the Berggren expansion, one calculates the root mean square deviation from the exact $1s_{1/2}$ halo wave function $u_{WS_2}(r)$ of $WS_2$ obtained by a direct integration of the Schrödinger equation. The root mean square deviations (see Eqs. (3.97) and (3.98)) are calculated on the real $r$-axis in the interval from $r=0$ to $r=15$ fm. Figure 3.6 shows the root mean square deviations (3.97) and (3.98) calculated in different Breggren bases. One can clearly see that the Berggren basis containing an antibound state (case (i) in Table 3.1) is less efficient in expanding the loosely bound $1s_{1/2}$ state. The number of discretized scattering states in this case must be two to four times bigger than that in cases (ii) and (iii) in order to attain the same precision for the real part of the wave function. For the imaginary part, the difference between results of these bases is even more pronounced.

Without an antibound state in the basis, 40–50 nonresonant scattering states are enough to obtain the precision of order $10^{-6}$ for the calculated $s_{1/2}$ wave function, whereas 150 nonresonant scattering states are necessary to reach the same precision with this state included. In cases (ii) and (iii), one finds similar root mean square deviations, because the $s_{1/2}$ basis wave functions are in both cases either bound or close to the real (positive) $k$-axis. Hence their contributions add up constructively. On the contrary, in case (i), the antibound state with exponentially increasing wave function interferes destructively with nonresonant scattering states in order to produce the halo state.

Inclusion of the antibound pole in the basis enormously enhances the role of the nonresonant continuum which has to efface the asymptotics of an antibound pole in order to create a bound state having the decaying asymptotics. This behavior is opposite to what is found when including a narrow resonance state in the Berggren ensemble which always concentrates a fairly large part of the expanded wave function [20, 72].

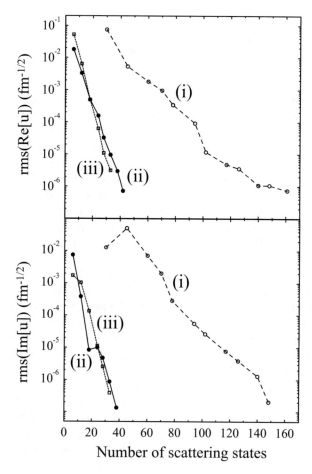

**Fig. 3.6** Real and imaginary parts of the rms deviations (3.97), (3.98) for the $1s_{1/2}$ halo wave function expanded in different Berggren bases as a function of the number of $s_{1/2}$ scattering wave functions on $L^+$. Cases (i)–(iii) of Table 3.1 are marked by dashed, solid, and dotted lines, respectively (adapted from Ref. [49])

## 3.8   Complex-Symmetric Operators and Matrices

In general, the Berggren basis contains complex-energy one-body states, so that the Hamiltonian matrix in this basis is complex symmetric. This is a fundamental difference with the hermitian Hamiltonian matrices in standard quantum mechanics, which are real symmetric and thus can be handled with standard diagonalization routines.

There are no standard libraries dedicated to the diagonalization of complex-symmetric matrices. Standard numerical routines are available for the general complex matrix, as in the LAPACK library for example, with which full numerical precision is attained, but at the price of abandoning complex symmetry. Moreover, no parallel version of the latter routine exists. Thus, it is the aim of this section to devise a diagonalization routine for complex-symmetric matrices which is both numerically precise and straightforward to parallelize.

The Hamiltonian matrix provided by a Berggren basis is a special type of complex symmetric matrix, as it is diagonally dominant for most of its rows and columns, and its imaginary parts are typically smaller than real parts. Consequently, it turns out to be possible to devise a stable method to diagonalize complex-symmetric matrices, based on a generalization of already existing methods for real symmetric matrices. Its first step consists to make the initial matrix tridiagonal, for example with orthogonal transformations using the standard Householder method [47]. The second step is to diagonalize the obtained tridiagonal matrix, by successively representing the matrix to be diagonalized by a product of an orthogonal matrix, commonly denoted as Q, times the matrix L, where only matrix elements below the diagonal are nonzero. This method is then deemed as the QL algorithm in numerical analysis [47].

Such an avenue has already been pursued in Refs. [73, 74], where the standard Householder and QL methods were extended to the complex case. However, the precision of eigenstates was assessed only for the largest eigenvalues in modulus. Moreover, as will be shown in the following, the Householder method becomes quickly imprecise when the dimension of the matrix increases. This phenomenon is due to the non-unitary character of orthogonal transformations in the complex case. Added to that, the parallelization of the Householder and QL methods were not considered. In order to clarify the situation a numerical study of several different methods to diagonalize complex-symmetric matrices will be effected in Sect. 3.8.3.

A simple and powerful method to transform a real symmetric matrix is the Lanczos method. It only demands matrix-vector multiplications and hence is widely used. However, it suffers from the well-known loss of orthogonality of Lanczos vectors when the number of Lanczos vectors increases. A full reorthogonalization of Lanczos vectors is an effective method to suppress this problem. Nevertheless, it makes the tridiagonalization process twice longer, so that it has been abandoned and replaced by the faster and stable real symmetric Householder method. However, as one will see in the following, the Lanczos method combined with full reorthogonalization extended to the complex symmetric case greatly alleviates the instabilities encountered in the complex symmetric Householder method. Moreover, parallelization of the Lanczos method is straightforward to implement using the open multiprocessing and message passing interface frameworks, contrary to that of the Householder method, where only open multiprocessing parallelization has been considered [74].

Nevertheless, the orthogonal transformation induced by the Lanczos method is not completely devoid of numerical inaccuracy, so that one cannot reach the full numerical precision of the LAPACK library routines diagonalizing the general complex matrix therein. Moreover, in the complex symmetric case, the so-called breakdown of Lanczos method can occur [73], which is the appearance of Lanczos vectors of vanishing norm. Indeed, the Berggren metric is not the hermitian metric, but the analytic continuation of the real symmetric scalar product. Consequently, the Berggren norm of a Gamow shell model vector can be in principle equal to zero. Such a vector cannot be normalized and the Lanczos method fails.

However, for this to occur, one would need matrix elements whose imaginary parts are of the same order of magnitude as their real parts, on the one hand, and rather large off-diagonal matrix elements, on the other hand. As the imaginary parts of matrix elements are always much smaller than real parts, and as diagonalized matrices are diagonally dominant, the breakdown of Lanczos method never occurs in practice. Indeed, as one will see in this section, the Lanczos method should be preferred to the Householder method when one diagonalizes Hamiltonian matrices.

### 3.8.1   Diagonalizability of Complex-Symmetric Matrices

Complex-symmetric matrices with nondegenerate eigenvalues can always be diagonalized. In Berggren basis applications, complex symmetric Hamiltonian matrices have typically different eigenvalues, because the continuum coupling lifts as a rule a possible degeneracy of the energies of basis states. Consequently, numerical diagonalization of these matrices poses no practical problem. If it occurs that two eigenvalues are equal, one can lift degeneracy by adding a small random matrix to the initial complex symmetric matrix $M$. As the latter matrix can be arbitrarily small, it cannot change the physical properties of $M$, known up to a given experimental precision.

The complex symmetric matrix with nondegenerate eigenvalues possesses very similar properties to hermitian matrices. The mathematical methods used to demonstrate this property are standard, but one has to pay attention therein that the non-hermitian norm is not positive definite. Conversely, complex-symmetric matrices with identical eigenvalues are, in general, non-diagonalizable. One will study in the following the case of $2 \times 2$ matrices to illustrate this fact.

### 3.8.2   The Two-Dimensional Complex Symmetric Matrix

In order to illustrate the properties of non-diagonalizable complex-symmetric matrices, one will consider the simplest case of $2 \times 2$ matrices. One will show that the general $2 \times 2$ complex symmetric matrix can be reduced to the study of a unique $2 \times 2$ complex symmetric matrix depending on one parameter.

The general $2 \times 2$ complex symmetric matrix

$$M = \begin{pmatrix} b & d \\ d & c \end{pmatrix}$$

is equal to

$$M = b \, I_2 + M'$$

where $I_2$ is the identity matrix, and:

$$M' = \begin{pmatrix} 0 & d \\ d & c - b \end{pmatrix}$$

Thus, the eigenvalues $\lambda$ of $M$ are equal to $b + \lambda'$, with $\lambda'$ the eigenvalues of $M'$. If $b = c$, then $M'$ is proportional to the real symmetric matrix:

$$P = \begin{pmatrix} 0 & 1 \\ 1 & 0 \end{pmatrix}$$

and is thus trivially diagonalizable.

If $b \neq c$, then it is convenient to introduce the following matrix:

$$A = \begin{pmatrix} 0 & a \\ a & 2 \end{pmatrix} .$$

Indeed

$$M = b \, I_2 + ((c - b)/2) \, A ,$$

with $a = (2d)/(c - b)$, so that $\lambda = b + ((c - b)/2) \, \lambda_A$, with $\lambda_A$ an eigenvalue of $A$. The characteristic polynomial of matrix $A$ is: $P(X) = X^2 - 2X - a^2$, whose roots are $\lambda_i = 1 - (-1)^i \sqrt{1 + a^2}, i \in 1, 2$.

The only $2 \times 2$ diagonalizable matrix with two identical eigenvalues is proportional to $I_2$. Hence, one supposes that $a \neq \pm i$. The normalized eigenvectors of $A$ thus read:

$$V_1 = \left(2(1 + a^2 + \sqrt{1 + a^2})\right)^{-1/2} \begin{pmatrix} 1 + \sqrt{1 + a^2} \\ -a \end{pmatrix}$$

and

$$V_2 = \left(2(1 + a^2 + \sqrt{1 + a^2})\right)^{-1/2} \begin{pmatrix} a \\ 1 + \sqrt{1 + a^2} \end{pmatrix} ,$$

$V_1$ and $V_2$ are always defined except when $a = \pm i$.

Matrix $A$ cannot be diagonalized only if $a = \pm i$. In this latter case, $|a| = 1$, thus of the same order of magnitude as the diagonal matrix element equal to 2 in $A$. In this case ($a = \pm i$), $A$ has only one eigenvector which up to a constant factor equals:

$$V_0 = \begin{pmatrix} \pm i \\ 1 \end{pmatrix} .$$

Its Berggren norm is equal to zero.

The study of exceptional points [25–30, 75] such as $V_0$ has practical applications in the domain of nuclear reactions. Indeed, double poles of the $S$-matrix can appear in the low-energy continuum and have an influence on phase shifts and elastic scattering cross sections [75, 76]. Moreover they are essential features of the configuration mixing in continuous phases in between the two successive branch points at the particle emission thresholds [76]. Eigenstates associated to the exceptional points can model various physical situations, such as the coalescence in Bose-Einstein condensates and the transition to quantum chaos in quantum billiards [77].

### 3.8.3 Numerical Studies of Complex-Symmetric Matrices

One will concentrate in this section on the numerical methods used to diagonalize complex-symmetric matrices. One will focus on the numerical precision of diagonalization algorithms, as well on their eventual parallelization, as diagonalization procedures can become very long even for matrices of moderate size.

A possible method to numerically diagonalize a complex symmetric matrix is to extend the Householder-QL method to the complex symmetric case, as was done in Refs. [73, 74]. The Householder method is an exact method to transform a full matrix into a tridiagonal matrix in $d - 2$ steps, with $d$ the dimension of the matrix. For this, a rotation is effected, which zeroes all the matrix elements of a column except its diagonal and the off-diagonal matrix element just below. As the matrix is symmetric, this is the same for its symmetric row. By repeating this action $d - 2$ times, one obtains a tridiagonal matrix similar to the initial one.

The use of Householder method for tridiagonalization of real symmetric matrices is preferred to the Lanczos method for two reasons. Firstly, the Householder method is very stable numerically and, secondly, the Lanczos method is more time consuming due to the reorthogonalization of Lanczos vectors. However, this is no longer the case in the complex symmetric case, where the Lanczos method becomes the method of choice for tridiagonalization.

While the Lanczos method had not been devised in the context of complex-symmetric matrices, it is straightforward to extend the hermitian Lanczos method to this case. Considering a symmetric matrix $M$, Lanczos vectors $|V_i\rangle$ for $i \in [1 : N]$ are defined by:

$$|V_2\rangle = M |V_1\rangle - \langle \widetilde{V}_1 | M | V_1 \rangle |V_1\rangle$$
$$|V_i\rangle = M |V_{i-1}\rangle - \langle \widetilde{V}_{i-1} | M | V_{i-1} \rangle |V_{i-1}\rangle - \langle \widetilde{V}_{i-1} | M | V_{i-2} \rangle |V_{i-2}\rangle, \ i \geq 3$$

$$(3.101)$$

The Lanczos procedure must be initialized with the vector $V_1$, which is called the pivot. Indeed, one can see from Eq. (3.101) that all Lanczos vectors with $i \geq 2$ can be generated iteratively from $V_1$. The pivot is typically a random vector, or an approximation of the vector one is looking for if the Lanczos method is used to find a few eigenvectors. Due to the symmetry of the matrix, the Lanczos method provides with a tridiagonal matrix, convenient to diagonalize. Theoretically, Eq. (3.101) provides with orthogonal $|V_i\rangle$ vectors with real or complex-symmetric matrices. However, numerically the orthogonality is quickly lost, so that a partial or full reorthogonalization of Lanczos vectors has to be done. For that reason, the Lanczos method has been abandoned for total diagonalization of real symmetric matrices, as other methods are more efficient therein.

The QL method is an iterative method applied to the tridiagonal matrix, obtained from either the Householder or Lanczos method. It consists of a sequence of orthogonal transformations applied to the tridiagonal matrix, which makes its off-diagonal matrix elements vanish very quickly. The whole process has a numerical cost of about $3d^3$ multiplications, so that it is tractable up to dimensions of few thousands typically. No loss of numerical precision has been noticed using the QL method with complex tridiagonal matrices. However, its convergence is much slower than in the real symmetric case when both eigenvalues and eigenvectors are computed (see Fig. 3.7). In fact, it is faster to calculate the eigenvalues of the tridiagonal complex symmetric matrix obtained using the QL method and their eigenvectors by inverse iteration if all eigenvalues differ by more than $10^{-5}$. One will denote this method by QL in the following.

It has been noticed as well that the tridiagonalization of the full complex symmetric matrix provided by the Lanczos + full reorthogonalization method,

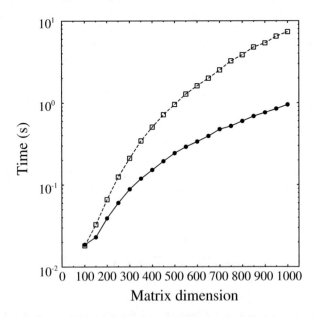

**Fig. 3.7** Time (in seconds) to diagonalize a random complex symmetric tridiagonal matrix whose elements are uniformly distributed in $[-0.5 : 0.5]$ for both real and imaginary part with the full QL method (dashed line with squares) and the QL method for eigenvalues and the shift-and-invert method for eigenvectors (solid line with filled circles)

followed by the QL method (stated as Lanczos-QL method in the following) is more precise than the complex symmetric Householder-QL method, so that the former is preferred to the latter when eigenvalues are well separated from one another. The presented scheme in the Lanczos-QL method is, however, more expensive in terms of storage, as one has to keep two matrices: the matrix to diagonalize and the matrix of eigenvectors. On the contrary, the Householder-QL method is "in place," that is, the initial matrix to diagonalize will contain its eigenvectors at the end of the calculation. This additional memory storage is however necessary in order to have a numerically precise algorithm for Lanczos-QL.

One will study the precision of eigenvalues and eigenvectors obtained with the Householder-QL, Lanczos-QL, and LAPACK routines. For this, for a given matrix $M$, one considers the maximal value of $||MX - EX||_\infty$ for all eigenvectors $X$, with $E$ the eigenvalue associated to $X$. The considered complex symmetric matrix has off-diagonal matrix elements whose real part is uniformly distributed in $[-0.01 : 0.01]$ and whose imaginary part is uniformly distributed in $[-0.001 : 0.001]$ in order to mimic the Hamiltonian interaction part in the continuum. Real numbers uniformly varying from 1 to 200 have been added to the diagonal of the considered matrix in order to simulate the one-body part of the Hamiltonian. Calculations have shown that the proposed method is very fast and precise, as a numerical precision of $10^{-10}$ or less is typically obtained with dimensions smaller than 1000, while the precision of the Householder-QL method is about $10^{-9}$ and can reach $10^{-8}$. The numerical precision of eigenvalues and eigenvectors becomes worse if their imaginary parts are of the same order of magnitude as real parts, that is, if imaginary part is uniformly distributed in $[-0.01 : 0.01]$ instead of $[-0.001 : 0.001]$. The precision of the Householder-QL method can become of the order of $10^{-6}$ for dimensions smaller than 1000, whereas the Lanczos-QL method which can be more precise by two orders of magnitude, reach the precision as large as $10^{-8}$. Maximal numerical precision, of about $10^{-12}$ is provided by the LAPACK library routine ZGEEV, whereas the overall precision of the Lanczos-QL method is close to $10^{-10}$ for considered dimensions. The precision of about $10^{-10}$ attained in Lanczos-QL method is sufficient in all practical cases, as the numerical error in calculated observables which arises from the necessary discretization of the Berggren basis contours is at best $10^{-6}$. Consequently, by using the Lanczos-QL method, the diagonalization of a dense complex symmetric Hamiltonian matrix as obtained in Gamow shell model, poses no problem, and the only numerical problem therein is the multiplication operation: Hamiltonian times vector, as in the standard shell model.

An example of the time taken for diagonalization for a few complex-symmetric matrices is shown in Fig. 3.8. The calculation time for all considered methods is comparable, as can be seen in Fig. 3.8. The simplicity of the Lanczos-QL method allows it to be easily parallelized, as the Lanczos method is the combination of matrix multiplications and classical Gram-Schmidt orthogonalizations. This is a very convenient aspect of the method, especially when one has to consider many complex-symmetric matrices of sizable dimension. On the contrary, the LAPACK library provides no parallel routine for the diagonalization of complex matrices.

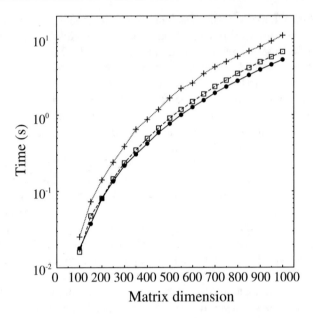

**Fig. 3.8** Time in seconds for the diagonalization of a random complex symmetric matrix $M$ with the Householder-QL method (dashed lines with squares), the Lanczos-QL method (solid lines with filled circles), and the ZGEEV routine of the LAPACK library (dotted line with pluses)

The Newton method applied to matrix diagonalization [78] can be generalized to complex-symmetric matrices, and parallelized as well as it involves only matrix multiplications. However, the better precision obtained comes at the price of slower calculations.

## Solutions to Exercises[1]

**Exercise I.** Equation (3.7) involves products of sine and cosine functions in its dominant term. It will then be rewritten as a sum of sine and cosine functions so that integrals can be conveniently evaluated:

$$I_{ab}(R_\delta)$$

$$= C_{k_a} C_{k_b} \frac{k_b \sin\left(k_a R_\delta - \eta_{k_a} \ln(2 k_a R_\delta) + \delta_{k_a}^{(\text{tot})}\right) \cos\left(k_b R_\delta - \eta_{k_b} \ln(2 k_b R_\delta) + \delta_{k_b}^{(\text{tot})}\right)}{(k_a - k_b)(k_a + k_b)}$$

$$- C_{k_a} C_{k_b} \frac{k_a \cos\left(k_a R_\delta - \eta_{k_a} \ln(2 k_a R_\delta) + \delta_{k_a}^{(\text{tot})}\right) \sin\left(k_b R_\delta - \eta_{k_b} \ln(2 k_b R_\delta) + \delta_{k_b}^{(\text{tot})}\right)}{(k_a - k_b)(k_a + k_b)}$$

---

[1] The input files, codes and code user manual associated to computer-based exercises can be found at https://github.com/GSMUTNSR.

$$+ \mathcal{O}(R_\delta^{-1})$$

$$= C_{k_a} C_{k_b} \frac{\sin\left((k_b - k_a)R_\delta - \eta_{k_b} \ln(2k_b R_\delta) + \eta_{k_a} \ln(2k_a R_\delta) + \delta_{k_b}^{(tot)} - \delta_{k_a}^{(tot)}\right)}{2(k_b - k_a)}$$

$$- C_{k_a} C_{k_b} \frac{\sin\left((k_a + k_b)R_\delta - \eta_{k_a} \ln(2k_a R_\delta) - \eta_{k_b} \ln(2k_b R_\delta) + \delta_{k_a}^{(tot)} + \delta_{k_b}^{(tot)}\right)}{2(k_a + k_b)}$$

$$+ \mathcal{O}(R_\delta^{-1}) = C_{k_a} C_{k_b} \frac{\sin\left(\Delta_k R_\delta + \beta_{ab} \, \Delta_k \, \ln(R_\delta) + f_-(k_a, k_b)\right)}{2\Delta_k}$$

$$- C_{k_a} C_{k_b} \frac{\sin\left((k_a + k_b)R_\delta - (\eta_{k_a} + \eta_{k_b}) \ln(R_\delta) + f_+(k_a, k_b)\right)}{2(k_a + k_b)} + \mathcal{O}(R_\delta^{-1})$$

$$= C_{k_a} C_{k_b} \frac{\sin\left(\Delta_k R_\delta + \beta_{ab} \, \Delta_k \, \ln(R_\delta)\right)}{2\Delta_k}$$

$$+ C_{k_a} C_{k_b} \sin\left(\Delta_k R_\delta + \beta_{ab} \, \Delta_k \, \ln(R_\delta)\right) \left(\frac{\cos\left(f_-(k_a, k_b)\right) - 1}{2\Delta_k}\right)$$

$$+ C_{k_a} C_{k_b} \cos\left(\Delta_k R_\delta + \beta_{ab} \, \Delta_k \, \ln(R_\delta)\right) \left(\frac{\sin\left(f_-(k_a, k_b)\right)}{2\Delta_k}\right)$$

$$- C_{k_a} C_{k_b} \frac{\sin\left((k_a + k_b)R_\delta - (\eta_{k_a} + \eta_{k_b}) \ln(R_\delta) + f_+(k_a, k_b)\right)}{2(k_a + k_b)} + \mathcal{O}(R_\delta^{-1}),$$

$$(3.102)$$

where $\Delta_k$, $\beta_{ab}$ and $f_\pm(k_a, k_b)$ are defined in Eqs. (3.8)–(3.10), respectively.

The interest of this calculation is that the first sine term of Eq. (3.102) resembles $\sin(\Delta_k R_\delta)/R_\delta$, whose weak limit is $\delta(\Delta_k)$ when $R_\delta \to +\infty$. On the other hand, the following terms do not exhibit any singularity in $\Delta_k = 0$, so that their rapidly oscillating character implies that they will weakly vanish when $R_\delta \to +\infty$. Hence, Eq. (3.102) allows to separate the main term responsible for the Dirac delta normalization from negligible terms when $R_\delta \to +\infty$.

**Exercise II.** $u(k, r)$ can be directly replaced by $C^+(k) \, u^+(k, r)$ in Eq. (3.33) as only bound states can give rise to residues. Moreover, the $S$-matrix appears explicitly in the obtained expression. Consequently, the demanded residue is straightforward to calculate from the residues of the $S$-matrix poles, evaluated in Exercise XIV of Chap. 2:

$$\text{Res}\left[u^+(k, r)u^+(k, r') \frac{C^+(k)}{2\pi \, C^-(k)}\right]_{k=k_n} = -\frac{1}{2\pi} \text{Res}\left[u^+(k, r)u^+(k, r') \, S(k)\right]_{k=k_n}$$

$$= -\frac{1}{2i\pi} \frac{u^+(k_n, r)u^+(k_n, r')}{\int_0^{+\infty} u^+(k_n, r)^2 \, dr}$$

$$= -\frac{1}{2i\pi} u_n(r)u_n(r'),$$

where the norm of the $u(k_n, r)$ bound state appears. The residue of Eq. (3.33) can then be concisely written in terms of normalized bound states $u_n(r)$ at $k = k_n$.

**Exercise III.** The considered integrals either weakly converge to a Dirac delta distribution or to zero when $K \to +\infty$:

$$-\frac{1}{2\pi} \int_{\mathscr{C}_K} e^{ikx} \, dk = \frac{\sin(Kx)}{\pi x} \to \delta(x) \tag{3.103}$$

$$\left| \int_0^{+\infty} f(x) \int_{\mathscr{C}_K} e^{ikx} \frac{a(k)}{k} \, dk \, dx \right|$$

$$\leq \left| \int_0^{K^{-1/2}} f(x) \int_{\mathscr{C}_K} e^{ikx} \frac{a(k)}{k} \, dk \, dx \right| + \left| \int_{K^{-1/2}}^{+\infty} f(x) \int_{\mathscr{C}_K} e^{ikx} \frac{a(k)}{k} \, dk \, dx \right|$$

$$\leq C \ln(K) \left( \pi \int_0^{K^{-1/2}} |f(x)| \, dx + \int_{K^{-1/2}}^{+\infty} |f(x)| \int_0^\pi e^{-Kx \sin(\theta)} \, d\theta \, dx \right)$$

$$\leq 2\pi \, C \, \ln(K) \, K^{-1/2} \int_0^{+\infty} |f(x)| \, dx \to 0 \,. \tag{3.104}$$

Equation (3.104) is valid for all smooth test functions $f(x)$. Therefore, it implies the following weak limit when $K \to +\infty$ and $x \geq 0$:

$$\int_{\mathscr{C}_K} e^{ikx} \frac{a(k)}{k} \, dk \to 0 \,. \tag{3.105}$$

**Exercise IV.** In this exercise, considered wave functions $u(k, r)$ can be either neutron $s$-states or Coulomb wave functions. Indeed, the presented methods are very close in both situations, so that they have been combined therein in a single exercise. Neutron $s$-states are considered in Sect. 3.2.1 and Coulomb wave functions in Sect. 3.2.2.

**A.** Here, one considers $0 < r' \leq r$. We also demand that $r, r' \leq R$ when discussing neutron $s$-states, while $r, r'$ can be chosen independently of $R$ in the case of point-particle Coulomb potential. For $K \to +\infty$, one has in this situation:

$$I(\mathscr{C}_K) = \frac{1}{2\pi} \int_{\mathscr{C}_K} \left( \frac{A^+(k) \, C_0 \, u_{\text{app}}^+(k, r) \, u_{\text{app}}^+(k, r')}{-2i \, C^-(k)} \right) \left( 1 + \mathscr{O}\left(k^{-2}\right) \right) dk$$

$$- \frac{1}{2\pi} \int_{\mathscr{C}_K} \left( \frac{B^+(k) \, C_0 \, u_{\text{app}}^-(k, r) \, u_{\text{app}}^-(k, r')}{-2i \, C^-(k)} \right) \left( 1 + \mathscr{O}\left(k^{-2}\right) \right) dk$$

$$-\frac{1}{2\pi} \int_{\mathscr{C}_K} \left( \frac{A^+(k)\, C_0\, u^+_{\mathrm{app}}(k,r)\, u^-_{\mathrm{app}}(k,r')}{-2i\, C^-(k)} \right) \left( 1 + \mathcal{O}\left(k^{-2}\right) \right) dk$$

$$+\frac{1}{2\pi} \int_{\mathscr{C}_K} \left( \frac{B^+(k)\, C_0\, u^-_{\mathrm{app}}(k,r)\, u^+_{\mathrm{app}}(k,r')}{-2i\, C^-(k)} \right) \left( 1 + \mathcal{O}\left(k^{-2}\right) \right) dk\ .$$

$$(3.106)$$

$\epsilon^\pm(k,r)$ and $\epsilon^\pm_u(k,r)$ (see Sect. 2.6.2) could both be replaced by $\mathcal{O}(k^{-2})$ because $|k| \to +\infty$ in Eq. (3.106).

**B.** Using the asymptotic expansions derived in Sect. 2.6.2, Eq. (3.106) reads:

$$I(\mathscr{C}_K) = \frac{1}{2\pi} \int_{\mathscr{C}_K} e^{ik(r+r')-i\ell_0\pi} \left( 1 + \alpha^+(r,r')\, k^{-1} + \beta^+(r,r')\, \ln(k)\, k^{-1} \right) dk$$

$$-\frac{1}{2\pi} \int_{\mathscr{C}_K} e^{ik(2R-r-r')+i\ell_0\pi} \left( \gamma^+(r,r')\, k^{-1} + \delta^+(r,r')\, \ln(k)\, k^{-1} \right) dk$$

$$-\frac{1}{2\pi} \int_{\mathscr{C}_K} e^{ik(r-r')} \left( 1 + \alpha^-(r,r')\, k^{-1} + \beta^-(r,r')\, \ln(k)\, k^{-1} \right) dk$$

$$+\frac{1}{2\pi} \int_{\mathscr{C}_K} e^{ik(2R-r+r')} \left( \gamma^-(r,r')\, k^{-1} + \delta^-(r,r')\, \ln(k)\, k^{-1} \right) dk$$

$$+\mathcal{O}\left( \frac{\ln^2(K)}{K} \right)\ .$$

$$(3.107)$$

Analyzing this result one finds the following:

- When considering neutron $s$-states, one can see that all exponential functions are bounded because $r \geq r'$ and $0 \leq r+r' \leq 2R$.
- The equivalents of values entering integrals when $|k| \to +\infty$ (see Sect. 2.6.2) allow to show that $\alpha^\pm(r,r')$, $\beta^\pm(r,r')$, $\gamma^\pm(r,r')$, and $\delta^\pm(r,r')$ are well defined and independent of $k$ (see Sect. 2.6.2). Their explicit calculation is then not necessary.
- Applying the formulas of Sect. 2.6.2 in the Coulomb wave function case, one has $\ell_0 = \ell$, $v_{c_0} = v_c$, $v_0(r) = 0$ (see Eqs. (2.2) and (2.144)), $C_0 = 1$, $A^+(k) = 1$, $B^+(k) = 0$, $C^- = -1/(2i)$ and $\gamma^\pm(r,r') = \delta^\pm(r,r') = 0$. Hence, all integrals are also well defined in the case of Coulomb wave function.

The dominant part of the first and third integrals of Eq. (3.107), that is, the term independent of $\alpha^\pm(r,r')$ and $\beta^\pm(r,r')$, becomes respectively proportional to $\delta(r+r') = 0$ and equal to $\delta(r-r')$ when $K \to +\infty$ (see Eq. (3.103)). From Eq. (3.105), one sees that all remaining terms in Eq. (3.107) weakly vanish. Therefore, Eq. (3.107) weakly converges to $\delta(r-r')$ when $K \to +\infty$.

**C.** When considering neutron $s$-states, one can check that if $0 < r' \leq R$ and $r \geq R$ then the situation is treated as in Eq. (3.107), with the exception of the following replacements: $u^+_{app}(k, r) \rightarrow H^+_{\ell,\eta}(kr)$, $u^-_{app}(k, r) \rightarrow 0$, $A^+(k) \rightarrow 1$ and $B^+(k) \rightarrow 0$, whereas all functions of $r'$ remain the same.

**D.** One still has to consider $r \geq r' \geq R$ in the case of neutron $s$-states. Note that all wave functions are provided by Eq. (2.7), so that the parameters entering the asymptotic expansions of $u(k, r)$ and $u^\pm(k, r)$ pertain to the Coulomb wave function case. The integral to evaluate then reads:

$$I(\mathscr{C}_K) = -\int_{\mathscr{C}_K} \frac{u^+(k, r)u^-(k, r')}{2\pi} \, dk - \int_{\mathscr{C}_K} \frac{C^+(k)u^+(k, r)u^+(k, r')}{2\pi \, C^-(k)} \, dk .$$

(3.108)

The first integral on the right-hand side of Eq. (3.108) can be shown to weakly converge to $\delta(r - r')$ similarly to the third integral of Eq. (3.107). To evaluate the second integral of Eq. (3.108), one will integrate separately the two contributions of $C^+(k)$ in Eq. (2.158). The integral part involving the first contribution of $C^+(k)$ on the right-hand side of Eq. (2.158) can be shown to be similar to the first integral of Eq. (3.107). Therefore, it becomes proportional to $\delta(r + r') = 0$ when $K \rightarrow +\infty$. The integral part involving the second contribution of $C^+(k)$ of the right-hand side of Eq. (2.158) has an integrand behaving as $\mathcal{O}(\exp(ik(r + r' - 2R)) \ln(k) k^{-1})$ when $K \rightarrow +\infty$. As $r + r' \geq 2R$, this integral weakly vanishes when $K \rightarrow +\infty$ (see Eq. (3.105)).

**Exercise V.** The first integral of Eq. (3.46) does not depend on $R^{(s)}$, so that the limit $R^{(s)} \rightarrow +\infty$ is trivial. One then has to show that the integral vanishes when $k_s \rightarrow 0$. Using Eqs. (2.141) and (3.20), one obtains that Dirac delta normalization is equivalent to the equality $\mathscr{N}_k^2(1 + A_k^2) = 2/\pi$. Moreover, $A_k \rightarrow 0$ in Eq. (2.141) when $k \rightarrow 0$ (see Sect. 2.6.2), so that $\mathscr{N}_k \sim \sqrt{2/\pi}$. Using Eq. (2.56), one obtains that $F_{\ell,\eta}(kr'') \rightarrow 0$ for $k \rightarrow 0$ when $r''$ is fixed. Therefore, using Eq. (2.142), $u(k, r)u(k, r') \rightarrow 0 \, \forall r, r'$ for $k \rightarrow 0$ (see Sect. 2.6.2), so that the considered integral vanishes when $k_s \rightarrow 0$.

Let us now consider the third integral of Eq. (3.46). Let us evaluate its integrand when $k \rightarrow +\infty$. As both functions $u(k, r'')$ and $u^{(s)}(k, r'')$ are normalized with a Dirac delta, their normalization constants $C_0$ and $C_0^{(s)}$ verify Eq. (3.25) when $k \rightarrow +\infty$. Note that $u(k, r'')$ and $u^{(s)}(k, r'')$ are bounded when $k \rightarrow +\infty$ (see Eqs. (2.145), (2.154), and (3.25)). Hence, the integrand of the third integral of Eq. (3.46) reads when $k \rightarrow +\infty$:

$$u(k, r)u(k, r') - u^{(s)}(k, r)u^{(s)}(k, r') = \mathcal{O}(\ln^2(k) k^{-2})$$

(3.109)

because $u(k, r) = C_0 \, f(k, r)$ and $u^{(s)}(k, r) = C_0^{(s)} \, f(k, r)$, with $f(k, r)$ a bounded function.

Consequently, one can determine the limit $R^{(s)} \to +\infty$ of the third integral of Eq. (3.46) by letting $R^{(s)} \to +\infty$ inside the integral. As $u^{(s)}(k, r'') \to u(k, r'')$ when $R^{(s)} \to +\infty$ (see Sect. 2.6.5), the limit is equal to zero.

**Exercise VI.** To show that the integral involving $u(k, r'')$ functions weakly converges to $\delta(r - r')$, one will rewrite Eq. (3.44) so that its $u(k, r'')$ and $u^{(s)}(k, r'')$ integrals are separated:

$$\sum_n u_n(r)u_n(r') + \int_0^K u(k, r)u(k, r')\, dk$$

$$= I_s(r, r', R^{(s)}) + \sum_n u_n^{(s)}(r)u_n^{(s)}(r') + \int_0^K u^{(s)}(k, r)u^{(s)}(k, r')\, dk + \epsilon(K),$$

$$(3.110)$$

where $K \to +\infty$, $\epsilon(K) \to 0$, and convergence of the integral of Eq. (3.44) (see Exercise V) has been used. Consequently, using the completeness of $u^{(s)}(k, r'')$ functions $\forall R^{(s)} > R$, one has:

$$\sum_n u_n(r)u_n(r') + \int_0^{+\infty} u(k, r)u(k, r')\, dk = I_s(r, r', R^{(s)}) + \delta(r - r'). \quad (3.111)$$

Hence, the completeness relation in the general case of Coulomb plus centrifugal potential is proved, with the limit $R^{(s)} \to +\infty$ as $I_s(r, r', R^{(s)}) \to 0$. One can also infer that $I_s(r, r', R^{(s)}) = 0$ $\forall R^{(s)} > R$ using a similar method as for $I_s^{(c)}(r, r', R^{(s)})$ in Eq. (3.48).

## Exercise VII.

A. Let us prove that the integration of $\exp(ikr)$ qualitatively corresponds to practical situations. In matrix elements, the product of operators and wave functions at large distance can be written as a linear combination of functions of the form $f(r)\exp(i S_k r)$, where $f(r)$ is a function of rational variations and $S_k$ is a linear combination of the linear momenta of involved one-body states. For a qualitative study of complex scaling, it is sufficient to take $f(r) = 1$. Thus, the integral of $\exp(ikr)$ reproduces the qualitative properties of matrix elements in the general case.

B. The equality is obtained by posing $z = R + xe^{i\theta}$, with $\theta$ fixed and integrating over $x \geq 0$.

C. The integral converges if $\exp(ikxe^{i\theta}) \to 0$ for $x \to +\infty$, which occurs if $\sin(\theta - \theta_k) > 0$. One has $\sin(\theta - \theta_k) > 0$ if $\theta \in ]\theta_k : \theta_k + \pi[$. Thus, the integral converges and is straightforward to calculate if $\theta \in ]\theta_k : \theta_k + \pi[$. Complex scaling is equivalent to the analytic continuation because it extends the domain of validity of a complex-valued function. As a consequence, no $\theta$ dependence can appear as the analytic continuation of a complex function is unique.

**Exercise VIII.**

**A.** The Gauss-Legendre quadrature can be applied only in finite intervals. Thus, it can be used if $x \in [0 : R^{-1/4}]$, but not if $x \geq 0$. In order to show that the proposed change of variable is efficient, let us consider $x \in [0 : R^{-1/4}]$. $|k|x^{-4}$ quickly increases when $x \to 0^+$, so that the integral converges rapidly even if $k$ is small. Indeed, if $\sin(\theta - \theta_k) = 0.1$, $|k| = 10^{-5}$ fm$^{-1}$, $x = 0.01$, $\exp(-|k|x^{-4}\sin(\theta - \theta_k))\, x^{-5} = 10^{10}\,\exp(-100) < 10^{-33}$. Consequently, the integrand goes quickly to zero when $x \to 0$ in all situations of practical interest. Thus, the proposed change of variable to calculate integrals initially defined on $[0 : +\infty)$ is efficient.

**B.** Even though theoretically the integral converges for $\theta \in ]\theta_k : \theta_k + \pi[$, nevertheless it converges very slowly for $\theta \sim \theta_k$ or $\theta \sim \theta_k + \pi$, so that the Gauss-Legendre integration cannot be precise therein. Conversely, $\exp(-|k|x^{-4}\sin(\theta - \theta_k))$ is smallest for $\theta$ in the middle of $]\theta_k : \theta_k + \pi[$, so that numerical integration is precise therein.

**C.** When one numerically calculates a resonance state, it is firstly necessary to fix its integration constant arbitrarily, as the norm of the resonance eigenstate is unknown. For example, one typically demands that $u(k, r) \sim r^{\ell+1}$ if $r \to 0$ (see Eq. (2.6)). Therefore, the norm of $u(k, r)$, calculated with complex scaling, will not be equal to one in general. $u(k, r)$ will be normalized to one after it is divided by the square root of the norm calculated with complex scaling.

**D.** Resonances states have a small width when their energy above the lowest decay threshold is close to zero. The particle-emission width quickly increases along with energy. As a consequence, the norm of resonance states has a very small imaginary part at small energies, which increases with increasing width of the Woods-Saxon unbound eigenstate.

**Exercise IX.**

(i) The $2p_{3/2}$ single-particle resonance ($E = 3.287$ MeV, $\Gamma = 931$ keV) in a Woods-Saxon potential of the depth $V_0 = 65$ MeV is expanded in the basis generated by the Woods-Saxon potential of the depth $V_0^{(B)} = 70$ MeV, where the $2p_{3/2}$ single-particle resonance has an energy $E = 1.905$ MeV and width $\Gamma = 61.89$ keV. After diagonalization in the discretized basis (3.87), one obtains $E = 3.287$ MeV and $\Gamma = 931$ keV for the $2p_{3/2}$ single-particle resonance, that is, the discretization error is negligible.

As both states are resonant, the squared amplitude of the $2p_{3/2}$ basis state is close to one. Nevertheless, the contribution from the nonresonant continuum is essential. It is due to the fact that the resonant state in the basis is very narrow, whereas the expanded resonant state is fairly broad. It is interesting to notice that the contribution from scattering states with energies smaller than that of the resonant state is practically negligible due to the confining effect of the Coulomb barrier.

(ii) One expands here a $2p_{3/2}$ state ($E = 1.905\,\text{MeV}$ and $\Gamma = 61.89\,\text{keV}$; $V_0 = 70\,\text{MeV}$) in a Woods-Saxon basis containing the bound $2p_{3/2}$ level ($E = -0.0923\,\text{MeV}$; $V_0^{(B)} = 75\,\text{MeV}$). As a consequence, the resonance width has to be brought by the scattering states. Nevertheless, the component of the $2p_{3/2}$ state of the basis is still close to one, whereas the continuum component plays a secondary role. Once again, one can see an effect of the Coulomb barrier: even if the expanded $2p_{3/2}$ state is unbound, its wave function is localized and has a large overlap with the bound $2p_{3/2}$ basis state. The diagonalization yields an almost exact result as well.

(iii) One now deals with the case of a $2p_{3/2}$ state that is bound in both potentials. Here $V_0 = 80\,\text{MeV}$ and $V_0^{(B)} = 75\,\text{MeV}$, and the $2p_{3/2}$ state lies at $E = -0.0923\,\text{MeV}$ and $E = -2.569\,\text{MeV}$, respectively. The scattering component in this case is almost negligible, which reflects the localized character of bound proton states. After the diagonalization, one obtains $E = -2.569\,\text{MeV}$ and a negligible value of $\Gamma$ for the $2p_{3/2}$ state, which is indeed very close to the exact result.

## Exercise X.

**A.** One notices from calculations that the bound eigenstates of Woods-Saxon potentials are well reproduced using a basis generated by a Pöschl-Teller-Ginocchio potential. Nevertheless, one cannot expand unbound states using the latter basis in practice. This arises because the Pöschl-Teller-Ginocchio potential bears neither centrifugal nor Coulomb barriers, necessary to generate resonance states of low energy. Thus, as the basis generated by a Pöschl-Teller-Ginocchio potential does not have physical asymptotic properties, it cannot provide with expansions of resonance states.

**B.** The set of bound and real-energy scattering eigenstates generated by the Pöschl-Teller-Ginocchio potential is complete in the domain of bound states. Hence, loosely bound states can also be expanded with this set of states. If one decreases the potential depth of the Woods-Saxon potential by a small amount, one can still expand the considered eigenstate of the Woods-Saxon potential with a Berggren basis generated by a Pöschl-Teller-Ginocchio potential. Therefore, the expansion of this loosely bound state with a basis of eigenstates generated by Pöschl-Teller-Ginocchio potential is still precise.

**C.** The single-particle basis generated by the Pöschl-Teller-Ginocchio contains bound states and real-energy scattering states. Consequently, the resonance states cannot be expanded therein. Moreover, even if the width of a resonance is narrow, a resonance still increases exponentially in modulus on the real axis. As the completeness relation generated by a Pöschl-Teller-Ginocchio potential can only expand bound states, it is impossible to diagonalize resonance states therein.

**Exercise XI.**

**A.** Radial integrals always converge with complex scaling when $k \neq k'$ (see Sect. 3.3). On the other hand, it is more complicated to calculate radial integrals for which $k = k'$ and $\omega_a \omega_b = -1$ with complex rotation (see Sect. 3.4). To show that these integrals converge, one considers $k' \sim k$ as

$$c_{k'} - c_k \sim \left( \frac{(k' - k)^n}{n!} \right) \frac{\partial^n c_\kappa}{\partial \kappa^n} \bigg|_{\kappa = k} ,$$

where $n \geq 1$ is the smallest integer so that the partial derivative is not equal to zero, as $c_k$ is an analytic function of $k$. The asymptote of $u_k^{\pm}(r)$ for $r \to +\infty$ is provided by Eq. (2.7) and one uses $2\pi \, C^+ C^- = 1$ (see Eq. (3.19)).

**B.** One firstly notices that when $k > k' > 0$, then:

$$\int_0^{+\infty} \frac{\sin(k'r)\sin(kr)}{r} \, dr = \int_0^{+\infty} \frac{\cos((k - k')r) - \cos((k + k')r)}{2r} \, dr$$

$$= \frac{\ln(k + k') - \ln(k - k')}{2} . \tag{3.112}$$

Equation (3.112) can be analytically continued to complex values of $k$ and $k'$ if one calculates the integral therein with complex rotation (see Sect. 3.3). The second integral on the right-hand side of Eq. (3.93) can then be evaluated analytically:

$$\int_{L^+} \text{Reg} \int_0^{+\infty} s(k', r) \, V_{\text{Coul}}(\Delta Z, r) \, s(k, r) \, dr \, dk'$$

$$= \frac{C_{\text{Coul}} \, \Delta Z}{\pi} \left[ \int_0^k [\ln(k + k') - \ln(k - k')] \, dk' \right.$$

$$\left. + \int_k^{k_{\max}} [\ln(k + k') - \ln(k' - k)] \, dk' \right]$$

$$= \frac{C_{\text{Coul}} \, \Delta Z}{\pi} \left[ (k_{\max} + k) \ln(k_{\max} + k) \right.$$

$$\left. - (k_{\max} - k) \ln(k_{\max} - k) - 2k \ln(k) \right] , \tag{3.113}$$

where one integrates along complex paths, as $k$ is complex, and where the fact that the complex linear momentum $k$ belongs to the $L^+$ contour has been taken into account.

**C.** The subtraction method thus leads to finite matrix elements in all situations, with divergences integrated analytically through the use of sine functions.

## Exercise XII.

**A.** Applying the trapezoidal rule to the considered integrals, while removing the infinite value at $k' = k$ arising from the end point contributions of the trapezoidal rule, one obtains:

$$\int_0^{k_{max}} \ln|k - k'| \, dk' = \int_0^k \ln(k - k') \, dk' + \int_k^{k_{max}} \ln(k' - k) \, dk'$$

$$\rightarrow \sum_{i=0}^{N-1} \ln(k - i\Delta k) \, \Delta k + \sum_{i=1}^{M} \ln(i\Delta k) \, \Delta k - \frac{\Delta k}{2} \ln(k) - \frac{\Delta k}{2} \ln(k_{max} - k)$$

$$= [N \ln(\Delta k) + \ln(N!) + M \ln(\Delta k) + \ln(M!)]\Delta k - \frac{\Delta k}{2} \ln(k) - \frac{\Delta k}{2} \ln(k_{max} - k)$$

$$= (k_{max} - k) \ln(k_{max} - k) - (k_{max} - k) + k \, \ln(k) - k - \Delta k \ln\left(\frac{\Delta k}{2\pi}\right)$$

$$+ \mathcal{O}(\Delta k^2), \quad \Delta k \rightarrow 0. \tag{3.114}$$

**B.** If one adds $\Delta k \ln[\Delta k/(2\pi)]$ to (3.114), then, with accuracy to error $\mathcal{O}(\Delta k^2)$, this expression becomes equal to the exact integral. This suggests to modify the range of $k, k'$ values in the trapezoidal rule so that $\ln|k - k'|$ is never infinite and generates an additional $\Delta k \ln[\Delta k/(2\pi)]$ term. One can check that the latter requirements are verified by applying the trapezoidal rule to $\ln|k - k'|$ for $k' \in [0 : k_{max}]$ using the same discretized $k$-points as in **A** and by replacing $k$ by $k - \Delta k/(4\pi)$ and $k'$ by $k' + \Delta k/(4\pi)$ if $k = k'$.

**C.** One can add a regular symmetric function $f(k, k')$ to $\ln|k - k'|$. Using the second-order Taylor expansion of $f(k, k')$, one finds $f(k - \Delta k/(4\pi), k + \Delta k/(4\pi)) = f(k, k) + \mathcal{O}(\Delta k^2)$. Consequently, the replacements $k \rightarrow k - \Delta k/(4\pi)$ and $k' \rightarrow k' + \Delta k/(4\pi)$ stated above have no sizable effect on the integral provided by the trapezoidal rule, as they only generate a $\mathcal{O}(\Delta k^2)$ term. One-body Coulomb potentials can always be written as arbitrary finite-range potential plus a point-particle Coulomb potential, proportional to $1/r$ in the asymptotic region. The finite-range part results in the appearance of a regular symmetric function $f(k, k')$ in $k$-representation, while the $1/r$ in the asymptotic region always generates a $\ln(k - k')$ function in $k$-representation. Consequently, the method hereby described covers all possible one-body Coulomb potentials bearing an asymptote proportional to $1/r$.

## Exercise XIII.
The precision of the described method is independent of $\Delta Z$, which is always a multiplicative factor.

All numerical errors are generated by the approximate treatment of the integration of $1/r$, independently of other parameters.

As $\Delta Z$ is in practice of the order of 10-100, it cannot create numerical instabilities.

# References

1. G. Birkhoff, G.C. Rota, Am. Math. Mon. **67**, 835 (1960)
2. E.C. Titchmarsh, *Eigenfunction Expansions Associated with Second-order Differential Equations* (Oxford University, Clarendon, 1958)
3. G. Fubini, Ann. Mat. Pura Appl. **20**, 217 (1913)
4. L. Lichtenstein, *Über eine Integro-Differentialgleichung und die Entwicklung willkürlicher Funktionen nach deren Eigenfunktionen* (Springer, Berlin, 1914)
5. H. Weyl, Math. Ann. **68**, 220 (1910)
6. N. Dunford, J.T. Schwartz, *Linear Operators* (Wiley Classics Library, New York, 1988)
7. T. Berggren, Nucl. Phys. A **109**, 265 (1968)
8. A. Messiah, *Quantum Mechanics*, vol. 1 and 2 (North Holland Publishing Co., Amsterdam, 1961)
9. R. Newton, *Scattering Theory of Waves and Particles*, 2nd edn. (Dover Publications, New York, 2013)
10. N. Michel, J. Math. Phys. **49**, 022109 (2008)
11. P.M. Morse, H. Feshbach, *Methods of Theoretical Physics* (Mc Graw-Hill, New York, 1953)
12. R. Newton, J. Math. Phys. **1**, 319 (1960)
13. N. Mukunda, Am. J. Phys. **49**, 910 (1978)
14. M.J. Seaton, Rep. Prog. Phys. **46**, 167 (1983)
15. Y.B. Zel'dovich, JETP (Sov. Phys.) **39**, 776 (1960)
16. N. Hokkyo, Prog. Theor. Phys. **33**, 1116 (1965)
17. W.J. Romo, Nucl. Phys. A **116**, 617 (1968)
18. B. Gyarmati, T. Vertse, Nucl. Phys. A **160**, 523 (1971)
19. G. Garcia-Calderon, R. Peierls, Nucl. Phys. A **265**, 443 (1976)
20. N. Michel, W. Nazarewicz, M. Płoszajczak, J. Okołowicz, Phys. Rev. C **67**, 054311 (2003)
21. V.I. Kukulin, V.M. Krasnopol'sky, J. Horáček, *Theory of Resonances* (Kluwer Academic Publishers, Dordrecht, 1989)
22. T. Vertse, P. Curutchet, R.J. Liotta, *Lecture Notes in Physics*, vol. 325 (Springer, Berlin, 1987), p. 179
23. P. Lind, Phys. Rev. C **47**, 1903 (1993)
24. B.F. Samsonov, J. Phys. A: Math. Gen. **38**, L571 (2005)
25. T. Kato, *Perturbation Theory for Linear Operators* (Springer, Berlin, 1995)
26. M.R. Zirnbauer, J.J.M. Verbaarschot, H.A. Weidenmüller, Nucl. Phys. A **411**, 161 (1983)
27. W.D. Heiss, W.H. Steeb, J. Math. Phys. **32**, 3003 (1991)
28. W.D. Heiss, A.L. Sannino, Phys. Rev. A **43**, 4159 (1991)
29. W. Heiss, J. Phys. A: Math. Gen. **37**, 2455 (2004)
30. J. Okołowicz, M. Płoszajczak, I. Rotter, Phys. Rep. **374**, 271 (2003)
31. M.V. Nikolaiev, V.S. Olkhovsky, Lett. Nuo. Cime. **8**, 703 (1973)
32. J. Bang, F.A. Gareev, M.H. Gizzatkulov, S.A. Gonchanov, Nucl. Phys. A **309**, 381 (1978)
33. W. Romo, Nucl. Phys. A **398**, 525 (1983)
34. P. Lind, Phys. Rev. C **47**, 1903 (1993)
35. P. Ring, P. Schuck, *The Nuclear Many-Body Problem*, 3rd ed. (Springer, Berlin, 2004)
36. D. Vautherin, M. Veneroni, Phys. Lett. B **25**, 175 (1967)
37. P. Fraser, K. Amos, S. Karataglidis, L. Canton, G. Pisent, J.P. Svenne, Eur. Phys. J. A **35**, 69 (2008)
38. E.I. Fredholm, Acta Math. **27**, 365 (1903)
39. R.J. Liotta, E. Maglione, N. Sandulescu, T. Vertse, Phys. Lett. B **367**, 1 (1996)
40. T. Vertse, R.J. Liotta, W. Nazarewicz, N. Sandulescu, A.T. Kruppa, Phys. Rev. C **57**, 3089 (1998)
41. P. Lind, R.J. Liotta, E. Maglione, T. Vertse, Z. Phys. A **347**, 231 (1994)
42. S. Saito, Suppl. Prog. Theor. Phys. **62**, 11 (1977)
43. T. Myo, A. Ohnishi, K. Katō, Prog. Theor. Phys. **99**, 801 (1998)

44. R.M. Id Betan, A.T. Kruppa, T. Vertse, Phys. Rev. C **78**, 044308 (2008)
45. N. Michel, W. Nazarewicz, M. Płoszajczak, Phys. Rev. C **82**, 044315 (2010)
46. B. Simon, Phys. Lett. A **71**, 211 (1979)
47. W.H. Press, S.A. Teukolsky, W.T. Vetterling, B.P. Flannery, *Numerical Recipes in C* (Cambridge University, Cambridge, 1988/1992)
48. N. Michel, Phys. Rev. C **83**, 034325 (2011)
49. N. Michel, W. Nazarewicz, M. Płoszajczak, J. Rotureau, Phys. Rev. C **74**, 054305 (2006)
50. H.M. Nussenzveig, *Causality and Dispersion Relations* (Academic, New York, 1972)
51. J.R. Taylor, *Scattering Theory* (Wiley, New York, 1972)
52. W. Domcke, J. Phys. B **14**, 4889 (1981)
53. A.B. Migdal, A.M. Perelomov, V.S. Popov, Yad. Fiz. **14**, 874 (1971)
54. A.B. Migdal, A.M. Perelomov, V.S. Popov, Sov. J. Nucl. Phys. **14**, 488 (1972)
55. L. Dubé, A. Herzenberg, Phys. Rev. Lett. **38**, 820 (1977)
56. M.A. Morrison, Phys. Rev. A **25**, 1445 (1982)
57. L.A. Morgan, Phys. Rev. Lett. **80**, 1873 (1998)
58. O.I. Tolstikhin, V.N. Ostrovsky, H. Nakamura, Phys. Rev. Lett. **79**, 2026 (1997)
59. B.A. Brown, in *Proceedings of the International School on Heavy Ion Physics, 4th Course: Exotic Nuclei*, ed. by R.A. Broglia, P.G. Hansen (World Scientific, Singapore, 1998), p. 1
60. I.J. Thompson, M.V. Zhukov, Phys. Rev. C **49**, 1094 (1994)
61. J. Wurzer, H.M. Hofmann, Z. Phys. A **135**, 354 (1996)
62. P. Descouvemont, Nucl. Phys. A **626**, 647 (1997)
63. S. Shimoura, T. Teranishi, Y. Ando, M. Hirai, N. Iwasa, T. Kikuchi, S. Moriya, T. Motobayashi, T. Murakami, T. Nakamura, T. Nishio, H. Sakurai, T. Uchibori, Y. Wabanabe, Y. Yanagisawa, M. Ishihara, Nucl. Phys. A **616**, 208 (1997)
64. M. Zinser, F. Humbert, T. Nilsson, W. Schwab, H. Simon, T. Aumann, M. Borge, L. Chulkov, J. Cub, T. Elze, H. Emling, H. Geissel, D. Guillemaud-Mueller, P. Hansen, R. Holzmann, H. Irnich, B. Jonson, J. Kratz, R. Kulessa, Y. Leifels, H. Lenske, A. Magel, A. Mueller, G. Münzenberg, F. Nickel, G. Nyman, A. Richter, K. Riisager, C. Scheidenberger, G. Schrieder, K. Stelzer, J. Stroth, A. Surowiec, O. Tengblad, E. Wajda, E. Zude, Nucl. Phys. A **619**, 151 (1997)
65. S. Shimoura, T. Teranishi, Y. Ando, M. Hirai, N. Iwasa, T. Kikuchi, S. Moriya, T. Motobayashi, T. Murakami, T. Nakamura, T. Nishio, H. Sakurai, T. Uchibori, Y. Wabanabe, Y. Yanagisawa, M. Ishihara, Nucl. Phys. A **630**, 387 (1998)
66. B.M. Young, W. Benenson, J.H. Kelley, N.A. Orr, R. Pfaff, B.M. Sherrill, M. Steiner, M. Thoennessen, J.S. Winfield, J.A. Winger, S.J. Yennello, A. Zeller, Phys. Rev. C **49**, 279 (1994)
67. T. Kobayashi, in *Proceedings of Radioactive Nuclear Beams III*, ed. by D.J. Morrisey (Editions Frontiers, Gif-sur-Yvette, France, 1993), p. 169
68. M. Zinser, F. Humbert, T. Nilsson, W. Schwab, T. Blaich, M.J.G. Borge, L.V. Chulkov, H. Eickhoff, T.W. Elze, H. Emling, B. Franzke, H. Freiesleben, H. Geissel, K. Grimm, D. Guillemaud-Mueller, P.G. Hansen, R. Holzmann, H. Irnich, B. Jonson, J.G. Keller, O. Klepper, H. Klingler, J.V. Kratz, R. Kulessa, D. Lambrecht, Y. Leifels, A. Magel, M. Mohar, A.C. Mueller, G. Münzenberg, F. Nickel, G. Nyman, A. Richter, K. Riisager, C. Scheidenberger, G. Schrieder, B.M. Sherrill, H. Simon, K. Stelzer, J. Stroth, O. Tengblad, W. Trautmann, E. Wajda, E. Zude, Phys. Rev. Lett. **75**, 1719 (1995)
69. H. Bohlen, W. von Oertzen, T. Stolla, R. Kalpakchieva, B. Gebauer, M. Wilpert, T. Wilpert, A. Ostrowski, S. Grimes, T. Massey, Nucl. Phys. A **616**, 254 (1997)
70. R.M. Id Betan, R.J. Liotta, N. Sandulescu, T. Vertse, Phys. Lett. B **584**, 48 (2004)
71. R.M. Id Betan, R.J. Liotta, N. Sandulescu, R. Wyss, Phys. Rev. C **72**, 054322 (2005)
72. N. Michel, W. Nazarewicz, M. Płoszajczak, Phys. Rev. C **70**, 064313 (2004)
73. I. Bar-on, V. Ryaboy, SIAM J. Sci. Comput. **18**, 1412 (1997)

74. J. Noble, M. Lubasch, J. Stevens, U. Jentschura, Comput. Phys. Commun. **221**, 304 (2017)
75. J. Okołowicz, M. Płoszajczak, Phys. Rev. C **80**, 034619 (2009)
76. J. Okołowicz, M. Płoszajczak, I. Rotter, Phys. Rep. **374**, 271 (2003)
77. R.G. Nazmitdinov, K.N. Pichugin, I. Rotter, P. Šeba, Phys. Rev. B **66**, 085322 (2002)
78. T. Ogita, K. Aishima, Jpn. J. Ind. Appl. Math. **35**, 1007 (2018)

# Two-Particle Systems in the Berggren Basis

**4**

The theoretical and numerical developments of Chap. 3 can be directly applied to the two-body problem which is solvable in terms of relative and center-of-mass coordinates and reduces to a one-body problem therein. Consequently, it is possible to study two-nucleon systems, consisting of diproton, dineutron, and deuteron, with one-body Schrödinger equations. These systems are of fundamental importance in nuclear physics as they provide most of the information gathered on the nucleon-nucleon interaction. Moreover, diproton, dineutron, and deuteron ground states are either loosely bound or unbound, so that their study with the Berggren basis is all the more justified.

Many complicated many-body systems can also be conveniently represented as two-body problems. This is particularly the case for weakly coupled subsystems, consisting of weakly bound or unbound nucleon (electron) interacting with a nuclear (molecular) core. Indeed, the action of the core on external particle can be modeled by the so-called pseudo-potentials, i.e. coupling potentials which mimic interactions in the many-body system. It will be shown in this chapter that in many cases, pseudo-potentials with a collective rotation of the core reproduce remarkably accurately the properties of many-body systems. As an example of such an approach, the particle-rotor model will be formulated and applied to weakly bound or unbound dipolar and quadrupolar anions, as well as to the one-neutron halo nucleus $^{11}$Be.

## 4.1 Exact Formulation of the Two-Particle Problem Using Relative Coordinates

As in classical mechanics, the two-body problem in quantum mechanics is solvable, as it reduces to two independent one-body problems. Let us introduce center-of-

© Springer International Publishing AG 2021
N. Michel, M. Płoszajczak, *Gamow Shell Model*, Lecture Notes in Physics 983,
https://doi.org/10.1007/978-3-030-69356-5_4

mass and relative coordinates of the two particles:

$$\mathbf{R}_{\mathrm{CM}} = \frac{m_1 \mathbf{r}_1 + m_2 \mathbf{r}_2}{m_1 + m_2} \tag{4.1}$$

$$\mathbf{P}_{\mathrm{CM}} = \mathbf{p}_1 + \mathbf{p}_2 \tag{4.2}$$

$$\mathbf{r}_{\mathrm{rel}} = \mathbf{r}_1 - \mathbf{r}_2 \tag{4.3}$$

$$\mathbf{p}_{\mathrm{rel}} = \frac{m_2 \mathbf{p}_1 - m_1 \mathbf{p}_2}{m_1 + m_2} , \tag{4.4}$$

where subscripts CM and rel stand for the center-of-mass and relative coordinates, respectively (see also Exercise I). As in classical mechanics, the center-of-mass and relative systems defined in Eqs. (4.1)–(4.4) do not correspond to physical systems but provide a convenient way of solving the two-body problem.

---

**Exercise I ⋆**
  Demonstrate that the commutators $[\mathbf{P}_{\mathrm{CM}}, \mathbf{R}_{\mathrm{CM}}]$ and $[\mathbf{p}_{\mathrm{rel}}, \mathbf{r}_{\mathrm{rel}}]$ are equal to $-i\hbar$. Compare these values to those of a one-particle system.

---

The two-particle Hamiltonian of a physical system reads:

$$H_2 = \frac{\mathbf{p}_1^2}{2m_1} + \frac{\mathbf{p}_2^2}{2m_2} + V(r_{\mathrm{rel}}) , \tag{4.5}$$

where $V(r_{\mathrm{rel}})$ is the interaction between the two particles, which depends on $\mathbf{r}_{rel}$ only as one supposes that the two-body system is not subject to an external potential. One can express Eq. (4.5) in terms of center-of-mass and relative coordinates:

$$H_2 = \frac{\mathbf{p}_{\mathrm{rel}}^2}{2m_{\mathrm{rel}}} + \frac{\mathbf{P}_{\mathrm{CM}}^2}{2M_{\mathrm{CM}}} + V(r_{\mathrm{rel}}) , \tag{4.6}$$

where one has introduced the masses of the center-of-mass and relative subsystems:

$$M_{\mathrm{CM}} = m_1 + m_2 \tag{4.7}$$

$$m_{\mathrm{rel}} = \frac{m_1 m_2}{m_1 + m_2} . \tag{4.8}$$

One can see that $H_2$ is the sum of two independent Hamiltonians, one depending on center-of-mass coordinates, denoted as $H_{\mathrm{CM}}$, and the other depending on relative coordinates, denoted as $H_{\mathrm{rel}}$ (see Exercise II):

$$H_{\mathrm{CM}} = \frac{\mathbf{P}_{\mathrm{CM}}^2}{2M_{\mathrm{CM}}} \tag{4.9}$$

$$h_{\mathrm{rel}} = \frac{\mathbf{p}_{\mathrm{rel}}^2}{2m_{\mathrm{rel}}} + V(r_{\mathrm{rel}}) . \tag{4.10}$$

> **Exercise II ⋆**
> Using Eqs. (4.1)–(4.4), show that $H_2 = H_{CM} + h_{rel}$.

Consequently, one can solve the two-body eigenproblem generated by Eq. (4.6) using the center-of-mass and relative eigenproblems with associated energies and wave functions:

$$H_2|\Psi\rangle = E|\Psi\rangle \tag{4.11}$$

$$|\Psi\rangle = |\Psi_{rel}\rangle|\Psi_{CM}\rangle \tag{4.12}$$

$$E = E_{CM} + e_{rel} \tag{4.13}$$

$$H_{CM}|\Psi_{CM}\rangle = E_{CM}|\Psi_{CM}\rangle \tag{4.14}$$

$$h_{rel}|\Psi_{rel}\rangle = e_{rel}|\Psi_{rel}\rangle . \tag{4.15}$$

Note that one can add a potential depending on center-of-mass coordinates without affecting previous results in Eq. (4.9). While this Hamiltonian is not physical, it is nevertheless very important theoretically when considered in the many-body case using a harmonic oscillator center-of-mass potential.

The fundamental importance of a harmonic oscillator center-of-mass Hamiltonian is that the decomposition on $|\Psi_{rel}\rangle$ and $|\Psi_{CM}\rangle$ in Eq. (4.12) can be imposed also in the many-body case if a basis of harmonic oscillator states is used. For this, one adds a center-of-mass Hamiltonian $H_{CM}$ to the initial Hamiltonian $H$, so that the center-of-mass part of the eigenstates of the Hamiltonian is the ground state of $H_{CM}$. Indeed, $H_{CM}$ and $H$ commute in a so-called $N\hbar\omega$ finite model space built from all Slater determinants of energy smaller than $N\hbar\omega$. Consequently, the separation of Eq. (4.12) can be also exactly obtained in $N\hbar\omega$-spaces by projecting $|\Psi_{CM}\rangle$ on the $0s$ harmonic oscillator ground state of $H_{CM}$. This is the object of the Lawson method [1] which is of fundamental importance in shell model and will be discussed in Chap. 5.

## 4.2 Study of Two-Nucleon Systems with Realistic Interactions

Physics of a two-particle system is described solely by Eq. (4.15) which describes internal structure of the system in relative variables. In the following, we will study three different two-nucleon systems: dineutron, diproton, and deuteron. These systems are of fundamental importance in nuclear physics. Indeed, they are the lightest nontrivial nucleonic systems whose binding energy depends on the nuclear interaction. Moreover, they are convenient to study experimentally via proton-proton and electron-deuteron scattering experiments. While proton-proton collisions give insight to the nuclear interaction when like particles are involved, the electron-deuteron scattering allows to investigate meson exchange process, which is the main component of the nuclear interaction. Despite complicated nature of the nuclear interaction, simple characteristics of various nucleon-nucleon systems could be experimentally discovered.

Two-nucleon systems have largest binding energy when the constituent nucleons form a symmetric radial wave function, which maximizes overlap of their wave functions and hence maximizes binding energy. When the two-nucleon wave function is antisymmmetric, the two nucleons cannot form a bound system. This fact is included in standard nucleon-nucleon interactions [2]. Indeed, due to the Pauli principle, two protons and two neutrons have to be antisymmetric in coordinate space, forming the $^1S_0$ singlet state ($S = 0$ and $T = 1$). Consequently, the nucleon-nucleon interaction in the $S = 0$ and $T = 1$ channel is constructed in such a way that diproton or dineutron are unbound. On the contrary, a proton-neutron system is not subject to this restriction, as proton and neutron are unlike particles, so that they can be in the $^3S_1$ triplet state ($S = 1$ and $T = 0$). The $T = 0$ nuclear interaction is symmetric in space, whereas the $T = 1$ nuclear interaction is antisymmetric therein. Therefore, the $T = 0$ proton-neutron interaction is stronger than that of the proton-proton or neutron-neutron interaction. Consequently, the $S = 1$, $T = 0$ channel possesses a bound state (deuteron ground state) close to the dissociation threshold, of energy equal to $-2.224$ MeV.

A fundamental nontrivial characteristic of deuteron is the relative admixture of $S$ and $D$ waves in the ground state, of respective weights close to 95% and 5% [2]. Indeed, a central and spin-orbit interaction would only generate $S$ components, so that the small but noticeable presence of $D$ waves implies that the nuclear interaction has also a tensor component.

The very different nature of bound deuteron compared to the unbound diproton and dineutron reflects one of the most important aspects of the nuclear interaction, which is the charge independence. Indeed, proton and neutron behave similarly in the presence of the nuclear interaction. In particular, the singlet state $S = 0$, $T = 1$ has almost the same binding energy in dineutron and deuteron. The charge independence of nuclear interaction is, however, not exact. Proton and neutron should have exactly the same mass in order to behave identically in the presence of nucleon-nucleon interaction. However, the masses of the up ($u$) and down ($d$) quarks are different, the down quark being about 2 MeV heavier than the up quark. Consequently, neutron of $udd$ structure is slightly heavier than proton of $uud$ structure and their masses equal 939.57 MeV/c$^2$ and 938.27 MeV/c$^2$2, respectively. While noticeable experimentally, this slight charge symmetry breaking is negligible as compared to that which is generated by the Coulomb interaction.

The approximate charge independence of the nuclear interaction has lead Heisenberg to introduce the concept of nuclear isospin [3]. Obviously, isospin symmetry is broken by the Coulomb interaction. Moreover, significant breaking of the isospin symmetry is also expected in loosely bound or resonance states, as in general the proton and neutron emission thresholds are different. All those effects can be clearly visible when considering the corresponding spectroscopic factors and radial overlap functions (see Chap. 7).

The Coulomb interaction strongly influences the nature of the singlet state of the diproton compared to that of the dineutron and of the $^1S_0$ singlet state of the deuteron. Even though both systems are unbound, the dineutron as well as the $^1S_0$ singlet state of the deuteron are antibound states of energy close to particle-emission threshold, whereas the S-matrix pole of the diproton has a low negative energy and a positive width. Thus, diproton, dineutron, and the $^1S_0$ singlet state of the deuteron are not resonances. Nevertheless, as they lie close to the real energy axis, their indirect effect on cross sections is experimentally visible. In fact, $np$ cross section at zero energy is close to 20 barns [4].

While the Lagrangian of quantum chromodynamics is well known [5–8], the nucleon-nucleon interaction cannot be directly calculated from quantum chromodynamics. This arises from the non-perturbative nature of quantum chromodynamics in the low energy region. In order to counteract this situation, Lagrangian of the quantum chromodynamics expressed in terms of quark and gluon fields is replaced by an effective interaction whose degrees of freedom are nucleons and mesons. Recent versions of this interaction bear the same chiral symmetry as the initial Lagrangian of quantum chromodynamics. Moreover, the interaction can be calculated perturbatively in the effective field theory which can be stated in one theorem [9, 10]:

*For a given set of asymptotic states, perturbation theory with the most general Lagrangian containing all terms allowed by the assumed symmetries will yield the most general S-matrix elements consistent with analyticity, perturbative unitarity, cluster decomposition and the assumed symmetries.*

Effective field theory [9–15] seems to be an attractive approach to describe physics of nucleons and nuclei at low energies, i.e., at energies $e$ below a certain energy scale $\Lambda$ imposed by the properties of a studied system and observables one wish to describe. This scale could be the nucleon mass or the pion mass in the chiral effective field theory, or the core excitation energy in halo effective field theory. In the effective field theory, relevant degrees of freedom at $e \ll \Lambda$ are explicitly taken into account whereas other degrees of freedom, related to energies $e \geq \Lambda$, are taken into account effectively in the renormalized coupling constants. If the interaction parameters converge to a finite value when $\Lambda \to +\infty$, the interaction is deemed as renormalizable. In the low-energy domain, the interactions stemming from the chiral effective theory are written as a perturbative expansion in a small parameter of the theory, such as $e/\Lambda$. In principle, effective field theory gives rise to an infinite number of terms in the $e/\Lambda$ expansion. For a fixed order in the expansion, the number of couplings is finite and one can renormalize the obtained interaction order by order.

Effective field theory can be applied to nuclei at low energy, where $e \leq 300$ MeV typically, whereas the nucleon mass $M$ is much heavier, as $M \sim 1$ GeV. Consequently, it is possible to devise quickly converging power series expansions in terms of $e/M$ ratios. The high energy degrees of freedom in this theory are indeed integrated out in order to suppress in particular the hard core nucleon-nucleon interaction which scatters nucleons away from the shell model space and is difficult to treat numerically. The low-momentum interactions obtained in this way can be conveniently applied in various basis expansion methods to solve the Schrödinger equation. As a matter of compromise, it has been advocated that the next-to-next-to-next-to-leading order ($N^3LO$) chiral interaction is sufficiently precise in practical applications. This standard approximation will be used in the following to calculate the ground states of the dineutron, diproton, and deuteron systems.

The theoretical errors made in the derivation of the nuclear forces have been thoroughly studied over last years. For example, the effects or fifth and sixth orders in chiral interactions have been considered in Refs. [16, 17], while the effect of regulators has been assessed in Ref. [18]. Statistical studies of the fit of nucleon-nucleon parameters have been done as well using Bayesian inference [19, 20] (see Sect. 5.10 for more details about the studies of the theoretical limitations of derived nucleon-nucleon forces).

### 4.2.1  Numerical Studies of Dineutron, Diproton, and Deuteron in Berggren Basis

In the following, we will calculate few observable quantities related to dineutron, diproton, and deuteron, namely the radius of deuteron and the scattering lengths extracted from nucleon-nucleon scattering cross sections (see Sect. 2.6.6). These studies are performed in Berggren basis using a variant of the $N^3LO$ chiral inter-action. Actually, the standard $N^3LO$ chiral interaction [21] cannot be diagonalized directly in the Berggren basis for two reasons.

Firstly, as it is the renormalized interaction, it depends on an energy cutoff parameter, beyond which all degrees of freedom are integrated out [22, 23]. The chiral symmetry breaking scale of the nuclear interaction, occurring for parameters $\Lambda \sim 5$ fm$^{-1}$, corresponding to energies close to 1 GeV, is often used as momentum cutoff [22–24]. In practice, a momentum cutoff of about 500 MeV is typically used in the chiral nuclear interaction expansion [24]. However, these energies are too large for Gamow shell model calculations, as the Berggren basis is suited for low energies, of the order of a few tens of MeV at most. Consequently, in the following calculations using Berggren basis, one will use $\Lambda = 1.9$ fm$^{-1}$, which is a standard value in many-body nuclear calculations, but where the cutoff dependence is still present.

Secondly, the $N^3LO$ interaction is defined in momentum space with propagators which decrease very slowly in coordinate space. Indeed, used regulators only suppress high energy components much larger than 1 GeV [24]. As a consequence, one cannot directly calculate the matrix elements of the Hamiltonian in the Berggren basis, so that it is necessary to expand the $N^3LO$ interaction in a basis of a few harmonic oscillator states, 5–10 typically per partial wave. This additional step generates basis dependence in Gamow shell model calculations of two-nucleon systems. Hence , in order to alleviate the latter problems, one has to choose multiply the $N^3LO$ interaction by a Fermi function, vanishing after 6–8 fm. Parameters of the Fermi function, its diffuseness and radius, have been fixed in order to reproduce as best as possible the basic properties of the dineutron, diproton, and deuteron, such as the binding energy, the scattering length, and the effective range (see Eqs. (2.185) and (2.186)). By taking a diffuseness of 0.65 fm, and a radius of 7.5 fm for dineutron and deuteron, and 6.85 fm for diproton, it has been possible to reproduce these observables with an error of 1–5% in comparison to $N^3LO$ calculations, where the Schrödinger equation is solved up to machine precision, in which $\Lambda = 4 - 6$ fm$^{-1}$ [24]. The Fermi radius for diproton has a different value compared to that of dineutron and deuteron due to the presence of the Coulomb interaction. Note that one takes the bare Coulomb interaction, in order to be consistent with the use of Coulomb wave functions to generate the diproton Berggren basis.

At most two channels have to be included to study eigenstates of the two-nucleon systems. These channels consist of the spin singlet ($S = 0$) and spin triplet ($S = 1$) channels. Due to the Pauli principle, $0^+$ states of two-nucleon systems have a $T = 1$ isospin and reduce to a single channel, which is the $^1S_0$ channel. Conversely, the orbital angular momenta $\ell = 0, 2$ are present in the partial decomposition of the $1^+$ state of two-nucleon systems, consisting in the $^3S_1$ and $^3D_1$ channels. The $1^+$ state has isospin $T = 0$ due to the Pauli principle.

Due to the strong binding character of the nucleon-nucleon $T = 0$ interaction, the nucleon-nucleon interaction is sufficiently strong to bind proton and neutron. The $1^+$ state in a deuteron is, in fact, the only bound state of two-nucleon systems. Conversely, the $T = 1$ interaction is weaker than the $T = 0$ interaction, so that the $0^+$ states of two-nucleon systems are always unbound. Due to the absence of Coulomb and centrifugal barrier for dineutron and deuteron in the $^1S_0$ channel, $0^+$ is an antibound state. However, even though the the $T = 1$ interaction cannot bind two nucleons, its attractive character is not negligible, as that the $0^+$ antibound state of dineutron and deuteron has a small binding energy. In the diproton case, the small Coulomb barrier implies, as stated before, that it is an $S$-matrix pole of negative energy but with a positive width.

One cannot generate experimentally an antibound state or an unbound state of negative energy and positive width, as they do not correspond to the physical states, which are either bound states, narrow resonances, or scattering states. Consequently, to infer the presence or absence of bound or unbound $S$-matrix poles close to the particle-emission threshold, one proceeds indirectly by considering the scattering length of the two-body system at zero energy (see Eqs. (2.185) and (2.186)). The measured scattering length is negative for the $0^+$ states and positive for the $1^+$ state, so that the unbound character of $0^+$ states and bound character of the $1^+$ state can be inferred from it.

One will now present results obtained with the Berggren basis expansion for the $S$-matrix poles of dineutron, diproton, and deuteron. The Berggren basis is generated by Woods-Saxon potentials with the diffusivity $d = 0.65$ fm, the radius $R_0 = 1.5$ fm, and the depth of central part $V_0 = 30$ MeV ($0^+$ states) or $V_0 = 40$ MeV ($1^+$ states). The $^1S_0$ channel for dineutron and diproton contains an antibound pole of the $S$-matrix with negative energy and positive width, while the $^3S_1$ and $^3D_1$ channels contain a loosely bound state, but no narrow resonance. The Berggren contours have to be complex to expand antibound states and $S$-matrix poles of negative energy and positive width in the $^1S_0$ channel. They consist of three segments defined by $k = 0$, $k = -0.1 - 0.1i$ fm$^{-1}$, $k = 0.2 - 0.1i$ fm$^{-1}$, and

**Table 4.1** Energies ($E$), widths ($\Gamma$), scattering lengths ($a$), and effective ranges ($r_0$) obtained with the Berggren basis diagonalization of dineutron ($nn$), diproton ($pp$), and deuteron ($pn$ singlet (s) and triplet (t) states)

| Observable | Theory | Experiment |
|---|---|---|
| $E_{nn}$ | $-0.099$ MeV | $-0.076 \pm 0.006$ MeV |
| $a_{nn}$ | $-19.180$ fm | $-18.9 \pm 0.4$ fm |
| $r_{0nn}$ | 2.859 fm | $2.75 \pm 0.11$ fm |
| $E_{pp}$ | $-0.089$ MeV | – |
| $\Gamma_{pp}$ | 814 keV | – |
| $a_{pp}$ | $-7.472$ fm | $-7.8196 \pm 0.0026$ fm |
| $r_{0pp}$ | 2.993 fm | $2.790 \pm 0.014$ fm |
| $E_{pn}(s)$ | $-0.0684$ MeV | $-0.080$ MeV |
| $a_{pn}(s)$ | $-23.421$ fm | $-23.740 \pm 0.020$ fm |
| $r_{0pn}(s)$ | 2.722 fm | $2.77 \pm 0.05$ fm |
| $E_{pn}(t)$ | $-2.136$ MeV | $-2.224$ MeV |
| $a_{pn}(t)$ | 5.579 fm | $5.419 \pm 0.007$ fm |
| $r_{0pn}(t)$ | 1.814 fm | $1.753 \pm 0.008$ fm |

Theoretical results are compared to available experimental data. Energies are given in MeV, widths in keV. Scattering lengths and effective ranges are given in fm

$k_{max} = 3$ fm$^{-1}$ for dineutron and $k = 0$, $k = 0.1 - 0.35i$ fm$^{-1}$, $k = 0.5 - 0.35i$ fm$^{-1}$, and $k_{max} = 3$fm$^{-1}$ for diproton. They have been discretized with 70–80 points to insure convergence.

For the calculation of the $1^+$ bound state of deuteron, one uses real contours up to $k_{max} = 3$ fm$^{-1}$. 10 and 9 harmonic oscillator shells have been used to expand the N$^3$LO interaction in the $S$ and $D$ channels, respectively. It has been checked that results do not significantly change when slightly changing basis parameters or increasing the number of harmonic oscillator shells. Energies, widths, scattering lengths, and effective ranges are presented in Table 4.1 for dineutron, diproton, and deuteron.

In fact, it is possible to reproduce the experimental data of dineutron, diproton, and deuteron almost exactly, but this demands to use methods proper to two-body systems. Indeed, the Hamiltonian of a two-body system is represented by a one-body integral equation in momentum space [25], so that its numerical resolution can be effected at machine precision. Moreover, no regularization at large distance is needed. On the contrary, a regularization of the nucleon-nucleon interaction is necessary in the Berggren basis calculation as Hamiltonian matrix elements must vary smoothly in the complex $k$-plane. To achieve this and show the efficiency of calculation using Berggren basis, we used a Fermi function regularization of the Hamiltonian matrix which slightly diminished the quality of reproduction of experimental data. Indeed, Berggren expansion method handles quickly decreasing nuclear interactions along the real coordinate axis very precisely, so that the discrepancy between these results and the results arising from standard approaches, with which experimental data are exactly reproduced, originates only from the use of a Fermi function regulator in the modified Hamiltonian.

The energies and widths obtained for the unbound $0^+$ state of dineutron, diproton, and deuteron are of particular interest, as they usually cannot be calculated with standard methods. One can see that one obtains these unbound states at energy close to zero, in accordance with experimental data. The real part of the energy of the diproton is nearly the same as that of the dineutron or the excited $T = 1$ deuteron state, whereas its imaginary part is close to 1 MeV in absolute value. This shows that the Coulomb interaction present in the Hamiltonian can drastically change the character of $S$-matrix pole of the two-body systems. In fact, the diproton ground state is not a resonance, but a virtual state, as it has a negative energy and a positive width.

An antibound or virtual state cannot be associated to a physical long-lived state, contrary to a resonance. Nevertheless, the presence of an antibound or virtual state at low energy has an important effect on the nucleon-nucleon cross section in the $^1S_0$ channel for very small scattering energies. Indeed, when $E \to 0$, the only important channel is the $^1S_0$ channel, as the $^3D_1$ channel wave function disappears due to its centrifugal barrier. Consequently, the knowledge of the effective range expansion is sufficient to determine the behavior of nucleon-nucleon cross section at low energy. As the scattering length is large and negative, one has to have an antibound state of energy close to zero, which has been verified numerically. The presence of an antibound state close to particle-emission threshold generates a more pronounced localization of scattering states in the nuclear zone. Consequently, the scattering cross section is much larger than in the case where no antibound state occurs.

The radial wave functions of the antibound state of dineutron and deuteron, and virtual state of diproton, are depicted in Fig. 4.1. The antibound characters of the dineutron and singlet deuteron states are immediately noticed as they increase exponentially along the real $r$-axis (see Fig. 4.1). Their small energy, however, makes the increase almost linear. The wave function of diproton virtual state has an imaginary part comparable to that of dineutron and singlet deuteron states. Indeed, the latter are purely imaginary, and all singlet systems have comparable energies. However, due to its positive width, the diproton wave function possesses a nonzero real part, of the same order of magnitude as its imaginary part already for small values of radius. This is in contrast with narrow resonance states, where real part is dominant and imaginary part is much smaller in the nuclear zone.

The $^3S_1$ and $^3D_1$ channel wave functions of the (bound) triplet state of a deuteron are shown in Fig. 4.2. As is the case of all nucleon-nucleon interactions, the $^3S_1$ channel is dominant, while the $^3D_1$ channel wave function is small but not negligible. Its content has been calculated and amounts to 4.24%, close to the exact value of 4.51% of the N$^3$LO interaction [24], which is obtained by solving the two-body relative Schrödinger equation in momentum space. It has been checked that the second maximum present in the $^3D_1$ channel wave function, and to a lesser extent in the $^3S_1$ channel wave function, is not due to numerical inaccuracy, but arises probably from the use of N$^3$LO interaction. Indeed, this minimum has been also noticed in Ref. [26]. The very smooth decrease of the bound $^3S_1$ and $^3D_1$ channel wave functions up to 25 fm indicates the completeness of Berggren basis

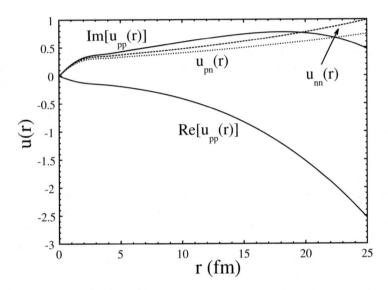

**Fig. 4.1** Wave functions of the $^1S_0$ antibound state of dineutron $u_{nn}(r)$ (dashed line), virtual state of diproton $u_{pp}(r)$ (solid lines), and antibound state of deuteron $u_{pn}(r)$ (dotted line). Dineutron and deuteron wave functions are purely imaginary as they are antibound. The real and imaginary parts of the diproton wave function are explicitly stated

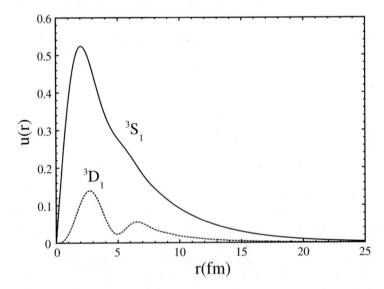

**Fig. 4.2** $^3S_1$ (solid line) and $^3D_1$ (dashed line) channel wave functions of the weakly bound $1^+$ deuteron ground state

wave functions, as the asymptote of two-body systems is built from scattering basis states, which are unbound, and one loosely bound basis state. Additional numerical illustrations of the two-body problem in the complex plane are done in Exercise III.

---

**Exercise III ★**

Using several sets of Hamiltonian parameters, we will study in a numerical example the physical observables associated to two-body systems presented in this section.

**A.** To calculate the observable quantities presented in this section, run the code dedicated to the two-body systems in intrinsic coordinates using default parameters for the three cases shown above, that is dineutron, diproton, and deuteron. Modify parameters in the Hamiltonian to check the precision of a numerical method, based on the diagonalization of two-body Hamiltonian in the Berggren basis. The set of parameters consists of the parameters of Woods-Saxon basis, the $k$ values defining Berggren basis contours, and the number of discretized scattering states. Modify each of these parameters by 10%. Notice that results are stable against these changes.

**B.** Explain why the use of a Fermi function is independent of the precision of the Berggren basis expansion.

---

## 4.3     The Particle-Rotor Model in the Berggren Basis

The main objective in physics of nonrelativistic complex systems is to solve the many-body Schrödinger equation as precisely as possible. While this is feasible when one has few nucleons or electrons, the numerical cost of calculations augments very rapidly with the number of active particles, so that approximation schemes become quickly necessary.

The simplest approach in this case is to separate particles of nucleus or molecules into inactive particles of the core and active particles of the valence part. The core has to possess the global features of the whole many-body system. It is modeled by a potential, whose parameters are typically fitted from experimental data. While in most applications the core can be spherical, there are situations where this assumption is no longer sufficient. Indeed, the multiple interactions existing at microscopic level may generate collective excitations, of vibrational and rotational nature, which implies deformation of the core. Even though this could be simply solved by the use of a nonspherical potential, a direct implementation of deformation in the core potential breaks rotational invariance of the Hamiltonian. In fact, the many-body wave function of a deformed nucleus is the linear combination of intrinsic wave functions of fixed deformation along a given axis, so that the full wave function fulfills the rotational invariance.

The restoration of rotational invariance can be effected in two ways. The first one is to directly project intrinsic wave function onto the space of rotationally invariant wave functions, where the total angular momentum $J$ is a good quantum number. While this approach is exact and treats the Pauli principle exactly, it is rather time-consuming even for well-bound systems, so that it would be very inefficient within the Berggren formalism.

The second method is to include the rotation of the core directly. Indeed, rotational invariance of the Hamiltonian is broken if one simply replaces a spherical core potential by a deformed core potential because the axes of the deformed potential are fixed. Consequently, if one allows the axes of the deformed potential to rotate freely in coordinate space, no direction of space is preferred and rotational invariance is restored. Hence, one generates in this method several rotational states for the core which are then coupled to the angular momentum of valence particles to form a many-body wave function of well-defined total angular momentum $J$.

Multipolar anions are among the most extended quantum systems. They have been measured experimentally and are being theoretically studied. In the following, one will study the physics of dipolar and quadrupolar anions using particle-rotor model formulated in the Berggren basis. The fundamental feature of anions is that they consist of a valence electron above a neutral core. Consequently, there is no long-range Coulomb potential acting on the valence electron, which can be bound to the core through the action of the dipolar and/or quadrupolar potentials, and higher multipoles of the Coulomb potential. Clearly, the valence electron can only be weakly bound or resonant, so that the Berggren basis is perfectly adapted for their study.

One of the most important characteristics of dipolar anions is their permanent dipole moment. The minimal value which allows for an electron to be bound to the molecule is called the critical dipole moment and is denoted by $\mu_c$. It has been first determined by Fermi and Teller [27] for a point-dipole ($\mu_c = 1.625\,\text{D}$) and then generalized to an extended dipole with an infinite moment of inertia [28]. However, high-resolution electron photodetachment experiments suggested a larger value for a critical moment [29–34], in accordance with nonadiabatic calculations ($\mu_c \sim 2.5\,\text{D}$) [35] which include the rotational degrees of freedom of the anion [36–44]. In this case, dipolar anions have only few bound states, and the value of $\mu_c$ depends on the moment of inertia of the neutral core. One should add that the molecular core possesses a permanent quadrupolar moment as well, whose effect has to be added to the Hamiltonian. The presence of an electron close to the polarized core also generates an additional induced dipole moment.

The particle-rotor model used along with the Berggren basis is also well adapted to the study of deformed nuclei in the vicinity of driplines. In this case, this model provides a simple theoretical formalism to describe wave functions of complex structure where both deformation and continuum degrees of freedom are coupled. The particle-rotor model is particularly useful for odd-mass nuclei, having one nucleon in a weakly bound or resonance state above a deformed well-bound core. As the nuclear interaction is short-ranged, it is sufficient to use a deformed Woods-Saxon potential to mimic the interaction between the core and the nucleon.

In the next subsection, we will provide the mathematical background of the particle-rotor model, in particular its coupled-channel equations which are generated by potentials coupling the valence particle and the core. Afterward, we will concentrate on the theory of weakly bound anions and discuss the use of pseudo-potentials, which effectively take into account the complex interactions occurring in the many-electron wave functions. Applications related to various molecular anions will be followed by the example of a one-neutron halo nucleus $^{11}$Be in which influence of continuum couplings on rotational bands will be studied.

### 4.3.1   Mathematical Formalism of the Particle-Rotor Model

The eigenfunction of the particle-rotor corresponding to the total angular momentum $J$ can be written as:

$$\Psi^J = \sum_c u_c^J(r)\Theta_{\ell_c j_c}^J . \tag{4.16}$$

In this expression, index $c$ labels the channel $(\ell, j)$, $u_c^J(r)$ is the radial wave function of the valence particle, $\Theta_{\ell_c j_c}^J$ is the channel function, and $j+\ell = J$. The eigenvalues are independent of the magnetic quantum number $M_J$ since the Hamiltonian is rotationally invariant. Hence, the magnetic quantum numbers will be omitted in the following.

The potential $V(r, \theta)$ between a valence particle and a core (see Eqs. (4.22)–(4.26)) is expanded in multipoles:

$$V(r, \theta) = \sum_\lambda V_\lambda(r) P_\lambda(\cos\theta) , \tag{4.17}$$

what allows to write the Schrödinger equation as a set of coupled-channel equations. In this expression, $V_\lambda(r)$ is the radial form factor and the angular part is given by:

$$P_\lambda(\cos\theta) = \left(\frac{4\pi}{2\lambda + 1}\right) Y_\lambda^{(mol)}(\hat{\mathbf{s}}) \cdot Y_\lambda^{(e)}(\hat{\mathbf{r}}) . \tag{4.18}$$

The matrix elements $\langle \Theta_{\ell_{c'} j_{c'}}^J | P_\lambda(\cos\theta) | \Theta_{\ell_c j_c}^J \rangle$ are obtained using the standard angular momentum algebra [44]. The radial wave functions $u_c^J(r)$ are obtained by solving the coupled-channel equations:

$$\left[\frac{d^2}{dr^2} - \frac{\ell_c(\ell_c + 1)}{r^2} - \frac{j_c(j_c + 1)}{I} + E_J\right] u_c^J(r) = \sum_{c'} V_{cc'}^J(r) u_{c'}^J(r) , \tag{4.19}$$

where $E_J$ is the energy of the system and

$$V_{cc'}^J(r) = \sum_{\lambda} \langle \Theta_{\ell' j_{c'}}^J | P_\lambda(\cos\theta) | \Theta_{\ell_c j_c}^J \rangle V_\lambda(r) \tag{4.20}$$

is the channel-channel coupling potential.

The standard method to solve Eq. (4.19) is to use the direct integration of coupled-channel equations. While it is apparently the most precise method in this case, it becomes unstable when the number of channels is large. Moreover, the presence of coupling potentials in the dipolar case, whose radial parts are proportional to $1/r^2$ for large $r$, prevents from having analytical asymptotes of channel wave functions. This is a problem for the direct integration as in this method one needs to impose asymptotic boundary conditions from an exact outgoing wave function which are unavailable hereby. However, direct integration can be used conveniently if the moment of inertia is infinite. In this case, as will be shown in Exercises IV and V, Eq. (4.19) possesses analytical solutions for large $r$, so that coupled-channel wave functions can be exactly matched to their asymptotic form and hence, the Berggren expansion method can be numerically tested against the direct integration method.

The Berggren expansion method for solving the particle-rotor model has been introduced in Ref. [44]. In this method, the Hamiltonian is diagonalized in a complete Berggren ensemble of single-particle states [45–47] which is generated by a finite-depth spherical one-body potential. While the finite-depth potential generating the Berggren ensemble can be chosen arbitrarily, to improve the convergence, one takes the diagonal part of the channel coupling potential $V_{cc'}(r)$. The basis states $\Phi_{k,c}(r)$ are eigenstates of the spherical potential $V_{cc}(r)$. These states are regular at the origin and meet outgoing boundary conditions for bound states ($b$), decaying states ($d$), or scattering states ($s$). Note that the wave number $k$ of eigenstates $\Phi_{k,c}(r)$ is in general complex. The normalization of the bound states is standard, while that for the decaying states involves the exterior complex scaling [44, 48, 49]. The scattering states are normalized to the Dirac delta function.

To determine Berggren ensemble, one calculates for all chosen partial waves first the single-particle bound and resonance states of the basis-generating potential. Then, for each channel $(\ell, j)$, one selects the contour $L_{\ell,j}^+$ in a fourth quadrant of the complex $k$-plane. All $(\ell, j)$-scattering states in this ensemble belong to the preselected contour $L_{\ell,j}^+$. The set of all resonant states and all scattering states on the contour $L_{\ell_c,j_c}^+$ form a complete single-particle basis for the channel $(\ell, j)$. Each contour $L_{\ell,j}^+$ is composed of three segments: the first one from the origin to $k_{\text{peak}}$ in the fourth quadrant of the complex $k$-plane, the second one from $k_{\text{peak}}$ to $k_{\text{middle}}$ on the real $k$-axis ($\mathcal{R}(k) > 0$), and the third one from $k_{\text{middle}}$ to $k_{\text{max}}$ also on the real $k$-axis. The segments are discretized with the Gauss-Legendre quadrature, which has been seen to be most precise numerically (see Sect. 3.7.1). The Berggren completeness relation in a coupled-channel formulation reads:

$$\sum_{i=1}^{N} |\Phi_{i,c}\rangle\langle\Phi_{i,c}| \simeq 1 \,, \tag{4.21}$$

where the $N$ basis states include bound, resonance, and discretized scattering states for each considered channel $c$.

### 4.3.1.1  Use of Pseudo-Potentials in Weakly Bound Anions

The main drawback of the particle-rotor model is that the Pauli principle is not exactly satisfied, as angular momentum coupling generates a partial occupation of one-body states in the core. To restore partially the Pauli principle, a short-range potential can be added to the Hamiltonian.

The interaction between the valence electron and the molecular core is modeled by a pseudo-potential, which is suppose to reproduce the main effects of the interactions between the valence electron and the electrons of the molecular core. It reads:

$$V(r,\theta) = V_{\text{dip}}(r,\theta) + V_\alpha(r,\theta) + V_{Q_{zz}}(r,\theta) + V_{\text{SR}}(r) \,, \tag{4.22}$$

where $\theta$ is the angle between the dipolar charge separation **s** and the electron coordinate. In this expression:

$$V_{\text{dip}}(r,\theta) = -\mu e \sum_{\lambda=1,3,\cdots} \left(\frac{r_<}{r_>}\right)^\lambda \frac{1}{s r_>} P_\lambda(\cos\theta) \tag{4.23}$$

is the electric dipole potential of the molecule, where $\mu$ is the dipole moment of the molecule;

$$V_\alpha(r,\theta) = -\frac{e^2}{2r^4}[\alpha_0 + \alpha_2 P_2(\cos\theta)] f(r) \tag{4.24}$$

is the induced dipole potential, where $\alpha_0$ and $\alpha_2$ are the spherical and quadrupole polarizabilities of the linear molecule;

$$V_{Q_{zz}}(r,\theta) = -\frac{e}{r^3} Q_{zz} P_2(\cos\theta) f(r) \tag{4.25}$$

is the potential due to the permanent quadrupole moment of the molecule; and

$$V_{\text{SR}}(r) = V_0 \exp\left[-(r/r_c)^6\right] \tag{4.26}$$

is the short-range potential, where $r_c$ is a radius defining the range of the potential. The short-range potential accounts for the exchange effects and compensates for spurious effects induced by the regulator:

$$f(r) = 1 - \exp\left[-(r/r_0)^6\right],\tag{4.27}$$

which was introduced in Eqs. (4.24) and (4.25) to avoid a singularity at $r \to 0$. The parameter $r_0$ in (4.27) defines an effective short-range scale for the regularization.

Quadrupolar anions are fascinating objects as well. Even less bound than dipolar anions, quadrupolar anions have been obtained experimentally only recently [50]. It is difficult to know whether an electron is bound through the quadrupole field or not [51–53]. Consequently, the accumulated knowledge about quadrupolar anions is mainly of theoretical origin. For example, ab initio calculations have indeed confirmed that quadrupole binding is much weaker than dipole binding [54, 55]. Moreover, due to the very small binding energies of quadrupolar anions, it is difficult to interpret experimental data, as in the case of the $CS_2^-$ molecule [56], where ab initio calculations have shown that it can exist in a weakly bound linear configuration.

Quadrupolar anions are also described in terms of pseudo-potentials. However, as its dipolar moment is hereby equal to zero, the pseudo-potentials will only have one term, generated by a linear charge distribution $(\pm q, \mp 2q, \pm q)$:

$$V_\lambda(r) = \left(\frac{e}{4\pi\varepsilon_0}\right)\left(\frac{Q^\pm}{s^2}\right)\begin{cases}\frac{1}{r_>} - \frac{1}{r} & \text{for } \lambda = 0 \\ \left(\frac{r_<}{r_>}\right)^\lambda \frac{1}{r_>} & \text{for } \lambda = 2, 4, 6\dots\end{cases}\tag{4.28}$$

with $r_> = \max(r, s)$ and $r_< = \min(r, s)$. For this simple geometry of the system and in the adiabatic limit, i.e., for an infinite moment of inertia of the neutral molecule, it is possible to calculate very precisely the positive and negative critical electric quadrupole moments of the core required to attach an excess electron in a $J^\pi = 0^+$ state [57]. In Ref. [57], using the finite-scaling method [58] recently introduced in atomic physics [59–61], the critical quadrupole moments have been expressed as a function of the scaled parameter $q_s = qs$, so that $Q_{zz}^\pm = \pm 2q_s s$. The critical values for the scaled parameter have been found analytically to be: $q_{s,c}^+ = 3.98251 \, (ea_0)$ and $q_{s,c}^- = 1.46970 \, (ea_0)$, for prolate and oblate critical quadrupole moments, respectively. These values are consistent with numerical calculations [62].

## 4.3.2   Dipolar Anions

Wave functions of an electron coupled to a neutral dipolar molecule [27, 28] are extreme examples of the giant quantum halos [63–68]. Resonance energies of dipolar anions, including those associated with rotational threshold states, can been determined in high-resolution electron photodetachment experiments [29, 31–34, 69]. Theoretically, however, the literature on the unbound part of the spectrum of dipole potentials, and multipolar anions in particular, are fairly limited [43, 70–77].

In this section, we will address the nature of the unbound part of the spectrum of dipolar anions. In particular, we will be interested in elucidating the transition from the rotational motion of weakly bound subthreshold states to the rotational-like behavior exhibited by resonances. The competition between continuum effects, collective rotation, and nonadiabatic aspects of the problem makes the description of threshold states in dipolar anions both interesting and challenging.

As an example, we will now discuss calculations for a rotational spectrum of the hydrogen cyanide $HCN^-$, which has long served as a prototype of the dipole-bound anions [44,78,79] and was a subject of experimental and theoretical studies [34,80, 81]. Here, one extends these previous studies of bound states of dipolar molecules to the unbound part of the spectrum. The parameters of the pseudo-potential for the $HCN^-$ anion are taken from Ref. [82]. One will also note that, since $V_{cc'}(r)$ decreases at least as fast as $r^{-2}$, all off-diagonal matrix elements of the coupling potential can be computed by means of the complex scaling.

To achieve stability of bound-state energies, the Berggren expansion method calculations were carried out by including all partial waves with $\ell \leq \ell_{max} = 9$ and taking the optimized number of points ($N = 165$) on the complex contour with $k_{max} = 6\,a_0^{-1}$ for each $J^\pi$. Detailed discussion can be found in Ref. [83].

The diagonalization of a complex-symmetric Hamiltonian matrix of the particle-rotor model in Berggren representation yields a set of eigenenergies which are the physical states ($S$-matrix poles of the Hamiltonian) and a large number of complex-energy scattering states. The resonances are thus embedded in a discretized continuum of scattering states and their identification is not trivial [84,85]. The eigenstates associated with resonances should be stable with respect to changes of the contour [84,85]. Moreover, their dominant channel wave functions should exhaust a large fraction of the real part of the norm. The norm of an eigenstate is given by:

$$\sum_c \sum_i \langle \Phi_{k,c}|u_c\rangle^2 = \sum_c n_c = 1\,, \tag{4.29}$$

where $n_c$ is the norm of the channel wave function.

In general, the norms of individual channel wave functions for resonances are complex numbers and their real parts are not necessarily positive definite. It may happen that if a large number of weak channels $\{c_i\}$ with small negative norms $\mathscr{R}(n_{c_i}) < 0$ contribute to the resonance wave function, then the dominant channel $c$ can have a norm $\mathscr{R}(n_c) > 1$. This does not come as a surprise as the channel wave functions have no obvious probabilistic interpretation. Stability of resonances with $\ell_{max}$ is excellent for $\ell_{max} > 6$. In general, $\mathscr{I}(E)$ is significantly more sensitive than $\mathscr{R}(E)$ with respect to the addition of channels with higher $\ell$- and $j$-values.

It is often instructive to present the density of the valence electron in the body-fixed frame. The $K$-representation, associated with the intrinsic frame, is useful to visualize wave functions, group the states with different $J$-values into the rotational bands, and interpret results in terms of the Coriolis mixing [86–91]. Even in the nonadiabatic case, the density representation (4.30) can be useful to assign members of rotational bands. In the body-fixed frame, the density of the valence particle (electron) in the state $J^\pi$ is axially symmetric and can be decomposed as:

$$\rho_J(r, \theta) = \sum_{K_J} \rho_{JK_J}(r, \theta) , \qquad (4.30)$$

where $(r, \theta)$ stand for the polar coordinates of the particle in the intrinsic frame, and the $K_J$-components of the density are:

$$\rho_{JK_J}(r, \theta) = \sum_{\ell,\ell'} \sum_j \left( \frac{2j+1}{2J+1} \right) \langle \ell K_J j0|J K_J \rangle \langle \ell' K_J j0|J K_J \rangle$$

$$\times \left( \frac{u^J_{\ell j}(r)^*}{r} \right) \left( \frac{u^J_{\ell' j}(r)}{r} \right) Y^{K_J*}_\ell(\theta, 0) Y^{K_J}_{\ell'}(\theta, 0) . \qquad (4.31)$$

If all $K_J$-components except one vanish in Eq. (4.30), the adiabatic strong-coupling limit is reached and $K_J$ becomes a good quantum number. In this particular case, $\rho_{JK_J}$ can be identified as the intrinsic density of a valence particle (electron) density in the body-fixed (dipole-fixed) reference frame. To quantify the degree of $K_J$-mixing, it is convenient to introduce the normalization amplitudes:

$$n_{JK_J} = \sum_{\ell,j} \left( \frac{2j+1}{2J+1} \right) \langle \ell K_J j0|J K_J \rangle^2 \int |u^J_{\ell j}(r)|^2 \, dr . \qquad (4.32)$$

Due to (4.29), the normalization amplitudes $n_{JK_J}$ fulfill the condition:

$$\sum_{K_J} n_{JK_J} = 1 . \qquad (4.33)$$

The excitation energies of the lowest-energy bound and resonance states are plotted in Fig. 4.3 as a function of $J(J + 1)$. The $J^\pi = 0^+, 1^-, 2^+$ bound states form a rotational band $K_J = 0$ build on the ground state $0^+_1$. Another $K_J = 0$ rotational band is built upon the $0^+_2$ resonance. Majority of the resonances are strongly $K_J$-mixed (see Fig. 4.3). Consequently, an identification of other rotational bands in the continuum, based on the concept of intrinsic density, is not obvious. As shown in Fig. 4.4, there appear clusters of resonances having the same total angular momentum $J$ within one group $g_i$. In each of those clusters, the dominant channel wave functions have the same orbital angular momentum of the valence electron $\ell$ but different rotational angular momenta of the molecule $j$.

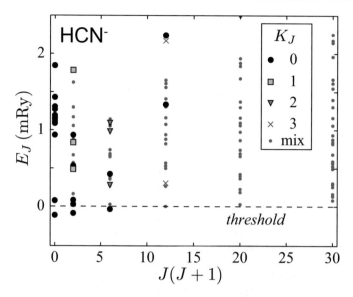

**Fig. 4.3** (Color online) Energy spectrum of the HCN$^-$ anion for $J^\pi = 0^+, 1^-, 2^+, 3^-, 4^+$, and $5^-$ is shown as a function of $J(J+1)$. The dominant $K_J$-component (4.32) is indicated. If several components are present, the state is marked as "mix" (from Ref. [83])

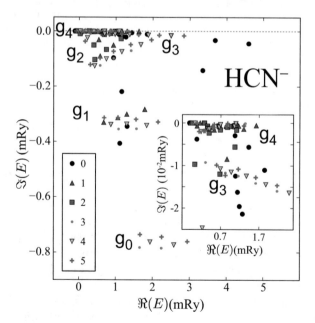

**Fig. 4.4** (Color online) Energies of the HCN$^-$ dipolar anion for $J^\pi = 0^+, 1^-, 2^+, 3^-, 4^+$, and $5^-$ states in the complex-energy plane. These states can be classified into five groups labeled $g_0$–$g_4$ based on their complex energies. Bound states and near-threshold resonances belonging to a $g_4$ group and narrow resonances in a group $g_3$ are shown in the insert (from Ref. [83])

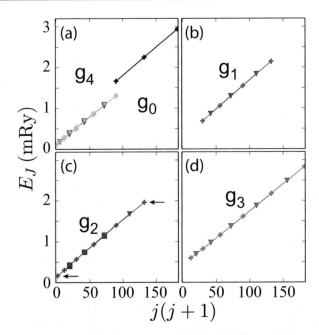

**Fig. 4.5** (Color online) Excitation energies of resonances in HCN⁻ dipolar anion are shown for various groups of states with the total angular momentum and parity $J^\pi$. Excitation energies are plotted as a function of $j(j+1)$, where $j$ is the rotational angular momentum of the molecule in the dominant channel wave function. The symbols ■, •, ▼, and + denote states with $J^\pi = 2^+$, $3^-$, $4^+$ and $5^-$, respectively (from Ref. [83])

**Exercise IV ★**

   One will compare the precision of the direct integration and Berggren expansion methods in numerical applications for the dipolar anions.

**A.** Run the particle-rotor code for the dipolar case with both direct integration and Berggren expansion methods.

   Compare energies and wave functions in these two methods and show that they provide with close, but not exactly the same values.

   Explain this difference in terms of the strengths of the $1/r^2$ coupling connecting different channels which is written on the output.

**B.** Run the particle-rotor code with the same input parameters as in **A** but with an infinite moment of inertia.

   Compare energies and wave functions obtained in direct integration and Berggren expansion methods and show that they provide with exact values this time.

   Explain this phenomenon by noticing that Eq. (4.19) is analytically solvable for $I = +\infty$ and $V_{cc'}^J(r) = a_{cc'}/r^2$, with $a_{cc'}$ a constant. For this, one will devise solutions of Eq. (4.19) in the asymptotic region of the form $u_c^J(r) = g_c\, H^+_{\ell^{(eff)},0}(kr)$, where $k$, $\ell^{(eff)}$, and $g_c$ are constants to determine.

**C.** Explain why the direct integration method is relatively precise here, and why the Berggren expansion method has to be preferred nevertheless for the study described in this section.

Excitation energies of resonances are plotted in Fig. 4.5 as a function of the molecular angular momentum $j$ for different groups of resonances of Fig. 4.4. It is seen that these states form very regular rotational band sequences in $j$ rather than in $J$. Different members of such bands lie close in the complex energy plane and have similar densities $\rho_{JK_J}(r,\theta)$. The rotational resonance structures are governed by a weak $\ell$-$j$ coupling, whereby the orbital motion of a valence electron is decoupled from the rotational motion of a dipolar neutral molecule.

At the detachment threshold, there appears a transition from the strong-coupling regime, in which the attached electron follows the rotational motion of the core, to the weak-coupling regime, where the electron's rotational motion is almost decoupled from that of the rotor.

### 4.3.3 Quadrupolar Anions

In order to benchmark the Berggren expansion method, results obtained in the adiabatic limit, i.e., for an infinite moment of inertia, are compared with the analytical results [57] for the quadrupolar anions with critical electric quadrupole moment $Q^{\pm}_{zz,c} = \pm 2q^{\pm}_{s,c}s$. The distance $s$ in Berggren calculations in the adiabatic limit is fixed at $1.6\,a_0$ [92], close to the value in $CS_2^-$ ($s = 1.554\,a_0$ [93]). The corresponding critical quadrupole moments are thus $Q^-_{zz,c} = -2.35152\,ea_0^2$ and $Q^+_{zz,c} = 6.372016\,ea_0^2$. In Berggren expansion method, in addition to $\ell_{max}$, the momentum cutoff $k_{max}$ has to be defined. By taking a real contour discretized with 80 points, and $k_{max} = 12\,a_0^{-1}$, one obtains $Q^-_{zz,c} = -2.35164\,ea_0^2$. The critical quadrupole moment is well described for the oblate configuration of an attached electron that is well localized around the two positive charges at the center of the molecule. For the prolate deformation, the situation is different. Indeed, as shown in Fig. 4.6, the results of the Berggren expansion method do not approach the analytical value as closely as for the oblate configuration. For $\ell_{max} = 14$ (and $k_{max} = 12\,a_0^{-1}$), one obtains $Q^+_{zz,c} = 6.3984\,ea_0^2$ with the Berggren expansion method. While the convergence of $Q^+_{zz,c}$ with $\ell_{max}$ (and $k_{max}$) is slower than for $Q^-_{zz,c}$, the Berggren expansion method results converges for $\ell_{max} = 14$ and $k_{max} = 12$, $a_0^{-1}$. The Pauli blocking at short distances [95–97] reduces the binding in the oblate configuration. Hence, in general, it is the prolate configuration that is more likely to bind electrons. Thus, while the oblate configuration results are useful for benchmarking purpose, their physical interpretation should be dealt with caution.

The analysis of the unbound states of quadrupolar anions is conveniently performed using the Berggren expansion method. The full excitation spectrum

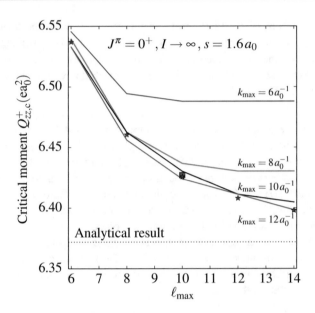

**Fig. 4.6** (Color online) Critical electric quadrupole moment for a prolate configuration as a function of the orbital angular moment cutoff in coupled-channel calculations in the adiabatic limit $(I \rightarrow \infty)$. The distance $s$ is fixed at $s = 1.6\,a_0$ and the corresponding value of $Q_{zz,c}^+ = 6.372016\,ea_0^2$ is indicated by the dotted line. The convergence of the Berggren expansion method results with respect to the momentum cutoff is shown for $\ell_{\max} = 6, 8, 10,$ and $12\,a_0^{-1}$. Calculations effected with a direct integration of Eq. (4.19) are indicated by stars. The result obtained with the direct integration method from Ref. [92] is denoted by a red square at $\ell_{\max} = 10$ (from Ref. [94])

is obtained in one diagonalization and the calculation remains tractable with the increased number of channels. As compared to the dipolar potential, the quadrupolar potential has a faster asymptotic falloff ( $\propto 1/r^3$ ) that may affect the structure of delocalized resonances. The impact on localized metastable states is less obvious.

In order to assess the importance of the fast falloff of the quadrupolar potential, the binding energy for $Q_{zz}^- = -2.42\,ea_0^2$ and $Q_{zz}^+ = +6.88\,ea_0^2$ is plotted in Figs. 4.7a,b, respectively, as a function of $J(J + 1)$. The contour $L_c^+$ is identical for all partial waves. It starts at zero and is defined by the three points: $(0.3, -10^{-5})$, $(0.6, 0)$, and $(6, 0)$ (all in $a_0^{-1}$). The three resulting segments are discretized with 30, 30, and 40 scattering states, respectively. The specific values of $Q_{zz}$ have been chosen so that the binding energy goes to zero for a total angular momentum $J \approx 2, 3$ at $\ell_{\max} = 4$.

A perfect rotational behavior is predicted for both prolate and oblate configurations, even above the detachment threshold. Moreover, Berggren expansion method and direct integration method give the same results. At the maximal orbital angular momentum cutoff $\ell_{\max}$, the states in the lowest-energy band are all dominated by the $\ell = 0$ channel at about 99.7% and 87.9%, for the oblate and prolate configuration,

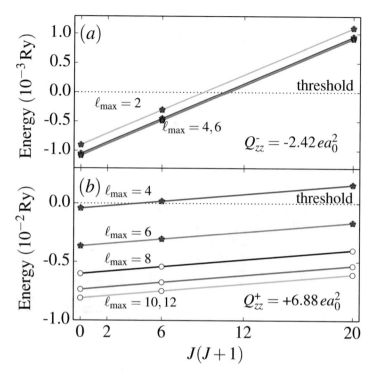

**Fig. 4.7** (Color online) $K_J = 0$ band of the lowest energy states for each given angular momentum (yrast band) in the quadrupolar anions is plotted as a function $J(J + 1)$ for different orbital angular momentum cutoff $\ell_{max}$. Quadrupolar anions are defined by a distance between charges $s = 1.6\,a_0$, a moment of inertia of $I = 10^4\,m_e a_0^2$, and quadrupole moments which equal $Q_{zz}^- = -2.42\,ea_0^2$ (panel (a)) and $Q_{zz}^+ = +6.88\,ea_0^2$ (panel (b)). Yrast band is calculated using both the Berggren expansion method (empty circles) and the direct integration method (stars). Results of these two methods are almost indistinguishable for all orbital angular momentum cutoffs (from Ref. [94])

respectively. Unlike in the dipolar case, rotational bands of quadrupolar anions persist in the continuum. The widths of unbound band members are very small ($\Gamma \sim 10^{-10}$ Ry).

Resonance spectrum calculations have been performed for the $J^\pi = 0^+$, $1^-$, and $2^+$ states in oblate and prolate configurations ($Q_{zz}^- = -2.42\,ea_0^2$, $Q_{zz}^+ = +6.88\,ea_0^2$) for $\ell_{max} = 8$, $s = 1.6\,a_0$, and $I = 10^4\,m_e a_0^2$. The contour $L_c^+$ for each partial wave starts at zero and is defined by the three points: $(0.3, -10^{-5})$, $(0.6, 0)$, and $(12, 0)$ (all in $a_0^{-1}$). The resulting segments have been discretized with 60, 40, and 100 points representing scattering states.

Calculations reveal the presence of families of narrow decaying resonances in the complex-energy plane as shown in Fig. 4.8 for oblate configurations. Remarkably similar resonance structures are obtained for prolate configurations, with widths

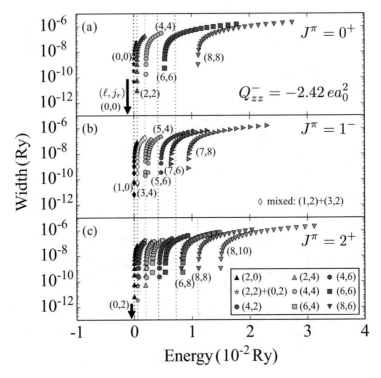

**Fig. 4.8** (Color online) The distribution of $J^\pi = 0^+$, $1^-$, and $2^+$ resonant states in the complex energy plane for the prolate configuration with $Q_{zz}^+ = +6.88\ ea_0^2$, $s = 1.6a_0$, and $I = 10^4 mea_0$ calculated with $\ell_{\max} = 8$. Bound states are marked by arrows. In most cases, families of resonances are characterized by one dominant channel. The corresponding labels $(\ell, j_r)$ are given. Mixed groups are represented by empty symbols. Dashed lines indicate rotational energies of the molecule, $E_{j_r} = \hbar^2 j_r(j_r + 1)/(2I)$ (from Ref. [94])

ranging from $10^{-10}$ Ry to $10^{-6}$ Ry corresponding to lifetimes in the range of $10^{-7} - 10^{-11}$ s. Each family of resonant states, marked by the same symbol and color in Fig. 4.8, is characterized by one dominant channel $(\ell, j_r)$ that represents about 99% of the total wave function, whereas mixed groups where two dominant channels exist are represented by empty symbols. Mixed groups in Fig. 4.8 are the (1,2) and (3,2) families for $J^\pi = 1^-$ and the (0,2) and (2,2) families for $J^\pi = 2^+$. Within each family, narrow resonances have a very diffuse dominant-channel wave function with a small number of nodes, while broader resonances tend to have wave functions peaked closer to the origin and having larger numbers of nodes. Dashed lines indicate rotational energies of the molecule, $E_{j_r} = \hbar^2 j_r(j_r + 1)/(2I)$. Overall, the resonance energies cluster close to the rotational states of the core, except for higher excitations where significant deviations can be seen. Indeed, higher-lying resonances have larger values of the orbital angular momentum $\ell$ in their dominant channel, which results in larger centrifugal barriers.

The pattern of families shown in Fig. 4.8 can be understood by considering angular momentum coupling. Since $\ell = j_r$ for $J^\pi = 0^+$ states, each family of resonances represents an electron anti-aligned with respect to the angular momentum of a rotor. The steadily increasing energy distance between groups is due to the centrifugal barrier that grows with $\ell$. The states within each family can be distinguished by the number of nodes in the radial wave function. For $J^\pi = 1^-$, the angular momentum selection rule becomes: $\ell = 1$ for $j_r = 0$ and $\ell = j_r \pm 1$ for $j_r = 2, 4, \ldots$. This yields 8 families (note that since $\ell_{\max} = 8$, there is only one channel with $j_r = 8$). The density of resonances in the complex energy plane is high because many resonances belonging to low-$\ell$ channels cluster around the rotational states of the molecule. This result in a multiple avoided crossings which imply the strong configuration mixing.

As demonstrated above for the supercritical quadrupolar molecules with $|Q_{zz}^\pm| > |Q_{zz,c}^\pm|$ (see Fig. 4.8), many resonances can exist in the vicinity of the rotor energies. Also the subcritical quadrupolar molecules with $|Q_{zz}^\pm| < |Q_{zz,c}^\pm|$ may accommodate resonances in spite of their weaker quadrupolar field [94].

The transition from a supercritical to subcritical regime in the quadrupolar anion is illustrated in Fig. 4.9 for $J^\pi = 0^+$ states. Figure 4.9a shows (real) energies of

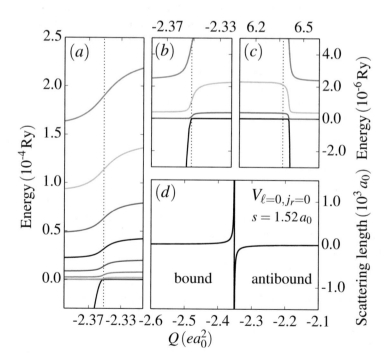

**Fig. 4.9** (Color online) The low-lying $J^\pi = 0^+$ eigenenergies (real parts) of a quadrupolar anion as a function of the electric quadrupole moment in the vicinity of $Q_{zz,c}^-$ (a,b) and $Q_{zz,c}^+$ (c). Panel (d) shows the scattering length of a scattering state at $E = 10^{-12}$ Ry, which is an eigenstate of the diagonal channel-channel coupling potential $V_{c,c}$ with $c = (\ell = 0, j_r = 0)$ for $s = 1.52\,a_0$ (from Ref. [94])

the lowest resonant states as a function of the quadrupole moment in an oblate setting. If $E_i$ is the energy of $i$th resonance of a supercritical molecule, then by changing the electric quadrupole moment continuously beyond $Q^-_{zz,c}$, one arrives at $E_i \rightarrow E'_i \approx E_{i+1}$. In the immediate vicinity of the critical quadrupole moment in Figs. 4.9b ($Q^-_{zz,c}$) and 4.9c ($Q^+_{zz,c}$), one can see the rearrangement of eigenvalues. Moreover, for $|Q^\pm_{zz}| << |Q^\pm_{zz,c}|$ the eigenvalues are almost equal for oblate and prolate configurations, as the corresponding wave functions are hardly sensitive to details of the potential.

Such a rearrangement of eigenvalues at the critical quadrupole moment can also be seen by considering the scattering length of the system at different values of $Q_{zz}$. At low energies ($k \rightarrow 0$), the scattering length is related to the $\ell = 0$ phase shift $\delta_0(k)$ of a scattering state through:

$$\lim_{k \to 0} \frac{k}{\tan \delta(k)} = -\frac{1}{a_0} . \tag{4.34}$$

To illustrate the criticality of the system, one may investigate a simple case where only $\ell = 0$ diagonal element of the channel-channel coupling potential is present (see panel (d) in Fig. 4.9). Indeed, the scaled parameter $q_{s,c} = qs$ reaches a constant at around the critical value of a quadrupole moment [57]. Therefore, by changing $s \rightarrow s'$ so that the $\ell = 0$ diagonal element of the potential becomes more and more important, one can effectively evolve the negative critical quadrupole moment $Q^-_{zz,c,0} = -2q_{s',c}s'$ for $\ell = 0$ to the value $Q^-_{zz,c}$ of the complete problem. To this end, one has to decrease $s$, or conversely increase $q_{s,c}$, to localize the electron in an almost pure $\ell = 0$ bound state.

---

**Exercise  V ★★**

One will consider the precision of the direct integration method and the Berggren expansion method for numerical studies of the quadrupolar anions, and compare results with those obtained in the dipolar anion case (see Exercise IV).

A. Run the particle-rotor code for the quadrupolar anion with both the direct integration method and the Berggren expansion method. Compare energies and wave functions obtained in these two methods and show that they provide with almost identical values.

   Explain why the situation is different as compared to the dipolar case.
B. By comparing the numerical stability of direct integration and Berggren expansion methods with the provided code, explain why it is preferable in practice to use the Berggren expansion method for the calculation of weakly bound and resonance states.

### 4.3.4 Weakly Bound and Unbound Atomic Nuclei

Studies of exotic nuclei far from the valley of beta-stability reveal novel features, such as the formation of nuclear halo [64, 98], near-threshold clustering [99–102], and the presence of new types of correlations [103]. In all these cases, atomic nucleus exhibits properties which are characteristic of the open quantum system whose properties depend on the coupling to reaction channels [104]. While the impact of reaction channels on nuclear structure has been recognized [49], the fact that some highly excited configurations can be interpreted in terms of nuclear clusters or nuclear molecules that experience collective motions such as rotations and vibrations is quite astonishing.

The success of the collective model of atomic molecules is based on the validity of the Born-Oppenheimer approximation [105] that relies on the huge difference of time scales for single-electron and ionic motions. The time-scale separation between single-nucleon and collective nuclear motion in atomic nuclei is small and, hence, the adiabatic approximation is expected to be badly violated [106]. One may ask, therefore, why a highly excited state, undergoing rapid particle emission, can be looked at using simple notions such as rotating or vibrating fields, or potentials, common for all nucleons. A good example is the spectrum of $^{12}$C which was discussed in the geometric/algebraic language of three $\alpha$-particle arrangements [107].

A related issue concerns the interpretation in terms of a nucleus, or a nuclear state of the broad resonances observed in scattering experiments. An excellent example is provided by the $0_1^+$, $2_1^+$, $4_1^+$ resonances of $^8$Be which have the width 5.57 eV, 1.15 MeV, and about 3.5 MeV, respectively [108]. Let us consider the single-particle time scale. The average time it takes for a nucleon to go across a light nucleus ($A \approx 10$) and come back can be roughly estimated at $T_{\text{s.p.}} \approx 1.3 \cdot 10^{-22}$ s [109], and corresponds to the time scale needed to form the nuclear mean field. Consequently, one is tempted to conclude that broad scattering features with $T_{1/2} < T_{\text{s.p.}}$ (or $\Gamma >$ 3.5 MeV for $A \approx 10$) can hardly be interpreted as the nuclear states [110].

Obviously, there is no sharp borderline that separates genuine nuclear states from broad structures seen in scattering experiments, and this often results in interpretational difficulties [111]. Moreover, a quantitative experimental characterization of broad resonances embedded in a large nonresonant background is not always possible. For instance, the extraction of experimental widths is usually model-dependent and relies on approximations [112, 113].

In order to study nuclear systems, one uses a particle-plus-core coupled-channel approach, based on the Berggren ensemble, which has been applied in Sects. 4.3.2 and 4.3.3 for the case of dipolar and quadrupolar anions. It has been shown in the dipolar case that below the ionization threshold, the motion of the valence electron is strongly coupled to the collective rotation of the molecule, forming rotational states. Above the ionization threshold, however, a rapid decoupling of the electron's motion from the rotational motion of the molecule takes place, thus leading to a disappearance of a collective rotational band. This observation brings the question whether such a suppression of collective rotation could also be expected in nuclear halo systems.

### 4.3.4.1 Deformed Proton Emitters: The Example of $^{141}$Ho

The particle-rotor model has many advantages for the study of deformed proton emitters, developed in the nonadiabatic approach and applied to the axially deformed case for that matter [88, 89, 114–116]. Firstly, medium-heavy and heavy nuclei, especially when they are deformed, cannot be studied with standard shell model due the very large dimensions of the matrix involved in their description. In fact, mean-field based methods are currently the only practical frameworks which can be applied in these regions of the nuclear chart. Moreover, as proton emitters typically consist of an unbound proton weakly coupled to a well bound core, the particle-rotor model recaptures the essential physical features of proton emitters and can be a good theoretical tool for their description. Indeed, deformation degrees of freedom in this model are exactly taken into account, and pairing correlations act only in well-bound core, so that they can be taken into account in the renormalized effective average potential acting on the emitted proton. Added to that, as said in Sect. 2.6.8, the proton emission widths of proton emitters are very small, of the order of $10^{-22}$ MeV, with associated lifetimes being of a few milliseconds. Consequently, the lifetimes of proton emitters are evaluated using Eq. (2.196), where the $u(k, r)$ state must be replaced by the amplitude of the considered channel. For this, the asymptotes of the wave functions of proton emitters must be very precisely evaluated, which is possible in the one-body picture provided by the particle-rotor model.

The Berggren basis has been applied to the description of the deformed proton emitter $^{141}$Ho (see Ref. [116] for details). For this, a Hamiltonian generated from a deformed Woods-Saxon potential, to which a Coulomb potential is added, was diagonalized in the Berggren basis. Proton emission widths were calculated either from the imaginary parts of eigenenergies, when width is larger than 1 keV typically, while the approximate expression of Eq. (2.196), generalized to the coupled-channel case, was used when widths were much smaller than 1 keV. Results are presented in Table 4.2.

When proton-emitting states have very narrow widths, coupled-channel calculations effected in coordinate space, diagonalization in harmonic oscillator basis, and Berggren basis diagonalization provide with similar results, except for a slightly smaller width for the $7/2^-$ ground state obtained in the Berggren basis diagonalization. Indeed, proton one-body states have a quasi-bound structure and emission widths are calculated from the continuity equation (see Eq. (2.196). The narrow character of proton one-body states implies that their wave functions vary smoothly on the real $r$-axis, so that all used methods are numerically stable. As could be expected, harmonic oscillator diagonalization can no longer provide with converged results once the width reaches 10–20 keV because wave functions start to significantly oscillate in the asymptotic region and, therefore, the wave function asymptotes cannot be calculated precisely. Consequently, Eq. (2.196) cannot be used to calculate the proton emission widths. Conversely, the energies and widths

**Table 4.2** Energies (in MeV) and widths (in brackets, in units of keV) of several proton-emitting states in $^{141}$Ho

| $\Omega^\pi$ | CC | HO diag. | Gamow diag. |
|---|---|---|---|
| | $0.756 \ (1.98 \cdot 10^{-20})$ | $0.756 \ (2.018 \cdot 10^{-20})$ | $0.758 \ (2.191 \cdot 10^{-20})$ |
| | 3.968 (0.053) | 3.968 (0.050) | 3.970 (0.051) |
| $1/2^+$ | 5.454 (0.035) | 5.454 (0.037) | 5.465 (0.032) |
| | 10.214 (833) | – | 10.534 (237) |
| | 11.686 (729) | – | 11.692 (651) |
| | 21.777 (527) | – | 21.809 (544) |
| | $1.190 \ (3.24 \cdot 10^{-16})$ | $1.190 \ (3.29 \cdot 10^{-16})$ | $1.194 \ (2.66 \cdot 10^{-16})$ |
| $7/2^-$ | 8.789 (17.51) | – | 8.790 (17.53) |
| | 9.933 (178) | – | 9.934 (178) |
| | 15.360 (104) | – | 15.375 (104) |

Their quantum numbers are $\Omega^\pi = 1/2^+$ (top) and $7/2^-$ (bottom). Theoretical frameworks consist of coupled-channel calculations effected in coordinate space (CC), harmonic oscillator diagonalization (HO diag.), and Berggren basis diagonalization (Gamow diag.) (Reproduced from Ref. [116] with the permission of AIP Publishing)

obtained from the coupled-channel calculations effected in coordinate space and Berggren basis diagonalization are close to one another, except for the fourth and fifth $1/2^+$ proton eigenstates. These discrepancies between the two approaches probably come from the fact that two broad proton states overlap in the same energy region.

Consequently, deformed proton emitters of very long lifetimes can be accurately described in the particle-rotor model and calculated very precisely using either coordinate space or basis expansion. As width is very narrow, even the real-energy formalism of harmonic oscillator diagonalization is well suited for that matter. However, broader proton emitters can be calculated only using the coordinate space integration or the Berggren basis diagonalization. Due to the possible overlapping of resonances, it is difficult to assess which of these two methods is more precise. The encountered case cannot be compared to those discussed in Sects. 4.3.2 and 4.3.3, because contrary to the case of multipolar anions, a repulsive Coulomb potential acting on the proton wave function is present in the asymptotic region. Nevertheless, the coordinate space integration and the Berggren basis diagonalization become equivalent when the resonances are well separated (see Table 4.2). Hence, while the harmonic oscillator diagonalization can be used in practice when width is but of a fraction of 1 keV, the Berggren basis diagonalization becomes a tool of choice when the resonances are broad. This is always the case when one considers neutron emitters, where the harmonic oscillator diagonalization can never be used in practice.

### 4.3.4.2 Rotational Bands in $^{11}$Be

In this section, we will investigate the existence of nuclear rotational states in the continuum of the deformed one-neutron halo nucleus $^{11}$Be. One assumes that the positive-parity ground-state band of $^{11}$Be can be viewed as a weakly bound/unbound neutron coupled to a deformed core of $^{10}$Be [117, 118]. The nonadiabatic approach developed in Refs. [88, 89, 114–116] (see also Sect. 4.3.4.1) in the context of proton emitters is now applied to the neutron + deformed core approach.

The exact form of the deformed pseudo-potential representing the nucleon-core interaction is not essential for the purpose of following discussion, as long as the one-nucleon threshold is correctly reproduced. One approximates the pseudo-potential by a deformed Woods-Saxon potential with a spherical spin-orbit term [115]. The total angular momentum of the system is $J = j + j_r$, where $j = \ell + s$ is the angular momentum of the valence nucleon and $j_r$ is that of the rotor. In the coupled-channel formalism, eigenstates of the decaying nucleus $|\Psi^{J^\pi}\rangle$ are expanded in the basis of channel wave functions labeled by channel quantum numbers $c = (\ell j j_r)$. Each channel state is given by the cluster radial wave function $u_c(r)/r$ representing the relative radial motion of the particle and the core, and the orbital-spin part $|j(\ell, s)j_r; JM_J\rangle$.

The rotational structure of nuclear states can be interpreted in terms of the intrinsic density of the valence particle in the core reference frame. One assumes that the core is associated with the rigid rotor axially deformed around the $z$-axis. When expressed in the deformed reference frame, the eigenstates $|\Psi^{J^\pi}\rangle$ can be expanded in the basis $|\Psi_K^{J^\pi}\rangle$, where $K$ is the projection of the total angular momentum $J$ on the symmetry axis in the core frame. The total density in intrinsic frame is obtained by summing up all the $K$-components of $\rho_{JK}(r, \theta)$ densities [83]. If only one $K$-component is nonzero, $K$ becomes a good quantum number and the strong coupling (adiabatic) limit is strictly obeyed [89, 90, 114, 115].

Coupled-channel equations of the particle-rotor model are solved up to a maximal radius of $R_{max} = 30$ fm and the rotation radius for the exterior complex scaling is $R_{rot} = 15$ fm. Since the studied rotational bands have positive parity, one takes partial waves with $\ell = 0, 2, 4, 6$ in the Berggren ensemble, and the maximum angular momentum of the core that is large enough to guarantee that the number of included states in the ground state band of the daughter nucleus does not impact the calculated widths [83, 115]. The neutron $0s_{1/2}$ shell in the core is excluded from the construction of the coupled-channel basis, in order that the Pauli principle between core and valence particles is approximately satisfied [120]. For the core energies $E_d^{j_r^\pi}$, one takes known experimental energies of $j_r^\pi = 0^+, 2^+$, and $4^+$ members of the ground-state band of $^{10}$Be [108]. The higher-lying band members are approximated by means of the rigid rotor expression with the moment of inertia corresponding to the $4^+$ level. The particle width of the core states with $j_r \geq 4$ will be ignored in the following. Experimentally, the $2_1^+$ state in $^{10}$Be is particle-bound, and the $4_1^+$ level has a fairly small width of 121 keV.

Parameters of the deformed pseudo-potential are fitted to the $1/2^+$ and $5/2^+$ members of the yrast band of $^{11}$Be [108]. These states collapse to the same band-head energy in the adiabatic limit ($I \to \infty$), i.e., they are members of the same rotational band. The experimental energy of higher-lying band members is uncertain, although candidates for the $3/2_1^+$, $7/2_1^+$, and $9/2_1^+$ states have been suggested

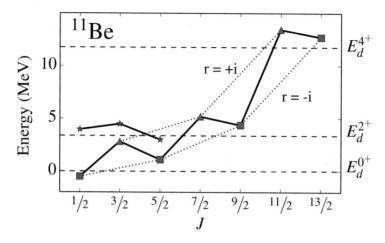

**Fig. 4.10** (Color online) Calculated lowest-energy bands of $^{11}$Be with $J \leq 13/2$ with signature $r = -i$ (squares) and $r = +i$ (triangles) are compared to the experimental ground-state band of $^{10}$Be core (horizontal dashed lines). Energies of some excited states of $^{11}$Be are marked by stars (from Ref. [119] )

[121, 122]. The calculated lowest energy states of $^{11}$Be are shown in Fig. 4.10. The large splitting between the favored signature band ($r = \exp(-i\pi J) = -i$) and the unfavored signature band ($r = +i$) is consistent with the results of a microscopic multicluster model [118, 123] and large-scale shell model [124, 125]. It is interesting to note that for yrast states $Q(J, j_r) < 0$ for $|J - j_r| = 1/2$. For instance, the $3/2_1^+$ and $5/2_1^+$ levels of $^{11}$Be are predicted to lie below the yrast $2^+$ state of $^{10}$Be. This means that the $\ell = 0$ neutron emission channel is blocked for both $r = -i$ and $r = +i$ bands.

The rotational structure of the ground-state band in $^{11}$Be is revealed by looking at the weights of individual $K$-components of valence neutron density. It turns out that the $K = 1/2$ component is dominant in most cases, in particular for the favored band. For the $7/2^+$, $11/2^+$, and $15/2^+$ states, the $K = 5/2$ and $3/2$ components dominate. In most cases, an appreciable degree of $K$-mixing is predicted. This suggests that a $K = 1/2$ label often attached to this band should be taken with a grain of salt.

Figure 4.11a shows the real part of the contributions of various partial waves ($\ell j$), denoted as $n_{\ell j}$, both for the ground-state band and the excited band in $^{11}$Be. One can see that the alignment pattern of the valence neutron is governed by a transition from the $s_{1/2}$ wave, which dominates at low angular momenta, to $d_{5/2}$, which governs the rotation at higher angular momenta. At high angular momenta, $J \geq 7/2$, the yrast line of $^{11}$Be can be associated with the weak coupling of neutron angular momentum $j = 5/2$ to the angular momentum $j_r$ of the core, resulting in the rotational alignment of $\boldsymbol{j}$ with $\boldsymbol{j_r}$. Indeed, as seen in Fig. 4.10, the computed energies $E^J$ of the $r = -i$ band are close to the energies of $^{10}$Be with $j_r = J - 5/2$.

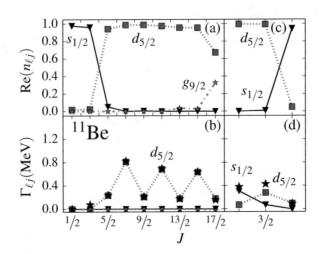

**Fig. 4.11** (Color online) Contribution from different partial waves ($\ell j$) to the norms (top) and widths $\Gamma_{\ell j} = \sum_{j_r} \Gamma_{\ell j j_r}$ (bottom) for different states of the ground-state band of $^{11}$Be (left) and for yrare states (right). The total widths are marked by stars (from Ref. [119])

The higher partial waves, such as $g_{9/2}$, are seen for $J > 15/2$. The structure of states in the excited band (see Fig. 4.11c) is dominated by $d_{5/2}$ at low angular momenta, and by $s_{1/2}$ at higher angular momenta.

To estimate one-neutron decay rates, one computes the current expression for the decay width (see Eq. (2.196) and Refs. [88,115,126]), which gives $\Gamma = \sum_c \Gamma_c(r)$ in terms of the partial widths of the channel wave functions. In every case considered here, the value of $\Gamma$ obtained from the current expression agrees with the eigenvalue estimate $-2\text{Im}(E^{J^\pi})$ and, moreover, the values of $\Gamma_c(r)$ are stable at $r = R_{\max}$. The calculated one-neutron widths corresponding to different partial waves are shown in panels (b) and (d) of Fig. 4.11.

As said before, the $s$-wave neutron decay is blocked in the ground band of $^{11}$Be. Consequently, the neutron decay widths of yrast states are primarily governed by $\ell = 2$ waves. Moreover, the $Q$-values for the neutron decay of the favored band are small in the $d_{5/2}$ channel because of the weak coupling of the valence neutron to the core states. As a result, the states of a favored yrast band of $^{11}$Be are predicted to have small neutron widths of the order of 200 keV.

In the case of the unfavored band, the neutron widths are larger ($\Gamma \sim 0.7$ MeV). This demonstrates that angular momentum alignment can stabilize collective behavior in highly excited yrast states of a neutron dripline system. Due to the Coriolis force, high-$\ell$ orbits which are responsible for the angular momentum alignment are occupied at high spins at the expense of low-$\ell$ states. The former are localized within the nuclear volume because of their large centrifugal barrier, whereas the latter ones determine halo properties and particle decay.

The above discussion does not hold for states $J = 1/2$ and $3/2$ of the excited band, for which the $\ell = 0$ channel is not blocked. As seen in Fig. 4.11d, the total width of those states is predicted to be around 0.4 MeV. For $J = 1/2$, the width is even totally dominated by $s_{1/2}$. For the $J = 5/2$ state of this band, the $\ell = 0$ channel is blocked again; hence, its width is small.

### 4.3.4.3 Effects of Deformation and Configuration Mixing on E1 Electromagnetic Transition in $^{11}$Be

One will now consider electromagnetic transitions in the particle-rotor model on the example of electric dipole transition between loosely bound $1/2^-$ and $1/2^+$ states in $^{11}$Be. In the study of this transition, energies of these states must be fitted to their experimental values. The $1/2^+$ and $1/2^-$ states are bound by about $-500$ keV and $-180$ keV, respectively, so the $1/2^- \rightarrow 1/2^+$ E1 transition will be enhanced due to the important nucleon density in the asymptotic zone for both states.

Nuclear interaction in the $N \sim 8$ region of nuclear chart leads to an inversion of the $0p_{1/2}$ and $1s_{1/2}$ shells in $^{11}$Be. This is directly seen in the $^{11}$Be spectrum, as the $1/2^+$ state, which is mainly a $^{10}$Be $+ 1s_{1/2}$ configuration, is more bound than the $1/2^-$ state, formed primarily by a $^{10}$Be $+ 0p_{1/2}$ configuration. Such shell inversions can be found in different regions of nuclear chart, e.g., inversion of $1s_{1/2}$ and $0d_{3/2}$ happens for $N \sim 14$, whereas the shells $0f_{7/2}$ and $1p_{3/2}$ are inverted at $N \sim 20$. Due to the simple Hamiltonian of the particle-rotor model, where most of nucleon-nucleon interactions are replaced by a deformed Woods-Saxon potential, it is necessary to fit the depth of Woods-Saxon potential independently in the $s_{1/2}$ and the $p_{1/2}$ neutron channels. In the Berggren ensemble, $1s_{1/2}$ single-particle state is weakly bound ($E = -0.00702$ MeV) and $0p_{1/2}$ is a broad single-particle resonance ($E = 0.00794$ MeV, $\Gamma = 0.725$ keV). These two single-particle states form an essential part of the $1/2^+$ and $1/2^-$ many-body states of $^{11}$Be, as they respectively correspond to the channels $[^{10}\mathrm{Be}(j_r = 0) \otimes \nu 1s_{1/2}]^{1/2^+}$ and $[^{10}\mathrm{Be}(j_r = 0) \otimes \nu 0p_{1/2}]^{1/2^-}$, where $j_r$ is the angular momentum of the $^{10}$Be rotor state (see Exercise VI for additional numerical applications). The calculated energies of the $1/2^+$ and $1/2^-$ many-body states with respect to one-neutron emission threshold are equal to $-0.503$ MeV and $-0.184$ MeV, respectively, and closely reproduce the experimental energies of these states, which are $-0.502$ MeV and $-0.182$ MeV, respectively [108].

---

**Exercise VI ★**

One will numerically calculate the B(E1) reduced transition probability between the ground state $1/2^+$ and the first excited state $1/2^-$ of $^{11}$Be, using different sets of Hamiltonian parameters.

Using the default parameters given in the input file, run the particle-rotor code in the case of $^{11}$Be to recalculate the binding energies and B(E1) transition probability discussed above.

Change slightly parameters of the Hamiltonian, e.g., by 10%, with respect to their default values. Notice that, while energies depend strongly on Hamiltonian parameters, B(E1) transition probability almost does not change in this case.

In order to calculate the B(E1) reduced transition probability, one has to use an effective charge for the neutron. The latter has two origins: one arising from the nuclear recoil and the other from the truncation of the configuration space. The recoil effective charge for the E1 transition is equal to $-Z/A$. Taking into account only the nuclear recoil, one obtains $B(E1) = 0.199$ e$^2$ fm$^2$, almost twice smaller than the experimental value $B(E1)_{exp} = 0.105$ e$^2$ fm$^2$. This indicates that the considered model space is too small and hence the configuration mixing, which reduces the transition probability, is insufficient. Indeed, the particle-hole excitations from $^{10}$Be are replaced in this model by a rotating core interacting with a neutron through a deformed Woods-Saxon potential, which can simulate only a fraction of the total configuration space. The experimental value of B(E1) reduced transition probability is recovered by taking an effective charge which is smaller by about 25% than the recoil effective charge.

## Solutions to Exercises[1]

**Exercise I.** The commutation relations: $[\mathbf{P}_{CM}, \mathbf{R}_{CM}] = -i\hbar$ and $[\mathbf{p}_{rel}, \mathbf{r}_{rel}] = -i\hbar$ are directly obtained using Eqs. (4.1)–(4.4), the standard commutators $[\mathbf{p}_1, \mathbf{r}_1] = [\mathbf{p}_2, \mathbf{r}_2] = -i\hbar$, and the fact that operators $\mathbf{p}_1$, $\mathbf{r}_1$ involving particle 1 commute with those involving particle 2: $\mathbf{p}_2$, $\mathbf{r}_2$.

**Exercise II.** One only has to consider the kinetic operators in $H_{CM} + h_{rel}$. Using Eqs. (4.1)–(4.4), (4.7), and (4.8), one directly obtains:

$$\frac{\mathbf{P}_{CM}^2}{2M_{CM}} + \frac{\mathbf{p}_{rel}^2}{2m_{rel}} = \frac{\mathbf{p}_1^2}{2m_1} + \frac{\mathbf{p}_2^2}{2m_2} . \tag{4.35}$$

**Exercise III.**

A. As the Berggren basis is complete, energies and observables are independent of the potential depth of the basis and of the form of the contour, as long as it encompasses all resonances of interest (see Sect. (3.5.1)). Consequently, a small change of parameters related to basis contours and potentials does not change results significantly in practice.

B. The use of a Fermi function changes the Hamiltonian as it removes unstable components at large distances. Consequently, results have to depend on the parameters of the Fermi function. Conversely, once the Fermi function is fixed, it generates moderate size matrix elements of the Hamiltonian. Such a Hamiltonian can then be efficiently diagonalized with the Berggren basis. As the latter is complete theoretically and in practice, results are independent of both basis potentials and contours.

---

[1] The input files, codes and code user manual associated to computer-based exercises can be found at https://github.com/GSMUTNSR.

## Exercise IV.

**A.** Direct integration and Berggren expansion method are not equivalent as $1/r^2$ coupling potentials have to be ignored in the asymptotic region. Indeed, one should match wave functions to analytical outgoing Hankel functions in the asymptotic region, but one cannot devise analytical asymptotic forms of wave functions when moment of inertia is finite. Thus, the direct integration method cannot provide with exact Hamiltonian eigenstates. As the Berggren basis is complete, this difficulty is absent in the Berggren expansion method.

**B.** Let us insert $u_c^J(r) = g_c \, H_{\ell(eff),0}^+(kr)$ in Eq. (4.19) in the asymptotic region to devise a solution of this equation. One has $k_c = k \equiv \sqrt{E_J} \; \forall c$ because $I = +\infty$. One then finds that the proposed ansatz is a solution of Eq. (4.19) at large radius if $\{g_c\}_c$ forms an eigenvector of the matrix $[\ell_c(\ell_c + 1)\delta_{cc'} + a_{cc'}]_{cc'}$, of associated eigenvalue $\alpha = \ell^{(eff)}(\ell^{(eff)} + 1)$, so that one can pose $\ell^{(eff)} = (2\alpha)/(1 + \sqrt{1 + 4\alpha})$ (see Sect. (2.3)). Hence, the asymptote of the wave functions of the particle-rotor system can be calculated exactly in this case, so that it is numerically precise to use direct integration to calculate the eigenstates of the particle-rotor Hamiltonian when $I = +\infty$.

**C.** As the calculated state is well bound, the direct integration is stable even if approximate for finite $\ell$. It becomes unstable for very weakly bound and resonance states of molecules. The numerical use of the Berggren expansion method is stable, provided that one uses appropriate contours.

## Exercise V.

**A.** As the potentials are proportional to $1/r^3$ for large $r$, they decrease quickly on the real $r$-axis, so that $u_c(r) \propto H_{\ell_c,0}^+(k_c r)$ in the asymptotic region.

Therefore, one can match wave functions to their exact analytical outgoing Hankel functions in the asymptotic region and the direct integration method is applicable.

Thus, both direct integration and Berggren expansion methods are equivalent in the quadrupolar case, contrary to the dipolar case (see Exercise IV).

**B.** Direct integration is hereby stable for well-bound states because of the fast decrease of wave functions on the real $r$-axis. However, as seen from effected numerical calculations, direct integration becomes unstable for very weakly bound and resonance states of molecules, similarly to the dipolar case (see Exercise IV). Consequently, the Berggren expansion method still has to be used for the calculation of weakly bound and resonance states in practice, even though direct integration and Berggren expansion methods are theoretically equivalent.

**Exercise VI.** Channel wave functions do not change much when varying Hamiltonian parameters, so that only energies follow this change, while the B(E1) reduced transition probabilities remain practically the same.

# References

1. R.D. Lawson, *Theory of the Nuclear Shell Model* (Clarendon, Oxford, 1980)
2. R.J.N. Phillips, Rep. Prog. Phys. **22**, 562 (1959)
3. W. Heisenberg, Z. Phys. **77**, 1 (1932)
4. M. Chadwick, P. Obložinský, M. Herman, N. Greene, R. McKnight, D. Smith, P. Young, R. MacFarlane, G. Hale, S. Frankle, A. Kahler, T. Kawano, R. Little, D. Madland, P. Moller, R. Mosteller, P. Page, P. Talou, H. Trellue, M. White, W. Wilson, R. Arcilla, C. Dunford, S. Mughabghab, B. Pritychenko, D. Rochman, A. Sonzogni, C. Lubitz, T. Trumbull, J. Weinman, D. Brown, D. Cullen, D. Heinrichs, D. McNabb, H. Derrien, M. Dunn, N. Larson, L. Leal, A. Carlson, R. Block, J. Briggs, E. Cheng, H. Huria, M. Zerkle, K. Kozier, A. Courcelle, V. Pronyaev, S. van der Marck, Nucl. Data Sheets **107**(12), 2931 (2006). Evaluated Nuclear Data File ENDF/B-VII.0
5. M.E. Peskin, D.V. Schroeder, *An Introduction to Quantum Field Theory* (CRC Press, Taylor and Francis Group, London/New York, 1995)
6. W. Greiner, S. Schramm, E. Stein, *Quantum Chromodynamics*, 3rd edn. (Springer, Berlin, 2007)
7. A.H. Mueller, *Perturbative QCD (Advanced Directions in High Energy Physics)* (World Scientific, Singapore, 1989)
8. R.D. Field, *Applications of Perturbative QCD*, 1st edn. (Basic Books, London, 1989)
9. S. Weinberg, Physica A **96**, 327 (1779)
10. H. Leutwyler, Ann. Phys. **235**, 165 (1994)
11. J.L. Friar, in *Few-Body Problems in Physics '95*, ed. by R. Guardiola (Springer, Vienna, 1996), p. 471
12. E. Epelbaum, H.W. Hammer, U.G. Meißner, Rev. Mod. Phys. **81**, 1773 (2009)
13. S. Weinberg, Ann. Rev. Nucl. Part. Sci. **43**, 209 (1993)
14. G. Ecker, Prog. Part. Nucl. Phys. **35**, 1 (1995)
15. A. Pich, Rep. Prog. Phys. **58**, 563 (1995)
16. D.R. Entem, N. Kaiser, R. Machleidt, Y. Nosyk, Phys. Rev. C **91**, 014002 (2015)
17. D.R. Entem, N. Kaiser, R. Machleidt, Y. Nosyk, Phys. Rev. C **92**, 064001 (2015)
18. J.E. Lynn, I. Tews, J. Carlson, S. Gandolfi, A. Gezerlis, K.E. Schmidt, A. Schwenk, Phys. Rev. Lett. **116**, 062501 (2016)
19. A. Ekström, C. Forssén, C. Dimitrakakis, D. Dubhashi, H.T. Johansson, A.S. Muhammad, H. Salomonsson, A. Schliep, J. Phys. G. Nucl. Part. Phys **46**, 095101 (2019)
20. S. Wesolowski, R.J. Furnstahl, J.A. Melendez, D.R. Phillips, J. Phys. G. Nucl. Part. Phys **46**(4), 045102 (2019)
21. R. Machleidt, D.R. Entem, Phys. Rep. **503**, 1 (2011)
22. S. Weinberg, Phys. Lett. B **251**, 288 (1990)
23. S. Weinberg, Nucl. Phys. B **363**, 3 (1991)
24. D.R. Entem, R. Machleidt, Phys. Rev. C **68**, 041001(R) (2003)
25. D.R. Entem, R. Machleidt, Phys. Lett. B **524**, 93 (2002)
26. T. Neff, H. Feldmeier, arXiv:1610.04066 [nucl-th] (2016)
27. E. Fermi, E. Teller, Phys. Rev. **72**, 399 (1947)
28. J.M. Lévy-Leblond, Phys. Rev. **153**, 1 (1967)
29. K.R. Lykke, R.D. Mead, W.C. Lineberger, Phys. Rev. Lett. **52**, 2221 (1984)
30. J. Marks, J.I. Brauman, R.D. Mead, K.R. Lykke, W.C. Lineberger, J. Chem. Phys. **88**, 6785 (1988)
31. T. Andersen, Phys. Scr. **1991**, 23 (1991)
32. E.A. Brinkman, S. Berger, J. Marks, J.I. Brauman, J. Chem. Phys. **99**, 7586 (1993)
33. A.S. Mullin, K.K. Murray, C.P. Schulz, W.C. Lineberger, J. Phys. Chem. **97**, 10281 (1993)
34. S. Ard, W.R. Garrett, R.N. Compton, A.L, S.G. Stepanian, Chem. Phys. Lett. **473**, 223 (2009)
35. J.E. Turner, Am. J. Phys. **45**, 758 (1977)
36. W.R. Garrett, Chem. Phys. Lett. **5**, 393 (1970)

37. W.R. Garrett, Phys. Rev. A **3**, 961 (1971)
38. H.O. Crawford, Mol. Phys. **20**, 585 (1971)
39. W.R. Garrett, J. Chem. Phys. **73**, 5721 (1980)
40. W.R. Garrett, Phys. Rev. A **22**, 1769 (1980)
41. W.R. Garrett, Phys. Rev. A **23**, 1737 (1981)
42. W.R. Garrett, J. Chem. Phys. **77**, 3666 (1982)
43. D.C. Clary, J. Phys. Chem. **92**, 3173 (1988)
44. K. Fossez, N. Michel, W. Nazarewicz, M. Płoszajczak, Phys. Rev. A **87**, 042515 (2013)
45. T. Berggren, Nucl. Phys. A **109**, 265 (1968)
46. T. Berggren, P. Lind, Phys. Rev. C **47**, 768 (1993)
47. P. Lind, Phys. Rev. C **47**, 1903 (1993)
48. B. Gyarmati, T. Vertse, Nucl. Phys. A **160**, 523 (1971)
49. N. Michel, W. Nazarewicz, M. Płoszajczak, T. Vertse, J. Phys. G. Nucl. Part. Phys. **36**, 013101 (2009)
50. C. Desfrançois, Y. Bouteiller, J.P. Schermann, D. Radisic, S.T. Stokes, K.H. Bowen, N.I. Hammer, R.N. Compton, Phys. Rev. Lett. **92**, 083003 (2004)
51. C. Desfrançois, V. Périquet, S. Carles, J.P. Schermann, L. Adamowicz, Chem. Phys. **239**, 475 (1998)
52. G.L. Gutsev, P. Jena, J. Chem. Phys. **111**, 504 (1999)
53. G.L. Gutsev, R.J. Bartlett, R.N. Compton, J. Chem. Phys. **108**, 6756 (1998)
54. T. Sommerfeld, J. Chem. Phys. **121**, 4097 (2004)
55. T. Sommerfeld, K.M. Dreux, R. Joshi, J. Phys. Chem. **118**, 7320 (2014)
56. M. Allan, J. Phys. B: At. Mol. Opt. Phys. **36**, 2489 (2003)
57. A. Ferron, P. Serra, S. Kais, J. Chem. Phys. **120**, 18 (2004)
58. V. Privman, *Finite Size Scaling and Numerical Simulation of Statistical Systems*, 1st edn. (World Scientific, Singapore, 1990)
59. J.P. Neirotti, P. Serra, S. Kais, Phys. Rev. Lett. **79**, 3142 (1997)
60. P. Serra, J.P. Neirotti, S. Kais, Phys. Rev. A **57**, R1481(R) (1998)
61. S. Kais, P. Serra, Int. Rev. Phys. Chem. **19**, 97 (2000)
62. V.I. Pupyshev, A.Y. Ermilov, Int. J. Quantum Chem. **96**, 185 (2004)
63. K. Riisager, D.V. Fedorov, A.S. Jensen, Europhys. Lett. **49**, 547 (2001)
64. A.S. Jensen, K. Riisager, D.V. Fedorov, E. Garrido, Rev. Mod. Phys. **76**, 215 (2004)
65. J. Mitroy, Phys. Rev. Lett. **94**, 033402 (2005)
66. D. Knoop, F. Ferlaino, M. Mark, M. Berninger, H. Schobel, H.C. Nagerl, R. Grimm, Nature **5**, 227 (2009)
67. H.W. Hammer, L. Platter, Ann. Rev. Nucl. Part. Sci. **60**, 207 (2010)
68. F. Ferlaino, R. Grimm, Physics **3**, 9 (2010)
69. J. Marks, D.M. Wetzel, P.B. Comita, J.I. Brauman, J. Chem. Phys. **84**, 5284 (1986)
70. T.F. O'Malley, Phys. Rev. **137**, A1668 (1965)
71. H. Estrada, W. Domcke, J. Phys. B: At. Mol. Phys. **17**, 279 (1984)
72. C.W. Clark, Phys. Rev. A **30**, 750 (1984)
73. I.I. Fabrikant, J. Phys. B **18**, 1873 (1985)
74. D.C. Clary, Phys. Rev. A **40**, 4392 (1989)
75. M. McCartney, P.G. Burke, L.A. Morgan, C.J. Gillan, J. Phys. B: Atom. Mol. Opt. Phys. **23**, L415 (1990)
76. H.R. Sadeghpour, J.L. Bohn, M.J. Cavagnero, B.D. Esry, I.I. Fabrikant, J.H. Macek, A.R.P. Rau, J. Phys. B. Atom. Mol. Opt. Phys. **33**(5), R93 (2000)
77. J. Martorell, J.G. Muga, D.W.L. Sprung, Phys. Rev. A **77**, 042719 (2008)
78. T. Klahn, P. Krebs, J. Chem. Phys. **109**, 531 (1998)
79. K.D. Jordan, F. Wang, Annu. Rev. Phys. Chem. **54**, 367 (2003)
80. P. Skurski, M. Gutowski, J. Simons, J. Chem. Phys. **114**, 7443 (2001)
81. K.A. Peterson, M. Gutowski, J. Chem. Phys. **11**, 3297 (2002)
82. W.R. Garrett, J. Chem. Phys. **133**, 224103 (2010)

83. F. Fossez, N. Michel, W. Nazarewicz, M. Płoszajczak, Y. Jaganathen, Phys. Rev. A **91**, 012503 (2015)
84. N. Michel, W. Nazarewicz, M. Płoszajczak, K. Bennaceur, Phys. Rev. Lett. **89**, 042502 (2002)
85. N. Michel, W. Nazarewicz, M. Płoszajczak, J. Okołowicz, Phys. Rev. C **67**, 054311 (2003)
86. A.T. Kruppa, W. Nazarewicz, P.B. Semmes, AIP Conf. Proc. **518**, 173 (1999)
87. A.T. Kruppa, W. Nazarewicz, P.B. Semmes, AIP Conf. Proc. **681**, 61 (2003)
88. A.T. Kruppa, W. Nazarewicz, Phys. Rev. C **69**, 054311 (2004)
89. H. Esbensen, C.N. Davids, Phys. Rev. C **63**, 014315 (2000)
90. C.N. Davids, H. Esbensen, Phys. Rev. C **69**, 034314 (2004)
91. V.E. Chernov, A.V. Dolgikh, B.A. Zon, Phys. Rev. A **72**, 052701 (2005)
92. W.R. Garrett, J. Chem. Phys. **128**, 194309 (2008)
93. V. Privman, *NIST Chemistry WebBook* (The National Institute of Standards and Technology, New York, 2016)
94. K. Fossez, X. Mao, W. Nazarewicz, N. Michel, W.R. Garrett, M. Płoszajczak, Phys. Rev. A **94**, 032511 (2016)
95. C. Desfrançois, H. Abdoul-Carime, J.P. Schermann, Eur. Phys. J. D **2**, 149 (1998)
96. H. Abdoul-Carime, J.P. Schermann, C. Desfrançois, Few-Body Syst. **31**, 183 (2002)
97. W.R. Garrett, J. Chem. Phys. **136**, 054116 (2012)
98. I. Tanihata, H. Savajols, R. Kanungo, Prog. Part. Nucl. Phys. **68**, 215 (2013)
99. W. von Oertzen, M. Freer, Y. Kanada-En'yo, Phys. Rep. **432**, 43 (2006)
100. M. Freer, Rep. Prog. Phys. **70**, 2149 (2007)
101. J. Okołowicz, W. Nazarewicz, M. Płoszajczak, Fortschr. Phys. **61**, 66 (2013)
102. J. Okołowicz, M. Płoszajczak, W. Nazarewicz, Prog. Theor. Phys. Supp. **196**, 230 (2012)
103. M. Matsuo, T. Nakatsukasa, J. Phys. G **37**, 064017 (2010)
104. N. Michel, W. Nazarewicz, M. Płoszajczak, Phys. Rev. C **82**, 044315 (2010)
105. M. Born, R. Oppenheimer, Ann. Physik (Leipzig) **389**, 457 (1927)
106. W. Nazarewicz, Nucl. Phys. A **557**, 489 (1993)
107. D.J. Marín-Lámbarri, R. Bijker, M. Freer, M. Gai, T. Kokalova, D.J. Parker, C. Wheldon, Phys. Rev. Lett. **113**, 012502 (2014)
108. http://www.nndc.bnl.gov/ensdf
109. V.I. Goldanskii, Ann. Rev. Nucl. Sci. **16**, 1 (1966)
110. M. Thoennessen, Rep. Prog. Phys. **67**, 1187 (2004)
111. E. Garrido, A.S. Jensen, D.V. Fedorov, Phys. Rev. C **88**, 024001 (2013)
112. H.O.U. Fynbo, R. Álvarez-Rodríguez, A.S. Jensen, O.S. Kirsebom, D.V. Fedorov, E. Garrido, Phys. Rev. C **79**, 054009 (2009)
113. K. Riisager, H. Fynbo, S. Hyldegaard, A. Jensen, Nucl. Phys. A **940**, 119 (2015)
114. A.T. Kruppa, B. Barmore, W. Nazarewicz, T. Vertse, Phys. Rev. Lett. **84**, 4549 (2000)
115. B. Barmore, A.T. Kruppa, W. Nazarewicz, T. Vertse, Phys. Rev. C **62**, 054315 (2000)
116. A.T. Kruppa, N. Michel, W. Nazarewicz, AIP Conf. Proc. **726**, 7 (2004)
117. F.M. Nunes, I.J. Thompson, R.C. Johnson, Nucl. Phys. A **596**, 171 (1996)
118. P. Descouvemont, Nucl. Phys. A **615**, 261 (1997)
119. K. Fossez, W. Nazarewicz, Y. Jaganathen, N. Michel, M. Płoszajczak, Phys. Rev. C **93**, 011305 (2016)
120. S. Saito, Prog. Theor. Phys. **41**, 705 (1969)
121. W. von Oertzen, Acta Phys. Pol. B **29**, 247 (1998)
122. H.G. Bohlen, W. von Oertzen, R. Kalpakchieva, T.N. Massey, T. Dorsch, M. Milin, C. Schulz, T. Kokalova, C. Wheldon, J. Phys. Conf. Ser. **111**, 012021 (2008)
123. P. Descouvemont, Nucl. Phys. A **699**, 463 (2002)
124. M.A. Caprio, P. Maris, J.P. Vary, Phys. Lett. B **719**, 179 (2013)
125. P. Maris, M.A. Caprio, J.P. Vary, Phys. Rev. C **91**, 014310 (2015)
126. J. Humblet, L. Rosenfeld, Nucl. Phys. **26**, 529 (1961)

# Formulation and Implementation of the Gamow Shell Model

**5**

## 5.1 Mathematical Foundation of the Gamow Shell Model

The Gamow shell model formulated in the Berggren basis [1] has all necessary features which allow to describe correlated $A$-body halo states and resonances. In contrast to the standard shell model, the Gamow shell model respects the flux conservation, i.e., provides unitary description of wave functions in the vicinity and above elastic reaction threshold. By using a configuration mixing framework, the Gamow shell model bears the same generality as the standard shell model, providing a microscopic description of all kinds of nuclei. However, due to the use of unbound states in the one-body basis, a rigorous mathematical foundation of the Gamow shell model is more complicated because resonances do not belong to the Hilbert space.

The Berggren basis has been introduced at the one-body level using the Newton completeness relation and Cauchy theorem. Completeness of this basis has then been demonstrated in a most general case using the analytical properties of Coulomb wave functions and the Cauchy theorem. However, if we intend to use the Berggren basis in shell model, then it becomes very cumbersome to continue using one-body completeness relations in the coordinate space ($r$-space), as shell model wave functions are multidimensional objects. Moreover, a feature as simple as the antisymmetry, which is straightforward to apply using Slater determinants, becomes very complicated to handle in coordinate space. Consequently, if one wants to keep the convenience of the shell model framework, it is necessary to use the abstract formulation of second quantization, based on the use of creation and annihilation operators on quantum bra and ket states.

Intuitively, one can already suppose that the algebra needed for the Gamow shell model will be very close to that used in the Hilbert space. Indeed, if one heuristically replaces the discrete completeness relation of bound states by a continuous one involving integration of bra and ket of unbound states, which one discretizes afterward, one obtains formulas which represent a multidimensional generalized Fourier transform. However, using Dirac notation with unbound states would remain

© Springer International Publishing AG 2021
N. Michel, M. Płoszajczak, *Gamow Shell Model*, Lecture Notes in Physics 983,
https://doi.org/10.1007/978-3-030-69356-5_5

purely formal and one would have no guarantee that they will provide with the proper eigenvectors of the considered Hamiltonian. Consequently, one has to leave the Hilbert space, where only square integrable states can be considered, and replace it by a larger space able to deal with resonance and scattering states. This is possible within the so-called rigged Hilbert space, at the price, however, of introducing a more complicated abstract framework involving different spaces for the states and for the linear functionals acting on them.

## 5.1.1  Rigged Hilbert Space Setting of the Gamow Shell Model

The Dirac notation, consisting in the standard bra and ket notation, was devised in the context of bound states. Its usefulness and mathematical rigor is based on the use of the Riesz representation theorem of the Hilbert space. Indeed, the latter theorem implies that the Hilbert space and its dual space are anti-isomorphic, so that the action of the dual of a state vector $|A\rangle$, i.e., the bra state $\langle A|$, on another ket state vector $|B\rangle$, provides with the same result as the scalar product between the two latter state vectors: $\langle A|(|B\rangle) = \langle A|B\rangle$.

This notation greatly simplifies equations, as all coordinate and momentum space dependences, as well as multidimensional integration, are embedded in an algebraic formalism of operators and vectors. It is not possible to use it rigorously if one includes scattering states, as they cannot be normalized except with a Dirac delta. Nevertheless, they can be formally handled with bra and ket states, as they correspond to well-defined wave functions. However, a rigorous mathematical framework with which bound, resonance, and scattering states can all be represented with the Dirac notation demands the introduction of rigged Hilbert spaces.

The rigged Hilbert space, or Gel'fand triple, is defined from three space inclusions [2–4]:

$$\Phi \subset \mathscr{H} \subset \Phi^{\times} . \tag{5.1}$$

$\Phi$ in this relation is the subspace of test functions, dense in the Hilbert space $\mathscr{H}$, whose elements can be seen as bounded operators bearing a fast decrease on the real axis. $\Phi^{\times}$ is the space of anti-linear functionals over $\Phi$, comprising in particular distributions, where all bound, resonance, and scattering states will be defined. Finally, $\Phi^{\times}$ is thus the rigged Hilbert space of interest.

The linear functionals over $\Phi$ are contained in another rigged Hilbert space, denoted as $\Phi'$:

$$\Phi \subset \mathscr{H} \subset \Phi' , \tag{5.2}$$

so that unbound bras and kets belong to $\Phi'$ and $\Phi^{\times}$, respectively.

The unbound states in $\Phi^{\times}$ will enter scalar products only with test functions of $\Phi$, so that they all are well defined even though $\Phi^{\times}$ contains non-integrable states. One may note that the Hilbert space $\mathscr{H}$ does not have to be included, as bound, resonance, and scattering states all belong to $\Phi^\times$. In fact, the inclusion of $\mathscr{H}$ is there to emphasize the more stringent properties which are demanded in $\Phi$. Indeed, the test functions of $\Phi$ have to decrease quickly on the $r$-axis, whereas $\mathscr{H}$ functions diverge in general for $r \rightarrow +\infty$.

For the case of bound states, $\Phi \sim \mathscr{H} \sim \Phi^\times$, as quantum states and their dual belong to anti-isomorphic spaces, so that their properties, i.e., those of bound states, are identical. On the other hand, the non-integrable character of $\Phi'$ elements, i.e., linear functionals on $\Phi$, which mathematically depict resonant and scattering states, imply that the $\Phi$ states must possess stronger regularity properties than those of the Hilbert space. In fact, $\Phi$ bears much resemblance with the Schwarz space, whose elements decrease on the real axis faster than any rational functions. One may note in passing that the Heisenberg uncertainty relations, involving position and impulsion which are unbounded operators, are properly defined on $\Phi$, $\Phi^\times$ and not on $\mathscr{H}$.

Rigged Hilbert spaces also provide with the mathematical setting to deal with time-dependent processes, such as particle capture or decay. Therefore, unbound states of complex eigenvalues, i.e., resonance or Gamow states, which are time-dependent, can be represented by a ket state of $\Phi^\times$ in rigged Hilbert spaces. It is thus possible to write the one-body Berggren completeness relation of Eq. (3.66) built from bound, resonance, and scattering state, demonstrated in Sect. 3.5, with the bra and ket states of the $\Phi'$ and $\Phi^\times$ spaces:

$$\sum_n |u_n\rangle \langle \widetilde{u_n}| + \int_{L^+} |u(k)\rangle \langle \widetilde{u(k)}| \; dk = \hat{\mathbf{1}} \,, \tag{5.3}$$

where $|u_n\rangle$ are resonant states and $|u(k)\rangle$ are scattering states of the $L^+$ contour, so that $\langle r|u_n\rangle = u_n(r)$ and $\langle r|u(k)\rangle = u(k, r)$ (see Sect. 3.5 for notations), and where the tilde sign indicates that one uses a different scalar product than in the Hilbert space [1]. Indeed, one has seen in Sect. 3.3 that matrix elements are defined from the analytic continuation of their real values obtained with bound states. This implies that no complex conjugation appears in the radial matrix elements, contrary to those defined in the Hilbert space. Eigenstates in Eq. (5.3) are biorthogonal [1]. This is effected with the time-reversal operator, represented by a tilde notation in Eq. (5.3). Physically, this means that a bra state must be a particle-capturing state, as a ket state is resonance and hence is a particle-emitting state. There is then no difference in the use of bound states, resonances and scattering states, which are all written using the same bra and ket notation.

The introduction of rigged Hilbert spaces has not however solved the problem of the interpretation of complex observables. Indeed, as resonances have a complex energy, and as one used the Berggren norm in which there is no complex conjugation, observables involving resonances become complex. The interpretation of complex observables then has to be done independently.

## 5.1.2   Complex Observables and Their Interpretation

The interpretation of a rigged Hilbert spaces formulation of quantum mechanics is still an open issue, and the relations between Berggren and rigged Hilbert spaces formulations are not fully solved [5]. In particular, the complex matrix elements of operators and the probabilistic interpretation of unbound wave function do not have a well-defined interpretation common to all quantum mechanics frameworks describing unbound states.

The interpretation of Berggren of the real and imaginary parts of the expectation value of an operator relies on the interference effects in reaction cross sections [6]. Berggren also studied observables which commute with Hamiltonian, which will be considered in this section. If one remains at the leading order, the rigged Hilbert spaces formulation of Bohm and Gadella [7, 8] and that of Berggren [9] are equivalent. When considering radial operators, expectation values have been studied using a two-channel model [10]. These expectation values follow Berggren's prescription and allow to identify resonances as long as the wavelength of the decaying wave is shorter than the radial extension of the resonance state.

One will extend here Berggren's prescription to the case of observables which do not commute with Hamiltonian. For this, one will consider explicitly the time dependence of resonance states in the quasi-stationary approximation and will connect complex observables to the latter. Indeed, Gamow states are supposed to be long-lived, so that their time-dependent wave function can be written in time-independent approach as the product of space and time components:

$$\Psi(x, t) = \Phi(x)e^{i\tilde{E}t/\hbar} \tag{5.4}$$

$$H|\Phi\rangle = \tilde{E}|\Phi\rangle , \tag{5.5}$$

where $\Psi(x, t)$ is a quasi-stationary time-dependent state, depending on $x$, representing all space coordinates, and time $t$. In this expression, $\Phi(x)$ is a spatial part of $\Psi(x, t)$ and an eigenstate of $\hat{H}$, and $e^{iEt/\hbar}$ is a time-dependent part characterized by a complex energy $\tilde{E}$. Clearly, $|\Psi(x, t)| = |\Phi(x)|$ if $\tilde{E}$ is real, so that complex energies are necessary if one wants to formally use the time-independent approach for resonant states. In this case, one has:

$$\tilde{E} = E - i\Gamma/2 , \tag{5.6}$$

where the real and imaginary parts of $\tilde{E}$ have been introduced. $\Psi(x, t)$ then exhibits the typical exponential decay law:

$$|\Psi(x, t)|^2 = e^{-\Gamma t/\hbar} . \tag{5.7}$$

Hence, the interpretation of $E$ and $\Gamma$ as energy and width of the state, respectively, is justified, and one can define the half-life of the state correspondingly:

$$T_{1/2} = \frac{\hbar}{\ln(2)\Gamma} . \tag{5.8}$$

The time-independent form of Eq. (5.4) implies that the probability density of $\Psi$ increases exponentially. Indeed, as $\Psi(x, t)$ is decaying, its wave function will locally vanish. In order for the number of particles to remain constant, the probability to find a particle has to increase in space, which can be understood semi-classically as a particle leaving the nucleus at a rate proportional to $\Gamma$. If $\Psi(x, t)$ had been calculated in a time-dependent approach, the probability of presence of the associated particle would be equal to zero beyond a given radius, which goes to infinity when $t \to +\infty$. Consequently, Eq. (5.4) is an idealization of the exact time-dependent state for $t \to +\infty$, so that a formal time independence implies that the memory of the formation of the composite has been lost, which is consistent with its long lifetime. Note that Eq. (5.4) implies above all that the spatial evolution of $\Psi(x, t)$ is very slow in a considered interval of time so that it can be neglected.

A system of a finite lifetime cannot have a well-defined energy. Hence, the fact that $\Psi(x, t)$ is an eigenstate of $\hat{H}$ (see Eq. (5.5)) seems then contradictory. This contradiction can be resolved by noting that the width is also the uncertainty of the resonance energy. Consequently, the eigenvalue equation of Eq. (5.5 has to be understood within the quasi-stationary approach, i.e., that $\Psi(x, t)$ has an expectation value of energy equal to $E$, while $\Gamma$ is the uncertainty of the measurement generated by its finite lifetime. Complex-valued observable quantities arise because Gamow states are normalized using the square of their wave function and not their modulus square (see Sect. 3.5). This demands the reinterpretation of expectation values as they have an additional imaginary part which is absent when dealing with states of the Hilbert space. We have shown that the use of complex energies in Eq. (5.6) has a well-defined physical interpretation. Berggren has extended this interpretation to the constants of motions, i.e., operators which commute with $\hat{H}$ [9]. Indeed, they can be expected to follow the same rules as for $\tilde{E}$, so that their real part is the average of all measurements, while their imaginary part is related to their imprecision, as will be seen in this section. This analysis can be applied to other observables if one makes a clear distinction between the uncertainties generated by the quantum nature of the state only and by its finite lifetime. To explain the situation, let us first consider the example of root-mean-square radius. The fundamental point therein is that it is well defined only for bound states. Indeed, as the wave function of a resonance state spreads in space, the value of the root-mean-square radius will change over time as well, and will eventually lose significance after decay.

Let us now consider the general case. It is useful therein to compare the same observable $\hat{A}$ calculated for a resonance state $\Psi(x, t)$ within a standard time-dependent formalism and Berggren formalism. In the standard time-dependent formalism, the expectation value of $\hat{A}$ is a real function of $t$:

$$A(t) = \langle \Psi | \hat{A} | \Psi \rangle = \langle \Phi | \hat{A} | \Phi \rangle \, e^{-\Gamma t / \hbar} \, , \qquad (5.9)$$

where one has assumed that Eq. (5.4), describing a quasi-stationary state, is valid. One also supposes that $A(t) \neq 0$, which implies that $\langle \Phi | \hat{A} | \Phi \rangle \neq 0$. The time-dispersion rate of $A(t)$ over a time interval $[0 : T_0]$ with $T_0 \ll T$ reads:

$$\Delta_t(A) = \frac{1}{T_0} \sqrt{\overline{A(t)^2} - \overline{A(t)}^2} \simeq \frac{\Gamma}{\hbar \sqrt{12}} \, | \langle \Phi | \hat{A} | \Phi \rangle | \, , \qquad (5.10)$$

where the time average of a function $f(t)$ is standard:

$$\overline{f(t)} = \frac{1}{T_0} \int_0^{T_0} f(t) \, dt \, , \qquad (5.11)$$

and where one has considered $\Gamma T_0 \ll 1$. It is important to state that Eq. (5.10) is different from the standard quantum dispersion of an operator, equal to:

$$\langle \Psi | \Delta \hat{A} | \Psi \rangle = \sqrt{\langle \Psi | \hat{A}^2 | \Psi \rangle - \langle \Psi | \hat{A} | \Psi \rangle^2} \, . \qquad (5.12)$$

Indeed, it is generally nonzero for bound states, contrary to $\Delta_t(A)$. In fact, one will see that it is comparable to $\Delta_t(A)$ only for constants of motion.

Let us now turn to the Berggren formalism. In this approach using the biorthogonal basis (see Sect. 5.1.1), Eq. (5.9) becomes:

$$\langle \widetilde{\Phi} | \hat{A} | \Phi \rangle = A_r + i A_i \, , \qquad (5.13)$$

where $A_r$ of $A_i$ are the real and imaginary parts of the resulting expectation value. When $\Gamma = 0$, the time-dependent and Berggren formalisms are identical if one considers the real part $A_r$ (see Eq. (5.13)) as the physical observable. For the particular case of constants of motion, where $[H, A] = 0$, then replacing complex expectation values as appearing in Eq. (5.13) by their real parts, one obtains [9]:

$$\langle \widetilde{\Phi} | \Delta \hat{A} | \Phi \rangle \simeq \sqrt{\Re[(A_r + i A_i)^2] - A_r^2} = i A_i \, , \qquad (5.14)$$

where the imaginary character of the expectation value only reflects the non-hermiticity of observables therein. $\langle \widetilde{\Phi} | \Delta \hat{A} | \Phi \rangle$ in (5.14) is comparable to $\Delta_t(A)$ (see Eq. (5.10)) in this case, as they become proportional when $\Gamma \ll E$.

There is no such simple relation between $A_i$ and $\hat{A}$ in the general case. However, $A_i$ is proportional to $\Gamma$ if $\Gamma \ll E$, which can be seen from the Taylor expansion of

$\langle \widetilde{\Phi} | \hat{A} | \Phi \rangle$ (see Eq. (5.13)):

$$\langle \widetilde{\Phi} | \hat{A} | \Phi \rangle = A(\tilde{E}) \simeq A(E) - i \frac{\Gamma}{2} A'(E) , \qquad (5.15)$$

so that $A_r \simeq A(E)$ and $A_i \simeq -(\Gamma/2) A'(E)$. Moreover, $\Delta_t(A) \propto \Gamma$ at first-order in $\Gamma$ if $\Gamma \ll E$ (see Eq. (5.10)). Consequently, at leading order, $A_r$ can be interpreted as the average value of measurements made at times $t \ll T$, whereas $A_i$ can be seen as its weighted dispersion rate over time:

$$A_r = \overline{A(t)} \qquad (5.16)$$

$$A_i = -\hbar\sqrt{3} \left( \frac{A'(E)}{|A(E)|} \right) \Delta_t(A) . \qquad (5.17)$$

Consequently, the interpretation of complex observables in a quasi-stationary formalism is formally clear. Indeed, both $A_r$ and $A_i$ can be in principle experimentally measured. However, as the values of cross sections involving unbound states are much smaller than those involving stable nuclei, a statistical analysis of observables related to resonance states seems very difficult to obtain. Moreover, electromagnetic and beta decays typically have time scales which are much longer than those of particle emission, so that they are very unlikely to precede particle decay. Thus, it seems that Eqs. (5.16) and (5.17) are primarily of theoretical interest, and that their experimental study would demand very precise experimental instruments, which are not available at present.

### 5.1.3 Many-Body Berggren Completeness Relation

The discretized basis (3.82) can be a starting point for establishing the completeness relation in the many-body case, in a full analogy with the standard shell model in a complete discrete basis, e.g., the harmonic oscillator basis. Indeed, the Berggren completeness relation is discretized with the Gauss-Legendre quadrature, so that it is formally identical to a discrete complete set of states (see Sect. 3.5). One can then derive the many-body completeness relation:

$$\sum_n |\Psi_n\rangle \langle \widetilde{\Psi}_n| \simeq \hat{1} . \qquad (5.18)$$

The $N$-body Slater determinants $|\Psi_n\rangle$ have the form $|\phi_1 \ldots \phi_N\rangle$, where $|\phi_k\rangle$ are resonant (bound and decaying) and scattering (contour) single-particle states. The approximate equality in (5.18) is an obvious consequence of the continuum discretization, similarly as in Eq. (3.82), so that the exact completeness relation is restored when the number of discretized basis scattering states becomes infinite. Like in the case of single-particle Gamow states, the normalization of the Gamow vectors in the configuration space is given by the squares of shell-model amplitudes:

$$\sum_n c_n^2 = 1 , \tag{5.19}$$

and not by the squares of their absolute values.

In the particular case of two-particle states, the completeness relation reads:

$$\sum_{i_1, i_2} |\phi_{i_1} \phi_{i_2}\rangle \langle \widetilde{\phi_{i_1} \phi_{i_2}}| \simeq \hat{1} . \tag{5.20}$$

This relation can be used to calculate two-body matrix elements (see Exercise I for a numerical study of the two-body Berggren completeness relation).

The domain of the completeness relation (5.20) is not trivial, as is already the case for the one-body completeness relation (see Sect. 3.5). Indeed, as one is no longer in the Hilbert space of integrable functions, the form of contours define the possible asymptotic behavior that many-body states can bear. As seen in Sect. 3.5, the one-body completeness relation allows to expand integrable one-body states and one-body-states of exponential asymptote $e^{ikr}$, with $k$ in the zone between the real axis and the $L^+$ contour. Consequently, Eq. (5.20) allows to expand many-body states where nucleon wave functions can have an asymptotic behavior of the form $e^{ikr} Y_{\ell jm}(\theta, \varphi)$, with $k$ belonging to the domain of applicability of the Berggren basis of the considered $(\ell, j)$ partial wave. This clearly limits the value of the width that a resonance can have. In particular, complex-energy neutron $\ell = 0$ scattering states can influence the value of the many-body width due to continuum coupling, even though no resonance state can exist in the neutron $\ell = 0$ partial wave.

Starting from this section, one will only use the Berggren formalism to calculate matrix elements. Consequently, in order to simplify notation, the tilde sign above bra states will no longer be written and will be implicit in the following.

---

**Exercise I ★**

One will study the numerical precision of the many-body Berggren completeness relation in practical calculations by considering a nucleus with two valence neutrons.

**A.** Run the two-nucleon Gamow shell model code using the modified surface Gaussian interaction (see Eq. (9.174)) for $^6$He to obtain its $0^+$ ground state and $2^+$ excited state. Check numerical accuracy by augmenting up to a factor 2 the number of discretized scattering states along the $p_{3/2}$ contour of the Berggren basis. Notice that energies and widths are stable with respect to this change.

**B.** Let us consider a $p_{3/2}$ contour peak very close to the $0p_{3/2}$ resonance state of basis, but still below it so that the Cauchy theorem is fulfilled. Run the code in these conditions and note the difference with initially obtained eigenstates. Redo the same calculation by considering now a $p_{3/2}$ contour peak further down in the complex plane by a factor larger than 10 compared to the initial peak.

Notice that the $^6$He eigenstates are poorly described with Berggren basis contours either too close or too far from the $0p_{3/2}$ basis resonance state for a fixed contour discretization, whereas the initial contour provided with very accurate results.

Show that increasing the number of discretized scattering states does not ameliorate the situation when the contour is too close to the $0p_{3/2}$ resonance state.

Alternatively, show that the number of discretized scattering states needed to obtain an acceptable precision becomes prohibitively large when the contour is too far from the $0p_{3/2}$ resonance state.

C. Discuss the imprecision induced by the discretization of Berggren scattering contours and phase shift variation along the $p_{3/2}$ contour.

Explain why optimal contours exist for the numerical implementation of the Berggren completeness relation, with which the number of discretized points needed to have precise results is minimized.

## 5.1.4 Complex Observables: The Generalized Variational Principle

The Gamow shell model Hamiltonian is represented by a complex symmetric matrix, which possess in general complex eigenvalues. The most important of them are those corresponding to resonance states, as the real part of the eigenvalue is interpreted as its energy and its imaginary part as minus half its width (see Sect. 2.6.6). However, one cannot assess for the moment the validity of the calculation of a many-body resonance state in a finite model space. Indeed, as its eigenvalue is complex, the variational principle cannot apply. Another method to state how precise the calculated eigenvector is compared to the exact eigenstate of the Hamiltonian is thus needed. For this, one relies on the generalized variational principle.

Indeed, the analytic continuation of the equations of the variational principle leads to the Gamow shell model eigenproblem in a finite model space. As one deals with complex energies, one will firstly vary their squared modulus in order to find the stationary point a real-valued function:

$$\delta(|E(\Psi)|^2) = 0$$
$$\Leftrightarrow E^*(\Psi)\delta E(\Psi) + E(\Psi)\delta E^*(\Psi) = 0 . \tag{5.21}$$

what implies

$$\delta E(\Psi) = 0$$

and therefore:

$$\delta \left( \frac{\langle \Psi | H | \Psi \rangle}{\langle \Psi | \Psi \rangle} \right) = 0 \, , \tag{5.22}$$

where $E(\Psi)$ is the expectation value of the Hamiltonian in $|\Psi\rangle$, which can always be nonzero, and where one has used the fact that $\delta E(\Psi)$ and $\delta E^*(\Psi)$ can be considered as independent.

As Eq. (5.22) is formally the same as for real-energy eigenstates, one can derive the equation verified by $|\Psi\rangle$ using the same technique as for real-energy ground states:

$$\delta \left( \frac{\langle \Psi | H | \Psi \rangle}{\langle \Psi | \Psi \rangle} \right) = 0$$

$$\Leftrightarrow \langle \delta \Psi | H | \Psi \rangle - E \langle \delta \Psi | \Psi \rangle = 0$$

$$\Leftrightarrow \langle \delta \Psi | H - E | \Psi \rangle = 0 \, , \tag{5.23}$$

where $E = E(\Psi)$.

---

**Exercise  II** ⋆⋆
One will illustrate now in a numerical example involving a nucleus with two valence nucleons that the generalized variational principle holds in practical calculations.

**A.** Run the two-body Gamow shell model code using the modified surface Gaussian interaction (see Eq. (9.174)) for $^{18}O$ to calculate the low energy observables: eigenstates, densities, and electromagnetic transitions. Notice how the generalized variational principle acts on the convergence of energies and widths of unbound states. Explain why energies do not converge in the same way as for bound states in the standard shell model, i.e., by increasing binding energy when the model space is enlarged by adding more Lanczos or Jacobi-Davidson vectors.

**B.** Notice in numerical studies that observables involving only bound states are always real, and that those involving at least one resonance state are complex. Explain from completeness arguments why the observables involving only bound states are not complex, even though basis states are complex in general.

**C.** Notice that narrow resonance states have observables whose imaginary part is very small, whereas broad resonance states bear observables with sizable imaginary parts. Explain qualitatively these values in terms of the Berggren interpretation of complex observables of Sect. 5.1.

---

Equation (5.23) is equivalent to the Hamiltonian matrix eigenproblem, as it must be fulfilled for all variations of $|\Psi\rangle$, denoted as $|\delta\Psi\rangle$. Consequently, even though $E$

is complex, $E$ is stationary with respect to small variations of $|\Psi\rangle$, so that it should not vary significantly in the vicinity of the exact energy of the eigenstate in the complex plane. As a consequence, the generalized variational principle allows to prove that the resonance states calculated from the diagonalization of Hamiltonian matrices in a finite model space are as precise as bound states (see Exercise II for a numerical illustration of the generalized variational principle).

## 5.2 Translationally Invariant Shell Model Scheme: The Cluster Orbital Shell Model

In the standard shell model, it is customary to remove center-of-mass excitations using the Lawson method [11]. In the Gamow shell model, however, this method is precluded because Berggren basis states are not eigenstates of the harmonic oscillator potential. Therefore, in order to eliminate center-of-mass excitations, while avoiding the numerical difficulties arising from the use of Jacobi coordinates in describing nuclei with many valence nucleons, one chose to work in the cluster orbital shell model framework [12]. In cluster orbital shell model, one considers a core plus valence particles, so that coordinates of valence particles are defined relatively to the center-of-mass of the core. This allows to work in a translationally invariant many-body framework.

Let us define the cluster orbital shell model coordinates:

$$\mathbf{r}_i = \mathbf{r}_{i,\text{lab}} - \mathbf{R}_{\text{CM,core}} \quad \text{if } i \in \text{val} \tag{5.24}$$

$$\mathbf{r}_i = \mathbf{r}_{i,\text{lab}} \quad \text{if } i \in \text{core} \tag{5.25}$$

where $\mathbf{r}_{i,\text{lab}}$ is the coordinate of a nucleon in the laboratory system and

$$\mathbf{R}_{\text{CM,core}} = \frac{1}{M_{\text{core}}} \sum_{i \in \text{core}} m_i \, \mathbf{r}_{i,\text{lab}} \tag{5.26}$$

is the coordinate of the center-of-mass of the core. In this expression, $m_i$ is the mass of the $i$th particle and $M_{\text{core}} = \sum_{i \in \text{core}} m_i$ is the mass of the core. The cluster orbital shell model momentum reads:

$$\mathbf{p}_i = -i\hbar \nabla_i , \tag{5.27}$$

where $\nabla_i$ is the gradient associated to $\mathbf{r}_i$.

It is possible to write the momentum in the laboratory frame $\mathbf{p}_{i,\text{lab}}$ as a function of the cluster orbital shell model operator $\mathbf{p}_i$ of Eq. (5.27). For this, the partial derivative operator $\dfrac{\partial}{\partial x_{i,\text{lab}}}$ needs to be expressed in cluster orbital shell model coordinates:

- For $i \in$ core:

$$\frac{\partial}{\partial x_{i,\text{lab}}} = \sum_j \frac{\partial x_j}{\partial x_{i,\text{lab}}} \frac{\partial}{\partial x_j} = \frac{\partial}{\partial x_i} - \sum_{j \in \text{val}} \left( \frac{m_i}{M_{\text{core}}} \right) \frac{\partial}{\partial x_j} \qquad (5.28)$$

- For $i \in$ val:

$$\frac{\partial}{\partial x_{i,\text{lab}}} = \sum_j \frac{\partial x_j}{\partial x_{i,\text{lab}}} \frac{\partial}{\partial x_j} = \frac{\partial}{\partial x_i} \qquad (5.29)$$

Equations for $\partial / \partial y_{i,\text{lab}}$ and $\partial / \partial z_{i,\text{lab}}$ are obtained by substituting $x$ by $y$ and $z$. As a consequence, the expression of $\mathbf{p}_i$ in cluster orbital shell model reads:

- For $i \in$ core:

$$\mathbf{p}_{i,\text{lab}} = \mathbf{p}_i - \sum_{j \in \text{val}} \left( \frac{m_i}{M_{\text{core}}} \right) \mathbf{p}_j \qquad (5.30)$$

- For $i \in$ val:

$$\mathbf{p}_{i,\text{lab}} = \mathbf{p}_i \qquad (5.31)$$

One can then calculate the center-of-mass linear momentum $\mathbf{P}_{\text{lab}}$ in cluster orbital shell model coordinates from Eq. (5.31):

$$\mathbf{P}_{\text{lab}} = \sum_{i \in \text{core}} \mathbf{p}_{i,\text{lab}} + \sum_{i \in \text{val}} \mathbf{p}_{i,\text{lab}} = \sum_{i \in \text{core}} \mathbf{p}_i - \sum_{\substack{i \in \text{core} \\ j \in \text{val}}} \left( \frac{m_i}{M_{\text{core}}} \right) \mathbf{p}_j + \sum_{i \in \text{val}} \mathbf{p}_i$$

$$= \sum_{i \in \text{core}} \mathbf{p}_i - \sum_{j \in \text{val}} \mathbf{p}_j + \sum_{i \in \text{val}} \mathbf{p}_i = \sum_{i \in \text{core}} \mathbf{p}_i \qquad (5.32)$$

Interestingly, $\mathbf{P}_{\text{lab}}$ is a function of core linear momenta only.

In order to deal with the kinetic energy operator, one will derive the Laplacian in cluster orbital shell model coordinates using Eqs. (5.30) and (5.31):

- For $i \in$ core:

$$\mathbf{p}_{i,\text{lab}}^2 = \mathbf{p}_i^2 + \sum_{j,j' \in \text{val}} \left( \frac{m_i}{M_{\text{core}}} \right)^2 \mathbf{p}_j \cdot \mathbf{p}_{j'} - 2 \sum_{j \in \text{val}} \left( \frac{m_i}{M_{\text{core}}} \right) \mathbf{p}_i \cdot \mathbf{p}_j \qquad (5.33)$$

- For $i \in$ val:

$$\mathbf{p}_{i,\text{lab}}^2 = \mathbf{p}_i^2 \qquad (5.34)$$

It is now possible to calculate the kinetic energy part of the Hamiltonian in the frame of the cluster orbital shell model, where the center-of-mass kinetic part is taken into account:

$$\sum_{i=1}^{A} \frac{\mathbf{p}_{i,\text{lab}}^2}{2m_i} - \frac{\mathbf{p}_{\text{lab}}^2}{2M} = \sum_{i \in \text{val}} \frac{\mathbf{p}_i^2}{2\mu_i} + \frac{1}{M_{\text{core}}} \sum_{i<j \in \text{val}} \mathbf{p}_i \cdot \mathbf{p}_j$$

$$+ \sum_{i \in \text{core}} \frac{\mathbf{p}_i^2}{2\mu_i'} - \frac{1}{M} \sum_{i<j \in \text{core}} \mathbf{p}_i \cdot \mathbf{p}_j - \frac{1}{M_{\text{core}}} \sum_{i \in \text{core}} \mathbf{p}_i \cdot \sum_{j \in \text{val}} \mathbf{p}_j \,.$$

$$(5.35)$$

In this expression, $M$ is the total mass, and $\mu_i$, $\mu'_i$ are reduced masses given by:

$$\frac{1}{\mu_i} = \frac{1}{m_i} + \frac{1}{M_{\text{core}}} \,, \tag{5.36}$$

and

$$\frac{1}{\mu_i'} = \frac{1}{m_i} - \frac{1}{M} \,. \tag{5.37}$$

The couplings between core and valence particles vanish because the core is coupled to $0^+$ and $\sum_{i \in \text{core}} \mathbf{p}_i$ is a spherical tensor of rank one.

Two-body matrix elements involving either only one-body states of the core, or only one-body states of the valence space, are straightforward to calculate in the cluster orbital shell model. Indeed, one obtains from Eqs. (5.24), (5.25), (5.33), and (5.34):

$$\mathbf{r}_{i,\text{lab}} - \mathbf{r}_{j,\text{lab}} = \mathbf{r}_i - \mathbf{r}_j \tag{5.38}$$

$$\mathbf{p}_{i,\text{lab}} - \mathbf{p}_{j,\text{lab}} = \mathbf{p}_i - \mathbf{p}_j \,, \tag{5.39}$$

so that standard shell model methods can be used to calculate associated two-body matrix elements.

However, the interaction matrix elements in which one valence state $|i\rangle$ and one core state $|j\rangle$ occur in both bra and ket states of the nuclear interaction are cumbersome to calculate. Indeed, in this case, Eqs. (5.24), (5.25), (5.33), and (5.34) provide with the following equalities:

$$\mathbf{r}_{i,\text{lab}} - \mathbf{r}_{j,\text{lab}} = \mathbf{r}_i - \mathbf{r}_j + \mathbf{R}_{\text{CM,core}} \tag{5.40}$$

$$\mathbf{p}_{i,\text{lab}} - \mathbf{p}_{j,\text{lab}} = \mathbf{p}_i - \mathbf{p}_j + \sum_{j' \in \text{val}} \left( \frac{m_j}{M_{\text{core}}} \right) \mathbf{p}_{j'} \,. \tag{5.41}$$

In fact, the two-body interaction in laboratory coordinates becomes the $A$-body interaction in cluster orbital shell model coordinates.

A solution to this problem can be found by demanding all couplings between core and valence spaces to vanish. In practice, in order to impose these conditions, it is necessary to define effective interactions directly with cluster orbital shell model coordinates. As realistic interactions are defined in a no-core picture, where all nucleons interact, the use of realistic interactions in the cluster orbital shell model seems to be impossible.

One can solve this problem by firstly defining an effective interaction derived from a realistic interaction, with core and valence parts calculated in the laboratory frame, so that couplings between core and valence spaces vanish. To illustrate this procedure, let us consider an effective Hamiltonian: $\hat{H} = T + U + V$, where $U$ is a one-body potential and $V$ is a two-body interaction, calculated from a realistic interaction in the laboratory frame. The core part of the Hamiltonian is not considered as the core is inert. In cluster orbital shell model coordinates, $\hat{H}$ is:

$$\hat{H} = \hat{T} + \hat{U} + \hat{V} = \sum_{i \in \text{val}} \left( \frac{\mathbf{p}_i^2}{2\mu_i} + \hat{U}_i(\mathbf{r}_i) \right) + \frac{1}{M_{\text{core}}} \sum_{(i<j) \in \text{val}} \mathbf{p}_i \cdot \mathbf{p}_j$$

$$+ \sum_{(i<j) \in \text{val}} \hat{V}_{i,j} \ . \tag{5.42}$$

The term involving $V_{i,j}$ is translationally invariant, so that using laboratory or cluster orbital shell model coordinates leads to the same results (see Eqs. (5.38) and (5.39)). The one-body part in Eq. (5.42) consists of $\hat{U}_i(\mathbf{r}_i)$ , which replaces the $\hat{U}_i(\mathbf{r}_{i,\text{lab}})$ operator defined with laboratory coordinates. The one-body part is indeed considered approximately, because the recoil terms induced by the change $\mathbf{r}_{i,\text{lab}} \to \mathbf{r}_i$ are neglected.

In fact, this approximation leads to a very small error. In order to show this, let us separate $\hat{U}_i$ in central and spin-orbit parts:

$$\hat{U}_i(\mathbf{r}_{i,\text{lab}}) = \hat{U}_i^{(C)}(\mathbf{r}_{i,\text{lab}}) + \hat{U}_i^{(LS)}(\mathbf{r}_{i,\text{lab}})(\mathbf{l}_{i,\text{lab}} \cdot \mathbf{s}_i) \ . \tag{5.43}$$

One has for the central part:

$$\hat{U}_i^{(C)}(\mathbf{r}_{i,\text{lab}}) = \hat{U}_i^{(C)}(\mathbf{r}_i) + \mathbf{R}_{\text{CM,core}} \cdot \nabla \hat{U}_i^{(C)}(\mathbf{r}_i)$$

$$+ \frac{1}{2} \mathbf{R}_{\text{CM,core}} \cdot \Delta \hat{U}_i^{(C)}(\mathbf{r}_i) \cdot \mathbf{R}_{\text{CM,core}} + \text{h.o.} , \tag{5.44}$$

where $\Delta \hat{U}_i^{(C)}$ is the Hessian matrix associated to $\hat{U}_i^{(C)}$ and higher order terms (h.o.) are neglected. The first-order term of Eq. (5.44) vanishes in many-body calculations because the core is coupled to $0^+$ and $\mathbf{R}_{\text{CM,core}}$ is of rank one. Similarly, the spin-orbit part of $U_i$ reads:

$$
\begin{aligned}
\hat{U}_i^{\text{(L S)}} & (\mathbf{r}_{i,\text{lab}})(\mathbf{l}_{i,\text{lab}} \cdot \mathbf{s}_i) \\
&= \hat{U}_i^{\text{(L S)}}(\mathbf{r}_{i,\text{lab}}) \left[ (\mathbf{r}_i + \mathbf{R}_{\text{CM,core}}) \times \mathbf{p}_i \right] \cdot \mathbf{s}_i \\
&= \hat{U}_i^{\text{(L S)}}(\mathbf{r}_i)(\mathbf{l}_i \cdot \mathbf{s}_i) \\
&\quad + (\mathbf{R}_{\text{CM,core}} \cdot \nabla \hat{U}_i^{\text{(L S)}}(\mathbf{r}_i))(\mathbf{l}_i \cdot \mathbf{s}_i) + \hat{U}_i^{\text{(L S)}}(\mathbf{r}_i) \left[ (\mathbf{R}_{\text{CM,core}} \times \mathbf{p}_i) \cdot \mathbf{s}_i \right] \\
&\quad + \frac{1}{2}(\mathbf{R}_{\text{CM,core}} \cdot \Delta \hat{U}_i^{\text{(L S)}}(\mathbf{r}_i) \cdot \mathbf{R}_{\text{CM,core}})(\mathbf{l}_i \cdot \mathbf{s}_i) \\
&\quad + (\mathbf{R}_{\text{CM,core}} \cdot \nabla \hat{U}_i^{\text{(L S)}}(\mathbf{r}_i)) \cdot \left[ (\mathbf{R}_{\text{CM,core}} \times \mathbf{p}_i) \cdot \mathbf{s}_i \right] + \text{h.o.} ,
\end{aligned} \tag{5.45}
$$

where the first-order terms of Eq. (5.45) vanishes for the same reason as for the central part and higher order terms are neglected as well.

The error made in the considered approximation is of the order of $\Delta U / M_{\text{core}}$. Indeed, the core matrix elements involving $\mathbf{R}_{\text{CM,core}}^2$ are of the order of $1/M_{\text{core}}$, which can be checked analytically using harmonic oscillator wave functions. Moreover, the Laplacian of a Woods-Saxon potential is about five times smaller than the potential itself, and the derivative of the spin-orbit part of the one-body potential is even smaller. The relative error on binding energy is of the order of 5% when the $\alpha$-particle core is used, which is the lightest core in the cluster orbital shell model applications [13]. Hence in practical calculations, the error resulting from neglecting the recoil term induced by the one-body term in Eq. (5.42) will not exceed few percents.

However, there remains an important point related to the use of cluster orbital shell model coordinates. As core and valence nucleons are not treated symmetrically, the antisymmetrization of many-body wave functions becomes difficult to impose. This problem can be solved by noting that couplings between core and valence nucleons always vanish using cluster orbital shell model. As core and valence nucleons do not interact, the core and valence parts of many-body wave functions do not have to be antisymmetrized. For this approximation to be consistent, it is sufficient to demand that core states are orthogonal to valence states. This procedure is called the orthogonality condition model [14]. In the calculation of an observable, the operator associated to this observable must be calculated with its recoil term included before being put to cluster orbital shell model form. Below, we will show that the recoil terms of the operators providing mean-square radius, electromagnetic, and beta transitions are either exactly treated or negligible.

Root-mean-square radius with the recoil correction included reads:

$$
R_{v;\text{rms}} = \left( \frac{1}{N_v} \sum_v (\mathbf{r}_{v,\text{lab}} - \mathbf{R}_{\text{CM}})^2 \right)^{1/2} , \tag{5.46}
$$

where $v$ is proton or neutron, $N_v$ is the number of protons $(Z)$ or neutrons $(N)$, $v$ runs over all protons or neutrons, and:

$$\mathbf{R}_{CM} = \frac{m_p}{M} \sum_p \mathbf{r}_{p,\text{lab}} + \frac{m_n}{M} \sum_n \mathbf{r}_{n,\text{lab}} . \tag{5.47}$$

is the center-of-mass of the nucleus. $\mathbf{R}_{CM}$ also bears a concise expression using cluster orbital shell model coordinates:

$$\mathbf{R}_{CM} = \mathbf{R}_{CM,\text{core}} + \frac{m_p}{M} \sum_{p \in \text{val}} \mathbf{r}_p + \frac{m_n}{M} \sum_{n \in \text{val}} \mathbf{r}_n , \tag{5.48}$$

where one has used Eqs. (5.24–5.26). The total root-mean-square radius is a simple function of the proton and neutron root-mean-square radii:

$$R_{\text{rms}} = \left( \frac{Z}{A} R_{\text{p;rms}}^2 + \frac{N}{A} R_{\text{n;rms}}^2 \right)^{1/2} . \tag{5.49}$$

Rewriting Eq. (5.46) in cluster orbital shell model coordinates, one obtains after a tedious but straightforward calculation:

$R_{v;\text{rms}}$

$$= \left( \frac{N_{v\,\text{core}}}{N_v} R_{\text{core}-v;\text{rms}}^2 + \sum_{v \in \text{val}} \left( \frac{1}{N_v} - \frac{2\,m_v}{M\,N_v} + \frac{m_v^2}{M^2} \right) r_v^2 + \sum_{\mu \in \text{val}} \left( \frac{m_\mu^2}{M^2} \right) r_\mu^2 \right.$$

$$+ \sum_{(v<v') \in \text{val}} \left( 2 \frac{m_v^2}{M^2} - \frac{4\,m_v}{M\,N_v} \right) \mathbf{r}_v \cdot \mathbf{r}_{v'} + \sum_{(\mu<\mu') \in \text{val}} \left( 2 \frac{m_\mu^2}{M^2} \right) \mathbf{r}_\mu \cdot \mathbf{r}_{\mu'}$$

$$\left. + \sum_{(v,\mu) \in \text{val}} \left( 2 \frac{m_v m_\mu}{M^2} - \frac{2\,m_\mu}{M\,N_v} \right) \mathbf{r}_v \cdot \mathbf{r}_\mu + \text{h.o.} \right)^{1/2} \tag{5.50}$$

where Eq. (5.48) has been used. In this expression, $\mu$ is neutron (proton) if $v$ is proton (neutron), $m_v$ stands for the mass of the $v$ particle (same for $\mu$), $M$ is the mass of the nucleus, and $R_{\text{core}-v;\text{rms}}^2$ is the proton or neutron root-mean-square radius of the core, whose value is taken from experimental data. The action of higher order terms vanishes because they are non-scalar operators in core space. Consequently, $R_{v;\text{rms}}$ can be exactly calculated in cluster orbital shell model coordinates.

---

**Exercise  III ⋆ ⋆ ⋆**

One will demonstrate that the recoil terms occurring in Eqs. (5.57–5.59) vanish when using cluster orbital shell model coordinates.

**A.** Explain why it is sufficient to consider $M_L = L$ in Eqs. (5.57–5.59).

**B.** Let us consider electric transitions. Show that Eq. (5.57) can be written as:

$$r_{\text{lab}}^L \, Y_{LL}(\Omega_{\text{lab}}) \propto (x_{\text{lab}} + i \, y_{\text{lab}})^L \propto \sum_{k=0}^{L} \binom{L}{k} (x + iy)^{L-k} (X_{\text{core}} + i \, Y_{\text{core}})^k$$

(5.51)

Explain why all recoil terms vanish in Eq. (5.51).

**C.** Let us now deal with magnetic transitions. Show that Eqs. (5.58) and (5.59) respectively read:

$$r_{\text{lab}}^{L-1} Y_{L-1,L-1}(\Omega_{\text{lab}}) \, l_{\text{lab}1}$$

(5.52)

$$r_{\text{lab}}^{L-1} Y_{L-1,L-1}(\Omega_{\text{lab}}) \, s_1$$

(5.53)

where $l_{\text{lab}1}$ and $s_1$ are the first covariant components of the spherical tensors $\mathbf{l}_{\text{lab}}$ and $\mathbf{s}$, respectively. Explain why Eq. (5.53) can be treated as Eq. (5.51).

**D.** One will show that the recoil terms obtained from Eq. (5.52) vanish identically. One recalls that $V_1 \propto V_x + iV_y$ for a spherical tensor $V$. Demonstrate the following expression:

$$l_{\text{lab}1} = l_1 + c(Z_{\text{core}} (p_x + ip_y) - (X_{\text{core}} + iY_{\text{core}}) p_z),$$

(5.54)

where $c$ is a constant. Deduce that Eq. (5.52) verifies:

$$r_{\text{lab}}^{L-1} Y_{L-1,L-1}(\Omega_{\text{lab}}) \, l_{\text{lab}1}$$

$$\propto \sum_{k=0}^{L-1} \binom{L-1}{k} (x + iy)^{L-1-k} (X_{\text{core}} + i \, Y_{\text{core}})^k \, l_1$$

$$- c \sum_{k=0}^{L-1} \binom{L-1}{k} (x + iy)^{L-1-k} (X_{\text{core}} + i \, Y_{\text{core}})^{k+1} \, p_z$$

$$+ c \sum_{k=0}^{L-1} \binom{L-1}{k} (x + iy)^{L-1-k} (X_{\text{core}} + i \, Y_{\text{core}})^k \, Z_{\text{core}} (p_x + ip_y).$$

(5.55)

Explain why the first and second term of Eq. (5.55) can be treated as in Eq. (5.51). To calculate the third term of Eq. (5.55, demonstrate the following relations:

$$(X_{\text{core}} + i \, Y_{\text{core}})^k \, Z_{\text{core}} \propto R_{\text{core}}^{k+1} \, Y_{kk}(\Omega_{\text{core}}) \, Y_{10}(\Omega_{\text{core}})$$

$$\propto R_{\text{core}}^{k+1} \sum_{\substack{k'=|k-1| \\ k'>k}}^{k+1} a_{kk'} \, Y_{kk'}(\Omega_{\text{core}})$$

(5.56)

where $a_{kk'}$ is a constant. Deduce that recoil terms in Eqs. (5.52) and (5.53)
cancel out.
E. Draw the conclusion that Eqs. (5.57–5.59) lead to the same results whether
laboratory or cluster orbital shell model coordinates are used, i.e., that recoil
corrections vanish.

Electromagnetic transitions of order $L$ in the long-wavelength approximation are
functions of the following operators:

$$r_{\text{lab}}^{L}\mathbf{Y}_L(\Omega_{\text{lab}}) \tag{5.57}$$

$$r_{\text{lab}}^{L-1}[\mathbf{Y}_{L-1}(\Omega_{\text{lab}}) \otimes \mathbf{l}_{\text{lab}}]^{L} \tag{5.58}$$

$$r_{\text{lab}}^{L-1}[\mathbf{Y}_{L-1}(\Omega_{\text{lab}}) \otimes \mathbf{s}]^{L} . \tag{5.59}$$

One will show in Exercise III that in this approximation, all recoil terms in
electromagnetic transitions vanish in cluster orbital shell model. This exercise is
difficult and can be omitted without creating problems of comprehension in the
following.

Without the long wavelength approximation, the recoil corrections lead in
general to the many-body operators which are cumbersome to calculate. It is,
however, possible to calculate the recoil corrections exactly if the wave function
of the core is built from harmonic oscillator states. In this case, core wave functions
separate in relative and center-of-mass coordinates, and the recoil operator becomes
a one-body operator when written in center-of-mass coordinates (see Sect. 9.2.10 for
the exact calculation of electromagnetic operator matrix elements with cluster wave
functions). As occupied core states are always well bound, the use of harmonic
oscillator states for their description covers in fact all practical situations.

Recoil corrections for beta transitions are also vanishing or negligible. This is
obvious for allowed beta transitions, whose operators are spin and isospin tensors
and hence are independent of space coordinates. The recoil corrections of the first-
forbidden beta decay operators without Coulomb corrections cancel out as their
spatial part is of rank one. It can be shown numerically that the Coulomb corrections
along with recoil correction in the first-forbidden beta decays are very small and
hence can be neglected.

Operators which might pose problem are the density operators. Indeed, they are
scalar operators so that the center-of-mass and valence parts lead to coupling matrix
between the intrinsic and center of mass parts of many-body wave functions. A
simple solution of this problem is to directly define density operators in cluster
orbital shell model coordinates, which is sound in practice as density is important
only in the valence space for both weakly bound and resonance nuclei.

## 5.3 Truncation of the Many-Body Berggren Basis in the Gamow Shell Model

The presence of numerous scattering states in the Berggren ensemble generates very large valence space dimensions in the Gamow shell model. Thus, the Hamiltonian matrix diagonalization cannot be handled with methods relying on full matrix times vector operations such as the Lanczos or Davidson methods [15, 16], unless drastic truncations are imposed.

In fact, the dimensionality problem in Gamow shell model is even more acute than in standard shell model [15], as one typically needs 30–50 Berggren basis states per partial wave to attain convergence, compared to the 5–10 harmonic states typically needed in no-core shell model. Consequently, in practice, one usually truncates the Gamow shell model space according to the number of occupied scattering states in Slater determinants, which is typically 2, 3, or 4 at most. By limiting the number of particles in the continuum, truncated Gamow shell model spaces are usually tractable when only a few partial waves are represented in the Berggren basis. Moreover, it is not necessary to use Berggren basis for the proton part of the wave function in neutron-rich nuclei, because protons are well bound and do not participate significantly in the asymptote of nuclear wave functions. This is also the case for a partial wave of large angular momentum, as its centrifugal barrier prevents nucleons from entering the asymptotic region. Thus, it is sufficient to use a basis of harmonic oscillator states for the latter cases, whereas the Berggren basis is utilized for $\ell \leq 1, 2$.

### 5.3.1 Natural Orbitals

Another powerful method to diminish the size of Gamow shell model space without loss of numerical precision could be implemented using the natural orbitals [17] defined as the eigenvectors of the scalar one-body density matrix:

$$\rho'_{mn} = \langle \Psi' | \left[ a_m^\dagger \tilde{a}_n \right]_0^0 | \Psi' \rangle, \tag{5.60}$$

where $m, n$ are indices of Berggren basis states, and $|\Psi'\rangle$ is an approximation of the final many-body state (see Exercise IV for numerical examples of the use of natural orbitals). Operator $\tilde{a}$ in (5.60) refers to the modified annihilation operator. Contrary to the creation operator $a^\dagger$, the annihilation operator $a_n$ is not a spherical tensor [18]. Therefore, it cannot be coupled with other creation and/or annihilation operators to a given value of the angular momentum. In fact, the annihilation operator $a$ must be multiplied by a phase factor of its angular part in order to become a spherical tensor [18]. It is standard to denote this newly defined modified annihilation operator by $\tilde{a}_n$ [18]. Note that this tilde has a different signification from that introduced in Sect. 5.1.1 to emphasize the biorthogonality of complex-energy states.

---

**Exercise  IV** ★★

In a numerical example involving a nucleus with three valence nucleons, one will show that the use of natural orbitals is efficient and precise in practical applications.

**A.** One will consider resonance of the $^7$He ground state in the approximation of an inert $^4$He core and 3 valence neutrons. The many-body basis generating the ground state wave function of $^7$He consists of the $p_{3/2}$ partial wave only.

Explain why it is sufficient from a physical point of view to have only basis states belonging to the neutron $p_{3/2}$ partial wave.

Run the many-body Gamow shell model code using the modified surface Gaussian interaction (see Eq. (9.174)) to calculate the $^7$He ground state wave function and its natural orbitals in the Berggren basis.

**B.** Run the Gamow shell model code in the same condition as in **A**, but with natural orbitals instead of the Berggren basis. Show that one obtains convergence using only a few natural orbitals, compared to the much larger number of demanded Berggren basis states.

**C.** Redo the same exercise with 2p-2h truncations for $^7$He and notice that using a non-truncated basis space with a few natural orbitals does not improve results. Explain the situation by comparison with the exact case considered in **A** and **B**.

---

Compared to the Hartree-Fock states, natural orbitals offer a more adapted basis for the diagonalization problem, on the assumption that $|\Psi'\rangle$ is close to the desired final many-body state.

Natural orbitals have been applied with success in the contexts of variational multiparticle-multihole configuration mixing method [19] and density matrix renormalization group approach [20]. In the present study, the approximate solution $|\Psi'\rangle$ is obtained in a smaller configuration space in which only two particles are allowed in the nonresonant continuum space. With the corresponding basis of natural orbitals, 5 to 7 states per partial wave offer results which are of a similar quality as those obtained in Berggren basis with about 30 states per partial wave. The difference of calculated Gamow shell model eigenergies using the Berggren basis or a basis of natural orbitals is in this situation smaller than 15 keV, so that both bases can be deemed as equivalent. In this way, one may reduce the sizes of matrices to be diagonalized by several orders of magnitude what enables large-space Gamow shell model calculations. However, if one aims at calculating nuclear states without truncating the Gamow shell model space, a basis of a few natural orbitals is no longer sufficient. For that matter, it is more efficient to use the density matrix renormalization group approach (see Sect. 5.9), based on a successive renormalization of nonresonant degrees of freedom within the used many-body basis.

## 5.4  Determination of Eigenvalues in Gamow Shell Model: The Overlap Method

Nuclear states in Gamow shell model are determined by the diagonalization of the Hamiltonian matrix expressed in a basis of Slater determinants, in a way formally identical to standard shell model. The fundamental difference is that the Slater determinants in Gamow shell model are built from the one-body states of the discretized completeness relation of Eq. (3.82). This implies that all the matrix elements of the Gamow shell model Hamiltonian must be calculated by direct integration, utilizing explicitly the radial wave functions of the Berggren basis states. The representation of the Gamow shell model Hamiltonian in a Berggren basis of discretized states is a complex symmetric matrix, as the basis states bear a complex energy.

In order to diagonalize the standard shell model matrices, the Lanczos method is widely employed, as it allows to determine low-energy nuclear states without having to fully diagonalize the Hamiltonian matrix. However, Lanczos method cannot be applied directly in Gamow shell model, because resonance $A$-body states are surrounded by many scattering $A$-body states. Hence, the low-energy spectrum obtained by the Lanczos method in Gamow shell model possesses bound, resonance, and scattering $A$-body states, where resonance and scattering states cannot be separated one from another. It is thus impossible to know if a Gamow shell model eigenstate is a resonance or scattering state considering only its eigenenergy. In fact, only bound states can be identified unambiguously, as they always bear negative eigenenergies.

The solution to this problem is the use of the overlap method [21]. It is a two-step method:

- The Hamiltonian is firstly diagonalized at the level of pole approximation, where the basis of Slater determinants is generated only by bound and resonance one-body states of the Berggren basis. The $|\Psi_0\rangle$ eigenvector, which is obtained from a full diagonalization of the Hamiltonian matrix at the level of pole approximation, is the zeroth-order approximation of the exact eigenvector.
- The Hamiltonian is then diagonalized by taking into account all one-body states of the Berggren basis, bound, resonance, and scattering, for which $|\Psi_0\rangle$ is the initial pivot. The Gamow shell model eigenstate $|\Psi\rangle$ corresponding to the sought bound or resonance $A$-body state is the one maximizing the $|\langle\Psi_0|\Psi\rangle|$ overlap.

Let us consider the example of $^{20}O$, where four valence neutrons interact above a $^{16}O$ core, modeled by a Woods-Saxon potential. The shell model space consists of the $0d_{5/2}$ and $1s_{1/2}$ bound states, $0d_{3/2}$ resonance of the Woods-Saxon potential, and of the nonresonant continuum of $d_{3/2}$ complex-energy scattering states. The eigenstates determined by the overlap method are enclosed by squares (see Fig. 5.1). Indeed, resonances should be stable with respect to the changes of the contour. As the Berggren sets involving scattering states on slightly modified contours are complete, the physical resonance states expanded using these different Berggren

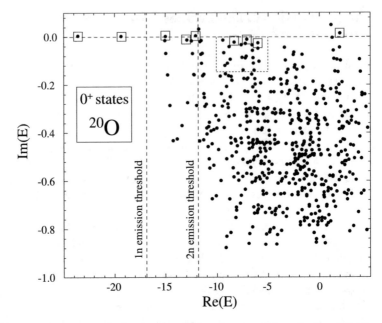

**Fig. 5.1** Complex energies of $0^+$ states of the $^{20}O$ nucleus issued from the diagonalization of the Gamow shell model Hamiltonian. The one-neutron (1n) and two-neutron (2n) emission thresholds are indicated. The bound states and physical resonances are marked by squares, while the other states form the nonresonant continuum. The zone enclosed by a dashed-line square is studied in more details in Fig. 5.2 (from Ref. [21])

basis contours of scattering states are theoretically equivalent. Consequently, one can verify that they are the sought resonance eigenstates by modifying the form of the $d_{3/2}$ complex contour of scattering states and noticing that they are independent of it (see Fig. 5.2). Hence, the overlap method allows to determine the $A$-body resonances, although their asymptotic behavior is not known explicitly. The practical efficiency of the overlap method is studied in Exercise V in a few numerical examples.

---

**Exercise   V ★**

In an example of nucleus with two valence neutrons, we will illustrate that the overlap method is efficient and precise in practical applications.

**A.** Run the two-particle Gamow shell model code for the $0^+$ eigenstates $^{18}O$ nucleus using the modified surface Gaussian interaction (see Eq. (9.174)) with full spectrum diagonalization. Plot obtained energies and widths similarly to Fig. 5.1. Discuss about the position of bound and resonance many-body states.

**B.** Run the Gamow shell model code to calculate $0^+$ pole states of $^{18}O$ only. Notice that the overlap between pole approximation and exact diagonalization is close to one for many-body bound and resonance states.

---

Run the Gamow shell model code with full spectrum diagonalization, but with slightly different Berggren basis contours. Plot results similarly to Fig. 5.2. Deduce that the energies and widths of pole states are independent of the used contours, contrary to that of scattering eigenstates.

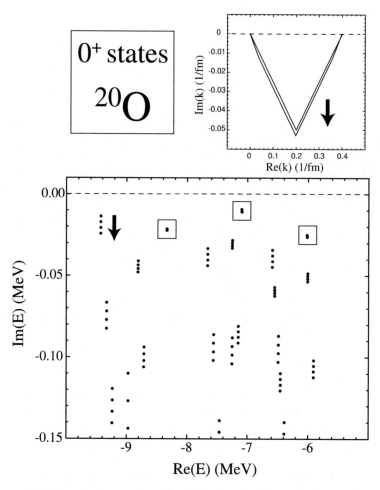

**Fig. 5.2** Effects of small changes of the $L^+$ contour on the stability of the energies of resonance and nonresonance $0^+$ states of the $^{20}$O nucleus. Upper panel: the $L^+$ contour of complex momenta corresponding to the $d_{3/2}$ continuum. The direction of deformation of the contour is indicated by an arrow. Calculations have been effected for four contours discretized with ten points. Only the two extremal contours are illustrated. Lower panel: Changes of the positions of the energies of the states close to the three resonances in the vicinity of the two-neutron emission threshold in Fig. 5.1 (from Ref. [21])

## 5.5    Optimization of the Gamow Shell Model One-Body Basis

While the Berggren basis is theoretically complete for any finite-range potential, one cannot use an arbitrary potential to generate the one-body basis of the Gamow shell model. A typical example of this problem is $^8$He, in a model space of $^4$He core plus four valence neutrons, as the $0p_{3/2}$ single-particle states of $^5$He have a sizable width, whereas $^8$He is weakly bound. As the single-particle approximation of the $^8$He ground state has a very large width, close to 500 keV, a very strong continuum coupling involving the four valence neutrons occurs to bind the $^8$He ground state. On the contrary, using a Berggren basis in which the $0p_{3/2}$ single-particle state is slightly bound leads to stable calculations, where particle-hole excitations in the continuum are small, so that truncating at 2p-2h level is generally sufficient.

This effect shows that the Berggren bases generated by different potentials are not equivalent in practical calculations, even though they are all theoretically complete. This arises from a necessary discretization of Berggren basis contours and truncation of Gamow shell model spaces. Indeed, the matrix elements provided by different Berggren basis sets of one-body states can vary significantly. For calculations to be most stable, it is then necessary to devise a Berggren basis with which 1p-1h excitations to the continuum are as small as possible. Moreover, while it is sufficient in principle to enclose all resonances of the basis between the $L^+$ contour of complex-energy scattering states and the real $k$-axis (see Sect. 3.5), the presence of a resonance too close to the $L^+$ contour creates numerical instabilities. Indeed, the rapid change of the phase shift of complex-energy scattering states close to the resonance generates large numerical cancellations between off-diagonal matrix elements involving these scattering states.

A simple and effective contour which is made of three segments can be devised to avoid this problem. The first segment starts from $k = 0$ and ends at $k_{peak}$, where $\Re(k_{peak}) \simeq \Re(k_{res})$ and $\Im(k_{peak}) \simeq -\Re(k_{res})$, with $k_{res}$ the complex linear momentum of the resonant state in the Berggren basis. The second segment starts from $k_{peak}$ and ends in $k_{middle} \simeq 0.5$–$1\,\mathrm{fm}^{-1}$, while the third segment starts from $k_{middle}$ and ends in $k_{max} \simeq 2$–$5\,\mathrm{fm}^{-1}$. The numerical advantage of this form of contour is twofold. On the one hand, it encompasses the resonance state of the Berggren basis and is sufficiently far from it not to generate numerical instabilities. On the other hand, as the phase shift of scattering states varies slowly at high energy, one can take fewer points on the third segment without inducing numerical inaccuracies.

Let us now consider the more difficult problem of optimization of the basis-generating potential. The Hartree-Fock potential provides with a very good single-particle basis in the case of closed-shell nuclei, as it is spherical and all 1p-1h excitations from the Hartree-Fock ground state vanish. However, it is no longer the case for open-shell nuclei, as its Hartree-Fock potential is no longer spherical. Deformation in Hartree-Fock potentials arises in open-shell nuclei because the one-body states of an open shell, of different angular momentum projections, are unevenly occupied, so that spherical symmetry is broken.

If one considers a Slater determinant of a configuration whose one-body states of maximal angular momentum projections are occupied, it is possible to build a Gamow-Hartree-Fock scheme, also called the $M$-potential, as the optimized Slater determinant is coupled to $J = M$ (see Ref. [22]). However, unless one considers single-particle or single-hole many-body states, the considered Slater determinant corresponds in general to an excited state, so that the basis-generating potential is not optimal. Moreover, partial averaging has to be done over the angular momentum projections of occupied one-body states. This procedure is called the fill-in approximation, as one considers therein that the single-particle states of the occupied shell are equally occupied.

While sphericity in the basis potential arises by construction in the fill-in approximation, the main drawback of this scheme is that it is not variational, so that its ability to provide with a well optimized basis potential is doubtful. Thus, it would be convenient to have a potential directly optimized at the ground state level, on the one hand, and which suppresses the most important 1p-1h excitations from the trial ground state, on the other hand. The optimized wave function can no longer be a Slater determinant, but it can be a linear combination of Slater determinants belonging to a given configuration and coupled to a given angular momentum $J$. Hence, the procedure just described will be called the multi-Slater determinant coupled Hartree-Fock method. We will see that the multi-Slater determinant coupled Hartree-Fock equations are very similar to Hartree-Fock equations, and that any nucleus can be treated therein.

The considered Hamiltonian consists of one-body and two-body parts. It is convenient to write it in the coupled representation:

$$
H = \sum_{\alpha\beta} \sqrt{2j_\alpha + 1} \, \langle \alpha|h|\beta \rangle \, [a_\alpha^\dagger \tilde{a}_\beta]_0^0
$$
$$
- \sum_{\alpha \leq \beta, \gamma \leq \delta, J'} \sqrt{2J' + 1} \langle \alpha\beta|V|\gamma\delta \rangle_{J'} [[a_\alpha^\dagger a_\beta^\dagger]^{J'} [\tilde{a}_\gamma \tilde{a}_\delta]^{J'}]_0^0 , \tag{5.61}
$$

where the tilde notation indicates the use of modified annihilation operators, which are necessary to couple creation/annihilation operators to a fixed angular momentum (see Sect. 5.3).

In order to determine the multi-Slater determinant coupled Hartree-Fock ground state, one considers the configuration of lowest energy provided by the basis potential. The Hamiltonian $\hat{H}$ of Eq. (5.61) is then diagonalized using the basis of the Slater determinants of the configuration of lowest energy so as to provide the multi-Slater determinant coupled Hartree-Fock ground state, denoted as $|\Psi_{GS}\rangle$. One then demands the most important 1p-1h excitations from $|\Psi_{GS}\rangle$ to vanish. As the basis potential must be spherical, 1p-1h excitations can only involve one-body states of the same $\ell$, $j$ quantum numbers. Moreover, the well-bound one-body occupied states do not have to be optimized, as their change through basis potential modification is minimal. Consequently, the most important 1p-1h excitations are those involving the resonant shell closest to the continuum, so that the multi-Slater determinant coupled Hartree-Fock potential can be defined from the variational principle, demanding that these excitations vanish, similarly to the Hartree-Fock procedure:

$$\langle \Psi_{\mathrm{GS}} \mid \hat{H} \mid \Psi_k \rangle = 0 \,, \tag{5.62}$$

where

$$|\Psi_k\rangle = [a^{\dagger}_{k\ell j}\tilde{a}_{n\ell j}]^0_0 \,|\Psi_{\mathrm{GS}}\rangle \,. \tag{5.63}$$

In this expression, $n\ell j$ is the last occupied shell of the multi-Slater determinant coupled Hartree-Fock configuration of $\ell$, $j$ quantum numbers, and $k\ell j$ is a scattering shell. $a^{\dagger}_{k\ell j}$ and $\tilde{a}_{n\ell j}$ are coupled to zero in order to have $|\Psi_k\rangle$ coupled to $J$. Using Eqs. (5.61) and (5.62), one obtains:

$$\sqrt{2j+1}\,\langle n\ell j | h | k\ell j\rangle \,\langle \Psi_{\mathrm{GS}} \mid [a^{\dagger}_{n\ell j}\,\tilde{a}_{k\ell j}]^0_0 \mid \Psi_k\rangle$$

$$-\sum_{\lambda_{\mathrm{occ}} J'} \sqrt{2J'+1}\langle \lambda_{\mathrm{occ}}\, n\ell j | V | \lambda_{\mathrm{occ}}\, k\ell j\rangle_{J'}$$

$$\times \langle \Psi_{\mathrm{GS}} \mid [[a^{\dagger}_{\lambda_{\mathrm{occ}}}\, a^{\dagger}_{n\ell j}]^{J'}\,[\tilde{a}_{\lambda_{\mathrm{occ}}}\, \tilde{a}_{k\ell j}]^{J'}]^0_0 \mid \Psi_k\rangle = 0 \,, \tag{5.64}$$

where

- $\alpha = n\ell j$ and $\beta = k\ell j$ in the one-body part of Eqs. (5.61) and (5.64) as $|\Psi_k\rangle$ is a 1p-1h excitation of $|\Psi_{\mathrm{GS}}\rangle$
- $\beta = n\ell j$, $\delta = k\ell j$ and $\alpha = \gamma = \lambda_{\mathrm{occ}}$ in the two-body part of Eqs. (5.61) and (5.64) because, on the one hand, $|\Psi_k\rangle$ is a 1p-1h excitation of $|\Psi_{\mathrm{GS}}\rangle$ and, on the other hand, the shell indices associated to $\alpha = n\ell j$ resonant states can be chosen so that they are always smaller than the shell indices associated to $\beta = k\ell j$ scattering states.

One can then define the multi-Slater determinant coupled Hartree-Fock potential from Eq. (5.64):

$$\langle \alpha | U_{\mathrm{MSDHF}} | \beta \rangle = \langle \alpha | h | \beta \rangle$$

$$-\frac{\displaystyle\sum_{\lambda_{\mathrm{occ}} J'} \sqrt{2J'+1}\langle \lambda_{\mathrm{occ}}\, \alpha | V | \lambda_{\mathrm{occ}}\, \beta\rangle_{J'} \,\langle \Psi_{\mathrm{GS}} \mid [[a^{\dagger}_{\lambda_{\mathrm{occ}}}\, a^{\dagger}_{n\ell j}]^{J'}\,[\tilde{a}_{\lambda_{\mathrm{occ}}}\, \tilde{a}_{k\ell j}]^{J'}]^0_0 \mid \Psi_k\rangle}{\sqrt{2j+1}\,\langle \Psi_{\mathrm{GS}} \mid [a^{\dagger}_{n\ell j}\,\tilde{a}_{k\ell j}]^0_0 \mid \Psi_k\rangle}$$

$$\tag{5.65}$$

$U_{\mathrm{MSDHF}}$ is spherical and independent of $k$ as the latter only induces angular momentum coupling and rearrangement phases (see Eq. (5.65)). It is solved self-consistently, diagonalizing $\hat{H}$ in the configuration of lowest energy to obtain

$|\Psi_{GS}\rangle$, and then calculating $U_{MSDHF}$ and its eigenstates similarly to a Hartree-Fock procedure (see Exercise VI for numerical illustrations).

---

**Exercise VI ★**

One will show in a numerical example that the multi-Slater determinant coupled Hartree-Fock method allows to generate a basis potential with which Gamow shell model calculations are stable and precise.

**A.** Run the two-particle Gamow shell model code with the modified surface Gaussian interaction (see Eq. (9.174)) for $^6$He, $^6$Be, and $^6$Li using bases generated either by a Woods-Saxon potential mimicking the core or by the multi-Slater determinant coupled Hartree-Fock potential. Notice that obtained energies and widths are almost the same in both cases for $^6$He and $^6$Be and explain why.

**B.** Show that calculations are problematic in the case of $^6$Li when using a basis generated by the Woods-Saxon potential of the core, especially for unbound states.

On the contrary, by using the multi-Slater determinant coupled Hartree-Fock potential, show that one obtains precise eigenenergies for the $^6$Li spectrum.

Explain this result by considering the overall behavior of Hamiltonian matrix elements with both basis-generating potentials.

Conclude by stating why optimization of the basis potential is necessary in the Gamow shell model.

---

$|\Psi_{GS}\rangle$ has to be redefined if no $\ell, j$ state is occupied in the multi-Slater determinant coupled Hartree-Fock ground state. Indeed, an $\ell, j$ state has to be occupied in $|\Psi_{GS}\rangle$ to be able to apply the multi-Slater determinant coupled Hartree-Fock method. This occurs, for example, for the neutron $p_{1/2}$ partial wave when considering the $^6$He nucleus, as the $0p_{1/2}$ neutron state is not occupied. In order to solve this problem, it is sufficient to have a 1p-1h excitation to the initial multi-Slater determinant coupled Hartree-Fock ground state. In the example of $^6$He with a $^4$He core, $|\Psi_{GS}\rangle = |[0p_{3/2} \, 0p_{1/2}]^{J=1}\rangle$ can be used to optimize the neutron $p_{1/2}$ partial wave.

If, however, the considered $\ell, j$ partial wave does not possess any pole in the basis, then its multi-Slater determinant coupled Hartree-Fock potential cannot be defined. But the $\ell, j$ partial wave only occurs through the continuum coupling, so that its effect will be much less important than the effect of pole states. It is then sufficient to use a Woods-Saxon or harmonic oscillator potential to generate the basis states of this particular partial wave.

## 5.6     Calculation of Two-Body Matrix Elements in the Gamow Shell Model

Realistic interactions are defined with the relative coordinate of the two nucleons. The shell model approaches, on the other side, demand the use of laboratory or cluster orbital shell model coordinates. The change from relative to laboratory coordinates is straightforward using a basis of harmonic oscillator states by the way of Talmi-Brody-Moshinsky coefficients [23]. Consequently, the calculation of the two-body matrix elements of realistic interactions poses no problem in standard shell model.

However, the Talmi-Brody-Moshinsky transformation cannot be directly applied in the Berggren basis. The extension of the Brody-Moshinsky transformation to arbitrary bases can be done, in fact, with vector brackets [24–26] at the price, however, of very time-consuming calculations. Indeed, the finite sums entering the Talmi-Brody-Moshinsky transformation become the two-dimensional integrals with vector brackets. Moreover, the presence in vector brackets of the Dirac delta and Heaviside function, which are not analytical in the complex plane, makes the use of complex-energy states difficult.

The solution to these theoretical and practical problems comes from the fact that the effect of nuclear interaction for bound and resonance states is localized in the vicinity of a nucleus. Consequently, one can assume that in Gamow shell model calculations a decomposition of the nuclear interaction in a basis of harmonic oscillator states will converge rapidly (see Sect. 2.2).

The used two-body nuclear interaction $\hat{V}$ can be written as [27]:

$$\langle ab|\hat{V}|cd\rangle = \sum_{\alpha\beta\gamma\delta}^{N_{max}} \langle \alpha\beta|\hat{V}|\gamma\delta\rangle \, \langle a|\alpha\rangle \, \langle b|\beta\rangle \, \langle c|\gamma\rangle \, \langle d|\delta\rangle \,, \qquad (5.66)$$

where $N_{max}$ is the number of harmonic oscillator states, and greek and latin letters refer respectively to harmonic oscillator states and Berggren basis states. As $\hat{V}$ appears only through matrix elements of harmonic oscillator states, the Talmi-Brody-Moshinsky transformation can be used to calculate them. Berggren basis states can be found only in overlaps of the form $\langle a|\alpha\rangle$. Thus, no complex scaling is necessary, as harmonic oscillator states always decrease like Gaussians for $r \to +\infty$ whereas Berggren basis states increase at most exponentially in modulus. Note that the one-body kinetic part of the Hamiltonian is directly expressed with the Berggren basis, hence without harmonic oscillator basis expansion, so that this calculation is not equivalent to a standard shell model calculation.

Equation (5.66) can be obviously extended to all types of operators. In particular, it is very convenient for the calculation of electromagnetic operators, whose radial part increases as $r^L$ for electric transitions and as $r^{L-1}$ for magnetic transitions, with $L$ the multipolarity of the considered transition. A direct computation of the matrix elements of electromagnetic operators with the Berggren basis would generate derivatives of the Dirac delta. Indeed, if one considers the important case of E2 transitions in a Berggren basis of Bessel functions for simplicity, the calculation of its matrix elements is equivalent to calculating the Fourier-Bessel transform of $r^2$, which is proportional to $\delta''(k - k')$. A direct numerical treatment of derivatives of the Dirac delta would be cumbersome as one would have to differentiate numerically the scattering components of the Gamow shell model eigenvector with respect to $k$. Conversely, Eq. (5.66) provides with well-defined matrix elements of electromagnetic operators. Moreover, it has been checked numerically that convergence with the number of used harmonic oscillator shells is rather fast in loosely bound or narrow resonance many-body states. Consequently, the harmonic oscillator expansion of Eq. (5.66) is the method of choice to deal with observables whose radial operator increases on the real axis.

This method is applied to the $^6$He and $^{18}$O nuclei, modeled by two neutrons above a core. One uses $^4$He core for $^6$He and $^{16}$O core for $^{18}$O. The N$^3$LO realistic interaction renormalized with the $V_{\text{low-k}}$ method [27] is used for nucleon-nucleon interaction. The convergence of the energies of ground and excited states of $^{18}$O nucleus is illustrated in Table 5.1. One may notice therein that convergence is attained for $n_{\max} \sim 10$, and that even for resonance states. Convergence of the density of the halo state $0_1^+$ of the $^6$He nucleus has been also studied in Ref. [27], where it was shown that convergence occurs for $n_{\max} \sim 5 - 10$.

The calculation of Coulomb two-body matrix elements in a Berggren basis follow the same method. Indeed, the Coulomb term $\hat{U}_{\text{Coul}}(Z_{\text{core}}) + \hat{V}_{\text{Coul}}$ must behave as $\hat{U}_{\text{Coul}}(Z - 1)(r)$ at large distances. Consequently, since $\hat{U}_{\text{Coul}}(Z)$ is additive in $Z$, one can rewrite the Coulomb interaction in the Gamow shell model Hamiltonian as $\hat{U}_{\text{Coul}}(Z - 1) + (\hat{V}_{\text{Coul}} - \hat{U}_{\text{Coul}}(Z_{\text{val}} - 1))$. The short-range character of the operator $\hat{V}_{\text{Coul}} - \hat{U}_{\text{Coul}}(Z_{\text{val}} - 1)$ also suggest to use the method described above:

$$\hat{V}_{\text{Coul}} - \hat{U}_{\text{Coul}}(Z_{\text{val}} - 1) = \sum_{\alpha\beta\gamma\delta}^{N} |\alpha\beta\rangle \langle\alpha\beta|\hat{V}_{\text{Coul}} - \hat{U}_{\text{Coul}}(Z_{\text{val}} - 1)|\gamma\delta\rangle \langle\gamma\delta|,$$

$$(5.67)$$

**Table 5.1** Convergence of the energies of the $0_1^+$, $0_2^+$, $2_1^+$, $2_2^+$, $4_1^+$, and $4_2^+$ states of the $^{18}$O nucleus. Energies are given as a function of the number of nodes of the harmonic oscillator states used in the expansion of the two-body interaction. The momentum cut is effected at $\Lambda = 1.9\,\text{fm}^{-1}$ in the space of relative coordinates for the N$^3$LO interaction. The harmonic oscillator parameter is equal to $b = 2\,\text{fm}$. Energies are given in MeV (adapted from Ref. [27])

| $n_{\max}$ | $J^\pi = 0_1^+$  E | $J^\pi = 0_2^+$  E | $J^\pi = 2_1^+$  E | $J^\pi = 2_2^+$  E | $J^\pi = 4_1^+$  E | $J^\pi = 4_2^+$ Re[E] | Im[E] |
|---|---|---|---|---|---|---|---|
| 4 | −12.225 | −8.438 | −12.1398 | −10.0488 | −11.0641 | −1.4373 | −0.8275 |
| 6 | −12.226 | −8.498 | −12.1465 | −10.0830 | −11.0907 | −1.4292 | −0.7600 |
| 8 | −12.228 | −8.499 | −12.1452 | −10.0853 | −11.0922 | −1.4380 | −0.7405 |
| 10 | −12.229 | −8.499 | −12.1450 | −10.0857 | −11.0921 | −1.4400 | −0.7390 |
| 12 | −12.228 | −8.499 | −12.1453 | −10.0858 | −11.0923 | −1.4393 | −0.7401 |
| 14 | −12.228 | −8.499 | −12.1453 | −10.0858 | −11.0923 | −1.4394 | −0.7401 |
| 16 | −12.228 | −8.499 | −12.1453 | −10.0858 | −11.0923 | −1.4394 | −0.7401 |
| 18 | −12.228 | −8.499 | −12.1453 | −10.0858 | −11.0923 | −1.4394 | −0.7401 |
| 20 | −12.228 | −8.499 | −12.1453 | −10.0858 | −11.0923 | −1.4394 | −0.7401 |

where Greek letters label harmonic oscillator states and $N$ is the number of harmonic oscillator states used in a given partial wave. This method allows to calculate the two-body part of the Coulomb interaction with the Talmi-Brody-Moshinsky transformation by separating its finite-range and infinite-range parts.

## 5.7    The Jacobi-Davidson Method in the Gamow Shell Model

The Lanczos method is the tool of choice in standard shell model as it makes use of matrix $\hat{H}$ times vector operations only, with which the lowest eigenvalues converge first. While the Lanczos method could be used for the search of bound states in Gamow shell model, the Lanczos method does not converge in general for resonance states. This arises from the presence of numerous scattering states surrounding any resonance state, with the same quantum numbers as the resonance. Consequently, it is necessary to use a diagonalization method which is targeting directly the resonant states. The Jacobi-Davidson method is the best method for that matter. Indeed, it only involves $\hat{H}$ times vector operations, as in the Lanczos method, while including an approximated shift-and-invert method, so that it quickly provides with eigenvalues of unbound states.

The algorithm of the Jacobi-Davidson method implemented in the Gamow shell model is shortly described in the following (see Ref. [16] for an introduction to this method):

- Start from an approximation of the Gamow shell model eigenvector $|\Psi_i\rangle$. The first eigenvector $|\Psi_0\rangle$ comes from pole approximation, where the Lanczos method can be used, or from another calculation.
- Calculate its residue $|R_i\rangle = \hat{H} |\Psi_i\rangle - E_i |\Psi_i\rangle$.
- Compute the new Gamow shell model Davidson vector by solving the linear system: $(\hat{H}_{app} - E_i) |\Phi_{i+1}\rangle = |R_i\rangle$, where $\hat{H}_{app}$ is a preconditioner of $\hat{H}$ (see explanations below).
- $|\Phi_{i+1}\rangle$ is orthogonalized with respect to all previous Jacobi-Davidson vectors $|\Phi_j\rangle$, $0 \leq j \leq i$, then projected on $J$ if necessary (see Sect. 5.8.4), and normalized.
- $\hat{H}$ is then diagonalized in the set of $|\Phi_j\rangle$, $0 \leq j \leq i + 1$, which provides with a new approximation of the Gamow shell model eigenvalue and eigenvector $E_{i+1}$ and $|\Psi_{i+1}\rangle$.
- Iterate until convergence is reached.

The fundamental problem of the Jacobi-Davidson method is to find a good preconditioner $\hat{H}_{app}$ for $\hat{H}$. A diagonal approximation for $\hat{H}_{app}$ is not sufficient therein, as off-diagonal matrix elements are large in the pole space. However, the diagonal of the matrix of $\hat{H}$ is dominant in the basis space spanned by scattering Slater determinants. Consequently, it is efficient to take $\hat{H}_{app}$ equal to $\hat{H}$ in the subspace generated by Slater determinants built from bound and resonance one-body states, and to take the diagonal of $\hat{H}$ for $\hat{H}_{\text{app}}$ in the rest of the Gamow shell model space. The diagonal of $\hat{H}_{\text{app}}$ must, in fact, be averaged on each configuration in order to have $[\hat{H}_{\text{app}}, \mathbf{J}] = 0$. Many applications of the $\hat{J}^2$ operator to project $|\Phi_i\rangle$ on $J$ (see Sect. 5.8.4) are then avoided. As the dimension of the pole space is small, the linear system involving $\hat{H}_{\text{app}}$ in the Jacobi-Davidson method is solved very quickly, while being sufficiently close to $\hat{H}$ so as to provide quick convergence.

As the main components of Gamow shell model eigenvectors are typically that of the pole space configurations, the Jacobi-Davidson method converges in typically 30 iterations, compared to the 50–100 typically necessary with the Lanczos method. Consequently, the $\hat{H}$ times vector operation in the Gamow shell model, which is slower than in standard shell model, is expected to be partially mitigated by the smaller number of Jacobi-Davidson iterations.

## 5.8  Memory Management and Two-Dimensional Partitioning

We will describe now an implementation of the Gamow shell model code which significantly improves both its performance and storage requirements for large-scale calculations [28].

From a computational point of view, the Gamow shell model code is formally a shell model code. As a consequence, in order to implementent the two-dimensional partitioning method in the Gamow shell model, one can reuse the techniques already developed in another shell model code, called the many-body fermion dynamics for nuclear structure, which has shown its efficiency when utilized on powerful parallel supercomputers [29–33]. In particular, one can directly implement the two-dimensional partitioning scheme of the many-body fermion dynamics code in the Gamow shell model code. The two-dimensional partitioning scheme is a powerful algorithm, taking advantage of the Hamiltonian matrix symmetry, and which reduces the overheads induced by message passing interface communication [34].

Nevertheless, even from pure computational considerations, the Gamow shell model code and the many-body fermion dynamics code are not identical. Firstly, the Gamow shell model uses the Berggren basis, which is in essence continuous, contrary to the discrete harmonic oscillator basis used in the many-body fermion dynamics code. Consequently, the Gamow shell model matrix is less sparse than the typical matrix generated in the many-body fermion dynamics code. Secondly, in standard shell model, it is customary to separate the large proton-neutron full model space into two small subspaces, built from only protons or neutrons, as it allows to greatly optimize standard shell model codes. On the contrary, the nuclei studied with the Gamow shell model typically possess a large asymmetry between the numbers of valence protons and neutrons. This implies that a memory storage optimization, which has no equivalent in the standard shell model, must be additionally devised in the Gamow shell model code. In particular, the storage of uncoupled two-body matrix elements and many-body matrix elements between Slater determinants built from only protons or only neutrons must be optimized. One also developed new features absent from the many-body fermion dynamics code, which deal with preconditioning (see Sect. 5.7) and angular momentum projection (see Sect. 5.8.4), in order to accelerate the convergence the used eigensolver.

### 5.8.1   Slater Determinants and Partitions in Shell Model Codes

Configurations (also called partitions) and Slater determinants form the many-body basic objects to build the Gamow shell model many-body basis. A Slater determinant is defined from the occupied one-body states by valence nucleons, while a configuration enumerates its occupied shells, independently of the $m$ quantum numbers of the one-body states occupied in Slater determinants. For example, if $|0s_{1/2}(-1/2)\, 0s_{1/2}(1/2)\, 0p_{3/2}(-3/2)\, 0p_{3/2}(3/2)\rangle$ is a Slater determinant, its associated configuration is $[0s_{1/2}^2\, 0p_{3/2}^2]$. One then has fewer configurations than Slater determinants.

This separation leads to many advantages. Firstly, configurations must be generated in a sequential manner due to the truncation imposed to the valence space. This step is quick as configurations are in small number. As Slater determinants of different configurations are independent and there are no truncations inside a configuration, it is straightforward to parallelize routines involving the construction of the Slater determinants of a fixed configuration. Looking for Slater determinant indices, which demands the use of binary search, is also much faster, as binary search is effected at the level of configuration first, and of Slater determinant afterward. This also allows to save memory, as all indices depending on shells only can be stored in arrays involving configurations only, while those depending on the $m$ quantum number only are handled in arrays dealing with Slater determinants. As a consequence, the consideration of configuration first and its Slater determinants afterward is always done when building the $\hat{H}$ matrix.

In the Gamow shell model, contrary to standard shell model, basis configurations and Slater determinants are built from few valence particles and many one-body shells and states. Therefore, one chose to use a different implementation to represent configurations and Slater determinants as in standard shell model. In the standard shell model, the method of choice is to store Slater determinants via the bit storage [35]. Indeed, due to the Pauli principle, a state can be occupied by one nucleon at most, so that one can build a one-to-one correspondence between Slater determinants and binary numbers. For example, $|1001100000\rangle$ is a Slater determinant where the states 1, 4, and 5 are occupied, while the states 2, 3, and 6, ..., 10 are not occupied. All operations on Slater determinants can then be reduced to operations on bits, so that the action of creation/annihilation operators becomes very fast. The downside of this method is that it becomes inefficient if many states are unoccupied. Indeed, as one integer possesses 32 bits, it is most efficient if one has at most than 32 basis one-body states.

Typically, the $sp$ partial waves are represented by Berggren basis contours in the Gamow shell model, discretized with 15–30 points each, whereas the $d$ partial waves are built from 10 states generated by a harmonic oscillator potential. As a whole, 340 one-body states are generated, so that one would need 11 integers in order to store one Slater determinant. This would be inefficient from both memory and speed points of view, as the computing efficiency of bit algebra would be hindered by the numerous look-ups of integers in arrays. Therefore, it has been preferred to store configurations and Slater determinants as integer arrays, so that the Slater determinant of the example considered in the previous paragraph becomes $\{1, 4, 5\}$. It has been noticed that calculations are nevertheless rather fast when using this storage scheme. Added to that, the parallelization of the routines using basis configurations and Slater determinants is straightforward.

One has seen that it would be impossible to store all the data related to the neutron space in the Gamow shell model if the considered nucleus has a large neutron-to-proton ratio (from symmetry arguments, the situation is analogous if one has more valence protons than valence neutrons). This arises due to the use of the Berggren basis, because it is customary to have 100–200 neutron valence shells on the discretized scattering contours, whereas the no-core shell model has, for example, 30 neutron shells in a $8 \, \hbar\omega$ space. The recalculation of many-body neutron matrix elements would also take too much time. Consequently, one devised memory optimization schemes in order to store many-body matrix elements between neutron Slater determinants. A short denomination of the latter matrix elements are the phases, because one has to store matrix elements of the form $\langle SD_f|a_\alpha^\dagger|SD_i\rangle$, $\langle SD_f|a_\alpha^\dagger a_\beta|SD_i\rangle$, equal to $\pm 1$. Clearly, the indices of involved one-body shells and states, as well as those of configurations and Slater determinants, must also be included in the storage.

Another useful memory optimization comes from $m$-reversal symmetry, where

$$\hat{\hat{T}} \, |j \, m\rangle = (-1)^{j-m} \, |j \, -m\rangle$$

by definition for a one-body state of total angular quantum number $j$ and angular quantum projection $m$. In the case of the storage of phases, if one applies $\hat{\hat{T}}$ to all states present in $\langle SD_f|a_\alpha^\dagger|SD_i\rangle$, $\langle SD_f|a_\alpha^\dagger a_\beta|SD_i\rangle$, ..., the obtained phase and associated indices can be easily recovered from the initial phase. This allows to provide an additional memory gain factor of about 2 for the storage of phases. One will also see in the following that $m$-reversal symmetry allows to halve the cost of computation of Hamiltonian times vector for even-even nuclei.

## 5.8.2 Memory Management of the Hamiltonian

Due to the rapid growth of matrix dimension with the number of valence nucleons, memory utilization must be carefully assessed. In order to compare standard shell model and Gamow shell model for that matter, one will consider $N_v$ valence nucleons in a many-body space generated by $N_s$ states. As one aims at giving an overall order of magnitude for used memory, antisymmetry, parity and $M$ projection will be neglected in the following discussion, as their impact is small when considering the Berggren basis. Identically, it is sufficient to consider the 2p-2h part of the $\hat{H}$ matrix, whereby $\langle SD_f|\hat{H}|SD_i\rangle = \pm\langle\alpha_f\,\beta_f|\hat{V}|\alpha_i\,\beta_i\rangle$, as they form the vast majority of nonzero matrix elements in the $\hat{H}$ matrix. In this approximate picture, the dimension of the $\hat{H}$ matrix in the Gamow shell model approach is proportional to $N_s^{N_v}$, so that the probability to have a nonzero matrix element is proportional to $(1/N_s)^{N_v-2}$, as all states in $|SD_i\rangle$ and $|SD_f\rangle$ must be equal, except for $|\alpha_i\,\beta_i\rangle \neq |\alpha_f\,\beta_f\rangle$. Hence, the total number of nonzeros matrix elements is of the order of $d^{1+2/N_v}$, which corresponds to $d^{1.67}$ and $d^{1.5}$ when one truncated the model space so that 3 or 4 valence nucleons are allowed in the continuum, respectively. In particular, for a dimension $d \sim 10^9$, one has typically $d^{1.4}$ nonzero matrix elements in standard shell model [29], so that the number of nonzero matrix elements in the Gamow shell model is typically one to two orders of magnitude larger that of the standard shell model for that same dimension. Consequently, computation is more expensive in Gamow shell model than in standard shell model, so that message passing interface communications are expected to play less important a role in Gamow shell model than in standard shell model.

Let us consider two many-body Gamow shell model vectors, i.e., the initial $|\Psi_i\rangle$ and final $|\Psi_f\rangle$ many-body states, verifying $\hat{H}\,|\Psi_i\rangle = |\Psi_f\rangle$. One can distribute the basis states of the model space over all processors, so that the parts of the basis model space on each processor are much smaller than the full space. In particular, the parts of the basis model space to which $|\Psi_i\rangle$ and $|\Psi_f\rangle$ belong are deemed as the initial and final spaces, respectively. They correspond to the number of rows and columns, respectively, similarly to the code of many-body fermion dynamics applied to nuclear structure [34]. Hence, one stores matrix elements of the form: $\langle SD_{int}|a_\alpha|SD_i\rangle$ and $\langle SD_f|a_\alpha^\dagger|SD_{int}\rangle$, $\langle SD_{int}|a_\alpha\,a_\beta|SD_i\rangle$, and $\langle SD_f|a_\alpha^\dagger\,a_\beta^\dagger|SD_{int}\rangle$, where $|SD_{int}\rangle$ is an intermediate Slater determinant, chosen so that the additional data to store are minimal. Any one-body or two-body observable can be calculated with this scheme. One has indeed:

$$\langle SD_f|a_\alpha^\dagger\,a_\beta|SD_i\rangle = \langle SD_f|a_\alpha^\dagger|SD_{int}\rangle\,\langle SD_{int}|a_\beta|SD_i\rangle \tag{5.68}$$

$$\langle SD_f|a_\alpha^\dagger\,a_\beta^\dagger\,a_\delta\,a_\gamma|SD_i\rangle = \langle SD_f|a_\alpha^\dagger\,a_\beta^\dagger|SD_{int}\rangle\,\langle SD_{int}|a_\delta\,a_\gamma|SD_i\rangle. \tag{5.69}$$

The phase matrix, containing by definition these matrix elements, is stored as a sparse matrix. For this, one fixes $|SD_i\rangle$, so that all the basis states $|SD_f\rangle$ which vary by one or two states are generated. Hence, the relative phase between $|SD_i\rangle$ and $|SD_f\rangle$, as well the indices of $|SD_f\rangle$ and of associated one-body states, denoted as $\alpha$, $\beta$, $\gamma$, and $\delta$ in Eq. (5.69), must be stored.

It is most efficient to firstly loop over configurations, so that only the shell indices associated to $\alpha$, $\beta$, $\gamma$, and $\delta$ and the configuration index associated to $|SD_f\rangle$ must be stored. The loop over Slater determinants is done afterward for fixed configurations, so that only $m$-dependent values are stored at this level. The phase matrix can then be recovered from this information. The number of phases associated to $|SD_{int}\rangle$ is proportional to $2N_v$ for one-body phases and $N_v(N_v - 1)$ for two-body phases, where $N_v$ is the number of valence nucleons. The factor 2 comes the fact that one has two arrays of one-body phases and two arrays of two-body phases. Memory storage can be decreased by demanding that $\alpha < \beta$ and $\gamma < \delta$, inequalities arising from antisymmetry requirements. By applying $m$-reversal symmetry, one can further divide the size of phase matrices by two. Indeed, the $\langle SD_{int} | a_\alpha | SD_i \rangle$ and $\langle SD_{int} | a_\alpha\, a_\beta | SD_i \rangle$ phases are straightforward to obtain from those where all $| j\, m \rangle$ one-body states are replaced by $| j - m \rangle$ (see Sect. 5.8.1).

Let us state that the presented memory optimization of the phase matrix is not applied to the angular momentum operator $\hat{J}^2$. It would be inefficient in this case, because the $\hat{J}^2$ operator is a function of $\hat{J}^\pm$. $\hat{J}^\pm$ is, in fact, a very sparse one-body operator, so that the number of phases occurring in $J^\pm$ is much smaller than those needed for the Hamiltonian matrix. Thus, the use of Eq. (5.68) when applying $\hat{J}^2$, where an intermediate Slater determinant would have to be introduced, is hereby not necessary.

The memory optimization discussed in this section pertained to the full storage scheme in the Gamow shell model code, in which all the many-body matrix elements of the Hamiltonian are stored. As the Hamiltonian matrix is sparse, only nonzero matrix elements and associated indices are retained in memory. Nevertheless, this scheme becomes quickly inapplicable as memory requirements in the Gamow shell model grow rapidly with the number of valence nucleons, In fact, this is basically the only issue related to memory, because the storage of Gamow shell model vectors, of a few hundreds at most, is negligible compared to the storage of nonzero matrix elements of the Hamiltonian.

As a matter of fact, the full storage scheme is limited by the aggregate memory space of used computer nodes on parallel machines. Therefore, other storage scheme have been developed in the Gamow shell model code as an alternative to the full storage scheme, which are the on-the-fly and partial storage schemes. In the on-the-fly scheme, the Hamiltonian matrix is not stored, but, on the contrary, is recalculated for each $\hat{H}$ times vector operation. While, on the one hand, the storage issues are of no importance in the on-the-fly scheme, the computation time, on the other hand, is maximal. This is due to the reconstruction of the Hamiltonian matrix at each eigensolver iteration.

In the partial storage scheme, one does not store the 2p-2h part of the Hamiltonian matrix, consisting of two-body matrix elements multiplied by a phase, which basically form the whole Hamiltonian matrix. Indeed, the remaining part of the Hamiltonian matrix is made of its diagonal and 1p-1h off-diagonal matrix elements, which is minute compared to the rest of the Hamiltonian matrix. Moreover, the number of two-body matrix elements is much smaller than the number of Hamiltonian matrix elements. Therefore, it is more efficient to store the index of the two-body matrix element and associated phase in a single integer, instead of storing the Hamiltonian matrix element itself. The Hamiltonian matrix element is recovered from a look-up into the array of two-body matrix elements, followed by a multiplication by its corresponding phase. However, this approach is more expensive than the full storage scheme because of the search of two-body matrix elements. Indeed, the array of two-body matrix elements is large and the two-body matrix elements of the look-up procedure are not contiguous in general. Therefore, the search process induces time lags absent from the full storage scheme.

Let us quantify the memory gain of the partial storage scheme compared to the full storage scheme. In the full storage scheme, one stores the index of Slater determinants $|SD_f\rangle$, which is an integer of 4 bytes, and the Hamiltonian matrix element, which is a complex number of 16 bytes. This adds up to 20 bytes. In the partial storage scheme, one still has to store the index of Slater determinants $|SD_f\rangle$, but, instead of the Hamiltonian matrix element, one stores an integer of 4 bytes, from which the index of the two-body matrix element in an array and associated phase can be recovered. Therefore, one stores 8 bytes in the partial storage scheme, as compared to the 20 bytes of the full storage scheme, hence obtaining the gain factor of 2.5. Consequently, the partial storage scheme is a compromise between full storage and on-the-fly schemes. Indeed, the partial storage scheme is faster than the on-the-fly scheme, as the Hamiltonian matrix is not reconstructed for every $\hat{H}$ times vector operation. However, the partial storage scheme still leads to significant memory requirements, albeit not as large as with the full storage method.

### 5.8.3  2D Partitioning Parallelization Scheme of the Hamiltonian

A 2D partitioning scheme has been implemented in Gamow shell model [28]. It allows in particular to distribute the Gamow shell model vectors among all nodes, so that memory usage is optimized for vectors as well. The symmetry of the Hamltonian $\hat{H}$ is included in this scheme. Moreover, the overlapping between message passing interface communications and calculations has been introduced therein. In this method, using a hybrid message passing interface/open multiprocessing scheme, a single thread takes care of message passing interface data transfer, while all the others are dedicated to matrix-vector multiplication. As message passing interface communications become of the same order of magnitude as matrix-vector multiplication when Gamow shell model dimensions increase, this scheme allows to save a significant amount of time.

In the 2D partitioning scheme, $\hat{H}$ is divided in $n_d(n_d + 1)/2$ squares, equal to $N$, where both $|SD_i\rangle$ and $|SD_f\rangle$ Slater determinants scale with the number of message passing interface nodes. The 1p-1h part of $\hat{H}$ is inexpensive in terms of both time and memory and hence will not be considered. Let us concentrate firstly on the neutron 2p-2h part of $\hat{H}$. For each Slater determinant $|SD_i\rangle$ present in the column, one loops over all intermediate configurations $C_{int}$ and Slater determinants $|SD_{int}\rangle$ (see Sect. 5.8.2). One then obtains the $\langle SD_{int}|a_\delta\, a_\gamma|SD_i\rangle$ phase and associated one-body states. Using the same procedure on $|SD_{int}\rangle$, one generates $\langle SD_f|a_\alpha^\dagger\, a_\beta^\dagger|SD_{int}\rangle$ phase and associated one-body states, so that the two-body phase $\langle SD_f|a_\alpha^\dagger\, a_\beta^\dagger\, a_\delta\, a_\gamma|SD_i\rangle$ is obtained with its one-body states.

The proton 2p-2h part of $\hat{H}$ writes evidently the same. The proton-neutron 2p-2h part of $\hat{H}$ is very similar, except that there is no intermediate Slater determinant:

$$\langle SD_f|\hat{H}|SD_i\rangle = \langle \alpha_p \, \beta_n|\hat{V}|\gamma_p \, \delta_n\rangle$$
$$\times \langle SD_{p\,(f)}|a^\dagger_{\alpha_p}|SD_{p\,(int)}\rangle \langle SD_{p\,(int)}|a_{\gamma_p}|SD_{p\,(i)}\rangle$$
$$\times \langle SD_{n\,(f)}|a^\dagger_{\beta_n}|SD_{n\,(int)}\rangle \langle SD_{n\,(int)}|a_{\delta_n}|SD_{n\,(i)}\rangle\,, \qquad (5.70)$$

where $p$ and $n$ refer to proton and neutron state, respectively. Otherwise, the computational method is the same as with the neutron 2p-2h part of the $\hat{H}$ matrix.

$m$-reversal symmetry is also taken into account for even-even nuclei, because $[\hat{H}, \hat{T}] = 0$ if $m = 0$, so that it allows to calculate only half of the output Gamow shell model vector when one multiplies $\hat{H}$ by an input Gamow shell model vector.

At this point, $\hat{H}$ has been stored in $N$ squares. There are no message passing interface communications because the phases needed in one node are already stored therein. For the $\hat{H}$ times vector operation, one uses row and column communicators, as in Ref. [34], which connect each $n_d$ nodes, so that message passing interface communications involve $n_d$ nodes within the 2D partitioning scheme. A simple example of matrix and vector distribution in the 2D scheme is depicted in Fig. 5.3.

Let us now describe the algorithm of matrix times vector when one applies $\hat{H}$ on a Gamow shell model vector within the 2D partitioning scheme, of dimension $d$:

- All occupied squares (see Fig. 5.3) are considered simultaneously.
- If one is on a diagonal square, whose associated node, i.e., the node where this square is stored, is the master node of the row and column communicators, one distributes its input Gamow shell model vector part to all the nodes of the row and column via the row and column communicators. For example, on Fig. 5.3, the nodes 2, 4, and 15 form a row communicator on the second row, whose master node is node 4, while the nodes 4, 5, and 6 form a column communicator. The

**Fig. 5.3** Example of the 2D partitioning scheme for the Hamiltonian matrix. Occupied squares are denoted by numbers and unoccupied squares by dashed lines (from Ref. [28])

| 1 | --- | --- | 12 | 14 |
|---|-----|-----|----|----|
| 2 | 4 | --- | --- | 15 |
| 3 | 5 | 7 | --- | --- |
| --- | 6 | 8 | 10 | --- |
| --- | --- | 9 | 11 | 13 |

row master node then casts the part 2 of the input Gamow shell model vector to its row and column.

- Using message passing interface overlapping, one multiplies at the same time the considered $\hat{H}$ square by the input Gamow shell model vector part to obtain a part of the output Gamow shell model vector. The latter is then reduced on the row via the row communicator.
- Using message passing interface overlapping, one multiplies at the same time the considered $\hat{H}$ square by the storage vector mentioned above, which takes care of the symmetric part of $\hat{H}$, not stored. A table of $n_c$ vectors is used therein for output, where $n_c$ is the number of threads per node, in order to avoid a race condition, i.e., a modification of the same variable by several threads at the same time, which leads to unpredictable results (see Ref. [34]). The race condition would occur because input and output indices are exchanged when one uses $\hat{H}$ symmetry.
- All resulting parts of $\hat{H}$ squares times input Gamow shell model vector parts are summed among the $n_c$ threads and then reduced on the column via the column communicator. The output Gamow shell model vector is then obtained.

The cost of message passing interface communications is then that of the transfer of 4 $d/n_{\mathrm{d}}$ complex numbers, with two transfers out of four which overlap with multiplications.

### 5.8.4   Treatment of the Angular Momentum Projection

The basis of Slater determinants is not rotationally invariant. While the total angular momentum projection $M$ is conserved, the total angular momentum $J$ is not. As explained in Ref. [36], both the requirements of rotational invariance and antisymmetry at basis level considerably increase complexity in building $\hat{H}$. Moreover, despite the smaller dimensions involved in a $J$-conserving basis, the relative number of nonzero matrix elements is much larger. Hence, the use of a basis of Slater determinants is more efficient computationally.

As $[H, \mathbf{J}] = 0$, applying $\hat{H}$ to a linear combination of Slater determinants coupled to $J$ still provides a vector coupled to $J$. However, in practice, $\hat{H}$ and $\mathbf{J}$ do not exactly commute due to numerical inaccuracy, so that the $J$ quantum number is eventually lost after several matrix-vector multiplications.

It is possible to recover rotational invariance by suppressing the Gamow shell model vector components with $J' \neq J$. For this, one uses the Löwdin operator [37]:

$$\hat{P}_J = \prod_{J' \neq J} \frac{\hat{J}^2 - J(J+1)}{J'(J'+1) - J(J+1)}, \qquad (5.71)$$

which removes all the angular momenta $J' \neq J$. The $J$ quantum number conservation has been checked to be numerically precise up to at least $10^{-10}$ in Gamow shell model applications.

The standard formulation $\hat{J}^2 = \hat{J}^- \hat{J}^+ + M(M + 1)$ is utilized to apply $\hat{J}^2$, where $\hat{J}^\pm$ is a sparse one-body operator. The computation of $\hat{J}^2$ times vector is thus fast. The main problem, however, is that the direct application of 2D partitioning is rather inefficient, that due to the block structure of the $\hat{J}^\pm$ operator matrix. Indeed, as $\hat{J}^\pm$ cannot connect two different configurations, it is entirely contained in a diagonal square of its matrix, except for the very few Slater determinants situated in two adjacent nodes and belonging to the same configurations. Therefore, a naive application of 2D partitioning to the $\hat{J}^\pm$ matrix would have the diagonal nodes associated to $\hat{J}^\pm$ to contain virtually all the $\hat{J}^\pm$ matrix, so that all the other nodes would be inactive. Hence, the $\hat{J}^\pm$ routines have been parallelized with a hybrid 1D/2D method, where a column of the $\hat{J}^\pm$ matrix is stored on one node. Gamow shell model vectors are then divided into $N$ parts, similarly to the 2D partitioning scheme, so that the output Gamow shell model vector can still be reduced on each node after application of the $\hat{J}^\pm$ operator. Considering that each node contains a part of the diagonal squares forming the $\hat{J}^\pm$ matrix, the memory distribution of the $\hat{J}^\pm$ matrix is well balanced.

$\hat{J}^\pm$ is not symmetric because it connects Slater determinants of total angular momentum projection $M$ and $M \pm 1$. Hence, one does not have to account for symmetry in the parallelization of $\hat{J}^\pm$ times vector, contrary to the 2D partitioning of the $\hat{H}$ matrix (see Sect. 5.8.3).

In fact, the fundamental problem here is message passing interface data transfer. Indeed, the number of angular momenta $J'$ to suppress (see Eq. (5.71)) is of the order of 50, so that, naively, one would have to do two message passing interface transfers of a full Gamow shell model vector per $\hat{J}^\pm$ application. This would lead to 200 message passing interface transfers of a full Gamow shell model vector per $\hat{P}_J$ application, which is prohibitively large. In order to solve this problem, one takes advantage of the block structure of the $\hat{J}^\pm$ matrix. Indeed, the necessary message passing interface transfers involve only neighboring nodes in Gamow shell model vectors. Thus, their total volume is that of the nonzero components of a configuration in between two nodes, and not the $d$ dimension, so that the amount of data to transfer is tractable. Moreover, the application of $\hat{P}_J$ occurs only a few times, because the $J$ quantum number is most often numerically conserved after a $\hat{H}$ application. Therefore, the projection on $J$ quantum number does not significantly slow down the implementation of eigenvectors.

In order to know whether or not it is necessary to apply $\hat{P}_J$, one has to test if a Gamow shell model vector is an eigenstate of the $\hat{J}^2$ operator. If $M = J$, the action of $\hat{J}^+$ on a $J$-coupled Gamow shell model vector provides zero, because one cannot generate a $J$-coupled many-body state whose angular momentum projection is $J + 1$. Hence, $\hat{P}_J$ has to be applied if the action of $\hat{J}^+$ on a Gamow shell model vector provides with a nonzero vector. This operation does not create any sizable computational burden because the $\hat{J}^+$ matrix is very sparse, so that the time taken by the $\hat{J}^+$ matrix times vector operator is negligible compared to a full $\hat{P}_J$ application. If $M \neq J$, it is convenient to apply $\hat{J}^2 - J(J + 1)$ to the considered Gamow shell model vector, because it is zero only if the considered Gamow shell model vector is coupled to $J$. Hence, as the application of $\hat{J}^\pm$ is done only once or twice per iteration, the previous test is very fast compared to the application of $\hat{H}$ or $\hat{P}_J$ operators.

## 5.9   Diagonalization of Very Large Gamow Shell Model Matrices with the Density Matrix Renormalization Group Approach

In order to be able to determine Gamow shell model eigenvalues almost exactly, the density matrix renormalization group approach has been developed and applied in the non-Hermitian case of the Gamow shell model [38, 39]. The density matrix renormalization group approach [40, 41] allows to calculate eigenstates of matrices whose dimensions are beyond capabilities of standard Krylov methods. The method has been tested in nuclear physics in the frame of standard shell model [42–45]. It was generalized to the non-Hermitian Gamow shell model several years ago, firstly for a space of valence neutrons [38], and then for a for space containing both protons and neutrons [39].

The application of density matrix renormalization group approach to Hermitian matrices of the nuclear shell model has shown that convergence to an exact result is difficult to attain due to the large matrix elements of inter-shell coupling of the nucleon-nucleon interaction [43–45]. In fact, the density matrix renormalization group is especially effective when the couplings induced by the used interactions are weak. This is particularly the case for spin chains [40, 41], where obtained binding energies are almost exact.

Initially, the density matrix renormalization group method had been developed in the context of real-energy physics only [40–42]. Consequently, the density matrix renormalization group method had firstly to be generalized to the use of non-Hermitian Hamiltonians in order to applied in the context of the Gamow shell model. While the construction of the complex symmetric matrices representing reduced density and Hamiltonian operators in the Berggren basis is straightfoward within the theory of rigged Hilbert spaces (see Sect. 5.1.1), the problem of the identification of resonance eigenstates among the numerous scattering eigenstates remains, as in the Gamow shell model (see Sect. 5.4). In order to solve this problem, one can note that continuum coupling typically generates small configuration mixing among scattering configurations. Hence, the use of the overlap method (see Sect. 5.4), devised in the Gamow shell model along with the Jacobi-Davidson method (see Sect. 5.7), can be applied as well in the density matrix renormalization group framework. This allows to efficiently calculate the eigenstates of the Hamiltonian arising from the density matrix renormalization group method. Moreover, due to the weak couplings occurring between the pole and scattering part of the Gamow shell model space, one can reasonably assume that the density matrix renormalization group used within the Gamow shell model will converge quickly. In fact, the application of the density matrix renormalization group method in the Gamow shell model is similar to that occurring in weakly coupled spin chains [40,41], in contrast to the situation encountered in standard shell model, where couplings are large and convergence slow [43–45].

The aim of the density matrix renormalization group method is to iteratively build bases becoming more and more correlated. In these correlated bases, Gamow shell model eigenstates can be expanded using a much smaller number of basis states, typically a few thousands. For this, the Gamow shell model valence space is separated in two spaces A and B. The space A is firstly constructed from the one-body resonant states of the Berggren basis. One also has to add to the space A a few one-body nonresonant states of quantum numbers $\ell, j$ different from those of resonant states. These additional nonresonant states are necessary in order to generate all possible couplings present in the Hamiltonian [39]. The space B is generated during the density matrix renormalization group process, by adding one by one the remaining scattering states of the Berggren basis (see Fig. 5.4). Its elements are numbered by $i_B = 0, 1, \cdots$.

One then starts the "warm-up" phase [38]. For this, the configurations $n_A$ and total angular momentum $j_A$, denoted as $|k_A\rangle$ (see Fig. 5.4), are built in the A space. Sub-operators of the Hamiltonian, as for example $(a^+ a^+)^K$, where $a^+$ is a creation operator and $K$ an angular momentum, are also calculated at this stage. Then, a new scattering shell of quantum numbers $(\ell, j)$ is added (see Fig. 5.4), and

**Fig. 5.4** Schematic illustration of the density matrix renormalization group procedure. States $\{k_A\}$ from the A reference space and $\alpha_B$ states from the B space are depicted. The newly added shell $(lj)_s$ to the B space generates the new basis states $\{k_A \otimes \{\alpha_B \otimes (lj)_s\}\}^J$ (from Ref. [38])

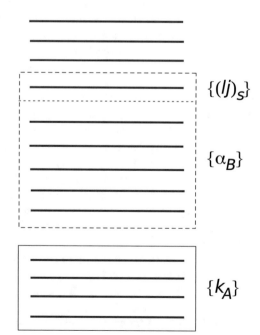

the configurations $|i_B\rangle$ for all possible number of particles $n_B$ and total angular momentum $j_B$ are constructed. This augments the size of the B space and generates a set of basis states $(|k_A(n_A, j_A)\rangle \otimes |i_B(n_B, j_B)\rangle)^J$. $n_{A(B)}$ and $j_{A(B)}$ are respectively the numbers of particles and angular momenta of the $k_A(i_B)$ states and $J$ is the total angular momentum of the calculated eigenstate.

When the number of configurations $i_B$ of the B space reaches $N_{opt}$, of the order of 10 to 100 typically, the Hamiltonian is diagonalized to provide an approximation of the calculated nuclear state:

$$|\Psi_J\rangle = \sum_{k_A, i_B} c_{i_B}^{k_A} \{|k_A\rangle \otimes |i_B\rangle\}^J . \tag{5.72}$$

The latter has been determined by the overlap method (see Sect. 5.4) [21]. The reduced density matrix $\hat{\rho}_{i_B i'_B}^B$ is thus calculated from Eq. (5.72):

$$\hat{\rho}_{i_B i'_B}^{j_B} = \sum_{k_A} c_{i_B}^{k_A} c_{i'_B}^{k_A} , \tag{5.73}$$

where the angular momentum $j_B$, which is the same for $|i_B\rangle$ and $|i_{B'}\rangle$, is fixed. The reduced density matrix is then diagonalized, and the resulting eigenvectors which possess the largest eigenvalues in modulus (at most $N_{opt}$ of them) are kept in the model space, the other ones being rejected. This method is indeed motivated by the variational principle, which states that the wave function bearing the largest overlap with the exact nuclear state is the one built from the aforementioned eigenvectors [40, 41]. The eigenvectors then generate the new $B$ space, and the sub-operators associated to the Hamiltonian are recalculated in this new space.

When all the shells of the Berggren basis have been taken into account, the "warm-up" phase terminates and the "sweep" phases start [38]. They are similar to the "warm-up" phase, except that the states of the space A are now replaced by the $(|k_A(n_A, j_A)\rangle \otimes |i_B(n_B, j_B)\rangle)^J$ states, issued from the previous iteration. The scattering states $(n, \ell, j)$ of the Berggren basis are then reintroduced one by one, with the index associated to $(n, \ell, j)$ firstly decreasing ("sweep down") and afterward increasing ("sweep up") [38,39]. Indeed, the truncation applied after each diagonalization of the reduced density matrix suppresses correlations, which must be recovered during sweeps. Convergence is obtained after a few "sweeps" [38,39]. Everytime a new scattering shell is added during a sweep, which translates into diagonalization of new reduced density matrice and recalculation of Hamiltonian sub-operators, by definition, a "step" occurs. The speed of convergence is measured via their number $N_{step}$.

The density matrix renormalization group approach has been tested by calculating the $3/2_1^-$ ground state and the $1/2_1^-$ first excited state of $^7$He [38] and the ground states of the $^7$Li and $^8$Li nuclei [39], using the same Hamiltonian as in Sect. 5.4. The rapid convergence of the density matrix renormalization group approach is illustrated in Fig. 5.5 for the $^7$He and $^7$Li nuclei.

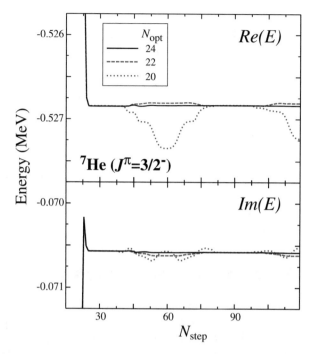

**Fig. 5.5** Convergence of the real (upper part) and imaginary (lower part) parts of the energy of $3/2_1^-$ ground state of $^7$He, as a function of $N_{step}$ for several values for $N_{opt}$ (from Ref. [38])

## 5.10   Statistical Evaluation of Modeling Errors and Quality of Predictions in the Gamow Shell Model

The Gamow shell model is used with effective nuclear interactions which are fitted to reproduce experimental data. These data usually consist of energies, but can also be phase shifts, as is the case to fix core potential parameters, or if one uses nucleon-nucleon interactions in the no-core Gamow shell model. Consequently, all calculated observables will contain errors arising from theoretical approximations, which must be quantitatively assessed if one aims at using the Gamow shell model to make sensible predictions. Experimental uncertainties are much smaller than theoretical uncertainties and can be neglected when modeling errors in the Gamow shell model [13].

The most common methods in the statistical error analysis related to theoretical calculations in nuclear physics involve linear regression and Bayesian analysis. The fundamental assumption of linear regression is that observable quantities vary linearly with parameters. Due to its simple formulas, hence convenient to implement, linear regression is used in the Gamow shell model. The Bayesian analysis is very promising to determine the plausibility of modeling assumptions when making predictions for observables. Moreover, it can be applied along with model averaging. In the context of Gamow shell model, model averaging consists of evaluating the same observable quantity with different sets of parameters in the interactions used so as to minimize the error made on the observables arising from all made calculations.

The Bayesian approach has been implemented in density functional theory, where it is used to assess the errors made on the parameters of used Skyrme interactions along with model averaging by comparing the results obtained in density functional theory with those from a finite-range liquid drop model [46]. The Bayesian analysis has also recently been applied to reaction theory [47, 48], where the theoretical uncertainties in calculated cross sections can be calculated from the used fitted optical potentials. While the Bayesian analysis has not been yet fully implemented in the Gamow shell model, introductory calculations have shown that it can provide with more efficient estimators of theoretical statistical errors of observables in the Gamow shell model.

## 5.10.1  Penalty Function and Linear Regression Approximation

The optimization of the interaction and the assessment of statistical uncertainties is performed according to Refs. [49, 50]. Given a model in which $N_p$ parameters $\mathbf{p} = \{p_1, ., p_{N_p}\}$ are adjusted to describe $N_d$ observables $\bar{O}_i$ $(i = 1, ., N_d)$, the optimization procedure is based on the minimization of the penalty function:

$$\chi^2(\mathbf{p}) = \frac{1}{2} \sum_{i=1}^{N_d} \left( \frac{\bar{O}_i(\mathbf{p}) - \bar{O}_i^{\text{exp}}}{\delta \bar{O}_i} \right)^2 , \qquad (5.74)$$

where $\bar{O}_i(\mathbf{p})$ are the calculated observables and $\bar{O}_i^{\text{exp}}$ are the experimental data (fit-observables) used to constrain the model. The adopted errors $\delta \bar{O}_i$ include different contributions stemming from experimental uncertainties, numerical inaccuracies, and theoretical errors due to model deficiency.

The choice for the theoretical error obviously involves a certain level of arbitrariness even when driven by physical considerations. Part of this arbitrariness can be removed by tuning the adopted errors so that they are consistent with the distribution of the residuals similarly to the case of a purely statistical distribution [50]. In particular, one requires the total penalty function to be normalized to the number of degrees of freedom $N_{\text{dof}} = N_d - N_p$ at the minimum $\mathbf{p}_0$ [51]:

$$\frac{\chi^2(\mathbf{p}_0)}{N_{\text{dof}}} \leftrightarrow 1 . \qquad (5.75)$$

In the case of a single type of data and on the assumption that experimental and numerical errors are negligible, the condition (5.75) can simply be achieved through a global scaling of the initial adopted errors:

$$\delta \bar{O}_i \rightarrow \delta \bar{O}_i \sqrt{\chi^2(\mathbf{p}_0)/N_{\mathrm{dof}}} \, . \tag{5.76}$$

With this choice of the normalization condition, one can apply the standard rules of linear regression in statistical analysis and assess quantities such as the covariance matrix and statistical uncertainties. The Jacobian $J$ associated to $\chi^2(\mathbf{p})$ and its Hessian matrix $\hat{H}$ are straightforward to calculate from Eq. (5.74):

$$\hat{\bar{J}} = \left\{ \frac{1}{\delta \bar{O}_i} \frac{\partial \bar{O}_i}{\partial p_j}(\mathbf{p}) \bigg|_{\mathbf{p}=\mathbf{p}_0} \right\}_{ij} \tag{5.77}$$

$$\hat{\bar{H}} = \left\{ \frac{\partial^2 \chi^2}{\partial p_i \, \partial p_j}(\mathbf{p}) \bigg|_{\mathbf{p}=\mathbf{p}_0} \right\}_{ij} = \hat{\bar{J}}^T \hat{\bar{J}} \, , \tag{5.78}$$

where the two first equations are valid in the general case whereas the last equation pertains only to the linear approximation (see Exercise VII).

---

**Exercise  VII** ⋆
Derive Eq. (5.78) by differentiating Eq. (5.74).

---

The probability distribution verified by $\mathbf{p}$ is somewhat arbitrary and has to be postulated based on physical grounds. One can demand that it has a Gaussian dependence with respect to $\chi^2(\mathbf{p})$ [50]:

$$\bar{P}(\mathbf{p}) \propto \exp(-\chi^2(\mathbf{p})) \, , \tag{5.79}$$

where its overall factor is determined by normalization.

Equation (5.79) can be understood intuitively: if $\chi^2(\mathbf{p})$ is very sharp around the minimum for some parameters, the latter are unlikely to be much different in the optimal set of parameters because modifying them would provide with results of poor quality for the already fitted observables. Conversely, if $\chi^2(\mathbf{p})$ has a flat bottom therein, these parameters are sloppy and can significantly vary if other observables are added to the fitting procedure, for example. The covariance matrix $\bar{C}(\mathbf{p}_0)$ can then be expressed in terms of the Hessian matrix $\hat{H}$ and hence Jacobian $J$ of Eqs. (5.77) and (5.78):

$$\bar{C}(\mathbf{p}_0) = \hat{\bar{H}}^{-1} = (\hat{\bar{J}}^T \hat{\bar{J}})^{-1} \, . \tag{5.80}$$

One will now implicitly assume that $\bar{C}$ is calculated in $\mathbf{p}_0$. The covariance between two observables $A$ and $B$ follows as:

$$\overline{\Delta A \, \Delta B} \simeq \sum_{\alpha,\beta=1}^{N_p} \frac{\partial A}{\partial p_\alpha}\bigg|_{\mathbf{p}_0} \bar{C}_{\alpha\beta} \frac{\partial B}{\partial p_\beta}\bigg|_{\mathbf{p}_0} . \tag{5.81}$$

In particular, for $A = B$, Eq. (5.81) gives the statistical uncertainty of the observable $A$:

$$\Delta A = \sqrt{\overline{\Delta A^2}} . \tag{5.82}$$

and the dimensionless correlation coefficient [49] is defined as:

$$c_{AB} = \frac{\overline{\Delta A \, \Delta B}}{\Delta A \, \Delta B} . \tag{5.83}$$

Applying Eqs. (5.81) and (5.83) to the model parameters, their statistical uncertainties reduce to $\Delta p_\alpha = \sqrt{\bar{C}_{\alpha\alpha}}$, and the correlation coefficients between two parameters $p_\alpha$ and $p_\beta$ are related to the covariance matrix elements by:

$$c_{\alpha\beta} = \frac{\bar{C}_{\alpha\beta}}{\sqrt{\bar{C}_{\alpha\alpha}\bar{C}_{\beta\beta}}} . \tag{5.84}$$

A demonstration of the validity of the fundamental equations used in linear regression is done in Exercise VIII. In principle, Eqs. (5.81) and (5.83) are valid for the model parameters and the observables to be predicted by the model. However, these do not hold for the observables defining parameters during the fitting procedures, as those are already well constrained around the minimum. The uncertainties on the adjusted observables can only be assessed through a full statistical analysis, such as the Bayesian inference.

---

**Exercise VIII ⋆⋆**

One will demonstrate the fundamental equations used in linear regression to calculate statistical errors.

**A.** Derive Eq. (5.80) directly from the assumption that the covariance matrix of differences

$$\left(\bar{\mathbf{O}}_\epsilon\right)_i = \frac{\bar{\mathbf{O}}_i(\mathbf{p}) - \bar{\mathbf{O}}_i^{\text{exp}}}{\delta \bar{O}_i},$$

is the identity matrix. For this, one may use the following identity, valid for $\mathbf{p} \sim \mathbf{p}_0$ in linear regression: $\hat{H}(\mathbf{p} - \mathbf{p}_0) = \hat{J}^T \bar{\mathbf{O}}_\epsilon$.

**B.** Explain why this assumption is consistent with the renormalization introduced in Eq. (5.76) when an overall scaling factor is used.

## 5.10.2 Bayesian Inference of Parameters

The probability distribution associated to the fitted parameters as in Eq. (5.79) has not yet been quantitatively assessed, even though it is of the undeniable interest. The standard procedure to calculate statistical errors in that framework is by using Bayesian inference. For this, a prior probability distribution of parameters is postulated, from which a posterior probability distribution is calculated, which can then confirm or infirm the hypothesis borne by the prior.

The fundamental equation determining the posterior is the Bayes theorem:

$$\bar{P}(\mathbf{p}|\bar{O}_{\text{pred}}) \propto \bar{P}(\mathbf{p})\bar{P}(\bar{\mathbf{O}}|\mathbf{p}) , \qquad (5.85)$$

where $\bar{O}_{\text{pred}}$ is a predicted observable, in general different from the $\bar{O}_i$ observables with which the model parameters are fitted. In this expression, $\bar{P}(\mathbf{p})$ is the prior probability distribution and $\bar{P}(\bar{\mathbf{O}}|\mathbf{p})$ is the so-called likelihood function. $\bar{P}(\mathbf{p}|\bar{O}_{\text{pred}})$ is the posterior probability distribution inferred by the Bayes theorem, where its overall factor is determined by normalization. The main difference with Eq. (5.79) is the appearance of the likelihood function, which has to be determined along with the prior. For this, one introduces the statistical model verified by the predicted observable $\bar{O}_{\text{pred}}$:

$$\bar{O}_{\text{pred}} = \bar{O}(\mathbf{p}) + \delta(\bar{\mathbf{O}}) + \epsilon , \qquad (5.86)$$

where $\delta(\bar{\mathbf{O}})$ and $\epsilon$ are the probability distributions of modeling and measurement errors, respectively, with $\delta(\bar{\mathbf{O}})$ a function of fitted observables $\bar{O}_i$. $\delta(\bar{\mathbf{O}})$ and $\epsilon$ are typically fitted with Gaussian processes [52]. Consequently, $\bar{P}(\mathbf{p})$ and $\bar{P}(\bar{\mathbf{O}}|\mathbf{p})$ can be calculated, as $\bar{O}(\mathbf{p})$, $\delta(\bar{\mathbf{O}})$, and $\epsilon$ are well defined.

Even though the Bayesian inference of parameters has not been implemented in the Gamow shell model code, preliminary calculations have been done and compared to the linear regression formulas. As could be expected, the errors provided by these two approaches for the well determined parameters are close, as they differ by a factor of about two at most. Thus, one can consider that the same amount of information is provided by linear regression and Bayesian inference formulas in this context and that the linear approximation is reliable for most important parameters.

On the contrary, errors can differ by up to two orders of magnitude for poorly defined parameters. In this case, the Bayesian inference might be necessary to precisely analyze the causes of the weak dependence of these parameters on observables in order to better constrain the model. Nevertheless, linear regression is qualitatively correct for that matter, as the parameters which are sloppy therein are also poorly defined using Bayesian inference.

Undeniably, a full statistical study of fitted interactions in the Gamow shell model will have to be effected with Bayesian statistics, as it will allow to precisely determine the quality of fitted parameters. One can consider adding additional observables for the fit of interaction parameters to better constrain them. Alternatively, as linear regression can both quantify errors made on well constrained parameters and identify poorly constrained parameters, it is a very reliable tool to test fitted interactions despite the simplicity of linear approximation. It can then be used to quantitatively analyze modeling errors in the Gamow shell model.

## Solutions to Exercises[1]

**Exercise I.**

**A.** The Berggren basis contours are discretized with the Gauss-Legendre quadrature. This method is very precise even with a small number of points as integrated functions in the $k$-space are smooth. Multiplying the number of points by two only slightly increases the numerical precision.

**B.** The calculation involving the Berggren basis contours either too close to the resonance in a basis or too deep in the complex plane are not precise, even with a large number of discretized scattering states. On the contrary, using a Berggren basis contour in between the latter contours lead to precise results, even if the number of discretized scattering states remains rather small.

**C.** If a Berggren basis contour is too close to a resonance of the basis, the phase shifts of scattering states vary very much in the vicinity of resonances. Consequently, the Gauss-Legendre discretization becomes imprecise. When basis scattering states are far away from the real axis, the imaginary parts of two-body matrix elements become very large, so that important numerical cancellations occur between Hamiltonian matrix elements involving scattering states. Thus, both these effects generate numerical inaccuracies.

Increasing the number of discretized scattering states when the $p_{3/2}$ contour is close from the $0p_{3/2}$ resonance does not ameliorate the situation as the change of phase shifts close to the $0p_{3/2}$ resonance is large for all discretization schemes. The numerical cancellations between large two-body matrix elements still exist even if one increases the number of discretized scattering states when

---

[1]The input files, codes and code user manual associated to computer-based exercises can be found at https://github.com/GSMUTNSR.

the $p_{3/2}$ contour is far from the $0p_{3/2}$ resonance. Consequently, there exists an optimal contour between these two extremes. It has to be sufficiently far from basis resonance states for scattering states to have smoothly varying phase shifts on the Berggren basis contour, and sufficiently close to the real axis for imaginary parts to remain small as well.

## Exercise II.

**A.** When one deals with bound states, adding more and more Lanczos or Jacobi-Davidson vectors enlarge the many-body model space in which the Hamiltonian matrix is diagonalized. Consequently, due to the variational principle, the eigenstate gains binding energy when the model space is enlarged. However, there is no such property for unbound states. Indeed, the generalized variational principle only states that eigenvalues are stationary when the eigenstate is slightly modified. Energies and widths converge to their exact values by adding more and more Lanczos or Jacobi-Davidson vectors. However, this convergence does not necessarily translate into increase of binding energy of the system, as is the case for bound states.

**B.** The Berggren basis is complete, so that taking a sufficiently large number of discretized scattering states insures that completeness will be numerically precise. Consequently, eigenvalues do not depend on the basis used. As the eigenenergies of bound states are real, the Hamiltonian matrix represented in the Berggren basis has real negative eigenvalues, even though it is complex symmetric. The fact that basis states are complex only implies that subtle numerical cancellations have to occur in order to obtain real eigenvalues.

**C.** Narrow resonance states are quasi-stationary, i.e., long lived. Hence, the statistical error on observables $A(t)$ (see Eq. (5.9)), denoted as $\Delta_t(A)$ in Eq. (5.10), is small. This explains the small imaginary part of the observables calculated in the Gamow shell model, which are proportional to statistical errors (see Eq. (5.17)). On the contrary, broad resonance states have a short lifetime. As a consequence, the physical observables $A(t)$ associated with broad resonances bear a large statistical error (see Eqs. (5.10) and (5.17)). This is visible on the complex values of observable quantities, where both real and imaginary parts are of the same order of magnitude (see Eqs. (5.16) and (5.17)).

## Exercise III.

**A.** Equations (5.57–5.59) contain only spherical tensor operators. The Wigner-Eckart theorem can then be used to calculate matrix elements. It is thus sufficient to consider $M_L = L$.

**B.** One has:

$$r_{lab}^L \, Y_{LL}(\Omega_{lab}) \propto r_{lab}^L \, \sin^L(\theta_{lab}) \, e^{iL\varphi_{lab}} \propto (x_{lab} + i \, y_{lab})^L \,.$$

Equation (5.51) is a direct consequence of Eq. (5.24) and of the binomial theorem.

Conversely, $(X_{\text{core}} + i\, Y_{\text{core}})^k \propto R_{\text{core}}^k\, Y_{kk}(\Omega_{\text{core}})$. Consequently, the core terms in Eq. (5.51) are scalar operators only for $k = 0$. As the core wave function is coupled to $0^+$, all recoil terms, bearing $k > 0$, vanish identically in Eq. (5.51).

C. $\mathbf{l}_{\text{lab}}$ and $\mathbf{s}$ in Eq. (5.58) are spherical tensors of order 1. As $M_L = L$, one necessarily has $M_L - 1$ for the $M$-projection of $Y_{L-1,M}$, and 1 for those of $\mathbf{l}_{\text{lab}}$ and $\mathbf{s}$. Equations (5.52) and (5.53) then follow from Eqs. (5.58) and (5.59). Equation (5.53) can be treated as Eq. (5.51) because spin is space-independent.

D. The orbital operator is handled from the equality:

$$\mathbf{l}_{\text{lab}} = \mathbf{r}_{\text{lab}} \times \mathbf{p}_{\text{lab}} = (\mathbf{r} + \mathbf{R}_{\text{core}}) \times \mathbf{p} = \mathbf{l} + \mathbf{R}_{\text{core}} \times \mathbf{p}\,,$$

where one used Eqs. (5.24) and (5.31). Equation (5.54) is then straightforward to deduce from Eqs. (5.24) and (5.31).

Equation (5.55) is obtained similarly to Eq. (5.51). $l_1$, $p_x$, $p_y$, and $p_z$ are operators involving valence nucleons only in Eq. (5.55). Consequently, the first and second terms of Eq. (5.55) can be treated as that of Eq. (5.51). $Z_{\text{core}}$ is proportional to $R_{\text{core}}\, Y_{10}(\Omega_{\text{core}})$ because $Z_{\text{core}} = R_{\text{core}}\,\cos(\theta_{\text{core}})$. Therefore, $(X_{\text{core}} + i\, Y_{\text{core}})^k\, Z_{\text{core}}$ is proportional to $R_{\text{core}}^{k+1}\, Y_{kk}(\Omega_{\text{core}})\, Y_{10}(\Omega_{\text{core}})$.

The right-hand side of Eq. (5.56) is obtained by using a standard property of spherical harmonics involving $Y_{kk}(\Omega_{\text{core}})\, Y_{10}(\Omega_{\text{core}})$. One notices that the sum in $k'$ of Eq. (5.56) possesses only non-scalar spherical tensors. Thus, they vanish identically when acted on the core wave function. Consequently, all recoil terms cancel out for magnetic transitions as well.

E. All the recoil terms induced by (5.51–5.53) vanish identically. As a consequence, laboratory coordinates can be formally replaced by cluster orbital shell model coordinates in Eqs. (5.57–5.59) without changing the results.

## Exercise IV.

A. The ground state of $^7$He mainly consists of the configuration having two protons and two neutrons in the $0s_{1/2}$ shells and three neutrons in the $0p_{3/2}$ resonance shell. The $p_{1/2}$ partial wave is not occupied in the ground state of $^7$He when considering a Hartree-Fock approximation of the many-body state. Consequently, a model space built from a $^4$He core and valence neutrons occupying the neutron $p_{3/2}$ partial wave only recaptures the main features of the ground state of $^7$He.

B. The natural orbitals have been calculated from the scalar density matrix of the exact ground state of $^7$He. Consequently, the nucleon occupation of these natural orbitals decreases very quickly when the number of natural orbitals increases. About five natural orbitals per partial wave are typically needed to obtain convergence. This number is much smaller than the tens of discretized scattering states needed to numerically obtain completeness when discretizing the Berggren basis contour.

**C.** In this situation, the natural orbitals have been calculated from the scalar density matrix of an approximate ground state of $^7$He. Thus, nucleon occupation is not negligible even on high lying natural orbitals. Consequently, the ground state of $^7$He is only approximately described using a few natural orbitals, even without the truncation of model space. Indeed, the 3p-3h configurations present in a non-truncated model space cannot compensate for the missing high lying natural orbitals.

## Exercise V.

**A.** One can clearly identify bound states in the negative energy region. On the contrary, unbound resonant states are embedded in the discretized continuum of many-body scattering states, making them impossible to identify. This situation is qualitatively the same as that of Fig. 5.2.
**B.** As continuum coupling is small for $^{18}$O, the overlap between the eigenstates obtained in pole approximation and in full space is close to one. As a consequence, the identification of resonances can be made without ambiguity. As the Berggren basis is complete, resonant eigenstates are stationary by slightly changing the Berggren basis contour, in full analogy with Fig. 5.2.

## Exercise VI.

**A.** The ground state and the first excited state of $^6$He and $^6$Be are either weakly bound or resonance states. As one has either two valence neutrons or two valence protons, the two-body interaction is in $T = 1$ channel. Hence, the nuclear interaction does not provide with much binding energy. Consequently, the use of Berggren basis generated by the Woods-Saxon potential of the core can provide with numerically stable results. This explains why the bases generated by the multi-Slater determinant coupled Hartree-Fock and Woods-Saxon potentials are equivalent.
**B.** In the case of $^6$Li, one has a valence proton and a valence neutron, so that both $T = 0$ and $T = 1$ components of the nuclear interaction are present. The $T = 0$ part of the nuclear interaction is responsible for the well-bound character of $^6$Li. Indeed, the ground state of $^5$Li is a broad resonance, whereas that of $^6$Li is bound by 3.7 MeV. Moreover, the spectrum of $^6$Li has unbound states built from the $0p_{1/2}$ one-body resonance states, whose widths are small compared to that of $^5$Li. $0p_{1/2}$ one-body state has an extremely large width of 5–7 MeV when it is generated by the Woods-Saxon potential of a core. Consequently, the Berggren basis produced by the Woods-Saxon potential of the core has very large two-body matrix elements. This makes the calculation imprecise.

On the contrary, the multi-Slater determinant coupled Hartree-Fock potential removes strong continuum coupling, so that calculations are precise using its Berggren basis. Indeed, the $0p_{3/2}$ and $0p_{1/2}$ one-body states are either bound or narrow resonances in the multi-Slater determinant coupled Hartree-Fock basis, so that Hamiltonian matrix elements involving unbound states are smaller. Let us compare the results provided by the basis generated by the Woods-Saxon potential of the core and that of the multi-Slater determinant coupled Hartree-Fock basis. One can see that, with the basis generated by the Woods-Saxon potential of the core, $^6$Li bound states are too unbound by about 200 keV. Moreover, high lying $T = 0$ unbound states have too large widths, of several MeVs, instead of a few hundreds of keV or less, as provided by the multi-Slater determinant coupled Hartree-Fock basis. This implies that basis potential optimization is necessary in the Gamow shell model, as calculations become unstable if one uses a basis-generating potential which is not tailored to the considered nucleus.

**Exercise VII.**

A straightforward differentiation of Eq. (5.74) provides with $\hat{\bar{H}} = \hat{\bar{J}}^T \hat{\bar{J}} + \hat{\bar{E}}$, where $\hat{\bar{E}}$ is a matrix whose elements arise from second derivatives of $\bar{O}_j$. In the linear regression approximation, the $\bar{O}_j$ observables are supposed to vary linearly with $\mathbf{p}$, which implies that $\hat{\bar{E}} \simeq 0$. The equality $\hat{\bar{H}} = \hat{\bar{J}}^T \hat{\bar{J}}$ at linear regression approximation then follows.

**Exercise VIII.**

**A.** Let us calculate the covariance matrix $\bar{C}(\mathbf{p}_0)$ in linear regression (see Eq. (5.74) and Exercise VII). One readily obtains for $\mathbf{p} \sim \mathbf{p}_0$ that $\mathbf{p}_0 = \mathbf{p} - \hat{\bar{H}}^{-1} \hat{\bar{J}}^T \bar{\mathbf{O}}_\epsilon$. As $\mathbf{p}$ is fixed, one has:

$$\bar{C}(\mathbf{p}_0) = \bar{C}(-\hat{\bar{H}}^{-1} \hat{\bar{J}}^T \bar{\mathbf{O}}_\epsilon) = \hat{\bar{H}}^{-1} \hat{\bar{J}}^T \bar{C}(\bar{\mathbf{O}}_\epsilon) \hat{\bar{J}} \hat{\bar{H}}^{-1} = \hat{\bar{H}}^{-1} (\hat{\bar{J}}^T \hat{\bar{J}}) \hat{\bar{H}}^{-1} = \hat{\bar{H}}^{-1},$$

where one has used standard properties of covariance matrices and the assumption $\bar{C}(\bar{\mathbf{O}}_\epsilon) = I$.

**B.** Demanding that $\bar{C}(\bar{\mathbf{O}}_\epsilon) = I$ is equivalent to demanding that the calculated statistical errors bear a value close to unity. This arises because the covariance matrix $\bar{C}(\bar{\mathbf{O}}_\epsilon)$ is directly related to statistical errors. Hence, it is consistent with Eq. (5.76), as the introduction of an overall scaling factor therein implies that $\chi^2 \sim 1$ in Eq. (5.74).

# References

1. T. Berggren, Nucl. Phys. A **109**, 265 (1968)
2. I.M. Gel'fand, N.Y. Vilenkin, *Generalized Functions*, vol. 4 (Academic Press, New York, 1961)
3. K. Maurin, *Generalized Eigenfunction Expansions and Unitary Representations of Topological Groups* (Polish Scientific Publishers, Warsaw, 1968)
4. A. Bohm, M. Gadella, S. Maxon, Comput. Math. Appl. **34**, 427 (1997)

5. O. Civitarese, M. Gadella, Phys. Rep. **396**, 41 (2004)
6. T. Berggren, Phys. Lett. B **73**, 389 (1978)
7. A. Bohm, Int. J. Theor. Phys. **42**, 2317 (2003)
8. A. Bohm, M. Gadella, *Lecture Notes in Physics*, vol. 348 (Springer, Berlin, 1989)
9. T. Berggren, Phys. Lett. B **373**, 1 (1996)
10. H. Friedrich, *Theoretical Atomic Physics*, Chap. 1 (Springer, New York, 1991)
11. R.D. Lawson, *Theory of the Nuclear Shell Model* (Clarendon-Press, Oxford, 1980)
12. Y. Suzuki, K. Ikeda, Phys. Rev. C **38**, 410 (1988)
13. Y. Jaganathen, R.M. Id Betan, N. Michel, W. Nazarewicz, M. Płoszajczak, Phys. Rev. C **96**, 054316 (2017)
14. S. Saito, Prog. Theor. Phys. **41**, 705 (1969)
15. E. Caurier, G. Martínez-Pinedo, F. Nowacki, A. Poves, A.P. Zuker, Rev. Mod. Phys. **77**, 427 (2005)
16. G.J.G. Sleijpen, H.A. van der Vorst, SIAM J. Matrix Anal. Appl. **17**, 401 (1996)
17. L. Brillouin, Act. Sci. Ind. **71**, 159 (1933)
18. K.L.G. Heyde, *The Nuclear Shell Model* (Springer, Berlin, 1994)
19. N. Pillet, J.F. Berger, E. Caurier, Phys. Rev. C **78**, 024305 (2008)
20. I.J. Shin, Y. Kim, P. Maris, J.P. Vary, C. Forssén, J. Rotureau, N. Michel, J. Phys. G. Nucl. Part. Phys. **44**, 075103 (2017)
21. N. Michel, W. Nazarewicz, M. Płoszajczak, J. Okołowicz, Phys. Rev. C **67**, 054311 (2003)
22. N. Michel, W. Nazarewicz, M. Płoszajczak, Phys. Rev. C **70**, 064313 (2004)
23. M. Moshinsky, Nucl. Phys. **13**, 104 (1959)
24. R. Balian, E. Brézin, Nuovo Cim. **61**, 403 (1969)
25. C.W. Wong, D.M. Clément, Nucl. Phys. A **183**, 210 (1972)
26. C.L. Kung, T.T.S. Kuo, K.F. Ratcliff, Phys. Rev. C **19**, 1063 (1979)
27. G. Hagen, M. Hjorth-Jensen, N. Michel, Phys. Rev. C **73**, 064307 (2006)
28. N. Michel, H. Aktulga, Y. Jaganathen, Comput. Phys. Commun. **247**, 106978 (2020)
29. J.P. Vary, P. Maris, E. Ng, Y. Yang, M. Sosonkina, J. Phys. Conf. Ser. **180**, 012083 (2009)
30. P. Maris, H.M. Aktulga, S. Binder, A. Calci, Ümit V Çatalyürek, J. Langhammer, E. Ng, E. Saule, R. Roth, J.P. Vary, C. Yang, J. Phys. Conf. Ser. **454**, 012063 (2013)
31. P. Maris, H.M. Aktulga, M.A. Caprio, Ümit V Çatalyürek, E.G. Ng, D. Oryspayev, H. Potter, E. Saule, M. Sosonkina, J.P. Vary, C. Yang, Z. Zhou, J. Phys. Conf. Ser. **403**, 012019 (2012)
32. H.M. Aktulga, M. Afibuzzaman, S. Williams, A. Buluç, M. Shao, C. Yang, E.G. Ng, P. Maris, J.P. Vary, IEEE Trans. Par. Dist. Sys. **28**, 1550 (2017)
33. M. Shao, H.M. Aktulga, C. Yang, E.G. Ng, P. Maris, J.P. Vary, Comp. Phys. Comm. **222**, 1 (2018)
34. H.M. Aktulga, C. Yang, E.G. Ng, P. Maris, J.P. Vary, Concurrency Computat. Pract. Exper. **26**, 2631 (2014)
35. C.W. Johnson, W.E. Ormand, P.G. Krastev, Comp. Phys. Comm. **184**, 2761 (2013)
36. R.R. Whitehead, A. Watt, B.J. Cole, I. Morrison, Adv. Nucl. Phys. **9**, 123 (1977)
37. P.O. Löwdin, Phys. Rev. **97**, 1509 (1955)
38. J. Rotureau, N. Michel, W. Nazarewicz, M. Płoszajczak, J. Dukelsky, Phys. Rev. Lett. **97**, 110603 (2006)
39. J. Rotureau, N. Michel, W. Nazarewicz, M. Płoszajczak, J. Dukelsky, Phys. Rev. C **79**, 014304 (2009)
40. S.R. White, Phys. Rev. Lett. **69**, 2863 (1992)
41. S.R. White, Phys. Rev. B **48**, 10345 (1993)
42. J. Dukelsky, S. Pittel, S.S. Dimitrova, M.V. Stoitsov, Phys. Rev. C **65**, 054319 (2002)
43. S. Pittel, N. Sandulescu, Phys. Rev. C **73**, 014301 (2006)
44. B. Thakur, S. Pittel, N. Sandulescu, Phys. Rev. C **78**, 041303 (2008)
45. O. Legeza, L. Veis, A. Poves, J. Dukelsky, Phys. Rev. C **92**, 051303 (2015)
46. L. Neufcourt, Y. Cao, W. Nazarewicz, F. Viens, Phys. Rev. C **98**, 034318 (2018)
47. A.E. Lovell, F.M. Nunes, J. Sarich, S.M. Wild, Phys. Rev. C **95**, 024611 (2017)

48. G.B. King, A.E. Lovell, F.M. Nunes, Phys. Rev. C **98**, 044623 (2018)
49. S. Brant, *Statistical and Computational Methods in Data Analysis* (Springer, New York, 1997)
50. J. Dobaczewski, W. Nazarewicz, P.G. Reinhard, J. Phys. G. Nucl. and Part. Phys. **41**, 074001 (2014)
51. R.T. Birge, Phys. Rev. **40**, 207 (1932)
52. M.C. Kennedy, A. O'Hagan, J. Roy. Stat. Soc. Ser. B (Stat. Meth.) **63**, 425 (2001)

# Physical Applications of the Gamow Shell Model

**6**

## 6.1 Effective Interaction in the Vicinity of the Particle-Emission Threshold

Structure of weakly bound/unbound nuclei close to particle driplines is different from that around the valley of beta-stability. There are many unique features of exotic nuclei that give prospects for entirely new phenomena likely to be different from anything we have observed to date. These phenomena can be described using the Gamow shell model. Certain specific features of interactions in weakly bound systems have been recently identified. In general terms, these interactions depend explicitly on the location of various emission thresholds and on the structure of $S$-matrix poles which dominate the corresponding decays.

For bound states appears a virtual scattering involving intermediate scattering states. Continuum coupling of this kind affects also the effective nucleon–nucleon interaction. For unbound states, the continuum structure appears explicitly in the properties of those states. In the mean-field picture, when approaching the one-nucleon dripline ($S_{n,p} \rightarrow 0$), the magnitude of the chemical potential approaches the pairing gap: $-\lambda_{n,p} - \Delta_{n,p} \rightarrow 0$, i.e. the pairing correlations become as important as the mean-field effects. Generalizing this observation, one expects that the dependence of weakly bound systems on effective nucleon–nucleon interaction is stronger than in "normal" nuclei.

One may ask how the effective nucleon–nucleon interaction is modified by the virtual scattering into the particle continuum. In shell model embedded in the continuum studies of binding energy systematics in $sd$ shell nuclei, it was found that in the nuclei close to the neutron dripline the neutron-proton $T=0$ interaction is significantly reduced with respect to the neutron-neutron $T=1$ interaction [1, 2]. Later, this observation has been confirmed in the study of lifetime of $^{16}$F [3]. The reduction of neutron-proton $T=0$ interaction in neutron-rich nuclei is a consequence of a decrease in the one neutron emission threshold when approaching the neutron dripline, i.e. it is a genuine continuum coupling effect which cannot be mimicked

© Springer International Publishing AG 2021
N. Michel, M. Płoszajczak, *Gamow Shell Model*, Lecture Notes in Physics 983,
https://doi.org/10.1007/978-3-030-69356-5_6

by the monopole components of the two-body shell model Hamiltonian [1, 3]. To account for this effect in the standard shell model, one would need to introduce a neutron-number dependence of the T=0 monopole terms [1] which appears if one includes the three-body interactions in the framework of a standard shell model [4]. Indeed, the nucleon–nucleon coupling via intermediate scattering states contributes effectively to three-body correlations which cannot be disentangled in the phenomenological analysis from the correlations generated by the genuine three-body force [1, 2, 5].

Even though the effective interaction theory for nuclear open quantum systems is not yet completed, nevertheless one can illustrate some of its aspects by analyzing continuum-coupling correction for the eigenenergy of nuclear closed quantum system (e.g. the standard shell model), using the real-energy continuum shell model [5–7]. In this model, the separation of (quasi-)bound ($Q$ subspace) and scattering ($P$ subspace) states (see Fig. 1.2) is achieved using the projection operator technique [8]. The total Hamiltonian $\hat{\mathscr{H}}$ in $Q + P$ can be formally divided into "unperturbed" Hamiltonians $\hat{H}_{QQ}$ and $\hat{H}_{PP}$ in $Q$ and $P$ subspaces, respectively, and the coupling terms $\hat{H}_{QP}$, $\hat{H}_{PQ}$ between these subspaces. In the $Q$ subspace, the closed quantum system approximation corresponds to replacing $\hat{\mathscr{H}}$ by the standard shell model Hamiltonian $\hat{H}_{QQ}$, whereas the open quantum system approximation is described by an energy-dependent effective Hamiltonian:

$$\hat{\mathscr{H}}_{QQ}(E) = H_{QQ} + H_{QP}G_P^{(+)}(E)\hat{H}_{PQ} , \qquad (6.1)$$

which includes coupling to the scattering continuum. In this expression, $G_P^{(+)}(E)$ is a Green's function for the motion of a single nucleon in $\mathscr{P}$ subspace. Hence the effective Hamiltonian $\hat{\mathscr{H}}_{QQ}$ is a complex-symmetric matrix for energies above the particle-emission threshold ($E^{(\text{thr})}$), and Hermitian below it. An eigenvalue

$$\mathscr{E}_i(E) = E_i + E_{\text{corr}}^{(i)}$$

of $\hat{\mathscr{H}}_{QQ}(E)$ can be written as a sum of a closed-system eigenenergy $E_i$ given by $\hat{H}_{QQ}$ and the correction due to the coupling to the decay channels. In particular, this correction depends on the distance of $E_i$ from the one-particle threshold and below the lowest-energy decay threshold ($E < 0$), i.e. for bound states, the energy correction is real.

Assuming the coupling to one decay channel, and neglecting the off-diagonal terms of $\hat{H}_{QP}G_P^{(+)}(E)\hat{H}_{PQ}$, the continuum-coupling correction to shell model energy $E_i$ becomes

$$E_{\text{corr}}^{(i)} = \langle \Phi_i| \hat{H}_{QP}G_P^{(+)}(E)\hat{H}_{PQ} |\Phi_i\rangle = \int_0^\infty dr\, \omega_i^{(+)}(r) \cdot w_i(r) , \qquad (6.2)$$

where $w_i = \hat{H}_{PQ} |\Phi_i\rangle$ is the source term, and $\omega_i^{(+)}(r) = G_P^{(+)} \hat{H}_{PQ} |\Phi_i\rangle$ is the solution of an inhomogeneous radial equation with the outgoing boundary condition [9, 10]:

$$\left( E - \hat{H}_{PP} \right) \omega_i^{(+)}(r) = w_i(r) . \tag{6.3}$$

In the following, we shall assume that the Hamiltonian in $P$-space $\hat{H}_{PP}$ consists of a kinetic energy operator and a finite-depth, square-well potential of radius $R_0$. The source term in Eq. (6.3) is modelled by $w(r) \propto r^\nu$ ($\nu \geq \ell + 1$) for $0 \leq r \leq R_0$ and $w(r) = 0$ for $r > R_0$, where $\ell$ is the orbital angular momentum of a particle. The assumed radial dependence of this localized source $w(r)$ for $r \to 0$ is consistent with the radial dependence of single-particle wave functions which enter in the microscopic calculation of this term [9, 10]. Under these assumptions, the continuum-coupling correction to the eigenenergy of the closed quantum system at the threshold ($E = 0$) can be calculated analytically as a function of the distance $\varepsilon$ of the eigenvalue of $\hat{H}_{PP}$ from the one-body continuum threshold:

$$E_{corr}^{(\ell)}(\varepsilon) = -\text{const}|\varepsilon|^{-1+\ell/2} + \mathcal{O}(|\varepsilon|^0) . \tag{6.4}$$

The continuum-coupling correction at the threshold becomes singular for $\ell = 0, 1$ states in the limit of $\varepsilon \to 0$, independently of the nature of the considered $S$-matrix pole. In the realistic case, the attenuation of this singular behavior is expected by the residual nucleon–nucleon interaction and the channel–channel couplings. For higher $\ell$-values ($\ell \geq 2$), a dependence of the continuum-coupling energy correction on the position of $S$-matrix pole is not singular and therefore, certain aspects of the continuum coupling can be imitated in standard shell model by introducing the particle number dependence in monopole terms of the effective interaction.

## 6.1.1   Monitoring Branch Points and Avoided Crossings in the Complex Plane

Monitoring effects of the configuration mixing in resonance wave function are an important issue in the studies of open quantum systems. Branch points associated with the reaction thresholds and avoided level crossings or exceptional points are essential elements of the configuration mixing in open quantum systems. A useful indicator of the effects caused by the continuum couplings is the phase rigidity [11–13]:

$$r_j = \frac{\langle \tilde{\Psi}_j | \Psi_j \rangle}{\langle \Psi_j | \Psi_j \rangle} , \tag{6.5}$$

given by the ratio of Berggren and Hermitian norms of an eigenfunction. The indicator (6.5) can also be written in a form [13]:

$$r_j = e^{2i\theta_j} \frac{\int dr \left(|\text{Re}\Psi_j(r)|^2 - |\text{Im}\Psi_j(r)|^2\right)}{\int dr \left(|\text{Re}\Psi_j(r)|^2 + |\text{Im}\Psi_j(r)|^2\right)} , \qquad (6.6)$$

which better illustrates its physical meaning. In this expression, the angle $\theta_j$ arises from a transformation of $\Psi_j$ so that $\text{Re}\Psi_j$ and $\text{Im}\Psi_j$ are orthogonal and it characterizes the degree to which the eigenfunction $\Psi_j$ is complex.

Phase rigidity varies between 1 and 0. It equals 1 for bound state eigenfunctions, whereas at the coalescence point of wave functions (the exceptional point) it equals 0. For unbound states, the condition $r_j=1$ means that the continuum couplings exert a negligible effect on the internal structure of an eigenfunction, i.e. its Berggren and Hermitian norms are identical. Abrupt variations of the phase rigidity in a certain interval of excitation energies are indicative of the instability of wave functions.

Another indicator of such instabilities could be the continuum-coupling energy correction $E_{\text{corr}}^{(i)}$ (Eq. (6.2)) which was studied extensively in the shell model embedded in the continuum [6, 7, 15, 16]. In Gamow shell model, one may use as an indicator the value of energy correction due to coupling to the nonresonant continuum background.

$$\Delta E_{\mathscr{B}-\text{space}} = \langle \Psi_i | \hat{H} | \Psi_i \rangle - \langle \Psi_{0;i} | \hat{H} | \Psi_{0;i} \rangle , \qquad (6.7)$$

where $\Psi_{0;i}$ is the pole space wave function corresponding to the Gamow shell model eigenfunction $\Psi_i$. This energy correction is a measure of continuum-induced instability of the pole space wave function and resulting rearrangement of occupancies of the resonant and nonresonant shells in Gamow shell model eigenfunctions, in particular, due to the proximity of the branch points associated with the particle-emission thresholds and the double-poles of the $S$-matrix (the exceptional points).

Figure 6.1 shows the difference $\Delta E_{\mathscr{B}-\text{space}}$ between Gamow shell model energies calculated in the full Berggren ensemble and in the pole space for the $0_1^+$ ground state of $^6$He, as a function of the one neutron separation energy $S_{1n}$ from $^5$He. In this calculation, the Hamiltonian consists of one-body potential generated by $^4$He core and the Furutani–Horiuchi–Tamagaki two-body interaction [17,18]. $S_{1n}$ of $^5$He is modified by changing the depth of the Woods–Saxon potential. One can see in Fig. 6.1 a non-analytic variation of $\Delta E_{\mathscr{B}-\text{space}}$ at around the threshold $S_{1n} = 0$. Further discussion of near-threshold singularities will be done in Sects. 7.4.1, 7.4.2, and 7.4.2.1.

## 6.1.2  Effect of Continuum Coupling on the Spin–Orbit Splitting

The coupling to the particle continuum in low-$\ell$ orbits may lead to the instability of the Hartree–Fock particle vacuum and the coexistence of local minima corresponding to different distribution of $S$-matrix poles around the threshold. In this

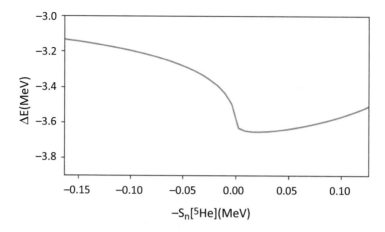

**Fig. 6.1** Continuum background energy correction to the pole space $0_1^+$ eigenvalue of Gamow shell model in $^6$He is plotted as a function of the neutron separation $S_{1n}$ energy in $^5$He [14]

coexistence zone, the many-body states cannot be generated by a limited number of configurations constructed from single-particle basis states corresponding to any of these minima separately, and a complete configuration mixing calculation is mandatory. This may have an influence on the spin–orbit splitting for $\ell = 1$ and 2 orbits. Below, we will demonstrate this effect on the example of the "spin–orbit splitting" of $3/2^-$ and $1/2^-$ states of $^7$He (see Fig. 6.2) calculated in Gamow shell model.

In this example, one uses the surface Gaussian two-body interaction [20]:

$$\hat{V}_{J,T}(\mathbf{r}_1, \mathbf{r}_2) = V_0(J, T) \cdot \exp\left[-\left(\frac{\mathbf{r}_1 - \mathbf{r}_2}{\mu}\right)^2\right] \cdot \delta(|\mathbf{r}_1| + |\mathbf{r}_2| - 2 \cdot R_0),$$

where $V_0(J, T)$ is the coupling constant which explicitly depends on the total angular momentum $J$ and the total isospin $T$ of the nucleonic pair and $\mu$ is the range of the interaction. The Gamow shell model results are generated using two different Gamow Hartree–Fock bases depending on the strength of the spin–orbit potential $V_{SO}$. For $V_{SO} \leq 6.5$ MeV, denoted as $GSM_1$ in Fig. 6.2, the optimal basis contains the $0p_{3/2}$ single-particle resonance. This pole becomes bound for $V_{so} \geq 8$ MeV (variant $GSM_2$ in Fig. 6.2). In both cases, the $0p_{1/2}$ pole is a broad resonance. In each discretized Gamow Hartree–Fock basis, one keeps the many-body Slater determinants with at most two particles in the nonresonant continuum.

In the intermediate range of the spin–orbit strength, around 8.5 MeV, the two Gamow Hartree–Fock minima coexist and the assumed truncation with up to two neutrons in the nonresonant continuum becomes insufficient because these two Gamow Hartree–Fock bases yield different physical results. To understand the underlying physics in this coexistence region, the full Gamow shell model

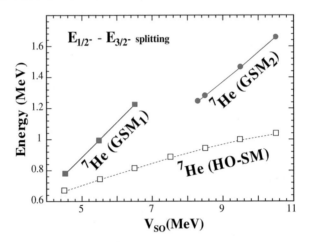

**Fig. 6.2** Energy splitting of the lowest $3/2^-$ and $1/2^-$ levels of $^7$He as a function of the strength of the spin–orbit potential $V_{SO}$. The dashed line represents the standard shell model results for $^7$He (HO-SM approximation). The Gamow shell model calculations are performed in the truncated many-body basis in which no more than two neutrons are allowed to occupy the nonresonant continuum. The optimal single-particle basis changes strongly with $V_{SO}$. For smaller values of the spin–orbit strength, the $0p_{3/2}$ state is a resonance (GSM$_1$) while at larger values of $V_{SO}$, the $0p_{3/2}$ state becomes bound (GSM$_2$) (adapted from Ref. [19])

calculation, including the three-neutron components in the nonresonant continuum, is mandatory.

To compare the Gamow shell model results with standard shell model, one calculates matrix elements of the surface Gaussian interaction in the harmonic oscillator basis. The single-particle energies in such "equivalent SM calculations," denoted as HO-SM approximation in Fig. 6.2, are given by real parts of $0p_{1/2}$ and $0p_{3/2}$ eigenvalues of the Woods–Saxon potential generating the Gamow shell model basis. The resulting "spin–orbit splittings" in $^7$He follow smoothly the spin–orbit strength of the one-body potential.

The behavior found in the truncated Gamow shell model calculations is more intricate. First of all, the splitting in $^7$He is enhanced with respect to the HO-SM variant. Secondly, the average slope of the energy difference $E_{1/2} - E_{3/2}$ vs $V_{SO}$ is increased as well, due to the two-particle correlations induced by the continuum coupling. One can see on this example that the continuum coupling can significantly modify the spin–orbit splitting, an effect similar to that produced by a genuine three-body force [21].

## 6.2    $T = 0, 1$ Nuclear Matrix Elements in the Berggren Basis

Matrix elements of the nuclear effective interaction have been extensively studied in the standard shell model. In fact, despite its complexity, the nuclear interaction gives rise to simple properties which can be understood from the Pauli principle or from the semi-classical arguments. Indeed, when one considers the one-body states of high orbital angular momenta, the geometrical features of their spherical harmonics prevail, whereas the different components of the nuclear interaction, of central, tensor, spin–orbit type for the most important ones, are integrated out and become unimportant as compared to the dependence both on the angular momenta of one-body states and on the total angular momentum of the pair of nucleons.

It is particularly interesting to study the generic features of nuclear matrix elements involving proton and neutron shells of a fixed angular momentum, and coupled to the total angular momentum $J$. Spherical harmonics are indeed simple when represented in coordinate space and their radial parts have very few nodes, so that the main features of nuclear matrix elements can be analyzed therein without problem. To study these matrix elements in well-bound nuclei, one can use a simple surface delta interaction [22] which captures main features of the realistic interactions and can be fitted to reproduce empirical nuclear matrix elements.

This interaction is not well suited in weakly bound and resonance nuclei due to its zero-range, which ignores the asymptotes of nuclear wave functions. Consequently, a finite-range interaction is better suited, as it depends on radial wave functions at all radii, and not only at the surface radius. In the following, we will use the Minnesota interaction [23] as nucleon–nucleon interaction for that matter. Minnesota interaction is a rather simple effective force, depending on a few parameters and bearing only central terms, so that the analysis of the its $T = 0, 1$ matrix elements in Gamow shell model can be made similarly as in the standard shell model [22].

In the study of generic properties of two-body matrix elements, one will consider various pairs of valence nucleons, consisting of either two neutrons, or one proton and one neutron. Ignoring the Coulomb interaction, i.e. assuming exact isospin symmetry, implies that pairs of protons or pairs of neutrons behave identically. The considered closed-shell cores, namely $^4$He, $^{16}$O, $^{40}$Ca, and $^{132}$Sn, are standard and are widely used in various spectroscopic studies. Bound and resonance valence proton and neutron shells consist of $0p_{3/2}$, $0p_{1/2}$ for $^4$He, $0d_{5/2}$, $1s_{1/2}$, and $0d_{3/2}$ for $^{16}$O, and $0f_{7/2}$, $1p_{3/2}$, $1p_{1/2}$, $0f_{5/2}$ for $^{40}$Ca. For nuclei above the core of $^{132}$Sn, one will consider only the shells of largest angular momenta, namely $0g_{7/2}$, $0h_{11/2}$ for protons and $0h_{9/2}$, $0i_{13/2}$, $1g_{9/2}$ for neutrons. The strengths of spin–orbit and Woods–Saxon potentials describing the used cores are first fitted to reproduce experimental single-particle energies. Then, in order to emphasize effects of the continuum coupling, the strengths of the Woods–Saxon core central potentials are modified keeping the spin–orbit strength unchanged so that the two-body states have a total binding energy of $-10$, $-1$, or $+1$ MeV with respect to the core. Coupling to the continuum states, which is rather negligible at $-10$ MeV, becomes visible at $-1$ MeV while the nuclear state is still bound. Finally, the continuum coupling becomes important for unbound states with energy $+1$ MeV above the core. These three examples then allow to assess the importance of continuum coupling on the considered nuclear matrix elements.

### 6.2.1  Dependence of Nuclear Matrix Elements on Angular Momentum and Continuum Coupling

When one considers the eigenstates of well-bound nuclei whose basis states belong to a single proton or neutron shell of fixed total angular momentum, binding energies are equal to the sum of core single-particle energies and nuclear two-body matrix element. Consequently, two-body matrix elements of the nuclear Hamiltonian can be directly extracted from binding energies by removing its one-body part. However, due to the configuration mixing induced by the presence of scattering states in the Berggren basis, this simple relation between binding energies and nuclear two-body matrix elements is no longer verified. One might argue that the previous situation would be restored by considering a unique proton or neutron resonance shell for unbound nuclear states. This is, however, not possible as in this case the completeness properties of the Berggren basis are lost. Indeed, for calculated nuclear states to bear positive widths if they are resonances, or to be real if they are bound, it is necessary to use a complete Berggren basis for the considered partial wave. Otherwise, while the real parts of binding energies might remain close to their exact values if a continuum coupling is not large, their imaginary parts would be uncontrolled and could no longer be associated with particle-emission widths. As a consequence, one has to include the scattering states of the same partial wave as that of the considered bound or resonance proton or neutron shell.

Clearly, calculation of nuclear Hamiltonian matrix elements in this case becomes not only more complicated but also undefined. Indeed, as a very large number of the nuclear matrix elements involving one-body scattering states enter the Hamiltonian matrix, their extraction from a binding energy becomes undefined. A solution to this problem is not to consider nuclear matrix elements *per se*, but to analyze the difference between binding energies and the sum of occupied core single-particle energies present in the Hamiltonian. This procedure allows to avoid the previous mentioned problem generated by the presence of the scattering states in the single-particle basis, while providing with energy differences from which nuclear matrix elements involving only resonant one-body states can be inferred. Indeed, the continuum coupling is in general not very large, so that the matrix elements of nuclear Hamiltonian involving resonance basis states and the aforementioned energy differences are very close. In order to simplify the following discussion, and to allow a direct comparison with the well-bound case, these aforementioned energy differences will be deemed as nuclear matrix elements even though, strictly speaking, they are not exactly equal to the nuclear two-body matrix elements.

In order to average out local effects due to the use of different nuclei, shells, and angular momenta, one does not consider directly nuclear matrix elements, but renormalized nuclear matrix elements [22]:

$$\Re(V_{12}) = \frac{E_n}{\bar{E}_n} \tag{6.8}$$

$$E_n = E - e_a - e_b \tag{6.9}$$

$$\bar{E}_n = \left| \frac{\sum_J (2J+1)(E - e_a - e_b)}{\sum_J (2J+1)} \right|, \tag{6.10}$$

where $\Re(V_{12})$ is the real part of the renormalized nuclear matrix element, $E_n$ is the interaction energy, i.e. a difference of the total energy $E$ of a two-nucleon system and the single-particle energies $e_a$ and $e_b$ of occupied levels, and $\bar{E}_n$ is the average real part of nuclear matrix elements, obtained by summing over all interaction energies $E$ for all possible two-body angular momenta $J$. Local effects are suppressed through the presence of $\bar{E}_n$ in Eq. (6.8) and the sign of nuclear matrix element remains unchanged by its renormalization as $\bar{E}_n > 0$. Consequently, one can plot the nuclear matrix elements coming from all considered cores and one-body shells on a single diagram, as in the standard shell model [22].

The imaginary part of nuclear matrix elements is renormalized analogously:

$$\Im(V_{12}) = \frac{\Gamma_n}{\bar{\Gamma}_n} \tag{6.11}$$

$$\Gamma_n = \Gamma - \gamma_a - \gamma_b \tag{6.12}$$

$$\bar{\Gamma}_n = \left| \frac{\sum_J (2J+1)(\Gamma - \gamma_a - \gamma_b)}{\sum_J (2J+1)} \right|. \tag{6.13}$$

$\Im(V_{12})$ is the difference between the width of a correlated two-nucleon system and the sum of one-body widths of single-particle levels $a, b$, denoted as $\gamma_a$ and $\gamma_b$. The latter sum is the width of two-body system in the independent-particle picture. $\bar{\Gamma}_n$ in Eq. 6.13 is the average width of the renormalized nuclear matrix element.

In order to emphasize the dependence of nuclear matrix elements on both angular momenta $j_a$ and $j_b$ of single-particle shells in Eqs. (6.8) and (6.11) and of the total angular momentum $J$ of the nuclear state, one defines the semi-classical angle $\theta$ between the vectors $\mathbf{j}_a$ and $\mathbf{j}_b$:

$$\cos(\theta) = \frac{J(J+1) - j_a(j_a+1) - j_b(j_b+1)}{2\sqrt{j_a(j_a+1)j_b(j_b+1)}}. \tag{6.14}$$

The renormalized two-body matrix elements (6.8), (6.11) are shown in Figs. 6.3 and 6.4 as a function of $\theta$.

It is clear that results shown in Figs. 6.3 and 6.4 follow two distinct trends according to the parity of $j_a + j_b + J$. If $j_a + j_b + J$ is even, $\Re(V_{12})$ and $\Im(V_{12})$ follow a curve resembling to a parabola, with a maximum around 90°. If $j_a + j_b + J$ is odd, nuclear matrix elements increase with decreasing $\theta$. These features can be semi-classically understood from the isospin $T$ of the two-nucleon wave function.

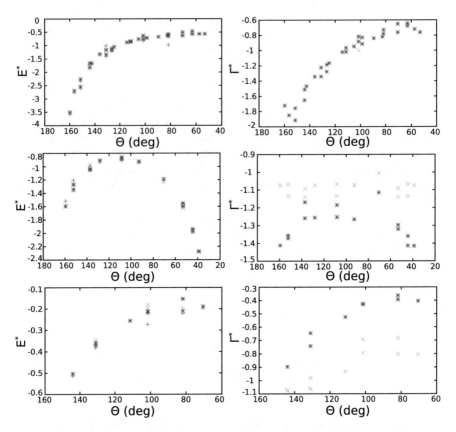

**Fig. 6.3** The real parts $\Re(V_{12})$ (left panels) and imaginary parts $\Im(V_{12})$ (right panels) of the renormalized nuclear matrix element (see Eqs. (6.8) and (6.11)) are plotted as a function of the semi-classical angle $\theta$ (see Eq. (6.14)). Protons and neutrons in this example occupy identical resonant orbits and scattering states of the same partial wave as resonant orbits. Upper panels are for neutron two-body systems for which $J + 2j$ is odd, with $j$ the total angular momentum of the neutron shell. Middle/lower panels are for proton–neutron two-body systems for which $J + j_p + j_n$ is even/odd, with $j_p$ the total angular momentum of the proton shell and $j_n$ that of the neutron shell. Red pluses, green crosses, and blue stars are used to denote nuclear state of energies $-10$, $-1$, and 1 MeV, respectively

If $j_a + j_b + J$ is even, one has an isospin $T = 0$, so that the two-nucleon wave function is symmetric in space and antisymmetric in isospace. Thus, nucleon–nucleon interaction is strongest when nucleons rotate on the same plane, i.e. where $\theta = 0°$ or $\theta = 180°$, and become weakest when they rotate on the orthogonal planes, i.e. where $\theta = 90°$. This results in a strongly attractive nuclear matrix element for $\theta = 0°$ or $\theta = 180°$ and a small or vanishing matrix element for $\theta = 90°$.

Conversely, if $j_a + j_b + J$ is odd, one has an isospin $T = 1$, so that the two-nucleon wave function is antisymmetric in space and symmetric in isospace. If $\theta = 0°$, $\mathbf{j}_a = \mathbf{j}_b$, so that the product of the one-body nucleon wave functions

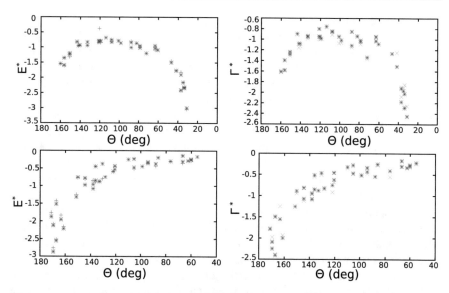

**Fig. 6.4**  Same as the middle and lower panels of Fig. 6.3, but for protons and neutrons occupying resonant orbits of different angular momenta. The upper part of Fig. 6.3 has no equivalent here

$\Phi_a(\mathbf{r}_a)\Phi_b(\mathbf{r}_b)$ is non-negligible only if $\mathbf{r}_a \simeq \mathbf{r}_b$. However, as the two-nucleon wave function is antisymmetric in space, i.e. proportional to $\Phi_a(\mathbf{r}_a)\Phi_b(\mathbf{r}_b) - \Phi_a(\mathbf{r}_b)\Phi_b(\mathbf{r}_a)$, the one-body nucleon wave functions therein partially cancel each other, so that the resulting two-nucleon wave function has very small values when $\mathbf{r}_a$ and $\mathbf{r}_b$ are not small in modulus. Consequently, it becomes possible to have an antisymmetric two-body wave function with sizable values close to the center of the nucleus. As the two-body wave function has a nonzero value close to the center of the nucleus, the strength of the nuclear interaction between the two nucleons is non-negligible. The product of the one-body nucleon wave functions $\Phi_a(\mathbf{r}_a)\Phi_b(\mathbf{r}_b)$ is also non-negligible when $\mathbf{r}_a$ and $\mathbf{r}_b$ are not almost identical if $\mathbf{j}_a \neq \mathbf{j}_b$. Thus, the nuclear matrix element becomes more and more negative when $\theta$ increases.

One can see that in the aforementioned approximation (see Eqs. (6.9) and (6.12), the averaged behavior of two-body matrix elements $\langle ab|V|ab\rangle_J$ in the Gamow shell model depends principally on $j_a$, $j_b$, and $J$. Let us note, however, that the discrepancy of $\Re(V_{12})$ and $\Im(V_{12})$ from the approximate smooth curve depicting their overall tendency is larger in Figs. 6.3 and 6.4 than seen in standard shell model calculations [22]. This is not related to the continuum coupling, as at the energies equal to $-10$ MeV the calculated reduced matrix elements deviate from the smooth shell model curves. The reason for this large dispersion lies in the use of the Minnesota interaction, which is more complex than the surface delta interaction. Indeed, Minnesota interaction generates different nuclear matrix elements according to the nature of used basis shells. Consequently, they cannot be perfectly renormalized using Eqs. (6.10) and (6.13) so as to form a smooth curve [22].

While the generic dependence of $\Re(V_{12})$ with respect to $\theta$ is the same as found in [22], it is remarkable that this is also true for $\Im(V_{12})$. This intriguing result can be understood if one recurs to the procedure of complex rotation, as is explained afterwards. Indeed, the complex nuclear matrix elements are analytic continuation of the real nuclear matrix elements involving bound single-particle states (see Sect. 3.3). Consequently, their analytic properties are analogous to that of one-body resonant states (see Sect. 3.5), whose width increases when energy increases. As widths are renormalized in the same manner as energies, it is better to consider $\Im(V_{12})$ as a function of the strength of the nuclear interaction. Indeed, the renormalization in Eqs. (6.11–6.13) removes the one-body widths, so that $\Im(V_{12})$ is identically zero without any nuclear interaction and becomes nonzero when the nuclear interaction is present.

It is important to state that $\Im(V_{12})$ does not have to be positive, as is the case for particle-emission width. Indeed, $\Im(V_{12})$ is the difference between the width of a correlated two-nucleon system and the sum of one-body widths, the latter sum being the two-body system width in an independent-particle picture (see Eqs. (6.11), (6.12), and (6.13)). In particular, one will typically have $\Im(V_{12}) < 0$ if the nucleon interaction is attractive. Indeed, the gain in energy arising from correlations will make the width of the two-nucleon system smaller than its value in the independent-particle case. Consequently, when the strength of the interaction augments, $\Im(V_{12})$ becomes more and more negative. This explains the qualitative similarity between the curves $\Re(V_{12})(\theta)$ and $\Im(V_{12})(\theta)$, even though the dispersion of calculated values for different shells and the two-body angular momentum $J$ is larger in $\Im(V_{12})(\theta)$ than in $\Re(V_{12})(\theta)$. Due to the renormalization of nuclear matrix elements (see Eqs. (6.8) and (6.11)), real $\Re(V_{12})$ and imaginary $\Im(V_{12})$ parts of two matrix elements are of the same order of magnitude and vary in the interval $[-4:0]$, independently of the very different initial values of these matrix elements in each studied two-nucleon system. Consequently, one can see that the two-body nuclear matrix elements involving resonance states possess similar properties to those pertaining to bound states, and that both the real and imaginary parts of two-body nuclear matrix elements behave similarly due to the analytic character of two-body wave functions in the complex-energy plane.

## 6.2.2   Influence of Continuum Coupling on Effective Nuclear Matrix Elements

The previous studies have been done in the context of renormalized nuclear matrix elements. One divided all initial nuclear matrix elements by their average value in order to exhibit their semi-classical behavior as a function of $\theta$ (see Figs. 6.3 and 6.4). This allowed to point out the common points between matrix elements for bound and resonance many-body states, as the overall behavior of renormalized nuclear matrix elements is similar independently of binding energy of the many-body states. However, the averaging procedure in Eqs. (6.10) and (6.13) masks differences between matrix elements that arise from the coupling to the continuum when one-body states are bound or resonance.

Consequently, it is interesting to discuss the absolute total binding energy of two-body states. For this, one fixes the total binding energy of the ground state of two-body systems at $-10$, $-1$, or $+1$ MeV with respect to the core, as in Sect. 6.2.1. One can then deduce the influence of continuum coupling on effective nuclear matrix elements from the energy of calculated two-body states.

The orbital angular momentum of involved partial waves cannot be too large, as otherwise continuum coupling becomes negligible. Alternatively, the total angular momentum of partial waves cannot be too small either, as otherwise too few two-body states are present and a global analysis can hardly be made. Thus, one only considers the $p_{3/2}$, $d_{5/2}$, and $f_{7/2}$ partial waves of both protons and neutrons, using, respectively, the cores of $^4$He, $^{16}$O, and $^{40}$Ca in model spaces. Indeed, as we will see below, effects of the continuum coupling are still visible when $\ell = 3$, whereas the number of possible total angular momentum $J$ of the two-body states ranges from 2 to 8, which allows extracting more general conclusions from the obtained results.

Results of calculations are shown in Tables 6.1, 6.2, and 6.3. One can see that the main effect of continuum coupling is to diminish the excitation energy of the two-body eigenstates of different $J$ values when one goes from well bound to unbound states. This means that the overall effects of the residual nucleon–nucleon interaction become weaker and weaker when eigenstates become less and less bound. This arises because unbound states, and to a lesser extent weakly bound states, are more spread in space than well-bound states. Moreover, the differences between excitation energies of the states with the same total angular momentum $J$ in well-bound, weakly bound, and unbound states can be as large as a few MeVs, and are typically a few hundreds of keV. Consequently, as the residual nucleon–nucleon interaction is mainly acting close to the core, the overall effect of the residual nucleon interaction decreases when two-body systems become more and more unbound. One can also notice that the relative spacings between the two-body states of different $J$ values in a given spectrum also decrease when total binding energy goes from well bound to unbound. This is also due to the fact that the residual nucleon–nucleon interaction is less and less binding in this situation, so that eigenstates, which would be degenerate in the absence of two-body interaction, become closer to each other.

These effects become less and less important when $\ell$ increases. The differences between the excitation energies of well-bound, weakly bound, and unbound states with the same $J$, which vary from 500 keV to 3 MeV when using $p_{3/2}$ partial waves, can reach up to 2 MeV for $d_{5/2}$ partial waves. Even for $f_{7/2}$, these differences of excitation energies may reach several hundreds of keV. This is a large effect with direct consequences on spectroscopy of weakly bound or resonance many-body states. The coupling to continuum, which is prominent in $p$ waves and $d$ waves and remains non-negligible in higher-$\ell$ waves, should be visible experimentally even in the medium-heavy nuclei.

A similar effect is seen in the particle-emission widths when $\ell = 1, 2$, whose variations are comparable to those of excitation energies (see Tables 6.1, 6.2,

**Table 6.1** Excitation energies (in MeV) and widths (in keV) of neutron-neutron (nn) and proton–neutron (pn) two-body states (second to seventh column) for different energies of the neutron-neutron and proton–neutron ground states. The energies of the ground states of neutron-neutron and proton–neutron systems, for which $J^\pi = 0^+$ and $J^\pi = 3^+$, respectively, are fixed at $-10$, $-1$, or 1 MeV. The total angular momentum of the considered two-body state is given in the first column. A $^4$He core is used and only $p_{3/2}$ partial waves are included in the model space. Widths are given inside round brackets

| $J^\pi$ | $-10$ MeV (nn) | $-1$ MeV (nn) | 1 MeV (nn) | $-10$ MeV (pn) | $-1$ MeV (pn) | 1 MeV (pn) |
|---|---|---|---|---|---|---|
| $3^+$ | – | – | – | 0 | 0 | 0 (78) |
| $1^+$ | – | – | – | 1.871 | 0.868 | 0.417 (368) |
| $0^+$ | 0 | 0 | 0 (149) | 2.672 | 1.102 (0.02) | 0.481 (467) |
| $2^+$ | 1.638 | 1.103 | 0.641 (1110) | 4.250 | 1.897 (322) | 1.077 (1794) |

**Table 6.2** Same as Table 6.1, except that a $^{16}$O core is used, so that only $d_{5/2}$ partial waves are included in the model space. The ground state of a proton–neutron pair has $J^\pi = 5^+$

| $J^\pi$ | $-10$ MeV (nn) | $-1$ MeV (nn) | 1 MeV (nn) | $-10$ MeV (pn) | $-1$ MeV (pn) | 1 MeV (pn) |
|---|---|---|---|---|---|---|
| $5^+$ | – | – | – | 0 | 0 | 0 (0.3) |
| $1^+$ | – | – | – | 1.435 | 0.896 | 0.687 (21) |
| $0^+$ | 0 | 0 | 0 (1) | 2.153 | 1.251 | 0.911 (60) |
| $3^+$ | – | – | – | 2.011 | 1.401 | 1.107 (110) |
| $2^+$ | 0.884 | 0.860 | 0.738 (113) | 3.051 | 2.046 (18) | 1.613 (369) |
| $4^+$ | 1.216 | 1.087 | 0.914 (187) | 3.566 | 2.235 (54) | 1.779 (491) |

**Table 6.3** Same as Table 6.1, except that a $^{40}$Ca core is used and only $f_{7/2}$ partial waves are included in the model space. The ground state of a proton–neutron pair has $J^\pi = 7^+$

| $J^\pi$ | $-10$ MeV (nn) | $-1$ MeV (nn) | 1 MeV (nn) | $-10$ MeV (pn) | $-1$ MeV (pn) | 1 MeV (pn) |
|---|---|---|---|---|---|---|
| $7^+$ | – | – | – | 0 | 0 | 0 |
| $1^+$ | – | – | – | 0.850 | 0.604 | 0.523 (0.2) |
| $0^+$ | 0 | 0 | 0 | 1.398 | 0.946 | 0.798 (3) |
| $5^+$ | – | – | – | 1.497 | 1.170 | 1.040 (10) |
| $3^+$ | – | – | – | 1.506 | 1.183 | 1.053 (11) |
| $2^+$ | 0.571 | 0.601 | 0.581 (6) | 1.977 | 1.543 | 1.362 (35) |
| $4^+$ | 0.869 | 0.833 | 0.784 (16) | 2.268 | 1.763 (0.4) | 1.552 (61) |
| $6^+$ | 0.936 | 0.892 | 0.837 (20) | 2.334 | 1.819 (0.8) | 1.603 (69) |

and 6.3). Indeed, the particle-emission widths of unbound two-body states are typically about 100 keV to 2 MeV when $p$ waves are occupied, and range from a few tens to few hundreds of keV when $d$ waves are occupied (see Tables 6.1 and 6.2). However, particle-emission widths no longer follow the trend seen with $p$ and $d$ partial waves when occupying the $f$ partial wave, as widths are only of a few tens of keV at most in the eigenstates of the $\ell = 3$ spectrum, and increase very mildly with excitation energy (see Table 6.3).

When comparing the results of neutron-neutron and proton–neutron states, one can see that continuum coupling acts essentially in the same manner in both systems, even though they are quantitatively different (see Tables 6.1, 6.2, and 6.3)). This could be expected as one has no Coulomb interaction in the Hamiltonian.

Effective nuclear matrix elements can then be expected to be different for well-bound, weakly bound, and unbound many-body states. Differences can be particularly large when $p$ and $d$ waves are occupied, and even for $f$ waves they remain of the order of a few hundreds of keV, which should still be possible to observe experimentally.

Obviously, the pure two-particle configuration is an idealization of the realistic situation. Nevertheless, these effects should be seen not only in the spectra of nuclei with two nucleons outside the closed (sub)shell but also in more complicate spectra of states when single-particle levels with low angular momenta are significantly occupied. The emission widths are sizable with $p$ and $d$ waves, as they are typically close to 500 keV, but become of the order of a few tens of keV when $f$ waves are occupied.

As a consequence, the derivation of effective nuclear matrix elements for nuclei close to driplines demands to include the continuum coupling. This is true for $p$ partial waves, but also when $\ell = 2, 3$, which is rather unexpected as these partial waves cannot give rise to halo configurations, for example, (see Sect. 7.2.1.1). In fact, only a model including both continuum coupling and inter-nucleon correlations, such as the Gamow shell model, can provide with definite conclusions for that matter. The inter-nucleon correlations in the continuum can generate a complex nuclear structure which is impossible to predict from an independent-particle model or from a shell model based on harmonic oscillator basis states.

## 6.3    Effective Nucleon–Nucleon Interactions for Gamow Shell Model Calculations in Light Nuclei

Traditionally, light nuclei have provided an excellent testing ground for microscopic nuclear structure models. The seminal work in the p-shell nuclei, effected in Refs. [24, 25], initiated the interacting shell model. This model became a pillar of nuclear structure theory and provided guidance to understand the data on energy levels, electromagnetic transitions, nuclear moments, and various particle decays. The identification of basic features and symmetries of the bare nucleon–nucleon interaction, and the continuous development of effective interactions in finite model spaces enabled this impressive progress in the description of nuclear spectra. Because of the harmonic oscillator basis used, only well-bound nuclear states can be reliably considered in the standard shell model. Continuum degrees of freedom, crucial in the proximity of particle-emission threshold in weakly bound and in resonance states, are neglected in this model. In fact, coupling to the scattering continuum and the decay channels is prominent close to driplines and in nuclear reactions [26–28].

Estimating the uncertainties of theoretical predictions is a prerequisite in physical models. Indeed, in the absence of the uncertainty quantification of theoretical results, it is impossible to assess the quality of predictions, thus precluding reliable extrapolations of the used models. The comparison of the quality of different models becomes also problematic (see Refs. [29–31]). In fact, the systematic calculation of the uncertainties of theoretical predictions has only started recently [32–39].

When considering light nuclei, it is convenient to separate the model space in a core and valence part, on the one hand, and, on the other hand, derive a Hamiltonian built from a one-body potential mimicking the effect of the core and an effective two-body interaction comprising central, spin–orbit, tensor, and Coulomb terms. This objective will be handled within two different model spaces. One will firstly study He, Li, and Be isotopes with $A \leq 9$ in a large (psdf)-model space, where both resonant and scattering basis states are included. The aim is to describe structure of nuclei in the $A \lesssim 12$ region (see Ref. [40] for the application of the Gamow shell model), with the inclusion of a statistical analysis of the model results. For this, singular value decomposition is used to estimate the indeterminacy of parameters. By using the covariance matrix obtained within the linear regression approach, the statistical uncertainties on the parameters and observables can be evaluated. One can then apply the Hamiltonian of Gamow shell model to predict two-nucleon correlation densities and excitation spectra with the quantified uncertainties. Secondly, one will study the proton-rich and neutron-rich nuclei with $A \approx 10$ nucleons.

To optimize calculations, the single-particle basis is adapted for each nucleus (see Sect. 5.5). However, it was noticed that the use of the multi-Slater determinant Hartree–Fock potential in light $p$-shell nuclei could become numerically imprecise in the presence of broad $0p_{3/2}$ and $0p_{1/2}$ resonance states. This arises because of the nonlocal character of the multi-Slater determinant Hartree–Fock potential. Indeed, resonance states generated by nonlocal potentials are typically less precise numerically than resonance states generated by local potentials. Consequently, it was preferred to replace the multi-Slater determinant coupled Hartree–Fock potential by a Woods–Saxon potential whose eigenstates have the same complex energy as obtained when using the multi-Slater determinant Hartree–Fock potential. As the Coulomb potential is directly included in the single-particle basis, one-body basis states possess the correct asymptotic behavior at infinity (see Sect. 3.7.1).

## 6.3.1  Study of the Helium, Lithium, and Beryllium Isotope Chains

In the studies of He, Li, and Be isotopes, one is using a core of $^4$He as its large binding energy implies that core excitations can be safely neglected. The Gamow shell model interaction has two components: the one-body potential of the core $\hat{U}_{core}$ and the two-body interaction $V$ between the valence nucleons. The core potential acting on valence particles is modelled by a Woods–Saxon potential with a Coulomb part, whose charge radius is taken from the experimental value of $^4$He: $R_{ch} = 1.681$ fm [41]. Different parameters are taken for proton and neutron Woods–Saxon potentials.

A general form of the two-body effective nuclear potential was derived in Refs. [42, 43]. In particular, a tensor part was added to central and two-body spin–

orbit components in order to reproduce the experimental quadrupole moment of the deuteron. An interaction of this type built from Gaussian form factors could reproduce nucleon–nucleon scattering data up to 300 MeV [44]. The nucleon–nucleon potential is a sum of central, spin–orbit, tensor, and Coulomb terms:

$$\hat{V} = \hat{V}_C + \hat{V}_{LS} + \hat{V}_T + \hat{V}_{Coul} . \tag{6.15}$$

The two-body Coulomb potential $V_{Coul}(r) = e^2/r$ between valence protons is treated exactly by incorporating its long-range part into the basis potential (see Ref. [45] for a detailed description of the method and Sect. 7.5 for a quantitative study of its precision).

The central, spin–orbit, and tensor part of the interaction are based on the Furutani–Horiuchi–Tamagaki-type interaction [17, 18]:

$$\hat{V}_C(r) = \sum_{n=1}^{3} V_C^n \left( W_C^n + B_C^n \hat{P}_\sigma - H_C^n \hat{P}_\tau - M_C^n \hat{P}_\sigma \hat{P}_\tau \right) e^{-\beta_C^n r^2} \tag{6.16}$$

$$\hat{V}_{LS}(r) = \boldsymbol{L} \cdot \boldsymbol{S} \sum_{n=1}^{2} V_{LS}^n \left( W_{LS}^n - H_{LS}^n \hat{P}_\tau \right) e^{-\beta_{LS}^n r^2} \tag{6.17}$$

$$\hat{V}_T(r) = S_{ij} \sum_{n=1}^{3} V_T^n \left( W_T^n - H_T^n \hat{P}_\tau \right) r^2 e^{-\beta_T^n r^2} , \tag{6.18}$$

where $r \equiv r_{ij}$ stands for the distance between the nucleons $i$ and $j$, $\boldsymbol{L}$ is the relative orbital angular momentum, $\boldsymbol{S} = (\boldsymbol{\sigma}_i + \boldsymbol{\sigma}_j)/2$, $S_{ij} = 3(\boldsymbol{\sigma}_i \cdot \hat{r})(\boldsymbol{\sigma}_j \cdot \hat{r}) - \boldsymbol{\sigma}_i \cdot \boldsymbol{\sigma}_j$, and $\hat{P}_\sigma$ and $\hat{P}_\tau$ are spin and isospin exchange operators, respectively. Each part of the interaction is the sum of two or three gaussians with different ranges: a short-range to account for the hard core, a long range to mimic the one-pion exchange potential, and an intermediate range.

The minimization of the penalty function (5.74) is central in the optimization process. An efficient minimization algorithm must solve two problems: a possible strong correlation between the adjusted observables and the sloppiness [46], or indeterminacy, among its parameters, i.e. the fact that parameters can be weakly constrained by the fit-observables [46–50]. For that matter, many minimization methods have been developed recently using Monte Carlo algorithms [33, 34] or the POUNDerS algorithm [51], which have been applied successfully to optimize the nuclear energy density functionals [52, 53] and chiral interactions [37] optimizations.

As the interaction used in this section is linear in strength parameters, the gradient of eigenenergies with respect to interaction parameters can be computed exactly using the Hellmann-Feynman theorem [54]. These derivatives are used in the Gauss-Newton method, which is combined with the singular value decomposition technique to optimize the parameters of the interaction so as to fit theoretical energies on experimental data, as is described in the following. The Gauss-Newton method is a variation of the standard Newton minimization algorithm for optimization problems. The singular value decomposition removes the instability of the Gauss-Newton method, appearing when the Jacobian matrix is non-invertible or has a very small determinant. This indeed happens when the fit-observables are highly correlated and/or some parameters are unconstrained (see Ref. [49] for a full description of the method).

The core-nucleon interaction of Eq. (3.83) was optimized to the experimental $p_{3/2}$, $p_{1/2}$, and $s_{1/2}$ nucleon—$^4$He scattering phase shifts up to 20 MeV [55–57]. The optimization procedure provided a well-converged result for both protons and neutrons corresponding to a precision of $\|\nabla\chi^2\|/N_{\text{dof}} \sim 10^{-12}$. In order to obtain a global minimum, the optimization is repeated several times starting from different initial points for the parameters of the fitted interaction.

Parameters of the optimized Woods–Saxon potentials and their statistical uncertainties are listed in Table 6.4. The corresponding phase shifts are shown in Fig. 6.5. As the $^4$He core at low excitation energies is approximately inert, its optical potential is well described by a Woods–Saxon model [58] which provides a very good description of experimental low-energy phase shifts. The small discrepancies seen at $E > 15$ MeV can be attributed to the virtual excitations to the excited states of $^4$He.

**Table 6.4** Parameters of the optimized core-nucleon interaction with associated statistical uncertainties. The charge radius $R_{\text{ch}}$ was set to the experimental value [41] and did not enter the optimization procedure (adapted from Ref. [40])

| Parameter | $V_0$ [MeV] | $V_{\ell s}$ [MeV fm$^2$] | $R_0$ [fm] | $a$ [fm] | $R_{\text{ch}}$ [fm] |
|---|---|---|---|---|---|
| Neutrons | 41.9 (10) | 7.2 (2) | 2.15 (4) | 0.63 (2) | – |
| Protons | 44.4 (11) | 7.2 (2) | 2.06 (4) | 0.64 (2) | 1.681 |

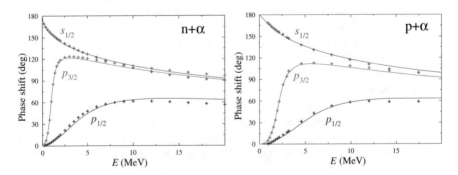

**Fig. 6.5** $s_{1/2}$, $p_{3/2}$, and $p_{1/2}$ neutron-$^4$He (proton-$^4$He) optimized phase shifts are plotted as functions of the neutron (proton) energy in the laboratory frame. Parameters of the Woods–Saxon potential are given in Table 6.4. The experimental values are represented by crosses. The neutron phase shifts are taken from Ref. [55]. The experimental data for proton phase shifts from 0 to 3 MeV are taken from Ref. [56] and the data from 3 to 20 MeV from Ref. [57] (adapted from Ref. [40])

**Table 6.5** Energies (in MeV) and widths (in keV) of the $3/2^-$ ground states of $^5$He and $^5$Li calculated using the optimized optical model with parameters listed in Table 6.4. The experimental values are taken from Refs. [59, 60] (adapted from Ref. [40])

| Nucleus | $E$ [MeV] | $E_{exp}$ [MeV] | $\Gamma$ [keV] | $\Gamma_{exp}$ [keV] |
|---------|-----------|-----------------|----------------|----------------------|
| $^5$He  | 0.755     | 0.798           | 651            | 648                  |
| $^5$Li  | 1.627     | 1.69            | 1351           | 1230                 |

**Table 6.6** Correlation matrix (5.84) of the optimized $^4$He-nucleon interaction for the protons (upper triangular matrix) and neutrons (lower triangular matrix). Interaction parameters $V_0$, $V_{\ell s}$ are in MeV, and $R_0$, $a$ in fm (adapted from Ref. [40])

| n/p          | $V_0$ | $V_{\ell s}$ | $R_0$ | $a$   |
|--------------|-------|--------------|-------|-------|
| $V_0$        | 1     | 0.62         | −0.95 | 0.59  |
| $V_{\ell s}$ | 0.55  | 1            | −0.78 | 0.81  |
| $R_0$        | −0.95 | −0.75        | 1     | −0.81 |
| $a$          | 0.52  | 0.84         | −0.75 | 1     |

The fact that parameters in Table 6.4 strongly differ in the proton and neutron cases, in particular for the Woods–Saxon depth $V_0$, has been well-documented in publications and is seen experimentally in the behavior of the proton and neutron $p_{1/2}$ phase shifts at $E < 1.5$ MeV (see Fig. 6.5) [58]. This consistency can be illustrated by calculating the energies and widths of the $3/2^-$ ground states of $^5$He and $^5$Li. Table 6.5 demonstrates a good agreement with experimental data, especially given the large widths of the resonance states.

Table 6.6 shows the correlation matrix (5.84) for the parameters of the optimized core potential. Together with the uncertainties on parameters given in Table 6.4, this information can be used to compute the uncertainties on observable quantities. One may notice that there is a strong correlation between the depth and radius of Woods–Saxon potential.

In the next step, the two-body nuclear interaction between valence nucleons is optimized. The calculations have been performed in the ($psdf$)-configuration space. The $0p_{3/2}$, $0p_{1/2}$ resonant states and the associated scattering continua were used as the Berggren basis for both protons and neutrons. Moreover, the $1s_{1/2}$ and $0d_{5/2}$ resonant states and the associated continua for neutrons were used to account for possible antibound shells and excited states of different parity.

As stated in Sect. 5.3, the basis potential that generates the Berggren basis was adapted for each nucleus. Consequently, the scattering continua were also chosen differently for all nuclei depending on the nature of the $0p_{3/2}$, $0p_{1/2}$, $1s_{1/2}$, and $0d_{5/2}$ poles. For example, if the considered pole and the searched many-body state were bound, the contour consisted of three segments on the real axis of the momentum plane defined by the points: $k_{peak} = (0.1, 0.0)$ fm$^{-1}$, $k_{mid} = (0.2, 0.0)$ fm$^{-1}$, and $k_{max} = (2.0, 0.0)$ fm$^{-1}$. In the case of unbound single-particle pole, $k_{peak}$ and $k_{mid}$ were moved into the complex momentum plane to encompass the resonance state. Finally, in the special casc of $^7$He, for which the $0p_{3/2}$ pole is bound but the many-body state is unbound, the corresponding contour was defined by the points $k_{\text{peak}} = (0.25, -0.24)\,\text{fm}^{-1}$, $k_{\text{mid}} = (0.5, 0.0)\,\text{fm}^{-1}$, and $k_{\text{max}} = (2.0, 0.0)\,\text{fm}^{-1}$ to generate a many-body configuration space which can describe an unbound many-body state. In all cases, the three segments were discretized with at least ten Gauss-Legendre points. The remaining higher-$\ell$ partial waves, and proton partial waves in neutron-rich nuclei, which play no role in the asymptotic region for the considered nuclei, were described using a harmonic oscillator basis. For that matter, and to reduce the size of the model space, the $sdf$ partial waves for protons and $df$ partial waves for neutrons were described by a harmonic oscillator basis with 11 shells ($n_{\text{max}}^{\text{HO}} = 10$). In this mixed basis, the natural orbitals were generated as discussed in Sect. 5.3 what allowed to have up to the four particles in the scattering continuum.

The two-body interaction was optimized to the experimental binding energies of the ground states and a few selected excited states of the helium, lithium, and beryllium isotopes. The binding energies with respect to $^4$He span a large energy range, from approximately $-30\,\text{MeV}$ to $+2\,\text{MeV}$, and different types of states are involved: bound states, resonances, and halo states, as for the ground state of $^6$He. The optimization provided a $\chi^2$ minimum with a precision $\|\nabla\chi^2\|/N_{\text{dof}} \sim 10^{-4}$ limited only by the singular value decomposition cutoff value. The optimized inter-action parameters are listed in Table 6.7 together with the associated uncertainties. As some parameters are weakly constrained, the singular value decomposition procedure played an important role in the optimization. To account for their different units and orders of magnitude, the parameters were normalized to the value of one during the singular value decomposition procedure, that is $p_\alpha \to \tilde{p}_\alpha = 1$, $J_{i\alpha} \to \tilde{J}_{i\alpha} = p_\alpha J_{i\alpha}$.

Table 6.8 lists the singular values (square roots of the eigenvalues of the Hessian matrix at the minimum together with the corresponding eigenvectors). The eigenvectors associated with large singular values define the directions along which the penalty function (see Eq. (5.74)) exhibits the largest variations. Following singular value decomposition, the parameter space is reduced to a smaller (relevant)

**Table 6.7** Optimized parameters of the two-body nuclear interaction together with their statistical uncertainties. As indicated in their superscript, parameters depend on the spin $S = 0, 1$ and isospin $T = 0, 1$ of the two nucleons, respectively (adapted from Ref. [40])

| Parameter | Value |
|---|---|
| $V_C^{11}$ [MeV] | $-3.2\ (220)$ |
| $V_C^{10}$ [MeV] | $-5.1\ (10)$ |
| $V_C^{00}$ [MeV] | $-21.3\ (66)$ |
| $V_C^{01}$ [MeV] | $-5.6\ (5)$ |
| $V_{LS}^{11}$ [MeV] | $-540\ (1240)$ |
| $V_T^{11}$ [MeV fm$^{-2}$] | $-12.1\ (795)$ |
| $V_T^{10}$ [MeV fm$^{-2}$] | $-14.2\ (71)$ |

**Table 6.8** Singular values $s_n$ and the corresponding eigenvectors of the normalized Hessian matrix, $\tilde{J}^T \tilde{J}$, with respect to the parameters at the minimum. The main components are written in boldface. The singular value decomposition cutoff separates the relevant space generated by the eigenvalues 1–4 from the irrelevant space. The displayed values of the interaction parameters (in units of MeV) are computed at the $\chi^2$ minimum but exhibit similar pattern during the optimization procedure (adapted from Ref. [40])

| $n$ | $s_n$ | $V_C^{11}$ | $V_C^{10}$ | $V_C^{00}$ | $V_C^{01}$ | $V_{LS}^{11}$ | $V_T^{11}$ | $V_T^{10}$ |
|---|---|---|---|---|---|---|---|---|
| 1 | 243 | 0.00 | **0.82** | −0.03 | **0.53** | 0.00 | 0.00 | 0.23 |
| 2 | 43.0 | 0.00 | **−0.49** | −0.02 | **0.85** | 0.00 | −0.01 | −0.19 |
| 3 | 7.06 | −0.04 | −0.16 | **0.79** | 0.05 | 0.04 | −0.07 | **0.58** |
| 4 | 3.94 | 0.02 | −0.25 | **−0.61** | 0.01 | −0.09 | −0.04 | **0.75** |
| 5 | 0.57 | −0.23 | −0.02 | −0.09 | 0.00 | **0.97** | −0.01 | 0.04 |
| 6 | 0.20 | **0.65** | −0.03 | 0.04 | 0.01 | 0.16 | **0.74** | 0.06 |
| 7 | 0.12 | **0.73** | 0.01 | 0.00 | 0.00 | 0.16 | **−0.66** | −0.04 |

space defined by the singular values greater than a given cutoff value $s_{min}$. In the case considered, a large value of $s_{min} = 1$ was needed for the optimization procedure to converge, reducing the parameter space to four main directions. Table 6.8 also shows that the two central-potential parameters $V_C^{10}$ and $V_C^{01}$ are the two parameters which primarily govern the optimization, as well as $V_C^{00}$ $V_T^{10}$ to a lesser extent. The three parameters with (ST) = (11) are poorly constrained by the experimental dataset chosen. More experimental data of different kinds, such as charge and matter radii and electromagnetic moments, will be useful in the future studies to constrain these parameters. At this point, the freedom on the sloppy parameters can be utilized to fine-tune the interaction to reproduce experimental reaction thresholds.

### 6.3.1.1 Energy Spectra

The results of the optimization of the Hamiltonian interaction parameters (see Eqs. (6.16–6.18)) are shown in Fig. 6.6. The overall quality of the optimization is excellent, with a root-mean-square deviation of 250 keV. The helium chain, where energies depend almost exclusively on a single parameter $V_C^{01}$, is well described with a root-mean-square deviation of 95 keV. The $T = 0$ nuclear interaction is responsible for clusterization effects and probably demands the inclusion of higher partial waves than $\ell = 3$ for a better description. This explains why the optimization slightly deteriorates for Li and Be isotopes. However, an overall agreement with experimental data over such a large range of energies is satisfactory and makes this interaction an excellent starting point for detailed structure and reaction studies across the $A \simeq 5 - 12$ nuclei. It is also worth noting that the widths of the unbound states, even if they do not enter the set of fit-observables and are extremely dependent on the threshold energies, are described fairly well.

The correlation coefficients (5.84) for the two-body interaction parameters are listed in Table 6.9. This table can be used to obtain the associate covariance matrix needed to assess the uncertainties on predicted observables. The two main interaction parameters $V_C^{10}$ and $V_C^{01}$ are strongly anti-correlated. The values that

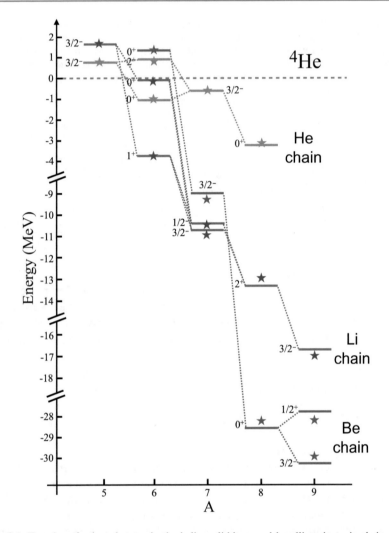

**Fig. 6.6** Energies of selected states in the helium, lithium, and beryllium isotopic chains with $A \leq 9$ are calculated using the optimized Gamow shell model interaction with parameters given in Tables 6.4 and 6.7. The experimental values shown by stars are taken from Ref. [60] (from Ref. [40])

are related to the sloppy parameters should not be taken too rigorously as they are computed within the linear regression framework. Only a fully consistent statistical study, based, e.g. on Bayesian techniques, can fully assess correlations related to these parameters.

Figure 6.7 shows predictions of the Gamow shell model using the optimized interaction (6.16–6.18) for several excited states of $^7$He, $^7$Be, and the ground state of $^7$B. The energy and widths are listed in Table 6.10 together with their uncertainties.

**Table 6.9** Correlation coefficients between the two-body interaction parameters. Parameters depend on the spin $S = 0, 1$ and isospin $T = 0, 1$ of the two nucleons, as indicated in their superscript (adapted from Ref. [40])

|            | $V_C^{11}$ | $V_C^{10}$ | $V_C^{00}$ | $V_C^{01}$ | $V_{LS}^{11}$ | $V_T^{11}$ | $V_T^{10}$ |
|------------|-----------|-----------|-----------|-----------|--------------|-----------|-----------|
| $V_C^{11}$ | 1 | 0.24 | 0.26 | −0.44 | 0.63 | −0.45 | −0.25 |
| $V_C^{10}$ | 0.24 | 1 | −0.22 | −0.92 | 0.01 | −0.89 | −0.99 |
| $V_C^{00}$ | 0.26 | −0.22 | 1 | 0.30 | −0.21 | 0.38 | 0.21 |
| $V_C^{01}$ | −0.44 | −0.92 | 0.30 | 1 | −0.17 | 0.96 | 0.89 |
| $V_{LS}^{11}$ | 0.63 | 0.01 | −0.21 | −0.17 | 1 | −0.28 | −0.04 |
| $V_T^{11}$ | −0.45 | −0.89 | 0.38 | 0.96 | −0.28 | 1 | 0.88 |
| $V_T^{10}$ | −0.25 | −0.99 | 0.21 | 0.89 | −0.04 | 0.88 | 1 |

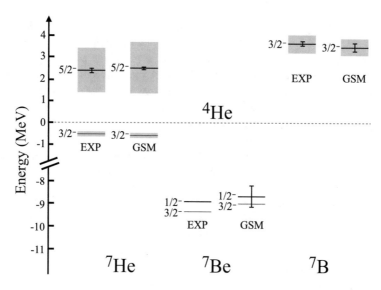

**Fig. 6.7** Spectra of $^7$He, $^7$Be, and $^7$B. The experimental values are taken from Ref. [60]. The ground state energies of $^7$He and $^7$Be were included in the optimization, hence are shown without uncertainties. The uncertainties on the widths, not shown in the figure, are listed in Table 6.10 (from Ref. [40])

Theoretical uncertainties have been calculated using the covariance matrices for the one-body (core-nucleon potential) and two-body potentials, and they can be expressed as $\Delta E = \sqrt{\Delta E_N^2 + \Delta E_{NN}^2}$. The main part of the uncertainty comes from the two-body interaction. For all $A < 8$ states, the contribution to the uncertainty $\Delta E_N < 0.07$ MeV coming from the core-nucleon potential is negligible. This is due to the fact that those states primarily involve the $p_{3/2}$ bound/resonance shell, which is well constrained by the core-nucleon potential optimization. The $p_{1/2}$ and $s_{1/2}$ pole shells are less constrained and result in larger values of $\Delta E_N$ for the $1/2^+$ and $1/2^-$ states of $^9$He (not shown in Fig. 6.7)

**Table 6.10** Calculated values of $A = 7$ nuclei and the helium nuclei. The experimental values come from [60, 61]. All energies are given with respect to the $^4$He core. The energies are given in MeV, the widths in keV. The theoretical and experimental uncertainties on the widths have different meanings and should not be compared (adapted from Ref. [40])

| State | $E$ [MeV] | $E_{\mathrm{exp}}$ [MeV] | $\Gamma$ [keV] | $\Gamma_{\mathrm{exp}}$ [keV] |
|---|---|---|---|---|
| $^7$He, $5/2^-$ | +2.50 (2) | +2.39 (9) | 2250 (280) | 1990 (170) |
| $^7$Be, $1/2^-$ | −8.67 (45) | −8.88 | | |
| $^7$B, $3/2^-$ | +3.42 (21) | +3.58 (7) | 740 (450) | 801 (20) |
| $^8$He, $2^+$ | −0.10 (75) | −0.41/+0.49 | 290 (1010) | 600 (200) |
| $^9$He, $1/2^+$ | −3.12 (31) | −2.93 (9) | ∼0 | 180 (160) |
| $^9$He, $1/2^-$ | −2.98 (102) | −1.88 (12) | 630 (330) | 130 (170) |

Both the energy and the width of the first excited $5/2^-$ state of $^7$He are well reproduced, with a small uncertainty. It is important to note that the uncertainties on calculated widths should be seen as the range of values where the parameters move within the interval of confidence [30]. Since the widths of many-body nuclear systems typically increase very quickly with energy when the state is unbound, the related uncertainties are usually large. Consequently, those values should not be directly compared to experimental uncertainties.

As expected, the $3/2^-$ ground state of $^7$B is well reproduced as its mirror state, the ground state of $^7$He, was included in the optimization. Finally, the agreement of the $1/2^-$ state of $^7$Be with the experimental data is also good within the uncertainty of 450 keV in calculated binding energies. This significant uncertainty comes from the fact that the less-constrained parameters of the interaction are important therein.

The energy of the $2^+$ of $^8$He is unresolved experimentally with the two values −0.41 MeV and +0.49 MeV [60]. The Gamow shell model prediction is consistent with both, due to the large statistical uncertainty of 750 keV.

The nucleus $^9$He is a difficult system to study experimentally and results of various experiments contradict each other. Experimental data in Table 6.10 for $^9$He is taken from Ref. [61]. The calculated uncertainties on predicted values are rather large, in particular for the $1/2^-$ state. This arises from the important occupation of the $p_{1/2}$ resonance shell, as the dominant configuration of $1/2^-$ state is that of a neutron $p_{1/2}$ resonance shell above a $^8$He core. The uncertainty of the $1/2^+$ state of $^9$He is somewhat smaller because $1/2^+$ state of $^9$Be was included in the optimization. Indeed, both these $1/2^+$ states mainly consist of one neutron occupying the weakly bound $1s_{1/s}$ shell, while other nucleons occupy the well bound $0p_{3/2}$ shells. Thus, the fit of the $1/2^+$ state of $^9$Be reduces the uncertainties present in the $1/2^+$ state of $^9$He. If one considers only the energies provided by the calculation, independently of statistical uncertainties, the $^9$He ground state is predicted to be the $1/2^+$ state. However, the energy difference between the $1/2^+$ and $1/2^-$ states of $^9$He is much smaller than the statistical uncertainty associated with the energies of these states.

The helium chain has been studied in Ref. [62] by reducing as much as possible the number of parameters needed to reproduce its main features. It was shown in this analysis that it is sufficient to adjust a single parameter, the spin-singlet central parameter denoted as $V_C^{01}$ in Table 6.7 to reproduce the experimental energies and widths of $^{5-8}$He up to a few tens of keV. A parity inversion of narrow resonances in $^9$He has also been predicted. The calculation of the ground state of $^{10}$He [62] provided with a wave function dominated by $s$-waves decaying predominantly by the two-neutron emission. The calculation of the ground state of $^{10}$He [62] provided with a wave function dominated by $s$-waves decaying predominantly by the two-neutron emission. For a complete description of the two-neutron emission of $^{10}$He it is necessary, however, to separate different emission channels in the Gamow shell model calculation. For this, one has to rely on the formulation of Gamow shell model in the representation of coupled-channels (see Sect. 9).

As we have discussed above, quite a few terms of the Furutani–Horiuchi–Tamagaki interaction are poorly constrained. In fact, only the central $V_C^{00}$, $V_C^{10}$, $V_C^{01}$, and tensor $V_T^{10}$ terms of this interaction are well fitted. This suggests another strategy which was put forward in Ref. [63]. In this strategy, one restricts the number of terms to those which are well constrained by the fitting procedure and the statistical analysis of the Furutani–Horiuchi–Tamagaki interaction. The reduction of the interaction terms can also be motived by arguments issued from the effective field theory [64–68], as the terms proportional to $V_C^{10}$, $V_C^{01}$, and $V_T^{10}$ appear at the leading order in the effective field theory expansion of the Furutani–Horiuchi–Tamagaki interaction [63]. Inclusion of the central term $V_C^{00}$, of higher order in this expansion, is motivated empirically as it improves the overall fit of energies. This simplified effective interaction was used to describe the chain of Li, Be isotopes [63] in a smaller model space including only the $s_{1/2}$, $p_{3/2}$, and $p_{1/2}$ partial waves. It was then possible to calculate the ground states of Li isotopes up to the neutron dripline, ending with the $^{11}$Li two-neutron halo state, as well as the isobaric analog states of the Li ground states.

Figure 6.8 shows the energies calculated in Gamow shell model for the ground states and selected excited states of $N = 3$ isotones from $^7$Be to $^{11}$O. As one can see, the devised interaction allows for a good reproduction of experimental energies. It is to be noted that the results for higher excited states, not included in the fit, are very satisfactory as well. The $5/2^-$ and $7/2^-$ excited states in $^7$Be are slightly above the corresponding experimental values, whereas the position of the resonant $3^+$ states in $^8$B and $5/2^-$ state in $^9$C are well reproduced, as well as the weakly bound ground states of $^8$B and $^9$C.

The spectrum of an unbound nucleus $^{10}$N is not experimentally known with certainty. Figure 6.8 shows the tentative level assignments given in http://www.nndc.bnl.gov/ensdf. According to Refs. [71, 72], the ground state of $^{10}$N is most likely a $1^-$ state with an excitation energy with respect to the core $E = (1.81 - 1.94)$ MeV. In a more recent work [73], two low-lying negative-parity states were observed but the spin assignment was not possible. The Gamow shell model calculation predicts $1^-$ ground state of $^{10}$N which is a resonance $(E, \Gamma) = (-8.93, 0.9)$ MeV that

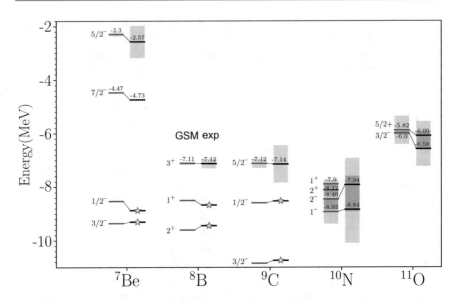

**Fig. 6.8** Level schemes of $N = 3$ isotones with respect to $^4$He calculated in Gamow shell model and compared to experiment. Width of resonances are marked by shaded boxes. The levels used in the Gamow shell model Hamiltonian optimization are marked by stars. Experimental energy of the $5/2^-$ resonance in $^9$C was taken from Ref. [69] and the data for $^{11}$O from Ref. [70] (from Ref. [63])

lies 1.92 MeV above the one-proton emission threshold. The first excited state is predicted to be a $2^-$ state with $\Gamma = 0.3$ MeV slightly below the value quoted in Ref. [73]. This result is consistent with the recent Gamow coupled-channel analysis [74]. One also predicts an excited $1^+$ state with $\Gamma = 0.3$ MeV, lying 2.9 MeV above the $^9$C+$p$ threshold, as well as a second positive-parity $2^+$ state with a width of 0.36 MeV.

### 6.3.1.2 Mirror Symmetry Breaking

The effect of different positions of particle-emission thresholds on spectra of mirror nuclei is shown in Fig. 6.9 which compares the level schemes of Li isotopes and their mirror partners. As expected, the proton-unbound states in proton-rich mirror nuclei are shifted down in energy as compared to the states in neutron-rich partners [75, 76], which lie below, or slightly above the one neutron emission threshold. The $^{10}$Li-$^{10}$N mirror pair is the most interesting one as both nuclei lie above the particle-emission thresholds. The effect of $^9$C+$p$ threshold in $^{10}$N on the negative-parity states $1^-$ and $2^-$ containing the $s$-wave proton partial wave is huge. It results in a large shift of both negative-parity states when going from $^{10}$Li to $^{10}$N that gives rise to a different structure of low-lying resonances in these nuclei.

The experimental situation in $^{10}$Li is still much debated. Several experimental [77–80] and theoretical studies [81, 82] indicated that the structure of the ground state in $^{10}$Li may correspond to a valence neutron in a virtual $s$-state. In a recent

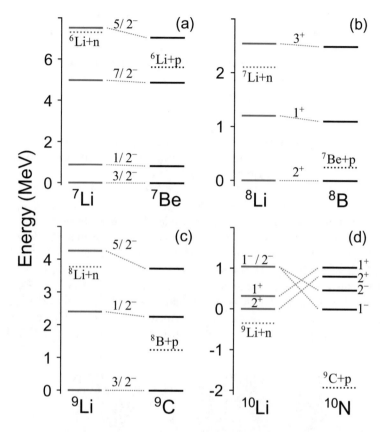

**Fig. 6.9** Level schemes of Li isotopes with (**a**) $A = 7$, (**b**) $A = 8$, (**c**) $A = 9$, (**d**) $A = 10$, and their mirror partner predicted in Gamow shell model calculations. The energies are plotted with respect to the ground state energy. The proton emission thresholds are denoted with a dashed line (from Ref. [63])

experiment [83], the presence of an appreciable low-energy $\ell = 0$ strength has not been confirmed. This conclusion was, however, challenged in theoretical studies [84, 85]. Gamow shell model calculations (see Fig. 6.9) predict the ground state of $^{10}$Li to be a $2^+$ state, about 0.35 MeV above the neutron emission threshold, in accordance with Ref. [86]. The lowest negative-parity state $1^-$ is predicted to lie $\sim$1.0 MeV higher, in agreement with Ref. [83].

The unbound $^{11}$O is the mirror partner of the two-neutron halo nucleus $^{11}$Li. The first observation of $^{11}$O was achieved recently [70]. A broad peak with a width of $\sim$3.4 MeV was observed. Gamow shell model calculations predict a $3/2_1^-$ ground state with a width of 0.13 MeV and the first excited $5/2_1^+$ state with $\Gamma \approx 1$ MeV, see Fig. 6.8. These predictions are consistent with the Gamow coupled-channel calculations of Ref. [74].

### 6.3.1.3 Pairing Correlations and Correlation Density in $^6$He and $^6$Li

Pairing correlations are important in nuclei close to the neutron dripline as they can stabilize weakly bound nuclei through the coupling to continuum states [87–91]. Two-nucleon correlations can be evaluated using correlation density [92–95]:

$$\rho_{NN}(r, \theta) = 8\pi^2 r_1^2 r_2^2 \sin(\theta)\rho(r_1, r_2) = \langle \Psi | \delta(r - r_1)\delta(r' - r_2)\delta(\theta - \theta_{12}) | \Psi \rangle, \tag{6.19}$$

in which $r_1$ and $r_2$ are the positions of nucleons and $\theta_{12}$ is the opening angle between the two nucleons. One follows the normalization convention of Ref. [94] in which the Jacobian $8\pi^2 r^2 r'^2 \sin\theta$ is implicitly included in $\rho_{NN}$.

Figure 6.10 shows the calculated pair correlation densities for the $2^+$ states of $^6$He and $^6$Li. As expected, the isobaric analog $0^+$ states shown in panels (a) and (c) are predicted to have similar correlation densities. Results are in agreement with the conclusions of Refs. [93–96], where dinucleon and cigar-like configurations coexist at small and large opening angles, respectively, and are radially extended. Such a behavior is absent for the $2^+$ resonance of $^6$He shown in panel (b), for which the valence neutrons are predicted to be weakly correlated [94]. As seen in Fig. 6.10d, the strong $T = 0$ interaction in the $1^+$ ground state of $^6$Li gives rise to a deuteron-like structure [95, 96]. This result is in agreement with models which describe this state as a deuteron orbiting in the potential generated by the alpha core [97, 98].

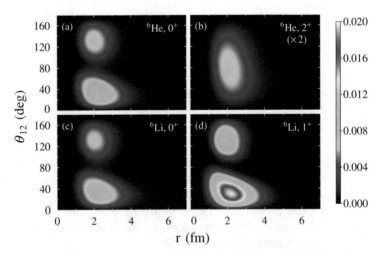

**Fig. 6.10** Two-nucleon correlation densities (in units of fm$^{-2}$) calculated for states in $^6$He and $^6$Li using the optimized Gamow shell model interaction. For the $2^+$ state in $^6$He the density was multiplied by a factor two to maintain the same scale as in other panels (from Ref. [40])

## 6.4 Three-Body Model in Berggren Basis

The cluster orbital shell model formalism has been employed in the Gamow shell model to remove spurious center-of-mass excitations (see Sect. 5.2). In this section, we shall compare Gamow shell model calculations realized with cluster orbital shell model coordinates and Jacobi coordinates in a three-body model. A three-body Hamiltonian can be conveniently solved with Jacobi coordinates using hyperspherical harmonics. As the latter framework is translationally invariant, both cluster orbital shell model coordinates and Jacobi coordinates are equivalent and hence, the eigenvalues of a Hamiltonian expressed using both approaches are formally identical. However, due to the different truncation schemes used in practical applications, the eigenvalues may not be equal in both frameworks.

### 6.4.1 Berggren Basis Expansion in the Three-Body Model

The three-body model discussed in this section describes nuclei in the approximation of a frozen core and two valence nucleons or clusters of nucleons interacting via a residual nucleon–nucleon interaction. The Hamiltonian can be written as:

$$\hat{H} = \sum_{i=1}^{3} \frac{\hat{p}_i^2}{2m_i} + \sum_{i>j=1}^{3} \hat{V}_{ij}(\boldsymbol{r}_{ij}) - \hat{T}_{\text{CM}} , \qquad (6.20)$$

where $V_{ij}$ is the interaction between clusters $i$ and $j$, including central, spin–orbit, and Coulomb terms, and $\hat{T}_{\text{CM}}$ is the kinetic energy of the center-of-mass.

The main drawback of three-body models is the appearance of Pauli-forbidden states arising from the lack of antisymmetrization between core and valence particles. In order to eliminate these states, one can implement the projection technique [99]:

$$\hat{H} \longrightarrow \hat{H} + \Lambda \sum_c |\varphi^{j_c m_c}\rangle \langle \varphi^{j_c m_c}| , \qquad (6.21)$$

where $\Lambda$ is a constant and $|\varphi^{j_c m_c}\rangle$ is a two-body state of the core with angular quantum numbers $j_c m_c$. At large values of $\Lambda$, Pauli-forbidden states appear at high energies, so that they are effectively suppressed.

When one uses cluster orbital shell model coordinates, the finite mass of the core induces a recoil term (see Eq. (5.42)) in the Hamiltonian. In order to describe asymptotics and to eliminate the spurious center-of-mass motion exactly, one expresses the three-body model in the Jacobi relative coordinates [100, 101]:

$$\boldsymbol{x} = \sqrt{\mu_{ij}}(\boldsymbol{r}_i - \boldsymbol{r}_j),$$

$$\boldsymbol{y} = \sqrt{\mu_{(ij)k}} \left( \boldsymbol{r}_k - \frac{A_i \boldsymbol{r}_i + A_j \boldsymbol{r}_j}{A_i + A_j} \right) , \qquad (6.22)$$

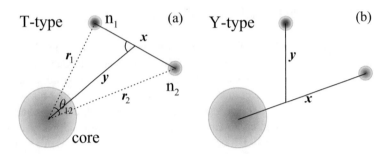

**Fig. 6.11** T-type (panel (**a**)) and Y-type (panel (**b**)) Jacobi coordinates in a three-body system (from Ref. [95])

where $r_i$ is the position vector of the i-th cluster, $A_i$ is the i-th cluster mass number, and $\mu_{ij}$ and $\mu_{(ij)k}$ are the reduced masses associated with $x$ and $y$, respectively:

$$\mu_{ij} = \frac{A_i A_j}{A_i + A_j}$$

$$\mu_{(ij)k} = \frac{(A_i + A_j)A_k}{A_i + A_j + A_k} .$$

(6.23)

As one can see in Fig. 6.11, the three-body system in Jacobi coordinates can be expressed in T- and Y-representations, both forming a complete basis. To describe the transformation between different types of Jacobi coordinates, it is convenient to introduce the basis of hyperspherical harmonics [102, 103]. The hyperspherical coordinates are constructed from a five-dimensional hyperangular part $\Omega_5$ and a hyperradial part $\rho = \sqrt{x^2 + y^2}$. The transformation between different types of Jacobi coordinates is then given by the Raynal–Revai coefficients [104].

The one-body completeness relation borne by the Berggren basis has been demonstrated in Sect. 3.5 and stated in Eq. (3.66). It is based on the Cauchy theorem applied to the real-energy completeness relation of Eq. (3.50), where the poles arising from complex contour integration become resonant states. The context in which it was demonstrated was that of a nucleon, hence of a physical particle. However, this is not a mathematical restriction, as Eq. (3.50) can be easily generalized to other systems. In fact, the only requirement for a Berggren completeness relation is the analyticity of used basis functions in the complex $k$-plane. Hence, one can formulate the complete Berggren basis of wave functions depending on hyper-radius $\rho$, thus associated with the relative motion of two particles above a core. Basis resonance states then physically correspond to unbound two-body systems where the two nucleons are either emitted as a cluster (small $\rho$), or separate from each other and move far away from the core in different directions (large $\rho$).

Besides this different interpretation of basis wave functions, the derivation of the Berggren basis using the hyper-radius $\rho$ instead of radius $r$ is formally identical. Indeed, the Coulomb potential defining the wave functions depending on the hyper-radius $\rho$ is formally the same as that used in Sect. 3.5. However, in order to emphasize the different physical content of calculated wave functions, one will use in this section the notation $\mathcal{B}(k, \rho)$ instead of $u(k, r)$. Let us note that the Berggren basis developed for particle states can also be generalized to pairs of nucleons. This will be the subject of Sect. 6.5, where the extension of Richardson pairing model to the continuum will be realized in the Berggren ensemble.

Similarly to Eq. (3.66), the Berggren completeness relation using the hyper-radius $\rho$ reads as:

$$\sum_{n \in b, d} \mathcal{B}_n(k_n, \rho) \mathcal{B}_n(k_n, \rho') + \int_{L^+} \mathcal{B}(k, \rho) \mathcal{B}(k, \rho') dk = \delta(\rho - \rho') , \qquad (6.24)$$

where $b$ and $d$ stand for bound and decaying states, respectively. For numerical purposes, the $L^+$ contour has to be discretized, e.g. by adopting the Gauss-Legendre quadrature [105]. Due to the symmetry between the second and fourth quadrants in complex momentum plane, only the contour in the fourth quadrant $L^+$ needs to be considered. If the contour $L^+$ is chosen along the real $k$-axis, the Berggren completeness relation reduces to the Newton completeness relation involving bound and real-energy scattering states.

The total wave function, whose quantum numbers are total angular momentum $J$, total angular momentum projection $M$, and parity $\pi$, can be written as a linear combination of basis states of coordinates $\Omega_5$ and $\rho$:

$$\Psi^{JM\pi}(\rho, \Omega_5) = \rho^{-5/2} \sum_{\gamma K} \psi_{\gamma K}^{J\pi}(\rho) \mathcal{Y}_{\gamma K}^{JM}(\Omega_5) , \qquad (6.25)$$

where $\gamma = \{s_1, s_2, s_3, S_{12}, S, \ell_x, \ell_y, L\}$ is a set of quantum numbers. $s$ and $l$ stand for spin and orbital angular momentum, respectively, $\psi_{\gamma K}^{J\pi}(\rho)$ is the hyperradial wavefunction, and the hyperangular basis state $\mathcal{Y}_{\gamma K}^{JM}(\Omega_5)$ is the hyperspherical harmonics [106].

In order to consider the continuum and resonances precisely, one uses the Berggren basis expansion for the hyperradial wavefunction:

$$\psi_{\gamma K}^{J\pi}(\rho) = \sum_n C_{\gamma n K}^{J\pi M} \mathcal{B}_{\gamma n}^{J\pi}(\rho) , \qquad (6.26)$$

where $\mathcal{B}_{\gamma n}^{J\pi}(\rho)$ is the Berggren basis and $C_{\gamma n K}^{J\pi M}$ is the expansion coefficient. As a result, the hyperradial Schrödinger equation can be written as a set of coupled-channel equations:

$$
\left[ -\frac{\hbar^2}{2m} \left( \frac{d^2}{d\rho^2} - \frac{(K+3/2)(K+5/2)}{\rho^2} \right) - \tilde{E} \right] \psi_{\gamma K}^{J\pi}(\rho)
$$

$$
+ \sum_{K'\gamma'} \hat{V}_{K'\gamma',K\gamma}^{J\pi}(\rho) \psi_{\gamma' K'}^{J\pi}(\rho) + \sum_{K'\gamma'} \int_0^{+\infty} \hat{W}_{K'\gamma',K\gamma}(\rho,\rho') \psi_{\gamma' K'}^{L\pi}(\rho') d\rho' = 0 ,
$$

$$(6.27)$$

where

$$
\hat{V}_{K'\gamma',K\gamma}^{L\pi}(\rho) = \langle \mathscr{Y}_{\gamma' K'}^{JM} | \sum_{i>j=1}^{3} \hat{V}_{ij}(r_{ij}) | \mathscr{Y}_{\gamma K}^{JM} \rangle
$$

$$
\hat{W}_{K'\gamma',K\gamma}(\rho,\rho') = \langle \mathscr{Y}_{\gamma' K'}^{JM} | \Lambda \sum_c | \varphi^{j_c m_c} \rangle \langle \varphi^{j_c m_c} | \mathscr{Y}_{\gamma K}^{JM} \rangle ,
$$

$$(6.28)$$

and where $\hat{W}_{K'\gamma',K\gamma}(\rho,\rho')$ is the nonlocal potential generated by the Pauli projection operator in Eq. (6.21). The radial matrix elements of the Hamiltonian in the Berggren basis are: $\langle \mathscr{B}_n(k_n,\rho)|V(\rho)|\mathscr{B}_m(k_m,\rho)\rangle$. The radial integral in these matrix elements can oscillate at large distances, especially for Coulomb and centrifugal potentials, so that it is numerically unstable. The nuclear potential in a hyperspherical harmonics expansion has also a long-range behavior in $\mathcal{O}(1/\rho^3)$ [107], so that it is difficult to handle numerically as well. Hence, one has to find an integral path where wavefunctions vanish in the asymptotic region. For potentials that decrease as $\mathcal{O}(1/\rho^2)$ (the centrifugal potential) or faster (the nuclear potential), one can use the exterior complex scaling [108], where integrals are calculated along a complex radial path:

$$
\langle \mathscr{B}_n|\hat{V}(\rho)|\mathscr{B}_m\rangle = \int_0^R \mathscr{B}_n(\rho)\hat{V}(\rho)\mathscr{B}_m(\rho)d\rho
$$

$$
+ \int_0^{+\infty} \mathscr{B}_n(R+\rho e^{i\theta})\hat{V}(R+\rho e^{i\theta})\mathscr{B}_m(R+\rho e^{i\theta})d\rho .
$$

$$(6.29)$$

In the above equation, $R$ is a radius taken sufficiently large to bypass all singularities, and $\theta$ is a rotation angle chosen so that the integral converges. $\theta$ can be positive or negative, according to the different behavior of outgoing ($H^+$) and incoming ($H^-$) wave functions. More details can be found in Ref. [109]. As the Coulomb potential is infinite-range, its diagonal matrix elements involving Berggren basis scattering states diverge even using exterior complex scaling. A practical solution of this problem is the off-diagonal method [110]. Basically, a small offset $\pm\delta k$ is added to the linear momenta $k_n$ and $k_m$ of involved scattering wave functions, so that the diverging diagonal Coulomb matrix element becomes a large but converging off-diagonal matrix element.

## 6.4.2 Convergence Properties of Eigenvalues Calculated in Jacobi and Cluster Orbital Shell Model Coordinates

Let us now compare Gamow shell model results obtained using Jacobi coordinates and cluster orbital shell model coordinates. In cluster orbital shell model coordinates, one defines a truncated model space according to a given maximum orbital angular momentum $\ell_{max}$ of particle orbital angular momenta $\ell_1, \ell_2$. In Jacobi coordinates, the model space is truncated according to the maximum value of $(\ell_x, \ell_y)$ (see Fig. 6.11).

In the cluster orbital shell model framework, one uses the Berggren basis for the $s, p, d$ orbits and harmonic oscillator basis states for partial waves bearing $\ell \geq 3$. Indeed, high orbital angular momentum components play a small role in wave function asymptotes so that they can be restricted to a few harmonic oscillator shells. The complex contours of the Berggren basis start from $k = 0\,\mathrm{fm}^{-1}$ and end in $k_{max} = 3\,\mathrm{fm}^{-1}$. In the framework using Jacobi coordinates, all calculations have been done taking into account the maximum hyperspherical quantum number $K_{max} = 20$. Similarly to cluster orbital shell model, one uses the Berggren basis for the $K \leq 6$ channels, where $K$ is the hyperspherical quantum number of the considered partial wave, while harmonic oscillator basis states are used when $K \geq 7$. Note that the energy range covered by the three-body model using Jacobi coordinates is about twice as large as that of cluster orbital shell model. In Jacobi coordinates, one has $k_\rho^2 = k_x^2 + k_y^2$, where $k_x$, $k_y$, and $k_\rho$ are the conjugate linear momenta of the hyperspherical coordinates $x$, $y$, and $\rho = \sqrt{x^2 + y^2}$, respectively (see Fig. 6.11). Let $k$ represents the linear momentum of a single nucleon in cluster orbital shell model. From Eq. (6.27), the energy of two-body cluster is $\hbar^2 k_\rho^2/2m$, while that of one nucleon is $\hbar^2 k^2/2m$ from Eq. (3.66). Consequently, at large kinetic energy, $k_\rho^2 \simeq 2k^2$, as the binding energy of the cluster can be hereby neglected. Maximal linear momenta thus have to be chosen differently for $k_\rho$ and $k$ if one wants model spaces to be as close as possible in both models.

As an example, let us consider $^6$He, $^6$Li, and $^6$Be. Their structure is relatively simple because they can be treated as an $\alpha$-particle core and two valence nucleons. For the calculation both in Jacobi coordinates and in cluster orbital shell model coordinates, details of the contours in the complex plane and their discretization are given in Ref. [95]. The original Minnesota interaction [111] is used to mimic the nuclear interaction between valence nucleons. Interaction between core and valence nucleons describes the Woods–Saxon potential whose parameters are fitted to the resonances of the core+n system. The Coulomb potential is added for charged particles. Parameters of the Woods–Saxon potential for $A = 6$ nuclei are detailed in Ref. [95]).

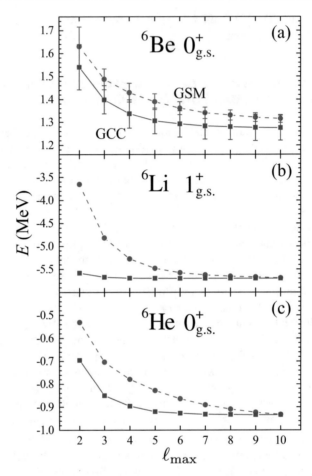

**Fig. 6.12** Comparison of convergence rates for ground state energies with respect to different maximum angular momenta $\ell_{max}$ in $A = 6$ nuclei: $^6$Be (panel (**a**)), $^6$Li (panel (**b**)), and $^6$He (panel (**c**)). Calculations are performed in the three-body model using Jacobi coordinates (denoted as GCC, for Gamow translationally invariant coupled channels) and cluster orbital shell model coordinates (denoted as GSM, for Gamow shell model). $\ell_{max}$ denotes the maximal single-particle orbital angular momentum for cluster orbital shell model and $\max(\ell_x, \ell_y)$ for the three-body model using Jacobi coordinates. The error bars stand for the decay widths (from Ref. [95])

Figure 6.12 illustrates the convergence rate for the ground state energies with respect to model spaces bearing different maximal angular momentum $\ell_{max}$ in the three-body model in Jacobi coordinates and cluster orbital shell model coordinates. The ground state energies of $^6$He and $^6$Be are in good agreement with experimental data. On the contrary, $^6$Li is overbound because the Minnesota interaction has no explicit dependence on the isospin $T$ of two-nucleon systems. The model spaces of cluster orbital shell model and Jacobi-coordinate framework are not exactly the same since different coordinate systems are used in these two approaches.

Nevertheless, obtained results become very close when $\ell_{max}$ reaches a value of 7–8 in both models.

One can see from Fig. 6.12 that all calculations done with Jacobi coordinates converge slightly faster than those done using cluster orbital shell model coordinates. The discrepancy between results of these two frameworks is largest when the cluster formed by two valence nucleons is well bound, as for $^6$Li. This comes from the attractive character of the proton–neutron interaction in the $T = 0$ channel. Indeed, the energy difference between ground states of $^6$Li in both approaches is $\sim$2 MeV for $\ell_{max} = 2$, whereas it is less than 200 keV for $^6$He and $^6$Be ground states for the same truncation.

The T-type Jacobi coordinates pertain to a cluster, as they consist of a center-of-mass coordinate and of a relative coordinate (see Fig. 6.11). In particular, a localized basis wave function in these coordinates represents a well-bound cluster close to the core. Consequently, inter-nucleon correlations are built in already in the basis wave functions, contrary to cluster orbital shell model, whose basis is made of Slater determinants, which possess no cluster structure. Therefore, three-body wave functions in Jacobi coordinates have to converge faster with the number of basis states than those calculated in cluster orbital shell model coordinates. This is in agreement with the conclusion of Ref. [112], where calculations using complex scaling method with cluster orbital shell model coordinates provide a slightly less bound wave function than that of Ref. [106], which used Jacobi coordinates.

This comparison demonstrates that one obtains very similar results for ground state energies when continuum effects and sufficiently large model spaces are considered. The typical energy differences between states in the three-body model using Jacobi and cluster orbital shell model coordinates are less than 100 keV. These results support also the argument [113] that cluster orbital shell model can successfully eliminate center-of-mass energy but at the price of a slower convergence due to the appearance of a recoil term.

Examples presented in Fig. 6.12 demonstrate that a standard three-body model in Jacobi coordinates formulated in Berggren formalism is an excellent theoretical tool to describe both weakly bound and resonance nuclei consisting of a two-body cluster above a well-bound core. The model is particularly advantageous for a calculation of emission of correlated two nucleons, e.g. two-proton emission in light and medium nuclei [74, 114].

### 6.4.3  Correlation Densities in a Three-Body Gamow Model

Correlations between the two valence neutrons can be conveniently studied using correlation density (see Eq. (6.19)) and Refs. [93, 94]: The two-nucleon angular correlation is obtained by integrating the correlation density of Eq. (6.19) over $r_1$ and $r_2$:

$$\rho(\theta) = \int_0^{+\infty} 8\pi^2 r_1^2 r_2^2 \sin(\theta)\rho(r_1, r_2)dr_1dr_2 . \tag{6.30}$$

While it is straightforward to calculate $\rho(\theta)$ with cluster orbital shell model coordinates, the two-nucleon correlation cannot be calculated directly in T-type coordinates, which are used to diagonalize the Hamiltonian of a three-body model in Jacobi coordinates. As a consequence, one can either calculate the density distribution $\rho_T(x, y, \varphi)$ in T-type variables and then using geometric relations (see the left panel of Fig. 6.11) transform it to the density $\rho(r_1, r_2, \theta)$ in cluster orbital shell model variables, or apply the coordinate transformation from T-type to V-type and then calculate the two-nucleon correlation directly using cluster orbital shell model coordinates. This transformation coefficient, which has been discussed in Ref. [104], provides with an analytical relation between hyperspherical harmonics in cluster orbital shell model coordinates $\mathscr{Y}_{\gamma'K'}^{JM}(r_1', r_2')$ and the T-type Jacobi coordinates $\mathscr{Y}_{\gamma K}^{JM}(x', y')$, where $r_1'$, $r_2'$, $x'$, and $y'$ are $r_1$, $r_2$, $x$, and $y$ multiplied by a mass-related constant:

$$
\begin{aligned}
r_1' &= \sqrt{A_i}\, r_1 \\
r_2' &= \sqrt{A_j}\, r_2 \\
x' &= x = \sqrt{\mu_{ij}}(r_1 - r_2) \\
y' &= \sqrt{\frac{A_i + A_j}{\mu_{(ij)k}}}\, y = \frac{A_i r_1 + A_j r_2}{\sqrt{A_i + A_j}} \,.
\end{aligned}
\tag{6.31}
$$

Angular correlation of the two valence neutrons in the ground state of $^6$He, calculated in the Gamow shell model using cluster orbital shell model coordinates and in the three-body model in Jacobi coordinates, is shown in Fig. 6.13 for model spaces which are defined by different values of $\ell_{max}$. The distribution $\rho(\theta)$ shows two maxima [93, 94, 112, 113, 115]. The higher peak, at a small opening angle, can be associated with a dineutron-like configuration. The second maximum, found in the region of large angles, represents the cigar-like configuration.

Results of the three-body model in Jacobi coordinates for $\ell_{max} = 2$ and $7$ are very close one to another. It is not the case for the Gamow shell model in cluster orbital shell model coordinates, which shows sensitivity to the cutoff value of $\ell_{max}$. However, as $\ell_{max}$ increases, the angular correlations obtained with the Gamow shell model and the three-body model in Jacobi coordinates become very similar. This shows that descriptions using either Jacobi or cluster orbital shell model variables are equivalent provided that the model space is sufficiently large.

Figure 6.14 compares two-nucleon angular correlations for $A = 6$ nuclei: $^6$He, $^6$Li, and $^6$Be, which are calculated with the Gamow shell model using cluster orbital shell model coordinates and with the three-body model defined with Jacobi coordinates. Similarly to Refs. [93, 94], one finds that the $T = 1$ configurations in $^6$He and $^6$Be have a dominant spin singlet ($S = 0$) component. The amplitude of the $S = 1$ component is small, i.e. the nuclear interaction tends to align nuclear spins in $^6$He and $^6$Be in the opposite directions. This is consistent with a cluster picture of the nucleon pair because, as discussed in Chap. 4, $S = 1$ for neutron-proton

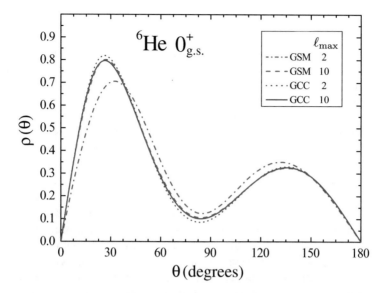

**Fig. 6.13** Comparison between the results for the neutron-neutron angular correlations in $^6$He, obtained in a three-body model for different model spaces defined by the cutoff values $\ell_{max}$. GSM and GCC denote, respectively, the results of Gamow shell model in cluster orbital shell model coordinates and three-body model in Jacobi coordinates (from Ref. [95])

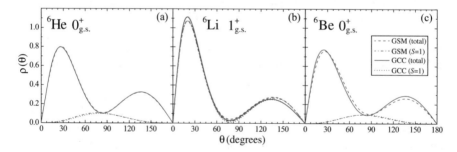

**Fig. 6.14** Two-nucleon angular correlation densities (the total density and its $S = 1$ component) in the ground states of $^6$He (**a**), $^6$Li (**b**), and $^6$Be (**c**). The model space is defined by the maximal angular momentum $\ell_{max} = 12$. For other information, see the caption of Fig. 6.13 (from Ref. [95])

(deuteron) ground state and $S = 0$ for dineutron and diproton ground states. Note, however, that the association of the valence nucleon pair in $A = 6$ nuclei to two-nucleon systems is only approximate due to the presence of a core. Indeed, while dineutron and diproton ground states are antibound and virtual states (see Sect. 4), respectively, they are loosely bound and weakly resonant in presence of a $^4$He core.

The fact that the ground states of $^6$He and $^6$Be are parts of the same $T = 1$ multiplet is also clearly seen in Fig. 6.14, as their two-nucleon angular correlations are very similar. One may, however, notice that the dineutron peak is slightly higher in $^6$He than in $^6$Be. This is due to the repulsive Coulomb interaction between valence protons in $^6$Be.

As for $^6$Li, the valence proton–neutron pair is very strongly correlated because the $T = 0$ interaction is much stronger than the $T = 1$ interaction. This is related to the structure of the ground state of $^6$Li, which mainly consists of a deuteron cluster orbiting around a $^4$He core. Due to the strong binding character of the $T = 0$ interaction, the cluster structure of the valence nucleons in $^6$Li is more prevalent than in $^6$He and $^6$Be (see Fig. 6.14). While the cluster shape component is about twice as large as the cigar shape component in $^6$He and $^6$Be, the cluster shape component of $^6$Li is three times larger than its cigar shape component. Consequently, the valence nucleon pair is more clusterized in $^6$Li than in $^6$He and $^6$Be, which is entirely due to the different isospin of their ground state wave functions.

For all nuclei, the two-nucleon angular correlations obtained with the Gamow shell model in cluster orbital shell model coordinates and with the three-body model in Jacobi coordinates are close one to another. All these different features seen in Figs. 6.13 and 6.14 demonstrate that the nucleon–nucleon angular correlations contain valuable information about the interaction of valence nucleons and their spatial configuration. Indeed, the cluster and cigar shapes directly appear from Fig. 6.14, as well as the averaged spin and isospin content of valence nucleons. Nevertheless, due to the integration over the radial $r$ coordinate, the asymptotic decrease of two-nucleon densities cannot be seen on two-nucleon angular correlations. To state the radial dependence of nucleon pairs in a nuclear wave function, it is necessary to consider the full correlation density, as done in Fig. 6.10.

## 6.5  Inclusion of Continuum Couplings in the Pairing Model

The pairing part of the nucleon–nucleon interaction is responsible for correlations and fluctuations related to the superfluid character of the nuclear medium, in both finite nuclei and neutron stars [116]. Therefore, Hamiltonians consisting of a one-body part and of a two-body part built from a pairing interaction are widely studied. Moreover, exact solutions of the pairing Hamiltonian for a constant pairing strength and a discrete set of single-particle levels are known since the seminal work of Richardson [117, 118].

It is possible to derive three classes of solvable models for fermions and bosons involving pairing Hamiltonians [119], if one combines the exact solution of the Richardson pairing model to that proposed by Gaudin for quantum spin systems [119]. The Richardson pairing model, i.e. that of a constant $g$ pairing Hamiltonian, is a particular case of the rational class of integrable models. Added to that, arbitrary combinations of the integrals of motion within each of the classes can generate more general exactly solvable pairing models. In particular, the pairing Hamiltonian encompassing the main features of heavy nuclei can be derived from the hyperbolic family of Gaudin models [120]. One can also note that the rational Gaudin model has been generalized to larger Lie algebras, with, for example, the $SO(5)$ for

$T = 1$ isovector pairing models [121] and the $SO(8)$ for $T = 0, 1$ spin-isospin pairing models [122], where proton–neutron pairing is treated exactly. The exercises proposed in this section are technical and can be omitted in the first reading.

The constant pairing Hamiltonian is given by

$$\hat{H} = \sum_{\alpha}^{D} \epsilon_\alpha c_\alpha^\dagger c_\alpha + G \sum_{\alpha,\beta}^{D} c_\alpha^\dagger c_{\bar\alpha}^\dagger c_{\bar\beta} c_\beta \,, \tag{6.32}$$

where $\epsilon_\alpha$ are the energies of bound single-particle levels, and $G$ is the pairing strength. Operators $c_\alpha^\dagger (c_\alpha)$ stand for the particle creation (annihilation) operators, and $\alpha \equiv \{a, m_\alpha\} = \{n_a, \ell_a, j_a, m_\alpha\}$, $\bar\alpha = \{a, \bar m_\alpha\}$. $c_{\bar\alpha}^\dagger$ is defined as $c_{\bar\alpha}^\dagger = (-)^{j_a - m_\alpha} c_{\alpha, -m_\alpha}^\dagger$. The degeneracy of a single-particle level $a$ is $\Omega_a = 2j_a + 1$. Let us define the particle number and pair creation operators:

$$\hat{n}_a = \sum_{m_\alpha = -j_a}^{j_a} c_\alpha^\dagger c_\alpha; \quad b_a^\dagger = \sum_{m_\alpha > 0} c_\alpha^\dagger c_{\bar\alpha}^\dagger = (b_a)^\dagger \,. \tag{6.33}$$

The operators of Eq. (6.33) obey the SU(2) commutator algebra:

$$\left[ \hat{n}_a, b_{a'}^\dagger \right] = 2\delta_{aa'} b_a^\dagger$$

$$\left[ b_a, b_{a'}^\dagger \right] = 2\delta_{aa'} \left( \frac{\Omega_a}{4} - \frac{\hat{n}_a}{2} \right)$$

$$\left[ b_a^\dagger, b_{a'}^\dagger \right] = 0 \,. \tag{6.34}$$

One can built the states of $N$ particles from the $\mathcal{N}$ single-particle states related to the operators $\hat{n}_a$, $b_a$ and $b_a^\dagger$:

$$|n_1, n_2, \cdots, n_{\mathcal{N}}, v\rangle = \frac{1}{\bar N} b_1^{\dagger n_1} b_2^{\dagger n_2} \cdots b_{\mathcal{N}}^{\dagger n_{\mathcal{N}}} |v\rangle \,, \tag{6.35}$$

where $|v\rangle = |v_1, v_2 \cdots v_{\mathcal{N}}\rangle$ is a state of the unpaired particles which satisfy

$$b_a |v\rangle = 0 \; ; \; \hat{n}_a |v\rangle = v_a |v\rangle \,. \tag{6.36}$$

$v = N - 2N_{\text{pair}}$ in Eq. (6.35) is the total number of the unpaired particles, with $N_{\text{pair}}$ the number of pairs, and $\bar N$ is the normalization constant. The pairing Hamiltonian (6.32) then reads

$$\hat{H} = \sum_a^{\mathcal{N}} \epsilon_a \hat{n}_a + G \sum_{a,a'}^{\mathcal{N}} b_a^\dagger b_{a'} \,, \tag{6.37}$$

where one has used the operators $\hat{n}_a$, $b_a$, $b_a^\dagger$. Richardson could derive the exact solution of the pairing Hamiltonian (see Eq. 6.37) with a discrete set of bound single-particle levels [117,118]. It was later demonstrated that the Richardson model can be solved by expressing the pairing Hamiltonian as a linear combination of integrals of motion [123].

The eigenvalue of the pairing Hamiltonian (6.37) for a given configuration of unpaired particles $v$, can be written as:

$$\tilde{\mathscr{E}}^{(K)} = \sum_{i=1}^{N_{\text{pair}}} E_i^{(K)} + \sum_{a=1}^{\mathscr{N}} \epsilon_a v_a \qquad K = 0, 1, \ldots, K_{\max} , \tag{6.38}$$

where index $K$ enumerates the eigenstates of increasing excitation energy and $K_{\max} + 1$ is the total number of eigenstates. $\tilde{\mathscr{E}}^{(K)}$ is complex in general. Thus, $\mathscr{R}(\tilde{\mathscr{E}}^{(K)}) = \mathscr{E}^{(K)}$ is the energy of the $K^{\text{th}}$ eigenstate, while $-2\mathscr{I}(\tilde{\mathscr{E}}^{(K)}) = \Gamma^{(K)}$ is the corresponding width.

The pair energies $E_i^{(K)}$ in Eq. (6.38) are solutions of $N_{\text{pair}}$ non-linear coupled equations:

$$1 - 2G \sum_a^{\mathscr{N}} \frac{d_a}{2\epsilon_a - E_i^{(K)}} + 2G \sum_{j \neq i}^{N_{\text{pair}}} \frac{1}{E_i^{(K)} - E_j^{(K)}} = 0 \qquad K = 0, 1, \ldots, K_{\max} , \tag{6.39}$$

where $d_a = v_a/2 - \Omega_a/4$.

Several authors attempted to derive an analytical solution to the pairing model including the continuum. Hasegawa and Kaneko studied effects of single-particle resonances on pairing correlations [124]. Id Betan attempted to solve Richardson equations with the real-energy continuum [125]. However, while an approximate solution of Richardson equations involving the continuum could be derived, no exact solution of the pairing problem was provided. Obviously, it is possible to diagonalize the pairing Hamiltonian including the continuum numerically using the Gamow shell model [28, 109, 126, 127]. However, one can only consider a small number of valence nucleons in this approach due to computer limitations.

One will formulate the rational Gaudin model in the presence of the continuum of single-particle states using the Berggren single-particle ensemble [128]. In this representation, the pairing Hamiltonian reads [129, 130]

$$\hat{H} = \sum_{i \in b, r} \epsilon_i \hat{n}_i + \sum_c \int_{L_c^+} \epsilon_{k_c} \hat{n}_{k_c} dk_c$$

$$+ G \sum_{i, i' \in b, r} b_i^\dagger b_{i'} + G \sum_{c, c'} \int_{L_c^+} b_{k_c}^\dagger b_{k'_{c'}} dk_c dk'_{c'}$$

$$+ G \sum_{(i \in b, r), c} \int_{L_c^+} \left( b_{k_c}^\dagger b_i + b_i^\dagger b_{k_c} \right) dk_c . \tag{6.40}$$

Sums over $c, c'$ denote summations over different partial waves, from $(\ell, j)$ to $(\ell_{\max}, j_{\max})$. The energy of a single-particle state $c$ in the nonresonant continuum reads: $\epsilon_c = \hbar^2 k_c^2 / 2m$, where $m$ is the particle mass, and $k_c$ is the associated linear momentum. The discrete sums run over the real-energy bound single-particle states and the complex-energy single-particle resonances situated between the contour $L_c^+$ and the real $k$-axis. The same contour $L_{c(\ell, j)}^+$ in the complex $k$-plane is used for all partial waves.

For discrete single-particle states (bound states and resonances), the pair creation (annihilation) operators satisfy the commutator relations in Eq. (6.34). Conversely, one has for the nonresonant scattering single-particle states [129, 130]:

$$\left[\hat{n}_{k_c}, b_{k'_{c'}}^\dagger\right] = 2\delta(k_c - k'_c)\delta_{cc'}b_{k_c}^\dagger$$

$$\left[b_{k_c}, b_{k'_{c'}}^\dagger\right] = \delta(k_c - k'_c)\delta_{cc'}\frac{\Omega_{k_c}}{2} - \delta_{k_c k'_c}\delta_{cc'}\hat{n}_{k_c}$$

$$\left[b_{k_c}^\dagger, b_{k'_{c'}}^\dagger\right] = 0 . \tag{6.41}$$

The continuum has to be discretized in practical applications. For this, it is convenient to define new number and pair operators:

$$\hat{\tilde{n}}_q = w_q \hat{n}_q \; ; \; \tilde{b}_q^\dagger = \sqrt{w_q} b_q^\dagger = (\tilde{b}_q)^\dagger, \tag{6.42}$$

where the index $q$ runs over all resonant and discretized scattering states of the Berggren basis, and $w_q$ is the Gaussian weight of the Gauss-Legendre quadrature. For resonant states, $w_q = 1$.

With this definition, all pair states are normalized to unity and treated in the same manner independently of their resonant or scattering character [129, 130]. The operators $\hat{\tilde{n}}_q, \tilde{b}_q, \tilde{b}_q^\dagger$ satisfy the same SU(2) commutation relations as the operators $\hat{n}_i, b_i, b_i^\dagger$ (see Eq. (6.34)):

$$\left[\hat{\tilde{n}}_q, \tilde{b}_{q'}^\dagger\right] = 2\delta_{qq'}\tilde{b}_q^\dagger$$

$$\left[\tilde{b}_q, \tilde{b}_{q'}^\dagger\right] = 2\delta_{qq'}\left(\frac{\Omega_q}{4} - \frac{\hat{\tilde{n}}_q}{2}\right)$$

$$\left[\tilde{b}_q^\dagger, \tilde{b}_{q'}^\dagger\right] = 0 . \tag{6.43}$$

The Hamiltonian of the generalized rational Gaudin model (6.40) can then be written in terms of the operators $\hat{\tilde{n}}_q, \tilde{b}_q, \tilde{b}_q^\dagger$:

$$\hat{H} = \sum_q^{\mathcal{N}} \epsilon_q \hat{\tilde{n}}_q + \sum_{q,q'}^{\mathcal{N}} G_{qq'} \tilde{b}_q^\dagger \tilde{b}_q \; ; \; G_{qq'} = \sqrt{w_q}\sqrt{w_{q'}}G , \tag{6.44}$$

where $\mathcal{N}$ is the total number of resonant and discretized continuum single-particle states. However, Eq. (6.44) cannot be analytically solved in the general case, even though it closely resembles Eq. (6.37). Indeed, state-dependent pairing Hamiltonians are not integrable in general. An exception is the hyperbolic model of Refs. [120, 131], where the Gaussian weights $w_q$ are linear functions of the single-particle energies $\epsilon_q$. Consequently, one must either approximately solve Eq. (6.44), or change the commutation relations of Eq. (6.41) for nonresonant scattering states, thus breaking the SU(2) commutator algebra, in order to obtain an ansatz for an exact eigenstate [129, 130].

The new normalized operators $\hat{\tilde{n}}_q$ and $\tilde{b}_q^\dagger$, $\tilde{b}_q$ must be applied when diagonalizing the Hamiltonian of Eq. (6.44). Indeed, the contour discretization leads not only to new normalized operators but also to new normalized Slater determinants, because the action of $\hat{\tilde{n}}_q$, $\tilde{b}_q^\dagger$, and $\tilde{b}_q$ is defined as in the discrete case. An approximate solution for the generalized rational pairing model (6.44) can be derived by replacing the Kronecker delta by the Dirac delta in the commutator of Eq. (6.41) for states in the nonresonant continuum [130]:

$$\left[ b_{k_c}, b_{k'_{c'}}^\dagger \right] = 2\delta(k_c - k'_c)\delta_{cc'} \left( \frac{\Omega_{k_c}}{4} - \frac{\hat{n}_{k_c}}{2} \right). \tag{6.45}$$

The pair operators $\tilde{b}_q^\dagger (\tilde{b}_q)$ for bound, resonance and discretized scattering states satisfy

$$\left[ \hat{\tilde{n}}_q, \tilde{b}_{q'}^\dagger \right] = 2\delta_{qq'}\tilde{b}_q^\dagger$$

$$\left[ \tilde{b}_q, \tilde{b}_{q'}^\dagger \right] = 2\delta_{qq'} \left( \frac{\Omega_q}{4} - \frac{\hat{\tilde{n}}_q}{2w_q} \right)$$

$$\left[ \tilde{b}_q^\dagger, \tilde{b}_{q'}^\dagger \right] = 0 . \tag{6.46}$$

Let us now derive the eigenvalues of the pairing Hamiltonian (6.44) in this approximation. For this, let us rewrite the eigenstate $K$ ($K = 1, 2, \ldots, K_{\max}$) as a product of the pair states, similarly to the discrete case:

$$|\Psi_{\text{norm}}\rangle = \prod_{\eta=1}^{N_{\text{pair}}} B_{\eta;\text{norm}}^\dagger |\nu\rangle , \tag{6.47}$$

where all considered eigenstates and operators implicitly depend on $K$. In Eq. (6.47), the pair operators read

$$B_{\eta;\text{norm}}^\dagger = c_\eta G \sum_q^{N_{\text{pair}}} \frac{\tilde{b}_q^\dagger \sqrt{w_q}}{2\epsilon_q - E_\eta} , \tag{6.48}$$

where $E_\eta$ are the pair energies in the eigenstate $K$. The normalization constants $c_\eta$ can be calculated from the following equation:

$$\frac{1}{(c_\eta G)^2} = \frac{1}{(C_\eta)^2} = \sum_q^{N_{\text{pair}}} \frac{w_q}{(2\epsilon_q - E_\eta)^2} . \tag{6.49}$$

In order to simplify notations, it is convenient to define

$$B_\eta^\dagger = B_{\eta;\text{norm}}^\dagger / C_\eta , \tag{6.50}$$

so that

$$|\Psi_{\text{norm}}\rangle = \prod_{\eta=1}^{N_{\text{pair}}} C_\eta B_\eta^\dagger |v\rangle = C|\Psi\rangle , \tag{6.51}$$

where

$$C = \prod_{\eta=1}^{N_{\text{pair}}} C_\eta \quad \text{and} \quad |\Psi\rangle = \prod_{\eta=1}^{N_{\text{pair}}} B_\eta^\dagger |v\rangle .$$

The operators $\hat{\tilde{n}}$, $B_\eta$, and $B_0$:

$$B_0^\dagger = \sum_q^{\mathscr{N}} \tilde{b}_q^\dagger \sqrt{w_q} \tag{6.52}$$

satisfy fundamental commutator relations borne by $\left[\hat{\tilde{n}}_q, B_\eta^\dagger\right]$, $\left[\hat{\tilde{n}}_q, B_0^\dagger\right]$, $\left[B_\eta, B_{\eta'}^\dagger\right]$, $\left[B_0, B_\eta^\dagger\right]$, and $\left[B_0^\dagger, B_\eta^\dagger\right]$ (see Exercise I).

---

**Exercise I** ⋆⋆
Calculate the commutators $\left[\hat{\tilde{n}}_q, B_\eta^\dagger\right]$, $\left[\hat{\tilde{n}}_q, B_0^\dagger\right]$, $\left[B_\eta, B_{\eta'}^\dagger\right]$, $\left[B_0, B_\eta^\dagger\right]$, and $\left[B_0^\dagger, B_\eta^\dagger\right]$ using the commutation relations for operators $\hat{\tilde{n}}_q, \tilde{b}_q^\dagger, \tilde{b}_q$ (see Eq. (6.46)).

---

The Hamiltonian of the generalized rational Gaudin model (6.44) expressed in these operators reads

$$\hat{H} = \sum_q^{\mathscr{N}} \epsilon_q \hat{\tilde{n}}_q + G B_0^\dagger B_0 . \tag{6.53}$$

The set of $N_{\text{pair}}$ non-linear coupled Richardson type equations associated with Eq. (6.53) then writes

$$1 - 2G \sum_{q}^{\mathcal{N}} \frac{w_q \left(v_q/2 - \Omega_q/4\right)}{2\epsilon_q - E_\eta^{(K)}} + 2G \sum_{\eta'=1; \neq \eta}^{N_{\text{pair}}} \frac{1}{E_\eta^{(K)} - E_{\eta'}^{(K)}} = 0$$

$$K = 0, 1, \ldots, K_{\text{max}} \,. \tag{6.54}$$

The first sum in these equations can be separated in resonant and discretized scattering parts. In the continuum limit, the generalized Richardson equations become

$$1 - 2G \left( \sum_{i \in b,r}^{\mathcal{N}} \frac{d_i}{2\epsilon_i - E_\eta^{(K)}} - \sum_{c}^{\ell_{\max}, j_{\max}} \int_{L_c^+} \frac{d_{k_c}}{\hbar^2 k_c^2 / m - E_\eta^{(K)}} dk_c \right.$$

$$\left. + \sum_{\eta'=1; \neq \nu}^{N_{\text{pair}}} \frac{1}{E_\eta^{(K)} - E_{\eta'}^{(K)}} \right) = 0 \,, \tag{6.55}$$

where $K = 0, 1, \ldots, K_{\text{max}}$. In this expression, $d_i = v_i/2 - \Omega_i/4$ and similarly for $d_{k_c}$.

By replacing the exact commutator relations (6.41) by the approximate commutator relations of Eq. (6.45), an approximate solution for the rational Gaudin model with the continuum (see Eq. (6.39)) could be obtained.

In certain limiting situations, however, this solution is exact. Equation (6.39) provides the exact solution of the rational Gaudin model [117, 118] when using a discrete set of bound single-particle levels, as $w_q = 1$ therein. Identically, Eq. (6.39) provides an exact solution of the pairing model with the continuum in the pole approximation, where the nonresonant continuum states are neglected. Interestingly, Eq. (6.39) exactly solves Eq. (6.53) if the Berggren ensemble contains only discretized states of the nonresonant continuum. Indeed, in this case, one can take the same weights $w_q \equiv w$ for all continuum states $q$, so that Eq. (6.39) becomes formally identical to the solvable discrete case by renormalizing the pairing strength so that $G' = Gw$. In this particular case, the third sum in Eq. (6.39) vanishes, so that one obtains

$$1 - 2G \sum_{c}^{\ell_{\max}, j_{\max}} \int \frac{d_{k_c}}{2\epsilon_{k_c} - E_\eta} dk_c = 0 \qquad K = 0, 1, \ldots, K_{\text{max}} \,. \tag{6.56}$$

A useful measure of pairing correlations in a given eigenstate $|\Psi^{(K)}\rangle$ is the canonical pairing gap:

$$\Delta^{(K)} = G \sum_{q}^{\mathcal{N}} \sqrt{n_q^{(K)} \left(1 - n_q^{(K)}\right)} \,, \tag{6.57}$$

where the sum runs over single-particle states, and $n_q^{(K)}$ is the occupation probability of the state $q$. Note that the canonical gap coincides with the BCS gap in the BCS approximation where the occupation probability $n_q^{(K)} = |v_q|^2$. The occupation probability $n_q^{(K)}$ can be calculated exactly by diagonalizing the Hamiltonian using the Gamow shell model.

Let us write the eigenstate $|\Psi^{(K)}\rangle$ of the pairing Hamiltonian (6.37) as an expansion in the basis of Slater determinants $|\Phi_\alpha\rangle$:

$$|\Psi^{(K)}\rangle = \sum_{\alpha} C_\alpha^{(K)} |\Phi_\alpha\rangle \,. \tag{6.58}$$

The expectation value of the particle number operator $\hat{N}$ in the eigenstate $|\Psi^{(K)}\rangle$ reads

$$N = \langle \Psi^{(K)}| \hat{N} |\Psi^{(K)}\rangle = \sum_{\alpha,\alpha'} C_\alpha^{(K)} C_{\alpha'}^{(K)} \langle \Phi_\alpha| \hat{N} |\Phi_{\alpha'}\rangle = \sum_{q} 2n_q^{(K)}. \tag{6.59}$$

The occupation probability can be evaluated numerically as:

$$n_q^{(K)} = \sum_{\alpha} g(\alpha, q; K) \, (C_\alpha^{(K)})^2. \tag{6.60}$$

In this expression, $g(\alpha, q; K) = 0$ or $1$ if the single-particle state $q$ is occupied or unoccupied in the Slater determinant $\alpha$ of an eigenstate $K$.

As one does not have access to the Slater determinant expansion of the many-body wave function when solving the Richardson equations, the particle numbers must be calculated differently. In the generalized Richardson equations, the single-particle occupation probabilities in an eigenstate $K$ are determined using the Hellmann-Feynman theorem [132, 133]:

$$n_q^{(K)} = \frac{\partial \tilde{\mathscr{E}}^{(K)}}{\partial \epsilon_q} \,, \tag{6.61}$$

where $\tilde{\mathscr{E}}^{(K)}$ is the total energy (6.38) of the eigenstate $K$.

### 6.5.1   Numerical Solution of the Generalized Richardson Equations

Equation (6.39) must be solved numerically due to its non-linear character. For this, a starting point is firstly devised in the weak coupling limit ($G \to 0$), which is then evolved iteratively by solving the generalized Richardson equations for increasing values of $G$. The solution for pair energies for a given $G$ is updated with the Newton-Raphson method using the solution of the previous step as the new starting point [134]. However, divergencies occur if at a given pairing constant $G$, two or more pair energies are equal to twice a single-particle energy. These divergencies exist only in numerical calculations, as infinities cancel exactly in Eq. (6.39). Hence, along with the Newton-Raphson procedure, a numerical method suppressing these numerical divergencies has to be applied as well.

The initial guess for a solution in the limit $G \to 0$ is determined by solving the generalized Richardson equations. The expression for pair energies $E_i$ in this limit reads

$$\lim_{G \to 0} E_i = 2\epsilon_q \tag{6.62}$$

with $i = 1, \cdots, N_{\text{pair}}$ and $q = 1, \cdots, \mathcal{N}$.

Analytical determination of the pair energies is more difficult if many pairs occupy the same single-particle level $q$. For $N_{\text{pair}}$ pairs occupying the same single-particle state of energy $\epsilon_q$, the starting pair energies $E_i$ are obtained by solving the set of $N_{\text{pair}}$ coupled equations:

$$1 - \frac{2Gd_q}{2\epsilon_q - E_i} + 2G \sum_{j \neq i}^{N_{\text{pair}}} \frac{1}{E_i - E_j} = 0 \qquad i = 1, \cdots, N_{\text{pair}} . \tag{6.63}$$

Notice that the nonresonant continuum states in the weak coupling limit $|G| \ll 1$ are not occupied and, hence, the corresponding terms in generalized Richardson equations are absent. Eq. (6.63) cannot be solved analytically in the general case. However, pair energies can be given in closed form in the one-pair and two-pair cases (see Exercise II). If in the limit $G \to 0$ two pairs occupy the same single-particle state $q$, then their energies are complex conjugate. For the spectrum of real single-particle energies, this symmetry of the pair energies at $G \to 0$ is preserved for any value of $G$ in an iterative procedure of solving equations of the generalized Richardson pairing model.

---

**Exercise II** ★ ★ ★
One will devise analytical solutions for pair energies in Eq. (6.63) in the one-pair and two-pair cases. Pair energies in Eq. (6.63) are no longer analytical when one has more than three pairs, so that pair energy solutions must be found numerically

using the Newton-Raphson method. One will show that one-pair and two-pair solutions can be used to devise a starting point for the Newton-Raphson method.

**A.** Assuming that the degeneracy of the single-particle level $q$ is $\Omega_q = 2$, show that Eq. (6.63) has a unique pair energy solution, which is

$$E_1 = 2\epsilon_q - 2Gd_q \, . \tag{6.64}$$

**B.** For a higher degeneracy of single-particle states $q$ ($\Omega_q \geq 4$), show that the analytical solution of Eq. (6.63) for two pairs of particles on the level $q$ is

$$E_1 = 2\epsilon_q - G(2d_q + 1) + iG\sqrt{|2d_q + 1|}$$
$$E_2 = 2\epsilon_q - G(2d_q + 1) - iG\sqrt{|2d_q + 1|} \, . \tag{6.65}$$

**C.** Let us now consider the case of three pairs occupying the same level $q$ for $|G| \ll 1$, hence with $\Omega_q \geq 6$. The solution of Eq. (6.63) can be found numerically with the Newton-Raphson method (see Ref. [132] for computational details).

  (i) Explain why a simple mean-field approximation cannot be used.
  (ii) Devise a starting point for the Newton-Raphson method using the results of **A** and **B**.
  (iii) Explain why the devised starting point leads in practice to a converging Newton-Raphson process.

This special symmetry of pair energies is broken if the nonresonant continuum states are included in the basis. Indeed, continuum states are absent in Eq. (6.63) but become occupied for finite values of the pairing strength $G$ and hence, the initial symmetry of pair energies is broken in the course of solving the generalized Richardson equations. The derivation of the starting solution of the Newton-Raphson procedure in the weak coupling limit is detailed in Exercise II.

The generalized seniority $v$ is not equal to zero for systems with an odd number of particles and/or broken pairs. Each configuration is defined by the seniorities $v_q$, equal to the number of unpaired particles occupying the level $q$. Equations (6.64) and (6.65) are then used to obtain an initial guess for the pair energies and initiate the iterative procedure.

As a first test of the approximate rational Gaudin model with the continuum, one will compare it with an exact Gamow shell model diagonalization of the pairing Hamiltonian (6.40). One discretizes the contour $L_c^+$ using the Gauss-Legendre quadrature method and builds the single-particle spectrum which is used both in the generalized Richardson equations (6.39) and in the Gamow shell model. Exact solutions of the constant pairing Hamiltonian (6.40) are obtained by diagonalizing the Hamiltonian matrix using the Jacobi–Davidson method (see Sect. 5.7). Figure 6.15 compares the approximate energies obtained by solving the generalized Richardson

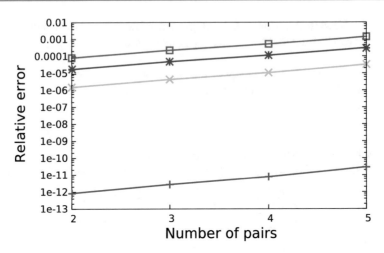

**Fig. 6.15** Comparison between exact Gamow shell model diagonalization of constant $G$ pairing Hamiltonian and Richardson calculation using Eqs. (6.39). The relative error of the total energy (6.38) calculated using Eqs. (6.39) is shown for various pairing strengths $G$, different numbers of fermion pairs and 45 discretization points along the real-energy contour. Results for pairing strengths $G$ equal to $-0.01$, $-0.3$, $-0.5$, $-0.7$ MeV are, respectively, depicted by pluses, crosses, stars and squares

equations (6.39) to the exact energies calculated in the Gamow shell model for a spectrum of well-bound single-particle levels: $\epsilon_i = \{-5, -4, -3, -2, -1\}$ MeV. For this, the relative error of the total energy $\mathscr{E}$, i.e. the real part of $\tilde{\mathscr{E}}$ in Eq. (6.38), calculated by using Eqs. (6.39) with respect to the exact Gamow shell model energy: $\delta(\mathscr{E}) = (\mathscr{E}_{GSM} - \mathscr{E})/\mathscr{E}_{GSM}$, is shown for different numbers of pairs and pairing constants $G$. Each level is doubly degenerate, i.e. there is only one pair of fermions per level. The set of single-particle states from the discretized real-energy contour is added, pertaining to the completeness of Berggren basis states. The contour is composed of three segments: $[k_0; k_1] = [0.0; 0.5]$, $[k_1; k_2] = [0.5; 1.0]$, and $[k_2; k_{max}] = [1.0; 2.0]$. Each segment of the contour $L_c^+$ is discretized with the same number of points. Different strengths $G$ of the pairing interaction: $G = 0.01$ MeV, $G = 0.3$ MeV, $G = 0.5$ MeV, and $G = 0.7$ MeV are used in calculations. The same set of single-particle levels and the corresponding Gaussian weights are then used to find the total energy of the system by solving both, the generalized Richardson equation (6.39) and the Gamow shell model. One can see that Eqs. (6.41) and (6.39), providing with approximate pair energies, do not accurately take the pair-pair interaction into account. Note that Eq. (6.39) is exact for a single pair case, so that solving Eq. (6.39) or diagonalizing the Hamiltonian with the exact Gamow shell model result gives the same energy.

Dependence of the relative error of the total energy $\mathscr{E}$ for weakly bound and resonance double degenerate single-particle levels is shown in Fig. 6.16 as a function of the pairing strength for two or three fermion pairs in three pole states $-1.5, -0.5, (0.5, -0.05)$ and corresponding discretized states of the nonresonant

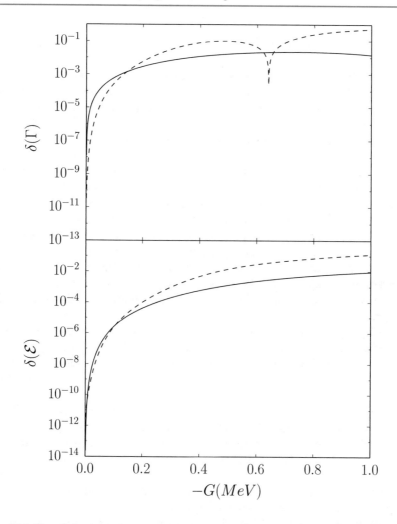

**Fig. 6.16** The relative error of the ground state energy $\delta(\mathscr{E})$ and width $\delta(\Gamma)$ which are calculated using the pair energies $E_i$ given by the generalized Richardson equations (6.39) and exact solution of the Gamow shell model. Results for 2 and 3 pairs of fermions are shown with solid and dashed line, respectively (from Ref. [130])

continuum for the considered resonance. The contour consists of the segments $k_0 = 0, k_1 = (0.1549, -0.14), k_2 = 1, k_{max} = 2$, in units of $\mathrm{fm}^{-1}$, and is discretized using the Gauss-Legendre quadrature procedure. One may notice a spike of the relative error for the ground state width at certain values of the pairing strength. At this discrete value of $G$, the imaginary part of the complex total energy (6.38) calculated using the generalized Richardson approach (6.39) is equal to the Gamow shell model energy. In general, both real and imaginary parts of the relative error of the complex energy exhibit spikes. These spikes in $\delta(\mathscr{E})$ and/or $\delta(\Gamma)$ are found only in the single-particle spectra with at least one resonance. As a rule, the relative error for the imaginary part of the total energy is larger than the corresponding error of the real part.

The Richardson-like solution for the generalized pairing problem including single-particle states of the discretized continuum becomes less accurate when the occupation of nonresonant continuum states is important, as in the case of strong pairing strength. In the weak pairing case, on the contrary, the relative error of the solutions of generalized Richardson equations remains small.

### 6.5.1.1  Physical Applications of the Generalized Richardson Equations

The generalized Richardson equations can find an application in the problem of ultra-small superconducting grains to understand the influence of continuum on pairing properties, in particular in the transitional region of the weak coupling limit [135–137]. To illustrate another possible application of the generalized Richardson equations, we will discuss the energy spectrum of $^{20}$C as an example.

The Hamiltonian (6.44) is parametrized using experimental energies of the spectrum of $^{13}$C and the binding energy of $^{14}$C. $^{12}$C is used as an inert core and the energies of all states in $^{14-20}$C are calculated with respect to the energy of the $^{12}$C core.

The Berggren basis consists of the pole single-particle states $0p_{1/2}$, $1s_{1/2}$, $0d_{5/2}$, $0d_{3/2}$, $0f_{7/2}$, and of the two nonresonant continua $\{d_{3/2}\}$, $\{f_{7/2}\}$. The single-particle energies of bound states $0p_{1/2}$, $1s_{1/2}$, $0d_{5/2}$ are given by the experimental energies of $1/2_1^-$, $1/2_1^+$ and $5/2_1^+$ states in $^{13}$C: $\epsilon_{0p_{1/2}} = -4.946\,\text{MeV}$, $\epsilon_{1s_{1/2}} = -1.857\,\text{MeV}$, and $\epsilon_{0d_{5/2}} = -1.093\,\text{MeV}$. The energy of resonances $0d_{3/2}$ and $0f_{7/2}$ are taken from Ref.[125]: $\epsilon_{0d_{3/2}} = (2.267\ \text{MeV};\ -0.416\,\text{MeV})$ and $\epsilon_{0f_{7/2}} = (9.288\ \text{MeV};\ -3.040\,\text{MeV})$. The complex contours $\{d_{3/2}\}$ and $\{f_{7/2}\}$, associated with the single-particle resonances $0d_{3/2}$ and $0f_{7/2}$, are discretized with ten points per segment, so that one has thirty points per contour. The pairing strength is given by $G = \chi/A$, where $\chi = -11.13\,\text{MeV}$, and $A$ is the number of nucleons. The constant $\chi$ is adjusted to reproduce the experimental binding energy of the ground state of $^{14}$C with respect to the $^{12}$C core.

To evaluate the role of the continuum in the spectra of carbon isotopes, the results of the generalized Richardson equations (6.39) are compared to results of the standard Richardson calculations (see Eq. (6.39)) without continuum couplings and with real single-particle energies. In the latter case, the single-particle energies of the bound states: $0p_{1/2}$, $1s_{1/2}$, $0d_{5/2}$, are the same as in the previous paragraph, and the energies of $0d_{3/2}$ and $0f_{7/2}$ resonances are real: $\epsilon_{0d_{3/2}} = 2.267\ \text{MeV}$ and $\epsilon_{0f_{7/2}} = 9.288\ \text{MeV}$. One uses the value $\chi = -15.064\,\text{MeV}$ in order to reproduce the experimental binding energy of $^{14}$C.

The spectrum of $^{20}$C, calculated using either the generalized Richardson equations with the continuum, or the standard Richardson equations without the continuum, is presented in Fig. 6.17. In both cases, the pairing strength $G$ is fitted to reproduce the experimental energy of the ground state of $^{14}$C with respect to $^{12}$C. One can see significant relative shifts of energy eigenvalues which are calculated in

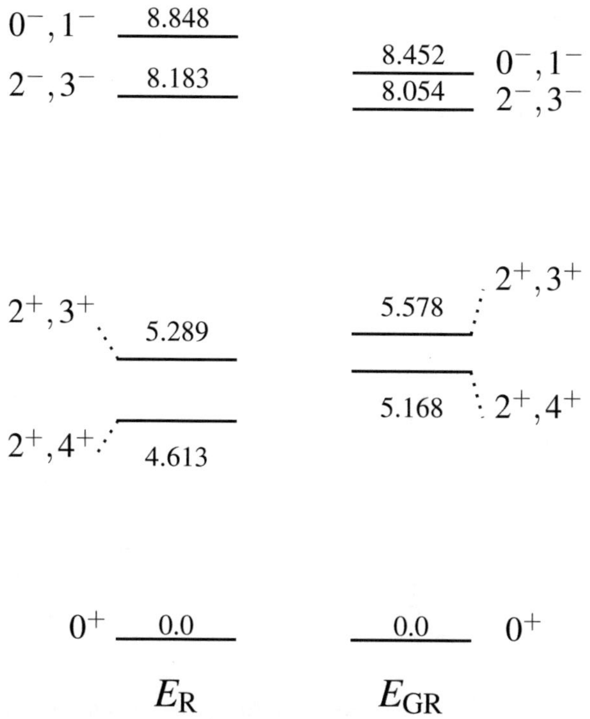

**Fig. 6.17** The energy spectrum of $^{20}$C, calculated using the standard Richardson equations (no continuum) ($E_R$), is compared with the spectrum obtained by solving the generalized Richardson equations ($E_{GR}$). No excited states are known experimentally for this nucleus (from Ref. [130])

these two approaches for solving the pairing Hamiltonian, which can be as large as 600 keV and depend strongly on the configuration of the considered state. Hence, the continuum couplings in pairing model have large and nontrivial effects on the energy spectra. It is correct in principle to adjust the parameters of the Hamiltonian in one nucleus in order to correct for missing continuum couplings. However, the problem of configuration-dependent energy shifts, arising from the continuum couplings in other isotopes of the same chain, remains. The $A$-dependence of the energies of the ground states of even-even carbon isotopes is illustrated in Fig. 6.19. All energies are given with respect to the $^{12}$C core. Experimental data are depicted with solid lines, whereas the dashed and dashed-dotted lines exhibit results of Richardson calculations with ($\mathscr{E}_{GR}$) and without ($\mathscr{E}_R$) continuum.

The evolution of pairing gap with the mass number in the ground states of even-even carbon isotopes is shown in Fig. 6.18. The pairing gap is calculated either by neglecting the continuum and solving the standard Richardson equations or by including the continuum and solving the generalized Richardson equations. These two procedures are indicated by $\Delta_R$ and $\Delta_{GR}$, respectively. One can see that the

**Fig. 6.18** Evolution of the
pairing gap with the mass
number in the ground state of
even-even carbon isotopes,
using either the standard
Richardson equations (no
continuum) ($\Delta_R$), or and
generalized Richardson
equations ($\Delta_{GR}$) (from
Ref. [138])

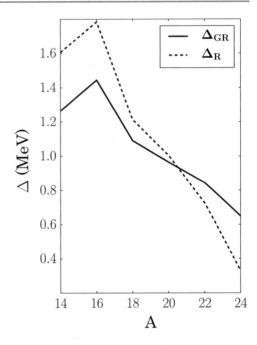

pairing gap in $^{14}$C and $^{16}$C is strongly reduced by the presence of $0d_{3/2}$ and $0f_{7/2}$ resonances and the associated $d_{3/2}$ and $f_{7/2}$ nonresonant scattering states. Note that the $A$-dependence of the pairing gap in both approaches differs significantly. At the $0d_{5/2}$ subshell closure in $^{24}$C, $\Delta_{GR}$ is almost twice larger than $\Delta_R$.

The $A$-dependence for $14 \leq A \leq 20$ is incorrect for both $\mathscr{E}_{GR}$ and $\mathscr{E}_R$ (Fig. 6.19), which is due to the absence of particle-hole components in the two-body interaction. It is astonishing that with a model as simple as the pairing Hamiltonian, $\mathscr{E}_{GR}$ reproduce both the binding energy of $^{20}$C, $^{22}$C isotopes and the experimental position of the neutron dripline. The fast increase of the energy, going from $A = 22$ to $A = 24$, depends on the $d_{5/2}$-$d_{3/2}$ spin–orbit splitting and the couplings to $d_{3/2}$ nonresonant continuum.

The pairing model is by no means a model realistically taking into account nucleon–nucleon correlations. Nevertheless, one can conclude that the coupling between discrete and continuum states cannot be mimicked by an adjustment of the parameters of an effective interaction in the standard shell model in order to reproduce observed levels in a given region of the periodic table. In fact, as shown in this section, such a procedure can lead to wrong conclusions about both the interactions and the structure of many-body states. This warning should be taken into account seriously if one aims at studying systematically the evolution of shell structure and spectra of excitations, going from the valley of stability to the vicinity of driplines.

Another observation concerns position of the neutron dripline. It seems that the main ingredients in a phenomenological description of driplines are: the correct

**Fig. 6.19** The $A$-dependence of the ground state energy in even-even carbon isotopes, calculated using either the standard Richardson equations (no continuum) ($\mathcal{E}_R$), or generalized Richardson equations ($\mathcal{E}_{GR}$), is compared with the experimental data (from Ref. [138])

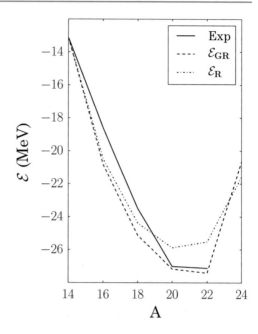

account of coupling to the scattering continuum, and the exact calculation of pairing energy, where the pairing strength is chosen to reproduce binding energy of the nucleus with two nucleons outside of the closed shell. Once these two elements are properly tuned, then other aspects of the problem like the mean-field or the three-body interaction play a secondary role. In that sense, the position of the dripline is a simple result of competing attractive pairing correlations and disruptive coupling to the scattering continuum. This observation also confirms the statement made in Sect. 6.1 that if neutron separation energy tends to zero, then pairing correlations become as important as the mean-field effects.

One should stress in passing that it is essential to calculate the evolution of pairing correlations with neutron number exactly, taking into account coupling to the continuum, since the change of pairing gap with the neutron number depends strongly on the coupling to nonresonant continuum.

## 6.6    Comparison of Gamow Shell Model and Hamiltonian Complex Scaling Formalisms

Calculation of the many-body resonant states in the Gamow shell model is done by the diagonalization of the Hamiltonian using the complex-energy Berggren basis (see Chap. 5). Many-body resonant states are therein linear combinations of bound, resonance, and scattering basis states, whereby complex rotation is used to normalize basis states and calculate matrix elements (see Sect. 3.3).

Alternatively, there exists a many-body formalism allowing to calculate the many-body resonances without recurring to the use of the Berggren basis. This formalism is based on the use of complex-scaled Hamiltonians [139, 140]. In the Hamiltonian complex scaling approach, the Hamiltonian is firstly transformed using a complex rotation similar to that described in Sect. 3.3 [139, 140]. The fundamental advantage of this transformation is that resonance states become integrable on the real $r$-axis, so that they can be expanded with a basis of bound states, as in standard shell model.

The fact that resonance many-body states, of unbound character, can be represented with bound states expansions, relies on Aguilar-Blaslev-Combes theorem, or shortly the ABC theorem [139, 140]. The ABC theorem states that resonance energies are independent of the complex scaling angle $\theta$ used in the transformation of the considered Hamiltonian, provided that $\theta$ is sufficiently large. Consequently, Hamiltonian complex scaling, which is based on the use of the ABC theorem, has been developed in order to calculate resonance energies of unbound quantum systems using the apparatus of bound state expansions. In fact, Hamiltonian complex scaling has been widely applied for many years to calculate the unbound spectrum of atomic systems and molecules [141–144] and light nuclei [145–154].

In the following, the ABC theorem is briefly studied in the one-body and many-body cases. From a practical point of view, it is of interest to compare the results obtained with the Gamow shell model and from the diagonalization of complex-scaled Hamiltonians. For that purpose, the Gamow shell model and Hamiltonian complex scaling method will be considered in a simple three-particle system. Due to the small model space dimensions involved, the un-truncated calculations can be performed to guarantee the convergence with respect to basis size at a numerical level. The overall numerical properties of Hamiltonian complex scaling in practical calculations will then be stated and compared to those of the Gamow shell model (see Chap. 5).

### 6.6.1   Complex-Scaled Hamiltonians and Their Eigenstates

The Hamiltonian complex scaling method is defined from the $\hat{U}_\theta$ operator, which applies a complex rotation to radial one-body wave functions [139]:

$$\langle r|\hat{U}_\theta|u\rangle = e^{i\theta/2}\left(\frac{u(r\,e^{i\theta})}{r}\right),\tag{6.66}$$

where $\theta$ is the rotation angle and $|u\rangle$ is a radial one-body state. Note that, similarly to Sect. 5.1.1, no complex conjugation enters the matrix elements of Eq. (6.66) and that angular matrix elements do not change when complex scaling is applied. This is the case because Hamiltonian complex scaling is defined from the analytic continuation of the real-axis dilation $r \to re^\theta$ [139].

The complex-scaled Hamiltonian arises from a similarity transformation using the $\hat{U}_\theta$ operator of Eq. (6.66):

$$\hat{H}_\theta = \hat{U}_\theta \hat{H} \hat{U}_\theta^{-1}. \tag{6.67}$$

The basic properties of $\hat{U}_\theta$ and $\hat{H}_\theta$ in Eqs. (6.66) and (6.67) in the one-body case, necessary in order to use complex-scaled Hamiltonians in practical applications, are studied in Exercise III.

One can write the general local one-body complex-scaled spherical Hamiltonian $\hat{H}_\theta$ (see Exercise III):

$$\hat{H}_\theta = e^{-2i\theta}\,\frac{\mathbf{p}^2}{2m} + U(r\,e^{i\theta})\,, \tag{6.68}$$

where $m$ is the mass of the nucleon and $\hat{U}(r)$ is a one-body potential. Let us consider the potentials which can give rise to complex-scaled Hamiltonians. The fundamental issue of Eq. (6.68) is evidently to be able to define $\hat{U}(r\,e^{i\theta})$ with complex values for the $\theta$ angles of interest. In fact, complex scaling of potentials is well defined only in the case of dilation analytic potentials, i.e. potentials which admit an analytic continuation for $|\theta| < \pi/4$, as demanded in the ABC theorem [139]. This is a rather restrictive condition, because potentials as pervasive as Woods–Saxon potentials do not belong to this category. Indeed, a Woods–Saxon potential possesses an infinity of poles in the complex $r$-plane. They are equal to

$$z_n = R_0 + i\pi d(2n + 1)\,,$$

where $R_0$ and $d$ are the radius and diffuseness of the potential, respectively, and $n$ is an integer. Thus, Hamiltonian complex scaling cannot be applied if $|\theta| \geq \arg(z_0)$ because of the appearance of poles. In particular, the requirements of the ABC theorem are not fulfilled by a Woods–Saxon potential bearing $R_0 = 3\,\text{fm}$ and $d = 0.65\,\text{fm}$, which are typical values for light nuclei, as one has $\arg(z_0) \sim 0.2$ in this case. Therefore, it is often preferred to use other potentials than Woods–Saxon potentials to mimic the effect of the core.

A popular potential is the Kanada–Kaneko–Nagata–Nomoto potential [155, 156], or shortly the KKNN potential [155]. The KKNN potential is built from Gaussian functions, so that it is dilation analytic, and is of physical interest as it resembles a Woods–Saxon potential on the real $r$-axis [155, 156]. Let us determine the spectrum of $\hat{H}_\theta$ (see Eq. (6.68)). For this, one will consider the operator $e^{2i\theta}\,\hat{H}_\theta$. Clearly, $e^{2i\theta}\,\hat{H}_\theta$ has the form of a non-rotated one-body Hamiltonian bearing a complex potential. Hence, the bound and scattering eigenstates of $e^{2i\theta}\,\hat{H}_\theta$ form a Newton completeness relation (see Sect. 3.5.2). Consequently, the resonant states of $e^{2i\theta}\,\hat{H}_\theta$ consist of localized wave functions on the real $r$-axis, hence whose linear momenta belong to the upper complex plane. The scattering eigenstates of $e^{2i\theta}\,\hat{H}_\theta$ bear real positive energies (see Sect. 3.5.2). Thus, the $\hat{H}_\theta$ operator of Eq. (6.68) has a resonant spectrum built from bound and resonance states, whose wave functions are integrable on the real $r$-axis and eigenenergies are independent of $\theta$ (see Exercise III and Ref. [139]). Obviously, the scattering spectrum of $\hat{H}_\theta$ is made of scattering eigenstates of energy $e^{-2i\theta} E$, with $E > 0$.

---

**Exercise III ⋆**
One will demonstrate the basic properties of $\hat{U}_\theta$ and $\hat{H}_\theta$ (see Eqs. (6.66) and 6.67)) in the one-body case.

**A.** Show that $\hat{U}_\theta^{-1} = \hat{U}_\theta^\dagger = \hat{U}_{-\theta}$ (see Eq. (6.66)). One recalls that complex conjugation does not enter adjoint definition in our case. Demonstrate that the application of $\hat{U}_\theta$ operators of Eq. (6.66) in Eq. (6.67) is equivalent to have $r \rightarrow r\,e^{i\theta}$ and $\nabla_r \rightarrow \nabla_r\,e^{-i\theta}$ in the matrix elements defining $\hat{H}$.

**B.** Let us consider the matrix element of $\hat{H}$ between two initial and final resonance states, denoted as $|u_i\rangle$ and $|u_f\rangle$, respectively.
   Write $\langle u_f|\hat{H}|u_i\rangle$ as a coordinate space integral in the complex plane using complex scaling of integral (see Eq. (3.56)) and $\hat{H}$ (see Eq. (6.67)), respectively.
   We will assume that one can integrate in the complex plane along the complex path defined by $z = r\,e^{i\theta}$. Deduce that both methods used to calculate $\langle u_f|\hat{H}|u_i\rangle$ are equivalent. Explain the presence of $e^{i\theta/2}$ in Eq. (6.66).

**C.** Explain why $\hat{H}$ and $\hat{H}_\theta$ have the same pole spectrum. Show how to determine the resonant energies of $\hat{H}$ by diagonalization of $\hat{H}_\theta$ in a basis of bound states (see Eq. (6.67)).

---

The generalization of Hamiltonian complex scaling method to the many-body case is formulated in Exercise IV. The complex-scaled many-body Hamiltonian, consisting of a one-body part $U$ and local two-body interaction $V$, can then be derived, and it takes the form similar to the one-body complex-scaled Hamiltonian (see Exercise IV):

$$\hat{H}_\theta = \sum_{i=1}^{A}\left(e^{-2i\theta}\frac{\mathbf{p}_i^2}{2m_i} + \hat{U}_i(r_i\,e^{i\theta})\right) + \sum_{i<j}^{A}\hat{V}_{ij}(r_i\,e^{i\theta}, r_j\,e^{i\theta}, \nabla_{r_i}\,e^{-i\theta}, \nabla_{r_j}\,e^{-i\theta}),$$

$$(6.69)$$

where $m_i$ is the mass of the $i$-th nucleon and angular dependence in $V_{ij}$ is not explicitly written as it is the same as in the absence of complex scaling.

An illustration of the position of eigenenergies, studied in Exercise IV, is depicted in the three-body case in Fig. 6.20. Similarly to the one-body case, the fact that the used interaction functionals must be dilation analytic is the main theoretical restriction for the use of many-body Hamiltonian complex scaling. Note that the two-body interaction arising from a harmonic oscillator basis expansion, studied in Sect. 5.6, is dilation analytic.

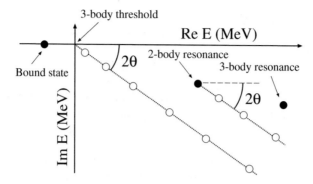

**Fig. 6.20** Complex-scaled eigenstates of the three-body Hamiltonian for the Borromean system. Solid circles are bound and resonance states, and open circles are continuum states (from Ref. [152])

In order to show this property, let us write the coordinate space matrix element of the complex scaled two-body interaction:

$$\langle r_1 r_2 | \hat{U}_\theta \hat{V} \hat{U}_\theta^{-1} | r_3 r_4 \rangle$$

$$= e^{2i\theta} \sum_{\alpha\beta\gamma\delta}^{N_{\max}} \langle \alpha\beta | \hat{V} | \gamma\delta \rangle \frac{u_\alpha(r_1\, e^{i\theta})\, u_\beta(r_2\, e^{i\theta})\, u_\gamma(r_3\, e^{i\theta})\, u_\delta(r_4\, e^{i\theta})}{r_1\, r_2\, r_3\, r_4},$$

(6.70)

where the notations introduced in Sect. 5.6 have been used here and where an additional $e^{2i\theta}$ factor appears due to the non-locality of the interaction. In this expression, $u_\alpha(r)$ is the radial part of the harmonic oscillator state $|\alpha\rangle$ (same for $\beta, \gamma, \delta$), and angular quantum numbers have not been written explicitly for simplicity. As the harmonic oscillator states are dilation analytic and bear decreasing Gaussian asymptotes for $\theta < \pi/4$ (see Eq. (2.17)), the complex-scaled matrix elements of Eq. (6.70) define a short-range dilation analytic two-body interaction. Consequently, one can include realistic interactions in complex-scaled Hamiltonians using a harmonic oscillator basis expansion, as it is done in the Gamow shell model (see Chap. 8 for calculations of the unbound states using the Gamow shell model with realistic interactions).

---

**Exercise IV** ★★

One will generalize the results demonstrated in the one-body case to the many-body case.

**A.** Explain how to formulate Hamiltonian complex scaling approach in the many-body case.

Formulate the dilation analytic properties of the interaction in this approach.

**B.** Write the many-body operator $\hat{H}_\theta$ for the case of non-interacting particles. Deduce that the eigenenergies of $\hat{H}_\theta$ can be written as

$$E_{res}(a) + e^{-2i\theta} E_{scat}(A - a),$$

where $a = 0, 1, 2, \ldots, A$, $E_{res}(a)$ is the energy of a resonant state of $a$ nucleons ($E_{res}(0) = 0$ by convention), and $E_{scat}(A - a) \geq 0$. Demonstrate that the associated $\hat{H}_\theta$ eigenstates of $\hat{H}_\theta$ form a complete set of states in the Hilbert space.

**C.** Show that the eigenenergies formulas of **B** are still valid when particles interact via a short-range force. For this, one will consider separately resonant and scattering eigenstates. Show that the case of resonant eigenstates can be treated as in the Gamow shell model. For scattering eigenstates, one will consider Hamiltonians consisting of several non-interacting parts and determine their eigenenergies $E$. The scattering eigenenergy of $\hat{H}_\theta$ will be shown to be equal to $E$ using the Lippmann–Schwinger equation. This approach is inspired by that developed in Ref. [140].

Note that potentials and interactions do not have to be dilation analytic in the Gamow shell model. Indeed, exterior complex scaling is used in the Gamow shell model to calculate matrix elements (see Sect. 3.3), i.e. complex scaling is used only in the asymptotic zone. As a consequence, dilation analyticity is trivially obtained, because only centrifugal and Coulomb potentials subsist at large distances.

## 6.6.2 Basis Dependence in Truncated Model Spaces

The theoretical results presented in Sect. 6.6.1 provide a simple and well defined method to determine the eigenstates of $\hat{H}$. For this, it is sufficient to diagonalize the complex-scaled operator $\hat{H}_\theta$ (see Eq. (6.67)) with a basis of bound states using a sufficiently large $\theta$ angle. However, in practice, it is necessary to truncate the basis, so that one has to assess the rate of convergence obtained with the number of basis states.

### 6.6.2.1 Basis Optimization

The fundamental problem is to precisely reproduce the many-body wave function in the asymptotic region. Let us firstly consider the case of halo states to ponder out this problem. It is straightforward to verify from Eq. (6.66) that the exponential decrease of weakly bound one-body states is slow $\forall \theta < \pi/4$. Consequently, one faces the same problem with the diagonalization of initial Hamiltonian or Hamiltonian transformed with complex scaling. Indeed, the most immediate choice for the basis of bound states is that of a basis of harmonic oscillator states, with which convergence when expanding weakly bound states is very slow due to the Gaussian fall-off of harmonic oscillator states. The situation is the same if one aims at calculating resonance states. Indeed, even though the application of the $\hat{U}_\theta$

operator of Eq. (6.66) makes resonance states integrable, the exponential decay rate of complex-rotated wave functions can be small. For example, in the one-body case, it is equal to $|k|$ at most, so that weakly unbound states decay slowly on the real $r$-axis. Consequently, convergence would also be very slow if one uses a basis of harmonic oscillator states to diagonalize $\hat{H}_\theta$ in order to evaluate weakly unbound resonance states (see Eq. (6.67)).

As a matter of fact, a basis of harmonic oscillator states, and in general a basis of bound states generated by a potential which goes to infinity with increasing $r$, is not suitable to calculate weakly bound and resonance many-body states. In principle, one could recur to the Newton completeness relation generated by a finite-range potential, as it has the same asymptotic properties as Fourier transform, so that slowly decreasing functions can be expanded (see Sect. 3.2.3). However, this would amount to utilize the Gamow shell model formalism, as one would have to discretize the continuum of the Newton completeness relation similarly to that of the Berggren basis (see Sect. 3.7). The whole point of using a complex-scaled Hamiltonian would then be forfeited, as its fundamental interest is to allow the use of discrete basis of integrable states for Hamiltonian diagonalization.

It is possible to solve this conundrum by using a non-orthogonal set of basis states. A widely used basis of this type is the Gaussian basis, where one uses Gaussian form factors of different lengths [147–149, 152]. The numerical advantage of Gaussian bases is twofold. On the one hand, convergence for weakly bound and resonance states is obtained with a fairly small number of basis states, because the Gaussian functions of the used basis have different spatial extensions. Moreover, the simple mathematical form of Gaussian basis states often allows to evaluate analytically two-body matrix elements, for example, if the used interaction is of Coulomb or Gaussian type. For nuclear applications, it is especially useful when the Hamiltonian consists of a KKNN potential mimicking a core and of a residual Gaussian force, as then Hamiltonian matrix elements can be calculated analytically.

The many-body Schrödinger equation becomes a generalized eigenvalue problem when representing the Hamiltonian with a non-orthogonal set of basis states, and is of the form:

$$\hat{H}_\theta \,|\Psi\rangle = E\hat{O}\,|\Psi\rangle ,$$

where $\hat{O}$ is the basis overlap matrix. The generalized eigenvalue problem is straightforward to solve when dimensions reach few thousands at most, as it can be reduced to the standard eigenvalue problem

$$\hat{O}^{-1/2}\hat{H}_\theta\hat{O}^{-1/2}\,|\Phi\rangle = E\,|\Phi\rangle ,$$

with $|\Psi\rangle = \hat{O}^{-1/2}|\Phi\rangle$. As the $\hat{O}^{-1/2}$ operator is straightforward to calculate from the eigenvalues and eigenvectors of $\hat{O}$, the numerical cost to solve a generalized eigenvalue problem is only a few times slower than that of a standard eigenvalue problem. Otherwise, it is possible to use extensions of the Lanczos or Jacobi–Davidson methods to the generalized eigenvalue problem, which have been developed to deal with matrices of large dimensions [157–159].

### 6.6.2.2 Dependence of Observables on the Rotation Angle in Practical Calculations

While eigenenergies are theoretically independent of $\theta$ due to the ABC theorem, a spurious $\theta$ dependence arises from the use of a finite model space. It is then necessary in practice to determine the optimal angle $\theta_{opt}$ with which truncation errors are minimal. For this, one uses the generalized variational principle, described in Sect. 5.1.4. Indeed, eigenenergies calculated with the rotation angle $\theta_{opt}$ are stationary with respect to small changes in $\theta$. The determination of $\theta_{opt}$ is straightforward, as it is sufficient for that matter to diagonalize $\hat{H}_\theta$ with $\theta_{min} < \theta < \pi/4$, where $\theta_{min}$ is the smallest rotation angle rendering the sought resonance state integrable on the real $r$-axis.

Another issue inherent to the use of Hamiltonian complex scaling method in a finite model space is the instability of wave function back-transformation, i.e. the application of $\hat{U}_\theta^{-1}$ (see Eq. (6.66)) on rotated wave functions to obtain the unbound eigenstates of $\hat{H}$ [112, 160, 161]. This problem occurs when one wants to calculate wave function observables which explicitly depend on $r$, such as density or correlation density [112, 160]. The situation is similar to inverse Fourier transform, which is notoriously known to be unstable numerically [162].

Indeed, the application of $\hat{U}_\theta^{-1}$ (see Eq. (6.66)) to the rotated resonance wave function, which is integrable, yields the physical resonance wave function on the real $r$-axis which diverges exponentially in modulus. Consequently, the small numerical inaccuracies present in the rotated resonance wave function are exponentially amplified in this process. In fact, it has been proved that the back-transformed resonance wave function calculated in a truncated space, in general does not converge to the exact resonance wave function at the limit of infinitely large model space [160]. Moreover, as could be expected, the numerical error increases along with $\theta$ [112].

As for inverse Fourier transform, this problem could be solved by using a priori knowledge on the eigenstate [112, 162]. Indeed, as one knows that the exact resonance wave function has smooth variations on the real $r$-axis, it can be reconstructed by suppressing spurious high frequency components. This has been realized in Ref. [112] using the Tikhonov regularization method [163]. It was shown that the density of the unbound $2^+$ first excited state of $^6$He, modelled by two valence neutrons above a $^4$He core, could be calculated accurately [112].

The Tikhonov regularization method depends on a regularization parameter, which is determined by way of a plateau condition [112]. For this, one determines the optimal regularization parameter for which the variations of density with respect to this parameter are minimal [112]. The precision of the Tikhonov regularization method could be assessed from Gamow shell model calculations, with which the density of resonance states can be almost exactly calculated [112]. Note that $r$-integrated observables, such as the angular correlation density, do not suffer from this instability and can be accurately calculated with the Hamiltonian complex scaling [112].

### 6.6.3   Numerical Comparison of Eigenenergies of Gamow Shell Model and Hamiltonian Complex Scaling Method

One will now compare Gamow shell model and Hamiltonian complex scaling frameworks in a numerical example. The considered model space must be sufficiently small to allow for exact calculations with both approaches, while the considered system must bear weakly bound or unbound eigenstates. The $^6$He and $^6$Be isotopes are ideal testing grounds for that matter [152]. Indeed, their $0^+$ ground state and $2^+$ first excited state are either halo or resonance states. Moreover, $^6$He and $^6$Be can be modelled as two valence nucleons above an inert $^4$He core, so that model spaces are not excessively large.

The used Hamiltonian consists of a KKNN potential mimicking a $^4$He core, to which a residual inter-neutron interaction is added. The cluster orbital shell model formalism is utilized (see Sect. 5.2), so that no center-of-mass excitation can occur. The used nuclear interaction consists of the Minnesota force [23], to which a phenomenological Gaussian interaction is added to account for missing three-body [152]. The Coulomb interaction is also included when considering $^6$Be. Hence, all radial form factors of the Hamiltonian are of Coulomb or Gaussian type, so that matrix elements can be calculated analytically in the basis of Gaussian functions [152].

Different model spaces have been considered to study $^6$He and $^6$Be eigenstates with the Gamow shell model and Hamiltonian complex scaling method (see Ref. [152] for details). In particular, the maximal orbital momentum of valence partial waves $l_{max}$ varies from 1 to 5. Results issued from the diagonalization in the Gamow shell model and Hamiltonian complex scaling frameworks are displayed in the upper and lower panels of Fig. 6.21 for $^6$He and $^6$Be, respectively.

As can be seen on Fig. 6.21), the eigenenergies provided by Gamow shell model and Hamiltonian complex scaling diagonalizations are close to each other. On the one hand, the halo $0^+$ ground state energy of $^6$He is almost exactly the same in the Gamow shell model and Hamiltonian complex scaling frameworks. On the other hand, the resonance energies of the ground state $0^+$ in $^6$Be, and the first excited states $2^+$ in $^6$He and $^6$Be, differ by about 10–20 keV in both approaches. As a consequence, the calculations depicted in Fig. 6.21 have proved that Gamow shell model and Hamiltonian complex scaling method are equivalent in practice. It is, nevertheless, important to understand the relatively larger discrepancy of resonance energies compared to bound state energies. This discrepancy has two main origins. The first one comes from the different methods used to calculate the two-body matrix elements. While a harmonic oscillator basis expansion is used in the Gamow shell model for that matter (see Sect. 5.6)), two-body matrix elements are analytical within the Hamiltonian complex scaling framework. Consequently, a slight dependence on the number of harmonic oscillator basis states $N_{max}$ arises in Gamow shell model calculations. Another reason is the dependence in practice on the used rotation angle $\theta$ for the diagonalization of complex-scaled Hamiltonians (see Sect. 6.6.2.2).

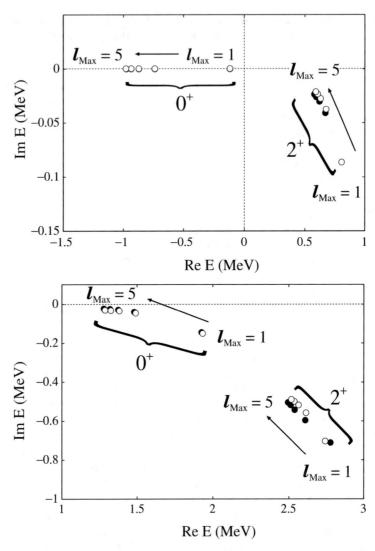

**Fig. 6.21** Poles of the ground $0_1^+$ and the first excited $2_1^+$ states of $^6$He (upper panel) and $^6$Be (lower panel), which are calculated using the Hamiltonian complex scaling approach and the Gamow shell model for $1 \leq \ell_{max} \leq 5$. Open and solid circles denote Hamiltonian complex scaling and Gamow shell model results, respectively (from Ref. [152])

An illustration of this dependence can be seen in Fig. 6.22 for the $2^+$ resonance eigenstate of $^6$He nucleus [152]. One can identify the $\theta_{opt}$ value (see Sect. 6.6.2.2) in Fig. 6.22, where the complex eigenenergy of the $2^+$ resonance has the smallest variations with respect to an angle $\theta$. As seen in Fig. 6.22, the eigenenergies provided by Gamow shell model and Hamiltonian complex scaling diagonalizations are almost the same when Hamiltonian complex scaling is applied with $\theta = \theta_{opt}$,

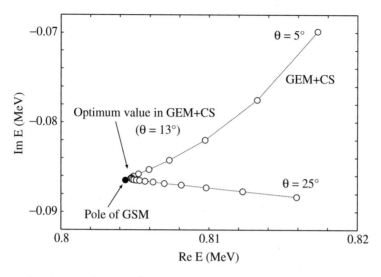

**Fig. 6.22** Complex energies of the $^6$He ($2^+$) resonance state as a function of the rotation angle $\theta$ for $l_{\max} = 1$. $\theta$ varies by $1°$ from one circle to another. The rotation angle $\theta$ varies from $5°$ to $25°$. The notation "GEM + CS" stands for Gaussian Expansion Method + Complex Scaling, and "GSM" for Gamow shell model (from Ref. [152])

so that the value $\theta = \theta_{\mathrm{opt}}$ is used by default in Hamiltonian complex scaling calculations. However, one can see in Fig. 6.22 that a sole variation of $1°$ of $\theta$ can induce a change of about $10\,\mathrm{keV}$. Consequently, the $\theta$-dependence can play a role in the energy differences noticed in the resonance eigenstates of $^6$He and $^6$Be when performing diagonalizations with Gamow shell model and Hamiltonian complex scaling frameworks.

---

## Solutions to Exercises[1]

### Exercise I.

Commutators are straightforward to calculate from Eqs. (6.41), (6.43), (6.45), (6.46), (6.50), and (6.52):

$$\left[\hat{\tilde{n}}_q, B_\eta^\dagger\right] = \frac{2\tilde{b}_q^\dagger \sqrt{w_q}}{2\epsilon_q - E_\eta}$$

$$\left[\hat{\tilde{n}}_q, B_0^\dagger\right] = 2\sqrt{w_q}\tilde{b}_q^\dagger$$

---

[1] The input files, codes and code user manual associated to computer-based exercises can be found at https://github.com/GSMUTNSR.

$$\left[B_\eta, B_{\eta'}^\dagger\right] = \sum_q^{\mathcal{N}} \frac{w_q \Omega_q/2 - \hat{\tilde{n}}_q}{(2\epsilon_q - E_\eta)(2\epsilon_q - E_{\eta'})}$$

$$\left[B_0, B_\eta^\dagger\right] = \sum_q^{\mathcal{N}} \frac{w_q \Omega_q/2 - \hat{\tilde{n}}_q}{2\epsilon_q - E_\eta}$$

$$\left[B_0^\dagger, B_\eta^\dagger\right] = 0 .$$                                    (6.71)

**Exercise II.**

**A.** As at most one pair of particles can occupy the level denoted as $q$, the sum in Eq. (6.63) disappears. Equation (6.64) is then immediately obtained.

**B.** Let us define $x_i = 2\epsilon_q - E_i$, where $i = 1, 2$. Eq. (6.63) then becomes:

$$(x_2 - x_1)x_1 - 2G \, d_q \, x_2 + 2G \, (d_q + 1) \, x_1 = 0$$
$$(x_1 - x_2)x_2 - 2G \, d_q \, x_1 + 2G \, (d_q + 1) \, x_2 = 0.$$

Let us define $S = x_1 + x_2$ and $D = x_2 - x_1$. The two previous equations translate into:

$$- D^2 + 2GS = 0$$
$$SD - 2G(2d_q + 1)D = 0.$$

As $G \neq 0$, $D$ cannot be equal to zero, as otherwise one obtains $S = 0$, and hence $x_1 = x_2 = 0$, so that Eq. (6.63) cannot hold. Thus, the two previous equations imply that:

$$S = 2G(2d_q + 1)$$
$$D^2 = 4G^2(2d_q + 1).$$

The equality $d_q = v_q/2 - \Omega_q/4 = -\Omega_q/4$ holds because the number of unpaired particles $v_q$ is equal to zero. As one has $\Omega_q \geq 4$, $2d_q + 1 \leq -1$, so that $D^2 < 0$ and one can pose $D = 2iG\sqrt{|2d_q + 1|}$. Consequently, the expression for $x_1$ and $x_2$ comes forward:

$$x_1 = G(2d_q + 1) - iG\sqrt{|2d_q + 1|}$$
$$x_2 = G(2d_q + 1) + iG\sqrt{|2d_q + 1|}.$$

Equation (6.65) is immediately obtained from the expression for $x_1$ and $x_2$.

**C.** (i) The mean-field approximation of the pairing model has all its energies equal to $2\epsilon_q - 2G \, d_q$. This leads to infinities in Eq. (6.63), so that a non-interacting system cannot be used as a starting point. Moreover, one knows from the two-pair case that the solutions of Eq. (6.63) are complex in general. Consequently, it is necessary to devise a complex-valued starting point. Indeed, the Newton-Raphson method applied to Eq. (6.63) cannot converge to a complex solution if its starting point is real.

(ii) One will then consider instead a system formed by two interacting pairs and by one non-interacting pair. Consequently, one can use a combination of the solutions (6.64) and (6.65), i.e. one pair is initiated with Eq. (6.64) while the two others are initiated with Eq. (6.65). This starting point is clearly complex.

(iii) One will now calculate the numerical value of the considered starting point in order to check that it can effectively lead to the convergence of the Newton-Raphson method. For the starting point of the Newton-Raphson method, one will use the values of $x_1$ and $x_2$ of the two-pair case devised in **B** for $x_1^{(0)}$ and $x_2^{(0)}$, respectively. Consequently, $x_3^{(0)} = 2\epsilon_q - E_3^{(0)} = 2G\, d_q$ from **A**. One can then evaluate Eq. (6.63) at the considered starting point:

$$\left| 1 - \frac{2G\, d_q}{x_1^{(0)}} + \frac{2G}{x_2^{(0)} - x_1^{(0)}} + \frac{2G}{x_3^{(0)} - x_1^{(0)}} \right| = \left| \frac{2G}{x_3^{(0)} - x_1^{(0)}} \right|$$

$$= \sqrt{\frac{4}{1 + |2d_q + 1|}}$$

$$\left| 1 - \frac{2G\, d_q}{x_2^{(0)}} + \frac{2G}{x_1^{(0)} - x_2^{(0)}} + \frac{2G}{x_3^{(0)} - x_2^{(0)}} \right| = \left| \frac{2G}{x_3^{(0)} - x_2^{(0)}} \right|$$

$$= \sqrt{\frac{4}{1 + |2d_q + 1|}}$$

$$\left| 1 - \frac{2G\, d_q}{x_3^{(0)}} + \frac{2G}{x_1^{(0)} - x_3^{(0)}} + \frac{2G}{x_2^{(0)} - x_3^{(0)}} \right| = \left| \frac{2G}{x_1^{(0)} - x_3^{(0)}} + \frac{2G}{x_2^{(0)} - x_3^{(0)}} \right|$$

$$= \frac{4}{1 + |2d_q + 1|} .$$

The devised starting point is well defined for $G > 0$. The Newton-Raphson method converges in practice as the interaction strength $G$ can be chosen to be arbitrarily small [130].

## Exercise III.

**A.** One will firstly calculate the matrix element $\langle r | \hat{U}_{-\theta} \hat{U}_\theta | u \rangle$ for arbitrary radius and radial wave function. Let us define for this the ket state $|u^{(\theta)}\rangle = \hat{U}_\theta |u\rangle$ for a given radial state $|u\rangle$. Thus, $u^{(\theta)}(r) = e^{i\theta/2} u(r \, e^{i\theta})$ from Eq. (6.66). One then has

$$\langle r|\hat{U}_{-\theta}\hat{U}_{\theta}|u\rangle = \langle r|\hat{U}_{-\theta}|u^{(\theta)}\rangle = e^{-i\theta/2} \left( \frac{u^{(\theta)}(r \, e^{-i\theta})}{r} \right) = e^{-i\theta/2} \, e^{i\theta/2} \left( \frac{u(r)}{r} \right)$$

$$= \frac{u(r)}{r} = \langle r|u\rangle \, . \tag{6.72}$$

The equality $\hat{U}_{\theta}^{-1} = \hat{U}_{-\theta}$ then follows.

In order to show that $\hat{U}_{\theta}^{\dagger} = \hat{U}_{-\theta}$, one will rewrite the matrix element $\langle u_f|\hat{U}_{\theta}|u_i\rangle$, where $|u_i\rangle$ and $|u_f\rangle$ are two bound states:

$$\langle u_f|\hat{U}_{\theta}|u_i\rangle = \int_0^{+\infty} u_f(r) \, u_i(r \, e^{i\theta}) \, e^{i\theta/2} \, dr$$

$$= \int_0^{+\infty} u_f(r \, e^{-i\theta}) \, u_i(r) \, e^{-i\theta/2} \, dr = \langle u_i|\hat{U}_{-\theta}|u_f\rangle \, , \tag{6.73}$$

where complex scaling of radial integrals has been used (see Sect. 3.3). As $\langle u_i|\hat{U}_{\theta}^{\dagger}|u_f\rangle = \langle u_f|\hat{U}_{\theta}|u_i\rangle$ by definition, one obtains $\hat{U}_{\theta}^{\dagger} = \hat{U}_{-\theta}$.

The general local one-body Hamiltonian can be written as $\hat{H}(r, \nabla_r)$. Let us then evaluate the matrix element $\langle u_f|\hat{H}_{\theta}|u_i\rangle$, with $|u_i\rangle$ and $|u_f\rangle$ being two bound states as well:

$$\langle u_f|\hat{H}_{\theta}|u_i\rangle = \langle u_f|\hat{U}_{\theta} \, \hat{H} \, \hat{U}_{\theta}^{-1}|u_i\rangle$$

$$= \int_0^{+\infty} u_f(r \, e^{-i\theta}) \, \hat{H}(r, \nabla_r) \, u_i(r \, e^{-i\theta}) \, e^{-i\theta} \, dr$$

$$= \int_0^{+\infty} u_f(r) \, H(r \, e^{i\theta}, \nabla_r \, e^{-i\theta}) \, u_i(r) \, dr \, , \tag{6.74}$$

where complex scaling of radial integrals has been used as well (see Sect. 3.3). Thus, applying $\hat{H}_{\theta}$ is equivalent to applying $\hat{H}$ by formally applying the replacements $r \to r \, e^{i\theta}$ and $\nabla_r \to \nabla_r \, e^{-i\theta}$ in $\hat{H}$.

**B.** Let us calculate $\langle u_f|\hat{H}|u_i\rangle$ using complex scaling of a radial integral on the complex path $L_z$ defined by $z = r \, e^{i\theta}$ for $r \geq 0$:

$$\langle u_f|\hat{H}|u_i\rangle = \int_{L_z} u_f(z) \, H(z, \nabla_z) \, u_i(z) \, dz$$

$$= e^{i\theta} \int_0^{+\infty} u_f(r \, e^{i\theta}) \, H(r \, e^{i\theta}, \nabla_r \, e^{-i\theta}) \, u_i(r \, e^{i\theta}) \, dr \, . \tag{6.75}$$

Let us now evaluate $\langle u_f | \hat{H} | u_i \rangle$ with complex scaling of Hamiltonian, i.e. by using Eqs. (6.66) and (6.67):

$$\langle u_f | \hat{H} | u_i \rangle = \langle u_f^{(\theta)} | \hat{H}_\theta | u_i^{(\theta)} \rangle$$

$$= \int_0^{+\infty} e^{i\theta/2} \, u_f(r \, e^{i\theta}) \, H(r \, e^{i\theta}, \nabla_r \, e^{-i\theta}) \, e^{i\theta/2} \, u_i(r \, e^{i\theta}) \, dr$$

$$= e^{i\theta} \int_0^{+\infty} u_f(r \, e^{i\theta}) \, H(r \, e^{i\theta}, \nabla_r \, e^{-i\theta}) \, u_i(r \, e^{i\theta}) \, dr , \quad (6.76)$$

where $|u_i^{(\theta)}\rangle = \hat{U}_\theta |u_i\rangle$ and $|u_f^{(\theta)}\rangle = \hat{U}_\theta |u_f\rangle$. The integrals converge because $u_i(r \, e^{i\theta})$ and $u_f(r \, e^{i\theta})$ exponentially vanish for $r \to +\infty$. As $\langle u_f | \hat{H} | u_i \rangle = \langle u_f^{(\theta)} | \hat{H}_\theta | u_i^{(\theta)} \rangle$, the application of complex scaling of integral with unbound resonance states or complex scaling of Hamiltonian with rotated resonance states are then equivalent. Previous results imply that the $e^{i\theta/2}$ factor in Eq. (6.66) arises from the Jacobian of the change of variable $r \to r \, e^{i\theta}$ in complex integration.

C. Let us consider a resonant state $|\Psi\rangle$ of $\hat{H}$, so that $\hat{H} |\Psi\rangle = E |\Psi\rangle$, with $E$ the energy of $|\Psi\rangle$. Thus, from Eq. (6.67), one has $\hat{H}_\theta |\Psi_\theta\rangle = E |\Psi_\theta\rangle$, with $|\Psi_\theta\rangle = \hat{U}_\theta |\Psi\rangle$. Consequently, $\hat{H}_\theta$ has the same resonant eigenenergies as $\hat{H}$, which are the poles of the $S$-matrix (see Sect. 2.6.6).

As the resonance eigenstates $|\Psi_\theta\rangle$ of $\hat{H}_\theta$ are integrable on the real $r$-axis (see Eq. (6.66)) for a sufficiently large angle $\theta$ in $[0 : \pi/4]$, they belong to the Hilbert space. Consequently, for $\theta$ sufficiently large, they can be obtained by diagonalizing $\hat{H}_\theta$ with a complete basis in the Hilbert space, hence with a basis of bound states.

## Exercise IV.

A. One aims at defining a generalization of the one-body $\hat{U}_\theta$ operator of Eq. (6.66) in the $A$-body case. For, let us consider a Slater determinant of the form $|s_1, \ldots, s_A\rangle$, with $|s_i\rangle$, $i \in [1 : A]$, a one-body state. One can then define an $A$-body operator $\hat{U}_\theta$ from the tensor product of $A$ one-body complex scaling operators of Eq. (6.66):

$$\hat{U}_\theta |u_1, \ldots, u_A\rangle = \prod_{i=1}^{A} U_\theta^{(i)} |u_i\rangle , \quad (6.77)$$

where $|u_i\rangle$ is the radial ket state associated with $|s_i\rangle$, and $U_\theta^{(i)}$ is the complex scaling operator associated with the $i$-th nucleon. The $\theta$ angle must be the same for all nucleon coordinates so that the $\hat{U}_\theta$ operator of Eq. (6.77) acts symmetrically on all nucleon states.

This implies that the complex scaled many-body Hamiltonian $\hat{U}_\theta H \hat{U}_\theta^{-1}$ is also symmetric with respect to the coordinates of the $A$ nucleons involved. Therefore, the many-body operator $\hat{H}_\theta$ has the same expression as in Eq. (6.67, provided that the complex scaling operator of Eq. (6.77) is used.

Similarly to the one-body case (see Eq. (6.68)), the used interaction in $\hat{H}_\theta$ must be dilatation analytic with respect to all rotated nucleon coordinates $r_i\, e^{i\theta}$.

**B.** The $\hat{H}_\theta$ operator in the non-interacting case is evidently the sum of $A$ one-body Hamiltonians of Eq. (6.68):

$$\hat{H}_\theta = \sum_{i=1}^{A} \left( e^{-2i\theta}\, \frac{\mathbf{p}_i^2}{2m_i} + \hat{U}_i(r_i\, e^{i\theta}) \right). \tag{6.78}$$

The eigenstates of $\hat{H}_\theta$ are clearly the Slater determinants $|s_1, \ldots, s_A\rangle$ generated by the eigenstates of the one-body Hamiltonians of the sum of Eq. (6.78).

One has $|s_1, \ldots, s_A\rangle = |s_1^{(\text{res})}, \ldots, s_a^{(\text{res})}, s_{a+1}^{(\text{scat})}, \ldots, s_A^{(\text{scat})}\rangle$, with $a = 0, 1, 2, \ldots, A$, where the resonant or scattering character of occupied states is explicitly written. Let us denote the eigenenergy of $|s_1, \ldots, s_A\rangle$ by $E$. From the discussion following Eq. (6.68), one deduces that

$$E = E_{\text{res}}(a) + e^{-2i\theta} E_{\text{scat}}(A - a) \, ,$$

where $E_{\text{res}}(a)$ is the sum of the resonant energies of $|s_1^{(\text{res})}\rangle, \ldots, |s_a^{(\text{res})}\rangle$ and $E_{\text{scat}}(A - a) \geq 0$.

As the one-body states generated by a complex-scaled one-body Hamiltonian build a completeness relation of Newton type, the Slater determinants generated by $\hat{H}_\theta$ of Eq. (6.78) form a many-body complete set of states in the Hilbert space.

**C.** The eigenenergies of many-body correlated resonant eigenstate are equal to $E_{\text{res}}(A)$ by definition. The $\theta$-independence of the many-body resonant eigenenergies of $\hat{H}_\theta$ can be demonstrated as in **C** of Exercise III, because the used argument in the one-body case clearly holds in the many-body case.

Let us now turn to the scattering eigenstates of the Hamiltonian $\hat{H}_\theta$ where nucleons interact. For this, one firstly constructs an approximate Hamiltonian $\hat{H}_\theta^{(\text{app})}$ from $\hat{H}_\theta$, by suppressing many-body interactions in $\hat{H}_\theta$ so that $A - a$ nucleons at least do not interact, with $a = 0, 1, 2, \ldots, A - 1$. Consequently, one can generate scattering eigenstates of $\hat{H}_\theta^{(\text{app})}$ of the form $|\Psi^{(\text{app})}\rangle = |\Psi_{\text{res}}^{(a)}\rangle \otimes |\Phi_{\text{scat}}^{(A-a)}\rangle$, where $|\Psi_{\text{res}}^{(a)}\rangle$ is a many-body resonant state of $a$ nucleons and $|\Phi_{\text{scat}}^{(A-a)}\rangle$ is a Slater determinant of $A - a$ nucleons, where only one-body scattering states are occupied. Thus, similarly to the non-interacting case, the eigenenergy of $|\Psi_{\text{res}}^{(a)}\rangle \otimes |\Phi_{\text{scat}}^{(A-a)}\rangle$ is equal to $E = E_{\text{res}}(a) + e^{-2i\theta} E_{\text{scat}}(A - a)$, where $E_{\text{res}}(a)$ is the eigenenergy of $|\Psi_{\text{res}}^{(a)}\rangle$ and $E_{\text{scat}}(A - a) \geq 0$.

One can now study the general case of correlated scattering eigenstates of $\hat{H}_\theta$. For this, one will suppose that there exists a scattering eigenstate $|\Psi\rangle$ of $\hat{H}_\theta$ of energy $E$ so that $E \neq E_{\text{res}}(a) + e^{-2i\theta} E_{\text{scat}}(A - a)\ \forall a \in 0, 1, 2, \ldots, A - 1$.

From the $|\Psi\rangle$ many-body state, one can build a scattering eigenstate $|\Psi^{(\mathrm{app})}\rangle$ of $\hat{H}_\theta^{(\mathrm{app})}$ of energy $E$ solution of the following Lippmann–Schwinger equation:

$$|\Psi^{(\mathrm{app})}\rangle = |\Psi\rangle - (\hat{H}_\theta^{(\mathrm{app})} - (E + i\eta))^{-1}(\hat{H}_\theta^{(\mathrm{app})} - \hat{H}_\theta)|\Psi\rangle, \qquad (6.79)$$

where $\eta \to 0$ is a regularizing parameter [164]. One has $|\Psi^{(\mathrm{app})}\rangle \neq |\Psi_{\mathrm{res}}^{(a)}\rangle \otimes |\Phi_{\mathrm{scat}}^{(A-a)}\rangle \ \forall a$, as otherwise the equation

$$E = E_{\mathrm{res}}(a) + e^{-2i\theta} E_{\mathrm{scat}}(A - a)$$

holds for a given number of nucleons $a$. Thus, $|\Psi^{(\mathrm{app})}\rangle = |\Psi_{\mathrm{scat}}^{(a)}\rangle \otimes |\Phi^{(A-a)}\rangle$, where $|\Psi_{\mathrm{scat}}^{(a)}\rangle$ is a correlated many-body scattering state of energy $E^{(a)}$ and $|\Phi^{(A-a)}\rangle$ is a Slater determinant, with $a \in 1, 2, \ldots, A - 1$.

Through the replacements in Eq. (6.79):

$$A \to a, \ |\Psi\rangle \to |\Psi_{\mathrm{scat}}^{(a)}\rangle, \ \hat{H}_\theta \to \hat{H}_\theta^{(\mathrm{app})}, \ \hat{H}_\theta^{(\mathrm{app})} \to \hat{H}_\theta^{(\mathrm{app}')},$$

$$|\Psi^{(\mathrm{app})}\rangle \to |\Psi_{\mathrm{scat}}^{(a)\,(\mathrm{app}')}\rangle, \ E \to E^{(a)}$$

with $\hat{H}_\theta^{(\mathrm{app}')}$ defined similarly to $\hat{H}_\theta^{(\mathrm{app})}$ with $a \to a' \in 1, 2, \ldots, a - 1$, one generates the decomposition $|\Psi_{\mathrm{scat}}^{(a)\,(\mathrm{app}')}\rangle = |\Psi_{\mathrm{scat}}^{(a')}\rangle \otimes |\Phi^{(a-a')}\rangle$, where the energy of $|\Psi_{\mathrm{scat}}^{(a)\,(\mathrm{app}')}\rangle \otimes |\Phi^{(A-a)}\rangle$ is equal to $E$. Similarly to $|\Psi^{(\mathrm{app})}\rangle$, $|\Psi_{\mathrm{scat}}^{(a')\,(\mathrm{app}')}\rangle \neq |\Psi_{\mathrm{res}}^{(a'')}\rangle \otimes |\Phi_{\mathrm{scat}}^{(a-a'')}\rangle \ \forall a''$.

Consequently, by recurrence on $a'$, one eventually obtains a many-body state $|\Psi_{\mathrm{scat}}^{(1)\,(\mathrm{app}')}\rangle \otimes |\Phi_{\mathrm{scat}}^{(A-1)}\rangle$. As $|\Psi_{\mathrm{scat}}^{(1)\,(\mathrm{app}')}\rangle$ is a one-body state and $|\Phi_{\mathrm{scat}}^{(A-1)}\rangle$ is a Slater determinant of $A - 1$ nucleons, $|\Psi_{\mathrm{scat}}^{(1)\,(\mathrm{app}')}\rangle \otimes |\Phi_{\mathrm{scat}}^{(A-1)}\rangle$ is a many-body non-interacting state $|\Phi_{\mathrm{scat}}^{(A)}\rangle$ of energy $E$. Therefore, its energy reads $E = e^{-2i\theta} E_{\mathrm{scat}}(A)$, which is impossible by assumption. As a consequence, all scattering eigenstates $|\Psi\rangle$ of $\hat{H}_\theta$ have an energy $E$ of the form $E_{\mathrm{res}}(a) + e^{-2i\theta} E_{\mathrm{scat}}(A - a)$, with $a \in 0, 1, 2, \ldots, A - 1$.

## References

1. Y.A. Luo, J. Okołowicz, M. Płoszajczak, N. Michel, arXiv nucl-th/0211068 [nucl-th] (2002)
2. N. Michel, W. Nazarewicz, J. Okołowicz, M. Płoszajczak, J. Rotureau, Acta Phys. Pol. B **35**, 1249 (2004)
3. R.J. Charity, K.W. Brown, J. Okołowicz, M. Płoszajczak, J.M. Elson, W. Reviol, L.G. Sobotka, W.W. Buhro, Z. Chajecki, W.G. Lynch, J. Manfredi, R. Shane, R.H. Showalter, M.B. Tsang, D. Weisshaar, J.R. Winkelbauer, S. Bedoor, A.H. Wuosmaa, Phys. Rev. C **97**, 054318 (2018)
4. A.P. Zuker, Phys. Rev. Lett. **90**, 042502 (2003)
5. J. Okołowicz, M. Płoszajczak, I. Rotter, Phys. Rep. **374**, 271 (2003)

6. J. Okołowicz, M. Płoszajczak, W. Nazarewicz, Prog. Theor. Phys. Supp. **196**, 230 (2012)
7. J. Okołowicz, W. Nazarewicz, M. Płoszajczak, Fortschr. Phys. **61**, 66 (2013)
8. H. Bartz, I. Rotter, J. Höhn, Nucl. Phys. A **275**, 111 (1977)
9. K. Bennaceur, F. Nowacki, J. Okołowicz, M. Płoszajczak, Nucl. Phys. A **651**, 289 (1999)
10. K. Bennaceur, F. Nowacki, J. Okołowicz, M. Płoszajczak, Nucl. Phys. A **671**, 203 (2000)
11. S.A. van Langen, P.W. Brouwer, C.W.J. Beenakker, Phys. Rev. E **55**, R1 (1997)
12. P.W. Brouwer, Phys. Rev. E **68**, 046205 (2003)
13. E.N. Bulgakov, I. Rotter, A.F. Sadreev, Phys. Rev. E **74**, 056204 (2006)
14. J-P. Linares, unpublished (2020)
15. J. Okołowicz, M. Płoszajczak, R.J. Charity, L.G. Sobotka, Phys. Rev. C **97**, 044303 (2018)
16. J. Okołowicz, M. Płoszajczak, W. Nazarewicz, Phys. Rev. Lett. **124**, 042502 (2020)
17. H. Furutani, H. Horiuchi, R. Tamagaki, Prog. Theor. Phys. **62**, 981 (1979)
18. H. Furutani, H. Kanada, T. Kaneko, S. Nagata, H. Nishioka, S. Okabe, S. Saito, T. Sakuda, M. Seya, Prog. Theor. Phys. Supp. **68**, 193 (1980)
19. N. Michel, W. Nazarewicz, J. Okołowicz, M. Płoszajczak, Nucl. Phys. A **752**, 335 (2005)
20. N. Michel, W. Nazarewicz, M. Płoszajczak, Phys. Rev. C **70**, 064313 (2004)
21. P. Navrátil, E. Ormand, Phys. Rev. C **68**, 034305 (2003)
22. P.J. Brussard, P.W.M. Glaudemans, *Shell Model Applications in Nuclear Spectroscopy* (North Holland Publishing Comp., Amsterdam, 1977)
23. D.R. Thompson, M. LeMere, Y.C. Tang, Nucl. Phys.A **268**, 53 (1977)
24. A.M. Lane, Proc. Phys. Soc. A **68**, 189 (1955)
25. D. Kurath, Phys. Rev. **101**, 216 (1956)
26. J. Dobaczewski, W. Nazarewicz, Phil. Trans. R. Soc. Lond. A **356**, 2007 (1998)
27. J. Dobaczewski, N. Michel, W. Nazarewicz, M. Płoszajczak, J. Rotureau, Prog. Part. Nucl. Phys. **59**, 432 (2007)
28. N. Michel, W. Nazarewicz, M. Płoszajczak, T. Vertse, J. Phys. G. Nucl. Part. Phys. **36**, 013101 (2009)
29. The Editors, Phys. Rev. A **83**, 040001 (2011)
30. J. Dobaczewski, W. Nazarewicz, P.G. Reinhard, J. Phys. G: Nucl. Part. Phys. **41**, 074001 (2014)
31. W. Nazarewicz, J. Phys. G **43**, 044002 (2016)
32. J.E. Amaro, R. Navarro Pérez, E. Ruiz Arriola, Few Body Syst. **55**, 977 (2014)
33. R. Navarro Pérez, J.E. Amaro, E. Ruiz Arriola, Phys. Rev. C **89**, 064006 (2014)
34. R. Navarro Pérez, J.E. Amaro, E. Ruiz Arriola, Phys. Lett. B **738**, 155 (2014)
35. M. Piarulli, L. Girlanda, R. Schiavilla, A. Kievsky, A. Lovato, L.E. Marcucci, S.C. Pieper, M. Viviani, R.B. Wiringa, Phys. Rev. C **94**, 054007 (2016)
36. M. Piarulli, L. Girlanda, R. Schiavilla, R.N. Pérez, J.E. Amaro, E.R. Arriola, Phys. Rev. C **91**, 024003 (2015)
37. A. Ekström, G. Baardsen, C. Forssén, G. Hagen, M. Hjorth-Jensen, G.R. Jansen, R. Machleidt, W. Nazarewicz, T. Papenbrock, J. Sarich, S.M. Wild, Phys. Rev. Lett. **110**, 192502 (2013)
38. A. Ekström, G.R. Jansen, K.A. Wendt, G. Hagen, T. Papenbrock, B.D. Carlsson, C. Forssén, M. Hjorth-Jensen, P. Navrátil, W. Nazarewicz, Phys. Rev. C **91**, 051301 (2015)
39. B.D. Carlsson, A. Ekström, C. Forssén, D.F. Strömberg, G.R. Jansen, O. Lilja, M. Lindby, B.A. Mattsson, K.A. Wendt, Phys. Rev. X **6**, 011019 (2016)
40. Y. Jaganathen, R.M. Id Betan, N. Michel, W. Nazarewicz, M. Płoszajczak, Phys. Rev. C **96**, 054316 (2017)
41. I. Sick, Phys. Rev. C **77**, 041302 (2008)
42. S. Okubo, R. Marshak, Ann. Phys. **4**, 166 (1958)
43. L. Eisenbud, E.P. Wigner, Proc. Nat. Acta. Sci. **27**, 281 (1941)
44. J.L. Gammel, R.M. Thaler, Phys. Rev. **107**, 291 (1957)
45. N. Michel, W. Nazarewicz, M. Płoszajczak, Phys. Rev. C **82**, 044315 (2010)
46. T. Nikšić, D. Vretenar, Phys. Rev. C **94**, 024333 (2016)
47. G.F. Bertsch, B. Sabbey, M. Uusnäkki, Phys. Rev. C **71**, 054311 (2005)

48. J. Toivanen, J. Dobaczewski, M. Kortelainen, K. Mizuyama, Phys. Rev. C **78**, 034306 (2008)
49. M. Kortelainen, J. Dobaczewski, K. Mizuyama, J. Toivanen, Phys. Rev. C **77**, 064307 (2008)
50. M. Stoitsov, M. Kortelainen, S.K. Bogner, T. Duguet, R.J. Furnstahl, B. Gebremariam, N. Schunck, Phys. Rev. C **82**, 054307 (2010)
51. T. Munson, J. Sarich, S.M. Wild, L.C. McInnes, Technical Memorandum ANL/MCS-TM-322, Argonne National Laboratory, Argonne (2012). http://www.mcs.anl.gov/tao
52. M. Kortelainen, T. Lesinski, J. Moré, W. Nazarewicz, J. Sarich, N. Schunck, M.V. Stoitsov, S.M. Wild, Phys. Rev. C **82**, 024313 (2010)
53. M. Kortelainen, J. McDonnell, W. Nazarewicz, P.G. Reinhard, J. Sarich, N. Schunck, M.V. Stoitsov, S.M. Wild, Phys. Rev. C **85**, 024304 (2012)
54. R.P. Feynman, Phys. Rev. **56**, 340 (1939)
55. J. Bond, F. Firk, Nucl. Phys. A **287**, 317 (1977)
56. L. Brown, W. Haeberli, W. Trächslin, Nucl. Phys. A **90**, 339 (1967)
57. P. Schwandt, T. Clegg, W. Haeberli, Nucl. Phys. A **163**, 432 (1971)
58. S. Sack, L.C. Biedenharn, G. Breit, Phys. Rev. **93**, 321 (1954)
59. D. Tilley, C. Cheves, J. Godwin, G. Hale, H. Hofmann, J. Kelley, C. Sheu, H. Weller, Nucl. Phys. A **708**, 3 (2002)
60. TUNL, Nuclear data evaluation project. http://www.tunl.duke.edu/nucldata
61. T. Al Kalanee, J. Gibelin, P. Roussel-Chomaz, N. Keeley, D. Beaumel, Y. Blumenfeld, B. Fernández-Domínguez, C. Force, L. Gaudefroy, A. Gillibert, J. Guillot, H. Iwasaki, S. Krupko, V. Lapoux, W. Mittig, X. Mougeot, L. Nalpas, E. Pollacco, K. Rusek, T. Roger, H. Savajols, N. de Séréville, S. Sidorchuk, D. Suzuki, I. Strojek, N.A. Orr, Phys. Rev. C **88**, 034301 (2013)
62. K. Fossez, J. Rotureau, W. Nazarewicz, Phys. Rev. C **98**, 061302 (2018)
63. X. Mao, J. Rotureau, W. Nazarewicz, N. Michel, R.M. Id Betan, Y. Jaganathen, Phys. Rev. C **102**, 024309 (2020)
64. C. Ordóñez, U. van Kolck, Phys. Lett. B **291**(4), 459 (1992)
65. P.F. Bedaque, U. van Kolck, Ann. Rev. Nucl. Part. Sci. **52**, 339 (2002)
66. P.F. Bedaque, H.W. Hammer, U. van Kolck, Phys.Lett. B **569**, 159 (2003)
67. I. Stetcu, J. Rotureau, B.R. Barrett, U. van Kolck, J. Phys. G **37**, 064033 (2010)
68. P. Capel, V. Durant, L. Huth, H.W. Hammer, D.R. Phillips, A. Schwenk, J. Phys. Conf. Ser. **1023**, 012010 (2018)
69. G.V. Rogachev, J.J. Kolata, A.S. Volya, F.D. Becchetti, Y. Chen, P.A. DeYoung, J. Lupton, Phys. Rev. C **75**, 014603 (2007)
70. T.B. Webb, S.M. Wang, K.W. Brown, R.J. Charity, J.M. Elson, J. Barney, G. Cerizza, Z. Chajecki, J. Estee, D.E.M. Hoff, S.A. Kuvin, W.G. Lynch, J. Manfredi, D. McNeel, P. Morfouace, W. Nazarewicz, C.D. Pruitt, C. Santamaria, J. Smith, L.G. Sobotka, S. Sweany, C.Y. Tsang, M.B. Tsang, A.H. Wuosmaa, Y. Zhang, K. Zhu, Phys. Rev. Lett. **122**, 122501 (2019)
71. R. Sherr, H.T. Fortune, Phys. Rev. C **87**, 054333 (2013)
72. H.T. Fortune, Phys. Rev. C **88**, 054623 (2013)
73. J. Hooker, G. Rogachev, V. Goldberg, E. Koshchiy, B. Roeder, H. Jayatissa, C. Hunt, C. Magana, S. Upadhyayula, E. Uberseder, A. Saastamoinen, Phys. Lett. B **769**, 62 (2017)
74. S.M. Wang, W. Nazarewicz, R.J. Charity, L.G. Sobotka, Phys. Rev. C **99**, 054302 (2019)
75. J.B. Ehrman, Phys. Rev. **81**, 412 (1951)
76. R.G. Thomas, Phys. Rev. **88**, 1109 (1952)
77. M. Zinser, F. Humbert, T. Nilsson, W. Schwab, T. Blaich, M.J.G. Borge, L.V. Chulkov, H. Eickhoff, T.W. Elze, H. Emling, B. Franzke, H. Freiesleben, H. Geissel, K. Grimm, D. Guillemaud-Mueller, P.G. Hansen, R. Holzmann, H. Irnich, B. Jonson, J.G. Keller, O. Klepper, H. Klingler, J.V. Kratz, R. Kulessa, D. Lambrecht, Y. Leifels, A. Magel, M. Mohar, A.C. Mueller, G. Münzenberg, F. Nickel, G. Nyman, A. Richter, K. Riisager, C. Scheidenberger, G. Schrieder, B.M. Sherrill, H. Simon, K. Stelzer, J. Stroth, O. Tengblad, W. Trautmann, E. Wajda, E. Zude, Phys. Rev. Lett. **75**, 1719 (1995)

78. M. Thoennessen, S. Yokoyama, A. Azhari, T. Baumann, J.A. Brown, A. Galonsky, P.G. Hansen, J.H. Kelley, R.A. Kryger, E. Ramakrishnan, P. Thirolf, Phys. Rev. C **59**, 111 (1999)
79. H. Jeppesen, A. Moro, U. Bergmann, M. Borge, J. Cederkäll, L. Fraile, H. Fynbo, J. Gómez-Camacho, H. Johansson, B. Jonson, M. Meister, T. Nilsson, G. Nyman, M. Pantea, K. Riisager, A. Richter, G. Schrieder, T. Sieber, O. Tengblad, E. Tengborn, M. Turrión, F. Wenander, Phys. Lett. B **642**, 449 (2006)
80. H. Simon, M. Meister, T. Aumann, M. Borge, L. Chulkov, U.D. Pramanik, T. Elze, H. Emling, C. Forssén, H. Geissel, M. Hellström, B. Jonson, J. Kratz, R. Kulessa, Y. Leifels, K. Markenroth, G. Münzenberg, F. Nickel, T. Nilsson, G. Nyman, A. Richter, K. Riisager, C. Scheidenberger, G. Schrieder, O. Tengblad, M. Zhukov, Nucl. Phys. A **791**, 267 (2007)
81. I.J. Thompson, M.V. Zhukov, Phys. Rev. C **49**, 1904 (1994)
82. R.M. Id Betan, R.J. Liotta, N. Sandulescu, T. Vertse, Phys. Lett. B **584**, 48 (2004)
83. M. Cavallaro, M. De Napoli, F. Cappuzzello, S.E.A. Orrigo, C. Agodi, M. Bondí, D. Carbone, A. Cunsolo, B. Davids, T. Davinson, A. Foti, N. Galinski, R. Kanungo, H. Lenske, C. Ruiz, A. Sanetullaev, Phys. Rev. Lett. **118**, 012701 (2017)
84. A. Moro, J. Casal, M. Gómez-Ramos, Phys. Lett. B **793**, 13 (2019)
85. F. Barranco, G. Potel, E. Vigezzi, R.A. Broglia, Phys. Rev. C **101**, 031305 (2020)
86. J. Smith, T. Baumann, J. Brown, P. DeYoung, N. Frank, J. Hinnefeld, Z. Kohley, B. Luther, B. Marks, A. Spyrou, S. Stephenson, M. Thoennessen, S. Williams, Nucl. Phys. A **940**, 235 (2015)
87. S.T. Belyaev, A.V. Smirnov, S.V. Tolokonnikov, S.A. Fayans, Sov. J. Nucl. Phys. **45**, 783 (1987)
88. J. Dobaczewski, W. Nazarewicz, T.R. Werner, J.F. Berger, C.R. Chinn, J. Dechargé, Phys. Rev. C **53**, 2809 (1996)
89. S. Fayans, S. Tolokonnikov, D. Zawischa, Phys. Lett. B **491**, 245 (2000)
90. M. Grasso, N. Sandulescu, N.V. Giai, R.J. Liotta, Phys. Rev. C **64**, 064321 (2001)
91. V. Rotival, T. Duguet, Phys. Rev. C **79**, 054308 (2009)
92. G. Bertsch, H. Esbensen, Ann. Phys. **209**, 327 (1991)
93. K. Hagino, H. Sagawa, Phys. Rev. C **72**, 044321 (2005)
94. G. Papadimitriou, A.T. Kruppa, N. Michel, W. Nazarewicz, M. Płoszajczak, J. Rotureau, Phys. Rev. C **84**, 051304(R) (2011)
95. S.M. Wang, N. Michel, W. Nazarewicz, F.R. Xu, Phys. Rev. C **96**, 044307 (2017)
96. W. Horiuchi, Y. Suzuki, Phys. Rev. C **76**, 024311 (2007)
97. T. Kopaleishvili, I. Vashakidze, V. Mamasakhlisov, G. Chilashvili, Nucl. Phys. **23**, 430 (1961)
98. P. Truoel, W. Bierter, Phys. Lett. B **29**, 21 (1969)
99. S. Saito, Prog. Theor. Phys. **41**, 705 (1969)
100. P. Navrátil, G.P. Kamuntavičius, B.R. Barrett, Phys. Rev. C **61**, 044001 (2000)
101. P. Navrátil, S. Quaglioni, G. Hupin, C. Romero-Redondo, A. Calci, Phys. Scr. **91**, 053002 (2016)
102. M. Fabre de la Ripelle, Ann. Phys. **147**, 281 (1983)
103. A. Kievsky, S. Rosati, M. Viviani, L.E. Marcucci, L. Girlanda, J. Phys. G **35**, 063101 (2008)
104. J. Raynal, J. Revai, Nuovo Cim. A **68**, 612 (1970)
105. G. Hagen, M. Hjorth-Jensen, N. Michel, Phys. Rev. C **73**, 064307 (2006)
106. P. Descouvemont, C. Daniel, D. Baye, Phys. Rev. C **67**, 044309 (2003)
107. P. Descouvemont, E. Tursunov, D. Baye, Nucl. Phys. A **765**, 370 (2006)
108. B. Gyarmati, T. Vertse, Nucl. Phys. A **160**, 523 (1971)
109. N. Michel, W. Nazarewicz, M. Płoszajczak, J. Okołowicz, Phys. Rev. C **67**, 054311 (2003)
110. N. Michel, Phys. Rev. C **83**, 034325 (2011)
111. D. Thompson, M. Lemere, Y. Tang, Nucl. Phys. A **286**, 53 (1977)
112. A.T. Kruppa, G. Papadimitriou, W. Nazarewicz, N. Michel, Phys. Rev. C **89**, 014330 (2014)
113. M.V. Zhukov, B.V. Danilin, D.V. Fedorov, J.M. Bang, I.J. Thompson, J.S. Vaagen, Phys. Rep. **231**, 151 (1993)
114. S.M. Wang, W. Nazarewicz, Phys. Rev. Lett. **120**, 212502 (2018)
115. Y. Kikuchi, K. Katō, T. Myo, M. Takashina, K. Ikeda, Phys. Rev. C **81**, 044308 (2010)

116. D. Dean, M. Hjorth-Jensen, Rev. Mod. Phys. **75**, 607 (2003)
117. R.W. Richardson, Phys. Lett. **3**, 277 (1963)
118. R.W. Richardson, N. Sherman, Nucl. Phys. **52**, 221 (1964)
119. J. Dukelsky, C. Esebbag, P. Schuck, Phys. Rev. Lett. **87**, 066403 (2001)
120. J. Dukelsky, S. Lerma, H.L.M. Robledo, R. Rodriguez-Guzman, S.M.A. Rombouts, Phys. Rev C **84**, 061301 (2011)
121. J. Dukelsky, V.G. Gueorguiev, P. Van Isacker, S. Dimitrova, H.B. Errea, S. Lerma, Phys. Rev. Lett. **96**, 072503 (2006)
122. S. Lerma, H.B. Errea, J. Dukelsky, W. Satuła, Phys. Rev. Lett. **99**, 032501 (2007)
123. M.C. Cambiaggio, A.M.F. Rivas, M. Saraceno, Nucl. Phys. A **624**, 157 (1997)
124. M. Hasegawa, K. Kaneko, Phys. Rev. C **67**, 024304 (2003)
125. R.M. Id Betan, Phys. Rev. C **85**, 064309 (2012)
126. N. Michel, W. Nazarewicz, M. Płoszajczak, K. Bennaceur, Phys. Rev. Lett. **89**, 042502 (2002)
127. R.M. Id Betan, R.J. Liotta, N. Sandulescu, T. Vertse, Phys. Rev. Lett. **89**, 042501 (2002)
128. T. Berggren, Nucl. Phys. A **109**, 265 (1968)
129. A. Mercenne, N. Michel, M. Płoszajczak, Acta Phys. Pol. B **47**, 967 (2016)
130. A. Mercenne, N. Michel, J. Dukelsky, M. Płoszajczak, Phys. Rev. C **95**, 024324 (2017)
131. S.M.A. Rombouts, J. Dukelsky, G. Ortiz, Phys. Rev. B **82**, 224510 (2010)
132. R.W. Richardson, J. Math. Phys. **6**, 1034 (1965)
133. J. Dukelsky, G.G. Dussel, J.G. Hirsch, P. Schuck, Nucl. Phys. A **714**, 63 (2003)
134. S.M.A. Rombouts, D. Van Neck, J. Dukelsky, Phys. Rev. C **69**, 061303(R) (2004)
135. I. Giaever, H.R. Zeller, Phys. Rev. Lett. **20**, 1504 (1968)
136. D.C. Ralph, C.T. Black, M. Tinkham, Phys. Rev. Lett. **74**, 3241 (1995)
137. C.T. Black, D.C. Ralph, M. Tinkham, Phys. Rev. Lett. **76**, 688 (1996)
138. A. Mercenne, Nuclear reactions in the Gamow shell model and solutions of the pairing Hamiltonian based on the rational Gaudin model. (Ph.D. Thesis, Université de Caen Normandie, 2017)
139. J. Aguilar, J. Combes, Commun. Math. Phys. **22**, 269 (1971)
140. E. Balslev, J. Combes, Commun. Math. Phys. **22**, 280 (1971)
141. N. Moiseyev, C. Corcoran, Phys. Rev. A **20**, 814 (1979)
142. J. Usukura, Y. Suzuki, Phys. Rev. A **66**, 010502 (2002)
143. S. Arias Laso, M. Horbatsch, Phys. Rev. A **94**, 053413 (2016)
144. K.B. Bravaya, D. Zuev, E. Epifanovsky, A.I. Krylov, J. Chem. Phys. **138**, 124106 (2013)
145. B. Gyarmati, A.T. Kruppa, Phys. Rev. C **34**, 95 (1986)
146. A. Csótó, Phys. Rev. C **49**, 2244 (1994)
147. S. Aoyama, K. Katō, K. Ikeda, Phys. Rev. C **55**, 2379 (1997)
148. H. Masui, K. Katō, K. Ikeda, Phys. Rev. C **73**, 034318 (2006)
149. H. Masui, K. Katō, K. Ikeda, Phys. Rev. C **75**, 034316 (2007)
150. T. Myo, K. Katō, K. Ikeda, Phys. Rev. C **76**, 054309 (2007)
151. T. Myo, Y. Kikuchi, K. Katō, Phys. Rev. C **85**, 034338 (2012)
152. H. Masui, K. Katō, N. Michel, M. Płoszajczak, Phys. Rev. C **89**, 044317 (2014)
153. A.T. Kruppa, G. Papadimitriou, W. Nazarewicz, N. Michel, Phys. Rev. C **89**, 014330 (2014)
154. E. Hiyama, R. Lazauskas, J. Carbonell, M. Kamimura, Phys. Rev. C **93**, 044004 (2016)
155. H. Kanada, T. Kaneko, S. Nagata, M. Nomoto, Prog. Theor. Phys. **61**, 1327 (1979)
156. S. Aoyama, S. Mukai, K. Katō, K. Ikeda, Prog. Theor. Phys. **93**, 99 (1995)
157. T. Kalamboukis, J. Comput. Phys. **53**(1), 82 (1984)
158. S. Sundar, B. Bhagavan, Comput. Math. Appl. **39**, 211 (2000)
159. G. Sleijpen, J. Booten, D. Fokkema, H. van der Vorst, BIT Num. Math. **36**, 595 (1996)
160. A. Csótó, B. Gyarmati, A.T. Kruppa, K.F. Pál, N. Moiseyev, Phys. Rev. A **41**, 3469 (1990)
161. R. Lefebvre, Phys. Rev. A **46**, 6071 (1992)
162. W.H. Press, S.A. Teukolsky, W.T. Vetterling, B.P. Flannery, *Numerical Recipes in C* (Cambridge University Press, 1988,1992)
163. A.N. Tikhonov, Sov. Math. Dokl. **4**, 1035 (1963)
164. P. Fröbrich, R. Lipperheide, *Theory of Nuclear Reactions* (Oxford Science Publications, Clarendon Press, Oxford, 1996)

# Wave Functions in the Vicinity of Particle-Emission Threshold

**7**

## 7.1 Configuration Mixing in the Vicinity of the Dissociation Threshold

What can be said about the properties of many-body states in the narrow range of energies around the reaction threshold? Are they universal, independent of any particular realization of the Hamiltonian? In other words, are the phenomena in the neighborhood of reaction thresholds sufficiently specific to be considered as a separate domain of interdisciplinary research, where nuclear phenomena share some properties with near-threshold phenomena in atomic physics, molecular physics or nanophysics?

The configuration mixing involving resonant states and background states of the nonresonant scattering continuum is a source of collective features such as, e.g. the resonance trapping [1–6] and super-radiance phenomenon [7, 8], multichannel coupling effects in reaction cross sections [9–11] and shell occupancies [12], modification of spectral fluctuations[13], and deviations from Porter-Thomas resonance widths distribution [6, 14, 15]. Another phenomenon which should be mentioned in this context is the formation of correlated structures (clusters) in states close to the cluster emission threshold. A comprehensive understanding of the universal occurrence of clustering is absent in the closed quantum system formulation of the nuclear many-body problem. The popular cluster model [16–22] is an a posteriori approach that assumes a pre-existence of the clusters. A priori type approaches, like the nuclear shell model, fail to predict the cluster states at observed low excitation energies around the cluster-decay thresholds. This failure to predict and describe cluster states in the closed quantum system framework is the central problem in nuclear theory.

Ikeda et al. [23] noticed that $\alpha$-cluster states can be found in the proximity of $\alpha$-particle decay thresholds. At that time, this important observation has not raised much interest in studies of the coupling to particle-decay channels as the general mechanism responsible for generating the correlations in near-threshold

© Springer International Publishing AG 2021
N. Michel, M. Płoszajczak, *Gamow Shell Model*, Lecture Notes in Physics 983,
https://doi.org/10.1007/978-3-030-69356-5_7

states which would resemble the degrees of freedom of the cluster model. Extensive studies in the shell model embedded in the continuum [24, 25] demonstrated that the troubling coexistence of cluster model and shell model wave functions in the description of low-energy excitations in the light nuclei, can be overcome in the open quantum system formulation of the shell model. These studies allowed to comprehend the phenomenological assertion of Ikeda et al. [23] and formulate its generalization, namely that the coupling to a nearby particle or cluster-decay channel induces particle or cluster correlations in shell model wave functions which are the imprint of this channel and do not originate from any particular property of the nuclear forces or any special symmetry of the nuclear many-body problem [24, 25]. The specific aspects of this generic phenomenon, depending, for example, on the interactions, are related to the energy and nature of various dissociation thresholds, and the stability of correlated multi-particle structures in the final state after the decay.

This generic phenomenon in open quantum systems should be seen also in atomic physics, molecular physics, or in nanoscience. However, it is in nuclear physics that those clustering effects are particularly interesting. This is due to the isospin structure of strong interactions which leads to a large number of different bound and unbound multi-nucleon systems and hence, to a rich spectrum of different particle or cluster emission thresholds. This in turn leads to a truly spectacular richness of correlations in near-threshold many-body states. Universality of the clustering phenomenon originates from basic properties of the scattering matrix in a multichannel system [26]. The decay threshold is a branching point of the scattering matrix. For energies below the lowest particle-decay threshold at energy $E_1$, one finds the analytic phase with a single solution which is regular in the entire space. Above $E_1$ and before the next decay channel opens at energy $E_2$, the new analytic phase corresponds to two regular solutions of the scattering problem, etc. Hence, one obtains a set of analytic phases, each one with a different number of regular scattering solutions. These phases are separated by decay thresholds (branching points) which are the non-analytic points of the scattering matrix.

Such a network is illustrated schematically in Fig. 7.1. For energies below the first decay threshold $E_{12_{C}\to3\alpha}$, the regular phase consists of elastic scattering solutions. For $E > E_{12_{C}\to3\alpha}$ and below the second threshold at energy $E_{12_{C}\to^{11}B+p}$, the regular phase contains two solutions coupled by unitarity, and the story repeats at higher excitation energies as more and more channels open up. The structure of any eigenfunction in this open quantum system depends on the total energy of the system which in turn determines the environment of decay channels accessible for couplings. With increasing excitation energy, the open quantum system network of couplings becomes more and more complex. The flux conservation (unitarity) implies that the mixing of shell model eigenfunctions changes if a new channel becomes available. The role of unitarity can also be seen on the example of spectroscopic factors [12] and radial overlap integrals [27] for near-threshold eigenfunctions (see a discussion in Sect. 7.4.2).

Microscopic description of states close to the branching point needs minimal assumptions about degrees of freedom involved but requires the unitary description

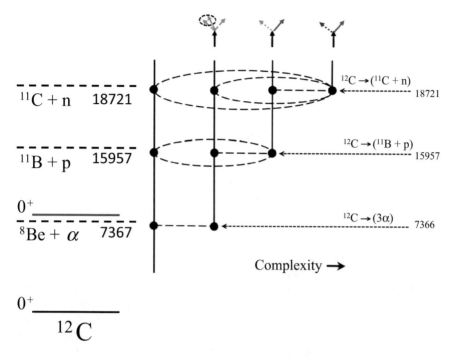

**Fig. 7.1** (Color online) The scheme of couplings in the multichannel representation of $^{12}$C. The structure of a near-threshold Hoyle resonance $0_2^+$ in $^{12}$C exhibits a strong imprint of the nearby $^{12}$C$\rightarrow^8$Be$+\alpha$ decay threshold. With the increasing excitation energy, subsequent decay channels open up, leading to an intricate multichannel network of couplings

of the transition between the two continuous phases. Proximity of the branching point at the threshold induces the collective mixing of shell model states, in which the essential role is played by a single eigenstate of the open quantum system Hamiltonian, the so-called aligned eigenstate. The strong configuration mixing, mediated by the aligned state which shares many features of the decay channel, is the essence of clustering phenomenon. It was conjectured in Refs. [24, 25] that this is a general mechanism by which clustering or correlated structures in the wave function appear in the vicinity of any charged or neutral particle-emission threshold. Hence, the presence of cluster states near their corresponding cluster emission thresholds is a signature of a profound change of the near-threshold shell model wave function and the direct manifestation of the continuum-coupling induced correlations. The mechanism responsible for the creation of an aligned state is mathematically similar to the formation mechanism of collective super-radiant or trapped states [1–5, 7, 8]. However, the domain of aligned states is not restricted to the region of large density of resonances in the continuum. It can even correspond to a bound state at energy below the lowest decay threshold. For example, neutral-cluster configurations are expected primarily below the threshold due to the rapid growth

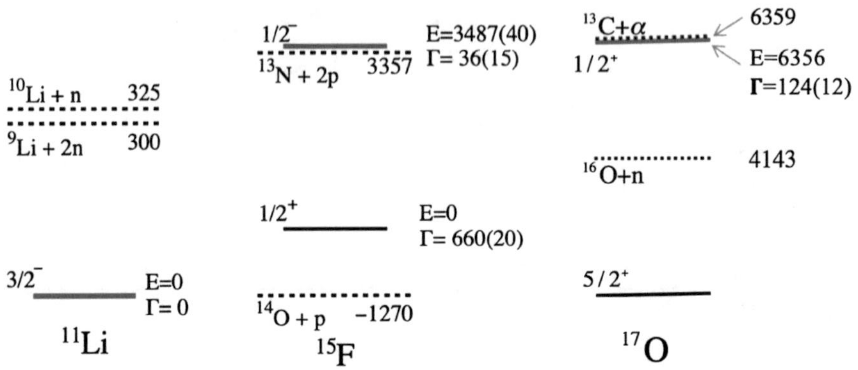

$^{13}$C+α .......... 6359

1/2⁻ ——— E=3487(40)    1/2⁺    E=6356
.............. 
$^{13}$N + 2p  3357  Γ= 36(15)       Γ=124(12)

$^{10}$Li + n    325
............
$^9$Li + 2n    300

$^{16}$O+n  ...............  4143

1/2⁺ ——— E=0
Γ= 660(20)

3/2⁻ ——— E=0       5/2⁺ ———
Γ= 0
............
$^{14}$O + p   −1270

$^{11}$Li       $^{15}$F       $^{17}$O

**Fig. 7.2** (Color online) Selected low-lying states and particle-decay thresholds (all in keV) in $^{11}$Li [28], $^{15}$F [29], and $^{17}$O [28]. The key near-threshold "aligned" states are indicated

of the decay width with energy. The spectacular examples of neutral clustering are one- and two-neutron halos in light nuclei. Favorable conditions for the appearance of charged-cluster configurations are above the charged-cluster emission threshold.

Few typical examples of near-threshold aligned states are shown in Fig. 7.2. The ground state of the Borromean halo nucleus $^{11}$Li (see Fig. 7.2) resembles the [$^9$Li$_{g.s.}$ ⊗ 2n] configuration of the nearest 2n-emission threshold rather that of the [$^{10}$Li$_{g.s.}$ ⊗n] state [30]. A similar situation happens in the mirror nucleus $^{11}$O, which is a 2p emitter [31]. The first $1/2^-$ state in $^{15}$F is a narrow resonance slightly above the 2p-emission threshold, well above the Coulomb barrier and the 1p-emission threshold. The dominant configuration in this state is [$^{13}$N$_{g.s.}$ ⊗ 2p] of the nearby 2p-emission channel and not that of [$^{14}$O$_{g.s.}$ ⊗ p], what explains its narrow width. More about this nucleus and the $1/2_1^-$ resonance can be found in Sect. 9.4.1.2. The last example in Fig. 7.2 is the narrow $1/2_2^+$ resonance situated 3 keV below the α-emission threshold. The alignment of this state with a nearby α-particle-emission threshold leads to a significant decrease of the neutron decay width in a channel $^{16}$O + n and enables slow neutron-capture process. Narrow resonances, both on and off the nucleosynthesis path are not fortuitously close to the particle-emission threshold but appear as a result of the generic in open quantum systems, mechanism of the collectivization of wave functions and the alignment with the nearby decay channel. Their frequent existence may significantly influence the nucleosynthesis path changing the branching ratios for capture and decay processes.

Excited new opportunity is a possible existence of exotic multi-neutron correlations in the states close to the corresponding multi-neutron emission thresholds. In Refs [32, 33] the observation of four correlated neutrons have been suggested. Subsequent theoretical studies concluded that narrow tetraneutron resonance is incompatible with the present understanding of nuclear forces [34–37]. Isospin structure of the nuclear force prevents that the tetraneutron separation energy $S_{4n}$ is less than one neutron $S_{1n}$ or two-neutron $S_{2n}$ separation energies. However, the nature of nuclear forces does not preclude the tetraneutron correlations in the vicinity of the 4n-emission threshold as a result of the collective coupling of many-body resonances via the 4n-emission channel. Tetraneutron correlations could exist, for example, in the ground states of $^{10}$He, $^{16}$Be, $^{19}$B, and in the low-energy continuum of $^{8}$He, $^{14}$Be, $^{22}$C. Manifestation of tetraneutron correlations in heavier nuclei is plausible in relatively high partial waves ($\ell \leq 3$) [38].

## 7.2  Halo Nuclei

As one approaches the particle driplines, the amount of spectroscopic information concerning bound states becomes scarce, and one relies on the analysis of decaying states and reaction dynamics to extract nuclear structure data. A halo in the spatial distribution of loosely bound neutron-rich nuclei appears [39, 40], and its radial extension for neutrons in low-$\ell$ orbits is limited only by the two-body correlations which cause an anti-halo effect [41, 42]. Similar phenomena related to the opening of quantum system appear, e.g. in $^{4}$He$_N$ atomic clusters where giant loosely bound clusters have been observed [43]. Theoretically, a proper microscopic description of both the many-body correlations and the continuum of positive energy states and decay channels is mandatory for these systems. Experimentally, a halo structure is noticed by an increase of the matter radius, larger by a sizable factor than the typical value of $1.25\,A^{1/3}$ fm encountered in well-bound nuclei [44]. The first halo nucleus to have been discovered is that of the $^{11}$Li ground state. $^{11}$Li has been studied at Lawrence Berkeley Laboratory by Tanihata et al. [39, 45], where a very large cross section involving neutron-rich helium and lithium isotopes was measured. This pointed out to the halo structure of the $^{11}$Li ground state, which, however, demanded another 20 years to be confirmed [46]. Interestingly, first halo nucleus $^{6}$He had been produced already in 1936, at the early age of quantum mechanics, using neutron beams on a $^{9}$Be target [47].

Halo states can exist only because of their quantum nature. Indeed, they consist of a few loosely bound nucleons, in most cases one or two, moving outside a well-bound core, and whose wave function extends in the classically forbidden region. The simplest example of a halo nucleus is that of a weakly bound neutron in $\ell = 0$ state moving in a spherical potential generated by the core. Due to the very slow exponential decay of the neutron wave function, the probability to find a neutron in the classically forbidden region will be large.

Several factors have to be gathered in order for halo to develop. Firstly, the considered nuclear state must be loosely bound, as otherwise all nucleons remain close to the center of the nucleus. Hence, halo is a phenomenon in the vicinity of particle-emission threshold. Moreover, the wave function of nucleons in the halo has to be formed mainly by partial waves of small orbital angular momentum. Indeed, partial waves of large orbital angular momentum have an important confining effect due to the centrifugal barrier, so that the least bound nucleons remain trapped, preventing the halo from appearing. Additionally, two-neutron halos have been noticed to arise in the so-called Borromean nuclei where all $A - 1$ subsystems are unbound but $A$ and $A - 2$ systems are bound because of the residual nucleon–

nucleon interaction in the continuum. Consequently, halo states in Borromean nuclei consist of a two-neutron halo surrounding a well-bound $A - 2$ core. Even more complicate halo configurations are possible, as can be seen in $^6$He and $^8$He. These two nuclei are Borromean, because they are bound, on the one hand, while the $^7$He and $^5$He nuclei and dineutron are neutron unbound, on the other hand. Moreover, the two-neutron separation energy is larger in $^8$He than in $^6$He, which is an abnormal situation. Indeed, neutron-rich isotopes are typically less and less bound when neutrons are added. In fact, the unusually large binding energy of $^8$He is due only to correlations. This arises because $^8$He consists mainly of a $^4$He $+0p_{3/2}^4$ neutron configuration, so that the neutrons in the $0p3/2$ shell strongly interact due to the closed-shell character of the $p_{3/2}$ shell at Hartree Fock level in the $^8$He ground state.

However, it is not sufficient to have either loosely bound nucleons occupying low-$\ell$ partial waves or the neutron emission thresholds set as in Borromean nuclei to form a halo state. In fact, it is also necessary for external nucleons to sufficiently decouple from the well-bound core, so that pairing correlations between nucleons inside and outside the core are significantly reduced. Otherwise, as one will see later in more details, the pairing anti-halo effect arise, which prevents formation of the giant halo in most states lying close to the particle-emission threshold.

Typical hint of the presence of halos for energies near particle-emission threshold are important correlations among nucleons in the asymptotic zone, as is the case for $^{11}$Li [48]. Large electromagnetic transitions are also typical features of halo nuclei, as they are enhanced by the unusually large density of nucleons in the asymptotic region [46]. A striking example is the strong dipole transition occurring between the first excited state $1/2^-$ and the ground state $1/2^+$ of $^{11}$Be [28]. A simple but accurate model to describe these states is that of a neutron interacting with a potential mimicking a $^{10}$Be core. The $1/2_1^+$ and $1/2_1^-$ states are essentially one neutron $1s_{1/2}$ and $0p_{1/2}$ states coupled to a $^{10}$Be core (see Sect. 4.3.4.3). Hence, the E1 electromagnetic transition is mainly proportional to the square of the integral of $r$ times the radial wave functions of the $1s_{1/2}$ and $0p_{1/2}$ states. Unusually large observed value of this transition implies that neutron one-body states extend significantly in space, i.e. $^{11}$Be is a halo nucleus.

Halo states have been discovered in the ground states of light nuclei, such as $^{6,8}$He, $^{11}$Be, and $^{11}$Li. Also $^{14}$Be, $^{14}$B, $^{15}$C, and $^{19}$C have been confirmed to be halo nuclei, whereas $^{15}$B, $^{17}$B, $^{19}$B, $^{22}$C, and $^{23}$O exhibit only certain features of halo configuration [46, 49]. Even though most halo nuclei are neutron halo states, proton halos also exist. One can cite examples of the one-proton halos in the ground states of $^8$B and $^{13}$N and of the first excited state of $^{17}$F, as well as the Borromean two-proton configuration in the ground state of $^{17}$Ne. However, due the presence of the Coulomb barrier, proton halo states are less probable to occur and bear a reduced spatial extension [46].

Halo states also exist in other quantum systems, outside the domain of nuclear physics. In atomic clusters, halo configurations have been observed in $^4$He$_2$ dimer [50, 51] and $^4$He$_2$–$^3$He trimer [50, 52]. The $^4$He$_2$ dimer has a very large root-mean square radius, of 50 Å [50], while that of the $^4$He$_2$–$^3$He trimer is relatively smaller, of 13 Å [50]. Spin-polarized triton is also suspected to form halos [53]. We have seen in Sect. 4.3 that electron halos can be found in multipolar anions, where extended electronic wave functions develop close to the detachment threshold.

### 7.2.1 Halo States in a Single-Particle Potential

The simplest model which can be used to describe halo configuration is that of a spherical potential, mimicking a well-bound core, with which a loosely bound proton or neutron interact. This model discards nucleon - nucleon correlations, but is useful to identify the partial waves which can give rise to the halo structure. Due to the confinement induced by the centrifugal barrier, a neutron halo can develop if the partial waves of lowest angular momentum are mainly occupied. Proton halos are also expected to occur in light nuclei. Indeed, the Coulomb barrier has to be sufficiently small not to confine protons inside the nuclear zone. Moreover, like in the case of neutron halo, low angular momentum partial waves should be mainly involved in proton halo.

The asymptotic behavior of single-particle states is obviously determinant for the description of one-particle halo states. Hence, for $r \geq R$ where $R$ is a radius outside the nucleus, one writes single-particle states of orbital angular momentum $\ell$ and total angular momentum $j$ the following way:

$$u_{\ell j}(r) = \beta_{\ell j} W_{-\eta,\ell+1/2}(2\kappa r) . \tag{7.1}$$

In this expression, $\beta_{\ell j}$ is the single-particle asymptotic normalization coefficient, $W_{-\eta,\ell+1/2}(2\kappa r)$ is the Whittaker function and $k = i\kappa$ (see Sect. 2.5). One does not use the outgoing Coulomb wave function $H^+_{\ell,\eta_k}(kr)$, as in Eq. (2.7) because it is more convenient to use the Whittaker function for bound states, as it is real for negative real energies:

$$W_{-\eta,\ell+1/2}(2\kappa r) = H^+_{\ell,\eta_k}(kr)e^{i\frac{\pi\eta}{2}+i\frac{\pi\ell}{2}-i\sigma_\ell(\eta)} , \tag{7.2}$$

where $\eta_k = -i\eta$ and $\sigma_\ell(\eta)$ is the Coulomb phase shift (see Sect. 2.3). and where one uses the notation $k = i\kappa$ (see Sect. 2.5), as $\kappa \geq 0$ for bound states.

#### 7.2.1.1 Neutron Halo States in a Spherical Potential

Let us consider neutron single-particle states. One will firstly show that neutron states of large angular momentum, i.e. for $\ell \geq 2$ typically, cannot generate a halo in a potential model. For simplicity, one assumes that the potential is of Woods–Saxon type and possesses a bound state at zero energy (see Sect. 2.5). Its wave function $u(k = 0, r)$ is then a very simple analytic function in the asymptotic region (see Eq. (2.128)):

$$u(k = 0, r) = A \, r^{-\ell} , \qquad r \geq R, \tag{7.3}$$

where $A$ is a normalization constant. The root-mean-square radius of $u(k = 0, r)$ is finite, as $r^2u(k = 0, r)^2 = \mathcal{O}(r^{-2\ell+2})$ for $r \rightarrow +\infty$, so that $r^2u(k = 0, r)^2$ is integrable.

A useful inequality verified by the root-mean-square radius when $\ell \geq 2$ can be obtained if one assumes that the Woods–Saxon potential mimicking the core potential vanishes after $R_0 + 2d$, with $R_0$ the radius of the Woods–Saxon potential and $d$ its diffuseness. Indeed, the Woods–Saxon potential is smaller than 12% of its maximal value for $r > R_0 + 2d$, which is sufficient to devise a qualitative estimate. Let us consider the expectation value of $r^2$ in this case:

$$\left( \int_0^{+\infty} r^2 u(k = 0, r)^2 \, dr \right) \left( \int_0^{+\infty} u(k = 0, r)^2 \, dr \right)^{-1}$$

$$= \left( \int_0^{R_0+2d} r^2 u(k = 0, r)^2 \, dr + A^2 \int_{R_0+2d}^{+\infty} r^{-2\ell+2} \, dr \right)$$

$$\times \left( \int_0^{R_0+2d} u(k = 0, r)^2 \, dr + A^2 \int_{R_0+2d}^{+\infty} r^{-2\ell} \, dr \right)^{-1}$$

$$\leq (R_0 + 2d)^2 \left( \int_0^{R_0+2d} u(k = 0, r)^2 \, dr + \frac{A^2(R_0 + 2d)^{-2\ell+1}}{2\ell - 3} \right)$$

$$\times \left( \int_0^{R_0+2d} u(k = 0, r)^2 \, dr + \frac{A^2(R_0 + 2d)^{-2\ell+1}}{2\ell - 1} \right)^{-1}$$

$$\leq \left( \frac{2\ell - 1}{2\ell - 3} \right) \left( R_0 + 2d \right)^2 , \tag{7.4}$$

where to suppress the $A$-dependence, the normalization integral of wave function $u(k = 0, r)$ has been explicitly considered. One readily obtains an inequality for the root-mean-square radius:

$$\langle r^2 \rangle^{1/2} \leq \sqrt{\frac{2\ell - 1}{2\ell - 3}} \left( R_0 + 2d \right) . \tag{7.5}$$

Thus, $\langle r^2 \rangle^{1/2}$ converges to a finite radius smaller or of the order of $R_0 + 2d$ if $\kappa \rightarrow 0$ in Eq. (7.5). Moreover, neglected short-range correlations may diminish nuclear root-mean-square radius even more, as will be discussed in Sect. 7.2.2. Consequently, in practice, a neutron halo cannot develop for angular momenta $\ell \geq 2$. Conversely, as we will now demonstrate, the root-mean-square radius of neutron states with $\ell \leq 1$ becomes infinite at the neutron dissociation threshold.

Let us firstly consider $\ell = 0$ partial wave. Using Eq. (7.1), that $u(k, r)$ for $\ell = 0$, $j = 1/2$ and $r \geq R$ reads

$$u(k, r) = \beta_{\ell j} \sqrt{\kappa} \exp(-\kappa r) , \tag{7.6}$$

where $\beta_{\ell j}$ is a normalization constant, chosen so that it remains finite and nonzero when $\kappa \to 0$. Indeed, as $\exp(-\kappa r) \to 1$ when $\kappa \to 0$, the part of the normalization integral involving $u(k, r)$ from 0 to $R$ becomes negligible, so that one obtains

$$\beta_{\ell j} \sim \sqrt{2}. \tag{7.7}$$

The expectation value of $r^2$ when $\kappa \to 0$ follows:

$$\int_0^{+\infty} r^2 u(k, r)^2 \, dr \sim 2\kappa \left( \int_R^{+\infty} r^2 \exp(-2\kappa r) \, dr \right) \sim \frac{1}{2\kappa^2}. \tag{7.8}$$

Similarly, an analogous relation is found for neutron $\ell = 1$ partial wave when $\kappa \to 0$. From Eq. (2.128, $u(k, r)$ for $\ell = 1$, $j = 1/2, 3/2$ and $r \geq R$ reads

$$u(k, r) = \beta_{\ell j} \exp(-\kappa r)(\kappa + 1/r), \qquad r \geq R, \tag{7.9}$$

where $\beta_{\ell j}$ is a normalization constant, which remains finite and nonzero when $\kappa \to 0$ as $u(k = 0, r)$ is integrable. Hence:

$$\beta_{\ell j} \to \beta_{\ell j}^{(0)}, \tag{7.10}$$

where $\beta_{\ell j}^{(0)}$ is the value of $\beta_{\ell j}$ when $\kappa = 0$. The expectation value of $r^2$ for $\kappa \to 0$ is straightforward to calculate

$$\int_0^{+\infty} r^2 u(k, r)^2 \, dr \sim \beta_{\ell j}^2 \int_R^{+\infty} \exp(-2\kappa r) \, (\kappa^2 r^2 + 2\kappa r + 1) \, dr \sim \left( \frac{5}{4\kappa} \right) \left( \beta_{\ell j}^{(0)} \right)^2, \tag{7.11}$$

where the finite part of the expectation value of $r^2$ from 0 to $R$, is neglected as it becomes negligible compared to its diverging asymptotic part. One then readily obtains asymptotic limits of root-mean-square radii of neutron whose orbital angular momentum verify $\ell \leq 1$:

$$\langle r^2 \rangle^{1/2} \sim \sqrt{\frac{1}{2}} \kappa^{-1}, \qquad \ell = 0, j = 1/2 \tag{7.12}$$

$$\langle r^2 \rangle^{1/2} \sim \sqrt{\frac{5}{4}} \left| \beta_{\ell j}^{(0)} \right| \kappa^{-1/2}, \qquad \ell = 1, j = 1/2, 3/2. \tag{7.13}$$

Contrary to the case for which $\ell \geq 2$, the root-mean-square radius of neutron level with $\ell \leq 1$ is diverging when $\kappa \to 0$. The effect of the centrifugal barrier for $\ell = 1$ is nevertheless visible, as the divergence of $\langle r^2 \rangle^{1/2}$, when $\kappa \to 0$, is slower than that occurring for $\ell = 0$. One can also note that the dependence on parameters of the potential disappears asymptotically for $\ell = 0$, whereas for $\ell = 1$

this dependence remains through the normalization constant $\beta_{\ell j}^{(0)}$. Many-body states whose dominant configurations possess weakly bound neutron bearing $\ell \leq 1$ are then good candidates for halo states.

### 7.2.1.2 Proton Halo States in a Spherical Potential

Even in the presence of a Coulomb barrier, it is possible for proton states to have a sizable probability distribution in the asymptotic, classically forbidden region. Due to the more complex structure of Coulomb wave functions compared to Hankel functions (see Sect. 2.3), it is necessary to consider different regimes dictated by the binding energy of the weakly bound proton state. For this, one will carefully state the transition from Sommerfeld parameter $\eta \sim 0$, for large binding energy, to $|\eta| \to +\infty$, close to proton emission threshold. Indeed, as one has seen in Sect. 2.3, the vicinity of $k = 0$ is an essential singularity in the complex plane.

For a discussion of the limiting cases, it is useful to introduce the complex turning point:

$$z_t = -i \left( \eta + \sqrt{\eta^2 + \ell(\ell + 1)} \right) . \tag{7.14}$$

At the turning point:

$$d^2 H^+(\ell, \eta_k, z)/dz^2 = 0 .$$

Using standard properties of the Coulomb wave functions (see Sect. 2.3.3 and Ref. [54]), one derives the asymptotic expressions:

$$W_{-\eta, \ell+1/2}(2\kappa r) \simeq \left( \frac{\kappa r}{2\eta} \right)^{1/4} \exp \left( \eta - \eta \ln(\eta) - 2\sqrt{2\eta\kappa r} \right)$$

$$\text{for} \quad \kappa r \ll |z_t| , \eta \to +\infty \tag{7.15}$$

$$W_{-\eta, \ell+1/2}(2\kappa r) \simeq \exp \left( -\kappa r - \eta \ln(2\kappa r) \right) \quad \text{for} \quad \kappa r \gg |z_t| . \tag{7.16}$$

Let us first consider the limit given in Eq. (7.15), which corresponds to very small separation energies. The asymptotic part of the single-particle wave function near the particle-emission threshold is

$$u_{\ell j}(r) \simeq \beta_{\ell j} \left( \frac{\kappa_0 r}{2\eta^2} \right)^{1/4} \exp \left( \eta - \eta \ln(\eta) - 2\sqrt{2\kappa_0 r} \right) , \tag{7.17}$$

where $\eta = \kappa_0/\kappa$ and $\kappa_0 = Z_b Z_c e^2 \mu/\hbar^2$. $\beta_{\ell j}$ in this expression has been defined in Eq. (7.1). As the external norm (7.39) must be finite, and $\exp[\eta - \eta \ln(\eta)]/\eta^{1/2} \to 0$ for $\eta \to +\infty$, therefore $\beta_{\ell j}$ in the limit of very weak binding must exhibit the universal $\eta$-dependence:

$$\tilde{\beta}_{\ell j}(\eta) = N_\beta(\ell, j)\eta^{1/2} \exp(\eta \ln \eta - \eta) . \tag{7.18}$$

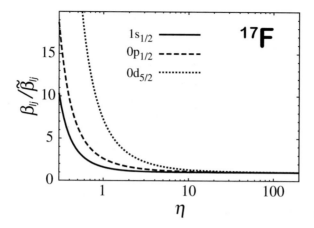

**Fig. 7.3** The dependence of $\beta_{\ell j}/\tilde{\beta}_{\ell j}$ on $\eta-$ for $1s_{1/2}, 0p_{1/2}$, and $0d_{5/2}$ single-particle proton wave functions in $^{17}F$ (from Ref. [55])

$N_\beta(\ell, j)$ in (7.18) is a pre-factor that depends on the structure of the single-particle state, in particular $\ell$.

To assess how quickly the limit (7.18) is reached, one will show results of calculations for the $1s_{1/2}$, $0p_{1/2}$, and $0d_{5/2}$ single-proton states in $^{17}F$. The single-particle radial wave function $u_{\ell j}$ is calculated using the Woods–Saxon potential. The Coulomb potential is given by a spherical uniform charge distribution with the radius $R_0$. For each $\ell$, $j$, the depth of the central potential is adjusted to the proton separation energy in $^{17}F$, which corresponds to a given value of $\eta$. The value of $N_\beta(\ell, j)$ can then be extracted from the calculated wave function at $\eta > 100$.

Figure 7.3 shows the $\eta$-dependence of the ratio $\beta_{\ell j}/\tilde{\beta}_{\ell j}$. One can see that the asymptotic behavior of this ratio is reached at $\eta \approx 1$ for $\ell = 0$ and $\eta \approx 10$ for $\ell = 2$. The ratio $\beta_{\ell j}/\tilde{\beta}_{\ell j}$ decreases smoothly with $\eta$ suggesting a polynomial dependence on the proton separation energy, or a Maclaurin series in $1/\eta$:

$$\beta_{n\ell j}(\eta) = \eta^{1/2} \exp(\eta \ln \eta - \eta) \sum_{i=0}^{\infty} \frac{a_i(n, \ell, j)}{\eta^i} . \tag{7.19}$$

The term $\eta^{1/2} \exp(\eta \ln \eta - \eta)$ which is responsible for the rapid growth of $\beta_{n\ell j}(\eta)$ around the threshold, is universal for all charged particles, independent of their quantum state. The dependence on quantum numbers $(n, \ell, j)$ of the bound state is contained in the coefficients $a_i$ of the Maclaurin series.

The limit (7.16) corresponds to finite separation energies that are sufficiently small so that one can neglect the $\eta$-variations of $\mathcal{N}_{\text{int}}$. In this case, the asymptotic part of the wave function:

$$u_{\ell j}(r) \simeq \beta_{\ell j} \exp\left(-\kappa_0 r/\eta - \eta \ln(2\kappa_0 r/\eta)\right) \tag{7.20}$$

shows the usual exponential decay. In order to keep $\mathcal{N}_{\text{ext}}$ finite when $\eta$ decreases, $\beta_{\ell j}$ has to increase. Consequently, one can expect a minimum of $\beta_{\ell j}$ as a function of $\eta$ going from small values of $\eta$ toward the threshold ($\eta = +\infty$).

Let us now discuss the behavior of $\beta_{\ell,j}$ in the full energy range. Figure 7.4 shows $\beta_{\ell j}(\eta)$ for the p+$^{16}$O capture reaction at $E_{\text{cm}} = 0$ in the channels where proton has the relative angular momentum $\ell = 0$ ($1s_{1/2}$) (panel (a)), $\ell = 1$ ($0p_{1/2}$ and $0p_{3/2}$) (panel (b)), or $\ell = 2$ ($0d_{5/2}$) (panel (c)). On may see that the magnitude of $\beta_{\ell,j}$ decreases with $\ell$. With increasing $\eta$, $\beta_{\ell,j}$ first decreases until a minimum value $\eta = \eta_{\text{crit}}$, and then increases again as the separation energy decreases towards the threshold ($S_{\text{p}}^{(a)} = 0$). One may thus conclude that $\eta_{\text{crit}}$ separates the regimes of strong ($\eta < \eta_{\text{crit}}$) and weak ($\eta > \eta_{\text{crit}}$) binding for a given partial wave and $Z_a$. In general, $\eta_{\text{crit}}$ scales almost linearly with $Z_b$ and strongly depends on the proton angular momentum $\ell$. The dependence of $\eta_{\text{crit}}$ on the channel angular momentum $j$ is weak.

**Fig. 7.4** Energy dependence of $\beta_{\ell j}$ in $^{17}$F for the bound state proton single-particle wave functions: (**a**) $1s_{1/2}$; (**b**) $0p_{1/2}$ and $0p_{3/2}$; and (**c**) $0d_{5/2}$ (from Ref. [55])

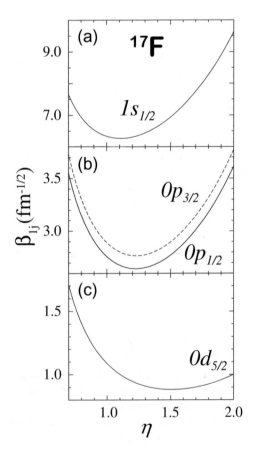

The case of $\ell = 0$ and large $\eta$ in Fig. 7.4a is characteristic of a proton halo. However, as seen in Fig. 7.4, the large values of $\beta_{\ell,j}$ are also expected for very bound states, so that a large $\beta_{n\ell j}(\eta)$ is not a sufficient indicator of a proton halo. Limiting regimes for $\beta_{n\ell j}(\eta)$ can be characterized by the complex turning point $z_t$ (see Eq. (7.14)). The low-binding regime, which is reached for $\kappa r \ll |z_t|$ and $\eta \to +\infty$, is characterized by the condition:

$$\left(\eta + \sqrt{\eta^2 + \ell(\ell+1)}\right)\eta \gg \kappa_0 R_f . \tag{7.21}$$

In this regime, the proton wave function behaves asymptotically as $u_{\ell j}(r) \propto r^{1/4} \exp(-2\sqrt{2\kappa_0 r})$, i.e. its decay is slower than exponential. The strong-binding regime is reached at $\kappa r \gg |z_t|$:

$$\left(\eta + \sqrt{\eta^2 + \ell(\ell+1)}\right)\eta \ll \kappa_0 R_f . \tag{7.22}$$

The proton wave function in the strong-binding regime shows the expected exponential decay (7.17).

The charged particle radiative capture cross sections can be discussed in the two limits: (i) $\eta_{CM} \to \infty$ ($E_{CM} \to 0$), and (ii) $\eta \to \infty$ ($S_c^{(a)} \to 0$). The center-of-mass energy of the system $\langle b + c \rangle$ can be varied experimentally. On the contrary, the charged particle separation energy is fixed for any state of the $a$-nucleus. Therefore, the limiting behavior of the radiative capture cross section when $\eta \to \infty$ cannot be studied experimentally in a single physical system. In the first limit ($\eta_{CM} \to \infty$), the radiative capture cross section $\sigma_{c\gamma}(\eta_{CM}, \eta)$ for a fixed value of $\eta$ is exponentially reduced :

$$\lim_{\eta_{CM} \to \infty} \sigma_{c\gamma} = \mathscr{S}_{bc}(\eta_{CM}, \eta) \frac{\exp(-2\pi\eta_{CM})}{E_{CM}} . \tag{7.23}$$

The second limit ($\eta \to \infty$), the radiative capture cross section $\sigma_{c\gamma}(\eta_{CM}, \eta)$ for a fixed value of $\eta_{CM}$ diverges as:

$$\lim_{\eta \to \infty} \sigma_{c\gamma} \propto \mathscr{S}_{bc}(\eta_{CM}, \eta) \propto \eta \exp(2\eta \ln(\eta) - 2\eta) . \tag{7.24}$$

Both asymptotic behaviors of $\sigma_{c\gamma}$ hold for any charged particle radiative capture reaction.

Figure 7.5 surveys the values of $\beta_{\ell,j}$ and $\eta$ for the proton emission channels with $\ell = 1$ ($p$-shell) and $\ell = 2$ ($d$-shell) in left and right panels, respectively. One can see that the $Z_a$-dependence of $\eta_{\text{crit}}$ is weak in both cases. All $p$-shell nuclei in Fig. 7.5 with the exception of the proton halo nucleus $^8$B, belong to the class $\eta < \eta_{\text{crit}}$. The ground state of $^{12}$N has $\eta = \eta_{\text{crit}}$. The neutron-rich nuclei have very small $\eta$-values in the ground state so their normalization constants $\beta_{\ell=1,j}(\eta)$ are large.

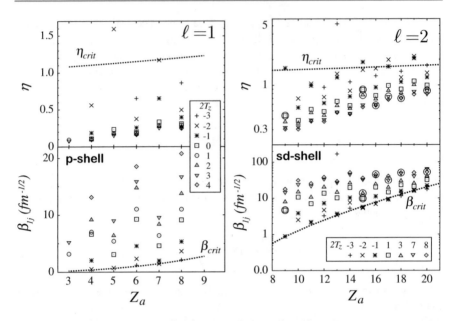

**Fig. 7.5** Experimental values of $\eta$ (top) and $\beta_{\ell,j}$ (bottom) for selected bound proton states. The left panel shows results for $\ell = 1$ and $j = 1/2, 3/2$ in various $p$-shell nuclei ($Z_a, T_z$). The right panel presents results for $\ell = 2$ bound proton wave functions in $sd$-shell nuclei. For certain nuclei, the radiative proton captured with the channel angular momentum $j = 3/2, 5/2$ cannot populate the ground state of a nucleus $a$. In such cases (encircled symbols), one plots $\eta$ and $\beta_{\ell,j}$ for the lowest-energy excited state that is populated by the capture with $\ell = 2$ protons. The dotted lines of $\eta_{crit}$ and $\beta_{crit}$ are calculated for $T_z = -1/2$ nuclei (from Ref. [55])

An odd-even staggering of $\eta$, due to pairing correlations, leads to an odd-even effect in $\beta_{\ell=1,j}(Z_a)$. The odd-even staggering of $\eta$ and $\beta_{\ell,j}$ is seen for both proton-rich and neutron-rich systems. The staggering in $\beta_{\ell j}$ is stronger for nuclei with $T_z \geq 0$ than for proton-rich nuclei with $T_z < 0$.

The data for $sd$-shell nuclei shown (see the right panel in Fig. 7.5) exhibit a similar behavior. Most of the particle-stable $sd$-shell nuclei are in the class $\eta < \eta_{crit}$. The only system that is in the regime of weak binding is the $5/2^+$ ground state of $^{23}$Al, which has a small proton separation energy ($S_p = 141$ keV) and a very large value of $\beta_{\ell,j}$ [56]. Other nuclei with $\eta > \eta_{crit}$ are $^{29}$P, $^{32}$Cl, and $^{36,37}$K, whereas $^{17}$F, $^{30}$S, $^{33}$Cl, and $^{37}$Ca are situated at the critical line $\eta_{crit}(Z_a)$.

The $\beta_{n\ell j}(\eta)$ values of single-particle asymptotic normalization coefficients for proton-rich nuclei remain close to $\beta_{crit}$. The transition from $\eta < \eta_{crit}$ to $\eta > \eta_{crit}$ can be further explored by studying in a given nucleus different particle-stable excited states populated in the capture reaction which are having different values of $\eta$.

### 7.2.1.3 Neutron Halo States in Deformed Nuclei

In the previous sections, we have discussed nuclear halo in spherical nuclei. However, as one saw in Sect. 4.3, deformed nuclei can also give rise to halo structures. The nature of halo states is more complicated in the deformed case than in the spherical case. This can be seen by expanding a one-body deformed state using a partial wave decomposition. Indeed, only partial waves of orbital momentum equal to $\ell = 0, 1$ can generate halos (see Sect. 7.2.1.1). Thus, a one-body deformed halo state is typically a mixture of partial wave components of halo and localized structure. It can also happen that the core and weakly bound nucleons are decoupled so that the deformation of the core and of the nucleons forming the halo can be different [57].

In order to illustrate the basic features of halos in deformed nuclei, we will consider in this section the main features of axially deformed neutron halo states (see also Ref. [57] for more detailed studies and numerical applications). As one has already considered the near-threshold properties of a nucleon in a spherical potential, it is convenient to expand deformed one-body states in a partial wave:

$$\Phi(k, r) = \sum_{\ell j} \left( \frac{u_{\ell j}(k, r)}{r} \right) \mathscr{Y}_m^{\ell j}(\theta, \phi) . \tag{7.25}$$

$u_{\ell j}(k, r)$ is the partial wave component of the nucleon deformed state $\Phi(k, r)$ and $\mathscr{Y}_m^{\ell j}(\theta, \phi)$ is the spin-coupled spherical harmonic of orbital angular momenta $\ell$, total angular momenta $j$ and projection $m$, the latter being fixed due to the axial symmetry of the Hamiltonian. One also assumes mirror symmetry, so that the parity $\pi$ is equal to 1 or -1, and only even or odd values of $\ell$ are present in the sum of Eq. (7.25), respectively.

---

**Exercise I** $\star\star$

The objective of this exercise is to deduce the asymptotical values of root-mean-square radii (see Eqs. (7.5), (7.12), (7.13) and Eqs. (7.27–7.30)) by means of a numerical example.

**A.** Run the one-body spherical Woods–Saxon code to calculate neutron states of orbital angular momentum $\ell = 0, 1$ close to the particle-emission threshold. Determine the depth of a Woods–Saxon central potential which provides a bound state for energies close to $-100\,\text{keV}$ and $-0.1\,\text{keV}$. The associated $\kappa$ values will be denoted as $\kappa_{max}$ and $\kappa_{min}$, respectively. Draw the root-mean-square radii for $\kappa_{min} < \kappa < \kappa_{max}$ values and show that $\langle r^2 \rangle^{1/2}$ verifies Eqs. (7.12), (7.13) when $\kappa \to 0^+$.

**B.** Plot the root-mean-square radius of a neutron of orbital angular momentum $\ell = 2, 3, 4$ in the same conditions as in **A**. Show that $\langle r^2 \rangle^{1/2}$ converges to a finite value when $\kappa \to 0^+$, as indicated by Eq. (7.5).

**C.** Generalize the study of **B** to the deformed case, by running the particle-rotor code in the static regime for neutron states bearing $m = 1/2, \ldots, 7/2 \, (\pi = 1)$ and $m = 1/2, \ldots, 5/2 \, (\pi = -1)$.

Explain why negative values of $m$ do not have to be considered.

Show that $\langle r^2 \rangle^{1/2}$ verifies Eqs. (7.27), (7.28), (7.29) when $\kappa \to 0^+$ for $m = 1/2$ $(\pi = 1)$ and $m = 1/2, 3/2$ $(\pi = -1)$.

Notice that $\langle r^2 \rangle^{1/2}$ converges for $\kappa \to 0^+$ when $m = 3/2, \ldots, 7/2$ $(\pi = 1)$ and $m = 5/2$ $(\pi = -1)$, in accordance with Eq. (7.30).

Explain the common features of $\langle r^2 \rangle^{1/2}$ in the spherical and deformed cases when $\kappa \to 0^+$.

One will consider the asymptotic behavior of the root-mean-square radius of $\Phi(k, r)$ for $\kappa \to 0$. For this, let us calculate the expectation value of $r^2$:

$$\langle \Phi | r^2 | \Phi \rangle = \sum_{\ell j} \int_0^{+\infty} r^2 u_{\ell j}(k, r)^2 \, dr \, . \tag{7.26}$$

Using results of Sect. 7.2.1.1, one arrives at the conclusion that the integrals in Eq. (7.26) either diverge if $\ell = 0, 1$, or are finite if $\ell \geq 2$ when $\kappa \to 0$. Hence, one obtains for the root-mean-square radius at particle-emission threshold:

$$\langle r^2 \rangle^{1/2} \sim \sqrt{\frac{1}{2}} \, \kappa^{-1} \, , \qquad |m| = 1/2 \, , \pi = 1 \tag{7.27}$$

$$\langle r^2 \rangle^{1/2} \sim \sqrt{\frac{5}{4}} \, \sqrt{\left( \beta^{(0)}_{\ell=1 j=1/2} \right)^2 + \left( \beta^{(0)}_{\ell=1 j=3/2} \right)^2} \, \kappa^{-1/2} \, , \qquad |m| = 1/2 \, , \pi = -1 \tag{7.28}$$

$$\langle r^2 \rangle^{1/2} \sim \sqrt{\frac{5}{4}} \, \left| \beta^{(0)}_{\ell=1 j=3/2} \right| \kappa^{-1/2} \, , \qquad |m| = 3/2 \, , \pi = -1 \tag{7.29}$$

$$\langle r^2 \rangle^{1/2} \to \text{const} \quad \text{otherwise} \tag{7.30}$$

where $\beta^{(0)}_{\ell j}$ has been defined in Eq. (7.10). Consequently, the near-threshold properties of deformed neutron states are analogous to that of spherical neutron states, as the $\ell = 0, 1$ condition must be replaced by the $\pi = \pm 1$ and $|m| = 1/2, 3/2$ conditions. Numerical examples showing the validity of Eqs. (7.5), (7.12), (7.13) and Eqs. (7.27–7.30) in practice, are done in Exercise I.

## 7.2.2   Effects of Nucleon–Nucleon Correlations on Halo States

The configuration mixing involving bound and unbound basis states may generate a weakly bound many-body state with the density distribution extending by tunnelling effect in the classically forbidden region. Indeed, even though halo states are bound, they possess sizable components of unbound basis states and in that sense are the states of *open quantum systems*. On the contrary, well-bound states whose density distribution is concentrated inside the nucleus and negligible in the classically forbidden region, represent states of *closed quantum systems*.

While pairing interaction provides with only a small additional binding in well-bound nuclei, it is responsible for binding of weakly bound nuclear states close to the particle-emission threshold. Indeed, in a sole presence of the mean-field, nuclei close to the particle-emission threshold would bear nucleons in unbound single-particle orbits. The residual interaction among these nucleons yields an additional binding so that the nucleus becomes weakly bound and nucleons in originally unbound orbits may form halo in the classically forbidden region.

The simplest systems in which pairing correlations are determinant for halo formation are the three-cluster systems. The asymptotic behavior of halo states can be analyzed therein under specific conditions [58, 59]. Rigorous results can be obtained if one lifts antisymmetrization in subsystems. The diverging character of neutron density, occurring when $\ell = 0, 1$ (see Sect. 7.2.1), is typically absent from the three-body systems. On the other hand, if one subsystem is bound and has an energy almost equal to the total energy of a three-body system, the bound subsystem behaves as one particle in the presence of the field generated by the other two subsystems [59]. No halo can occur if two subsystems of the three-body system are bound as then the root-mean-square radius of the ground state is always finite at zero binding energy.

When considering two nucleons interacting with a core, the exponential decay of the one- and two-body densities of valence neutrons is mainly a function of the energy of the two correlated valence nucleons, while there is basically no dependence on single-particle energies of the combined system. This model is, however, oversimplified as the pairing interaction is not treated self-consistently. The antisymmetrization in the total system can be restored at the price of adding complexity in the initially concise cluster description [60]. The main result obtained in Ref. [60] using fully antisymmetrized wave functions is that the exponential decay constant of the total density is equal to $\kappa = \sqrt{2mE_3}/\hbar$, where $E_3$ is the binding energy of the three-cluster system relatively to the three-body break-up threshold. The asymptotic behavior of the three-particle density is of two-body character if a two-body subsystem is bound therein,

The previous discussion applies to nuclei having two- or three-cluster structure. Therefore, it would be desirable to devise a model which can be applied in various mass regions in order to numerically assess if a ground state wave function can give rise to a halo. In such an approach, it is necessary to incorporate both pairing correlations and the proper asymptotic behavior of single-particle wave functions. Moreover, the resulting model should be applicable in the whole nuclear chart. Therefore, one has to recur to mean-field methods. A convenient tool for that matter is the Hartree–Fock-Bogolyubov approach in coordinate space [61]. For even particle number N, where N is the number of protons or neutrons, the asymptote of neutron density in the Hartree–Fock-Bogolyubov approach is provided by the total binding energy relative to the two-body break-up threshold $E_2$. As a consequence, the aforementioned decay constant becomes $\kappa = \sqrt{2mE_2}/\hbar$, where

$E_2 = \min[E_\mu] - \lambda$, with $\min[E_\mu]$ being the lowest discrete quasi-particle energy verifying $E_\mu < \lambda$, and $\lambda$ is the chemical potential. The quasi-particle energies $E_\mu$ are rather close to the canonical quasi-particle energies [62]:

$$E_\mu \simeq E_\mu^{(can)} = \sqrt{(\epsilon_\mu^{(can)})^2 + (\Delta_\mu^{(can)})^2} , \qquad (7.31)$$

where $\epsilon_\mu^{(can)}$ and $\Delta_\mu^{(can)}$ are the diagonal matrix elements of the particle-hole and particle-particle mean fields in the canonical basis. Hence, a canonical state at the Fermi surface, for which $\epsilon_\mu^{(can)} \simeq \lambda$, provides with the decay constant of the asymptotic density with $E_2 = \Delta_\mu^{(can)} - \epsilon_\mu^{(can)}$. One may note the presence of the pairing gap $\Delta_\mu^{(can)}$, which vanishes in the absence of pairing correlations. This additional term prevents a halo from appearing in even-N nuclei. This salient pairing effect involving all nucleons of the nucleus is called pairing anti-halo effect [63]. It has significant effect on low-energy collective excitations at around driplines [42].

The situation is different in odd-N nuclei. The neutron densities arise from one-quasi-particle states [64], so that the decay constant governing their asymptote corresponds to $E_2 = -\min[E_\mu] - \lambda$. At the Fermi surface, $E_2 \simeq 0$ when $\lambda \simeq -\Delta_\mu^{(can)}$, hence the root-mean-square radius is diverging when $\epsilon_\mu^{(can)} \to -\Delta_\mu^{(can)}$.

Consequently, while a large occupation of $s$ and $p$ waves in nuclei close to particle-emission threshold typically generates halo states (see Sect. 7.2.1), the nucleon–nucleon correlations induced by the residual nuclear interaction act against the halo formation through the pairing anti-halo effect. Therefore, the nucleus must bear very weakly bound or unbound subsystems for pairing anti-halo effect to be alleviated. Indeed, the pairing gap $\Delta_\mu^{(can)}$ is non-negligible in even-N nuclei if the nuclear ground state is well described by a generalized Slater determinant, i.e. by a product of quasi-particles generated by Hartree–Fock-Bogolyubov equations, which is not the case if the nuclear wave function is composed of clusters.

Such a situation is, for example, in the two-neutron halo configurations of $^{11}$Li and $^6$He which are the Borromean systems. More precisely, the two-body subsystems present in $^{11}$Li and $^6$He, are either dineutron, $^9$Li + neutron or $^4$He + neutron, which are all unbound. This emphasizes the strong coupling induced by the nuclear interaction among subsystems, because in the mean-field picture $^5$He + neutron or $^{10}$Li + neutron two-body systems would be even more unbound than $^5$He or $^{10}$Li.

Pairing correlations are also responsible for the odd-even staggering of one-nucleon separation energies. In the vicinity of neutron dripline, typically, odd-N systems are unbound and even-N systems are bound. A striking example of this odd-even staggering effect is the helium chain, as $^{5,7,9}$He are neutron unbound, whereas $^{6,8}$He are weakly bound halo states. One can also mention boron, carbon, and fluorine isotope chains at the neutron dripline, as $^{16,18,20}$B, $^{21}$C and $^{28,30}$F are all unbound, whereas $^{15,17,19}$B, $^{20,22}$C, and $^{27,29,31}$F are bound. The two-neutron halo states are often found in even-N nuclei neighboring unbound odd-N nuclei. For example, the ground states of $^{15,17,19}$B [46] and $^{31}$F [65] are suspected to be two-neutron halos. On the other hand, $^{10}$He and $^{26,28}$O are weakly unbound. Pairing correlations in this case are not sufficiently strong to bind these nuclei mitigating only their resonance character.

It is clear that halo nuclei have to be described within a quantum framework comprising both nucleon–nucleon correlations and continuum coupling. The Gamow shell model is thus the tool of choice, as continuum coupling is present at the one-body level through the use of the Berggren basis, and nucleon–nucleon correlations are exactly taken into account by configuration mixing. Moreover, as the wave functions are explicitly calculated, one can directly assess the structure of halo nuclear states with overlap functions and spectroscopic factors, which, as one will see, provide with the weights and radial formfactors of two-cluster configurations, such as $^5$He + neutron in $^6$He. The exact consideration of the Coulomb interaction in the Gamow shell model also allows to precisely determine the isospin mixing features of the nuclear Hamiltonian, prominent to understand physics at particle-emission threshold, and thus halo nuclei.

## 7.3   Asymptotic Normalization Coefficient in Mirror Nuclei

The one-nucleon overlap functions and the resulting spectroscopic factors are standard tools in the theory of direct reactions [66–71]. The fundamental interest of asymptotic normalization coefficients is that they are not sensitive to the changes of short-range physics (see Ref. [72], the discussion in Ref. [73] about observables and non-observables, and also Ref. [74] for an overview of unitary transformations and their impact on nuclear structure operators). Indeed, asymptotic normalization coefficients are invariant under unitary transformations, so that they are observables, contrary to spectroscopic factors. This arises because a unitary transformation of the Hamiltonian only affects the interior region of the wave function, so that asymptotic normalization coefficients remain unchanged.

Experimentally, asymptotic normalization coefficients have been measured for $s$-shell and $p$-shell nuclei (see Ref. [75] and the references therein). Theoretical calculations of asymptotic normalization coefficients have been performed using the hyperspherical harmonics approach [76], the Gamow shell model and the shell model embedded in the continuum [55], the Green's function Monte Carlo [77–79] (see also Ref. [80]).

The spectroscopic factors are deduced from experimental cross sections and measure configuration mixing in the $A$-body wave functions. However, the determination of spectroscopic factors depends on the models used, which implies problems in the experimental determination of spectroscopic factors. The common assumptions in standard shell model, i.e. that nucleons occupy localized states, and degrees of freedom associated with the continuum are neglected, can been tested using the Gamow shell model [12, 81]. In the following, we shall discuss the radiative capture reaction $b + c \rightarrow a + \gamma$, and define the radial overlap function $I^a_{bc}$ for a process $a \rightarrow b + c$. The radial overlap function $I^a_{bc}$ can be written in the asymptotic region as:

$$I^a_{bc;\ell j}(r) \sim \left(\frac{1}{r}\right) C_{\ell j} W_{-\eta,\ell+1/2}(2\kappa r) , \tag{7.32}$$

where $W_{-\eta,\ell+1/2}$ is the Whittaker function. In the above expression, $C_{\ell j}$ is the asymptotic normalization coefficient which characterizes the virtual decay of a nucleus into two particles $b$ and $c$, $r$ is the relative distance between particles $b$ and $c$, $\kappa = \sqrt{2\mu S^{(a)}_c/\hbar^2}$ where $S^{(a)}_c$ is the separation energy of particle $c$ in the nucleus $a$, and $\eta = Z_b Z_c e^2 \mu/\hbar^2 \kappa$ where $\mu$ is the reduced mass of $b + c$. The quantum numbers $\ell$ and $j$ are the orbital angular momentum and the total angular momentum, respectively. We shall assume that the particle $c$ is a nucleon (proton or neutron); hence, $S^{(a)}_n$ ($S^{(a)}_p$) is the one neutron (one-proton) separation energy. In this case, the radial overlap integral can be written as:

$$I^a_{bc;\ell j}(r) = \frac{1}{\sqrt{2J_a+1}} \sum_{\mathscr{B}} \langle \Psi^{J_a}_a || a^{\dagger}_{\ell j}(\mathscr{B}) || \Psi^{J_b}_b \rangle \, \langle r\ell j | u_{\mathscr{B}} \rangle , \tag{7.33}$$

where $a^{+}_{\ell j}(\mathscr{B})$ is a creation operator associated with the single-particle basis state $|u_{\mathscr{B}}\rangle$ and $\langle r\ell j | u_{\mathscr{B}} \rangle$ is the radial single-particle wave function. The sum runs over the complete single-particle basis. The spectroscopic factor $S_{\ell j}$ is defined as the squared norm of the radial overlap integral (7.33). In the general case of multichannel coupling, the asymptotic normalization coefficient for a radiative capture reaction is defined in terms of the Hermitian norm of all the contributions corresponding to different couplings of the target state and the state in a parent nucleus:

$$|C| = \sqrt{\sum_{\ell,j} |C_{\ell j}|^2} \tag{7.34}$$

For bound states, the radial overlap function $I^a_{bc;\ell j}$ is well approximated by the product of the spectroscopic amplitude $S^{1/2}_{\ell j}$ and the single-particle radial wave function $u_{\ell j}/r$ at a single-particle energy $-S^{(a)}_c$:

$$I^a_{bc;\ell j}(r) \sim \left(\frac{1}{r}\right) S^{1/2}_{\ell j} u_{\ell j}(r) . \tag{7.35}$$

Therefore, $I^a_{bc;\ell j}$ behaves as:

$$I^a_{bc;\ell j}(r) \sim \left(\frac{1}{r}\right) S^{1/2}_{\ell j} \beta_{\ell j} W_{-\eta,\ell+1/2}(2\kappa r) , \tag{7.36}$$

far from the region of nuclear interaction. $\beta_{\ell j} = \beta_{\ell j}(S_c^{(a)})$ is the single-particle asymptotic normalization coefficient of Eq. (7.1), where $\beta_{\ell j}$ is directly related to the many-body asymptotic normalization coefficient [82, 83]:

$$C_{\ell j} = S_{\ell j}^{1/2} \beta_{\ell j} \; . \tag{7.37}$$

The relations (7.35–7.37) are satisfied also for many-body resonances. Indeed, it was demonstrated in Gamow shell model [27] that the overlap function $I_{bc;\ell j}^a$ is well approximated by a product of the spectroscopic amplitude $S_{\ell j}$ and the single-particle resonance wave function which reproduces the $Q$-value of the reaction studied. Let us also remind that the astrophysical factor $\mathscr{S}_{bc}$ in the limit of zero center-of-mass energy, $E_{\mathrm{cm}} = 0$, is proportional to $\beta_{\ell j}^2$ [84]:

$$\mathscr{S}_{bc}(0) \sim \beta_{\ell j}^2 \equiv \beta_{\ell j}^2(\eta) \; . \tag{7.38}$$

Let us now calculate the norm of bound state in the external region where the nuclear part of the potential is practically zero. In the region $(r \geq R_f)$, where $R_f$ $(R_f \gg R)$ is the radius in the external region where the nuclear part of the potential vanishes, the single-particle wave function $u_{\ell j}$ is described by Eq. (7.1), i.e. is given by $W_{-\eta, \ell+1/2}(2\kappa r)$. For the normalized bound state wave function: $\mathscr{N}_{\mathrm{int}} + \mathscr{N}_{\mathrm{ext}} = 1$, where $\mathscr{N}_{\mathrm{int}}$ is the norm of the internal part of the wave function, and

$$\mathscr{N}_{\mathrm{ext}} = \beta_{\ell j}^2 \int_{R_f}^{+\infty} |W_{-\eta, \ell+1/2}(2\kappa r)|^2 \, dr \tag{7.39}$$

is the norm of the external part. On the condition that the energy dependence of $\mathscr{N}_{\mathrm{int}}$ is weak, which is a reasonable assumption even if the separation energy is close to zero, $\beta_{\ell j}$ should be strongly depending on the value of $\mathscr{N}_{\mathrm{ext}}$.

An extensive analysis of proton and neutron asymptotic normalization coefficients for light mirror nuclei has been performed using standard shell model, cluster model, and various effective $NN$ interactions [85–90]. Calculations of the overlap functions $I_{bc;\ell j}^a(r)$ in the Gamow shell model has been for the $p$-shell nuclei: $^7$Li/$^8$Li and $^7$Be/$^8$B [55]. The scattering contours in these calculations have been discretized utilizing a Gauss-Legendre quadrature. The number of discretization points have been chosen in order to maintain a correct asymptotic behavior of $I_{bc;\ell j}^a(r)$ up to at least $\sim$8 fm. Gamow Hartree–Fock ensemble [91] corresponding to a parent nucleus has been taken for the single-particle basis.

One will firstly discuss the asymptotic normalization coefficients corresponding to single-nucleon capture reactions involving only bound states. The asymptotic normalization coefficients in Gamow shell model can be directly extracted from the calculated radial overlap integrals by fitting their tail to Whittaker functions at large The radial overlap integrals are calculated using Eq. (7.33), where the sum over $\mathscr{B}$ states runs over the complete Berggren ensemble; hence, the result is independent

**Fig. 7.6** Radial overlap integrals $I^a_{bc;\ell j}(r)$ calculated in Gamow shell model for the $J^\pi = 2^+_1$ ground state: $^7Be_{3/2^-} + p \to {}^8B_{2+}$ (**a**) $^7Li_{3/2^-} + n \to {}^8Li_{2+}$ (**b**) for $\ell = 1$ and $j = 3/2$. The tail of the radial overlap integral is fitted by the Whittaker function (dashed line) to extract asymptotic normalization coefficient (from Ref. [55])

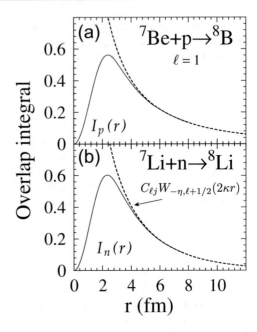

of the single-particle basis representation. The asymptotic normalization coefficient is then obtained directly from Eq. (7.32).

Figure 7.6 shows how this procedure works for the radial overlap integrals corresponding to $\ell = 1$ ($0p_{1/2}$ and $0p_{3/2}$) protons and neutrons in the $2^+$ ground states of the mirror nuclei $^8B$, $^8Li$. This example is not trivial as the configuration mixing is appreciable, and $^8B$ is a proton halo as seen from the extended tail of the overlap function (see the upper panel in Fig. 7.6). Nevertheless, the extraction of the asymptotic normalization coefficients does not cause any problem.

One now considers the more interesting case of overlap functions and asymptotic normalization coefficients involving unbound many-body states. This demands a slight modification of the formalism used to calculate the asymptotic normalization coefficients for bound states. Indeed, the definition of asymptotic normalization coefficients via Eqs. (7.36), (7.37) is no longer appropriate for negative separation energies, i.e. when the state of a nucleus $a$ ($A$-particle system) is unbound with respect to the nucleus $b$ ($A - 1$-particle system). In this case, $\kappa$ becomes complex and the Whittaker function becomes complex as well. The imaginary part of the Whittaker function is not vanishing even at the limit of vanishing width and the associated asymptotic normalization coefficients are complex.

A suitable definition of the asymptotic normalization coefficient for negative separation energies involves the outgoing Coulomb wave function $H^+_{\ell,\eta}(kr)$, where $k = (-2\mu S^2_a/\hbar^2)^{1/2}$ and $\eta = Z_b Z_c e^2 \mu / \hbar^2 k$:

$$I^a_{bc;\ell j}(r) \sim \left(\frac{1}{r}\right) C_{\ell j} H^+_{\ell,\eta}(kr) . \qquad (7.40)$$

Note that as the state in $A$-particle system is unbound, $k$ and $\eta$ are complex [27]. In the limit of vanishing resonance width, $k$, $I^a_{bc;\ell j}$ and $H^+_{\ell,\eta}(kr)$ become real and, hence, $C_{\ell j}$ given by Eq. (7.40) becomes real as well. Numerical examples showing that the ansatz expressed Eqs. (7.32), (7.40)) are valid in practice in the asymptotic zone are presented in Exercise II.

---

**Exercise  II** ⋆⋆

We will show in this numerical exercise that overlap functions can be effectively fitted by Hankel or Coulomb wave functions (see Eqs. (7.32), (7.40).

   While it is not strictly necessary, it is preferrable to run the many-body Gamow shell model code using a parallel machine with at least 5–10 cores and a total computer memory of about 30 Gb, as calculations can take several hours otherwise.

**A.** Run the many-body Gamow shell model code using the Furutani–Horiuchi–Tamagaki interaction (see Ref. [92, 93] and Eq. (6.18)) to calculate energies and overlap functions of $^8\text{Li}/^7\text{Li}$ and $^8\text{B}/^7\text{Be}$, as shown above. Fit the obtained overlap functions in the asymptotic region by using the code extracting asymptotic normalization coefficients from overlap functions.
Notice that they are well reproduced by Hankel or Coulomb wave functions using Eqs. (7.32) and (7.40).

**B.** Change the interaction parameters to have slightly different binding energies for $^8\text{Li}/^7\text{Li}$ and $^8\text{B}/^7\text{Be}$. Show that the overlap functions obtained in this way can be fitted with Hankel or Coulomb wave functions by changing the $\kappa$ or $k$ parameter of Eqs. (7.32) and (7.40) accordingly.

---

The overlap function $I^a_{bc;\ell j}(r)$ defined in Eq. (7.32) obeys a Schrödinger-like equation, albeit inhomogeneous [66]. A source term in this equation, however, is generated by the two-body interaction. However, separating the full interaction into a one-body term and a two-body residual interaction, the source term can be decomposed into a dominant homogeneous part and a residual inhomogeneous part. Indeed, the full interaction can be written as a mean-field one-body part, to which a small two-body residual part is added. One-body parts generate homogeneous terms, whereas inhomogeneous terms arise from the two-body part of the interaction. The inhomogeneous part can then be absorbed into the homogeneous part. As only bound states or narrow resonance states are involved, one can also assume that the potential entering the equation defining $I^a_{bc;\ell j}(r)$ is real. These approximations have been tested successfully in Ref. [27] for both bound and unbound states.

   Under these assumptions, one can easily derive the relation between asymptotic normalization coefficient and partial width [94], which reads mutatis mutandis as Eq. (2.197). The total asymptotic normalization coefficient constant $C$ (see Eq. (7.34)) can then be derived by expressing the total width $\Gamma$ in terms of the sum of partial widths $\Gamma_{\ell j}$:

$$C = \sqrt{\sum_{\ell j} |C_{\ell j}|^2} = \sqrt{\Gamma\left(\frac{\mu}{\hbar^2 \Re(k)}\right)}. \tag{7.41}$$

In this expression:

$$\Gamma_{\ell j} = \left(\frac{\hbar^2}{\mu}\right) |C_{\ell j}|^2 \Re(k) \tag{7.42}$$

is given by the same formula as that obtained in Ref. [95], even though approximations and boundary conditions are different in the real-energy $R$-matrix approach and in the complex-energy Gamow-state formalism [96, 97].

Figure 7.7 shows the radial overlap integrals for a $3_1^+$ broad resonance in $^8$B and a narrow mirror resonance in $^8$Li. The tails of real and imaginary parts of radial overlap integrals can be seen to be well fitted with the outgoing Coulomb wave functions of a complex argument $k$.

**Fig. 7.7** Radial overlap integrals $I_{bc;\ell j}^a(r)$ calculated in Gamow shell model for the $J^\pi = 3_1^+$ resonance: $^7$Be$_{3/2-}$ + $p \to$ $^8$B$_{3+}$ (upper) and $^7$Li$_{3/2-}$ + $n \to$ $^8$Li$_{3+}$ (lower). $^8$B in the $3_1^+$ state is a broad one-proton resonance which results in a large imaginary part of radial overlap integral. The tail of the radial overlap integral is fitted by the Coulomb wave function $H_\ell^+(\ell, \eta)(kr)$ to extract asymptotic normalization coefficient as $J^\pi = 3_1^+$ states are unbound (from Ref. [55])

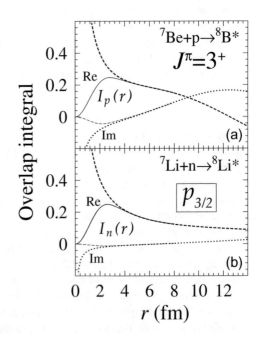

## 7.4    Near-Threshold Behavior of Wave Functions

Due to the unitarity of the scattering matrix and the resulting flux conservation, an opening reaction channel can generate the so-called Wigner cusps in other open channels with lower thresholds. Wigner cusps are the non-analyticities appearing in cross sections at energies when a new particle-emission channel opens. They translate into very sharp changes of cross sections, discontinuities of the derivative of cross sections or of the cross sections themselves, depending on the angular momentum content of the wave functions involved in cross sections.

In order to study Wigner cusps, one will evaluate spectroscopic factors of weakly bound and resonance systems, whose energy will vary smoothly by changing the parameters of the Hamiltonian. Contrary to the evaluation of energies, a precise calculation of spectroscopic factors with the Gamow shell model demands a very large number of discretized scattering basis states. In the following analysis, one will concentrate on systems, such as $^{5-7}$He and $^{17-18}$O isotopes, as they one can be described with only a few valence nucleons in the presence of a core of $^4$He or $^{16}$O core, respectively. In these systems, the Gamow shell model calculations can be realized without any model space truncation. It will be shown that the spectroscopic factors exhibit qualitatively different features according to the bound or unbound nature of involved nuclear states. Moreover, due to the different partial waves involved in the studied isotopes, $p$ shells for $^{5-7}$He isotopes and $sd$ shells for $^{17-18}$O isotopes, spectroscopic factors will exhibit Wigner cusps of different nature. The radial overlap functions associated with the calculated spectroscopic factors will also be evaluated. In particular, it will be shown that the calculated radial overlap functions resemble one-body states generated by Woods–Saxon potential, whose depth parameter is real or complex.

Another aspect of near-threshold properties of many-body wave functions is the isospin mixing which has been estimated to be very small, of the order of a few percent at most in well-bound light nuclei [98,99]. This arises because the Coulomb interaction has an infinite range, so that its main effect is to change the energy of valence shells by a nearly constant shift. Consequently, well-bound nuclear states possess an almost exact isospin, the Coulomb Hamiltonian translating only into a charge dependence of binding energies of nuclear states.

Conversely, in the vicinity of particle-emission threshold, the asymptotic behavior of the wave functions of mirror nuclei is very different. Neutron wave functions are only subject to the centrifugal barrier at large distance, while proton wave functions also depend on the Coulomb barrier. While this effect is negligible in well-bound states, it is prominent in near-threshold wave functions, whose asymptote extends far away from the nuclear zone. In particular, the isobaric analog states of mirror nuclei near particle-emission threshold bear different excitation energies and widths, whereas they should be identical in the presence of exact isospin symmetry. In order to exhibit isospin mixing close to particle-emission threshold, one will consider the isobaric analogue states of $^{5-6}$He, $^{5-6}$Li, and $^6$Be. Indeed, these states lie close to the particle-emission threshold. Isospin symmetry breaking will be shown to be particularly visible in spectroscopic factors. It will be directly quantified through the calculation of the expectation value of the isospin operator, denoted as $\hat{T}^2$.

### 7.4.1  Two-Particle Reaction Cross Sections and Wigner Cusps

In a closed quantum system framework, the cross section is often modelled by the product of the cross sections coming from a one-body potential and the associated spectroscopic factors:

$$\sigma = \sum_{n\ell jm} S_{n\ell jm}\, \sigma_{s.p.}^{n\ell jm} , \tag{7.43}$$

where $S_{n\ell jm}$ is the spectroscopic factor associated with the $n\ell jm$ single-particle state, and $\sigma_{s.p.}^{n\ell jm}$ contains all the kinetic dependence related to the transfer of the nucleon of the projectile. Single-particle cross sections $\sigma_{s.p.}^{n\ell jm}$ in Eq. (7.43) are derived from a one-body calculation so that they do not include the effects of inter-nucleon correlations. Information about the nucleon–nucleon correlations is contained in spectroscopic factors which typically arise from a shell model calculation. The sum in Eq. (7.43) is complete in a closed quantum system framework if one sums over all basis states, i.e. taking into account all the possible values of $n, \ell, j,$ and $m$ quantum numbers therein. Nevertheless, it is customary to take only one major harmonic oscillator shell in standard shell model spaces, thus creating a spurious basis dependence.

Equation (7.43) has to be generalized to open quantum systems. For this, one will rely on the completeness of the Berggren basis to define spectroscopic factors. Indeed, they correspond to the norm $S$ of the overlap functions $I_{\ell j}(r)$ of Eq. (7.33):

$$S = \frac{1}{2J_A + 1} \sum_{\mathscr{B}} \left( \langle \widetilde{\Psi_A^{J_A}} || a_{\ell j}^+(\mathscr{B}) || \Psi_{A-1}^{J_{A-1}} \rangle \right)^2 , \tag{7.44}$$

where $\mathscr{B}$ runs over all the Berggren basis states of quantum numbers $(\ell, j)$. Since the partial wave represented by the Berggren basis states is complete, Eqs. (7.33), (7.44) are independent of the choice of the basis, contrary to standard shell model. Moreover, as the scattering states are taken into account in the Berggren basis, the continuum degrees of freedom are included without any approximation.

Wigner cusp [100] appears in spectroscopic factors when a many-body state crosses particle-emission threshold. This phenomenon obeys a simple law in the vicinity of threshold energy and appears in experimental cross sections involving neutral particles [101]:

$$\sigma \sim k^{2\ell-1} \tag{7.45}$$

$$\sigma \sim k^{2\ell+1} , \tag{7.46}$$

where $\sigma$ is the cross section, $k \sim 0$ is the transferred momentum, and $\ell$ is the transferred orbital angular momentum. Equation (7.45) is for a capture process and Eq. (7.46) for an emission process. The generalization of the Wigner threshold law to three particles have been studied in Refs. [102–104]

Wigner cusps are particularly well visible for neutron $\ell = 0, 1$ waves, while their manifestation is subtle in neutron waves with $\ell \geq 2$, as we will see in the next section. One can notice in Eqs. (7.45), (7.46) that the cross sections at the particle-emission threshold are mainly different in partial waves of low angular momentum. No Wigner cusp appears for charged particles, as in this case Eqs. (7.45), (7.46) become

$$\sigma \sim \exp\left(-2\pi\,\eta\right)\,k^{-2} \tag{7.47}$$

$$\sigma \sim \exp\left(-2\pi\,\eta\right)\,. \tag{7.48}$$

Here, Eq. (7.47) describes the capture process and Eq. (7.48) the emission process.

One can notice that the energy dependence in Eq. (7.46) is identical to that of Eq. (2.199). This feature is not accidental and arises from the peripheral character of cross sections when $k \sim 0$. Indeed, one-body bound or resonance states are extended in space close to the particle-emission threshold, so that cross sections are proportional to the modulus square of their asymptotic wave function, equal to $|C^+\,H^+_{\ell,\eta}(kR)|^2$. Nuclear structure is entirely contained in $C^+$, which is the asymptotic normalization coefficient of Eq. (7.32). Consequently, Eqs. (7.45), (7.46) are obtained similarly to Eq. (2.199), by separating $k$-dependence for $k \sim 0$ from factors depending on Hamiltonian parameters and kinetic operator only.

Charged particle cross sections are proportional to the Gamow factor (see Eq. (2.31)), which strongly suppresses the cross section at low energies, hence preventing from a visible Wigner cusp to develop at $k = 0$. One can see as well the analogy between Eqs. (7.48) and (2.201), arising for the same reason as for neutral particles.

### 7.4.2 Spectroscopic Factors and Radial Overlap Integrals

One neutron spectroscopic factors in the ground state of $^6$He, $^7$He, and excited state of $^{18}$O* are shown in Figs. 7.8 and 7.9. In the oxygen case, it is the excited state of dominant configuration $\langle^{17}$O $\otimes (0d_{3/2})^2\rangle$ coupled to $J = 0$ which will be studied. The involved partial waves in these spectroscopic factors are $(\ell, j) = p_{3/2}$ for helium isotopes and $(\ell, j) = d_{3/2}$ for oxygen isotopes. The Hamiltonian consists of the Woods–Saxon potential mimicking the $^4$He or $^{16}$O core (see Eq. (3.83)) and the finite-range modified surface Gaussian interaction (see Eq. (9.174)). The Hamiltonian parameters are varied in such a way that the ground states of the $^6$He nucleus (upper and lower panels of Fig. 7.8), and $^5$He nucleus (middle panel of Fig. 7.8), as well as the $3/2^+$ excited state of $^{17}$O, denoted as $^{17}$O*, i.e. the $d_{3/2}$ Woods–Saxon states, (see Fig. 7.9) vary from bound to unbound, continuously. This

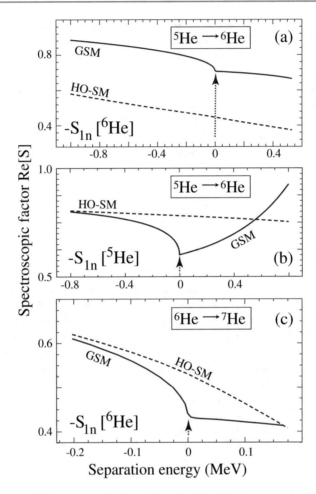

**Fig. 7.8** Real part of the spectroscopic factor $S$ as a function of the one neutron separation energy $S_{1n}$. Upper (**a**) and middle (**b**) panels display: $\langle {}^6\mathrm{He}(\text{g.s.})|[{}^5\mathrm{He}(\text{g.s.}) \otimes p_{3/2}]^{0^+}\rangle^2$, whereas lower panel (**c**) shows: $\langle {}^7\mathrm{He}(\text{g.s.})|[{}^6\mathrm{He}(\text{g.s.}) \otimes p_{3/2}]^{0^+}\rangle^2$. The solid line represents results issued from Gamow shell model calculations while the dotted line depicts standard shell model approximation (HO-SM) where the surface Gaussian interaction matrix elements are calculated in the basis of harmonic oscillator states $\{0p_{3/2}, 0p_{1/2}\}$. The neutron emission threshold in ${}^6\mathrm{He}$ (upper (**a**) and lower (**c**) panels) and ${}^5\mathrm{He}$ (middle panel (**b**)) are indicated by arrows. The results illustrated in the middle part (**b**) of the figure are given as function of the energy of the resonance $0p_{3/2}$, which is equal to $-S_{1n}[{}^5\mathrm{He}]$ in the model used (from Ref. [12])

simulates formation of a composite system at different excitation energies [12, 81]. The other nucleus of the pair, i.e. ${}^5\mathrm{He}(\text{g.s.})$ for ${}^6\mathrm{He}$ nucleus (upper panel of Fig. 7.8), ${}^6\mathrm{He}(\text{g.s.})$ for ${}^5\mathrm{He}$ nucleus (middle panel of Fig. 7.8), ${}^7\mathrm{He}(\text{g.s.})$ for ${}^6\mathrm{He}$ nucleus (lower panel of Fig. 7.8) and ${}^{18}\mathrm{O}(0^+_3)$ state for ${}^{17}\mathrm{O}^*$ nuclear state (Fig. 7.9), is always bound.

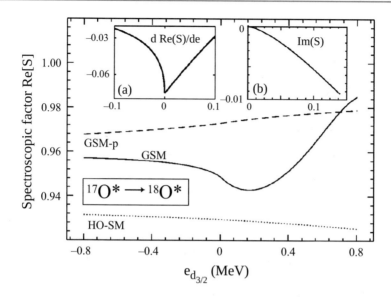

**Fig. 7.9** Real part of the spectroscopic factor $S$ as a function of the energy $e_{d_{3/2}}$ of the one-body state $0d_{3/2}$, equal in the model used to $-S_{1n}[^{17}O^*]$. One considers: $\langle ^{18}O(0_3^+)|[^{17}O(3/2_1^+)\otimes d_{3/2}]^{0^+}\rangle^2$. The first derivative of Re($S$) in the vicinity of particle-emission threshold $e_{d_{3/2}}=0$ is depicted in the insert (**a**), while insert (**b**) depicts the imaginary part of $S$ (from Ref. [27]).

The Wigner cusp is clearly noticeable in Figs. 7.8 and 7.9. One can see that the spectroscopic factor $S$ exhibits a cusp at $k = 0$ which is similar to that of the transfer cross section of Eqs. (7.45), (7.46). The Wigner cusp can be seen to originate uniquely from coupling to the nonresonant continuum, as it disappears in standard shell model calculations utilizing a basis of harmonic oscillator states or in the pole approximation of Gamow shell model (see Sect. 5.4), where only bound and resonance states $0d_{5/2}$, $1s_{1/2}$, and $0d_{3/2}$ are present in the valence space (see Fig. 7.9).

Coupling to the nonresonant scattering continuum is essential to describe correctly the spectroscopic factors close to particle-emission threshold. This coupling is also crucial to preserve unitarity at particle-emission threshold in the Gamow shell model. Therefore, the salient features of cross sections seen in Eqs. (7.45–7.48) are the direct consequences of respecting unitarity in near-threshold wave functions.

The radial overlap functions connected to the spectroscopic factors of the $^6$He/$^5$He pair are studied by fixing the real one neutron separation energy $S_{1n} = \Re[E_{^5\mathrm{He}}] - \Re[E_{^6\mathrm{He}}]$ and complex one neutron separation energy $\tilde{S}_{1n} = E_{^5\mathrm{He}} - E_{^6\mathrm{He}}$ for bound and resonance ground states of $^5$He and $^6$He, respectively [81] (see Fig. 7.10). The complex one neutron separation energy:

$$\tilde{S}_{1n} \equiv E(N-1) - E(N) = S_{1n} - \frac{i}{2}[\Gamma(N-1) - \Gamma(N)] \qquad (7.49)$$

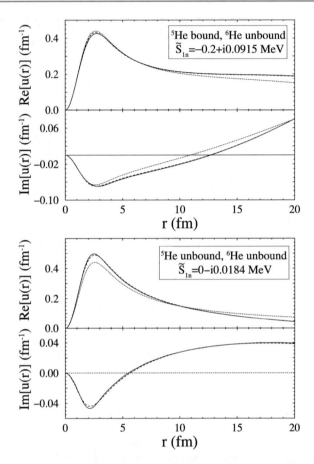

**Fig. 7.10** The solid line represents the radial overlap function $I(r)$ associated with the spectroscopic factor $\left(\langle {}^6\text{He(g.s.)}|[{}^5\text{He(g.s.)} \otimes p_{3/2}]^{0^+}\rangle\right)^2$ which is calculated in Gamow shell model for the bound and resonance states in ${}^5\text{He}$, and the resonance state in ${}^6\text{He}$ ($S_{1n}$ is negative). The dotted line shows the radial wave function of the $0p_{3/2}$ resonance generated by a real Woods–Saxon potential such that the real part of the energy of $0p_{3/2}$ single-particle state equals $-S_{1n}$. The dashed line depicts the radial wave function of the $0p_{3/2}$ resonance generated by a complex Woods–Saxon potential such that the complex energy of $0p_{3/2}$ single-particle state equals $-\tilde{S}_{1n}$. All wave functions are normalized to unity to be able to compare $u(r)$ and radial wave functions of $0p_{3/2}$ resonance states. The real (imaginary) parts of wave functions are shown in the upper (lower) part of each panel (adapted from Ref. [27])

replaces the real separation energy $S_{1n}$ when unbound states enter in the calculation of the spectroscopic factor, because $\tilde{S}_{1n}$ and not $S_{1n}$ determines the asymptotic behavior of radial overlap integral:

$$I_{\ell j}(r) \sim \exp(\kappa r) \,,$$

where the decay constant is given by

$$\kappa = \sqrt{2m\tilde{S}_{1n}/\hbar} \,. \tag{7.50}$$

One should add in passing that if the unbound states are involved in the extraction of the effective matrix elements of nucleon–nucleon interaction from experimental binding energies of neighboring nuclei, one should use complex and not real binding energies because in this case the two-body matrix elements calculated in Gamow shell model become complex as well.

For bound states, the real overlap function can be modelled by a Woods–Saxon single-particle state of energy $-S_{1n}$ [27]. However, a real Woods–Saxon potential cannot model complex one neutron separation energy $\tilde{S}_{1n}$. Hence, the radial overlap functions of Fig. 7.10 have been compared to one-body states generated by a real Woods–Saxon potential whose depth fixes the real-energy $-S_{1n}$, or by a Woods–Saxon potential whose depth is complex in order to reproduce the complex-energy $-\tilde{S}_{1n}$. One can see that only a complex potential can reproduce almost exactly the radial overlap function if $S_{1n}$ is complex, as a real potential cannot provide, for example, a state of zero or negative energy and nonzero width, as is the case in Figs. 7.10.

A numerical example showing that cusps occur in the context of the spectroscopic factors is done in Exercise III for the pair of nuclei $^9$He/$^{10}$He.

---

**Exercise III ★**

In this numerical example, one will consider the spectroscopic factors of the pair of nuclei $^9$He/$^{10}$He and one will show that a cusp occurs at particle-emission threshold.

A. We will use a core of $^8$He. The Hamiltonian is otherwise the same as discussed in this section.

   Explain why the use of a core of $^8$He to describe the unbound eigenstates of $^9$He and $^{10}$He is sound from both physical and numerical perspective.
   Considering the small widths of the $1/2^-$ state of $^9$He and $0^+$ state of $^{10}$He, explain why the study of the $p_{1/2}$ spectroscopic factors of $^9$He/$^{10}$He is meaningful.

B. Run the two-body Gamow shell model code using cluster orbital shell model coordinates to calculate $p_{1/2}$ spectroscopic factors of $^9$He/$^{10}$He.
   Plot the spectroscopic factors as a function of the energy, similarly to Fig. 7.8.
   Notice the Wigner cusp at $S_{1n} = 0$ MeV occurring using full Berggren basis and compare results with the $^5$He/$^6$He case.

### 7.4.2.1 Spectroscopic Factors and Pairing Correlations in $^{11}$Li

Pairing correlations are an essential ingredient of the two-neutron halos. In the last section, one considered the case of the ground state of $^6$He, which was bound by about 1 MeV, while the two subsystems in its partition into $^5$He and dineutron are unbound by about 900 keV and 50 keV, respectively. By considering $^6$He composite system close to neutron emission threshold, one could point out to the importance of continuum via the study of spectroscopic factors of the form $[^5\text{He(g.s.)} \otimes \nu p_{3/2}]^{0^+}$. Indeed, large Wigner cusps appear in spectroscopic factors, indicating an important role of coupling to the nonresonant continuum occurring in the structure of $^6$He composite at low energy. In this study, one could only obtain information about correlations in neutron $\ell = 1$ partial wave. It would be interesting to determine also the effects of neutron $\ell = 0$ partial wave at neutron emission threshold. Indeed, as clearly shown by the Wigner power laws of Eqs. (7.45), (7.46), both $\ell = 0$ ($s$ wave) and $\ell = 1$ ($p$ wave) partial waves are likely to create cusps in transfer cross sections, as they induce an inverse-square-root- or square-root-dependence of transfer cross sections, respectively, as a function of linear momentum $k$ for $k \sim 0$.

A good candidate for the study of near-threshold properties involving neutron $\ell = 0$ partial wave is $^{11}$Li. The ground state of $^{11}$Li is a two-neutron halo state of two-neutron separation energy close to 350 keV [28]. Hence, it can be modelled by a $^9$Li core surrounded by two valence neutrons. $^{11}$Li can be considered as a Borromean system because all subsystems, namely dineutron in the partition $[2n \oplus ^9\text{Li}]$, or $^{10}$Li in the partition $[n \oplus ^{10}\text{Li}]$ are unbound.

The wave function of the ground state of $^{10}$Li in the three-body model is an antibound $\ell = 0$ neutron state $1s_{1/2}$ coupled to the $^9$Li core. Consequently, the ground state of $^{11}$Li possesses large components of neutron-$s$ basis states. Theoretical calculations typically predict a many-body wave function of $^{11}$Li made of about 50% neutron-$s$ states, 50% neutron-$p$ states, possibly along with a minor part built from neutron-$d$ states [105–108], in accordance with experimental studies [109]. The wave function components involving $\ell = 1$ neutron basis states can be explained from the vicinity of a one-body $0p_{1/2}$ neutron state lying close to neutron emission threshold. Consequently, besides the $3/2_1^-$ ground state of $^{11}$Li, one can also expect a low-lying $3/2_1^+$ excited state of $^{11}$Li, as suggested in Ref. [110]. One can reasonably assume that the $3/2_1^+$ state of $^{11}$Li would then mainly consist of two neutrons occupying $1s_{1/2}$ and $0p_{1/2}$ one-body states and coupled to $0^-$.

One will represent the $^9$Li core by a Woods–Saxon potential, whose $1s_{1/2}$ and $0p_{1/2}$ one-body states have an energy of $-0.1$ MeV and $0.236$ MeV, respectively. The energy of the $0p_{1/2}$ one-body state is fitted to that of the $1_1^+$ excited state of $^{10}$Li. This state can be approximated by the $^9$Li core coupled to a neutron in a $0p_{1/2}$ shell. One can also note that the width of the calculated $0p_{1/2}$ one-body state is 122 keV, very close to the experimental value of 100 keV. Consequently, the $1_1^+$ excited state of $^{10}$Li is well described in a single-particle model, as it mainly consists of a neutron occupying a $0p_{1/2}$ resonance state, generated by a well-bound $^9$Li core. The ground state of $^{10}$Li is unbound, and is very close to a $^9$Li core coupled to an antibound $1s_{1/2}$ neutron state, of energy around $-25$ keV. It would then seem logical to consider an antibound $1s_{1/2}$ neutron state in the model used. However, the spectroscopic factor loses its physical meaning in this situation.

Indeed, as demonstrated in Sect. 3.7.2, the inclusion of an antibound state in the Berggren basis generates a very large coupling to continuum. In particular, the pole approximation (see Sect. 5.4) including antibound states is not appropriate when considering many-body bound states. In this case, the scattering components of many-body bound states would have to suppress the antibound character of pole components, and generate the bound structure of the considered many-body states. As the result, the many-body wave function calculated in full model space would have a very small overlap with that devised in pole approximation. Therefore, spectroscopic factors which are a measure of correlations in many-body wave functions, lose their significance in this situation.

Spectroscopic factors are close to one when a resonant valence nucleon is decoupled from the rest of the nucleus. On the contrary, if the neutron occupies an antibound $1s_{1/2}$ state, its associated spectroscopic factor is arbitrary and does not provide with any useful information. Therefore, in the example discussed below it has been chosen to include a loosely bound $1s_{1/2}$ neutron state instead of an antibound state. We will see that even if energy of $s_{1/2}$ neutron state is changed by about 100 keV, this choice generates unique features in the pairing correlations involving the $s_{1/2}$ neutron partial wave. In the following, we will use the Berggren basis generated by the $^9$Li core, as in this case the evaluation of $^{11}$Li wave functions is numerically precise. For the nucleon–nucleon interaction, one will use the Minnesota interaction [111] whose parameters are varied to achieve either weakly bound or unbound wave functions for $^{11}$Li. The Berggren basis consist of the $s_{1/2}$ and $p_{1/2}$ neutron scattering states, along with the resonant $1s_{1/2}$ and $0p_{1/2}$ states, to which $p_{3/2}$ scattering states are added. The $0s_{1/2}$ and $0p_{3/2}$ single-particle states are not taken into account as they are occupied in the $^9$Li core.

Let us first consider the spectroscopic factor $S = \langle ^{11}\text{Li}(3/2_1^-) | [^{10}\text{Li}(2_1^-) \otimes s_{1/2}]^{3/2^+} \rangle^2$, where $^{10}\text{Li}(2_1^-)$ consists of the weakly bound $1s_{1/2}$ neutron single-particle state of the Berggren basis coupled to the $^9$Li ground state. In order to study the behavior of the $^{11}$Li wave function at small neutron separation energy, the parameters of Minnesota interaction are adjusted so that the $^{11}$Li ground state has a binding energy varying from $-1$ MeV to $-0.1$ MeV. Results are shown in Fig. 7.11. One can see that the energy dependence of spectroscopic factors consists of two different regions. Firstly, the spectroscopic factor decreases smoothly from 2 at $-0.2$ MeV to a value close to 1 at $-1$ MeV. Indeed, if $^{11}$Li ground state is bound, it has a sizable $p$ components and its spectroscopic factor is smaller than the maximal value of $S = 2$. When the interaction becomes less attractive, the coupling to $p$ waves decreases and the $^{11}$Li wave function becomes very close to the single-particle $[1s_{1/2}]^2$ configuration. The spectroscopic factor in this case decreases abruptly to 1. Moreover, the energy of the $^{11}$Li ground state remains smaller than $-0.1$ MeV, which is the energy of the $1s_{1/2}$ neutron state, when the nucleon–nucleon interaction becomes infinitely repulsive.

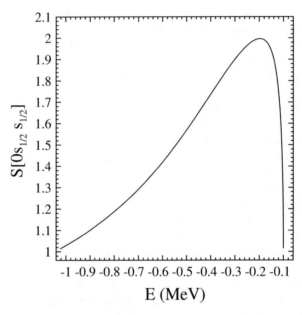

**Fig. 7.11** Real part of the spectroscopic factor $S = \langle {}^{11}\text{Li}(3/2_1^-)|[{}^{10}\text{Li}(2_1^-) \otimes s_{1/2}]^{3/2^+}\rangle^2$ is shown as a function of the energy $E$ of the $3/2_1^-$ state of ${}^{11}\text{Li}$. In this example, the $3/2_1^-$ state is built from a two-neutron wave function, of quantum number $0_1^+$, coupled to the $3/2_1^-$ ground state of the ${}^9\text{Li}$ core. $2_1^-$ state of ${}^{10}\text{Li}$ is a $1s_{1/2}$ neutron single-particle state coupled to the $3/2_1^-$ ground state of the ${}^9\text{Li}$ core

The phenomenon of a bound ground state whose energy converges to a finite value when the interaction strength becomes infinite, has not been encountered in previous studies. It is, in fact, relating to a wave function possessing almost exclusively neutron $s$ waves. As the neutron separation energy goes to zero, the wave function of ${}^{11}\text{Li}$ becomes almost equal to one neutron occupying a weakly bound $1s_{1/2}$ state, and a second neutron occupying $s_{1/2}$ scattering states. As the energy of the $1s_{1/2}$ neutron state is fixed, the extension of the two-neutron wave function of ${}^{11}\text{Li}$ at neutron emission threshold is independent of this state. On the contrary, the $s_{1/2}$ scattering states occupied by the second neutron generate a weakly bound wave function, which spreads more and more in space as the nucleon–nucleon interaction becomes infinitely repulsive. The extension of the two-neutron wave function of ${}^{11}\text{Li}$ is then only governed by the second neutron therein.

The wave function of a second neutron is thus qualitatively similar to the $s_{1/2}$ neutron one-body state whose energy goes to zero (see the discussion in Sect. 7.2.1.1 of neutron $s$ states with $\kappa \rightarrow 0$). Consequently, the probability to find the second neutron in the nuclear zone vanishes when the neutron separation energy goes to zero. As the nuclear interaction is short-ranged, the two-neutron wave function is no longer depending on nucleon–nucleon interaction when this interaction becomes infinitely repulsive. Thus, the two-neutron wave function of ${}^{11}\text{Li}$ in the limit of an infinitely repulsive neutron-neutron interaction becomes that of one neutron occupying the $1s_{1/2}$ single-particle state with an energy of $-0.1\,$MeV, and of a second neutron, of almost zero energy with vanishing probability in the nuclear zone (see the above discussion about the replacement of a neutron antibound state $1s_{1/2}$ by a weakly bound $1s_{1/2}$ neutron state). Consequently, energy of the $^{11}$Li ground state converges to $-0.1\,$MeV and the spectroscopic factor tends to half its maximal value, i.e. becomes equal to 1.

This unusual situation occurs because there is no centrifugal barrier in a neutron-$s$ partial wave, so that a neutron can spread throughout all space. Consequently, it is impossible for a correlated two-body state to be unbound and have a decay width generated by neutron $s$ states only. Indeed, the zero neutron separation energy corresponds to the energy of the weakly bound $1s_{1/2}$ shell for the $^{11}$Li wave function, so that having an energy larger than this one is impossible even with an infinitely repulsive interaction.

If a wave function of one neutron is negligible in the nuclear zone, then also the matrix elements of interaction involving both valence neutrons are negligible. This occurs if the valence neutrons occupy only the neutron $s$ states. Thus, the pairing correlation vanishes in a vicinity of particle-emission threshold if the many-body wave function mainly consist of neutron $s$ basis states. Note that due to the disappearance of pairing correlations close to particle-emission threshold, the pairing anti-halo effect [63] cannot develop. Therefore, the wave function generates a halo of neutron $s$ waves at low energy, which fills all space at the limit of zero neutron separation energy.

The last study concerns the spectroscopic factors where the $^{11}$Li$(3/2_1^+)$ excited state is involved. Due to the parity conservation, the two valence neutrons of $^{11}$Li$(3/2_1^+)$ in a pole approximation are in the $[0p_{1/2}\,1s_{1/2}]$ configuration and are coupled to $0^-$. Consequently, two different pole states are occupied and therefore two different one neutron spectroscopic factors are important in this configuration. In the first spectroscopic factor, equal to $\langle^{11}\text{Li}(3/2_1^+)|[^{10}\text{Li}(2_1^-)\otimes p_{1/2}]^{3/2^+}\rangle^2$, the weakly bound $1s_{1/2}$ basis neutron state is coupled to the $^9$Li ground state (see Fig. 7.12). In the second spectroscopic factor, equal to $\langle^{11}\text{Li}(3/2_1^+)|[^{10}\text{Li}(2_1^+)\otimes s_{1/2}]^{3/2^+}\rangle^2$, the $0p_{1/2}$ neutron state is coupled to the $^9$Li ground state (see Fig. 7.13). Due to the presence of the $0p_{1/2}$ state, the many-body wave function of $^{11}$Li$(3/2_1^+)$ has components depending on $\ell = 1$ partial waves, so that the current situation is much closer to that of Fig. 7.8.

Indeed, one can clearly see a Wigner cusp in Fig. 7.12 at the energy of $-0.1\,$MeV, when the state $^{11}$Li$(3/2_1^+)$ becomes unbound. The dependencies seen in Wigner expressions of Eqs. (7.45), (7.46) are also visible. Indeed, denoting as $S_{1n}$ the separation energy of one neutron, the spectroscopic factor behaves as $S_{1n}^{1/2}$ for $S_{1n} > 0$ and has an infinite derivative for $S_{1n} \to 0^+$, whereas it becomes regular for $S_{1n} < 0$, in accordance with the $S_{1n}^{3/2}$ law. In Fig. 7.12, one can see the opening of a $p_{1/2}$ neutron emission channel, occurring at $E = 0.236\,$MeV. The spectroscopic factor $\langle^{11}\text{Li}(3/2_1^+)|[^{10}\text{Li}(2_1^-)\otimes p_{1/2}]^{3/2^+}\rangle^2$ starts to increase quickly

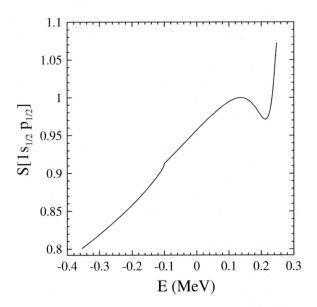

**Fig. 7.12** Real part of the spectroscopic factor $S = \langle {}^{11}\text{Li}(3/2_1^+)|[{}^{10}\text{Li}(2_1^-) \otimes p_{1/2}]^{3/2^+}\rangle^2$ is plotted as a function of the energy $E$ of the $3/2_1^+$ state of ${}^{11}\text{Li}$. The ${}^{11}\text{Li}(3/2_1^+)$ state is built from a two-neutron wave function, of quantum number $0_1^-$, coupled to the ground state of the ${}^{9}\text{Li}$ core, of quantum number $3/2_1^-$. The ${}^{10}\text{Li}(2_1^-)$ state is a $1s_{1/2}$ neutron one-body state coupled to the $3/2_1^-$ ground state of the ${}^{9}\text{Li}$ core

at $E = 0.236\,\text{MeV}$, whereas it was smoothly decreasing between $E = 0.1\,\text{MeV}$ and $E = 0.2\,\text{MeV}$. For energies above this channel threshold, the ${}^{11}\text{Li}(3/2_1^+)$ wave function can decay either to the $2_1^-$ ground state or to the $1_1^+$ excited state of ${}^{10}\text{Li}$. However, as the $0p_{1/2}$ state is unbound, the spectroscopic factor at $E = 0.236\,\text{MeV}$ behaves differently than at $E = -0.1\,\text{MeV}$.

Indeed, one can see that there is no cusp at $0.236\,\text{MeV}$, even though the spectroscopic factor varies abruptly in this region. One can explain the absence of cusp by the fact that the ${}^{11}\text{Li}$ wave function was already unbound for $E < 0.236\,\text{MeV}$. Consequently, the change of its asymptote is not as important as at $E = -0.1\,\text{MeV}$, and the spectroscopic factor has a smooth behavior even though a new channel opens.

The absence of cusp at $E = 0.236\,\text{MeV}$ in Fig. 7.12 is not inconsistent. Indeed, the opening of a channel in an already unbound many-body state is not described by Eqs. (7.45), (7.46), where an emission channel opens in a bound many-body nuclear wave function. Added to that, two phenomena occur around $E = 0.236\,\text{MeV}$, denoted as P1 and P2. Firstly, the nucleon–nucleon interaction strength is minimal for $E \simeq 0.135\,\text{MeV}$, so that the spectroscopic factor reaches a maximum therein (P1). Secondly, the spectroscopic factor decreases for $E$ slightly above $0.135\,\text{MeV}$ as the interaction strength starts increasing in absolute value (P2).

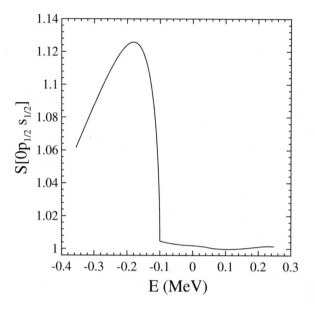

**Fig. 7.13** Real part of the spectroscopic factor $S = \langle{}^{11}\text{Li}(3/2_1^+)|[{}^{10}\text{Li}(2_1^+) \otimes s_{1/2}]^{3/2^+}\rangle^2$ as a function of the energy $E$ of the $3/2_1^+$ state of ${}^{11}\text{Li}$. The ${}^{11}\text{Li}(3/2_1^+)$ ground state is built from a $0_1^-$ two-neutron wave function, coupled to the $3/2_1^-$ ground state of the ${}^{9}\text{Li}$ core. The ${}^{10}\text{Li}(2_1^+)$ state is a $0p_{1/2}$ neutron one-body state coupled to the $3/2_1^-$ ground state of the ${}^{9}\text{Li}$ core

This effect is peculiar, because the components associated with the $[p_{1/2}\, s_{1/2}]$ basis configurations augment for $E > 0.236$ MeV. This arises because the basis components associated with the $[0p_{1/2}\, s_{1/2}^{(\text{scat})}]$ basis configurations[1] firstly have a positive real part which decreases, and eventually become negative. Therefore, the spectroscopic factor decreases, and then increases, instead of exhibiting the expected decrease occurring when a nuclear system becomes more correlated. The $[p_{1/2}^{(\text{scat})}\, 1s_{1/2}]$ basis state components have been checked to be negligible. The combination of the P1 and P2 phenomena (see above) thus explains the minimum found at $E = 0.236$ MeV, and also the presence of a spectroscopic factor larger than 1 for $E > 0.236$ MeV.

Let us now consider the spectroscopic factor $S = \langle{}^{11}\text{Li}(3/2_1^+)|[{}^{10}\text{Li}(2_1^+) \otimes s_{1/2}]^{3/2^+}\rangle^2$, depicted in Fig. 7.13. The form of the spectroscopic factor for $S_{1n} > 0$ is similar to that of Fig. 7.11. Indeed, as the interaction strength become less and less attractive, the coupling between $s$ and $p$ waves decreases, so that the spectroscopic factor reaches a maximum around $E = -0.2$ MeV. When the neutron separation energy in ${}^{11}\text{Li}(3/2_1^+)$ approaches zero, the wave function spreads more in space

---

[1] The superscript "*scat*" indicates that only scattering states are considered in the associated partial wave.

because relative weight of the $p_{1/2}$ scattering components in the wave function of $^{11}$Li($3/2^+_1$) increases. The short-ranged nuclear interaction is then less and less effective, which leads to a decrease of the spectroscopic factor. However, that effect is not as pronounced as in Fig. 7.11 due to the presence of a centrifugal barrier for $\ell = 1$ basis states. As in Fig. 7.11, the $p_{1/2}$ components always augment when the binding energy of $^{11}$Li($3/2^+_1$) becomes smaller, so that the corresponding spectroscopic factor also decreases (see Fig. 7.13).

A Wigner cusp for $\ell = 1$ neutron states also develops at $S_{1n} = 0$, as imposed by the Wigner laws (7.45), (7.46). However, the spectroscopic factor $\langle ^{11}\text{Li}(3/2^+_1)|[^{10}\text{Li}(2^+_1) \otimes s_{1/2}]^{3/2^+}\rangle^2$ remains almost constant for $S_{1n} < 0$ MeV, even at the threshold of the $0p_{1/2}$ emission channel ($E = 0.236$ MeV), where a cusp might a priori be expected to occur. This arises due to a destructive interference which occurs between the $[0p_{1/2}\, 1s_{1/2}]$ and $[0p_{1/2}\, s^{(\text{scat})}_{1/2}]$ basis states components. Indeed, the contribution of the $[0p_{1/2}\, s^{(\text{scat})}_{1/2}]$ basis states is negative, while that of $[0p_{1/2}\, 1s_{1/2}]$ is larger than 1 by about the same amount. Hence, the spectroscopic factor does not necessarily change significantly when a new decay channel opens, as pairing correlations therein can have an enhancing or mitigating effect at particle-emission threshold.

The presence of neutron-$s$ waves in many-body wave functions is especially important when other partial waves play a negligible role. In the example depicted in Fig. 7.11, one has seen that it is impossible to obtain unbound resonance states, even though the nucleon–nucleon interaction becomes infinitely repulsive. In this situation, bound states become more extended in space due to the absence of a centrifugal barrier, similarly to the one-body case (see Sect. 3.7.2). If one has both $s$ and $p$ waves in the many-body wave function, Wigner cusps do not radically change compared to the case where only $p$ waves play a significant role (see Figs. 7.12 and 7.13). Indeed, the passage from bound to resonance many-body state is only due to the $p$-wave components, so that Wigner cusps obey the power laws of Eqs. (7.45), (7.46) for $\ell = 1$.

The situation changes if the many-body state of $A - 1$ nucleons is a resonance (see Fig. 7.12). In this case, the spectroscopic factor does not follow the power-law expressions of Eqs. (7.45), (7.46), because the $^{11}$Li($3/2^+_1$) nuclear many-body state was already unbound for smaller energies. However, that emission channel can still generate important change of the neutron spectroscopic factor. This change is nevertheless dependent on the pairing correlations, which can provide with constructive or destructive interferences when a channel associated with a many-body resonance state of $A - 1$ nucleons opens.

## 7.5    Mirror Nuclei and Isospin Symmetry Breaking

The very different asymptotic behavior of proton and neutron wave functions at the particle-emission threshold induces breaking of isospin symmetry. Moreover, in loosely bound and resonance many-body states, a strong coupling to the scattering continuum is present. If one uses realistic nucleon–nucleon interactions, the continuum coupling may indeed lead to isospin breaking even in the absence of Coulomb interaction. Indeed, the charge dependence of realistic nucleon–nucleon interactions might lead to sizable isospin breaking. Consequently, we will study the spectroscopic factors of isobaric analog states in mirror nuclei using a Hamiltonian where isospin is broken solely by the Coulomb interaction. This will allow to disentangle the isospin-breaking effects of the nuclear and Coulomb parts in the Hamiltonian, on the one hand, and to quantify the importance of the Coulomb interaction at particle-emission threshold, on the other hand.

Light nuclei are good candidates for that matter, as they are experimentally accessible at both proton and neutron driplines. With this in mind, we will study the spectroscopic factors $[^5\text{He(g.s.)} \otimes \nu p_{3/2}]^{0^+}$, $[^5\text{Li(g.s.)} \otimes \pi p_{3/2}]^{0^+}$, $[^5\text{Li(g.s.)} \otimes \nu p_{3/2}]^{0^+}$ and $[^5\text{He(g.s.)} \otimes \pi p_{3/2}]^{0^+}$, Corresponding, respectively, to the $^6\text{He}/^5\text{He}$, $^6\text{Be}/^5\text{Li}$, $^6\text{Li}/^5\text{Li}$ and $^6\text{Li}/^5\text{He}$ pairs. In order to isolate the effect of Coulomb interaction, the Hamiltonian parameters have been fitted so that the ground state energies of the $A = 5$ mirror nuclei of a given pair have the same value. Consequently, the differences between proton and neutron spectroscopic factors arise solely from the different asymptotics of the corresponding wave functions.

The Hamiltonian of He and Li isotopes consists of a Woods–Saxon potential representing the core (see Eq. (3.83)), a finite range modified surface Gaussian interaction (see Eq. (9.174)), and the recoil term which arises from the use of cluster orbital shell model coordinates (see Sect. 5.2). The Coulomb potential is generated by a proton density of a Gaussian form. This form for the Coulomb potential is convenient as it is analytical, hence its matrix elements can be calculated with a Berggren basis numerically, as discussed in Sect. 3.7.1. The Coulomb Hamiltonian is separated in one- and two-body parts in order to treat its infinite-range exactly (see Eq. (5.67)). As the $^6\text{Be}$ nucleus has two valence protons, the two-body Coulomb interaction, denoted as $V_{\text{Coul}}$, must be taken into account. Due to the high precision needed to calculate spectroscopic factors, it is necessary to check that the expansion of Eq. (5.67) converges rapidly.

The convergence of the energy and width of $^6\text{He}$ and $^6\text{Be}$ ground states, as well as that of the spectroscopic factor associated with the pairs of nuclear ground states $^6\text{He}/^5\text{He}$ and $^6\text{Be}/^5\text{Li}$, are shown in Fig. 7.14. Even though recoil is automatically included using cluster orbital shell model coordinates, calculations have been done also without recoil to emphasize the infinite-range character of the recoil term. One can see in Fig. 7.14 that the convergence is slower when recoil is taken into account. In order to explain this fact, let us consider the point Coulomb potential and recoil in the momentum space. The matrix elements of the Coulomb potential in the Berggren basis bear a logarithmic divergence in $\ln(|k - k'|)$ for $k \sim k'$, where $k$ and $k'$ are the linear momenta of two scattering one-body states (see Sect. 3.7.1). The Fourier transform of the recoil term is analytical. It consists of the matrix elements $\langle \mathbf{k}_1' \mathbf{k}_2' | (\mathbf{p}_1 \cdot \mathbf{p}_2)/M_{\text{core}} | \mathbf{k}_1 \mathbf{k}_2 \rangle$, where $|\mathbf{k}_{1,2}\rangle$ and $|\mathbf{k}_{1,2}'\rangle$ are states of plane waves of linear momenta $\mathbf{k}_{1,2}$ and $\mathbf{k}_{1,2}'$, respectively, and where $M_{\text{core}}$ is the mass of the core. These recoil matrix elements are clearly equal

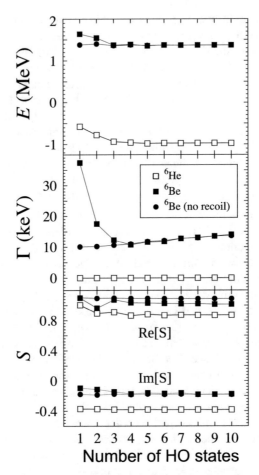

**Fig. 7.14** Numerical test of the expansion of the Coulomb interaction ($^6$Be) and recoil term ($^6$He and $^6$Be) in a basis of harmonic oscillator states. The Hamiltonian consists of the core Woods–Saxon potential, the residual two-body nuclear interaction, the Coulomb interaction (see Eq. (5.67) and related discussion for its treatment) and the two-body recoil term issued from the use of cluster orbital shell model coordinates (see Sect. 5.2). Upper panel represents energy of the ground states of $^6$He and $^6$Be. Width of $^6$He and $^6$Be ground states is shown in the middle panel. Lower panel depicts the spectroscopic factors $[^5$He(g.s.)$\otimes \nu p_{3/2}]^{0^+}$ and $[^5$Li(g.s.)$\otimes \pi p_{3/2}]^{0^+}$. The data related to the $^6$Be ground state are represented with full symbols. One uses circles or squares for the data involving the $^6$Be ground state when the recoil term of the Hamiltonian is removed or included, respectively. Results depicting the convergence pattern of the $^6$He ground state are shown with empty squares (from Ref. [112])

to $(\mathbf{k}_1 \cdot \mathbf{k}_2)/M_{\text{core}}\, \delta(\mathbf{k}'_1 - \mathbf{k}_1)\, \delta(\mathbf{k}'_2 - \mathbf{k}_2)$. To be properly expanded, the Dirac delta function, which is singular, demands more harmonic oscillator states than the $\ln(|k - k'|)$ function, which is integrable for $k \sim k'$. This explains the slower convergence encountered in Fig. 7.14 when recoil is included. One can see that

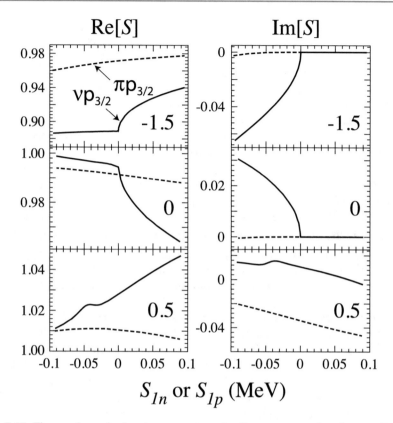

**Fig. 7.15** The real and imaginary parts of the spectroscopic factor $S$ for $\langle {}^6\mathrm{He(g.s.)}|[{}^5\mathrm{He(g.s.)} \otimes \nu p_{3/2}]^{0^+}\rangle$ (solid line) and $\langle {}^6\mathrm{Be(g.s.)}|[{}^5\mathrm{Li(g.s.)} \otimes \pi p_{3/2}]^{0^+}\rangle$ (dotted line) as a function of one-nucleon separation energy ($S_{1n}$ for ${}^6\mathrm{He}$ and $S_{1p}$ for ${}^6\mathrm{Be}$) for three different values of the one-nucleon threshold energy in the respective $A = 5$ nucleon systems: $E_T = -1.5, 0$ and $0.5\,\mathrm{MeV}$, indicated on the right of each panel (from Ref. [112])

the convergence is attained for both energies and spectroscopic factors with nine harmonic oscillator states. The width is obtained with a tiny error of $\sim$2 keV which is unimportant compared to other theoretical uncertainties of the model.

In the following, one will compare properties of mirror nuclei: a Borromean halo nucleus ${}^6\mathrm{He}$ and ${}^6\mathrm{Be}$ which is a 2p emitter in the ground state. The spectroscopic factors for ${}^6\mathrm{He}$ and ${}^6\mathrm{Be}$ are plotted in Fig. 7.15 as a function of the one-nucleon separation energy in $A = 6$ nuclei ($S_{1n}$ for ${}^6\mathrm{He}$ and $S_{1p}$ for ${}^6\mathrm{Be}$) for three different values of the one-nucleon threshold energy ($-S_{1n}$ for ${}^5\mathrm{He}$ and for $-S_{1p}$ ${}^5\mathrm{Li}$) in ${}^5\mathrm{He}$ and ${}^5\mathrm{Li}$ nuclei, respectively. For the bound $A = 5$ nuclei ($E_T = -1.5\,\mathrm{MeV}$), the spectroscopic factors of neutron and proton differ for all separation energies $S_{1n}$ and $S_{1p}$ in ${}^6\mathrm{He}$ and ${}^6\mathrm{Be}$, respectively. The difference between these spectroscopic factors is maximal at one-nucleon emission threshold. When $S_{1n}/S_{1p}$ increases, spectroscopic factors become close to one which is the value predicted by simple standard shell model considerations [12]. The Wigner cusp, characteristic of the $\ell = 1$ neutron partial wave, is visible for $^6$He. The absence of the cusp in the spectroscopic factors of the $^6$Be nucleus, originates in the different asymptotic between proton and neutron wave functions [100].

The energy dependence of spectroscopic factors changes if the $A = 5$ particle system is exactly at one-nucleon threshold ($E_T = 0$) or is unbound ($E_T = 0.5\,\text{MeV}$). In these two situations, the spectroscopic factors of bound $A = 6$ particle systems ($S_{1n}$ or $S_{1p}$ positive) slightly differ. Wigner cusp disappears in neutron spectroscopic factor when $^5$He becomes unbound (see the $E_T = 0.5\,\text{MeV}$ case in Fig. 7.15). In fact, one can only see a bump developing close to $S_{1n} \simeq -0.05\,\text{MeV}$ when $E_T = 0.5\,\text{MeV}$, whose origin cannot be related to the particle-emission threshold, as neutron spectroscopic factors do not change significantly at $S_{1n} = 0\,\text{MeV}$. For the experimental values of $S_{1n}$, $S_{2n}$ and $S_{1p}$, $S_{2p}$ in $^6$He and $^6$Be, respectively, the spectroscopic factors equal: $\langle^6\text{He(g.s.)}|[^5\text{He(g.s.)} \otimes \nu p_{3/2}]^{0^+}\rangle = 0.88 - i0.386$ and $\langle^6\text{Be(g.s.)}|[^5\text{Li(g.s.)} \otimes \pi p_{3/2}]^{0^+}\rangle = 1.057 - i0.181$. One sees that $^6$He in the ground state is significantly more collective than its mirror partner $^6$Be which can be viewed as a single proton attached to the proton-unbound core of $^5$Li.

In some cases, the real part of spectroscopic factors can be larger than one (see $\Re[S]$ of the $^6$He nucleus for $E_T = 0.5\,\text{MeV}$ in Fig. 7.15). This non-intuitive behavior may have several origins. Firstly, the imaginary part of the average value of an operator acting on resonance state can be interpreted as an uncertainty on the physical value of the associated observable induced by the possibility of decay during the measuring process (see Sect. 5.1.2). The right column of Fig. 7.15 illustrates the uncertainty $\Im[S]$ of the spectroscopic factors shown in the left column. The uncertainty disappears when both $A = 6$ and $A = 5$ nuclei are bound. However, one may notice in Fig. 7.15, that $\Re[S] > 1$ is not totally explained by a large value of $\Im[S]$. Indeed, $\Re[S] > 1$ corresponds to $\Im[S] \simeq 0$ in $^6$He at $E_T = 0.5\,\text{MeV}$ and $0.05 < S_{1n} < 0.1\,\text{MeV}$.

Secondly, $\Re[S] > 1$ may originate from the coupling between $\Psi_{A-1}^{J_{A-1}} \equiv \Psi_{A-1;R}^{J_{A-1}}$, where R stands for resonance, and the scattering states of the nonresonant continuum $\{\Psi_{A-1;c}^{J_{A-1}}\}$ of energies close to that of the resonance [113]. Let us remind here the interference phenomenon between resonant and nonresonant continuum, which explains the appearance of a cusp. Indeed, neglecting coupling to nonresonant continuum destroys the unitarity of the formalism, so that no redistribution of flux can occur. The contributions of the spectroscopic factors $S_c$ originating from the non-resonant continuum of $A - 1$ particles are depicted in Fig. 7.16) (upper and lower parts for its real and imaginary part, respectively), as well as those corresponding to resonant states. In all the situations presented in Fig. 7.16), the contribution of the one-body resonance state to $\Re[S]$ is dominant.

The impact of the nonresonant continuum depends on $E_T$ and $S_{1p}/S_{1n}$. For $E_T \leq 0$ and $S_{1p}/S_{1n} \geq 0$, i.e. for bound ground states of $A = 5$ and $A = 6$ nucleons, the nonresonant continuum is practically negligible. The same phenomenon is present for $^6$Be with $E_T \leq 0$ and $S_{1p} < 0$, hence for a bound $^5$Li but for an

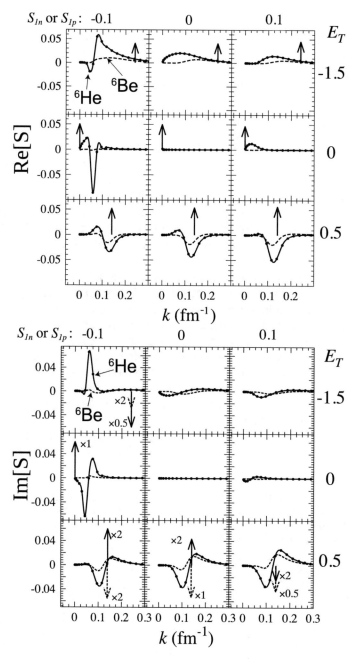

**Fig. 7.16** Distributions of real (upper) and imaginary (lower) parts of $S_c$ for $^6$He (solid line) and $^6$Be (dotted line) as a function of the Re[$k$] value of scattering states $3/2^-$ of the $^5$He and $^5$Li nuclei. Calculations have been done for three different values of the threshold energy $E_T$ of the $A = 5$ particle nuclei, which is $-1.5$ MeV (upper panel), 0 (middle panel), and 0.5 MeV (lower panel). Arrows indicate the contributions of resonant state, multiplied by a scaling factor equal to 0.05 unless precised differently on the figure (from Ref. [112])

unbound $^6$Be. However, when the $A = 5$ nucleon ground state is unbound ($E_T > 0$), or when $E_T \leq 0$ and $S_{1n} < 0$ for $^6$He, i.e. when $^6$He is unbound with respect to a bound $^5$He, the nonresonant continuum plays an important role for both real and imaginary parts of spectroscopic factors. In particular, when $E_T > 0$ and $S_{1p}/S_{1n} \geq 0$, $\Re[S_c]$ is negative, which implies that $\Re[S] > 1$. Moreover, $\Im[S_c]$ and $\Re[S_c]$ are comparable therein, although this does not occur every time that $\Re[S] > 1$. As a result, spectroscopic factors obtained solely via $A$-body resonances cannot fully reproduce the experimental spectroscopic factor, which can contain an important nonresonant contribution. The requirement of unitarity oblige us to keep both resonant and nonresonant contributions, as otherwise their artificial separation leads to errors.

Let us now consider a physical situation of $0^+$ and $2^+$ states of $^6$He, $^6$Li, and $^6$Be, which belong to the $T = 1$ triplet in the absence of Coulomb interaction [112]. The $0^+$ and $2^+$ states of $^6$He are fitted to their experimental value. This fixes the nuclear part of the Hamiltonian. Consequently, the Hamiltonian differs only by the way of its Coulomb part for $^6$Li and $^6$Be.

Energies, widths, and spectroscopic factors are given in Table 7.1. Due to the very simple nuclear interaction and the restricted model space, the energy of $^6$Be is 300 keV away from its experimental value. In particular, $s$ and $d$ partial waves are not included in the model space, so that one cannot exactly reproduce results obtained with the Furutani–Horiuchi–Tamagaki interaction in Sect. 6.3.1, where the fit of neutron-rich nuclei guarantees a good reproduction of their mirror proton-rich partners. One can see that the Coulomb Hamiltonian also induces differences between proton and neutron spectroscopic factors therein. Nevertheless, considering that their imaginary parts are related to the uncertainty of the real parts, it is difficult to differentiate between the values of the $0^+$ states.

One may note the important imaginary part of the spectroscopic factor of $^6$He/$^5$He. This arises because $^5$He is unbound and $^6$He is bound, so that the overlap

**Table 7.1** Energies ($E$ in MeV), widths ($\Gamma$ in keV) and spectroscopic factors ($S$) associated with the $T = 1$ triplet of $0^+$ and $2^+$ states of $^6$He, $^6$Li, and $^6$Be. Experimental energies and widths are denoted by $E_{exp}$ and $\Gamma_{exp}$, respectively. Proton and neutron spectroscopic factors of $^6$Li (see text) are indicated by "(p)" and "(n)" (see Ref. [112] for details)

| | $E$ | $E_{exp}$ | $\Gamma$ | $\Gamma_{exp}$ | |
|---|---|---|---|---|---|
| | (MeV) | (MeV) | (keV) | (keV) | $S$ |
| $^6$He $0^+$ | −0.974 | −0.973 | 0 | 0 | $0.8 - i0.383$ |
| $^6$He $2^+$ | 0.823 | 0.824 | 89 | 113 | $1.061 + i0.0011$ |
| $^6$Be $0^+$ | 1.653 | 1.371 | 41 | 92 | $1.015 - i0.0147$ |
| $^6$Be $2^+$ | 2.887 | – | 986 | – | $0.973 - i0.0142$ |
| $^6$Li $0^+$ (p) | 0.0866 | 0.136 | $8.85 \times 10^{-3}$ | $8.2 \times 10^{-3}$ | $1.061 - i0.280$ |
| $^6$Li $0^+$ (n) | 0.0866 | 0.136 | $8.85 \times 10^{-3}$ | $8.2 \times 10^{-3}$ | $0.911 - i0.361$ |
| $^6$Li $2^+$ (p) | 1.667 | 1.667 | 329 | 541 | $0.987 - i0.00326$ |
| $^6$Li $2^+$ (n) | 1.667 | 1.667 | 329 | 541 | $1.034 - i0.0235$ |

between $^5$He and $^6$He wave functions is not very large. Indeed, the bound asymptote of $^6$He is entirely due to continuum coupling, whereas its pole approximation, consisting of two $0p_{3/2}$ resonant neutron states, is unbound. This decreases the probability to find a resonance $^5$He ground state in the wave function of the $^6$He ground state. On the contrary, the ground states of $^5$Li and $^6$Be are both unbound, so that their associated spectroscopic factor is closer to one.

---

**Exercise IV ★★**

One will illustrate Wigner cusps in spectroscopic factors in different partial waves by considering a numerical example involving $s$, $d$ shells in the pairs $^{13}$C/$^{14}$C and $^{13}$N/$^{14}$O. Even though these nuclei are $p$-shell nuclei, one will concentrate on their excited states, where $p$-shells are unimportant. In this study, one will use a core of $^{12}$C.

**A.** Point out the advantages and disadvantages of the use of a $^{12}$C core in a shell model. For this, one will take into account the closed-shell character of the $^{12}$C ground state at the Hartree–Fock level, as well as the oblate deformation of the $^{12}$C ground state.

**B.** In the following, we will concentrate on the spectroscopic factors involving $s$ and $d$ shells. Explain why the $0p_{1/2}$ valence shells are unimportant for this study. Present the current situation when isospin symmetry is exact and explain why the isospin symmetry breaking is important in this example.

**C.** Run the two-body Gamow shell model code in cluster orbital shell model coordinates to calculate spectroscopic factors $^{13}$C/$^{14}$C and $^{13}$N/$^{14}$O.

Plot the spectroscopic factors and their derivatives as a function of energy, and explain the presence or absence of Wigner cusps at $S_{1n/1p} = 0$ MeV.

---

One can also study the impact of continuum couplings on isospin symmetry breaking by considering overlaps between the $A = 6$ nuclear ground states and the isomultiplet generated by the $0^+$ ground state of $^6$He [112]. For this, one needs to properly define the isospin $\hat{T}^-$ operator:

$$
\hat{T}^- = \sum_{\mathscr{B}} a^+_{\ell\,j\,t_z=-1/2}(\mathscr{B})\, a_{\ell\,j\,t_z=1/2}(\mathscr{B}) , \tag{7.51}
$$

where the proton and neutron $\mathscr{B}$ states of the same coordinate and spin space quantum numbers have to bear the same radial dependence. In order to fulfill the latter requirement, the neutron potential is used to generate both proton and neutron basis states. This demands to diagonalize the one-body Coulomb potential in the Gamow shell model basis, i.e. the calculation of matrix elements of Coulomb potential in a basis of neutron states. This is done using the method described in Sect. 3.7.1.

Acting on the ground state of $^6$He with an operator $\hat{T}^-$, respectively, once and twice, and normalizing accordingly, one obtains the $T = 1$ isomultiplet of this state, which will be denoted with a $T$ subscript. The overlap squared $\langle ^6\text{Li}|^6\text{Li}\rangle_T^2$ is equal to 0.995, which signifies that $|^6\text{Li}\rangle \simeq |^6\text{Li}\rangle_T$, so that isospin symmetry is almost exact in this case. This is an example of partial dynamical symmetry, where a symmetry is verified by some eigenstates of Hamiltonian but not by the Hamiltonian itself [114]. The overlap squared $\langle ^6\text{Be}|^6\text{Be}\rangle_T^2$ is equal to $0.951 - i0.050$, so that there is a reduction compared to the value of 1 of exact isospin symmetry. Uncertainties being smaller for $2^+$ states, one can conclude that the mirror symmetry breaking of order 5% is present in the eigenvectors calculated with the Gamow shell model. The isospin-breaking properties of the spectroscopic factors of the pairs $^{13}$C/$^{14}$N and $^{13}$N/$^{14}$O are studied as a numerical example in Exercise IV.

The explanation of isospin breaking in the framework of standard shell model demands the use of nuclear interactions which explicitly break isospin symmetry [115, 116]. We have seen, that the correct treatment of the continuum coupling can also generate isospin symmetry breaking effects, even for the isospin invariant interactions and, hence, is capable to explain at least a part of the isospin breaking effects in atomic nucleus.

## Solutions to Exercises[2]

### Exercise I

**A.** One obtains the root-mean-square radius limits (see Eqs. (7.12), (7.13)) when $\kappa \to 0$ by fitting the radii plots obtained with the attached computer code for small values of $\kappa$. As mentioned in Sect. 7.2.1.3, the root-mean-square radius value for $\ell = 0$ depends only on $\kappa$, while that for $\ell = 1$ depends on $\kappa$ and on the potential depth.

**B.** When $\ell \geq 2$, the root-mean-square radius converges to a finite value when $\kappa \to 0$, whose limit at $\kappa = 0$ is a function of the potential depth. Note that Eq. (7.5) does not strictly hold in the general case, as one has assumed that the Woods–Saxon potential vanishes after $R_0 + 2d$. However, the root-mean-square radius should converge to a value which is smaller or close to that given by Eq. (7.5), so that Eq. (7.5) must be seen as a rough estimate of that limit $\kappa \to 0$.

**C.** Negative values of $m$ do not have to be considered because one considers axially symmetric potentials. Axial symmetry implies that energies of eigenstates depend only on $|m|$. In the spherical case, root-mean-square radius limits are obtained similarly as in **A** and **B**. Eqs. (7.27–7.30) can be directly compared to Eqs. (7.5), (7.12), (7.13). One has $m = 1/2$ with $\pi = \pm 1$ which corresponds, respectively, to $\ell = 0, 1$. The case $m = 3/2$ with $\pi = -1$ is associated with

---

[2]The input files, codes and code user manual associated to computer-based exercises can be found at https://github.com/GSMUTNSR.

$\ell = 1$. All other cases provide with finite root-mean-square radius limits when $\kappa \to 0$, similarly to the $\ell \geq 2$ case of spherical potentials. The plots of radii issued from the calculations involving a spherical or deformed potential then have the same qualitative properties.

## Exercise II

A. The overlap asymptote of Eqs. (7.32) and (7.40) for the bound and unbound cases, respectively, is quickly reached. The overlap functions become equal to Eq. (7.32) or Eq. (7.40) around 10–15 fm. It is then straightforward to determine the asymptotic normalization coefficient by fitting the overlap function both in the case of bound and unbound states.

B. Eqs. (7.32), (7.40) depend on the $\kappa$ or $k$ value associated with one neutron separation energy, so that changing the interaction parameters also changes $\kappa$ or $k$. Consequently, the asymptotic normalization coefficient is modified, as the structure of involved many-body states changes along with interaction parameters.

## Exercise III

A. $^8$He is a bound nucleus, where proton $0s_{1/2}$ and neutron $0s_{1/2}$ and $0p_{3/2}$ shells are fully occupied. Added to that, the neutron $0p_{1/2}$ shell is unbound. Consequently, the use of $^8$He as a core to describe $^9$He and $^{10}$He allows to recapture the main features of the eigenstates of $^9$He and $^{10}$He, while having small Gamow shell model space dimensions.

The $1/2^-$ state of $^9$He and $0^+$ state of $^{10}$He have a small width, close to 100 keV. Therefore, similarly to the study of the $p_{3/2}$ spectroscopic factors of $^5$He/$^6$He, one can assume that the ground state of $^{10}$He is bound and have the energy of the $1/2^-$ state of $^9$He vary so that it goes from bound to unbound. A composite system of $2n+^8$He is then simulated for different excitation energies.

B. As one considers neutron states, the non-analytic cusps of Eqs. (7.45), (7.46) must occur at zero neutron separation energy for $\ell = 0, 1$ only. Moreover, as wave functions are built from the neutron $p_{1/2}$ partial wave, the overall behavior of spectroscopic factors is similar to the $^5$He/$^6$He case.

## Exercise IV

A. $^{12}$C has a deformed ground state, so that excitations from $0p_{3/2}$ shells are sizable. The $^{12}$C ground state wave function thus bears an important configuration mixing in a no-core shell model picture [117]. However, $^{12}$C is a closed-shell nucleus at Hartree–Fock level, and there is a sizeable gap between $0p_{3/2}$ and $0p_{1/2}$, of about 5 MeV. Thus, it is reasonable to assume that effective interactions can be generated by using a $^{12}$C core. Consequently, one can assume that taking $^{12}$C as a core can provide with an effective interaction able to reproduce the main features of $^{13,14}$C, $^{13}$N, and $^{14}$O. In fact, the Zuker-Buck-McGrory interaction

[118], an effective interaction based on the use of a $^{12}C$ core, can properly describe the energy spectra and observables of $^{16,17}O$ and $^{17}F$ [118–120].

**B.** As one considers spectroscopic factors involving $(sd)$ shells, the considered eigenstates of $^{13}C$, $^{13}N$, and $^{14}O$ will have energies close to particle-emission threshold. As $0p_{1/2}$ shells are well bound, their effect on spectroscopic factors at $S_{1n/1p} \sim 0$ will be minimal, contrary to the $1s_{1/2}$, $0d_{5/2}$, and $0d_{3/2}$ shells, which will be loosely bound or narrow resonances. Thus, $0p_{1/2}$ shells will not significantly impart spectroscopic factors.

Isospin symmetry is exact in the absence of Coulomb Hamiltonian. In this case, the pairs $^{13}C/^{14}C$ and $^{13}N/^{14}O$ behave identically. In the studied example, the Coulomb interaction is included in the Hamiltonian. Consequently, one can expect different behaviors of spectroscopic factors as the asymptotes of nuclear wave functions in the pairs $^{13}C/^{14}C$ and $^{13}N/^{14}O$ will be built from Hankel and Coulomb wave functions, respectively.

**C.** One does not obtain Wigner cusps at $S_{1n/1p} = 0\,MeV$ because the partial wave present in the model space bears $\ell = 0$ or $\ell = 2$. One has seen in Sect. 7.4.2.1 that $s$-waves cannot give rise to Wigner cusps, as one cannot have unbound states mainly occupied by $s$-waves. Moreover, neutron $d$-waves cannot generate Wigner cusps on spectroscopic factors, as they become visible only on the derivative of spectroscopic factors (see Sect. 7.4.2).

Thus, the studied spectroscopic factors involving neutrons, i.e. for the pair $^{13}C/^{14}C$, are similar to the spectroscopic factors $\langle ^{11}Li(3/2_1^-)|[^{10}Li(2_1^-) \otimes s_{1/2}]^{3/2^+}\rangle$ and $\langle ^{18}O(0_3^+)|[^{17}O(3/2_1^+) \otimes d_{3/2}]^{0^+}\rangle$ of Sect. 7.4.2 for $s$ and $d$ waves, respectively. Values of neutron $d$-waves spectroscopic factors form a continuous and differentiable function, but its derivative presents a non-analyticity at $S_{1n} = 0\,MeV$. Spectroscopic factors involving protons only, that is for the pair $^{13}N/^{14}O$, cannot generate Wigner cusps due to the presence the Coulomb barrier, as can be seen in Eqs. (7.47), (7.48).

## References

1. P. Kleinwächter, I. Rotter, Phys. Rev. C **32**, 1742 (1985)
2. I. Rotter, Rep. Prog. Phys. **54**, 635 (1991)
3. V. Sokolov, V. Zelevinsky, Phys. Lett. B **202**, 10 (1988)
4. V. Sokolov, V. Zelevinsky, Nucl. Phys. A **504**, 562 (1989)
5. V. Sokolov, V. Zelevinsky, Phys. Lett. B **202**, 10 (1988)
6. S. Drożdż, J. Okołowicz, M. Płoszajczak, I. Rotter, Phys. Rev. C **62**, 24313 (2000)
7. R. Dicke, Phys. Rev. **93**, 99 (1954)
8. N. Auerbach, V. Zelevinsky, Rep. Prog. Phys. **74**, 106301 (2011)
9. A. Baz, Soviet Phys. JETP **6**, 709 (1957)
10. R. Newton, Phys. Rev. **114**, 1611 (1959)
11. C. Hategan, Ann. Phys. (NY) **116**, 77 (1978)
12. N. Michel, W. Nazarewicz, M. Płoszajczak, Phys. Rev. C **75**, 031301 (2007)
13. Y. Fyodorov, B. Khoruzhenko, Phys. Rev. Lett. **83**, 65 (1999)
14. P. Kohler, F. Bečvář, M. Krtička, J. Harvey, K. Guber, Phys. Rev. Lett. **105**, 072502 (2010)
15. G. Celardo, N. Auerbach, F. Izrailev, G. Berman, Phys. Rev. Lett. **106**, 042501 (2011)

16. J. Wheeler, Phys. Rev. **52**, 1083 (1937)
17. C. von Weizsäcker, Naturwiss. **26**, 209 (1938)
18. W. Wefelmeier, Z. Phys. **107**, 332 (1937)
19. J. Hiura, I. Shimodaya, Prog. Theor. Phys. **30**, 585 (1963)
20. J. Hiura, I. Shimodaya, Prog. Theor. Phys. **36**, 977 (1966)
21. R. Tamagaki, H. Tanaka, Prog. Theor. Phys. **34**, 191 (1965)
22. S. Drożdż, J. Okołowicz, M. Płoszajczak, Phys. Lett. B **128**, 5 (1983)
23. K. Ikeda, N. Takigawa, H. Horiuchi, Prog. Theor. Phys. Suppl., Extra Number 464 (1968)
24. J. Okołowicz, M. Płoszajczak, W. Nazarewicz, Prog. Theor. Phys. Supp. **196**, 230 (2012)
25. J. Okołowicz, W. Nazarewicz, M. Płoszajczak, Fortschr. Phys. **61**, 66 (2013)
26. A. Baz, Y. Zel'dovich, A. Perelomov, *Scattering Reactions and Decay in Nonrelativistic Quantum Mechanics* (Israel Program for Scientific Translations Jerusalem, 1969)
27. N. Michel, W. Nazarewicz, M. Płoszajczak, Nucl. Phys. A **794**, 29 (2007)
28. http://www.nndc.bnl.gov/ensdf
29. F. de Grancey, A. Mercenne, F. de Oliveira Santos, T. Davinson, O. Sorlin, J. Angélique, M. Assié, E. Berthoumieux, R. Borcea, A. Buta, I. Celikovic, V. Chudoba, J. Daugas, G. Dumitru, M. Fadil, S. Grévy, J. Kiener, A. Lefebvre-Schuhl, N. Michel, J. Mrazek, F. Negoita, J. Okołowicz, D. Pantelica, M. Pellegriti, L. Perrot, M. Płoszajczak, G. Randisi, I. Ray, O. Roig, F. Rotaru, M. Saint Laurent, N. Smirnova, M. Stanoiu, I. Stefan, C. Stodel, K. Subotic, V. Tatischeff, J. Thomas, P. Ujić, R. Wolski, Phys. Lett. B **758**, 26 (2016)
30. E. Garrido, D. Fedorov, A. Jensen, Nucl. Phys. A **617**, 153 (1997)
31. T.B. Webb, S.M. Wang, K.W. Brown, R.J. Charity, J.M. Elson, J. Barney, G. Cerizza, Z. Chajecki, J. Estee, D.E.M. Hoff, S.A. Kuvin, W.G. Lynch, J. Manfredi, D. McNeel, P. Morfouace, W. Nazarewicz, C.D. Pruitt, C. Santamaria, J. Smith, L.G. Sobotka, S. Sweany, C.Y. Tsang, M.B. Tsang, A.H. Wuosmaa, Y. Zhang, K. Zhu, Phys. Rev. Lett. **122**, 122501 (2019)
32. F.M. Marqués, M. Labiche, N.A. Orr, J.C. Angélique, L. Axelsson, B. Benoit, U.C. Bergmann, M.J.G. Borge, W.N. Catford, S.P.G. Chappell, N.M. Clarke, G. Costa, N. Curtis, A. D'Arrigo, E. de Góes Brennand, F. de Oliveira Santos, O. Dorvaux, G. Fazio, M. Freer, B.R. Fulton, G. Giardina, S. Grévy, D. Guillemaud-Mueller, F. Hanappe, B. Heusch, B. Jonson, C. Le Brun, S. Leenhardt, M. Lewitowicz, M.J. López, K. Markenroth, A.C. Mueller, T. Nilsson, A. Ninane, G. Nyman, I. Piqueras, K. Riisager, M.G.S. Laurent, F. Sarazin, S.M. Singer, O. Sorlin, L. Stuttgé, Phys. Rev. C **65**, 044006 (2002)
33. K. Kisamori, S. Shimoura, H. Miya, S. Michimasa, S. Ota, M. Assie, H. Baba, T. Baba, D. Beaumel, M. Dozono, T. Fujii, N. Fukuda, S. Go, F. Hammache, E. Ideguchi, N. Inabe, M. Itoh, D. Kameda, S. Kawase, T. Kawabata, M. Kobayashi, Y. Kondo, T. Kubo, Y. Kubota, M. Kurata-Nishimura, C.S. Lee, Y. Maeda, H. Matsubara, K. Miki, T. Nishi, S. Noji, S. Sakaguchi, H. Sakai, Y. Sasamoto, M. Sasano, H. Sato, Y. Shimizu, A. Stolz, H. Suzuki, M. Takaki, H. Takeda, S. Takeuchi, A. Tamii, L. Tang, H. Tokieda, M. Tsumura, T. Uesaka, K. Yako, Y. Yanagisawa, R. Yokoyama, K. Yoshida, Phys. Rev. Lett. **116**, 052501 (2016)
34. C.A. Bertulani, V. Zelevinsky, J. Phys. G. Nucl. Part. Phys **29**, 2431 (2003)
35. S.C. Pieper, Phys. Rev. Lett. **90**, 252501 (2003)
36. E. Hiyama, R. Lazauskas, J. Carbonell, M. Kamimura, Phys. Rev. C **93**, 044004 (2016)
37. K. Fossez, J. Rotureau, N. Michel, M. Płoszajczak, Phys. Rev. Lett. **119**, 032501 (2017)
38. J. Okołowicz, M. Płoszajczak, W. Nazarewicz, Acta Phys. Pol. B **45**, 331 (2014)
39. I. Tanihata, H. Hamagaki, O. Hashimoto, Y. Shida, N. Yoshikawa, K. Sugimoto, O. Yamakawa, T. Kobayashi, N. Takahashi, Phys. Rev. Lett. **55**, 2676 (1985)
40. Y.E. Penionzhkevich, Nucl. Phys. A **616**, 247 (1997)
41. S.A. Fayans, O.M. Knyazkov, I.N. Kuchtina, Y.E. Penionzhkevich, N.K. Skobelev, Phys. Lett. B **357**, 509 (1995)
42. M. Yamagami, Eur. Phys. J. A **25**, 569 (2005)
43. R.E. Grisenti, W. Schöllkopf, J.P. Toennies, G.C. Hegerfeldt, T. Köhler, M. Stoll, Phys. Rev. Lett. **85**, 2284 (2000)
44. J.S. Al-Khalili, J.A. Tostevin, Phys. Rev. Lett. **76**, 3903 (1996)

45. I. Tanihata, Phys. Lett. B **160**, 380 (1985)
46. J. Al-Khalili, *The Euroschool Lectures on Physics with Exotic Beams*, vol. I (Springer, Berlin, 2004)
47. T. Bjerge, K.J. Borgström, Nature **138**, 400 (1936)
48. A.M. Poskanzer, S.W. Cosper, E.K. Hyde, J. Cerny, Phys. Rev. Lett. **17**, 1276 (1966)
49. P. Mueller, I.A. Sulai, A.C.C. Villari, J.A. Alcántara-Núñez, R. Alves-Condé, K. Bailey, G.W.F. Drake, M. Dubois, C. Eléon, G. Gaubert, R.J. Holt, R.V.F. Janssens, N. Lecesne, Z.T. Lu, T.P. O'Connor, M.G. Saint-Laurent, J.C. Thomas, L.B. Wang, Phys. Rev. Lett. **99**, 252501 (2007)
50. A.S. Jensen, K. Riisager, D.V. Fedorov, E. Garrido, Rev. Mod. Phys. **76**, 215 (2004)
51. W. Schöllkopf, J.P. Toennies, Science **266**, 1345 (1994)
52. A. Kalinin, O. Kornilov, W. Schöllkopf, J.P. Toennies, Phys. Rev. Lett. **95**, 113402 (2005)
53. P. Stipanović, L.V. Markić, J. Boronat, B. Kežić, J. Chem. Phys. **134**, 054509 (2011)
54. M. Abramowitz, *Handbook of Mathematical Functions, National Bureau of Standards, Applied Mathematics*, vol. 55, ed. by M. Abramowitz, I.A. Stegun, (National Bureau of Standards, Gaithersburg, 1972)
55. J. Okołowicz, N. Michel, W. Nazarewicz, M. Płoszajczak, Phys. Rev. C **85**, 064320 (2012)
56. A. Banu, L. Trache, F. Carstoiu, N.L. Achouri, A. Bonaccorso, W.N. Catford, M. Chartier, M. Dimmock, B. Fernández-Domínguez, M. Freer, L. Gaudefroy, M. Horoi, M. Labiche, B. Laurent, R.C. Lemmon, F. Negoita, N.A. Orr, S. Paschalis, N. Patterson, E.S. Paul, M. Petri, B. Pietras, B.T. Roeder, F. Rotaru, P. Roussel-Chomaz, E. Simmons, J.S. Thomas, R.E. Tribble, Phys. Rev. C **84**, 015803 (2011)
57. T. Misu, W. Nazarewicz, S. Åberg, Nucl. Phys. A **614**, 44 (1997)
58. D.V. Fedorov, A.S. Jensen, K. Riisager, Phys. Rev. C **49**, 201 (1994)
59. D.V. Fedorov, A.S. Jensen, K. Riisager, Phys. Rev. C **50**, 2372 (1994)
60. I.J. Thompson, B.V. Danilin, V.D. Efros, J.S. Vaagen, J.M. Bang, M.V. Zhukov, Phys. Rev. C **61**, 024318 (2000)
61. J. Dobaczewski, H. Flocard, J. Treiner, Nucl. Phys. A **422**, 103 (1984)
62. J. Dobaczewski, W. Nazarewicz, T.R. Werner, J.F. Berger, C.R. Chinn, J. Dechargé, Phys. Rev. C **53**, 2809 (1996)
63. K. Bennaceur, J. Dobaczewski, M. Płoszajczak, Phys. Lett. B **496**, 154 (2000)
64. P. Ring, P. Schuck, *The Nuclear Many-Body Problem*, 3rd edn. (Springer, Berlin, Heidelberg, New York, Hong King, London, Milan, Paris, Tokyo, 2004)
65. Y. Utsuno, T. Otsuka, T. Mizusaki, M. Honma, Phys. Rev. C **64**, 011301(R) (2001)
66. G.R. Satchler, *Direct Nuclear Reactions* (Clarendon Press, Oxford, 1983)
67. W.T. Pinkston, G.R. Satchler, Nucl. Phys. **72**, 641 (1965)
68. M.H. Macfarlane, J.B. French, Rev. Mod. Phys. **32**, 567 (1960)
69. N.K. Glendenning, Ann. Rev. Nucl. Sci. **13**, 191 (1963)
70. N.K. Glendenning, *Direct Nuclear Reactions* (Academic Press, London, 1983)
71. P. Fröbrich, R. Lipperheide, *Theory of Nuclear Reactions* (Oxford Science Publications, Clarendon Press, Oxford, 1996)
72. A.M. Mukhamedzhanov, A.S. Kadyrov, Phys. Rev. C **82**, 051601 (2010)
73. R.J. Furnstahl, A. Schwenk, J. Phys. G: Nucl. Part. Phys. **37**, 064005 (2010)
74. S.K. Bogner, R.J. Furnstahl, A. Schwenk, Prog. Part. Nucl. Phys. **65**, 94 (2010)
75. K.M. Nollett, R.B. Wiringa, Phys. Rev. C **83**, 041001 (2011)
76. M. Viviani, A. Kievsky, S. Rosati, Phys. Rev. C **71**, 024006 (2005)
77. K.M. Nollett, Phys. Rev. C **86**, 044330 (2012)
78. D.R. Entem, R. Machleidt, Phys. Rev. C **68**, 041001(R) (2003)
79. I. Brida, S.C. Pieper, R.B. Wiringa, Phys. Rev. C **84**, 024319 (2011)
80. A.M. Mukhamedzhanov, Phys. Rev. C **86**, 044615 (2012)
81. N. Michel, W. Nazarewicz, J. Okołowicz, M. Płoszajczak, Nucl. Phys. A **752**, 335 (2005)
82. A.M. Mukhamedzhanov, F.M. Nunes, Phys. Rev. C **72**, 017602 (2005)
83. D.Y. Pang, F.M. Nunes, A.M. Mukhamedzhanov, Phys. Rev. C **75**, 024601 (2007)
84. H.M. Xu, C.A. Gagliardi, R.E. Tribble, A.M. Mukhamedzhanov, N.K. Timofeyuk, Phys. Rev. Lett. **73**, 2027 (1994)

85. N.K. Timofeyuk, P. Descouvemont, Phys. Rev. C **72**, 064324 (2005)
86. L.J. Titus, P. Capel, F.M. Nunes, Phys. Rev. C **84**, 035805 (2011)
87. N.K. Timofeyuk, R.C. Johnson, A.M. Mukhamedzhanov, Phys. Rev. Lett. **97**, 069904 (2006)
88. N.K. Timofeyuk, Nucl. Phys. A **632**, 38 (1998)
89. N.K. Timofeyuk, R.C. Johnson, A.M. Mukhamedzhanov, Phys. Rev. Lett. **91**, 232501 (2003)
90. N.K. Timofeyuk, P. Descouvemont, R.C. Johnson, Phys. Rev. C **75**, 034302 (2007)
91. N. Michel, W. Nazarewicz, M. Płoszajczak, Phys. Rev. C **70**, 064313 (2004)
92. H. Furutani, H. Horiuchi, R. Tamagaki, Prog. Theor. Phys. **62**, 981 (1979)
93. H. Furutani, H. Kanada, T. Kaneko, S. Nagata, H. Nishioka, S. Okabe, S. Saito, T. Sakuda, M. Seya, Prog. Theor. Phys. Supp. **68**, 193 (1980)
94. J. Humblet, L. Rosenfeld, Nucl. Phys. **26**, 529 (1961)
95. A.M. Mukhamedzhanov, R.E. Tribble, Phys. Rev. C **59**, 3418 (1999)
96. B. Barmore, A.T. Kruppa, W. Nazarewicz, T. Vertse, Phys. Rev. C **62**, 054315 (2000)
97. A.T. Kruppa, W. Nazarewicz, Phys. Rev. C **69**, 054311 (2004)
98. W.M. MacDonald, Phys. Rev. **100**, 51 (1955)
99. W.M. MacDonald, Phys. Rev. **101**, 271 (1956)
100. E.P. Wigner, Phys. Rev. **73**, 1002 (1948)
101. J.T. Wells, A.B. Tucker, W.E. Meyerhof, Phys. Rev. **131**, 1644 (1963)
102. H.R. Sadeghpour, J.L. Bohn, M.J. Cavagnero, B.D. Esry, I.I. Fabrikant, J.H. Macek, A.R.P. Rau, J. Phys. B. Atom. Mol. Opt. Phys. **33**(5), R93 (2000)
103. E. Nielsen, J.H. Macek, Phys. Rev. Lett. **83**, 1566 (1999)
104. B.D. Esry, C.H. Greene, H. Suno, Phys. Rev. A **65**, 010705 (2001)
105. I.J. Thompson, M.V. Zhukov, Phys. Rev. C **49**, 1094 (1994)
106. J. Meng, P. Ring, Phys. Rev. Lett. **77**, 3963 (1996)
107. T. Myo, K. Katō, H. Toki, K. Ikeda, Phys. Rev. C **76**, 024305 (2007)
108. N. Michel, W. Nazarewicz, M. Płoszajczak, J. Rotureau, Phys. Rev. C **74**, 054305 (2006)
109. H. Simon, D. Aleksandrov, T. Aumann, L. Axelsson, T. Baumann, M.J.G. Borge, L.V. Chulkov, R. Collatz, J. Cub, W. Dostal, B. Eberlein, T.W. Elze, H. Emling, H. Geissel, A. Grünschloss, M. Hellström, J. Holeczek, R. Holzmann, B. Jonson, J.V. Kratz, G. Kraus, R. Kulessa, Y. Leifels, A. Leistenschneider, T. Leth, I. Mukha, G. Münzenberg, F. Nickel, T. Nilsson, G. Nyman, B. Petersen, M. Pfützner, A. Richter, K. Riisager, C. Scheidenberger, G. Schrieder, W. Schwab, M.H. Smedberg, J. Stroth, A. Surowiec, O. Tengblad, M.V. Zhukov, Phys. Rev. Lett. **83**, 496 (1999)
110. S. Karataglidis, P.G. Hansen, B.A. Brown, K. Amos, P.J. Dortmans, Phys. Rev. Lett. **79**, 1447 (1997)
111. D.R. Thompson, M. LeMere, Y.C. Tang, Nucl. Phys.A **268**, 53 (1977)
112. N. Michel, W. Nazarewicz, M. Płoszajczak, Phys. Rev. C **82**, 044315 (2010)
113. T. Berggren, Phys. Lett. B **373**, 1 (1996)
114. A. Leviatan, P. Van Isacker, Phys. Rev. Lett. **89**, 222501 (2002)
115. M.A. Bentley, S.M. Lenzi, Prog. Part. Nucl. Phys. **59**, 497 (2007)
116. A.P. Zuker, S.M. Lenzi, G. Martínez-Pinedo, A. Poves, Phys. Rev. Lett. **89**, 142502 (2002)
117. R. Roth, Phys. Rev. C **79**, 064324 (2009)
118. A.P. Zuker, B. Buck, J.B. McGrory, Phys. Rev. Lett. **21**, 39 (1968)
119. N. Michel, J. Okołowicz, F. Nowacki, M. Płoszajczak, Nucl. Phys. A **703**, 202 (2002)
120. K. Bennaceur, N. Michel, F. Nowacki, J. Okołowicz, M. Płoszajczak, Phys. Lett. B **488**, 75 (2000)

# No-Core Gamow Shell Model

**8**

The most fundamental approach to describe nuclear states is to use a fully microscopic Hamiltonian where all nucleons are active. It is the subject of ab initio methods, where the only input is the nucleon–nucleon interaction and nuclear states arise from an exact solution of the many-body Schrödinger equation. Indeed, the nucleon–nucleon interaction is now well understood and several two-body versions of the nuclear interaction exist, which reproduce the deuteron binding energy and radius, as well as cross sections of nucleon–nucleon scattering. One can cite the Argonne $v_{18}$ interaction [1], defined in coordinate space with several combinations of spin and isospin operators and fitted in two-nucleon properties, the CD Bonn interaction [2], following the same approach in momentum space, as well as chiral interactions, which possess the same symmetries as the chiral Lagrangian of quantum chromodynamics while using nucleons as degrees of freedom.

While theoretically sound and intuitive, the ab initio approaches are nevertheless very difficult to apply in practice. Its first caveat is the hard core repulsion of nucleons at short distance. Indeed, the latter generates very important high momentum components, preventing the simple picture of independent particles to be used even as a starting point. In order to avoid this problem, equivalent low-energy interactions have been devised, which do not bear repulsion at low energy, but are built to reproduce the physical properties of two-nucleon systems for all physical energies of interest, typically up to 300 MeV [1, 2].

A widely used method to devise low-energy interactions is based on the renormalization group method and is deemed $V_{\text{low-k}}$. It relies on a Feshbach separation of the model space, in which the components of high momentum in the interaction are renormalized by a standard decoupling method between low and high momentum space. The latter method has been successfully applied to the phenomenological Argonne $v_{18}$ and CD Bonn interactions. When considered with chiral interactions, it leads to the most fundamental approach to treat nucleons starting from quantum chromodynamics, called the effective field theory [3]. It is based on the perturbative expansion of the nucleon–nucleon interaction in powers of $m_\pi/\Lambda$, where $m_\pi$ is the

© Springer International Publishing AG 2021
N. Michel, M. Płoszajczak, *Gamow Shell Model*, Lecture Notes in Physics 983,
https://doi.org/10.1007/978-3-030-69356-5_8

pion mass and $\Lambda$ the breakdown scale, i.e. the energy after which the model can no longer be used. It then provides a hierarchy of two-body, three-body, four-body, ..., operators, whose contribution decreases very quickly. In practice, the two-body and three-body terms have been seen to be the most important. Note that this method also renormalizes the quark and gluonic effects present in nucleons in the parameters of the many-body operators of effective field theory.

The first applications of interest have been the $N^3LO$ (next-to-next-to-next-to-leading-order) interaction, which could provide a realistic interaction [4]. Nevertheless, it came at the price of an important three-body term, difficult to take into account in practice. Hence, it is most of the time approximated by a nucleus-dependent two-body Hamiltonian, calculated from the normal ordering of the Hamiltonian in second quantization. The $N^3LO$ interaction, however, is problematic when applied to light nuclei and cannot reproduce the binding energies of the oxygen isotopes, for example.

Another option was the $N^2LO_{opt}$ interaction, whose parameters are fitted directly on nucleon–nucleon phase shifts only. This interaction has the advantage of having very small three-body forces and can reproduce the binding energies of the oxygen isotopes. One can also cite the JISP16 ($J$-matrix inversion of scattering process) interaction, an effective interaction based on the fit of nucleon–nucleon phase shifts and nuclear energies up to $^{16}O$, which has no three-body force as well [5].

The three-body force in nuclear effective interactions contains both "initial" and "induced" components. The "initial" component denotes the three-body force which appears due to the neglect of quark degrees of freedom and arise from the expansion of the nucleon–nucleon interaction in the framework of the (chiral) effective field theory. Indeed, the chiral interaction is written as a power series in $p/M$, with $p$ the typical momentum transfer between nucleons and $M$ the mass of a nucleon. Consequently, high energy degrees of freedom are integrated out in the first terms of the expansion provided by the effective field theory. This results in the appearance of many-body forces [6–10]. Added to that, the renormalization such as the $V_{low-k}$ or the similarity renormalization group (see Refs. [11–14] and discussion afterwards) also generate the three-nucleon forces. These "induced" three-nucleon forces arise from the components of the interaction beyond the cutoff $\Lambda$ and are uniquely related to the renormalization [11–14].

Typically, nucleon–nucleon interactions are fitted on the properties of nuclei having $A \leq 4$ nucleons. While these very light nuclei are well described therein, precision deteriorates quickly when considering heavier nuclei [15, 16]. In order to counter this tendency, the $N^2LO_{sat}$ (next-to-next-to-leading order with saturation) interaction has been introduced [17]. The nucleon–nucleon chiral interaction is considered at $N^2LO$ level, to which a fit of carbon and oxygen isotopes binding energies and radii of $A = 3, 4$ bound nuclei are added. Two and three-body forces are taken into account in the determination of $N^2LO_{sat}$ interaction. As a result, the error made in heavier nuclei is much reduced, as the $N^2LO_{sat}$ interaction is precise up to $^{40}Ca$ [17].

The study of nucleon–nucleon chiral interactions is still subject to development. Indeed, one can continue the chiral expansion beyond fourth order, i.e. that of the

$N^3LO$ interaction. The fifth and sixth orders have been considered in the $N^4LO$ and $N^5LO$ interactions [18, 19]. It has been noticed that the inclusion of the fifth order is not negligible, as it compensates for the energy overbinding provided by the fourth order [18]. However, the sixth order of chiral interactions is rather small, which suggests that convergence could be attained at this level [19]. Further calculations must be done in order to confirm this assertion.

The chiral effective field theory is in general not renormalizable, even though some effective field theories can be, as is the case for the pionless effective field theory [20]. In general, the chiral effective field theory exhibits not only the regularization scale dependence but also unwanted dependence on the particular form of regulators [21]. Added to that, it is not clear which three-body and four-body terms are really important in the chiral expansion of the nuclear interaction [22].

A pionless chiral effective field theory, satisfying the criteria of a physically correct theory, in particular the independence on the regularization scale, has been developed in Ref. [20]. In this approach, the three-body interaction appears at leading order, but four-body terms might be necessary at higher order [23]. Moreover, the dependence on the number of particles remains to be investigated [23]. This dependence of chiral effective field theory expansions is not understood in all presently discussed formulations (see also Ref. [22]).

As the calculation of realistic interactions is based on the fit of a restricted set of parameters, a statistical analysis of errors made in that process is important to check the physical content of parameters. The use of Bayesian inference is well suited for that purpose [24]. Indeed, this approach can identify interesting parameter domains and better fitting functionals. For example, redundancy has been found in determining the $s$-wave phase shifts through the use of Bayesian inference [25]. Using this approach, the range of data, such as energies, can be improved [25, 26], and the truncation and parameter errors can be combined to make better predictions [26].

The naturalness priors of the Bayesian formalism also allow to examine the influence of higher order terms in the expansions provided by the effective field theory [26]. Those terms can be included in the fit, using an already known functional or an effective field theory expansion for observables. The estimation of parameters can account for sources of uncertainty in the extraction of low-energy constants. They include data uncertainties issued from experiment or numerical calculations, systematic errors arising from the truncation of effective field theory, and errors induced by theoretical frameworks considered to calculate observables. Overfitting and underfitting are also mitigated by Bayesian inference.

A prior based on the naturalness of parameters suppresses spurious linear combinations of parameters. It also makes sure that low-energy constants do not over-reproduce experimental data in a necessarily limited theoretical framework. This occurs through the calculations of posterior probability distributions of low-energy constants, naturalness being explicitly taken into account. The amount of information found in the fitted low-energy constants is thus heavily increased.

Posterior probability distributions also test the dependence on prior distributions. Error propagation on predictions of effective field theory shows the dependence on the uncertainties of the experimental data used to fit the low-energy constants.

In the domain of the light nuclei, with a number of nucleons typically smaller than about 15, the most fundamental approaches are that of the no-core shell model [27, 28] and of the Green-Function Monte Carlo approach [6]. They are based on the quasi-exact resolution of the many-body Schrödinger equation. The Green-Function Monte Carlo approach starts from an approximate many-body wave function of Jastrow type provided by the variational principle, which is then evolved in configuration space via stochastic paths generated by a Monte Carlo process [6]. The no-core shell model makes use of a large basis of Slater determinants, generated by harmonic oscillator states, with which completeness is achieved [28]. Both approaches are rather successful in describing binding energies and spectra of light nuclei [29, 30].

The no-core shell model can also calculate particle-emission widths and reaction cross sections by including continuum degrees of freedom with the resonating group method approach [31–40]. This approach is very close to that of the shell model embedded in the continuum, developed about 20 years ago in the context of nucleon and cluster direct reactions [41–44]. The resulting no-core shell model with continuum was applied to calculate the $S$-matrix of reactions populating $^{11}$Be [45], $(d, t)$ fusion reactions [46], the calculation of $^6$He and $^9$He nuclear states [34, 47], elastic scattering of protons on $^{10}$C [48] and the $^7$Li(d,p)$^8$Li transfer reaction [36].

While very powerful, the no-core shell model and the Green-Function Monte Carlo approach are, however, hindered by computational limitations when the number of nucleons increases, upon which their numerical cost scales exponentially [6, 28]. The often necessary use of three-body forces in the no-core shell model worsens the situation, as the latter greatly augments the number of nonzero matrix elements of the Hamiltonian, so that nuclei with more than 10 nucleons can hardly be considered therein. Moreover, the Green-Function Monte Carlo approach can only provide with bound states, so that resonance states cannot be calculated in principle, except through extrapolation methods [49]. Hence, different methods have to be used to consider other nuclei of the nuclear chart.

For this, the coupled-cluster and Gorkov–Green-function frameworks have been developed, where the many-body wave function is no longer calculated, but is replaced by an exponential ansatz of a simple many-body wave function in coupled-cluster approach [50–54], and by many-body particle and pairing densities in Gorkov–Green-function framework [55–57], which iteratively converge to their exact value starting from a Hartree–Fock Bogoliubov many-body state via the use of Gorkov propagators. Consequently, medium and heavy nuclei can be treated, with the examples of Sn and Ni isotopes in the coupled-cluster framework [53,58] and of the Ca, Ti, and Ni isotopes in the Gorkov–Green-function framework [56,57]. The neutron dripline can also be reached by the coupled-cluster method and unbound states can be calculated therein using the Berggren basis to include continuum coupling [51].

The fundamental limitation of the coupled-cluster and Gorkov–Green-function methods is that they cannot deal with all nuclei. For example, the coupled-cluster method can handle in practice only closed-shell nuclei or nuclei with one or two particles or holes outside closed shells [59]. Moreover, the Gorkov–Green-function method can only be used for the calculation of even nuclei whose proton or neutron part must consist of closed shells, and immediate neighbors of even number of nucleons [55–57]. With above restrictions, the calcium isotopes could be calculated in the coupled-cluster approach up to the neutron dripline. Proton-rich tin isotopes, in the vicinity of the doubly-closed shell nucleus $^{100}$Sn, could also be calculated in the coupled-cluster framework [53]. However, the fact that the coupled-cluster method is restricted to closed-shell nuclei $\pm 2$ nucleons can be partly mitigated by calculating an equivalent interaction using a core and a restricted valence space [60]. Such an interaction can then be used in the framework of standard shell model. The advantage of such a two step approach is that dimensions of the Hamiltonian matrix remain tractable and, moreover all nuclei in a chosen subspace can be described.

It appears then that a profitable strategy is not to solve the initial Hamiltonian using an exact resolution method, but to transform it so that it can be used further within shell model using a number of configurations sufficiently small to allow for numerical diagonalization. Such a numerical strategy can be implemented via the in-medium similarity renormalization group [61–65] or the many-body perturbation theory [66, 67]. In the in-medium similarity renormalization group method, a unitary transformation depending on a continuous parameter is applied to the initial Hamiltonian, so that its low and high energy excitations are increasingly decoupled [61–65]. Consequently, the eigenstates of a truncated shell model calculation converge to their exact value, with a few percent precision. At present, when using a basis of harmonic oscillator states, the in-medium similarity renormalization group method can only be applied in a shell model space generated by a single harmonic oscillator major shell [61–65]. Consequently, certain aspects of nuclear deformation cannot be described in this method, as multishell valence spaces would be needed therein [68]. Nevertheless, axial nuclear deformation could be recently included in the $fp$ shell by using the generator coordinate method, where Hartree–Fock–Bogoliubov solutions of several different deformations are used as basis states (see Ref. [69] where neutrinoless beta decay is studied in $A = 48$ deformed nuclei.)

The in-medium similarity renormalization group method has been applied in light and medium-heavy well-bound nuclei, such as the isotopes of oxygen, calcium, and nickel [61, 62, 65]. The method allows also to evaluate the electromagnetic transitions [68]. Recently, the in-medium similarity renormalization group method has been considered in the context of the Gamow shell model to calculate the spectra of $^{22}$C and $^{24}$O [70].

States of deformed nuclei can be calculated in an ab initio framework if the basis states are generated by a Hamiltonian depending on deformation degrees of freedom. For example, the Monte Carlo shell model [71–74] makes use of Slater determinants generated by a deformed potential, where the rotational invariance of the many-body Hamiltonian is restored with projection techniques [75]. Alternatively, a new ab initio shell model framework has been recently introduced to describe deformed nuclei, where basis states are issued from solvable Hamiltonians bearing SU(3) symmetry, so that deformation is included in basis states [76–78]. SU(3) symmetry arises from the fact that basis states are described in terms of irreducible representations of the Sp(3,ℝ) symplectic group [76]. As a consequence, the shell model framework based on the use of basis states bearing SU(3) symmetry is also called symplectic shell model. The symplectic symmetry borne by SU(3) basis states allows to classify the shell model space in different irreducible representations labeled by two collective quantum numbers [79,80]. The fundamental interest of symplectic irreducible representations is that the dominant representations of nuclear states bear small quantum numbers, so that symplectic shell model space dimensions are greatly reduced as compared to standard shell model [76]. Usefulness of the symplectic symmetry has been confirmed in many ab initio calculations [80–83]. A successful description of light nuclei, including in particular the Hoyle state of $^{12}$C, can then be obtained with very small model space dimensions [78].

Another class of models which also allow to include the continuum couplings is based on the solution of Faddeev and Faddeev–Yakubovsky equations [84, 85]. These models are, however, typically restricted to $A \leq 5$ nuclei due to computer limitations. The only general *ab initio* approach including continuum couplings is that of the no-core shell model with continuum [31–40], where the degrees of freedom of unbound states are accounted for by coupled channel equations. While theoretically justified, it nevertheless becomes quickly inapplicable in practice due to the growing dimensions of model spaces. Moreover, particle-emission channels are non-orthogonal in general, which can generate problems due to overcomplete model spaces. Up to now, the heaviest nuclear system considered in the no-core shell model with continuum is that of $^{11}$Be [45].

An alternative ab initio theoretical approach, which also includes the continuum coupling, is based on the generalization of Gamow shell model. This no-core Gamow shell model, which can describe weakly bound and unbound states of light nuclei using realistic interactions and allowing all nucleons to be active, will be discussed in next sections. This fully microscopic model is using the Berggren ensemble of single-particle states to build Slater determinants.

## 8.1   Formulation of the No-Core Gamow Shell Model

One will solve the $A$-body Schrödinger equation with $H$ the intrinsic Hamiltonian:

$$\hat{H} = \frac{1}{A} \sum_{i<j}^{A} \frac{(\mathbf{p}_i - \mathbf{p}_j)^2}{2m} + \sum_{i<j}^{A} \hat{V}_{ij}^{NN} + \sum_{i<j<k}^{A} \hat{V}_{ijk}^{NNN} + \dots , \qquad (8.1)$$

where $\hat{V}^{NN}$, $\hat{V}^{NNN}$, … are realistic NN, NNN, … interactions. Typically, $\hat{H}$ is restricted to the two and three-body interactions due to computer limitations. In the following, one will consider only two-body interaction.

The following identities are useful to exhibit the center-of-mass degrees of freedom in Eq. (8.1):

$$\mathbf{P} = \sum_{i=1}^{A} \mathbf{p}_i , \tag{8.2}$$

where $\mathbf{P}$ is the center-of-mass momentum, and

$$\sum_{i=1}^{A} \mathbf{p}_i^2 = \frac{1}{A} \left[ \mathbf{P}^2 + \sum_{i<j} \left( \mathbf{p}_i - \mathbf{p}_j \right)^2 \right] , \tag{8.3}$$

resulting in:

$$\sum_{i=1}^{A} \frac{\mathbf{p}_i^2}{2m} - \frac{\mathbf{P}^2}{2mA} = \frac{1}{2mA} \sum_{i<j} \left( \mathbf{p}_i - \mathbf{p}_j \right)^2 . \tag{8.4}$$

In the no-core Gamow shell model, there is no restriction on the type of the NN interaction, contrary to the Green's Function Monte Carlo approach, for example, where difficulties arise for nonlocal potentials. One can hence use a local interaction, such as the Argonne $v_{18}$ potential [1], or a nonlocal interaction, such as the CD-Bonn 2000 [2] and various chiral interactions. There is also the possibility to use renormalized versions of the aforementioned forces, by applying techniques such as $V_{\text{low-k}}$ [86], the similarity renormalization group approach [12, 13], or the $G$-matrix [66, 87].

In the examples discussed in this chapter, one employs the phenomenological Argonne $v_{18}$ potential and the specific chiral interactions [4]. Both potentials are renormalized via the $V_{\text{low-k}}$ technique with a sharp momentum cutoff $\Lambda = 1.9\,\text{fm}^{-1}$ to decouple high- from low-momentum degrees of freedom and, henceforth, improving the convergence of nuclear structure calculations [14]. The single-particle basis is generated by the realistic two-body interaction, solving the integro-differential Schrödinger equation which contains both local and nonlocal parts [88–90], i.e. with the multi-Slater determinant coupled Hartree–Fock method (see Sect. 5). The one-body self-consistent potential $U_{\text{MSDHF}}(r)$ is then used to solve the one-body Schrödinger equation:

$$u_k''(r) = \left[ \frac{\ell(\ell+1)}{r^2} + \frac{2m}{\hbar^2} U_{\text{MSDHF}}(r) + V_{\text{Coul}}(Z, r) - k^2 \right] u_k(r) , \tag{8.5}$$

where $V_{\text{Coul}}(Z, r)$ is the one-body Coulomb potential (see Eq. (3.88)).

By adopting a single-particle basis upon which one builds many-body basis states, one effectively localizes the nucleus in space and, hence, one breaks the translational invariance of the Hamiltonian. Moreover, to describe the $A$ particle system, one needs $3A - 3$ coordinates, whereas the constructed wave function, depends on $3A$ coordinates. The redundant degrees of freedom are responsible for the spurious center-of-mass excitations. Plane wave single-particle states are eigenstates of the momentum operator and, hence, preserve the translational invariance but, unfortunately, cannot be used to describe a localized system. The standard method to suppress spurious center-of-mass excitations is to solve the many-body problem using relative coordinates, e.g. Jacobi coordinates. This ensures that the translational invariance of the system is preserved. However, antisymmetrization of states becomes unfeasible for $A \geq 6$, typically, due to the current limitations of computing resources. Consequently, an alternative strategy is considered for practical applications. For this, one uses the unique analytical properties of the harmonic oscillator single-particle basis in a full $N\hbar\omega$ space, in which the total wave function factorizes into $|\psi_{rel}\rangle \otimes |\psi_{CM}\rangle$. However, due to the confining character of the harmonic oscillator basis states, this limits the application of this method to well-bound systems only.

In the case of the no-core Gamow shell model, the latter factorization is not guaranteed. Since the Hamiltonian of Eq. (8.1) is expressed in relative coordinates of nucleons, one is expecting that eigenstates would not contain spurious center-of-mass excitations in an infinite space. However, because one is working in a finite space, it is necessary to demonstrate numerically this condition. Assuming that the factorization into a center-of-mass and a relative wave functions is valid and also the center-of-mass wave function has a Gaussian shape, the expectation value of the center-of-mass operator is [91]

$$\hat{H}_{CM} = \frac{1}{2mA}\mathbf{P}_{CM}^2 + \frac{mA\omega^2}{2}\mathbf{R}_{CM}^2 - \frac{3}{2}\hbar\omega , \qquad (8.6)$$

where $\hbar\omega$ is the parameter that characterizes the Gaussian wave function. The matrix elements of (8.6) are calculated with the harmonic oscillator expansion method (see Sect. 5.6 and Ref. [92]).

Following the assertion of Ref. [93], if $\langle \hat{H}_{CM}\rangle \sim 0$ then the factorization is valid (see Sect. 8.2.1 for a detailed study of the center-of-mass correction in the no-core Gamow shell model). However, this condition is useful only for bound states. Indeed, it has been noticed that the use of the Lawson method for many-body resonances may bind these states. Moreover, (8.6) is the only two-body center-of-mass Hamiltonian that can be applied in shell model applications with the Lawson method. The only other available two-body center-of-mass Hamiltonian is $\mathbf{P}_{CM}^2$, which possesses no ground state as all its eigenstates are center-of-mass Bessel functions. Consequently, there is a priori no method to fix the center-of-mass of an unbound state.

## 8.2 Benchmarking of No-Core Gamow Shell Model Against Other Methods

To benchmark the no-core Gamow shell model, one will compare results for the ground state energy of $^3$H and $^4$He obtained using the Jacobi–Davidson method with results issued from the harmonic oscillator no-core shell model using the Lanczos method. Both $^3$H and $^4$He are well-bound systems so that results of the no-core Gamow shell model can be compared against other well-known bound states methods. The numerical task in solving a three and four nucleon system is relatively easy so that one can check the convergence with respect to the maximal angular momentum of the model space.

### 8.2.1 Binding Energies and Center-of-Mass Excitations

One will demonstrate in this section that the generalized variational principle can be applied with the Berggren basis, *i.e.* that the binding energies of ground states in truncated shell model spaces converge to their exact value in infinite space. One will also study the center-of-mass properties of no-core Gamow shell model many-body states in details.

In Fig. 8.1, one compares results of the no-core Gamow shell model with results of the Faddeev calculations obtained using the Argonne $v_{18}$. Calculations are performed using $s_{1/2}$, $p_{3/2}$, $p_{1/2}$, $d_{3/2}$, $d_{5/2}$, $f_{7/2}$, $f_{5/2}$, $g_{9/2}$, $g_{7/2}$ partial

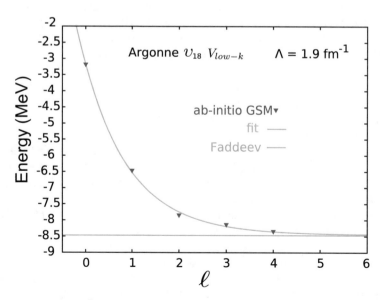

**Fig. 8.1** (Color online) $^3$H results as a function of the (maximum) angular momentum, in the interval from $\ell = 0$ to $\ell = 4$ using the Argonne $v_{18}$ interaction (from Ref. [94])

waves for protons and neutrons. The basis-generating potential is that of the multi-Slater determinant coupled Hartree–Fock which binds the proton and neutron $0s_{1/2}$ states with energies $-10.417$ MeV and $-11.982$ MeV, respectively. The scattering continua $i\{s_{1/2}\}, i\{p_{3/2}\}, i\{p_{1/2}\}$ are chosen along the real $k$-axis and each of them is discretized with 18 points. Here $i$ denotes the number of discretization points which ranges from 1 to 18 for the $s$-wave and from 0 to 18 for the $p$-waves. The harmonic oscillator length is fixed at $b = 1.5$ fm in order to provide with energies closest to experimental values. $\hbar\omega$ is then equal to 18.5 MeV. $d$-waves are represented by five harmonic oscillator states, which from now on are denoted as $5d$. The physical argument behind this choice lies in the fact that for $\ell > 1$ the centrifugal barrier is large enough to confine the single-particle states in the potential well and, hence, the harmonic oscillator basis becomes a realistic alternative. For $f$ and $g$ partial waves, one used three harmonic oscillator states. In total, the single-particle basis consists of 154 partial waves. The no-core Gamow shell model energy of $^3$H is then: $E_{NCGSM} = -8.39$ MeV, whereas the result of Fadeev calculation [95] is: $E_{Fadeev} = -8.47$ MeV.

Figure 8.1 serves as an illustration of the rate of convergence of the energy with the angular momentum of basis states. An exponential fit of the no-core Gamow shell model results for different $\ell$ values of the single-particle basis allows to extrapolate energies in the non-truncated model space. The extrapolated energy is: $E_{extrp} = -8.449 \pm 0.087$ MeV, where the fit function is: $E = E_{extrp} + b \times e^{-c\ell}$. One sees that the inclusion of partial waves with angular momentum larger than $\ell = 4$, should have a very small contribution, of the order of 50 keV, to the eigenenergy.

In the above no-core Gamow shell model calculation, the features related to the center-of mass have not been assessed. Thus, we will now calculate the expectation value of the center-of-mass $\hat{H}_{CM}$ operator (see Eq. (8.6)) of the $^3$H ground state. Following the assertion of Ref. [93], one assumes that the no-core Gamow shell model wave function in a sufficiently large model space factorizes and the center-of-mass wave function is that a harmonic oscillator $0s$ state. In the largest model space used in Fig. 8.1, the expectation value of the $\hat{H}_{CM}$ is approximately 7 keV. Consequently, one can confirm that in this case the ground state wave function of $^3$H indeed separates into an intrinsic wave function and harmonic oscillator center-of-mass $0s$ wave function.

However, although this example has proven effectiveness of Berggren basis in the no-core Gamow shell model applications, it cannot be considered as representative of the general case. Indeed, while the ground state binding energy of $^3$H calculated using the local Argonne $v_{18}$ interaction converged fast to its exact value given by a Faddeev calculation, that convergence is much slower using the nonlocal N$^3$LO interaction. Consequently, we will use the N$^3$LO interaction to quantify the center-of-mass properties in the ground states $J^\pi = 1/2^+$ in $^3$H and $J^\pi = 0^+$ in $^4$He. The momentum cutoff of the N$^3$LO interaction in this calculation is $\Lambda = 1.9$ fm$^{-1}$. In this study, one uses three different methods of calculation. The first method is that of the standard shell model, as the use of the harmonic oscillator basis states in this model allows to separate exactly many-body wave functions in intrinsic and center-of-mass parts. The second method is that of the no-core Gamow shell model using the Lawson method. One considers therein the $s_{1/2}$, $p_{1/2}$ and $p_{3/2}$ proton and neutron partial waves as Berggren basis states, whereas all other partial waves, for which $2 \leq \ell \leq 10$, are described by the harmonic oscillator functions. The Berggren basis in $^3$H is generated by a Woods–Saxon potential [94]. The single-particle pole states for $^3$H in this basis are $0s_{1/2}$ for protons and neutrons. Bound $0p_{3/2}$ proton and neutron pole states are added for the $^4$He calculations. All poles of the Berggen basis are bound, as the ground states of $^3$H and $^4$He are well bound so that the continuum couplings are negligible. The resonances appearing in the partial waves of Berggren basis for $\ell \geq 2$ are very high in energy, so that it is sufficiently precise to take real-energy contours for the Berggren basis.

In standard shell model using a basis of harmonic oscillator states, the $N\hbar\omega$ many-body basis space is defined by taking all the Slater determinants having a harmonic oscillator energy of $N\hbar\omega$ or less. In order to effectively fulfill this condition in the Berggren basis calculation, one assigns equivalent harmonic oscillator energies to valence Berggren shells. For this, the bound $0s_{1/2}$ and $0p_{3/2}$ one-body states have an equivalent harmonic oscillator energy of $0\hbar\omega$ and $1\hbar\omega$, respectively. Scattering $s_{1/2}$ Berggren basis shells are assigned a harmonic oscillator energy of $2\hbar\omega$. As there are no Berggren basis poles besides $0s_{1/2}$ states for $^3$H, the scattering $p_{3/2}$ and $p_{1/2}$ Berggren basis contours are given an energy of $1\hbar\omega$. One does not have any $0p_{1/2}$ Berggren basis pole states for $^4$He as well. Hence, the $p_{1/2}$ Berggren basis scattering states are also assigned an energy of $1\hbar\omega$. $0p_{3/2}$ bound states which have an energy of $1\hbar\omega$ are present in the Berggren basis of $^4$He, so that the scattering $p_{3/2}$ Berggren basis contour are given an energy of $3\hbar\omega$. By assigning the $\hbar\omega$-dependent values to scattering one-body basis states, one insures that $N\hbar\omega$ spaces are implicitly included in the Berggren basis.

Let us consider the $1s_{1/2}$ harmonic oscillator state as an example. Due to the completeness of the Berggren basis, it can be expanded using basis states on the $s_{1/2}$ contour. As the $s_{1/2}$ scattering states have an energy of $2\hbar\omega$, the $1s_{1/2}$ harmonic oscillator state is then always implicitly included in the Gamow shell model basis space by demanding $N\hbar\omega \geq 2\hbar\omega$.

For $^3$H, the contours in Berggren basis are truncated at $k_{\max} = 4\,\text{fm}^{-1}$ and discretized with 22 states per contour. For $^4$He, $k_{\max} = 3\,\text{fm}^{-1}$ and one takes 15 states per contour. As the Berggren basis is complete, the harmonic oscillator $N\hbar\omega$ many-body states are effectively included in the Berggren many-body basis spanned by Slater determinants of equivalent harmonic oscillator energy smaller or equal to $N\hbar\omega$. Consequently, one can expect the Lawson method to converge if a sufficiently large constant $\lambda$ is used in the Hamiltonian $\hat{H} + \lambda\hat{H}_{CM}$ defined in Eqs. (8.1) and (8.6). For example, the convergence of the expectation value of $H_{CM}$ to zero is achieved for $\lambda \geq 5$. The harmonic oscillator length has been adjusted so as to provide the smallest expectation value of $\hat{H}_{CM}$ and is, respectively, equal to 1.6 fm and 1.5 fm for the ground states of $^3$H and $^4$He. The $N\hbar\omega$ spaces used run from $N = 2$ to $N = 8$ what is sufficient to quantify the center-of-mass properties.

The ground state energies of $^3$H (left panel) and $^4$He (right panel) are depicted in Fig. 8.2, while the expectation values of $\hat{H}_{CM}$, associated with these energies, are shown in Fig. 8.3. One can see that the ground state energies, which are calculated

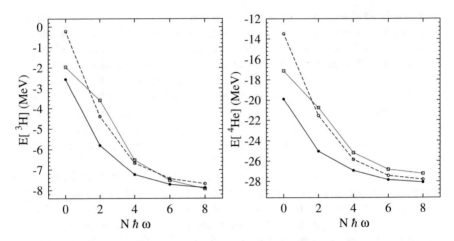

**Fig. 8.2** Energy of the ground state of $^3$H (left panel) and $^4$He (right panel) as a function of $N\hbar\omega$. Dotted lines with squares depict energies issued from harmonic oscillator shell model, dashed lines with circles show results arising from no-core Gamow shell model, in which the Lawson method has been applied, and solid lines with bullets represent energies obtained from the no-core Gamow shell model without the center-of-mass correction

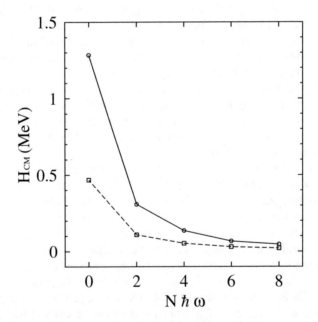

**Fig. 8.3** Expectation values of $H_{CM}$ (see Eq. (8.6)) calculated using the Lawson method in no-core Gamow shell model are plotted as a function of $N\hbar\omega$ for the ground state wave function of $^3$H (dashed line with squares) and $^4$He (solid line with circles). Expectation value of $H_{CM}$ is always equal to zero in the harmonic oscillator no-core shell model so that it is omitted in the figure

using the Lawson method, with either the harmonic oscillator basis or the Berggren basis, or without any center-of-mass correction using the Berggren basis only, are similar and converge to their exact value. Typically, the largest ground state binding energy is obtained using the no-core Gamow shell model without the center-of-mass correction. Indeed, using the Lawson method in the no-core Gamow shell model wave functions, one effectively projects ground state wave functions in the largest $N\hbar\omega$ model space included in the Gamow shell model space. Therefore, the ground state binding energy is smaller than in the absence of the Lawson method.

The ground state of $^4$He in the no-core Gamow shell model is also slightly more bound than obtained in the no-core shell model with harmonic oscillator basis. This comes from the use of a Woods–Saxon potential which is very close to the Hartree–Fock potential for the $^4$He ground state, so that the no-core Gamow shell model results converge faster when $N$ increases, as compared to the no-core shell model in the harmonic oscillator basis. As can be seen from Fig. 8.3, expectation values of $\hat{H}_{CM}$ are already very small in $2\hbar\omega$ spaces and eventually become smaller than 50 keV in $8\hbar\omega$ spaces when the Lawson method is applied to no-core Gamow shell model. This implies that it is possible to have a separation of intrinsic and center-of-mass parts in many-body wave functions, as long as the model space is sufficiently large.

As already seen from Fig. 8.2, the ground state energy becomes closer and closer to its exact value even when the Lawson method is not applied. Indeed, due to the generalized variational principle, ground state eigenenergy is stationary when enlarging Gamow shell model basis spaces, so that it has to converge to the exact eigenenergy at the limit of infinite space, where eigenstates are separated in center-of-mass and intrinsic parts. Hence, the spuriosity of eigenstates decreases when the size of Gamow shell model space increases, to vanish identically at the limit of infinite space. It is then not necessary to apply the Lawson method in order to compute the ground state energy. In fact, one can obtain a ground state energy close to the exact energy if the model space used is sufficiently large.

However, the generalized variational principle in general does not apply to excited states. In order to demonstrate this statement, let us consider the case of the first two eigenstates of $^4$He, which are $0^+$ states. If one diagonalizes the Gamow shell model Hamiltonian without consideration of the center-of-mass properties of the calculated $0^+$ states, it is possible that the two first $0^+$ states of $^4$He converge to the same $0^+$ ground state. As the Hamiltonian is intrinsic, both $0^+$ states can indeed become degenerate at the limit of infinite space, with their intrinsic parts being that of the ground state, whereas their center-of-mass parts are orthogonal. Consequently, the generalized variational principle can only be applied to ground states of subspaces with fixed angular momentum and parity. For example, one can calculate the binding energies of the $3/2_1^-$ and $1/2_1^-$ eigenstates of $^5$He using the generalized variational principle, because both these states are lowest energy states for the considered quantum numbers.

Another method to solve this problem would be to work in a fully intrinsic ab initio coupled-channel formalism, as in the no-core shell model with continuum [31–40]. In this approach, the cluster + target composite is built with a harmonic oscillator basis and hence can be separated in center-of-mass and relative coordinates. Continuum coupling is included because of the presence of coupled-channel basis states, so that one can expand resonance states. However, in this case, the resolution of continuum in practical applications depends on the choice of reaction channels.

The Lawson method can be applied in the no-core Gamow shell model as long as the Gamow shell model space is sufficiently large to generate a many-body ground state wave function separable in intrinsic and center-of-mass parts. As bound many-body states are well localized, it is possible to project their center-of-mass part on a harmonic oscillator center-of-mass $0s$ state, similarly to no-core shell model using a basis of harmonic oscillator states. In fact, the use of harmonic oscillator and Berggren bases is equivalent for that matter in the domain of well-bound many-body states. However, because of the very large model spaces involved in no-core Gamow shell model, it is hardly possible to apply the Lawson method beyond $^4$He. Consequently, it is more convenient to rely on the generalized variational principle (see Sect. 5.1.4), with which no projection on center-of-mass harmonic oscillator ground state is needed. In other words, one does not apply the Lawson method, but uses the fact that the eigenenergy is stationary when enlarging the Gamow shell model space. The eigenenergy then converges to the exact nuclear binding energy at the limit of infinite Gamow shell model space because of the intrinsic character of the Hamiltonian (see Sect. 5.1.4). The generalized variational principle can be also applied to unbound many-body states, contrary to the Lawson method, where the projection on a harmonic oscillator center-of-mass $0s$ state automatically binds the many-nucleon system. This will be the method of choice for the next applications, concerning the resonance ground states of $^5$He and $^4$n.

### 8.2.2   $^5$He Unbound Nucleus

The unbound character of $^5$He makes its description in a no-core picture challenging. Indeed, both its ground and first excited states are many-body resonances with sizable widths. The complex-energy formulation of the no-core Gamow shell model using the Berggren ensemble is then suitable for the calculation of $^5$He many-body states. Indeed, in the Gamow shell model formalism, the energy and width of $^5$He will be obtained from the eigenvalues of the complex-symmetric Hamiltonian matrix. Contrary to the previous applications, where one used a real-energy basis, one will employ a basis made of complex-energy basis states, which also includes the $0p_{3/2}$ resonance. The one-body states consist of the bound $0s_{1/2}$ neutron state with an energy $-23.290$ MeV, the bound $0s_{1/2}$ proton state with an energy $-23.999$ MeV and the $0p_{3/2}$ resonance with a real part of energy $1.193$ MeV and a width $1267$ keV. The position of this resonance in the complex-energy plane is: $k = (0.277, -0.068)$ fm$^{-1}$. The single-particle basis for protons and neutrons is generated by a Gamow Hartree–Fock calculation using the N$^3$LO $V_{\text{low-k}}$ interaction with $\Lambda = 1.9$ fm$^{-1}$. The $p_{3/2}$ contour is complex and encompasses the $0p_{3/2}$

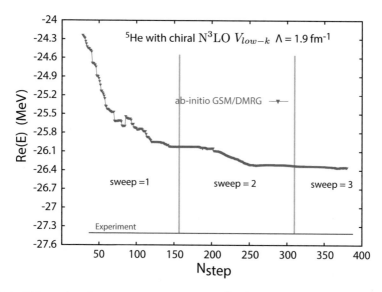

**Fig. 8.4** (Color online) Convergence of the real part of $^5$He ground state energy, calculated with the density matrix renormalization group, is shown as a function of the number of iterations, also called sweeps. Experimental data are also provided [96] (from Ref. [94])

resonance, whereas the $s_{1/2}$ and $p_{1/2}$ contours are chosen along the real-$k$ axis. For states with $\ell > 1$, one uses harmonic oscillator basis functions, denoted as $(5d, 3f, 3g)$, similarly to previous cases.

Figure 8.4 shows the convergence of the real part of the ground state energy in $^5$He, calculated using the density matrix renormalization group method. Results of calculation are presented starting from the fortieth shell in the sweep-down phase. The converged energy: $\Re(E_{\text{NCGSM}}) = -26.31$ MeV, is about 1 MeV higher than the experimental binding energy [96]. The truncation error in this calculation is $\epsilon = 10^{-6}$ and the maximum number of eigenvectors kept in the density matrix renormalization group many-body basis space (see Sect. 5.9) is $N_{\text{opt}} \sim 300$. The corresponding dimension of the matrix is $D_{\text{max}} \sim 10^5$, whereas the equivalent dimension of Slater determinant basis in the Gamow shell model is about $3 \times 10^9$.

The imaginary part of the $^5$He ground state energy is shown in Fig. 8.5. The converged value obtained with the density matrix renormalization group method is: $\Im(E_{\text{NCGSM}}) = -0.2$ MeV, i.e. $\Gamma_{\text{NCGSM}} = 400$ keV. From the calculated binding energies of the ground states $^4$He and $^5$He, one obtains the position of the $^5$He resonance at energy 1.17 MeV above the $\alpha + n$ threshold.

Experimentally, resonance parameters can be extracted in two ways [97]. One may apply the conventional $R$-matrix approach on the real axis and use the Lane and Thomas prescription for the extraction of positions and widths of a state [98]. The alternative method is the extended $R$-matrix approach [99, 100], which associates the resonance parameters with the complex poles of the $S$-matrix. This is the most appropriate prescription to compare with the no-core Gamow shell model results,

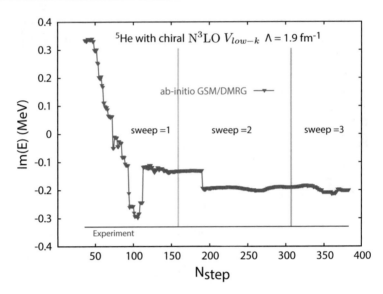

**Fig. 8.5** (Color online) Same as in Fig. 8.4 but for the imaginary part of $^5$He ground state energy (from Ref. [94])

because complex eigenvalues correspond to the poles of $S$-matrix. The extended $R$-matrix approach determines the position of the $(3/2)^-$ state at an energy 798 keV above the $\alpha + n$ threshold and with a width $\Gamma = 648$ keV. The difference between no-core Gamow shell model and experimental data is probably due to the used chiral interaction and/or to missing many-body forces.

One can verify these findings in the no-core Gamow shell model. For this, one compares results obtained with two different renormalization scale parameters: $\Lambda = 1.9 \,\text{fm}^{-1}$ and $2.1 \,\text{fm}^{-1}$. The calculations are performed using the Jacobi–Davidson diagonalization method [101]. The model space in these calculations is truncated and only configurations with no more than four particles in the nonresonant continuum are included. Moreover, the single-particle basis is reduced by taking only the neutron $p_{3/2}$ states as Berggren states, whereas remaining single-particle basis states with angular momenta up to $\ell = 4$ are the harmonic oscillator basis functions with an oscillator length equal to $b = 1.5 \,\text{fm}$.

The energies and widths of $^4$He and $^5$He in this calculation are close to the exact no-core Gamow shell model results obtained with the density matrix renormalization group method. Using the $V_{\text{low-k}}$ N$^3$LO force with $\Lambda = 1.9 \,\text{fm}^{-1}$, one obtains $E_{\text{NCGSM}}^{(4\text{p}4\text{h})} = -27.386 \,\text{MeV}$ for the energy of $^4$He. For $^5$He, one finds an energy $\Re\left(E_{\text{NCGSM}}^{(4\text{p}4\text{h})}\right) = -25.825 \,\text{MeV}$ and a width $\Gamma_{\text{NCGSM}}^{(4\text{p}4\text{h})} = 370 \,\text{keV}$, i.e. the $^5$He ground state is 1.56 MeV above the $\alpha$+n decay threshold. For $\Lambda = 2.1 \,\text{fm}^{-1}$, the $^4$He energy equals $E_{\text{NCGSM}}^{(4\text{p}4\text{h})} = -26.060 \,\text{MeV}$, whereas for $^5$He one finds $\Re\left(E_{\text{NCGSM}}^{(4\text{p}4\text{h})}\right) = -23.903 \,\text{MeV}$ and $\Gamma = 591 \,\text{keV}$. Hence, the $^5$He ground state for

**Table 8.1** Density matrix renormalization group result of the no-core Gamow shell model is compared to experimental energy and width of the $^5$He ground state deduced by various $R$-matrix procedures. Energies are given with respect to the $\alpha +$ n threshold (adapted from Ref. [94])

| Method | Energy (MeV) | $\Gamma$ (MeV) |
|---|---|---|
| Density Matrix Renormalization Group | 1.17 | 0.400 |
| Extended R-matrix [97] | 0.798 | 0.648 |
| R-matrix [97] | 0.963 | 0.985 |
| R-matrix [102] | 0.771 | 0.644 |
| NUBASE evaluation [103] | 0.890 | 0.651 |
| $^3$He + t [104] | 0.79 | 0.525 |

$\Lambda = 2.1\,\mathrm{fm}^{-1}$ lies $2.15\,\mathrm{MeV}$ above the $\alpha +$ n decay threshold. As can be expected, the unbound state of $^5$He is seen to be very sensitive to the choice of the nucleon–nucleon interaction.

Several extracted experimental data for the position and width of $^5$He ground state can be found in Table 8.1. Let us note that, because of the broad width of $^5$He ground state and the proximity to the $\alpha +$ n decay threshold, the extracted values depend on the algorithm used to extract the resonance parameters from the experimental data.

### 8.2.3  Radial Overlap Integral and Asymptotic Normalization Coefficient

One will now study the asymptotic normalization coefficients in $^5$He (see Sect. 7.3 for a definition and an application of asymptotic normalization coefficients in light nuclei). To obtain accurate asymptotic normalization coefficients, it is mandatory to describe correctly the tail of the radial overlap integrals. This requirement is fulfilled in the no-core Gamow shell model for either bound, weakly bound or unbound states due to the completeness properties of the Berggren basis. Moreover, the one neutron separation energy $S_{1n}$ must be precisely reproduced because the tail of the radial overlap integral is sensitive to the $S_{1n}$. This is particularly problematic in *ab initio* calculations with no adjustable parameters since even small differences of $S_{1n}$ for different interactions can cause large differences in the asymptotic normalization coefficients. A method to alleviate this problem was proposed in the Green's function Monte Carlo approach, where the calculations could be performed for experimental separation energies [105, 106]. For this, radial overlap integrals are written as a linear combination of Coulomb wave functions. Their coefficients are given by matrix elements involving many-body wave functions [105, 106]. These coefficients are analytic with respect to the complex linear momentum associated with one-nucleon separation energy [105, 106], This guarantees a physical asymptotic behavior of the radial overlap integral when the many-body state is unbound, even though the asymptotes of the initial many-body wave functions used in the matrix elements providing with Coulomb wave functions coefficients are not correct [105, 106]

The calculation of the asymptotic normalization coefficient is based on the tail of radial overlap integral (see Eq. (7.33) and Sect. 7.3 for a general definition of the asymptotic normalization coefficient). In the case of reaction: $^4\text{He}_{0+} + \text{n} \rightarrow$ $^5\text{He}_{3/2-}$, the operator $a^\dagger_{\ell j}(\mathcal{B})$ creates a neutron in any $p_{3/2}$ single-particle state of the Berggren basis and $\langle r\ell j | u_\mathcal{B}\rangle$ is the $p_{3/2}$ radial wave function. The radial overlap integral then reads

$$I_{\ell=1, j=3/2}(r) = \frac{1}{2}\sum_i \langle \widetilde{^5\text{He}_{3/2-}} ||a^\dagger_{i,\ell=1,j=3/2}||^4\text{He}_{0+} u_{i,\ell=1,j=3/2}(r)\rangle . \qquad (8.7)$$

The model space and $V_{\text{low-k}}$ momentum cutoff used in this calculation are the same as the one at the end of Sect. 8.2.2. One uses a $p_{3/2}$ Berggren basis partial wave, where $0p_{3/2}$ is a resonance, while the $s_{1/2}$ and $p_{1/2}$ neutron partial waves consist of real-energy Berggren basis contours. Partial waves with $\ell \leq 2$ are generated by harmonic oscillator basis functions. The no-core Gamow shell model calculation is performed for chiral N$^3$LO interaction (see Ref. [4]) with a cutoff parameter $\Lambda = 1.9\,\text{fm}^{-1}$.

The radial overlap integral for $^4\text{He}_{0+} + \text{n} \rightarrow\ ^5\text{He}_{3/2-}$ is shown in Fig. 8.6. The asymptotic normalization coefficient $C = 0.197\,\text{fm}^{-1/2}$ is extracted by fitting both the real and imaginary parts of the radial overlap integral with a Hankel function in the asymptotic region. Results are very sensitive to the value of the

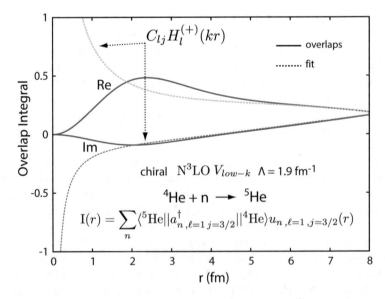

**Fig. 8.6** (Color online) The radial overlap integral for $^4\text{He}_{0+} + \text{n} \rightarrow\ ^5\text{He}_{3/2-}$. The asymptotic region is fitted by the Hankel function (from Ref. [94])

**Table 8.2** No-core Gamow shell model results for the dependence of one neutron separation energy $S_{1n}$ and spectroscopic amplitude $S_{p3/2}^{1/2}$ on the $V_{\text{low}-k}$ cutoff parameter $\Lambda$. In this calculation, number of particles in the scattering continuum is limited to four particles (adapted from Ref. [94])

| $\Lambda$ fm$^{-1}$ | $S_{1n}$ (MeV) | $S_{p3/2}^{1/2}$ |
|---|---|---|
| 2.1 | −2.15 | 0.812 |
| 1.9 | −1.56 | 0.787 |
| 1.5 | −1.38 | 0.774 |

cutoff parameter $\Lambda$. For $\Lambda = 1.9\,\text{fm}^{-1}$ and $2.1\,\text{fm}^{-1}$, neutron separation energy equals $S_{1n} = -1.56\,\text{MeV}$ and $S_{1n} = -2.15\,\text{MeV}$, respectively (see Table 8.2). Different neutron separation energies have direct impact on asymptotic behavior of radial overlaps, and hence on asymptotic normalization coefficients which equal $0.197\,\text{fm}^{-1/2}$ for $\Lambda = 1.9\,\text{fm}^{-1}$, and $0.255\,\text{fm}^{-1/2}$ for $\Lambda = 2.1\,\text{fm}^{-1}$.

In fact, the asymptotic normalization coefficients are meaningful only if the corresponding experimental separation energy is well reproduced. For different $\Lambda$, however, the particle separation energies change also because the transformation which is used to alter the short-range physics is not unitary. The unitarity is violated because the induced many-body forces are neglected in practical applications. The separation energies and the asymptotic normalization coefficients are independent of $\Lambda$ if all many-body interactions induced by the $V_{\text{low}-k}$ renormalization scheme are taken into account, what never happens in practice. However, even in the restricted, unitarity violating scheme, the calculated asymptotic normalization coefficient is a consistency check of the used many-body approach. Indeed, for narrow resonances, the resonance width $\Gamma$ and the asymptotic normalization coefficient $C$ are related together by an expression (see Eq. (7.42), Refs. [106–108] and references cited therein):

$$C = \sqrt{\frac{\Gamma \mu}{\hbar^2 \Re(k)}}, \qquad (8.8)$$

where $\mu$ is the effective mass. Equation (8.8) describes the outgoing flux probability for a decaying resonance. In a process: $^{4}\text{He}_{0^+} + n \rightarrow {}^{5}\text{He}_{3/2^-}$, $k$ is the linear momentum $\left(k = \sqrt{2mS_{1n}/\hbar^2}\right)$ associated with the one neutron separation energy of $^{5}\text{He}$, and $\Gamma$ is the one neutron emission width of $^{5}\text{He}$ ground state. In general, $k$ is a complex number:

$$k \sim \sqrt{\Re(S_{1n}) \left(1 - i\frac{\Gamma}{2\Re(S_{1n})}\right)}. \qquad (8.9)$$

The imaginary part of $k$ can be neglected if:

$$\frac{\Gamma}{2S_{1n}} \to 0 . \tag{8.10}$$

Hence, the validity of Eq. (8.8) does not necessarily mean that the resonance width is small, but that the ratio of the resonance width and the separation energy is small.

Using the relation (8.8), one obtains the one neutron emission width $\Gamma_{ANC} = 311\,\text{keV}$ and $\Gamma_{ANC} = 570\,\text{keV}$ for $\Lambda = 1.9\,\text{fm}^{-1}$ and $2.1\,\text{fm}^{-1}$, respectively. These values for the widths are consistent with those obtained in the no-core Gamow shell model, in spite of the fact that the ground state width of $^5$He is broad. The ratio $\Gamma/(2S_{1n})$ varies in the range from 10% to 15% in the considered interval of $\Lambda$ cutoff parameters. The corresponding error is compatible with the deviation between the width obtained from the no-core Gamow shell model diagonalization and its extraction from the asymptotic normalization coefficient formula (8.8).

The norm of the radial overlap integral squared $I_{\ell j}(r)^2$ (see Eqs. (7.33), (7.44)):

$$S_{\ell j} = \int_0^{+\infty} I_{\ell j}(r)^2 \, dr \tag{8.11}$$

defines the spectroscopic factor. It is solely a theoretical construction and should not be considered as a measurable quantity [109]. The value of spectroscopic factor $S$ depends on the representation of a Hamiltonian, and hence on the interior of the wave function. Indeed, one has the freedom to change the off-shell behavior of the potential in any possible manner, while maintaining the NN phase shifts. In this way, families of phase shift equivalent potentials are obtained. Each method of changing the off-shell behavior is then considered as a different representation or scheme. In spite of this shortcoming, the spectroscopic factors can provide useful information about shell occupancies and hence the nucleon–nucleon correlations, i.e. how much the nucleus in a given state deviates from the single-particle picture. Such spectroscopic factors have been calculated by several methods, either in *ab initio* or more phenomenological approaches [105, 110–114].

Using no-core Gamow shell model solutions for $^4$He/$^5$He in the truncated model space with at most four nucleons in the nonresonant continuum, one finds that the spectroscopic amplitude of $^5$He corresponding to the channel $\left[^4\text{He(g.s.)} \otimes p_{3/2}\right]^{3/2^-}$ is 0.787 ($S_{p_{3/2}} = 0.62$) for $\Lambda = 1.9\,\text{fm}^{-1}$ and 0.812 ($S_{p_{3/2}} = 0.66$) for $\Lambda = 2.1\,\text{fm}^{-1}$. By decreasing the renormalization scale parameter $\Lambda$, i.e. by "softening" the interaction, the spectroscopic factor deviates further from unity. This tendency is seen also at $\Lambda = 1.5\,\text{fm}^{-1}$ (see Table 8.2) which yields the energies $-28.670\,\text{MeV}$ and $-27.285\,\text{MeV}$ for $^4$He and $^5$He, respectively.

## 8.3 Multi-Neutron Resonances

Whether the four-neutron system exists or not is a long-standing question which rests on the hypothesis that such a system could be formed as a result of a subtle interplay between the many-body components of the nuclear interaction, the Pauli principle and the coupling to the neutron continuum. Tremendous efforts have been made during the last few decades to understand few-nucleon systems, in particular few-neutron systems [115–117]. Earlier experiments failed to find positive evidence for the existence of multi-neutron systems (see a discussion in Ref. [115]). In 2002, Marqués et al. reported a possible observation of a tetraneutron in a breakup reaction for the $^{14}Be \rightarrow {}^{10}Be+4n$ channel [117]. After that experiment, several theoretical attempts were performed to examine the possible existence of bound tetraneutron, but all calculations failed to reproduce experimental data [118, 119].

The experimental hints of the tetraneutron resonance provided by the recent measurement [116] exacerbate the need for reliable nuclear interactions and *ab initio* methods able to cope with couplings to the continuum. According to those experimental results, if such a state exists it would have an energy $E = 0.83 \pm 0.65$ (stat) $\pm$ 1.25 (syst) MeV above the tetraneutron threshold and a maximal width $\Gamma = 2.6$ MeV. This large width is unlikely to correspond to a nuclear state [120].

In Ref. [121], the possibility to form the tetraneutron resonance by adding a phenomenological $T = 3/2$ three-body force to a realistic two-body interaction was investigated. In this study, the continuum was included and it was shown that the unrealistic modifications of the nuclear interaction would be necessary to obtain the tetraneutron. On the contrary, *ab initio* calculation in the harmonic oscillator basis [122] using the two-body JISP16 interaction [123] obtained the energy and the width of the tetraneutron ($E = 0.8$ MeV and $\Gamma = 1.4$ MeV) from extrapolation into the continuum of the phase shifts of artificially bound tetraneutron, close to the values published in Ref. [116]. However, couplings to the continuum at low energy below and above the particle-emission thresholds may strongly affect the configuration mixing and make such extrapolations misleading for broad resonances. Results of Ref. [121] based on the uniform complex scaling method [124, 125] are not affected by this issue and suggest that if the tetraneutron system exists its energy and width should be larger than in Ref. [122]. Below, one will investigate the conditions of existence of the tetraneutron system using different *ab initio* methods and various two-body chiral interactions while including the continuum coupling.

The quantum Monte Carlo framework has been used to calculate multi-neutron systems based on local chiral interactions via an extrapolation of bound states, obtained with a confining auxiliary potential, to the physical domain of unbound multi-neutron systems. These calculations predicted that the trineutron is located at 1.11 MeV, below the energy of tetraneutron, at 2.12 MeV [126, 127]. Calculations in this approach suggested that the existence of multi-neutron resonances can only be obtained by strongly modifying standard nuclear forces, which is incompatible with the current understanding of nucleon–nucleon interactions [121].

In the no-core Gamow shell model, the Hamiltonian matrix is either diagonalized with the Jacobi–Davidson method or with the density matrix renormalization group method. In order to reduce the computational challenge that are configuration interaction calculations in the continuum, the density matrix renormalization group method in no-core Gamow shell model has been supplemented with natural orbitals (see Sect. 5.3).

The density matrix renormalization group calculations are performed as follows. First the standard density matrix renormalization group run is performed, with some given truncations, to approach the final eigenstate of the Hamiltonian. Then, the corresponding natural orbitals are calculated and a new density matrix renormalization group run is performed where the natural orbitals replace the Berggren basis states and the truncations are removed. This technique leads to an impressive gain in computational time, while efficiently incorporating many-body correlations with the successive sweeps of the density matrix renormalization group procedure. In the calculation of tetraneutron, natural orbitals were generated from a truncated space which is large enough to have a decent approximation of the targeted Hamiltonian eigenstate. A small basis of natural orbitals can be built, by including only those whose associated eigenvalue $\eta$ in the density matrix is larger than some chosen value $0 < \eta_{min} < 1$. It is then possible to do a many-body calculation of the tetraneutron without model space truncation when Berggren basis states are replaced by a few natural orbitals.

### 8.3.1   Tetraneutron Energy from the Extrapolation of Overbinding Interactions

In the exploratory no-core Gamow shell model study of a tetraneutron [128], the model space consisted of the $0s_{1/2}$ and $0p_{3/2}$ resonant shells (pole space), and associated nonresonant partial waves whose energies are selected along the contours in the complex-momentum plane. It has been checked that the positions of the poles are unimportant as long as the poles are embedded in their associated complex continua. This model space is then augmented by 30 nonresonant $p_{1/2}$ partial waves along a real-momentum contour, and seven harmonic oscillator shells for each of the partial waves $d$, $f$ and $g$. Details of the calculation can be found in Ref. [128].

When allowing only two neutrons into the continuum, the energies and widths of the $J^\pi = 0^+$ tetraneutron, calculated for various two-body chiral interactions ($N^3LO$ [4], $N^2LO_{opt}$ [129] and $N^2LO_{sat}$ [17]) with different renormalization cutoffs are all mutually consistent as shown in Table 8.3. For a qualitative comparison, in the $N^2LO_{sat}$ interaction one only used its two-body part. Similar simplification has been done in the $N^3LO$ interaction in previous sections. The small effect of the $\Lambda$-cutoff variations on the calculated results indicates a weak influence of the missing induced three- and four-body forces on the tetraneutron properties.

In the following, the $N^3LO$ two-body interaction with a renormalization cutoff $\Lambda = 1.9\,\text{fm}^{-1}$ is used. The role of the continuum in shaping the tetraneutron is illustrated by scaling the interaction by a factor $f = 2.0$ so that the system is

**Table 8.3** Energies and widths (in brackets) of the $J^\pi = 0^+$ four-neutron pole (in MeV) for various two-body interactions. The asterisk means that only the two-body part of the interaction was considered (adapted from Ref. [128])

| | $\Lambda = 1.7\,\text{fm}^{-1}$ | $\Lambda = 1.9\,\text{fm}^{-1}$ | $\Lambda = 2.1\,\text{fm}^{-1}$ |
|---|---|---|---|
| N³LO | 7.27 (3.69) | 7.28 (3.67) | 7.28 (3.69) |
| N²LO$_\text{opt}$ | 7.32 (3.74) | 7.33 (3.78) | 7.34 (3.95) |
| N²LO$_\text{sat}$ * | 7.24 (3.48) | 7.22 (3.58) | 7.27 (3.55) |
| JISP16 | 7.00 (3.72) | | |

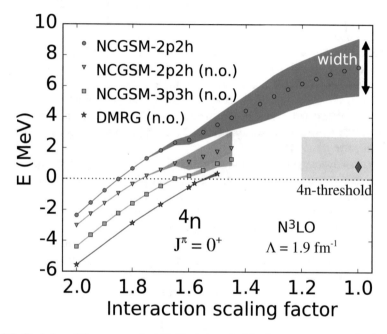

**Fig. 8.7** Evolution of the energy and width (shaded area) of the four-neutron system is plotted as a function of the scaling factor $f$ multiplying the N³LO interaction. $f$ changes in the interval from 2.0 to 1.0 by step of 0.05 and for a total of 20 points. The circles represent the no-core Gamow shell model results with two neutrons in the continuum, which is used to generate the no-core Gamow shell model results based on natural orbitals with two (triangles) and three (squares) neutrons in the continuum. The density matrix renormalization group results without truncations are represented by stars. The experimental energy is indicated by a diamond and the gray area shows the maximal experimental uncertainties [116] (from Ref. [128])

artificially bound, and then one follows the evolution of the energy and width of the $J^\pi = 0^+$ state when $f \to 1.0$ as shown in Fig. 8.7.

The first set of results denoted NCGSM-2p2h corresponds to the no-core Gamow shell model calculations with only two neutrons in the continuum and shows a rapid increase of the width when the scaling of the interaction is gradually decreased. The small variation of the energy with the scaling factor around $f = 1.6$ is due to the use of different single-particle bases where the $0p_{3/2}$ shell is bound for $f \geq 1.6$

and unbound for $f < 1.6$, to acknowledge the opening of the $^4n \rightarrow {}^3n + n$ channel [128]. For each scaling factor, one uses results of the no-core Gamow shell model with two neutrons in the continuum to generate natural orbitals and keeps only those natural orbitals whose associated eigenvalues in the scalar density matrix are sufficiently large ($\eta > 10^{-7}$ in the density matrix), therefore reducing the size of the basis by a factor $\approx 3$. The first $s_{1/2}$ and $p_{3/2}$ natural orbitals are treated as pole states while remaining orbitals are considered as continuum shells. Then in the results denoted NCGSM-2p2h (n.o.) and NCGSM-3p3h (n.o.), one considers at most two and three neutrons in the natural orbitals continuum shells, respectively. This technique allows to reduce significantly the computational cost and to include additional many-body correlations in a more efficient way.

The results denoted NCGSM-2p2h (n.o.), shown in Fig. 8.7, illustrate this point clearly, as they are all lower in energy than the results of initial no-core Gamow shell model with at most two neutrons in the continuum shells (NCGSM-2p2h) obtained in the Berggren basis. However, the quality of natural orbitals for the description of targeted state depends on how close the generating eigenstate is to the final eigenstate. For that reason, the NCGSM-2p2h (n.o.) calculations could only be performed for $f > 1.45$.

Another advantage of the natural orbitals is the possibility to remove some of the truncations as compared to NCGSM-2p2h calculations. However, the NCGSM-2p2h calculations need to include enough of correlations in the continuum for the removal of truncations in the calculations with natural orbitals is meaningful. The NCGSM-3p3h (n.o.) calculations which allow for three neutrons in the natural orbitals outside the pole space, contain more correlations than the NCGSM-2p2h (n.o.). However, also these calculations are limited to $f > 1.45$. It was not possible to completely remove the truncations in the no-core Gamow shell model and hence one had to rely on the density matrix renormalization group method to calculate binding energies.

The density matrix renormalization group applied to no-core Gamow shell model allows the calculation for a tetraneutron without truncations on the number of particles in the continuum. The same shells are used as in the no-core Gamow shell model using natural orbitals and the convergence criterion of the density matrix renormalization group method is fixed by the parameter $\varepsilon = 10^{-8}$ [130, 131] (see Sect. 5.9). At the scaling factor $f = 2.0$, these results are about 1 MeV lower than results of the no-core Gamow shell model with three particles at most in the continuum (NCGSM-3p3h (n.o.)), indicating important missing correlations in the NCGSM-3p3h (n.o.). This shows that configurations with four neutrons in the continuum shells have a large contribution to the wave function even when the system is artificially bound. In fact, the opening of new decay channels and the presence of continuum states in the configuration mixing above the threshold is expected to make the width explode when $f \rightarrow 1$, especially in the density matrix renormalization group results where all decay channels are open. This is in qualitative agreement with the results in Ref. [121] which show a rapid increase of the width when the strength of the phenomenological $T = 3/2$ three-body force decreases. This explosion of the width is already visible in the no-core Gamow shell model results with natural orbitals where the width increases faster than in the NCGSM-2p2h calculations. Another hint for the explosion of the width is the impossibility to perform the density matrix renormalization group calculations far above the four-neutron threshold, even when using the improved identification technique for broad resonances. This is due to the strong couplings to the continuum, resulting in large overlaps between complex-energy scattering states and the targeted decaying resonance, making them indistinguishable. Finally, while the energy of the four-neutron system may be compatible with the experimental value when $f \rightarrow 1$, calculations including more than two particles in the continuum as in Table 8.3 suggest that the width of the tetraneutron is larger than $\Gamma \approx 3.7\,\mathrm{MeV}$, as can be seen from the NCGSM-2p2h calculation in Fig. 8.7.

This first attempt to determine the binding energy of the tetraneutron has shown that the no-core Gamow shell model is accurate enough to deal with this difficult theoretical problem. The consideration of model spaces of different levels of truncation, as well as the use of the density matrix renormalization group in a non-truncated model space, proved that the tetraneutron energy can be reasonably estimated by extrapolating no-core Gamow shell model results in truncated spaces into an infinite space and the physical nucleon–nucleon interaction. One should stress, however, that the density matrix renormalization group cannot provide with definitive information about the width of the tetraneutron, as its domain of applicability is where the tetraneutron is either bound or has a relatively small width, i.e. in the overbinding regime of nucleon–nucleon interactions.

### 8.3.2 Trineutron and Tetraneutron in the Auxiliary Potential Method

Another method to identify physical resonant states of multi-neutron systems has been introduced in the quantum Monte Carlo framework [118, 126]. It relies on the use of potentials confining the valence neutrons in an external trap. In this method, one adds to Eq. (8.1) an auxiliary Woods–Saxon potential of the form:

$$\hat{V}_{\text{WS-aux}}(r) = V_{\text{aux}}/[1 + \exp((r - R_{\text{aux}})/a_{\text{aux}})], \qquad (8.12)$$

where the depth $V_{\text{aux}}$ varies from a finite negative value to zero. Consequently, multi-neutron systems become well bound for sufficiently large negative values of $V_{\text{aux}}$, so that the calculation of Hamiltonian eigenstates is stable therein.

The first calculation is done with the auxiliary potential strength $V_{\text{aux}}^{(0)}$, for which the eigenstate is bound. One then slowly decreases the absolute value of $V_{\text{aux}}$, i.e. $|V_{\text{aux}}| < |V_{\text{aux}}^{(0)}|$. At each iteration, the overlap method is applied in order to identify the eigenstate generated using the auxiliary potential (see Sect. 5.4). The $|\Psi_0\rangle$ eigenvector employed for that matter (see Sect. 5.4 for notation) is that of the previous iteration, i.e. $|\Psi_0\rangle$ is the eigenstate generated using the $V_{\text{aux}}^{(0)}$ auxiliary potential. The method is iterated until convergence, by replacing $V_{\text{aux}}^{(0)}$ by $V_{\text{aux}}$ and restarting the previously described procedure, up to the physical value of $V_{\text{aux}} = 0$. This allows to follow Hamiltonian eigenstates adiabatically through the region of very strong continuum couplings. The precision of this adiabatic identification procedure has been checked to be nearly exact for a dineutron system [127].

One will calculate the ground states of trineutron and tetraneutron using the auxiliary potential method. In these calculations, both resonance and continuum basis states are included, so that one can calculate energies and widths of multi-neutron systems. Model spaces are defined by 3p-3h and 4p-4h truncations, as in Sect. 8.3.1. One uses the so-called large model space, where Slater determinants are built from $s_{1/2}, p_{3/2}$ Berggren basis partial waves and $p_{1/2}, f_{5/2}, f_{7/2}, g_{9/2}$ harmonic oscillator partial waves. In the small model space, one builds Slater determinants using the same partial waves as in the large model space, but the $p_{1/2}, d_{3/2}, d_{5/2}$ and $f_{5/2}, f_{7/2}, g_{9/2}$ partial waves are represented only with two and one harmonic shells, respectively. Numerical tests showed that the dineutrons channel does not open when decreasing $V_{\text{aux}}$, so that eigenenergies can be expected to smoothly vary when $V_{\text{aux}} \to 0$.

The Gamow shell model code based on the two-dimensional partitioning method (see Sect. 5.8) has been employed to calculate the Hamiltonian ground state. The single-particle Berggren basis is generated by the finite-depth Woods–Saxon potential including spin–orbit coupling and the chiral two-body interaction N$^3$LO [132], as in Sect. 8.3.1.

As can be seen in Fig. 8.8, it was not possible to calculate tetraneutron energies in the large model space beyond the value of $V_{\text{aux}} = -1.5$ MeV. The physical region ($V_{\text{aux}} = 0$) could be reached only in truncated spaces, either using a reduced number of shells, denoted as "small" in Fig. 8.8, or using 3p-3h truncations. Consequently, one still does not have a definitive result for the tetraneutron energy with realistic interactions, and one has to rely on extrapolations, as it was done in Sect. 8.3.1.

Both the auxiliary potential method and the density matrix renormalization group approach (Sect. 8.3.1), predict similar tetraneutron energy, close to 2.6 MeV, which is within the range of the experimental error [116]. The calculation of the trineutron ground state is converged in a full space, i.e. without truncation on the number of neutrons in the continuum, where one obtains $E_{3n} = 1.29$ MeV and $\Gamma_{3n} = 0.91$ MeV. The energy and width of the trineutron, obtained with the confining potential, is smaller than that of the tetraneutron, while being larger than that of the dineutron. Consequently, it is plausible that the trineutron resonance lives longer than the tetraneutron resonance and, hence, the chance to observe it experimentally are bigger.

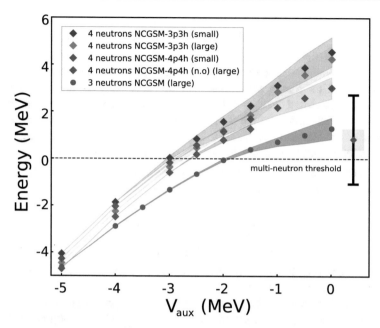

**Fig. 8.8** Evolution of the energies and widths of trineutron and tetraneutron as function of the depth of the auxiliary Woods–Saxon potential. Small and large no-core Gamow shell model spaces with three particles at most in the continuum and without model space truncation are considered. Depending on the value of $V_{aux}$, either Berggren basis or natural orbital basis is used. To help readability, the scale of a tetraneutron width is decreased by a factor 3. Trineutron results are obtained without any truncations. The experimental data [116] for tetraneutron are also shown (adapted from Ref. [132])

## 8.4   Renormalization of Realistic Interactions in a Finite Model Space Using $\hat{Q}$-Box Method

The main reason of interest in no-core shell model approaches is that they allow to provide with almost exact eigenenergies of realistic Hamiltonians. However, as one saw earlier, it is possible to derive equivalent Hamiltonians in smaller model spaces using unitary transformations of the in-medium similarity renormalization group [61–65], or the many-body perturbation theory [66, 67]. Consequently, it would be useful in practical applications to use such transformations in the framework of no-core Gamow shell model to reduce the dimensionality of the model space while working with realistic Hamiltonians. With this objective in mind, the $\hat{Q}$-box method has been recently applied in Gamow shell model [70]. As in standard shell model, it consists in deriving an effective interaction in a core + valence particle picture from the in-medium similarity renormalization group method [61–65] or many-body perturbation theory [66, 67].

The main problem arising in Gamow shell model and absent in standard shell model is the optimization of the single-particle basis, as discussed in Sect. 5.5. The first calculations involving the $\hat{\bar{Q}}$-box method in Gamow shell model were done using a Woods–Saxon potential to generate the single-particle basis. This was not an optimal solution, however, because the many-body perturbation expansion poorly converges when the basis potential is of Woods–Saxon type [70]. Much better convergence was obtained if the Berggren basis was generated by the Hartree–Fock potential of the Hamiltonian. Indeed, the use of the Hartree–Fock potential allows to cancel many diagrams of the many-body perturbation theory.

In order to apply the $\hat{\bar{Q}}$-box method, one separates the many-body Hamiltonian (8.1) into a zeroth-order part $\hat{H}_0$ and a perturbation $\hat{V}$:

$$\hat{H} = \hat{H}_0 + (\hat{H} - \hat{H}_0) = \hat{H}_0 + \hat{V} \,. \tag{8.13}$$

The Hartree–Fock Hamiltonian is used for $\hat{H}_0$, and the $\hat{\bar{Q}}$-box folded-diagram calculation are performed in the basis generated by the multi-Slater determinant coupled Hartree–Fock method.

A powerful renormalization method of many-body perturbation theory is that of the extended Kuo–Krenciglowa method [133] which was originally derived with degenerate basis states. However, one-body states of Berggren ensemble have different energies, so it is preferable to use the extended Kuo–Krenciglowa method with nondegenerate basis states to build the effective Hamiltonian $\hat{H}_{\text{eff}}$. It is done by way of an iterative process [67, 134, 135]:

$$\hat{H}_{\text{eff}}^{(n)} = \hat{P}\hat{H}_0\hat{P} + \hat{\bar{Q}}(E) + \sum_{k=1}^{\infty} \frac{1}{k!} \frac{d^k \hat{\bar{Q}}(E)}{dE^k} \{\hat{H}_{\text{eff}}^{(n-1)} - E\}^k \,, \tag{8.14}$$

where $k$ indicates $k$-th derivative, $E$ is the starting energy, and $\hat{P}$, $\hat{Q}$ represent the model space and the excluded space projectors, respectively. The $\hat{\bar{Q}}$-box is then defined as:

$$\hat{\bar{Q}}(E) = \hat{P}\hat{V}\hat{P} + \hat{P}\hat{V}\hat{Q} \frac{1}{E - \hat{Q}\hat{H}\hat{Q}} \hat{Q}\hat{V}\hat{P} \,. \tag{8.15}$$

In practical applications, the $\hat{\bar{Q}}$-box is calculated perturbatively in the power series in $\hat{V}$ [66, 136, 137]. The one-body part of the $\hat{\bar{Q}}$-box Hamiltonian, i.e. the shell model single-particle energies, must be constructed from the realistic force similarly to its two-body part. For that, one performs the so-called $\hat{S}$-box folded-diagram renormalization, i.e. the renormalization method applied to the one-body part of the $\hat{\bar{Q}}$-box, to generate shell model single-particle energies [138]. In this way, one arrives at a totally microscopic approach, with which one can define a valence space Hamiltonian for the Gamow shell model calculations.

### 8.4.1 Application of the $\hat{Q}$-Box Method in Oxygen and Fluorine Chains of Isotopes

The first application of $\hat{Q}$-box method for Gamow shell model calculations can be found in Ref. [139]. Below, we will discuss briefly most important features of these calculation for the isotopes of oxygen and fluorine. The considered nucleon–nucleon interaction is the optimized chiral interaction $NNLO_{opt}$, which offers a good description of nuclear structure, i.e. binding energies, excitation spectra, positions of driplines and neutron matter equation of state, without the need of three-body forces [129]. Figure 8.9 compares the Gamow shell model ground state energies ($E_{gs}$) and single-neutron separation energies ($S_n$) in the oxygen and fluorine isotopic chains with experimental data and other theoretical calculations. The neutron dripline of oxygen isotopes, situated at $^{24}O$, has been reproduced in shell model calculations [144], thanks to the repulsive character of three-nucleon forces (see curves for shell model with two- and three-nucleon forces in Fig. 8.9). All calculations which include the effects of three-nucleon forces predict the correct position of the neutron dripline, as shown in Fig. 8.9. However, quantitative results vary widely, which is the case, in particular, for nuclei beyond the neutron dripline. This might arise from the fact that open decay channels and particle continuum are not always included, as it occurs in standard shell model with USDB interaction [145], shell model with two-nucleon and three-nucleon forces [144], or valence space in-medium similarity renormalization group, also with two-nucleon forces and three-nucleon forces [143]. Indeed, in the shell model embedded in the continuum studies of oxygen and fluorine chains of isotopes, it was shown that the position of driplines can be described with standard two-body interaction when the coupling to the decay channels is included [147, 148]. Hence, the effects of three-nucleon forces and the coupling to the scattering continuum can hardly be disentangled. The Gamow shell model approach, hereby based on many-body perturbation theory using two-body interactions, provides with the correct location for the neutron dripline of oxygen isotopes. One also obtains a good description of the $^{25,26}O$ isotopes, which are situated beyond the neutron dripline (see the left panel of Fig. 8.9).

Let us consider the two other calculations employing the Berggren representation presented in Fig. 8.9. The first calculation corresponds to the coupled-cluster model using Berggren basis [141] and a schematic three-nucleon force. In this case the neutron dripline is reproduced, but results deviate from experimental data. The second calculation is that of the Gamow shell model [142], where an effective two-body interaction fitted to the bound states and resonances of $^{23-26}O$ is utilized. One assumes that the core of $^{22}O$ is inert. A proper description of energies is obtained therein. It is also predicted that the $^{28}O$ ground state is very loosely bound or weakly unbound. Moreover, the $^{28}O$ isotope is still more bound than $^{27}O$, while being slightly unbound with respect to $^{26}O$.

The unbound character of the $^{25-28}O$ isotopes is clearly seen in the one neutron separation energies (see Fig. 8.9). The issue of tetraneutron radioactivity in $^{28}O$ was raised in Ref. [149]. A direct tetraneutron decay could occur if $S_{4n} < 0$ and

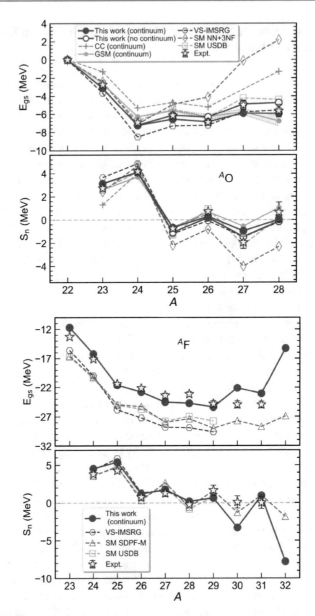

**Fig. 8.9** Ground state energies with respect to the $^{22}$O core and one neutron separation energies for oxygen (upper panel) and fluorine (lower panel) isotopes are compared with experimental data [140] (the AME2016 extrapolated values are taken for $^{27,28}$O and $^{30,31}$F) and other calculations: coupled-cluster approach with continuum (CC) [141], Gamow shell model (GSM) [142], valence space in-medium similarity renormalization group (VM-IMSRG) [143], shell model with two-nucleon forces and three-nucleon forces (SM NN+3NF) [144], shell model with the USDB interaction (SM USDB) [145] and shell model with the SDPF-M interaction (SM SDPF-M) [146] (from Ref. [139])

$\{S_n, S_{2n}, S_{3n}\} > 0$. Gamow shell model calculations with the $\hat{Q}$-box interactions exclude a direct tetraneutron decay because $S_{2n} < 0$, though a sequential emission of two dineutrons is still possible.

The main configuration in the ground state of $^{28}O$ is $(\nu 1s_{1/2})^2(\nu 0d_{3/2})^4$. The single-particle resonance $\nu 0d_{3/2}$ is occupied by four neutrons in the neutron-unstable ground state of $^{28}O$. The decay width of $^{28}O$ is predicted to be several tens of keV. In order to exhibit the influence of the $\nu 0d_{3/2}$ resonance and the scattering contours, one did calculation with and without coupling to the continuum (see Fig. 8.9). The gained binding energy arising from the coupling to the continuum is about 0.5 MeV for $^{26}O$ and 1 MeV for $^{28}O$.

Results for fluorine isotopes obtained with the Gamow shell model and NNLO$_{opt}$ interaction are shown on the lower panel of Fig. 8.9. Energies are given with respect to that of the $^{22}O$ ground state. One obtains a good description of ground state energies for the $^{23-28}F$ isotopes. While experimentally $^{31}F$ is located at the neutron dripline [150, 151], the Gamow shell model calculation yields the energy of $^{31}F$ lower than that of $^{30}F$. Nevertheless, $^{31}F$ remains unbound with respect to $^{29}F$. This discrepancy originates from the truncations which had to be imposed because of the rapid increase of Hamiltonian matrix dimensions with the number of valence nucleons. For example, the matrix dimension in the calculation of $^{29}F$ ground state in $(sdpf_{7/2})$ valence space without truncations, is $\sim 1.95 \times 10^{10}$. Gamow shell model matrices of this size cannot be handled at present so in the calculations of $^{29-32}F$ isotopes, at most two valence particles in the nonresonant continuum were allowed.

A partial explanation of the difference between positions of driplines for oxygen and fluorine isotopes can be put forward by analyzing the effective single-particle energies. At the end of the isotopic chains of oxygen and fluorine, $\nu 0d_{3/2}$ is unbound in the oxygen isotopes, whereas it is bound in fluorine isotopes where about 1 MeV of binding energy is gained due to the proton–neutron interaction.

## References

1. R.B. Wiringa, V.G.J. Stoks, R. Schiavilla, Phys. Rev. C **51**, 38 (1995)
2. R. Machleidt, Phys. Rev. C **63**, 024001 (2001)
3. H. Leutwyler, Ann. Phys. **235**, 165 (1994)
4. D.R. Entem, R. Machleidt, Phys. Rev. C **68**, 041001(R) (2003)
5. R. Skibiński, J. Golak, K. Topolnicki, H. Witała, Y. Volkotrub, H. Kamada, A.M. Shirokov, R. Okamoto, K. Suzuki, J.P. Vary, Phys. Rev. C **97**, 014002 (2018)
6. S.C. Pieper, R.B. Wiringa, Annu. Rev. Nucl. Part. Sci. **51**, 53 (2001)
7. S. Coon, M. Scadron, P. McNamee, B. Barrett, D. Blatt, B. McKellar, Nucl. Phys. A **317**, 242 (1979)
8. U. van Kolck, Prog. Part. Nucl. Phys. **43**, 337 (1999)
9. A. Nogga, H. Kamada, W. Glöckle, Phys. Rev. Lett. **85**, 944 (2000)
10. P. Navrátil, W.E. Ormand, Phys. Rev. C **68**, 034305 (2003)
11. E.D. Jurgenson, P. Navrátil, R.J. Furnstahl, Phys. Rev. C **83**, 034301 (2011)
12. E.D. Jurgenson, P. Navrátil, R.J. Furnstahl, Phys. Rev. Lett. **103**, 082501 (2009)
13. R. Roth, J. Langhammer, A. Calci, S. Binder, P. Navrátil, Phys. Rev. Lett. **107**, 072501 (2011)
14. S.K. Bogner, R.J. Furnstahl, A. Schwenk, Prog. Part. Nucl. Phys. **65**, 94 (2010)

15. S. Binder, J. Langhammer, A. Calci, R. Roth, Phys. Lett. B **736**, 119 (2014)
16. T.A. Lähde, E. Epelbaum, H. Krebs, D. Lee, U.G. Meißner, G. Rupak, Phys. Lett. B **732**, 110 (2014)
17. A. Ekström, G.R. Jansen, K.A. Wendt, G. Hagen, T. Papenbrock, B.D. Carlsson, C. Forssén, M. Hjorth-Jensen, P. Navrátil, W. Nazarewicz, Phys. Rev. C **91**, 051301 (2015)
18. D.R. Entem, N. Kaiser, R. Machleidt, Y. Nosyk, Phys. Rev. C **91**, 014002 (2015)
19. D.R. Entem, N. Kaiser, R. Machleidt, Y. Nosyk, Phys. Rev. C **92**, 064001 (2015)
20. S. König, H.W. Grießhammer, H.W. Hammer, U. van Kolck, J. Phys. G: Nucl. Part. Phys. **43**, 055106 (2016)
21. J.E. Lynn, I. Tews, J. Carlson, S. Gandolfi, A. Gezerlis, K.E. Schmidt, A. Schwenk, Phys. Rev. Lett. **116**, 062501 (2016)
22. R. Machleidt, D.R. Entem, Phys. Rep. **503**, 1 (2011)
23. H.W. Hammer, A. Nogga, A. Schwenk, Rev. Mod. Phys. **85**, 197 (2013)
24. A. Ekström, C. Forssén, C. Dimitrakakis, D. Dubhashi, H.T. Johansson, A.S. Muhammad, H. Salomonsson, A. Schliep, J. Phys. G. Nucl. Part. Phys **46**, 095101 (2019)
25. S. Wesolowski, R.J. Furnstahl, J.A. Melendez, D.R. Phillips, J. Phys. G. Nucl. Part. Phys **46**(4), 045102 (2019)
26. S. Wesolowski, N. Klco, R.J. Furnstahl, D.R. Phillips, A. Thapaliya, J. Phys. G. Nucl. Part. Phys **43**, 074001 (2016)
27. P. Navrátil, S. Quaglioni, I. Stetcu, B.R. Barrett, J. Phys. G: Nucl. Part. Phys. **36**, 083101 (2009)
28. B.R. Barrett, P. Navrátil, J.P. Vary, Prog. Part. Nucl. Phys. **69**, 131 (2013)
29. T. Fukui, L. De Angelis, Y.Z. Ma, L. Coraggio, A. Gargano, N. Itaco, F.R. Xu, Phys. Rev. C **98**, 044305 (2018)
30. M. Piarulli, A. Baroni, L. Girlanda, A. Kievsky, A. Lovato, E. Lusk, L.E. Marcucci, S.C. Pieper, R. Schiavilla, M. Viviani, R.B. Wiringa, Phys. Rev. Lett. **120**, 052503 (2018)
31. S. Quaglioni, P. Navrátil, Phys. Rev. C **79**, 044606 (2009)
32. S. Baroni, P. Navrátil, S. Quaglioni, Phys. Rev. Lett. **110**, 022505 (2013)
33. G. Hupin, S. Quaglioni, P. Navrátil, Phys. Rev. Lett. **114**, 212502 (2015)
34. C. Romero-Redondo, S. Quaglioni, P. Navrátil, G. Hupin, Phys. Rev. Lett. **117**, 222501 (2016)
35. P. Navrátil, S. Quaglioni, G. Hupin, C. Romero-Redondo, A. Calci, Phys. Scr. **91**, 053002 (2016)
36. F. Raimondi, G. Hupin, P. Navrátil, S. Quaglioni, Phys. Rev. C **93**, 054606 (2016)
37. G. Hupin, J. Langhammer, P. Navrátil, S. Quaglioni, A. Calci, R. Roth, Phys. Rev. C **88**, 054622 (2013)
38. P. Navrátil, R. Roth, S. Quaglioni, Phys. Rev. C **82**, 034609 (2010)
39. S. Baroni, P. Navrátil, S. Quaglioni, Phys. Rev. C **87**, 034326 (2013)
40. S. Quaglioni, P. Navrátil, Phys. Rev. Lett. **101**, 092501 (2008)
41. J. Okołowicz, M. Płoszajczak, I. Rotter, Phys. Rep. **374**, 271 (2003)
42. K. Bennaceur, N. Michel, F. Nowacki, J. Okołowicz, M. Płoszajczak, Phys. Lett. B **488**, 75 (2000)
43. N. Michel, J. Okołowicz, F. Nowacki, M. Płoszajczak, Nucl. Phys. A **703**, 202 (2002)
44. J. Rotureau, J. Okołowicz, M. Płoszajczak, Nucl. Phys. A **767**, 13 (2006)
45. A. Bonaccorso, F. Cappuzzello, D. Carbone, M. Cavallaro, G. Hupin, P. Navrátil, S. Quaglioni, Phys. Rev. C **100**, 024617 (2019)
46. G. Hupin, S. Quaglioni, P. Navrátil, Nat. Commun. **10**, 351 (2019)
47. M. Vorabbi, A. Calci, P. Navrátil, M.K.G. Kruse, S. Quaglioni, G. Hupin, Phys. Rev. C **97**, 034314 (2018)
48. A. Kumar, R. Kanungo, A. Calci, P. Navrátil, A. Sanetullaev, M. Alcorta, V. Bildstein, G. Christian, B. Davids, J. Dohet-Eraly, J. Fallis, A.T. Gallant, G. Hackman, B. Hadinia, G. Hupin, S. Ishimoto, R. Krücken, A.T. Laffoley, J. Lighthall, D. Miller, S. Quaglioni, J.S. Randhawa, E.T. Rand, A. Rojas, R. Roth, A. Shotter, J. Tanaka, I. Tanihata, C. Unsworth, Phys. Rev. Lett. **118**, 262502 (2017)

49. K.M. Nollett, S.C. Pieper, R.B. Wiringa, J. Carlson, G.M. Hale, Phys. Rev. Lett. **99**, 022502 (2007)
50. G. Hagen, T. Papenbrock, M. Hjorth-Jensen, D.J. Dean, Rep. Prog. Phys. **77**, 096302 (2014)
51. G. Hagen, T. Papenbrock, M. Hjorth-Jensen, Phys. Rev. Lett. **104**, 182501 (2010)
52. G. Hagen, M. Hjorth-Jensen, G.R. Jansen, T. Papenbrock, Phys. Scr. **91**, 063006 (2016)
53. T.D. Morris, J. Simonis, S.R. Stroberg, C. Stumpf, G. Hagen, J.D. Holt, G.R. Jansen, T. Papenbrock, R. Roth, A. Schwenk, Phys. Rev. Lett. **120**, 152503 (2018)
54. G. Hagen, T. Papenbrock, D.J. Dean, A. Schwenk, A. Nogga, M. Włoch, P. Piecuch, Phys. Rev. C **76**, 034302 (2007)
55. V. Somà, T. Duguet, C. Barbieri, Phys. Rev. C **84**, 064317 (2011)
56. V. Somà, C. Barbieri, T. Duguet, Phys. Rev. C **87**, 011303 (2013)
57. V. Somà, C. Barbieri, T. Duguet, Phys. Rev. C **89**, 024323 (2014)
58. G. Hagen, T. Papenbrock, D.J. Dean, M. Hjorth-Jensen, Phys. Rev. Lett. **101**, 092502 (2008)
59. G.R. Jansen, M. Hjorth-Jensen, G. Hagen, T. Papenbrock, Phys. Rev. C **83**, 054306 (2011)
60. Z.H. Sun, T.D. Morris, G. Hagen, G.R. Jansen, T. Papenbrock, Phys. Rev. C **98**, 054320 (2018)
61. H. Hergert, S. Binder, A. Calci, J. Langhammer, R. Roth, Phys. Rev. Lett. **110**, 242501 (2013)
62. H. Hergert, S.K. Bogner, T.D. Morris, S. Binder, A. Calci, J. Langhammer, R. Roth, Phys. Rev. C **90**, 041302 (2014)
63. S.K. Bogner, H. Hergert, J.D. Holt, A. Schwenk, S. Binder, A. Calci, J. Langhammer, R. Roth, Phys. Rev. Lett. **113**, 142501 (2014)
64. H. Hergert, S.K. Bogner, T.D. Morris, A. Schwenk, K. Tsukiyama, Phys. Rep. **621**, 165 (2016). Memorial Volume in Honor of Gerald E. Brown
65. H. Hergert, S.K. Bogner, S. Binder, A. Calci, J. Langhammer, R. Roth, A. Schwenk, Phys. Rev. C **87**, 034307 (2013)
66. M. Hjorth-Jensen, T.T.S. Kuo, E. Osnes, Phys. Rep. **261**, 125 (1995)
67. N. Tsunoda, T. Otsuka, N. Shimizu, M. Hjorth-Jensen, K. Takayanagi, T. Suzuki, Phys. Rev. C **95**, 021304 (2017)
68. N.M. Parzuchowski, S.R. Stroberg, P. Navrátil, H. Hergert, S.K. Bogner, Phys. Rev. C **96**, 034324 (2017)
69. J.M. Yao, B. Bally, J. Engel, R. Wirth, T.R. Rodríguez, H. Hergert, Phys. Rev. Lett. **124**, 232501 (2020)
70. B.S. Hu, Q. Wu, Z.H. Sun, F.R. Xu, Phys. Rev. C **99**, 061302(R) (2019)
71. M. Honma, T. Mizusaki, T. Otsuka, Phys. Rev. Lett. **75**, 1284 (1995)
72. M. Honma, T. Mizusaki, T. Otsuka, Phys. Rev. Lett. **77**, 3315 (1996)
73. T. Otsuka, M. Honma, T. Mizusaki, Phys. Rev. Lett. **81**, 1588 (1998)
74. T. Abe, P. Maris, T. Otsuka, N. Shimizu, Y. Utsuno, J.P. Vary, Phys. Rev. C **86**, 054301 (2012)
75. P. Ring, P. Schuck, *The Nuclear Many-Body Problem*, 3rd edn. (Springer, Berlin, Heidelberg, New York, Hong King, London, Milan, Paris, Tokyo, 2004)
76. G.K. Tobin, M.C. Ferriss, K.D. Launey, T. Dytrych, J.P. Draayer, A.C. Dreyfuss, C. Bahri, Phys. Rev. C **89**, 034312 (2014)
77. T. Dytrych, A.C. Hayes, K.D. Launey, J.P. Draayer, P. Maris, J.P. Vary, D. Langr, T. Oberhuber, Phys. Rev. C **91**, 024326 (2015)
78. A.C. Dreyfuss, K.D. Launey, T. Dytrych, J.P. Draayer, R.B. Baker, C.M. Deibel, C. Bahri, Phys. Rev. C **95**, 044312 (2017)
79. D.J. Rowe, Rep. Prog. Phys. **48**, 1419 (1985)
80. T. Dytrych, K.D. Sviratcheva, J.P. Draayer, C. Bahri, J.P. Vary, J. Phys. G. Nucl. Part. Phys **35**, 123101 (2008)
81. T.c.v. Dytrych, K.D. Sviratcheva, C. Bahri, J.P. Draayer, J.P. Vary, Phys. Rev. Lett. **98**, 162503 (2007)
82. T. Dytrych, K.D. Sviratcheva, C. Bahri, J.P. Draayer, J.P. Vary, Phys. Rev. C **76**, 014315 (2007)
83. T. Dytrych, K.D. Launey, J.P. Draayer, P. Maris, J.P. Vary, E. Saule, U. Catalyurek, M. Sosonkina, D. Langr, M.A. Caprio, Phys. Rev. Lett. **111**, 252501 (2013)

84. R. Lazauskas, J. Carbonell, Phys. Rev. C **71**, 044004 (2005)
85. R. Lazauskas, J. Carbonell, Phys. Rev. C **70**, 044002 (2004)
86. S.K. Bogner, T.T.S. Kuo, A. Schwenk, Phys. Rep. **386**, 1 (2003)
87. Hjorth-Jensen M., Jansen G., CENS: A computational environment for nuclear structure. http://www.fys.uio.no/compphys/software.html
88. N. Michel, W. Nazarewicz, M. Płoszajczak, Phys. Rev. C **70**, 064313 (2004)
89. N. Michel, Eur. Phys. J. A **73**, 523 (2009)
90. D. Vautherin, M. Veneroni, Phys. Lett. B **25**, 175 (1967)
91. F. Palumbo, Nucl. Phys. A **99**, 100 (1967)
92. R.D. Lawson, *Theory of the Nuclear Shell Model* (Clarendon Press, Oxford, 1980)
93. G. Hagen, T. Papenbrock, D.J. Dean, Phys. Rev. Lett. **103**, 062503 (2009)
94. G. Papadimitriou, J. Rotureau, N. Michel, M. Płoszajczak, B.R. Barrett, Phys. Rev. C **88**, 044318 (2013)
95. A. Nogga, S.K. Bogner, A. Schwenk, Phys. Rev. C **70**, 061002 (2004)
96. G. Audi, A.H. Wapstra, C. Thibault, Nucl. Phys. A **729**, 337 (2003)
97. D. Tilley, C. Cheves, J. Godwin, G. Hale, H. Hofmann, J. Kelley, C. Sheu, H. Weller, Nucl. Phys. A **708**, 3 (2002)
98. A.M. Lane, R.G. Thomas, Rev. Mod. Phys. **30**, 257 (1958)
99. G.M. Hale, R.E. Brown, N. Jarmie, Phys. Rev. Lett. **59**, 763 (1987)
100. A. Csótó, G.M. Hale, Phys. Rev. C **55**, 536 (1997)
101. N. Michel, W. Nazarewicz, M. Płoszajczak, T. Vertse, J. Phys. G. Nucl. Part. Phys. **36**, 013101 (2009)
102. J. Bond, F. Firk, Nucl. Phys. A **287**, 317 (1977)
103. G. Audi, O. Bersillon, J. Blachot, A.H. Wapstra, Nucl. Phys. A **279**, 3 (2003)
104. D.B. Smith, N. Jarmie, A.M. Lockett, Phys. Rev. **129**, 785 (1963)
105. K.M. Nollett, R.B. Wiringa, Phys. Rev. C **83**, 041001 (2011)
106. K.M. Nollett, Phys. Rev. C **86**, 044330 (2012)
107. A.M. Mukhamedzhanov, R.E. Tribble, Phys. Rev. C **59**, 3418 (1999)
108. J. Okołowicz, N. Michel, W. Nazarewicz, M. Płoszajczak, Phys. Rev. C **85**, 064320 (2012)
109. B.K. Jennings. Non-observability of spectroscopic factors (2011)
110. N. Michel, W. Nazarewicz, M. Płoszajczak, Phys. Rev. C **75**, 031301 (2007)
111. P. Navrátil, C.A. Bertulani, E. Caurier, Phys. Rev. C **73**, 065801 (2006)
112. P. Navrátil, Phys. Rev. C **70**, 054324 (2004)
113. N.K. Timofeyuk, Phys. Rev. Lett. **103**, 242501 (2009)
114. O. Jensen, G. Hagen, M. Hjorth-Jensen, B.A. Brown, A. Gade, Phys. Rev. Lett. **107**, 032501 (2011)
115. R.Y. Kezerashvili, in *Proceedings of the Sixth International Conference on Fission and Properties of Neutron-Rich Nuclei (ICFN6): Fission and Properties of Neutron-Rich Nuclei*, ed. by J.H. Hamilton, A.V. Ramayya, P. Talou (World Scientific, Singapore, 1998), p. 403
116. K. Kisamori, S. Shimoura, H. Miya, S. Michimasa, S. Ota, M. Assie, H. Baba, T. Baba, D. Beaumel, M. Dozono, T. Fujii, N. Fukuda, S. Go, F. Hammache, E. Ideguchi, N. Inabe, M. Itoh, D. Kameda, S. Kawase, T. Kawabata, M. Kobayashi, Y. Kondo, T. Kubo, Y. Kubota, M. Kurata-Nishimura, C.S. Lee, Y. Maeda, H. Matsubara, K. Miki, T. Nishi, S. Noji, S. Sakaguchi, H. Sakai, Y. Sasamoto, M. Sasano, H. Sato, Y. Shimizu, A. Stolz, H. Suzuki, M. Takaki, H. Takeda, S. Takeuchi, A. Tamii, L. Tang, H. Tokieda, M. Tsumura, T. Uesaka, K. Yako, Y. Yanagisawa, R. Yokoyama, K. Yoshida, Phys. Rev. Lett. **116**, 052501 (2016)
117. F.M. Marqués, M. Labiche, N.A. Orr, J.C. Angélique, L. Axelsson, B. Benoit, U.C. Bergmann, M.J.G. Borge, W.N. Catford, S.P.G. Chappell, N.M. Clarke, G. Costa, N. Curtis, A. D'Arrigo, E. de Góes Brennand, F. de Oliveira Santos, O. Dorvaux, G. Fazio, M. Freer, B.R. Fulton, G. Giardina, S. Grévy, D. Guillemaud-Mueller, F. Hanappe, B. Heusch, B. Jonson, C. Le Brun, S. Leenhardt, M. Lewitowicz, M.J. López, K. Markenroth, A.C. Mueller, T. Nilsson, A. Ninane, G. Nyman, I. Piqueras, K. Riisager, M.G.S. Laurent, F. Sarazin, S.M. Singer, O. Sorlin, L. Stuttgé, Phys. Rev. C **65**, 044006 (2002)
118. S.C. Pieper, Phys. Rev. Lett. **90**, 252501 (2003)

119. C.A. Bertulani, V. Zelevinsky, J. Phys. G. Nucl. Part. Phys **29**, 2431 (2003)
120. K. Fossez, W. Nazarewicz, Y. Jaganathen, N. Michel, M. Płoszajczak, Phys. Rev. C **93**, 011305 (2016)
121. E. Hiyama, R. Lazauskas, J. Carbonell, M. Kamimura, Phys. Rev. C **93**, 044004 (2016)
122. A.M. Shirokov, G. Papadimitriou, A.I. Mazur, I.A. Mazur, R. Roth, J.P. Vary, Phys. Rev. Lett. **117**, 182502 (2016)
123. A.M. Shirokov, J.P. Vary, A.I. Mazur, T.A. Weber, Phys. Lett. B **644**, 33 (2007)
124. S. Aoyama, T. Myo, K. Katō, K. Ikeda, Prog. Theor. Phys. **116**, 1 (2006)
125. T. Myo, Y. Kikuchi, H. Masui, K. Katō, Prog. Part. Nucl. Phys. **79**, 1 (2014)
126. S. Gandolfi, H.W. Hammer, P. Klos, J.E. Lynn, A. Schwenk, Phys. Rev. Lett. **118**, 232501 (2017)
127. S. Gandolfi, H.W. Hammer, P. Klos, J.E. Lynn, A. Schwenk, Phys. Rev. Lett. **123**, 069202 (2019)
128. K. Fossez, J. Rotureau, N. Michel, M. Płoszajczak, Phys. Rev. Lett. **119**, 032501 (2017)
129. A. Ekström, G. Baardsen, C. Forssén, G. Hagen, M. Hjorth-Jensen, G.R. Jansen, R. Machleidt, W. Nazarewicz, T. Papenbrock, J. Sarich, S.M. Wild, Phys. Rev. Lett. **110**, 192502 (2013)
130. J. Rotureau, N. Michel, W. Nazarewicz, M. Płoszajczak, J. Dukelsky, Phys. Rev. C **79**, 014304 (2009)
131. J. Rotureau, N. Michel, W. Nazarewicz, M. Płoszajczak, J. Dukelsky, Phys. Rev. Lett. **97**, 110603 (2006)
132. J.G. Li, N. Michel, B.S. Hu, W. Zuo, F.R. Xu, Phys. Rev. C **100**, 054313 (2019)
133. A.M. Krenciglowa, T.T.S. Kuo, Nucl. Phys. A **235**, 171 (1974)
134. K. Takayanagi, Nucl. Phys. A **852**, 61 (2011)
135. N. Tsunoda, K. Takayanagi, M. Hjorth-Jensen, T. Otsuka, Phys. Rev. C **89**, 024313 (2014)
136. L. Coraggio, A. Covello, A. Gargano, N. Itaco, T. Kuo, Ann. Phys. **327**, 2125 (2012)
137. Q. Wu, F.R. Xu, B.S. Hu, J.G. Li, J. Phys. G: Nucl. Part. Phys. **46**, 055104 (2019)
138. N. Itaco, L. Coraggio, Phys. Lett. B **616**, 43 (2005)
139. B. Hu, Q. Wu, J. Li, Y. Ma, Z. Sun, N. Michel, F. Xu, Phys. Lett. B **802**, 135206 (2020)
140. M. Wang, G. Audi, F. Kondev, W. Huang, S. Naimi, X. Xu, Chin. Phys. C **41**, 030003 (2017)
141. G. Hagen, M. Hjorth-Jensen, G.R. Jansen, R. Machleidt, T. Papenbrock, Phys. Rev. Lett. **108**, 242501 (2012)
142. K. Fossez, J. Rotureau, N. Michel, W. Nazarewicz, Phys. Rev. C **96**, 024308 (2017)
143. S.R. Stroberg, H. Hergert, J.D. Holt, S.K. Bogner, A. Schwenk, Phys. Rev. C **93**, 051301(R) (2016)
144. T. Otsuka, T. Suzuki, J.D. Holt, A. Schwenk, Y. Akaishi, Phys. Rev. Lett. **105**, 032501 (2010)
145. B.A. Brown, W.A. Richter, Phys. Rev. C **74**, 034315 (2006)
146. Y. Utsuno, T. Otsuka, T. Mizusaki, M. Honma, Phys. Rev. C **60**, 054315 (1999)
147. Y.A. Luo, J. Okołowicz, M. Płoszajczak, N. Michel, arXiv nucl-th/0211068 [nucl-th] (2002)
148. N. Michel, W. Nazarewicz, J. Okołowicz, M. Płoszajczak, J. Rotureau, Acta Phys. Pol. B **35**, 1249 (2004)
149. L.V. Grigorenko, I.G. Mukha, C. Scheidenberger, M.V. Zhukov, Phys. Rev. C **84**, 021303(R) (2011)
150. H. Sakurai, S. Lukyanov, M. Notani, N. Aoi, D. Beaumel, N. Fukuda, M. Hirai, E. Ideguchi, N. Imai, M. Ishihara, H. Iwasaki, T. Kubo, K. Kusaka, H. Kumagai, T. Nakamura, H. Ogawa, Y. Penionzhkevich, T. Teranishi, Y. Watanabe, K. Yoneda, A. Yoshida, Phys. Lett. B **448**, 180 (1999)
151. S.M. Lukyanov, Y.E. Penionzhkevich, R. Astabatyan, S. Lobastov, Y. Sobolev, D. Guillemaud-Mueller, G. Faivre, F. Ibrahim, A.C. Mueller, F. Pougheon, O. Perru, O. Sorlin, I. Matea, R. Anne, C. Cauvin, R. Hue, G. Georgiev, M. Lewitowicz, F. de Oliveira Santos, D. Verney, Z. Dlouh, J.M. zek, D. Baiborodin, F. Negoita, C. Borcea, A. Buta, I. Stefan, S. Grevy, J. Phys. G. Nucl. Part. Phys **28**, L41 (2002)

# The Unification of Structure and Reaction Frameworks

9

It has become customary to separate nuclear structure from nuclear reactions in the development of theoretical models, even though both of them arise from the same Hamiltonian [1–7]. In standard reaction frameworks [3, 4, 6], the structure aspects of many-body nuclear states have been only indirectly included. For example, it is customary to use coupled-channel equations and one-body optical potentials to calculate reaction cross sections, whereby the parameters entering one-body optical potentials are directly fitted from experimental data [8–11]. Consequently, the correlations present in many-body nuclear states are considered only from an effective point of view, as they only enter the fitted parameters of reaction potentials [8–11]. It is also common to separate cross sections into two parts, one coming from a calculation where the effect of nuclear structure degrees of freedom is averaged within a mean-field potential, such as the distorted wave Born approximation, and the other being a spectroscopic factor calculated in the standard shell model. In fact, nuclear structure and reaction degrees of freedom are decoupled in this approach, so there is no consistency between the Hamiltonians used for the calculation of nuclear structure and nuclear reaction observables.

Cross sections can be usually written as the product of two factors, one containing the kinetic dependence of the considered reaction and the other its correlation part (see Sect. 7.4.1). However, even if the results obtained in this approach compare well with experimental data in well-bound nuclei, they are not satisfactory when considering weakly bound and resonance many-body states. Indeed, in a vicinity of the driplines, nuclei exhibit halos or are unbound even in the ground state. Moreover, many-body wave functions strongly depend on separation energies and on the partial waves present in the model (see Sect. 7.4.2). Therefore, in the description of nuclear reactions, aspects of the structure and the reaction can no longer be approximated by the aforementioned factorization and are, on the opposite, intertwined. Hence, it is necessary to devise a theoretical approach for nuclear reactions treating inter-nucleon correlations and continuum degrees of freedom on the same footing.

© Springer International Publishing AG 2021
N. Michel, M. Płoszajczak, *Gamow Shell Model*, Lecture Notes in Physics 983,
https://doi.org/10.1007/978-3-030-69356-5_9

A framework going along these lines has been formulated in the context of the continuum shell model [12–14], where the Hilbert space is decomposed following the Feshbach projection technique. In this model, the Hilbert space is divided into orthogonal subspaces $\mathcal{Q}_0$, $\mathcal{Q}_1$, ... containing 0, 1, ... particles in the scattering continuum, respectively. All basis states have real energies, contrary to the Gamow shell model. Despite the real-energy character of the basis of the continuum shell model, unbound states can still be calculated in this framework. Indeed, coupling to the continuum generates non-Hermitian components, which are added to the initial Hermitian Hamiltonian describing discrete states of $\mathcal{Q}_0$ subspace in the closed quantum system approximation. Consequently, an open quantum system description of states in $\mathcal{Q}_0$ includes couplings to the environment of decay channels through the energy-dependent and in general non-Hermitian effective Hamiltonian (see Ref. [15] and references cited therein). The eigenvalues of the effective Hamiltonian are real at energies below all decay thresholds and complex above the lowest decay threshold.

In the past, the continuum shell model was applied with schematic Hamiltonians [13,14]. In such a formulation, the continuum shell model cannot provide a realistic description of complex structures in atomic nuclei. To remedy this problem, the shell model embedded in the continuum has been developed [16–21], which derives from the continuum shell model. In the shell model embedded in the continuum, the subspace of discrete states is identified with a standard shell model space, i.e. one utilizes a standard shell model interaction to generate configuration mixing in $\mathcal{Q}_0$. Furthermore, a zero-range interaction is used for the effective nucleon–nucleon interaction pertaining to continuum coupling [15]. This allowed to reproduce nuclear spectra of light nuclei, which is impossible using a schematic nuclear interaction [16–23]. Moreover, due to the coupled-channel representation of the Hamiltonian in the shell model embedded in the continuum, one can conveniently calculate various cross sections, such as the elastic/inelastic scattering, radiative capture, and Coulomb dissociation cross sections, using the $S$-matrix formalism [16–20]. Up to now, the shell model embedded in the continuum studies has been performed in a core + valence nucleons approach, but the model can be easily generalized into an ab initio no-core formulation as well. Attempts in this direction have been made by including continuum degrees of freedom on top of the no-core shell model using the resonating group method [24–33]. Another yet unexplored path would be the formulation of no-core Gamow shell model in the coupled-channel representation (see Sect. 8).

The calculation of reaction cross sections in an ab initio framework has been developed in the context of the no-core shell model with the resonating group method [24, 28–31, 33] and, more recently, in the no-core shell model with continuum [26–29]. In the no-core shell model with the resonating group method, target and projectile wave functions are calculated in the no-core shell model framework in a basis of harmonic oscillator states. In the no-core shell model with continuum, these wave functions are complemented with a continuum part, taking into account asymptotic degrees of freedom. The Hamiltonian can then be written in a coupled-channel representation with a basis of target–projectile reaction channels within the resonating group method.

These theoretical frameworks are exact if all target and projectile many-body states, as well as all emission channels, are included. However, due to the current numerical limitations, it is impossible to attain full convergence. Reasonable convergence can be obtained in lightest many-body systems, but the no-core shell model, with or without the inclusion of continuum degrees of freedom, to which the resonating group method is added, cannot treat nuclei in the $sd$ shell [34]. Moreover, the inclusion of many-nucleon projectiles is very difficult therein, due to the complex character of target–projectile channels. This is due to the necessary use of Jacobi coordinates to properly describe target–projectile composites in the asymptotic region.

A similar problem is encountered in standard reaction frameworks, where it is customary to treat reactions in the frame of three-body or four-body problems, i.e. with three or four subsystems in the asymptotic region. Indeed, while solving the five-body and six-body problems is still possible [35, 36], they are at the limit of current computing capabilities. Hence, the no-core shell model used along the resonating group method, with or without the inclusion of continuum degrees of freedom, is very limited and cannot be systematically used to consider reactions involving nuclei beyond $s$- and $p$-shell nuclei.

Another possibility is to use the coupled-cluster method to calculate optical potentials and cross sections [37–39]. In Ref. [37], composite states, formed by target state and a one-proton projectile state, are expanded with a hybrid basis, formed by the Berggren basis for the most important partial waves occurring in the reaction process, and by a basis of harmonic oscillator states for the other partial waves. This allows to precisely calculate proton overlap functions and asymptotic normalization coefficients (see Sect. 7.3), from which phase shifts and hence reaction cross sections can be calculated [37]. This framework was successfully applied to the $^{40}$Ca(p,p) elastic scattering reaction at low energies [37]. Another formalism has been followed in Refs. [38, 39]. In this approach, targets are still expanded in a hybrid basis formed by harmonic oscillator and Berggren basis states in order to account for continuum coupling, whereas optical potentials are calculated from the Green's function representation of the $A$-nucleon system [38, 39]. These optical potentials then provide with scattering states, from which phase shifts are extracted and reaction cross sections implemented. First applications consisted of the examples of neutron scattering on $^{16}$O and $^{40,48}$Ca [38, 39]. While the devised numerical method is stable, the obtained optical potential lacks the absorption, as its imaginary part is very small. The absence of imaginary potential was attributed to missing higher order correlation effects in the coupled-cluster formalism using the singles-and-doubles approximation [38, 39]. Optical potentials describing one-nucleon scattering on oxygen and calcium isotopes have recently been calculated in the context of the Gorkov–Green-function formalism [40]. However, as one has already seen in Chap. 8, the coupled cluster or the Gorkov–Green-function methods are also restricted to a small number of targets, near closed-shell nuclei typically.

As the Gamow shell model can be defined accurately in a core + valence particle approach, it can be considered in wider sections of the nuclear chart. Moreover, realistic nucleon–nucleon forces issued from many-body perturbation theory [41, 42] or the in-medium similarity renormalization group method [43–47] can be included in this framework as well [48]. However, the Gamow shell model methods presented in previous chapters cannot be used directly to calculate reaction cross sections. Indeed, the many-body scattering states arising from the diagonalization of the Gamow shell model Hamiltonian matrix do not have a well defined asymptotic behavior so the many-body eigenstates are generally a linear combination of many reaction channels. Hence, the scattering eigenstates of the Gamow shell model cannot be identified, contrary to resonant states, for which the overlap method can be utilized (see Sect. 5.4).

Therefore, one has to build a new theoretical framework to be able to use the Gamow shell model therein and obtain in this way a unified theory of nuclear structure and reactions, which remains valid in both bound and unbound atomic nuclei. To achieve this objective, a route similar to that used in the no-core shell model with the resonating group method [24, 28–31, 33] has been followed. One constructs target–projectile reaction channels using the resonating group method, where target and projectile wave functions are issued from the diagonalization of the Gamow shell model Hamiltonian. In this way, the considered targets and projectiles can be weakly bound or resonance nuclei, and nevertheless, the reaction channels are well defined. Many-body scattering states are obtained from the solution of a Gamow shell model Hamiltonian expressed in the coupled-channel representation. Reaction observables, such as, for example, the elastic/inelastic scattering and radiative capture cross sections, can then be calculated.

In this approach, the Gamow shell model unifies structure and reaction degrees of freedom as both spectra and reaction cross sections can be calculated using the same Hamiltonian. In particular, the calculation of the coupling potentials entering coupled-channel equations includes complete information about structure of target and projectile, and the wave functions of target and projectile are expanded on a basis of Berggren many-body states. Added to that, continuum degrees of freedom are taken into account to calculate spectra, as resonant states in the coupled-channel Gamow shell model are linear combination of reaction channels. As all degrees of freedom are included in the coupled-channel Gamow shell model, it is similar to no-core shell model with continuum used with the resonating group method [26–29], or the shell model embedded in the continuum in $S$-matrix formulation [15, 19–21].

The fact that target nuclei can be loosely bound or even unbound is a very interesting feature of the coupled-channel Gamow shell model. As targets are directly expanded with the Berggren basis, it is not necessary to expand their asymptotes with reaction channels, as is necessary in the no-core shell model with resonating group method [26–29]. Therefore, a smaller number of reaction channels are needed to represent the same many-body scattering wave functions. For example, the description of scattering reaction p + $^{11}$Li in the no-core shell model with continuum demands three nucleons in the asymptotic region, whereas it is a two-body reaction using coupled-channel Gamow shell model, as the two-neutron halo in $^{11}$Li is already included in the many-body wave function arising from the Gamow shell model.

## 9.1   Nucleon–Nucleus Reactions

One will firstly consider elastic/inelastic scattering and radiative capture cross sections involving one-nucleon projectiles. They are of important experimental interest, and many theoretical aspects of the presented formalism are common to the reactions making use of one-nucleon and many-nucleon projectiles.

The $A$-body system representing a composite system, i.e. the many-nucleon system formed by the projectile and the target, is described by the Hamiltonian (5.42) given in the cluster orbital shell model coordinates. To determine reaction cross sections, one has to calculate the scattering state $|\Psi_{M_A}^{J_A}\rangle$ of the $A$-body system, which verifies the Schrödinger equation:

$$\hat{H} |\Psi_{M_A}^{J_A}\rangle = E |\Psi_{M_A}^{J_A}\rangle \ . \tag{9.1}$$

Here, due to the energy conservation, one has: $E = E_T + E_P$, where $E_P$ and $E_T$ are, respectively, the total energy of the projectile and the total energy of the target. To calculate the cross section with a given entrance channel, one has to determine the $A$-body scattering state $|\Psi_{M_A}^{J_A}\rangle$

$$\hat{\mathscr{A}} \{|\Psi_T^{J_T}\rangle \otimes |\Psi_P^{J_P}\rangle\}_{M_A}^{J_A} \ . \tag{9.2}$$

Here $|\Psi_T^{J_T}\rangle$ and $|\Psi_P^{J_P}\rangle$ are, respectively, the target and projectile states with their momentum $J_T$ and $J_P$ so that the total angular momentum is: $\mathbf{J_A} = \mathbf{J_T} + \mathbf{J_P}$. All $A$-body wave functions are fully antisymmetrized, as emphasized by the $\mathscr{A}$ symbol.

The first step is to build the channel basis, and the second step is to determine the expansion of the $A$-body scattering state $|\Psi_{M_A}^{J_A}\rangle$ in this channel basis. In order to consider reactions involving one-proton or neutron scattering processes, one defines the projectile state as

$$|\Psi_P^j\rangle = |r \ell j \tau\rangle \ , \tag{9.3}$$

where the angular momentum notation $J_P$ introduced in Eq. (9.2) has been replaced by $j$ for convenience. Here, $r$ is the radial coordinate, $\ell$, $j$ are the angular quantum numbers, and $\tau$ is the isospin quantum number.

To express antisymmetry in a convenient way, the $|r \ell j \tau\rangle$ channel is firstly expanded with the single-particle basis of Gamow shell model wave functions $u_n(r)$

$$|r \ell j \tau\rangle = \sum_n \left(\frac{u_n(r)}{r}\right) |u_n \ell j \tau\rangle \ , \tag{9.4}$$

where

$$\langle r\ell j\tau | u_n \ell j\tau \rangle = \frac{u_n(r)}{r} ,$$

which implies for its associated particle creation operator

$$a^{\dagger}_{r\ell j\tau} = \sum_n \left( \frac{u_n(r)}{r} \right) a^{\dagger}_{n\ell j\tau} . \tag{9.5}$$

Hence, the expression of the $A$-body wave function becomes

$$\hat{\mathscr{A}} \, |\{ |\Psi_T^{J_T} \rangle \otimes |r\ell j\tau \rangle \}_{M_A}^{J_A} \rangle = \sum_n \left( \frac{u_n(r)}{r} \right) \hat{\mathscr{A}} \, |\{ |\Psi_T^{J_T} \rangle \otimes |u_n \ell j\tau \rangle \}_{M_A}^{J_A} \rangle$$

$$= \sum_n \left( \frac{u_n(r)}{r} \right) \{ a^{\dagger}_{n\ell j\tau} |\Psi_T^{J_T} \rangle \}_{M_A}^{J_A} . \tag{9.6}$$

One will now expand the $A$-body wave function $\{ a^{\dagger}_{n\ell j\tau} |\Psi_T^{J_T} \rangle \}_{M_A}^{J_A}$, appearing in Eq. (9.6), in a complete basis of $A$-body Slater determinants

$$\{ a^{\dagger}_{n\ell j\tau} |\Psi_T^{J_T} \rangle \}_{M_A}^{J_A} = \sum_{\alpha} c_{\alpha} |SD_{\alpha} \rangle . \tag{9.7}$$

For that, one uses the expansion of $|\Psi_T^{J_T, M_T} \rangle$ in $(A-1)$-body Slater determinants, obtained from the diagonalization of the Gamow shell model Hamiltonian $H$

$$|\Psi_T^{J_T, M_T} \rangle = \sum_{\beta} b_{\beta}^{A-1} |SD_{\beta} \rangle^{A-1} , \tag{9.8}$$

where the $(A-1)$-body character of expansion coefficients and Slater determinants is explicitly written. Applying the creation operator defined in Eq. (9.5) on Eq. (9.8), one obtains the uncoupled fully antisymmetrized $A$-body wave function:

$$\{ a^{\dagger}_{n\ell jm\tau} |\Psi_T^{J_T, M_T} \rangle \} = \sum_{\beta} b_{\beta}^{A-1} a^{\dagger}_{n\ell jm\tau} |SD_{\beta} \rangle^{A-1} = \sum_{\alpha} (-1)^{\varphi_{\alpha}} b_{\alpha} |SD_{\alpha} \rangle , \tag{9.9}$$

where $m$ is the angular momentum projection of the projectile state, verifying $M_T + m = M_A$, and the Slater determinants $|SD_{\alpha} \rangle = (-1)^{\varphi_{\alpha}} a^{\dagger}_{n\ell jm\tau} |SD_{\beta} \rangle^{A-1}$ are $A$-body basis functions, in which a $(-1)^{\varphi_{\alpha}}$ rearrangement phase appears. Note that all $A$-body states, $a^{\dagger}_{n\ell jm\tau} |SD_{\beta} \rangle^{A-1}$, for which $|u_n \ell jm\tau \rangle \in |SD_{\beta} \rangle^{A-1}$, vanish identically due to the Pauli principle. The $A$-body wave function expansion of Eq. (9.7) can be determined by coupling the wave function defined in Eq. (9.9) to a given angular momentum using Clebsch–Gordan coefficients.

### 9.1.1 Coupled-Channel Hamiltonian for Structure and Reaction Observables

One will now express the Gamow shell model Hamiltonian in a coupled-channel formalism. The Hamiltonian is denoted as: $\hat{H} = \hat{T} + \hat{U}_{\text{core}} + \hat{V}_{\text{res}}$, where $\hat{U}_{\text{core}} = \sum_{i \in \text{val}} \hat{U}_i^{(\text{core})}$ and $\hat{V}_{\text{res}} = \sum_{i < j \in \text{val}} \hat{V}_{ij}$. One will evaluate the matrix elements of $\hat{H}$ (multiplied by the radii $r_i$ and $r_f$ for convenience), denoted by $H_{if}$:

$$H_{if} = r_i\, r_f\, \langle \hat{\mathscr{A}}\{\langle \Psi_{T_f}^{J_{T_f}} | \otimes \langle r_f \ell_f j_f \tau_f |\}_{M_A}^{J_A} | \hat{H} | \hat{\mathscr{A}}\{| \Psi_{T_i}^{J_{T_i}} \rangle \otimes | r_i \ell_i j_i \tau_i \rangle\}_{M_A}^{J_A} \rangle . \tag{9.10}$$

For this, one separates the Hamiltonian $\hat{H}$ into basis and residual parts:

$$\hat{H} = \hat{T} + \hat{U}_{\text{basis}} + (\hat{V}_{\text{res}} - \hat{U}_0) , \tag{9.11}$$

where $\hat{U}_{\text{basis}}$ is a basis potential recapturing the bulk properties of the $A$-particle system and $\hat{U}_0 = \hat{U}_{\text{basis}} - \hat{U}_{\text{core}}$. The advantage of this decomposition is that $\hat{V}_{\text{res}} - \hat{U}_0$ is a finite-range operator.

In order to derive analytical expressions for the infinite-range part of $\hat{H}$, one will suppose that only a finite number of Slater determinants appear in Eq. (9.10). This approximation is justified because Gamow shell model eigenvector expansion coefficients decrease exponentially with the energy of basis scattering Slater determinants, so that they can be approximated with a finite expansion of Slater determinants for any arbitrarily small precision. Hence, there exists $n_{\max}$ for which $\langle s | u_n \ell j \tau \rangle = 0, \forall |s\rangle \in |SD_\beta\rangle^{A-1}$ with $\beta \leq \beta_{\max}$ and $\forall n > n_{\max}$. Therefore, antisymmetry no longer plays any role in $|u_n \ell j \tau\rangle$ and $|\Psi_T^{J_T}\rangle$ (see Eq. (9.7)) for $n > n_{\max}$.

---

**Exercise I ★ ★ ★**
One will derive the expression for the matrix elements of the Gamow shell model Hamiltonian in a coupled-channel formalism (see Eq. (9.10)).

**A.** Expand the $H_{if}$ matrix element of Eq. (9.10) using the Berggren basis representation of channels. One will denote as $|n_i \ell j \tau\rangle$ and $|n_f \ell j \tau\rangle$ the Berggren basis channels associated to the $|r_i \ell j \tau\rangle$ input channel and $|r_f \ell j \tau\rangle$ output channel. Separate the obtained expression for $H_{if}$ in different parts where $n_i \leq n_{\max}$, $n_i > n_{\max}$, $n_f \leq n_{\max}$, $n_f > n_{\max}$, $n_i = n_f$, or $n_i \neq n_f$. Show that $\hat{H}$ reduces to $\hat{H}_T + \hat{H}_P$ and that antisymmetrizers $\mathscr{A}$ can be removed if $n_i > n_{\max}$ or $n_f > n_{\max}$.

**B.** Show that the sums for which $n_i \neq n_f$ and where $n_i > n_{\max}$, or $n_f > n_{\max}$ vanish in the $H_{if}$ matrix element. Demonstrate that the $H_{if}$ matrix element only possesses finite sums if $|\ell_i j_i \tau_i\rangle \neq |\ell_f j_f \tau_f\rangle$ in Eq. (9.10).

**C.** One will now consider that $|\ell_i j_i \tau_i\rangle = |\ell_f j_f \tau_f\rangle = |\ell j \tau\rangle$ in Eq. (9.10). Write the $H_{if}$ matrix element as the sum of one off-diagonal part involving $n_i \neq n_f$

and of two diagonal parts in which $n \geq 0$ and $n \leq n_{\max}$, with $n = n_{\mathrm{i}} = n_{\mathrm{f}}$. Explain why it is convenient to have an infinite sum starting from $n = 0$. Calculate the infinite sum of $H_{\mathrm{if}}$ using the completeness properties of the Berggren basis.

**D.** Using all previous results, write $H_{\mathrm{if}}$ as a function of target states energies, single-particle energies of Berggren basis states, and Slater determinant expansions of channel states. Conclude that $H_{\mathrm{if}}$ can be calculated numerically using the Gamow shell model formalism.

A convenient way to remove antisymmetry when $n > n_{\max}$ is to have matrix elements $\langle \alpha \beta | V_{\mathrm{res}} - U_0 | \gamma \delta \rangle$ vanished when $n_\alpha > n_{\max}$, $n_\beta > n_{\max}$, $n_\gamma > n_{\max}$, or $n_\delta > n_{\max}$. These assumptions are always present in Gamow shell model calculations as one always uses a finite model space. In the absence of antisymmetry if $n > n_{\max}$, it is convenient to rewrite the Hamiltonian $\hat{H}$ of Eq. (9.11) introducing an operator acting only on the target

$$\hat{H} = \hat{T} + \hat{U}_{\mathrm{basis}} + (\hat{V}_{\mathrm{res}} - \hat{U}_0)^{A-1} + [(\hat{V}_{\mathrm{res}} - \hat{U}_0) - (\hat{V}_{\mathrm{res}} - \hat{U}_0)^{A-1}] , \qquad (9.12)$$

where one defines $(\hat{V}_{\mathrm{res}} - \hat{U}_0)^{A-1}$ by its action on the non-antisymmetrized $A$-body states $|\Psi_{\mathrm{T}}^{J_{\mathrm{T}}}\rangle \otimes |u_{n\ell j m \tau}\rangle$

$$(\hat{V}_{\mathrm{res}} - \hat{U}_0)^{A-1}(|\Psi_{\mathrm{T}}^{J_{\mathrm{T}}}\rangle \otimes |u_{n\ell j m \tau}\rangle) = [(\hat{V}_{\mathrm{res}} - \hat{U}_0)|\Psi_{\mathrm{T}}^{J_{\mathrm{T}}}\rangle] \otimes |u_{n\ell j m \tau}\rangle . \qquad (9.13)$$

In this expression, $(\hat{V}_{\mathrm{res}} - \hat{U}_0)^{A-1}$ is the finite-range part of $\hat{V}_{\mathrm{res}} - \hat{U}_0$ restricted to act on the $(A-1)$-body states only. Consequently, one can evaluate matrix elements of $\hat{T} + \hat{U}_{\mathrm{basis}} + (\hat{V}_{\mathrm{res}} - U_0)^{A-1}$ (see Eqs. (9.12), (9.13)) without recurring to the Slater determinant expansion of Eq. (9.10). Indeed, $\hat{T} + \hat{U}_{\mathrm{basis}} + (\hat{V}_{\mathrm{res}} - \hat{U}_0)^{A-1}$ can be written as a sum of target and projectile Hamiltonians, $\hat{H}_{\mathrm{T}} + \hat{H}_{\mathrm{P}}$, with $\hat{H}_{\mathrm{T}}$ acting only on $|\Psi_{\mathrm{T}}^{J_{\mathrm{T}}}\rangle$ and $\hat{H}_{\mathrm{P}}$ acting only on $|u_{n\ell j \tau}\rangle$ (see Eq. (9.6))

$$\hat{H}_{\mathrm{T}} = \hat{T}^{A-1} + (\hat{U}_{\mathrm{basis}})^{A-1} + (\hat{V}_{\mathrm{res}} - \hat{U}_0)^{A-1} \qquad (9.14)$$

$$\hat{H}_{\mathrm{P}} = \hat{t} + \hat{U}_{\mathrm{basis}} . \qquad (9.15)$$

In these expressions, $\hat{T}$ and $\hat{U}_{\mathrm{basis}}$ operators have been separated in target ($A-1$ superscript) and projectile parts. Eq. (9.10) is calculated in Exercise I. This exercise is technical, however, and can be omitted in a first reading.

### 9.1.1.1 Orthogonality of the Composite States with Respect to the Core States

As the cluster orbital shell model formulation is used in coupled-channel Gamow shell model, the valence states have to be orthogonal to the core states. For this, one introduces the projectors "in" and "out" of the core

$$\hat{Q} = \sum_{s \in \text{core}} |s\rangle \langle s| \tag{9.16}$$

$$\hat{P} = \hat{1} - \hat{Q}, \tag{9.17}$$

where $|s\rangle$ is a one-body state occupied in the core.

---

**Exercise II** ⋆

One will study the use of projectors within the cluster orbital shell model.

**A.** Show that $\hat{Q}$ and $\hat{P}$ are orthogonal Hermitian projectors and state their eigenvalues and eigenvectors.
**B.** Explain how to project an arbitrary operator on the $P$ space with projectors $\hat{Q}$ and $\hat{P}$.

Show that the projection of a Hermitian operator on the $P$ space remains Hermitian.
**C.** State the physical significance of the use of Hermitian projectors $\hat{Q}$ and $\hat{P}$ within the cluster orbital shell model.

Explain why considering complex-symmetric projectors and Hamiltonians is not sufficient for practical applications.

---

The projectile Hamiltonian $\hat{H}_P$ (see Eq. (9.15)) must be replaced by $\hat{P}\hat{H}_P\hat{P}$ in order to have valence states orthogonal to core states (see Exercise II). Thus, one obtains from Eqs. (9.16, 9.17)

$$\hat{H}_P \to \hat{P}\hat{H}_P\hat{P} = \hat{H}_P - \hat{Q}\hat{H}_P - \hat{H}_P\hat{Q} + \hat{Q}\hat{H}_P\hat{Q}. \tag{9.18}$$

The $|s\rangle$ states entering the $\hat{Q}$ projector in Eq. (9.16) belong to the core, so that they are well bound. As a consequence, the $\hat{Q}$ projector takes a form of the potential localized in the nuclear zone when represented in coordinate space. Therefore, $\hat{P}\hat{H}_P\hat{P}$ induces an additional nonlocal potential in $\hat{U}_{\text{basis}}$ of Eq. (9.15). With this modification of $\hat{U}_{\text{basis}}$, all previously derived equations remain the same, as long as core states $|u_{n\ell j\tau}\rangle$ are suppressed in all expansions involving Berggren basis states. Therefore, composite states in the Gamow coupled-channel formalism, consisting of target many-body states, provided by the Gamow shell model, and of the one-body state of the projectile, are orthogonal to the core occupied states.

### 9.1.1.2 Coupled-Channel Equations

As one has defined the channel basis and the matrix elements of the Hamiltonian in a basis of channel many-body states, one can now determine the $A$-body scattering states solutions of coupled-channel equations. Let us define the $A$-body scattering state in a coupled-channel representation:

$$|\Psi_{M_A}^{J_A}\rangle = \sum_c \int_0^{+\infty} \left(\frac{u_c(r)}{r}\right) |(c, r)\rangle \, r^2 dr \,, \tag{9.19}$$

where

$$|(c, r)\rangle = \hat{\mathscr{A}}\{|\Psi_T^{J_T}\rangle \otimes |r, \ell, j, \tau\rangle\}_{M_A}^{J_A} \,.$$

The index $c$ stands for the $\{J_T, \ell, j, \tau\}$ quantum numbers. Here $u_c(r)$ is the radial amplitude of channel $c$, which has to be determined. The coupled-channel equations derive formally from the Schrödinger equation $\hat{H} |\Psi_{M_A}^{J_A}\rangle = E |\Psi_{M_A}^{J_A}\rangle$ as

$$\sum_c \int_0^{+\infty} \left(H_{cc'}(r, r') - E \, N_{cc'}(r, r')\right) u_c(r') dr' = 0 \,, \tag{9.20}$$

with

$$H_{cc'}(r, r') = rr' \langle (c, r)| \hat{H} |(c', r')\rangle \tag{9.21}$$

$$N_{cc'}(r, r') = rr' \langle (c, r)|(c', r')\rangle \,. \tag{9.22}$$

The calculation of $H_{cc'}(r, r')$

$$H_{cc'}(r, r') = \left(\frac{\hbar^2}{2m} \left(-\frac{d^2}{dr^2} + \frac{\ell(\ell+1)}{r^2}\right) + E_T\right) \delta(r - r') \delta_{cc'}$$

$$+ U_{\text{basis}}(r, r') \delta_{cc'} + V_{cc'}(r, r') \tag{9.23}$$

is detailed in Exercise I. In this expression $E_T$ is the energy of $|\Psi_T^{J_T}\rangle$ and $V_{cc'}(r, r')$ is the channel coupling potential. Matrix elements of $V_{cc'}$ are expanded in a basis of harmonic oscillator states. Even though the use of the Berggren basis is theoretically correct for that matter, it would lead to unstable calculations in practice. Indeed, discretization effects in this case would generate spurious components, so that calculated potentials would diverge at large distances instead of vanishing. Moreover, it would be numerically prohibitive to calculate and store potentials for every $(r, r')$ coordinate. Therefore, as it is already the case for the basis potential $U_{\text{basis}}$, one calculates the coupled-channel potentials of Sect. 9.1.1.2 in a basis of harmonic oscillator states. For that, due to the completeness properties of both coordinate and harmonic oscillator basis expansions, it is sufficient to replace $u_n(r)$ by the overlap $\langle n|n^{\text{HO}}\rangle$, where $|u_n^{\text{HO}} \ell j \tau\rangle$ is a harmonic oscillator basis state. The expression of the $V_{cc'}$ matrix element using a basis of harmonic oscillator states reads from the results of Exercise I

$$
V_{cc'}(n_i{}^{HO}, n_f{}^{HO}) = \sum_{n_i n_f \alpha_i \alpha_f}^{n_{max}} c_{\alpha_i}^{(i)} c_{\alpha_f}^{(f)} \langle SD_{\alpha_f} | \hat{H} | SD_{\alpha_i} \rangle \langle n_i{}^{HO} | n_i \rangle \langle n_f{}^{HO} | n_f \rangle
$$

$$
- \delta_{cc'} \sum_{n \leq n_{max}} (E_{T_i} + e_n) \langle n_i{}^{HO} | n \rangle \langle n_f{}^{HO} | n \rangle , \tag{9.24}
$$

where $|n_i{}^{HO} \ell j \tau\rangle$ and $|n_f{}^{HO} \ell j \tau\rangle$ are one-body harmonic oscillator states, $|n_i \ell j \tau\rangle, |n_f \ell j \tau\rangle$, and $|n \ell j \tau\rangle$ are one-body Berggren basis states, $E_{T_i}$ is the energy of $|\Psi_{T_i}^{J_{T_i}}\rangle$, $e_n$ is the single-particle energy of $|n \ell j \tau\rangle$, and the expansion coefficients $c_\alpha^{(i)}$ and $c_\alpha^{(f)}$ are those of Eq. (9.7) for the incoming and outgoing channels, respectively.

The expression for matrix element of the $N_{cc'}$ is similar to that of $V_{cc'}$ and can be calculated from Eqs. (9.23) and (9.24)

$$
N_{cc'}(r, r') = \delta(r - r') \delta_{cc'} + \Delta N_{cc'}(r, r') , \tag{9.25}
$$

where

$$
\Delta N_{cc'}(n_i{}^{HO}, n_f{}^{HO}) = \sum_{n_i n_f \alpha}^{n_{max}} c_\alpha^{(i)} c_\alpha^{(f)} \langle n_i{}^{HO} | n_i \rangle \langle n_f{}^{HO} | n_f \rangle - \delta_{cc'} \sum_{n \leq n_{max}} \langle n_i{}^{HO} | n \rangle \langle n_f{}^{HO} | n \rangle .
$$

$$
\tag{9.26}
$$

A fundamental property of the coupled-channel Gamow shell model Hamiltonian of Eq. (9.23) is that both resonant and scattering states can be calculated by solving this equation. Resonant states are indeed eigenstates of Eq. (9.23) by imposing an outgoing wave condition therein, whereas scattering states are solution of Eq. (9.23) for all positive energies. Therefore, Eq. (9.23) unifies structure and reaction formalisms. Hence, it is possible to study the mutual dependence between structure observables and reaction cross sections with the coupled-channel Gamow shell model. For example, a sudden increase of reaction cross section at a given energy can be explained by the presence of a narrow resonance state in the energy spectrum. Conversely, one can assess the single-particle or collective nature of weakly bound and resonance states, as they are linear combination of Berggren basis reaction channels.

## 9.1.2 Electromagnetic Transitions in a Gamow Coupled-Channel Formalism

Radiative capture is a very important reaction process in nuclear and astrophysical applications. Indeed, a very large number of isotopes are generated in stars from radiative nucleon capture. For example, the reactions entering the CNO cycles consist to a large extent of radiative proton capture. Nuclei $^7$Li and $^7$Be were also generated during the big bang nucleosynthesis by the $\alpha$-particle capture on $^3$H and $^3$He, respectively. Finally, heavy nuclei are created in stars by nucleon capture, with the so-called p (proton), r (rapid neutron), and s (slow neutron) capture processes.

Radiative capture reactions in stars occur at very low energies and are peripheral, so that they mainly depend on the asymptotes of many-body wave functions involved in these reactions. Many-body wave functions give a correct asymptotic behavior in the Gamow coupled-channel framework due to the use of the Berggren basis to expand cluster scattering wave functions. As a consequence, the coupled-channel Gamow shell model is an appropriate theoretical tool to study radiative capture processes.

Let us now discuss how one can calculate proton and neutron radiative capture cross sections in the coupled-channel Gamow shell model approach. For this, one has to calculate the matrix elements of the electromagnetic operators (see Eqs. (9.53–9.55)). The presence of resonance and scattering states of the Berggren basis in Eq. (9.5) generates new problems, inexistent in standard reaction models. A direct calculation of matrix elements involving one-body scattering states of the Berggren ensemble is indeed not possible. This arises because of the infinite-range of electromagnetic operators, which induces infinite matrix elements even when exterior complex scaling is applied. However, if one neglects the antisymmetrization in $|(c, r)\rangle = \hat{\mathscr{A}}\{|\Psi_T^{J_T}\rangle \otimes |\Psi_P^{J_P}\rangle\}_{M_A}^{J_A}$, then the calculation of these matrix elements involves only overlaps of one resonant state and one scattering state, so that they converge using complex scaling. Note that the antisymmetry plays a significant role only close to the target. Indeed, if one considers a projectile scattering state orthogonal to all the occupied states inside the target, antisymmetry disappears between target and projectile as the index of the projectile state can be taken to be larger than the indices of the bound states present in the target. This technique has been successfully implemented in the shell model embedded in the continuum (see Refs. [19–21]).

In order to separate target from projectile for a given operator $\hat{O}_{M_L}^L$, it is sufficient to assume that all $A$-nucleons of the target are distinguishable from those of the projectile:

$$\hat{O}_{M_L}^L = \sum_{i \in A} \hat{O}_{M_L}^L(i) + \hat{O}_{M_L}^L(\mathrm{p}) \,. \tag{9.27}$$

In this expression, $\hat{O}_{M_L}^L(i)$ is the part of the many-body operator $\hat{O}_{M_L}^L$ associated to the $i$-th nucleon of the target and $\hat{O}_{M_L}^L(\mathrm{p})$ is the operator associated with the one-nucleon projectile. The operator $\hat{O}_{M_L}^L(i)$ acts only on the target nucleus, which has $A$-nucleons. In this case, channels are not antisymmetrized, and matrix elements can be calculated in the following way:

$$\langle J_f||\hat{O}^L||J_i\rangle = \langle J_f||\hat{O}^L||J_i\rangle_{\mathrm{nas}} + [\langle J_f||\hat{O}^L||J_i\rangle - \langle J_f||\hat{O}^L||J_i\rangle_{\mathrm{nas}}]^{\mathrm{HO}} \,, \tag{9.28}$$

where "nas" means that channels are not antisymmetrized in the considered matrix element and HO means that operators and scattering states $u_c(r)$ are expanded in harmonic oscillator basis. One can use an expansion in terms of the harmonic oscillator states for the difference between antisymmetrized and non-antisymmetrized matrix elements in Eq. (9.28) because antisymmetry is important only inside of the nucleus. Thus, divergence problems are avoided, as there is no need of complex scaling when using an expansion of harmonic oscillator states.

Let us now calculate the matrix elements involving reaction channels, where both target and projectile are present. The reaction channel used is

$$|JM\rangle_c = \int_0^{+\infty} \left(\frac{u_c(r)}{r}\right)|c, r\rangle r^2 \, dr \,. \tag{9.29}$$

One will denote initial and final channels as $c_i$ and $c_f$, respectively.

Let us consider firstly the matrix element $\langle J_f||\hat{O}^L||J_i\rangle_{\text{nas}}$ in Eq. (9.28). As one does not impose antisymmetry in this matrix element, $\hat{O}^L = \hat{O}^L(\text{T}) + \hat{O}^L(\text{P})$, where $\hat{O}^L(\text{T})$ acts only on the target and $\hat{O}^L(\text{P})$ only on the projectile. Therefore, using the Wigner–Eckart theorem, one obtains

$$[c_f\langle J_f||\hat{O}^L(\text{T})||J_i\rangle_{c_i}]_{\text{nas}} = \langle[J_{T_f} \otimes u_{c_f}\ell_f j_f]^{J_f}||\hat{O}^L(\text{T})||[J_{T_i} \otimes u_{c_i}\ell_i j_i]^{J_i}\rangle$$

$$= (-1)^{J_{T_f}+j_f+J_i+L}\sqrt{(2J_i+1)(2J_f+1)}$$

$$\times \left\{\begin{matrix} J_{T_f} & j_f & J_f \\ J_i & L & J_{T_i} \end{matrix}\right\} \langle J_{T_f}||\hat{O}^L||J_{T_i}\rangle$$

$$\times \delta_{\ell_i\ell_f}\delta_{j_i j_f} \int_0^{+\infty} u_{c_i}(r)u_{c_f}(r) \, dr \tag{9.30}$$

$$[c_f\langle J_f||\hat{O}^L(\text{p})||J_i\rangle_{c_i}]_{\text{nas}} = \langle[J_{T_f} \otimes u_{c_f}\ell_f j_f]^{J_f}||\hat{O}^L(\text{p})||[J_{T_i} \otimes u_{c_i}\ell_i j_i]^{J_i}\rangle$$

$$= \delta_{T_i T_f}(-1)^{J_{T_i}+j_i+J_f+L}\sqrt{(2J_i+1)(2J_f+1)} \left\{\begin{matrix} J_{T_f} & j_f & J_f \\ L & J_i & j_i \end{matrix}\right\}$$

$$\times \langle u_{c_f}\ell_f j_f||\hat{O}^L||u_{c_i}\ell_i j_i\rangle \,. \tag{9.31}$$

Complex scaling is used to calculate the overlap in Eq. (9.30), as $u_{c_i}(r)$ is scattering of real energy and $u_{c_f}(r)$ is a resonant state.

One cannot expand a scattering wave function $u_c(r)$ in a Berggren basis. As a consequence, it is necessary to expand $u_c(r)$ with harmonic oscillator basis states

$$u_c(r) \rightarrow u_c^{\text{HO}}(r) = \sum_{n^{\text{HO}}} \langle u_c|u_n^{\text{HO}}\rangle u_n^{\text{HO}}(r) \,, \tag{9.32}$$

where $|u_n^{HO}\rangle$ is a harmonic oscillator state. One thus has to consider the following channel wave functions:

$$|JM\rangle_c^{HO} = \int_0^{+\infty} \left( \frac{u_c^{HO}(r)}{r} \right) |c,r\rangle\, r^2\, dr = \sum_n \langle u_c^{HO}|u_n\rangle\, [a_{n\ell_c j_c}^\dagger |JT_c\rangle]_M^J ,$$

(9.33)

where the Berggren basis expansion of the $|c,r\rangle$ channel of Eq. (9.5)) has been used. The expansion in terms of harmonic oscillator states is hereby justified by the fact that $|J_{T_{i(f)}}\rangle$ targets are localized. The electromagnetic matrix elements $[_{cf}\langle J_f||\hat{O}^L(p)||J_i\rangle_{c_i}]^{HO}$ are calculated using the standard Gamow shell model formulas as the $|JM\rangle_c^{HO}$ states are expanded in a Berggren many-body basis of Slater determinants.

The last many-body matrix element of Eq. (9.28), equal to $\langle J_f||\hat{O}^L||J_i\rangle_{\text{nas}}^{HO}$, is calculated using Eqs. (9.30), (9.31), and replacing $u_{c_{i(f)}}(r)$ by $u_{c_{i(f)}}^{HO}(r)$ in Eq. (9.32). The total many-body matrix element $\langle J_f||\hat{O}^L||J_i\rangle$ of Eq. (9.28) is then obtained by summing over all $c_i$, $c_f$ channels.

## 9.1.3   Cross Sections

Once the solutions for many-body wave function of the Gamow shell model Hamiltonian in the coupled-channel representation have been calculated, one can directly apply the standard formulas of reaction theory providing with reactions cross sections. As the coupled-channel equations of the Gamow shell model Hamiltonian are defined with cluster orbital shell model coordinates, whose utilization is not standard in reaction theory, so one will express the reaction cross section formulas in these new coordinates for clarity.

### 9.1.3.1   Cross Sections of Direct Reactions

In order to calculate elastic/inelastic scattering cross sections, one has to state the asymptotic form of many-body scattering wave functions. Indeed, these cross sections are direct function of the phase shifts of many-body scattering wave functions.

Let us consider for this a scattering reaction from a given entrance channel denoted as "$e$." All the other $c$ indices stand for the exit channels. Let us rewrite Eq. (9.19) for a given entrance channel

$$|\Psi_{M_A}^{e,J_A}\rangle = \sum_c \int_0^{+\infty} \left( \frac{u_c^{e,J_A}(r)}{r} \right) |(c,r)\rangle\, r^2 dr ,$$

(9.34)

where the radial amplitude $u_c^{e,J_A}(r)$ is explicitly written for a given composite state $J_A$ and entrance channel $e$. The $|(c,r)\rangle$ channel state is here written as

$$|(c,r)\rangle = \hat{\mathcal{A}}\{|\Psi_T^{J_c^\pi}\rangle \otimes |r \ \ell_c \ j_c \ m_c \ \tau_c)\}_{M_A}^{J_A} .$$

The asymptotic behavior of the radial amplitude associated to the channel $c$ is

$$u_c^{e,J_A^\pi}(r) \longrightarrow -\frac{1}{2i}\left[\delta_{ce}H_{\ell_e,\eta_e}^-(k_e r) - S_{ec}^{J_A}H_{\ell_c,\eta_c}^+(k_c r)\right]$$

$$= \delta_{ce}F_{\ell_e,\eta_e}(k_e r) + T_{ec}^{J_A}H_{\ell_c,\eta_c}^+(k_c r) , \tag{9.35}$$

where $F_{\ell_e,\eta_e}(k_e r)$ and $H_{\ell_c,\eta_c}^\pm(k_c r)$ are the regular and outgoing/incoming Coulomb wave functions, respectively (see Sec.(2.3)), and $T_{ec}^{J_A}$ is the $T$-matrix element [3, 4, 6]. $S$-matrix (see Sect. 2.6.6) and $T$-matrix elements are related by the expression:

$$T_{ec}^{J_A} = \frac{S_{ec}^{J_A} - \delta_{ce}}{2i} . \tag{9.36}$$

Let us denote by $\tilde{c}$ the set of quantum numbers that completely defines a target state; the spherical channel $c$ can then be formally rewritten as $(\tilde{c}, \ell_c, j_c)$. Hence, the physical scattering wave function, characterized by the entrance state of the target $\tilde{e}$ (usually the ground state), the magnetic quantum numbers $m_{\tilde{e}}$ and $M_{\tilde{e}}$, respectively of the nucleon and the target, reads [3,4,6]

$$|\varphi_{m_{\tilde{e}}M_T^{\tilde{e}}}^{\tilde{e}}\rangle = \sum_{\ell_e j_e J_A} \frac{A_{s_{\tilde{e}}M_{\tilde{e}}}^{\ell_e j_e J_T^{\tilde{e}} J_A}}{k_{\tilde{e}}} |\Psi_{m_{\tilde{e}}+M_{\tilde{e}}}^{\tilde{e}\ell_e j_e J_A}\rangle , \tag{9.37}$$

where the coefficients are given by

$$A_{m_{\tilde{e}}M_T^{\tilde{e}}}^{\ell_e j_e J_T^{\tilde{e}} J_A} = \langle j_e m_{\tilde{e}}|\ell_e 0 \ \tfrac{1}{2}m_{\tilde{e}}\rangle \ \langle J_A m_{\tilde{e}}+M_T^{\tilde{e}}|j_e m_{\tilde{e}} \ J_T^{\tilde{e}} M_T^{\tilde{e}}\rangle \ i^{\ell_e}\sqrt{4\pi(2\ell_e+1)}e^{i\sigma_{\ell_e}} , \tag{9.38}$$

and $\sigma_\ell = \arg(\ell+1+i\eta)$ is the Coulomb phase shift in the $\ell$-partial wave.

This particular partial wave decomposition in terms of the $A$-body scattering states (see Eq. (9.34)) confers to the physical scattering state its correct asymptotic behavior, that is [3, 4, 6]

$$
\langle \mathbf{r} \, | \varphi^{\tilde{e}}_{m_{\tilde{e}} M^{\tilde{e}}_{\mathrm{T}}} \rangle \longrightarrow \exp\left(i \, [kz + \eta \, \ln(k(r-z))]\right) \cdot \left( |\chi_{1/2}\rangle \otimes |\Psi^{J^{\tilde{e}}_{\mathrm{T}}}_{\mathrm{T}}\rangle \right)
$$

$$
+ \sum_{\tilde{c} m_{\tilde{c}} M^{\tilde{c}}_{\mathrm{T}}} f_{\tilde{e} m_{\tilde{e}} M^{\tilde{e}}_{\mathrm{T}} \to \tilde{c} m_{\tilde{c}} M^{\tilde{c}}_{\mathrm{T}}}(\theta, \phi) \left( \frac{\exp\left(i \, [kr - \eta \, \ln(2kr)]\right)}{r} \right)
$$

$$
\cdot \left( |\chi_{1/2}\rangle \otimes |\Psi^{J^{\tilde{c}}_{\mathrm{T}}}_{\mathrm{T}}\rangle \right) , \tag{9.39}
$$

in which $|\chi_{1/2}\rangle$ stands for the intrinsic spin wave function of the projectile in the channel $\tilde{c}$. Indeed, the state (9.39) exhibits at large distance an incoming plane wave (modified by the infinite-range Coulomb potential) in the entrance channel $\tilde{e}$ and outgoing spherical waves in each outgoing channel $\tilde{c}$.

Using Eqs. (9.34), (9.35), (9.37), and (9.38), one can express the scattering amplitudes in terms of the T-matrix elements [3, 4, 6]:

$$
f_{\tilde{e} m_{\tilde{e}} M^{\tilde{e}}_{\mathrm{T}} \to \tilde{c} m_{\tilde{c}} M^{\tilde{c}}_{\mathrm{T}}}(\theta, \phi) = \delta_{\tilde{c}\tilde{e}} \, \delta_{m_{\tilde{c}} m_{\tilde{e}}} \delta_{M^{\tilde{c}}_{\mathrm{T}} M^{\tilde{e}}_{\mathrm{T}}} \, f_C(\theta)
$$

$$
+ \sum_{\ell_{\tilde{e}} j_{\tilde{e}} \ell_{\tilde{c}} j_{\tilde{c}} J_{\mathrm{A}}} \frac{c^{\ell_{\tilde{e}} j_{\tilde{e}} J^{\tilde{e}}_{\mathrm{T}} \ell_{\tilde{c}} j_{\tilde{c}} J^{\tilde{c}}_{\mathrm{T}} J_{\mathrm{A}}}_{m_{\tilde{e}} M^{\tilde{e}}_{\mathrm{T}} m_{\tilde{c}} M^{\tilde{c}}_{\mathrm{T}}}}{k_{\tilde{e}}} \, T^{J_{\mathrm{A}}}_{\tilde{e}\ell_{\tilde{e}} j_{\tilde{e}}, \tilde{c}\ell_{\tilde{c}} j_{\tilde{c}}} \, Y^{\ell_{\tilde{c}}}_{m_{\tilde{e}} + M^{\tilde{e}}_{\mathrm{T}} - m_{\tilde{c}} - M^{\tilde{c}}_{\mathrm{T}}}(\theta, \phi),
$$

$$
\tag{9.40}
$$

where $f_C(\theta)$ is the Coulomb scattering amplitude given by the relation [3]:

$$
f_C(\theta) = -\frac{\eta}{2k \sin^2(\theta/2)} e^{-i[\eta \ln \sin^2(\theta/2) - 2\sigma_0]} \tag{9.41}
$$

and:

$$
c^{\ell_{\tilde{e}} j_{\tilde{e}} J^{\tilde{e}}_{\mathrm{T}} \ell_{\tilde{c}} j_{\tilde{c}} J^{\tilde{c}}_{\mathrm{T}} J_{\mathrm{A}}}_{m_{\tilde{e}} M^{\tilde{e}}_{\mathrm{T}} m_{\tilde{c}} M^{\tilde{c}}_{\mathrm{T}}} = \langle j_{\tilde{e}} m_{\tilde{e}} | \ell_{\tilde{e}} 0 \, \tfrac{1}{2} m_{\tilde{e}} \rangle \, \langle j_{\tilde{c}} \, m_{\tilde{e}} + M^{\tilde{e}}_{\mathrm{T}} - M^{\tilde{c}}_{\mathrm{T}} | \ell_{\tilde{c}} \, m_{\tilde{e}} + M^{\tilde{e}}_{\mathrm{T}} - m_{\tilde{c}} - M^{\tilde{c}}_{\mathrm{T}} \, \tfrac{1}{2} m_{\tilde{c}} \rangle
$$

$$
\times \langle J_{\mathrm{A}} \, m_{\tilde{e}} + M^{\tilde{e}}_{\mathrm{T}} | j_{\tilde{e}} m_{\tilde{e}} \, J^{\tilde{e}}_{\mathrm{T}} M^{\tilde{e}}_{\mathrm{T}} \rangle \, \langle J_{\mathrm{A}} \, m_{\tilde{e}} + M^{\tilde{e}}_{\mathrm{T}} | j_{\tilde{c}} \, m_{\tilde{e}} + M^{\tilde{e}}_{\mathrm{T}} - M^{\tilde{c}}_{\mathrm{T}} \, J^{\tilde{c}}_{\mathrm{T}} M^{\tilde{c}}_{\mathrm{T}} \rangle
$$

$$
\times i^{(\ell_{\tilde{e}} - \ell_{\tilde{c}})} \sqrt{4\pi(2\ell_{\tilde{e}} + 1)} e^{i(\sigma_{\ell_{\tilde{e}}} + \sigma_{\ell_{\tilde{c}}})} . \tag{9.42}
$$

From the expression (9.39) of the physical scattering state, the differential cross section for the scattering process to the channel $(\tilde{c}, m_{\tilde{c}}, M^{\tilde{c}}_{\mathrm{T}})$ at a given angle $\theta$ reads

$$
\frac{d\sigma_{\tilde{e} m_{\tilde{e}} M^{\tilde{e}}_{\mathrm{T}} \to \tilde{c} m_{\tilde{c}} M^{\tilde{c}}_{\mathrm{T}}}}{d\Omega}(\theta) = \frac{k_{\tilde{c}}}{k_{\tilde{e}}} \left| f_{\tilde{e} m_{\tilde{e}} M^{\tilde{e}}_{\mathrm{T}} \to \tilde{c} m_{\tilde{c}} M^{\tilde{c}}_{\mathrm{T}}}(\theta, \phi) \right|^2 . \tag{9.43}
$$

Note that the $\phi$-dependence disappears from scattering amplitudes, as one would expect in this axially symmetric process.

In general, reaction experiments involve target nuclei and beams of projectile particles that are both unpolarized. The differential cross sections of the scattering process to a given state $\tilde{c}$ of a target are thus equal to the average value over the entrance magnetic quantum numbers $m_{\tilde{e}}$ and $M_{\tilde{e}}$ and the sum over all final possible orientations $m_{\tilde{c}}$ and $M_{\tilde{c}}$:

$$\frac{d\sigma_{\tilde{e}\to\tilde{c}}}{d\Omega}(\theta) = \frac{1}{2(2J_{\mathrm{T}}^{\tilde{e}}+1)} \sum_{m_{\tilde{e}}M_{\mathrm{T}}^{\tilde{e}}M_{\mathrm{T}}^{\tilde{c}}m_{\tilde{c}}} \frac{k_{\tilde{c}}}{k_{\tilde{e}}} \left| f_{\tilde{e}m_{\tilde{e}}M_{\mathrm{T}}^{\tilde{e}}\to\tilde{c}m_{\tilde{c}}M_{\mathrm{T}}^{\tilde{c}}}(\theta,\phi) \right|^2 . \tag{9.44}$$

### 9.1.3.2 Projectile Energies in the Different Reference Frames

Cross sections are typically measured in the center-of-mass frame of a target–projectile composite system, whereas in the coupled-channel Gamow shell model they are calculated with cluster orbital shell model. Therefore, one has to derive the transformation formulae from one reference frame to another. For this, one has to express the projectile energies in laboratory, center-of-mass, and cluster orbital shell model frames.

In the cluster orbital shell model, all coordinates of particles are defined with respect to the center-of-mass of the inert core. The motion of the center-of-mass of the target core is negligible compared to that of the light projectile in a target–projectile reaction, so that one considers that the target is at rest in the cluster orbital shell model framework. Consequently, in a target–projectile reaction, the total kinetic energy of the composite is then equal to the kinetic energy of the projectile:

$$E^{(\mathrm{COSM})} = \frac{\mathbf{p}^2}{2\mu} , \tag{9.45}$$

where $\mu$ is the reduced mass of the projectile in the cluster orbital shell model (see Eq. (5.36)), $\mathbf{p} = \mu\mathbf{v}$, and $\mathbf{v}$ is the velocity of the projectile. Using the standard definition $E_{\mathrm{lab}} = m\mathbf{v}^2/2$ of the composite energy in the laboratory frame, with $m$ the mass of the projectile, Eq. (9.45) can be expressed as

$$E^{(\mathrm{COSM})} = \frac{M_{\mathrm{core}}}{m + M_{\mathrm{core}}} E_{\mathrm{lab}} . \tag{9.46}$$

The energy in the center-of-mass frame $E_{\mathrm{CM}}$ is obtained from the standard formulas arising from the two-body problem:

$$E_{\mathrm{CM}} = \frac{M_{\mathrm{T}}}{m + M_{\mathrm{T}}} E_{\mathrm{lab}} , \tag{9.47}$$

where $M_{\mathrm{T}}$ is the mass of the target.

### 9.1.3.3 Cross Sections of Radiative Capture Reactions

One now presents the calculation of radiative capture reaction cross sections in coupled-channel Gamow shell model. The differential cross section for one-nucleon radiative capture reaction from an initial $A$-body state $|E_i\,M_i\rangle$, where the projectile is a plane wave of energy $E_i$, to a final $A$-body state $|J_f M_f\rangle$ of energy $E_f$ reads [49–51]

$$\frac{d\sigma}{d\Omega_\gamma} = \left(\frac{q}{8\pi k}\right)\left(\frac{e^2}{\hbar c}\right)\left(\frac{\mu_u c^2}{\hbar c}\right)\left(\frac{1}{2(2J_T+1)}\right)$$

$$\times \sum_{\substack{M_i,M_f,\\ M_T,M_L,\\ P_\gamma,m_s}}\left|\sum_L i^L\sqrt{\frac{2\pi(2L+1)(L+1)}{L}}\left(\frac{q^L}{k}\right)\left(\frac{P_\gamma}{(2L+1)!!}\right)\right.$$

$$\left.\times D^L_{M_L P}(\varphi_\gamma,\theta_\gamma,0)\,\langle J_f M_f|\mathscr{M}_{L,M_L}|E_i\,M_i\rangle\right|^2$$

$$= \left(\frac{q}{8\pi k}\right)\left(\frac{e^2}{\hbar c}\right)\left(\frac{\mu_u c^2}{\hbar c}\right)\left(\frac{1}{2(2J_T+1)}\right)$$

$$\times \sum_{\substack{M_i,M_f,\\ M_T,M_L,\\ P_\gamma,m_s}}\left|\sum_L g^L_{M_L,P}(k,q,\varphi_\gamma,\theta_\gamma)\,\langle J_f M_f|\mathscr{M}_{L,M_L}|E_i\,M_i\rangle\right|^2, \tag{9.48}$$

with

$$g^L_{M_L,P_\gamma}(k,q,\varphi_\gamma,\theta_\gamma) = i^L\sqrt{\frac{2\pi(2L+1)(L+1)}{L}}\left(\frac{q^L}{k}\right)$$

$$\times \left(\frac{P_\gamma}{(2L+1)!!}\right)D^L_{M_L P_\gamma}(\varphi_\gamma,\theta_\gamma,0), \tag{9.49}$$

and

$$|E_i\,M_i\rangle = \sum_{J_i,c_e} |(J_i M_i)_{c_e}\rangle\,\langle \ell_{c_e} 0\,s m_s|j_{c_e} m_s\rangle\,\langle j_{c_e} m_s\,J_T M_T|J_i M_i\rangle, \tag{9.50}$$

where one sums over all entrance channels $c_e$ and

$$|(J_i M_i)_{c_e}\rangle = \sum_c |(J_i M_i)_{c_e}\rangle_c. \tag{9.51}$$

Thus, the differential cross section writes

$$\frac{d\sigma}{d\Omega_\gamma} = \left(\frac{q}{8\pi k}\right)\left(\frac{e^2}{\hbar c}\right)\left(\frac{\mu_u c^2}{\hbar c}\right)\left(\frac{1}{2(2J_T+1)}\right)$$

$$\times \sum_{J_i,c_e,L} \langle J_f||\mathscr{M}_L||(J_i)_{c_e}\rangle \sum_{J_i',c_e',L'} \langle J_f||\mathscr{M}_{L'}||(J_i')_{c_e'}\rangle$$

$$\times \sum_{M_L,P_\gamma} [g^L_{M_L,P_\gamma}(k,q,\varphi_\gamma,\theta_\gamma)]^* g^{L'}_{M_L,P_\gamma}(k,q,\varphi_\gamma,\theta_\gamma)$$

$$\times \sum_{M_i,M_f} \begin{pmatrix} J_f & L & J_i \\ -M_f & M_L & M_i \end{pmatrix} \begin{pmatrix} J_f & L' & J_i' \\ -M_f & M_L & M_i \end{pmatrix}$$

$$\times \sum_{m_s} \langle \ell_{c_e} 0\, s m_s | j_{c_e} m_s\rangle \langle \ell_{c_e'} 0\, s m_s | j_{c_e'} m_s\rangle$$

$$\times \sum_{M_T} \langle j_{c_e} m_s J_T M_T | J_i M_i\rangle \langle j_{c_e'} m_s J_T M_T | J_i' M_i\rangle, \tag{9.52}$$

where

- $q$, in $\text{fm}^{-1}$, is the linear momentum of the emitted photon: $q = (E_f - E_i)/(\hbar c)$;
- $e^2/(\hbar c)$ has no units and is the electromagnetic coupling constant ($\simeq 1/137$);
- $k$, in $\text{fm}^{-1}$, is the linear momentum of the incoming nucleon in the center-of-mass frame;
- $\mu_u c^2$, in MeV, is the reduced mass of the system $p+A$;
- $s$ is the spin of the nucleon, and $J_T$ is the total spin of the target;
- $P_\gamma = \pm 1$ is the polarization of the photon;
- $L$ and $M_L$ are the multipoles and multipole projections of the photon;
- $D^L_{M_L P_\gamma}(\varphi_\gamma, \theta_\gamma, 0)$ is the Wigner $D$-matrix, function of the angular variables $\theta_\gamma$ and $\varphi_\gamma$ of the photon;
- $\mathscr{M}_{L,M_L}$ is the electromagnetic transition operator.

Note that the differential cross section is in unit of $\text{fm}^2$.

The electromagnetic transition operators $\mathscr{M}_{L,M_L}$ separate into an electric part, denoted as $\mathscr{M}^E_{L,M_L}$, and a magnetic part denoted as $\mathscr{M}^M_{L,M_L}$, which read [52]

$$\mathscr{M}^E_{L,M_L} = \left(\frac{(2L+1)!!}{(L+1)\,q^L}\right) \sum_i e_i \left(\hat{j}'_L(qr_i) + \left(\frac{qr_i}{2}\right)\hat{j}_L(qr_i)\right) Y_{LM_L}(\Omega_i)$$

$$+ \left(\frac{(2L+1)!!}{(L+1)\,q^L}\right)\left(\frac{\hbar c}{2mc^2}\right) \sum_i g^s_i \left(\frac{\hat{j}_L(qr_i)}{r_i}\right) (\mathbf{l}_i \cdot \mathbf{s}_i) Y_{LM_L}(\Omega_i)$$

$$\tag{9.53}$$

$$\mathscr{M}_{L,M_L}^M = \left(\frac{\hbar c}{2mc^2}\right)\left(\frac{(2L+1)!!}{(L+1)\,q^L}\right)\sum_i \left(g_i^\ell \nabla_i \left(\frac{\hat{j}_L(qr_i)}{qr_i}Y_{LM_L}(\Omega_i)\right)\cdot \mathbf{l}_i\right.$$

$$\left.+ g_i^s \nabla_i \left(\hat{j}_L'(qr_i)Y_{LM_L}(\Omega_i)\right)\cdot \mathbf{s}_i\right)$$

$$+\left(\frac{\hbar c}{2mc^2}\right)\left(\frac{(2L+1)!!}{(L+1)\,q^L}\right)\sum_i g_i^s\, q\,\hat{j}_L(qr_i)\,(\mathbf{s}_i\cdot \mathbf{e}_i)\,Y_{LM_L}(\Omega_i)\,. \tag{9.54}$$

Consequently,

$$\mathscr{M}_{L,M_L}^M = \left(\frac{\hbar c}{2mc^2}\right)\left(\frac{(2L+1)!!}{(L+1)\,q^L}\right)\sum_i \left(-\left(\frac{g_i^\ell}{r_i}\right)\sqrt{\frac{L+1}{2L+1}}\right.$$

$$\times\left(\hat{j}_L'(qr_i) - (L+1)\left(\frac{\hat{j}_L(qr_i)}{qr_i}\right)\right)[Y_{L+1}(\Omega_i)\otimes \mathbf{l}_i]_{M_L}^L$$

$$+\left(\frac{g_i^\ell}{r_i}\right)\sqrt{\frac{L}{2L+1}}\left(\hat{j}_L'(qr_i) + L\left(\frac{\hat{j}_L(qr_i)}{qr_i}\right)\right)[Y_{L-1}(\Omega_i)\otimes \mathbf{l}_i]_{M_L}^L$$

$$-\left(\frac{g_i^s}{r_i}\right)\sqrt{\frac{L+1}{2L+1}}\left[\left(\frac{L(L+1)}{qr_i} - qr_i\right)\hat{j}_L(qr_i) - L\hat{j}_L'(qr_i)\right)$$

$$[Y_{L+1}(\Omega_i)\otimes \mathbf{s}_i]_{M_L}^L$$

$$+\left(\frac{g_i^a}{r_i}\right)\sqrt{\frac{L}{2L+1}}\left(\left(\frac{L(L+1)}{qr_i} - qr_i\right)\hat{j}_L(qr_i) + (L+1)\hat{j}_L'(qr_i)\right)$$

$$[Y_{L-1}(\Omega_i)\otimes \mathbf{s}_i]_{M_L}^L$$

$$+\, g_i^s\, q\,\hat{j}_L(qr_i)\,(\mathbf{s}_i\cdot \mathbf{e}_i)\,Y_{LM_L}(\Omega_i)\bigg)\quad, \tag{9.55}$$

where

- $i$ runs over all considered nucleons, with $\mathbf{l}_i$ and $\mathbf{s}_i$ its orbital and spin angular momenta, respectively;
- $Y_{LM_L}(\Omega)$ is a spherical harmonic of order $L$ and projection $M_L$;
- $\hat{j}_L$ is the Ricati–Bessel function of order $L$: $\hat{j}_L(qr) = qr j_L(qr)$;
- $r_i$, $\Omega_i$ are the radial and angular coordinates of the nucleon $i$, respectively, and $\mathbf{e}_i = \mathbf{r}_i/r_i$;
- $e_i$ is the dimensionless effective charge of the nucleon $i$, so that bare charge is 1 for proton and 0 for neutron;
- $g_i^s$ is the dimensionless magnetic spin moment of the nucleon $i$ and is 5.5857 for proton and $-3.8263$ for neutron;

- $mc^2$, in MeV, is the mass of the nucleon;
- $g_i^\ell$ is the dimensionless magnetic orbital momentum of the nucleon $i$ multiplied by $L + 1$, so that it is 2 for proton and 0 for neutron.

Corresponding formulas in the long wavelength approximation are obtained from Eqs. (9.53 and 9.55) by letting $q \to 0$ therein:

$$\mathcal{M}_{L,M_L}^E = \sum_i e_i \, r_i^L Y_{LM_L}(\Omega_i) \tag{9.56}$$

$$\mathcal{M}_{L,M_L}^M = \left( \frac{\hbar c}{2mc^2} \right) \sqrt{L(2L+1)} \sum_i \left( \left[ r_i^{L-1} Y_{L-1}(\Omega_i) \otimes \left( \left( \frac{g_i^\ell}{L+1} \right) \mathbf{l}_i + g_i^s \, \mathbf{s}_i \right) \right]_{M_L}^L \right). \tag{9.57}$$

Coupled-channel Gamow shell model calculations are done in cluster orbital shell model coordinates, but the radiative capture cross section is expressed in the center-of-mass reference frame. The initial energy is

$$E_i^{(\mathrm{COSM})} = E_{\mathrm{proj}}^{(\mathrm{COSM})} + E_T^{(\mathrm{COSM})}, \tag{9.58}$$

where $E_i^{(\mathrm{COSM})}$, $E_{\mathrm{proj}}^{(\mathrm{COSM})}$, and $E_T^{(\mathrm{COSM})}$ are the total energy, the projectile energy, and the Gamow shell model target binding energy, respectively. All energies are calculated in the cluster orbital shell model coordinate system.

The link between the projectile energies in cluster orbital shell model and center-of-mass reference frames is provided by Eqs. (9.45–9.47). Energy conservation implies that the final energy is

$$E_i^{(\mathrm{COSM})} = E_f^{(\mathrm{COSM})} + E_\gamma, \tag{9.59}$$

where $E_f^{(\mathrm{COSM})}$ is the compound system binding energy in the cluster orbital shell model frame of reference, and $E_\gamma$ is the photon energy, which does not depend on the chosen reference frame.

Resonances in the spectrum of a composite $A$-nucleon system correspond to the peaks in the radiative capture cross section at the center-of-mass energy.

The cross section for a final state of the total angular momentum $J_f$ is

$$\sigma_{J_f}(E_{\mathrm{CM}}) = \int_0^{2\pi} \left( \int_0^\pi \frac{d\sigma_{J_f}(E_{\mathrm{CM}}, \theta_\gamma, \varphi_\gamma)}{d\Omega_\gamma} \sin \theta_\gamma \, d\theta_\gamma \right) d\varphi_\gamma, \tag{9.60}$$

and the total cross section is thus

$$\sigma(E_{\mathrm{CM}}) = \sum_{J_f} \sigma_{J_f}(E_{\mathrm{CM}}). \tag{9.61}$$

In practice, one often shows the astrophysical factor, which removes the exponential dependence of the cross section at low energies due to the Coulomb potential barrier

$$\mathscr{S}(E_{CM}) = \sigma(E_{CM}) \, E_{CM} \, e^{2\pi \eta_{CM}} , \qquad (9.62)$$

where $\eta_{CM}$ is the Sommerfeld parameter (see Sect. 2.1).

## 9.2    Generalization to Many-Nucleon Projectiles

In this section, one will discuss the coupled-channel Gamow shell model formulation for the case of a complex projectile composed of two or more nucleons. In principle, all nucleons of the projectile can be scattered in the reaction, so that one would need to impose three-, four-body, etc., outgoing wave function conditions in the asymptotic zone. However, reactions involving many-body asymptotes are very seldom. Indeed, in most cases, the projectile is either the same in incoming and outgoing channels, as in elastic scattering reactions, or gains/loses one nucleon through pick-up/stripping reactions, as in $(d, p)$ or $(p, d)$ reactions. Therefore, all projectiles behave as point particles in the asymptotic region, so that reactions reduce to a two-body problem therein. This reduction of the number of degrees of freedom in the asymptotic region, called cluster approximation, considerably simplifies the theoretical formulation of the reaction framework. The internal structure of the projectile is nevertheless considered in its entirety in the nuclear zone, where all the nucleons of the target–projectile composite interact.

In order to treat projectiles in the asymptotic region, it is necessary to separate their center-of-mass and intrinsic degrees of freedom. Indeed, the trajectory of the projectile is a function of its center-of-mass coordinates, while its intrinsic part intervenes in reaction channels and Hamiltonian matrix elements. In order to do this separation rigorously, one must either use a basis of harmonic oscillator states, due to the exact factorization of many-body wave functions in center-of-mass and intrinsic parts therein, or Jacobi coordinates, which are intrinsic by construction. Both methods have their advantages and disadvantages. On the one hand, one can conveniently use the basis of harmonic oscillator states in a shell model framework. Due to the localized character of the basis of harmonic oscillator states, however, one can only expand well-bound clusters using this method. Hence, it is most useful for projectiles such as $^{3}$H, $^{3}$He, and especially $^{4}$He particles, which are well bound, by about 6 MeV for the former and 20 MeV for the latter. On the other hand, it is not appropriate for deuteron projectiles, which is prone to breakup due to its small binding energy, of about 2 MeV. Deuteron can be directly treated in an intrinsic framework, as it has only two particles. Therefore, it will be handled numerically by diagonalization of a realistic Hamiltonian with the Berggren basis using relative coordinates. Hence, break-up effects can be included as one can form scattering deuteron projectile states near the target. Due to their complementary natures, both aforementioned methods to generate projectile wave functions will be presented.

### 9.2.1 Clusters in the Berggren Basis Formalism

Let us define cluster states:

$$|\Psi_P^{J_P}\rangle = [|K_{CM}, L_{CM}\rangle \otimes |K_{int}, J_{int}\rangle]_{M_P}^{J_P} , \qquad (9.63)$$

where $K_{CM}$ and $L_{CM}$ are, respectively, the linear momentum and angular momentum of the center-of-mass, $K_{int}$ is the intrinsic linear momentum, and $J_{int}$ represents the intrinsic angular momentum, so that one has $\mathbf{J_P} = \mathbf{J_{int}} + \mathbf{L_{CM}}$.

Composite states are then built from the antisymmetrized tensor product of target and projectile states. For antisymmetry to be fulfilled, one expands both target and projectile in the same complete basis of Slater determinants. By construction, as the target state was generated by a Gamow shell model calculation, it is expanded with Slater determinants. Eq. (9.63) must be then expanded in the basis of Slater determinants used for the target:

$$|\Psi_P^{J_P}\rangle = \sum_n C_n |SD_n\rangle , \qquad (9.64)$$

where the Slater determinants are constructed from the single-particle states of the Berggren ensemble. However, the overlap $\langle\Psi_P^{J_P}|\Psi_{P'}^{J_{P'}}\rangle = \delta(K_{CM} - K'_{CM})$ is difficult to treat numerically because the treatment of the Dirac delta function would require an extremely fine discretization of the continuum for the center-of-mass/intrinsic separation of $|\Psi_P^{J_P}\rangle$ in Eq. (9.63) at large distances. Consequently, one has to proceed indirectly.

As reactions are localized close to the target, the wave function of the projectile can be approximated by a bound state wave function, so that one can use the harmonic oscillator basis instead. Let us define the harmonic oscillator projectile state as

$$|\Psi_P^{J_P}\rangle^{HO} = \left[|N_{CM}, L_{CM}\rangle^{HO} \otimes |K_{int}, J_{int}\rangle^{HO}\right]_{M_P}^{J_P} , \qquad (9.65)$$

where $|N_{CM}, L_{CM}\rangle^{HO}$ is a center-of-mass harmonic oscillator state and $|K_{int}, J_{int}\rangle^{HO}$ is an intrinsic projectile cluster state expanded on a basis of harmonic oscillator states.

### 9.2.2 Numerical Treatment of Well-Bound Projectiles

The wave function of a bound projectile is approximated by a bound state wave function, represented by harmonic oscillator basis states. One has to calculate the many-body state $|N_{CM}, L_{CM}, J_{int}\rangle_{M_P}^{J_P}$ of Eq. (9.65). For this, one starts by calculating the ground state of the cluster with a $0s$ center-of-mass part, which is standard with the Lawson method

$$|N_{CM} = 0, \ L_{CM} = 0, \ M_{CM} = 0\rangle \, |J_{int}, M_{int}\rangle \ . \tag{9.66}$$

In order to calculate center-of-mass harmonic oscillator many-body states efficiently, one introduces the ladder operator:

$$A^{\dagger}_{\mu, CM} = \sqrt{\frac{M\omega}{2\hbar}} R^{(1)}_{\mu} - i \sqrt{\frac{1}{2M\hbar\omega}} P^{(1)}_{\mu} \ , \tag{9.67}$$

where $\mu = -1, 0, +1$, and $R^{(1)}_{\mu}$, $P^{(1)}_{\mu}$ represent the position and momentum of the projectile in center-of-mass system, respectively,

$$R^{(1)}_{\mu} = \frac{1}{a} \sum_{i=1}^{a} r^{(1)}_{i,\mu} \quad , \quad P^{(1)}_{\mu} = \sum_{i=1}^{a} p^{(1)}_{i,\mu} \ . \tag{9.68}$$

$r^{(1)}_{i,\mu}$, and $p^{(1)}_{i,\mu}$ are, respectively, the position and impulsion of each nucleon of the projectile in the center-of-mass system, and $a$ is the number of nucleons in the projectile. When one applies the $A^{\dagger}_{\mu, CM}$ tensor operator on the $|N_{CM} \, L_{CM} \, M_{CM}\rangle$ state, one increases the $2N_{CM} + L_{CM}$ by one harmonic oscillator center-of-mass quantum:

$$2N_{CM} + L_{CM} \rightarrow 2N_{CM} + L_{CM} + 1 \equiv 2N'_{CM} + L'_{CM} \ . \tag{9.69}$$

At this point, one couples the tensor operator $A^{\dagger}_{CM}$ acting on $|N_{CM} \, L_{CM} \, J_{int}, M_{int}\rangle$ to $L'_{CM} \, M'_{L'_{CM}}$

$$\left[ A^{\dagger}_{CM} |N_{CM} \, L_{CM} \, J_{int} \, M_{int}\rangle \right]^{L'_{CM}}_{M'_{L'_{CM}}} \propto |N'_{CM} \, L'_{CM} \, M'_{L'_{CM}} \, J_{int} \, M_{int}\rangle \ . \tag{9.70}$$

Note that $A^{\dagger}_{CM}$ does not act on $|J_{int} \, M_{int}\rangle$. Therefore, one can build the set of states:

$$\{|N_{CM} \, L_{CM} \, M_{L_{CM}} \, J_{int} \, M_{int}\rangle\} \ , \tag{9.71}$$

because the harmonic oscillator degeneracy induced by $A^{\dagger}_{\mu, CM}$ is lifted by the $L_{CM}$ coupling. Then, using Clebsch–Gordan coefficients $\langle L_{CM} M_{L_{CM}} \, J_{int} M_{int} | J_P M_P \rangle$, one couples the many-body state of Eq. (9.70) to $J_P$. $|N_{CM} \, L_{CM} \, M_{L_{CM}} \, J_{int}\rangle^{J_P}_{M_P}$ states are calculated within harmonic oscillator shell model, so that they read

$$|N_{CM} \, L_{CM} \, M_{L_{CM}} \, J_{int}\rangle^{J_P}_{M_P} = \sum_{N} c^{HO}_{N} |SD_{N}\rangle^{HO} \ . \tag{9.72}$$

One can now express the projectile state in the Berggren basis:

$$|\Psi_P^{J_P}\rangle = \sum_n c_n |SD_n\rangle . \tag{9.73}$$

$c_n$ coefficients are readily obtained from Eqs. (9.72 and 9.73):

$$c_n = \langle SD_n|N_{CM} \, L_{CM} \, J_{int}\rangle_{M_P}^{J_P} = \sum_N c_N^{HO}\langle SD_n|SD_N\rangle^{HO} . \tag{9.74}$$

As $|SD_n\rangle$ and $|SD_N\rangle^{HO}$ are built from different one-body basis states, the overlap $\langle SD_n|SD_N\rangle^{HO}$ must be calculated using the definition of Slater determinants as a linear combination of non-antisymmetrized tensor products:

$$\langle SD_n|SD_N\rangle^{HO} = \sum_k (-1)^{\varphi_k} \langle s_1 \ldots s_a| P_k |\sigma_1 \ldots \sigma_a\rangle , \tag{9.75}$$

where $P_k$ is a permutation operator, which is labeled by its $k$ index, enumerating all possible permutations, $|s_i\rangle$ are the single-particle states of the Berggren basis occupied in $|SD_n\rangle$, the $|\sigma_i\rangle$ are the single-particle states from the harmonic oscillator basis occupied in $|SD_N\rangle^{HO}$, and $(-1)^{\varphi_k}$ is the reordering phase associated to the permutation operator $P_k$.

### 9.2.3 Inclusion of Break-up Reactions

One will concentrate in this section on the treatment of weakly bound projectiles, which can break up during a reaction. One will deal for this with the deuteron projectile, which is widely used in experiment and has a low binding energy, of about $-2$ MeV, so that it easily breaks up during a reaction. Moreover, as the deuteron forms a two-body system, it is convenient to treat numerically. Generalization to projectiles having more than two nucleons will be discussed at the end of the section.

In order to calculate deuteron projectile states, which are prone to breakup, one will use a direct Berggren basis diagonalization of the relative two-body Hamiltonian. The intrinsic part of $|\Psi_P^{J_P}\rangle^{HO}$ in Eq. (9.65) is provided by the relative Hamiltonian $H_{int}$, which is firstly diagonalized in a Berggren basis of relative bound and scattering states. This guarantees a good asymptotic behavior of relative scattering deuteron states. The associated Schrödinger equation is that of a two-body problem, so that it can be exactly solved when written in center-of-mass and relative cluster coordinates (see Sect. 4.1). The relative Hamiltonian $H_{int}$ then possesses only one degree of freedom (see Eq. (4.5)). Its matrix representation using a relative Berggren basis thus bears a small dimension, so that $H_{int}$ can be exactly diagonalized. This provides with the relative deutcron eigenstates $|K_{int} \, J_{int}\rangle$

generated by $H_{\text{int}}$ in the relative Berggren basis. They are projected on a basis of harmonic oscillator states in order to provide with the $|K_{\text{int}} \, J_{\text{int}}\rangle^{\text{HO}}$ states.

The $|\Psi_{\text{P}}^{J_{\text{P}}}\rangle^{\text{HO}}$ states of Eq. (9.65) have to be expressed in cluster orbital shell model coordinates to be able to use them in a shell model formalism. For this, one can apply the Talmi–Brody–Moshinsky transformation (see Ref.[53, 54] and Sect. 2.2) to the relative + center-of-mass two-body wave functions formed by the deuteron states of Eq. (9.65). Consequently, the coefficients of the Slater determinant expansion of $|\Psi_{\text{P}}^{J_{\text{P}}}\rangle^{\text{HO}}$ in Eq. (9.65) are straightforward to obtain

$$|\Psi_{\text{P}}^{J_{\text{P}}}\rangle^{\text{HO}} = \sum_N c_N^{\text{HO}} |SD_N\rangle^{\text{HO}} . \tag{9.76}$$

Let us write the expansion of the projectile wave function in the Berggren basis:

$$|\Psi_{\text{P}}^{J_{\text{P}}}\rangle^{\text{HO}} = \sum_n c_n \, |SD_n\rangle , \tag{9.77}$$

where the $c_n$ coefficients come from a direct expansion of the harmonic oscillator basis in the Berggren basis:

$$c_n = \langle SD_n | N_{\text{CM}} \, L_{\text{CM}} \, K_{\text{int}} \, J_{\text{int}}\rangle_{M_{\text{P}}}^{J_{\text{P}}} = \sum_N c_N^{\text{HO}} \langle SD_n | SD_N\rangle^{\text{HO}} . \tag{9.78}$$

Note that the $\langle SD_n | SD_N\rangle^{\text{HO}}$ overlap is easily computed even though $|SD_n\rangle$ and $|SD_N\rangle^{\text{HO}}$ are built from different one-body basis states. Indeed, the two latter Slater determinants bear only one proton and one neutron state, so that $\langle SD | SD\rangle^{\text{HO}}$ is a product of proton and neutron overlaps:

$$\langle SD | SD\rangle^{\text{HO}} = \langle \phi^{(\text{p})} | \phi^{(\text{p})}\rangle^{\text{HO}} \langle \phi^{(\text{n})} | \phi^{(\text{n})}\rangle^{\text{HO}} , \tag{9.79}$$

where $|\phi^{(\text{p})}\rangle \, \left(|\phi^{(\text{n})}\rangle\right)$ and $|\phi^{(\text{p})}\rangle^{\text{HO}} \, \left(|\phi^{(\text{n})}\rangle^{\text{HO}}\right)$ denote the proton (neutron) one-body states of the $|SD\rangle$ and $|SD\rangle^{\text{HO}}$ Slater determinants, respectively.

Aside from the harmonic oscillator projection of the intrinsic state $|K_{\text{int}} \, J_{\text{int}}\rangle$, Eqs. (9.64) and (9.77) differ by their center-of-mass part, as it is a Berggren center-of-mass state in Eq. (9.63) and a harmonic oscillator state in Eq. (9.65). As the center-of-mass asymptote of composite states will be provided by the integration of coupled-channel equations, defined in coordinate space, having a localized center-of-mass state in Eq. (9.77) creates no problem.

The case of three-body projectiles using the Berggren basis can be treated similarly to the three-body model written in terms of Jacobi coordinates (see Sect. 6.4). $H_{\text{int}}$ is therein diagonalized using a Berggren basis defined with Jacobi coordinates, while the $c_N^{HO}$ coefficients of Eq. (9.76) are determined using the Raynal–Revai coefficients (see Ref.[55] and Sect. 6.4). However, the presented method becomes not applicable when the number of nucleons increases, as the use of Jacobi coordinates becomes cumbersome in this situation [35, 36]. In practice, this method might be useful for projectiles bearing two or three nucleons only.

### 9.2.4  Berggren Basis Expansion of Center-of-Mass States

In order to compute the center-of-mass part of the projectile state $|K_{CM}\, L_{CM}\, J_{int}\rangle$, one derives a Hamiltonian generating these states from the composite Hamiltonian written in laboratory coordinates. The composite cluster orbital shell model Hamiltonian cannot be used for the projectile, as it is defined for a wave function where core particles are present. The Hamiltonian in laboratory coordinates reads

$$\hat{H} = \sum_i \frac{\mathbf{p}_{i,lab}^2}{2m_i} + \sum_{i<j} \hat{V}_{ij} \,, \tag{9.80}$$

where $i, j$ cover all nucleons and $\hat{V}_{ij}$ is the nucleon–nucleon interaction in laboratory coordinates. Let $\hat{U}_i^{(T)}$ be the mean field created by all target nucleons:

$$\sum_{j\in T} \hat{V}_{ij} \rightarrow \hat{U}_i^{(T)} \,, \tag{9.81}$$

where $i$ is a particle of the projectile and averagings have been done so that $\hat{U}_i^{(T)}$ bears spherical symmetry. Thus, neglecting couplings between target and projectile, the projectile Hamiltonian $\hat{H}_P$ can be defined as

$$\hat{H}_P = \sum_{i\in P} \left( \frac{\mathbf{p}_{i,lab}^2}{2m} + \hat{U}_i^{(T)} \right) + \sum_{i<j\in P} \hat{V}_{ij} \,. \tag{9.82}$$

Linear momenta of valence particles are identical in laboratory and cluster orbital shell model coordinates (see Eq. (5.31)). Moreover, $R_{core}$ corrections are of the order of 5% at most when one goes from laboratory to cluster orbital shell model coordinates (see Sect. 5.2). Therefore, up to these small corrections, which can be neglected, $\hat{H}_P$ in cluster orbital shell model coordinates reads the same as in Eq. (9.82). Consequently, Eq. (9.82) in cluster orbital shell model coordinates takes a form:

$$\hat{H}_P = \sum_{i\in P} \frac{\left( \mathbf{p}_i - \frac{1}{a}\mathbf{P}_{CM} \right)^2}{2m} + \sum_{i<j\in P} \hat{V}_{ij} + \frac{\mathbf{P}_{CM}^2}{2M_P} + \sum_{i\in P} \hat{U}_i^{(T)} \,, \tag{9.83}$$

with $\mathbf{P}_{CM} = \sum_{i \in P} \mathbf{p}_i$, $a$ the number of nucleons in the projectile, $m$ the nucleon mass, and $M_P$ the mass of the projectile.

Assuming the cluster approximation, one has $\mathbf{r}_i \simeq \mathbf{R}_{CM}$, so that the central part of the mean field $\hat{U}_{CM;C}(\mathbf{R}_{CM})$ created by all target nucleons can be approximated by

$$\hat{U}_{CM,C}(\mathbf{R}_{CM}) = \sum_{i \in P} \hat{U}_{i,C}^{(T)}(\mathbf{R}_{CM}) = a_p \, \hat{U}_{p,C}^{(T)}(\mathbf{R}_{CM}) + a_n \, \hat{U}_{n,C}^{(T)}(\mathbf{R}_{CM}) , \tag{9.84}$$

where $a_p$ and $a_n$ are the number of protons and neutrons of the projectile, respectively. The spin–orbit part of the mean field $\hat{U}_{CM;SO}(\mathbf{R}_{CM})$ is calculated through a similar averaging procedure, as the cluster approximation also implies that $\mathbf{l}_i \simeq \mathbf{L}_{CM}/a$ and $\sum_{i \in P} \mathbf{s}_i \simeq \mathbf{J}_{int}$

$$\sum_{i \in P} \hat{U}_{i,SO}^{(T)}(\mathbf{R}_{CM}) \, (\mathbf{l}_i \cdot \mathbf{s}_i) = \hat{U}_{CM,SO}(\mathbf{R}_{CM})$$

$$\sum_{i \in P} \mathbf{l}_i \cdot \mathbf{s}_i = \frac{1}{a} \, \hat{U}_{CM,SO}(\mathbf{R}_{CM}) \, (\mathbf{L}_{CM} \cdot \mathbf{J}_{int}) , \tag{9.85}$$

where $\hat{U}_{CM,SO}$ is the average of all $\hat{U}_{i,SO}^{(T)}$ potentials

$$\hat{U}_{CM,SO}(\mathbf{R}_{CM}) = \frac{a_p}{a} \, \hat{U}_{p,SO}^{(T)}(\mathbf{R}_{CM}) + \frac{a_n}{a} \, \hat{U}_{n,SO}^{(T)}(\mathbf{R}_{CM}) . \tag{9.86}$$

From Eqs. (9.84) and (9.85), one finds the potential $\hat{U}_{CM}(\mathbf{R}_{CM})$ that generates the center-of-mass states $|K_{CM}, L_{CM}\rangle$

$$\hat{U}_{CM}(\mathbf{R}_{CM}) = \hat{U}_{CM,C}(\mathbf{R}_{CM}) + \frac{1}{a} \, \hat{U}_{CM,SO}(\mathbf{R}_{CM}) \, (\mathbf{L}_{CM} \cdot \mathbf{J}_{int}) . \tag{9.87}$$

Consequently, $\hat{H}_P$ reads

$$\hat{H}_P = \hat{H}_{int} + \hat{H}_{CM}, \tag{9.88}$$

with

$$\hat{H}_{int} = \sum_{i \in P} \frac{\left(\mathbf{p}_i - \frac{1}{a}\mathbf{P}_{CM}\right)^2}{2m} + \sum_{i < j \in P} \hat{V}_{ij}, \tag{9.89}$$

and

$$\hat{H}_{CM} = \frac{\mathbf{P}_{CM}^2}{2M_P} + \hat{U}_{CM} . \tag{9.90}$$

The mean-field approximation described in Eqs. (9.84) and (9.85) insures that $\hat{U}_{CM}(\mathbf{R}_{CM})$ recaptures the features of the Hamiltonian of Eq. (9.82) for the $|K_{CM}, L_{CM}\rangle$ center-of-mass states at the level of cluster approximation. It is thus possible to calculate a Berggren basis of center-of-mass $|K_{CM}, L_{CM}\rangle$ states, as these states are formally identical to one-body states. Moreover, even though cluster approximation is no longer valid in break-up reactions, the latter potential can still be used therein as it provides with a complete set of $|K_{CM}, L_{CM}\rangle$ center-of-mass states.

### 9.2.5  Orthogonalization of Composites with Respect to the Core

In the case of a one-nucleon projectile, all calculated wave functions must be orthogonal to the occupied states in the core, as demanded by the orthogonalization condition model (see Sect. 5.2). Both target and projectile wave functions of the composite state in this case could be in principle diagonalized using the same Berggren basis, from which the occupied core one-body states are removed. Consequently, the basis configurations expanding target and projectile wave functions are all orthogonal to the occupied states in the core, so that the orthogonalization condition model would be a priori exactly verified.

However, we have seen that projectile wave functions, describing a cluster of a few nucleons, must be written as the product of center-of-mass and relative parts (see Eq. (9.63)). As the occupied states in the core are written in cluster orbital shell model coordinates, the orthogonalization condition model would become very cumbersome to apply. Indeed, as said before, the transformation from cluster orbital shell model coordinates to center-of-mass and relative coordinates is conveniently treated only through the use of a basis of harmonic oscillator states or Jacobi coordinates. This would a priori preclude the inclusion of the Berggren basis to describe cluster wave functions.

Thus, one has to introduce an alternative procedure to impose the orthogonalization condition model for composite eigenstates whose projectile is a cluster and whose wave function is a product of center-of-mass and relative wave functions (see Eq. (9.63)). In order to avoid the problems related to the transformation from cluster orbital shell model coordinates to center-of-mass and relative coordinates, one will define projectors in terms of center-of-mass and relative coordinates:

$$\hat{Q}_{CM} = \sum_{N_{CM} \leq N_{CM\,min}} |N_{CM}\, L_{CM}\, J_{int}\, J_P\, M_P\rangle \langle N_{CM}\, L_{CM}\, J_{int}\, J_P\, M_P| \tag{9.91}$$

$$\hat{P}_{CM} = \hat{1} - \hat{Q}_{CM} , \tag{9.92}$$

where $N_{\text{CMmin}}$ is chosen so as to include the cluster eigenstates of $\hat{H}_P$ sizably occupying the core. $\hat{H}_{\text{CM}}$ is then redefined similarly to $H$ in the one-nucleon projectile case (see Sect. 9.1.1.1):

$$\hat{H}_{\text{CM}} \to \hat{P}_{\text{CM}} \hat{H}_{\text{CM}} \hat{P}_{\text{CM}} = \hat{H}_{\text{CM}} - \hat{Q}_{\text{CM}} \hat{H}_{\text{CM}} - \hat{H}_{\text{CM}} \hat{Q}_{\text{CM}} + \hat{Q}_{\text{CM}} \hat{H}_{\text{CM}} \hat{Q}_{\text{CM}} , \tag{9.93}$$

which thereby generates an additional short-range interaction to be added to $\hat{H}_P$ in Eq. (9.15).

As the $\hat{Q}_{\text{CM}}$ and $\hat{P}_{\text{CM}}$ projectors (see Eqs. (9.91) and (9.92)) are written in center-of-mass and relative coordinates, they cannot be exactly equal to $\hat{Q}$ and $\hat{P}$ written in orbital shell model coordinates (see Eqs. (9.16) and (9.17)). However, as it has been checked in practical applications, $\hat{Q} \simeq \hat{Q}_{\text{CM}}$ and $\hat{P}_{\text{CM}} \simeq \hat{P}$, so that the use of $\hat{Q}_{\text{CM}}$ and $\hat{P}_{\text{CM}}$ allows the orthogonalization condition model to be verified. Hence, in practical calculations, similarly to the one-nucleon case, composite states can be considered to be orthogonal to the core occupied states, even though a strict orthogonality is no longer exactly fulfilled with many-body projectiles.

## 9.2.6   Asymptotic Cancellation of Correlations at High Projectile Energy

Contrary to the one-nucleon projectile case, the nucleon–nucleon interaction between target and projectile does not become negligible when the energy of the projectile nucleons becomes very large (see Sect. 9.1.1). Indeed, even though the nucleons inside a rapidly moving cluster do not interact with the nucleons of the target, they still strongly interact with each other. Thus, the description of a many-nucleon projectile of high velocity will demand to include two-body matrix elements involving one-body states of high energy. Nevertheless, as the projectile almost does not interact with the target, the Hamiltonian describing target–projectile channels of high energy will greatly simplify. In fact, it will decouple similarly to the one-nucleon case (see Sect. 9.1.1), with a low-energy part, containing target–projectile correlations, and a high energy part, where target and projectile are decoupled.

Let us define the projectile state as a product of low-energy pairs of particles, which embodies its compact cluster structure:

$$|\Psi_P^{J_P}\rangle = \left( \mathscr{A} \prod_{k_{\text{rel}} \leq k_{\text{rel}}^{(\text{max})}} |k_{\text{rel}}\rangle \right) |K_{\text{CM}}\rangle , \tag{9.94}$$

where $|k_{\text{rel}}\rangle$ stands for a particle pair, $k_{\text{rel}}^{(\text{max})}$ is a maximal linear momentum allowed for pairs, related to the average relative velocity of nucleons inside the cluster, and the $|K_{CM}\rangle$ state describes the center-of-mass of the projectile. The Hamiltonian is denoted as $\hat{H} = \hat{T} + \hat{U}_{core} + \hat{V}_{res}$, where $\hat{U}_{core}$ is the potential of the core and $\hat{V}_{res}$ is the two-body residual interaction.

In order to evaluate the following matrix elements:

$$\langle \hat{\mathscr{A}} \{ \langle \Psi_{T_f}^{J_{T_f}} | \otimes \langle R_{CM,f} \, L_{CM,f} \, J_{int,f} \, J_{P,f} | \}_{M_A}^{J_A} |$$

$$\hat{H} \, | \hat{\mathscr{A}} \{ | \Psi_{T_i}^{J_{T_i}} \rangle \otimes | R_{CM,i} \, L_{CM,i} \, J_{int,i} \, J_{P,i} \rangle \}_{M_A}^{J_A} \rangle \,, \qquad (9.95)$$

one uses the separation of the Hamiltonian $\hat{H}$ into basis and residual parts (see Eq. (9.11)):

$$\hat{H} = \hat{T} + \hat{U}_{basis} + \hat{V}_{res} - \hat{U}_0 \,.$$

We are going to prove that the residual matrix elements $\langle ab | \hat{V}_{res} - \hat{U}_0 | cd \rangle$ involved in target–projectile coupling at high energy can be neglected

$$\langle ab | \hat{V}_{res} - \hat{U}_0 | cd \rangle \simeq 0 \ \text{if} \ \exists (i,j) \in \{a,b,c,d\} \mid k_i \ll k_{max} \ \text{and} \ k_j \gtrsim k_{max} \,, \qquad (9.96)$$

where $|a\rangle$, $|b\rangle$, $|c\rangle$, and $|d\rangle$ are the Berggren states, of momentum $k_a$, $k_b$, $k_c$, and $k_d$, respectively, and where $k_{max}$ is an arbitrarily large one-body state momentum. Clearly, the property embedded in Eq. (9.96) implies that the nucleons of the target and projectile no longer interact at high projectile energy.

Let us quantify the properties verified by linear momenta and matrix element in Eq. (9.96). For this, one introduces the maximal linear momentum $k_{max}^{(H)} \ll k_{max}$, which embodies the finite-range character of $\hat{U}_0$ and $\hat{V}_{res}$ in $\hat{H}$. Indeed, the matrix elements of $\hat{U}_0$ can be made arbitrarily small if one involved linear momentum is larger than $k_{max}^{(H)}$, with $k_{max}^{(H)}$ sufficiently large. Similarly, due to the translational invariance of $\hat{V}_{res}$, the relative matrix elements of $\hat{V}_{res}$ verify the same property if one relative linear momentum is larger than $k_{max}^{(H)}$. This requirement is clearly consistent with the cluster definition of Eq. (9.94) if $k_{rel}^{(max)}$ and $k_{max}^{(H)}$ have the same order of magnitude.

The conditions $k_i \ll k_{max}$ and $k_j \gtrsim k_{max}$ in Eq. (9.96) must then be understood as short-hand notations for $k_i < k_{max}^{(H)}$ and $k_j > k_{max} - k_{max}^{(H)}$, respectively. Moreover, the notation $\langle ab | \hat{V}_{res} - \hat{U}_0 | cd \rangle \simeq 0$ signifies that this matrix element can be made arbitrarily small for sufficiently large values of $k_{max}^{(H)}$ and $k_{max}$.

### 9.2.6.1 Cancellation of High Energy Matrix Elements with Plane Waves

Neutron wave functions can always be expanded in a basis of plane waves. This is not true for proton wave functions because the matrix elements of Coulomb potential in the basis of plane waves diverge in general (see Sect. 3.7.1). Plane waves will be denoted as $|\mathbf{k}\rangle$, and one will perform a decomposition into relative and center-of-mass parts:

$$\langle \mathbf{r}_a \, \mathbf{r}_b | \mathbf{k}_a \, \mathbf{k}_b \rangle = e^{i\mathbf{k}_a \cdot \mathbf{r}_a} e^{i\mathbf{k}_b \cdot \mathbf{r}_b} = e^{i\mathbf{k}_{rel} \cdot \mathbf{r}_{rel}} e^{i\mathbf{K}_{CM} \cdot \mathbf{R}_{CM}} , \tag{9.97}$$

where

$$\mathbf{r}_{rel} = \mathbf{r}_a - \mathbf{r}_b, \quad \mathbf{R}_{CM} = \frac{1}{2} (\mathbf{r}_a + \mathbf{r}_b), \quad \mathbf{k}_{rel} = \frac{1}{2} (\mathbf{k}_a - \mathbf{k}_b)$$

$$\text{and} \quad \mathbf{K}_{CM} = \mathbf{k}_a + \mathbf{k}_b . \tag{9.98}$$

Antisymmetry of two-body states is not considered in Eq. (9.97) for simplicity. There is no loss of generality to do so because non-antisymmetrized two-body states form a complete set of states.

Let us consider firstly the part of Eq. (9.96) depending on $\hat{U}_0$. One has immediately

$$\langle \mathbf{k}_a \, \mathbf{k}_b | \hat{U}_0 | \mathbf{k}_c \, \mathbf{k}_d \rangle = \langle \mathbf{k}_a | \hat{U}_0 | \mathbf{k}_c \rangle \, \delta(\mathbf{k}_b - \mathbf{k}_d) + \langle \mathbf{k}_b | \hat{U}_0 | \mathbf{k}_d \rangle \, \delta(\mathbf{k}_a - \mathbf{k}_c) . \tag{9.99}$$

As $\hat{U}_0$ is a finite-range operator, one has $\langle \mathbf{k} | \hat{U}_0 | \mathbf{k}' \rangle \simeq 0$ unless $k \ll k_{max}$ and $k' \ll k_{max}$ (see Sect. 9.2.6). As a consequence, the two-body matrix element of Eq. (9.99) is not negligible only if $k_i \ll k_{max}$ $\forall i \in \{a, b, c, d\}$.

Secondly, the translationally invariant character of $\hat{V}_{res}$ implies that

$$\langle \mathbf{k}_a \, \mathbf{k}_b | \hat{V}_{res} | \mathbf{k}_c \, \mathbf{k}_d \rangle = \langle \mathbf{k}_{rel} \, \mathbf{K}_{CM} | \hat{V}_{res} | \mathbf{k}'_{rel} \, \mathbf{k}'_{CM} \rangle = \delta(\mathbf{K}_{CM} - \mathbf{K}'_{CM}) \, \langle \mathbf{k}_{rel} | \hat{V}_{res} | \mathbf{k}'_{rel} \rangle , \tag{9.100}$$

where Eq. (9.98) has been used, with $\mathbf{k}_{rel}$, $\mathbf{K}_{CM}$ the relative and center-of-mass linear momenta arising from $\mathbf{k}_a$, $\mathbf{k}_b$, and $\mathbf{k}'_{rel}$, $\mathbf{K}'_{CM}$ the relative and center-of-mass linear momenta issued from $\mathbf{k}_c$ and $\mathbf{k}_d$, respectively.

Let us write the equation of energy conservation arising from Eq. (9.98):

$$K_{CM}^2 + 4k_{rel}^2 = 2(k_a^2 + k_b^2) . \tag{9.101}$$

Due to cluster approximation, having $k_a \gtrsim k_{max}$, $k_b \ll k_{max}$, or $k_a \ll k_{max}$, $k_b \gtrsim k_{max}$ is impossible as they correspond to configurations in which the nucleons of the cluster separate from each other. Consequently, as the considered cluster has a large kinetic energy and as one demands that $k_i \ll k_{max}$ or $k_i \gtrsim k_{max}$ in Eq. (9.96) $\forall i \in \{a, b, c, d\}$, one has only one possibility for the matrix element of $\hat{V}_{res} - \hat{U}_0$ to be non-negligible

$$k_a, k_b \gtrsim k_{max} \Rightarrow K_{CM} \simeq \sqrt{2(k_a^2 + k_b^2)} \gtrsim 2 \, k_{max} , \tag{9.102}$$

where one has used the fact that $k_{rel}^{(max)} \ll k_{max}$.

Therefore, one has $\langle \mathbf{k}_a | \hat{U}_0 | \mathbf{k}_c \rangle \simeq 0$ and $\langle \mathbf{k}_b | \hat{U}_0 | \mathbf{k}_d \rangle \simeq 0$ in Eq. (9.99). As one has $k_a, k_b \gtrsim k_{max}$, one can have $k_c \ll k_{max}$ or $k_c \gtrsim k_{max}$ (same for $k_d$) in Eq. (9.96). Let us then consider the different values that $k_c$ and $k_d$ can bear. On the one hand, one obtains from Eq. (9.101)

$$k_c, k_d \ll k_{max} \Rightarrow K'_{CM} \ll 2 \, k_{max} \ . \tag{9.103}$$

Consequently, $\delta(K_{CM} - K'_{CM}) = 0$ in Eq. (9.100) in this situation. On the other hand, having $k_c \gtrsim k_{max}$, $k_d \ll k_{max}$, or $k_c \ll k_{max}$, $k_d \gtrsim k_{max}$ leads to vanishing matrix elements of $\hat{V}_{res}$ due to cluster approximation. Therefore, the matrix elements $\langle \mathbf{k}_a \, \mathbf{k}_b | \hat{V}_{res} - \hat{U}_0 | \mathbf{k}_c \, \mathbf{k}_d \rangle$ are negligible if $\exists (i, j) \in \{a, b, c, d\} \ | \ k_i \ll k_{max}$ and $k_j \gtrsim k_{max}$. Eq. (9.96) is then fulfilled in a basis of plane waves.

### 9.2.6.2 Cancellation of High Energy Matrix Elements with Real-Energy Neutron States

Let us consider a one-body neutron state of real energy, of orbital angular momentum $\ell$, and linear momentum $k$, denoted by $|i\rangle$. The asymptote of $|i\rangle$ is that of Eq. (2.7), so that $|i\rangle$ can be written as

$$|i\rangle = C |k\ell\rangle_{\mathscr{B}} + |i^+\rangle \ , \tag{9.104}$$

where $|k\ell\rangle_{\mathscr{B}}$ is a state of Bessel function, $C$ is a constant and $|i^+\rangle$ is a state satisfying an outgoing wave condition. As $|i\rangle$ is generated by a smooth finite-range potential, one can assume that the basis-generating potential vanishes identically if $k \geq k_{max}^{(H)}$ (see Sect. 9.2.6), as results will be the same up to an arbitrarily small precision. Equation (2.2) expressed in momentum space [56] then implies that $\langle i^+ | k'\ell \rangle_{\mathscr{B}} = 0$ for $k' \geq k_{max}^{(H)}$.

Let us write the matrix elements of $\hat{V}_{res} - \hat{U}_0$ using Eq. (9.104):

$$\langle ab | \hat{V}_{res} - \hat{U}_0 | cd \rangle = \sum_{f_a, f_b, f_c, f_d} \langle f_a \, f_b | \hat{V}_{res} - \hat{U}_0 | f_c \, f_d \rangle \ , \tag{9.105}$$

where $|f_i\rangle = \{|k_i \ell_i m_i\rangle_{\mathscr{B}}, |i^+\rangle\}$ with $i = \{a, b, c, d\}$. The matrix elements $\langle f_a \, f_b | \hat{V}_{res} - \hat{U}_0 | f_c \, f_d \rangle$ are expressed using the Bessel function expansion of $|f_i\rangle$ states:

$$\langle f_a \, f_b | \hat{V}_{res} - \hat{U}_0 | f_c \, f_d \rangle = \int \langle k'_a \ell_a m_a k'_b \ell_b m_b | \hat{V}_{res} - \hat{U}_0 | k'_c \ell_c m_c k'_d \ell_d m_d \rangle_{\mathscr{B}}$$
$$\times \langle f_a | k'_a \ell_a m_a \rangle_{\mathscr{B}} \langle f_b | k'_b \ell_b m_b \rangle_{\mathscr{B}}$$
$$\times \langle f_c | k'_c \ell_c m_c \rangle_{\mathscr{B}} \langle f_d | k'_d \ell_d m_d \rangle_{\mathscr{B}} dk'_a dk'_b dk'_c dk'_d \ , \tag{9.106}$$

where (9.96) is supposed to be verified.

All functions are bounded in Eq. (9.106), due to the normalizations of basis states and the low-energy character of $\hat{V}_{res} - \hat{U}_0$. Therefore, in order to show that $\langle f_a \; f_b | \hat{V}_{res} - \hat{U}_0 | f_c \; f_d \rangle \simeq 0$, it is sufficient to find one negligible factor in the integrand of Eq. (9.106). One will show in Exercise III that $\langle ab | \hat{V}_{res} - \hat{U}_0 | cd \rangle \simeq 0$ when Eq. (9.96) is fulfilled using Eq. (9.106). This exercise is technical, so that readers can admit its main results if they find the exercise too difficult.

---

**Exercise III $\star\star\star$**

One will demonstrate that the matrix elements $\langle ab | \hat{V}_{res} - \hat{U}_0 | cd \rangle$ vanish when Eq. (9.96) is verified. This will be necessary to derive the coupled-channel Hamiltonian equations for many-body projectiles.

**A.** Let us firstly consider that the two conditions $k'_i \ll k_{max}$ and $k'_j \gtrsim k_{max}$ are not verified $\forall (i,j) \in \{a, b, c, d\}$. Express these conditions in terms of $k_{max}$ and $k_{max}^{(H)}$. Show that the integrand of Eq. (9.106) vanishes in this case.

**B.** Let us now suppose that $\exists (i,j) \in \{a, b, c, d\}$ so that $k'_i \ll k_{max}$ and $k'_j \gtrsim k_{max}$. Expand the matrix element $\langle k'_a \ell_a m_a \; k'_b \ell_b m_b | \hat{V}_{res} - \hat{U}_0 | k'_c \ell_c m_c \; k'_d \ell_d m_d \rangle_{\mathscr{B}}$ using plane waves. Deduce from the plane wave expansion that Eq. (9.106) vanishes in this situation.

**C.** Conclude that Eq. (9.96) is fulfilled by real-energy neutron Berggren basis states.

---

### 9.2.6.3 Cancellation of High Energy Matrix Elements with Real-Energy Proton States

For the cases involving proton states, one can consider the screening method in order to avoid divergences due to the infinite range of the Coulomb potential. This is justified by the fact that reactions occurring far from the target are only sensitive to the Coulomb interaction. Hence, one can consider that the Coulomb Hamiltonian vanishes after a radius $R$. Proton states for which $k \sim 0$ can be neglected as well, as they can only induce Rutherford scattering at large distances. Consequently, as proton states generated by a screened Coulomb Hamiltonian can be expanded with planes waves, one obtains (see Sect. 2.6.5)

$$\langle ab | \hat{V}_{res} - \hat{U}_0 | cd \rangle = \langle a_s \; b_s | \hat{V}_{res} - \hat{U}_0 | c_s \; d_s \rangle + \mathscr{O}(R^{-1}) , \qquad (9.107)$$

where the index $s$ stands for "screening." Consequently, as $R$ can be arbitrarily large, one can consider that $\langle ab | \hat{V}_{res} - \hat{U}_0 | cd \rangle \simeq 0$ in Eq. (9.107) with both proton and neutron states when the conditions in Eq. (9.96) are verified.

### 9.2.6.4 Suppression of Antisymmetry at High Projectile Energy

The results previously obtained imply that the only significant two-body matrix elements $\langle ab | \hat{V}_{res} - \hat{U}_0 | cd \rangle$ verify $k_i \gtrsim k_{max} \; \forall i \in \{a, b, c, d\}$, assuming $K_{CM} \gtrsim K_{CM\,max}$, with $K_{CM\,max} = 2k_{max}$. As the target is always in a low-energy eigenstate, one has $k_i \ll k_{max}$ $\forall i \in \{a, b, c, d\}$ in the two-body matrix elements entering the target Hamiltonian. Therefore, the antisymmetry between projectile and target wave functions for large $K_{CM}$ can be suppressed, and methods used in Sect. 9.1.1 can also be utilized for many-nucleon projectiles in the frame of cluster approximation.

### 9.2.6.5 Generalization to Complex-Energy States

The Berggren basis possesses complex-energy states, so that it might seem that Eq. (9.96) has to be extended to this case. However, realistic interactions cannot be directly analytically continued to the complex plane [57]. Indeed, it has been seen that the use of a harmonic oscillator expansion of two-body matrix elements is necessary to make the realistic interactions suitable for applications using the Berggren basis (see Sect. 5.6). Nevertheless, one cannot derive an equivalent of Eq. (9.101) with harmonic oscillator states. This is due to the center-of-mass part in Eq. (9.63) because Eq. (9.101) relating the linear momenta of one-body states and projectile center-of-mass states is valid only for plane waves. As the plane wave expansion of Bessel functions is done at fixed energy, the $K_{CM}$-dependent relations derived for plane waves can also be used with Bessel functions, and thus with real-energy Berggren basis functions (Eq. (9.106)), but not with harmonic oscillator states.

A solution to this conundrum can be found by using the analytic properties of the wave functions of the solutions of the coupled-channel equations of the Gamow shell model. For this, one firstly derives the coupled-channel equations of the Gamow shell model within the framework of resonating group method with real energies, so that Eq. (9.96) is valid. Then, once the coupled-channel equations of the Gamow shell model are obtained on the real-energy axis, the real-energy solutions of the coupled-channel equations can be analytically continued to the complex-energy plane. Consequently, one can use the coupled-channel Hamiltonian of the resonating group method with complex-energy states, even though the coupled-channel equations were derived using real energies.

## 9.2.7 Coupled-Channel Equations and Wave Functions for Structure and Reactions with Multi-Nucleon Projectiles

Let us now derive the representation of Hamiltonian in coupled-channel equations for multi-nucleon projectiles. The procedure we follow in this derivation is similar to the one-nucleon case. The multi-nucleon composite is expressed in a basis of Slater determinants in order to be able to fulfill the antisymmetry requirement. Then, by separating the Hamiltonian in low- and high-energy parts, one is able to determine the coupling potentials entering the coupled-channel equations of the Gamow shell model.

### 9.2.7.1 Many-Body Wave Functions

Let us calculate the expansion of the $A$-body wave function $\{A^\dagger_{|N_{\mathrm{CM}},L_{\mathrm{CM}},J_{\mathrm{int}}\rangle^{J_\mathrm{P}}_{M_\mathrm{P}}}$ $|\Psi_\mathrm{T}^{J_\mathrm{T}}\rangle\}^{J_\mathrm{A}}_{M_\mathrm{A}}$, which will be used to evaluate the matrix elements in a complete basis of $A$-body Slater determinants:

$$\{A^\dagger_{|N_{\mathrm{CM}},L_{\mathrm{CM}},J_{\mathrm{int}}\rangle^{J_\mathrm{P}}_{M_\mathrm{P}}}\,|\Psi_\mathrm{T}^{J_\mathrm{T}}\rangle\}^{J_\mathrm{A}}_{M_\mathrm{A}} = \sum_\alpha c_\alpha\,|SD_\alpha\rangle\;. \tag{9.108}$$

For that, one uses the expansion of $|\Psi_\mathrm{T}\rangle^{J_\mathrm{T}}_{M_\mathrm{T}}$ in $(A - a)$-body Slater determinants and $|N_{\mathrm{CM}}, L_{\mathrm{CM}}, J_{\mathrm{int}}\rangle^{J_\mathrm{P}}_{M_\mathrm{P}}$ in $a$-body Slater determinants, where $a$ is the number of nucleons of the projectile, obtained from the diagonalization of the Gamow shell model Hamiltonian $\hat{H}$:

$$|\Psi_{M_\mathrm{T}}^{J_\mathrm{T}}\rangle = \sum_\beta b_\beta^{\mathrm{A\text{-}a}}|SD_\beta\rangle^{\mathrm{A\text{-}a}} \tag{9.109}$$

$$|N_{\mathrm{CM}}, L_{\mathrm{CM}}, J_{\mathrm{int}}\rangle^{J_\mathrm{P}}_{M_\mathrm{P}} = \sum_\gamma b_\gamma^{\mathrm{a}}|SD_\gamma\rangle^{\mathrm{a}}\,, \tag{9.110}$$

where the $a$ and $(A - a)$-body characters of expansion coefficients and Slater determinants are explicitly written. Applying the creation operator of Eq. (9.108) on Eq. (9.109) with a given angular momentum projection $M_\mathrm{P}$, and using Eq. (9.110), the following uncoupled fully antisymmetrized $A$-body wave function comes forward:

$$\begin{aligned}
\{A^\dagger_{|N_{\mathrm{CM}},L_{\mathrm{CM}},J_{\mathrm{int}}\rangle^{J_\mathrm{P}}_{M_\mathrm{P}}}\,|\Psi_\mathrm{T}\rangle^{J_\mathrm{T}}_{M_\mathrm{T}}\} &= \sum_\beta b_\beta^{\mathrm{A\text{-}a}} A^\dagger_{|N_{\mathrm{CM}},L_{\mathrm{CM}},J_{\mathrm{int}}\rangle^{J_\mathrm{P}}_{M_\mathrm{P}}}\,|SD_\beta\rangle^{\mathrm{A\text{-}a}} \\
&= \sum_{\beta\gamma} b_\beta^{\mathrm{A\text{-}a}} b_\gamma^{\mathrm{a}}\,\mathscr{A}\{|SD_\beta\rangle^{\mathrm{A\text{-}a}}|SD_\gamma\rangle^{\mathrm{a}}\} \\
&= \sum_\alpha b_\alpha\,|SD_\alpha\rangle\,, \tag{9.111}
\end{aligned}$$

where $|SD_\alpha\rangle = \mathscr{A}\{|SD_\beta\rangle^{A-a}|SD_\gamma\rangle^{a}\}$ and $b_\alpha = \pm b_\beta^{A-a} b_\gamma^{a}$, $\pm 1$ being a rearrangement phase. The $A$-body wave function expansion of Eq. (9.108) can then be obtained by coupling the wave function defined in Eq. (9.111) to a given angular momentum using Clebsch–Gordan coefficients.

### 9.2.7.2 Hamiltonian Coupled-Channel Equations

Similarly to Sect. 9.1.1, let us develop the coupled-channel equations for multi-nucleon projectiles. One considers a scattering $A$-body state decomposed in reaction channels, denoted by $c$,

$$|\Psi_{M_A}^{J_A}\rangle = \sum_c \int_0^{+\infty} \left(\frac{u_c(R_{CM})}{R_{CM}}\right) |(c, R_{CM})_{M_A}^{J_A}\rangle R_{CM}^2 \, dR_{CM} , \qquad (9.112)$$

with

$$|(c, R_{CM})\rangle = \hat{\mathcal{A}} |\{|\Psi_T^{J_T}\rangle \otimes |R_{CM} \, L_{CM} \, J_{int} \, J_P \, M_P\rangle\}\rangle_{M_A}^{J_A} ,$$

where the spin target $J$ and the spin projectile $J_P$ are coupled to $J_A$. The channel index $c$ stands for the $\{A - a, J_T; a, L_{CM}, J_{int}, J_P\}$ quantum numbers. $u_c(R_{CM})$ is the radial amplitude of the $c$ channel to be determined. The coupled-channel equations can then be formally derived from the Schrödinger equation: $H |\Psi_{M_A}^{J_A}\rangle = E |\Psi_{M_A}^{J_A}\rangle$, as

$$\sum_c \int_0^{+\infty} \left(H_{cc'}(R_{CM}, R'_{CM}) - E N_{cc'}(R_{CM}, R'_{CM})\right) u_c(R_{CM}) = 0 , \qquad (9.113)$$

with

$$H_{cc'}(R_{CM}, R'_{CM}) = R_{CM} \, R'_{CM} \, \langle(c, R_{CM})|\hat{H}|(c', R'_{CM})\rangle \qquad (9.114)$$

$$N_{cc'}(R_{CM}, R'_{CM}) = R_{CM} \, R'_{CM} \, \langle(c, R_{CM})|(c', R'_{CM})\rangle . \qquad (9.115)$$

---

**Exercise  IV ★ ★ ★**
One will derive the coupled-channel Hamiltonian matrix elements involving cluster projectiles (see Eq. (9.120)).

**A.** Expand $H_{cc'}^{J_A M_A}(R_{CM}, R'_{CM})$ using the eigenbasis of $\hat{H}_{CM}$. One will use the notation $|N_{CM} \, L_{CM} \, J_{int} \, J_P \, M_P\rangle$ and $|N'_{CM} \, L_{CM} \, J_{int} \, J_P \, M_P\rangle$ for the eigenbasis expanding the $R_{CM}$-dependent and $R'_{CM}$-dependent channels, respectively. Write the obtained expansion as four different sums, involving $N_{CM} \leq N_{CM max}$, $N_{CM} > N_{CM max}$, $N'_{CM} \leq N_{CM max}$ and $N'_{CM} > N_{CM max}$.
**B.** Notice that the sum of the expansion function of $N_{CM}$, $N'_{CM} \leq N_{CM max}$ can be conveniently calculated. Show that $\hat{H} = \hat{H}_T + \hat{H}_{CM} + \hat{H}_{int} + \hat{\mathcal{A}} \hat{H}_{TP} \hat{\mathcal{A}}$ if $N_{CM} > N_{CM max}$ or $N'_{CM} > N_{CM max}$. Deduce that the non-vanishing sums of the expansion are those for which $N_{CM}, N'_{CM} \leq N_{CM max}$ and $N_{CM}, N'_{CM} > N_{CM max}$.
**C.** One will now consider input and output channels whose projectiles can bear different numbers of nucleons, denoted as $a$ for the input channel and $a'$ for the output channel. They will be deemed as identical/different partitions if the number of nucleons of these channels is identical/different. Calculate the sum

of the $H_{cc'}^{J_A M_A}(R_{CM}, R'_{CM})$ expansion for identical partitions and $N_{CM}, N'_{CM} > N_{CM\,max}$. Show that this sum vanishes for different partitions.

**D.** Write $H_{cc'}^{J_A M_A}(R_{CM}, R'_{CM})$ as a function of target states, center-of-mass, and projectile intrinsic energies using the completeness properties of the eigenstates of $\hat{H}_{CM}$.

Due to the decoupling of the target and projectile at high energy, it is more convenient to rewrite the Hamiltonian $\hat{H}$ by introducing the target Hamiltonian $\hat{H}_T$:

$$\hat{H}_T = \hat{T}_T + \hat{U}_{T;\,basis} + \left(\hat{V}_{res} - \hat{U}_0\right)^{A-a}, \qquad (9.116)$$

where $\left(\hat{V}_{res} - \hat{U}_0\right)^{A-a}$ is the part of $\hat{V}_{res} - \hat{U}_0$ acting on the $(A-a)$-body target state, and where $\hat{T}_T$ and $\hat{U}_{basis}^T$ are the target kinetic and potential parts of the Hamiltonian $\hat{H}$, respectively. The action of target and projectile Hamiltonians $\hat{H}_T$ and $\hat{H}_P$ (see Eq. (9.88)) on $A$-body states is performed by considering non-fully antisymmetrized $A$-body states:

$$\hat{H}_T \left(|\Psi_T\rangle \otimes |\Psi_P\rangle\right) = \left(\hat{H}_T |\Psi_T\rangle \otimes |\Psi_P\rangle\right) \qquad (9.117)$$

$$\hat{H}_P \left(|\Psi_T\rangle \otimes |\Psi_P\rangle\right) = \left(|\Psi_T\rangle \otimes \hat{H}_P |\Psi_P\rangle\right). \qquad (9.118)$$

Thus, one can write the Hamiltonian as

$$\hat{H} = \hat{H}_T + \hat{H}_P + \hat{H}_{TP}, \qquad (9.119)$$

where $\hat{H}_{TP} = \hat{H} - \hat{H}_T - \hat{H}_P$ by definition.

Similarly to the one-nucleon projectile, the matrix elements $H_{cc'}^{J_A M_A}(R_{CM}, R'_{CM})$ are expanded in a basis of center-of-mass harmonic oscillator states (see Sect. 9.1.1.2):

$$\begin{aligned}
H_{cc'}^{J_A M_A}(N_{CM\,i}^{HO}, N_{CM\,f}^{HO}) = &\sum_{N_{CM} N_{CM\,f} \alpha_i \alpha_f}^{N_{CM\,max}} c_{\alpha_i}^{(i)} c_{\alpha_f}^{(f)} \langle SD_{\alpha_f} | \hat{H} | SD_{\alpha_i}\rangle \\
&\times \langle N_{CM\,f}^{HO} | N_{CM\,f}\rangle \langle N_{CM\,i}^{HO} | N_{CM\,i}\rangle \\
&- \sum_{N_{CM} \leq N_{CM\,max}} (E_T + E_{int} + E_{CM}) \\
&\times \langle N_{CM\,f}^{HO} | N_{CM}\rangle \langle N_{CM\,i}^{HO} | N_{CM}\rangle, \qquad (9.120)
\end{aligned}$$

where $E_T$ is the energy of $|\Psi_T^{J_T}\rangle$, $E_{int}$ and $E_{CM}$ are the intrinsic and center-of-mass energies of $|N_{CM} L_{CM} J_{int} J_P M_P\rangle$, respectively, and the expansion coefficients $c_\alpha^{(i)}$ and $c_\alpha^{(f)}$ are those of Eq. (9.108) for the input and output channels, respectively. The derivation of Eq. (9.120) is analogous to the one-nucleon case and is the subject of Exercise IV. This exercise is difficult, but it resembles Exercise I. Hence, readers can admit results therein if they did not complete Exercise I, but should be able to solve Exercise IV without too many problems if they have already finished Exercise I.

One can then write the coupled-channel representation of $\hat{H}$, denoted as $H_{cc'}^{J_A M_A}(R_{CM}, R'_{CM})$, using the results of Exercise IV and the methods presented in Sect. 9.1.1.2:

$$
H_{cc'}^{J_A M_A}(R_{CM}, R'_{CM})
$$

$$
= \left( \frac{\hbar^2}{2M_P} \left( -\frac{d^2}{d R_{CM}^2} + \frac{\ell(\ell+1)}{R_{CM}^2} \right) + E_T + E_{int} \right) \delta(R_{CM} - R'_{CM}) \, \delta_{cc'}
$$

$$
+ U_{CM}^{L_{CM}}(R_{CM}, R'_{CM}) \, \delta_{cc'} + V_{cc'}^{J_A M_A}(R_{CM}, R'_{CM}) , \tag{9.121}
$$

where $V_{cc'}^{J_A M_A}$ includes the remaining short-range potential terms of the Hamiltonian kernels (see Exercise IV).

As seen in Sect. 9.1.1.2, it is necessary to use a basis of harmonic oscillator states to calculate coupled-channel potentials. As in the one-nucleon projectile case, the use of a basis of harmonic oscillator states is also more convenient and stable numerically. In order to express the matrix elements of the coupled-channel potentials $V_{cc'}^{J_A M_A}$ of the Hamiltonian in a basis of harmonic oscillator states in Eq. (9.121), one replaces the $|R_{CM}L_{CM}\rangle$ states of Eq. (9.121) by the $|N_{CM}^{HO}L_{CM}\rangle$ states. Indeed, the $|N_{CM}^{HO} L_{CM} J_{int} J_P M_P\rangle$ harmonic oscillator states form a complete set and $H_{TP}$ is finite-ranged, so that $V_{cc'}^{J_A M_A}$ can be expected to converge quickly with $N_{CM\,max}^{HO}$.

Similarly to the one-nucleon case (see Sect. 9.1.1.2), Eq. (9.121) is used to either calculate resonant many-body eigenstates or scattering many-body states. Structure and reaction frameworks are then unified in Eq. (9.121). One uses the same Hamiltonian to calculate resonant and scattering many-body states, so that the cross sections and spectra involving many-nucleon projectiles are inter-dependent. Thus, the changes induced by nuclear structure in cross sections can be directly associated to the spectrum calculated with the same Hamiltonian.

A novel aspect of Eq. (9.121) is that it allows to calculate the cluster emission widths in resonance states, which could not be effected in the Gamow shell model. Indeed, it would be necessary therein to expand cluster wave functions with Slater determinants. Even though it is theoretically possible, this cannot be done in practice due to the large number of Slater determinants basis states needed to reconstruct the cluster wave function in the asymptotic region. On the contrary, cluster emission widths are immediately obtained using Eq. (9.121), as the channels used are directly built from target-cluster two-body systems. In particular, the deuteron emission widths of the low-lying resonance states of $^6$Li can be obtained in a small model space from Eq. (9.121) (see Sect. 9.4.3).

## 9.2.8    Treatment of Faddeev Equations with the Berggren Basis

As one could see in Sect. 6.4, the three-body approach to nuclear systems is both conceptually simple and convenient in practical applications. While the approach detailed in Sect. 6.4 is devoted to nuclear structure, the three-body models are also widely used for nuclear reaction applications. For example, the three-body Schrödinger equation can be numerically solved with Lagrange mesh and continuum discretized coupled-channel techniques [58–62]. However, the latter methods are limited to real-energy states. Consequently, one cannot directly calculate resonances with these techniques.

It would be convenient to solve the three-body Schrödinger equation for both real and complex energies, so that the structure and reaction aspects of coupled-channel Hamiltonians can be unified therein. Therefore, one will present how to solve the three-body Schrödinger equation with the Berggren basis.

Let us write the Faddeev equations describing an interacting three-body system, formed by a target, seen as a one-body object, and a projectile formed by two subsystems. Hamiltonian and Schrödinger equations read

$$\hat{H} |\Psi\rangle = E |\Psi\rangle \tag{9.122}$$

$$\hat{H} = \hat{H}_P + \frac{\mathbf{P}^2}{2M_{CM}} + \hat{V}_T(\mathbf{r}, \mathbf{R}) , \tag{9.123}$$

where $\mathbf{P}^2/(2M_{CM})$ is the kinetic energy operator of the center-of-mass of the projectile, $\mathbf{P}$ is the linear momentum of the center-of-mass of the projectile with respect to that of the target, and $M_{CM}$ is the mass of the projectile. $\hat{H}_P$ is the intrinsic Hamiltonian of the projectile, taking into account nucleon–nucleon correlations in the projectile, $\hat{V}_T(\mathbf{r}, \mathbf{R})$ is the potential depending on the intrinsic spatial coordinate of the projectile $\mathbf{r}$ and the coordinate $\mathbf{R}$ between the center-of-mass of the target and that of the projectile, generated by target, which describes the scattering of the projectile.

In Eqs. (9.122) and (9.123), the target is structureless as its effect on the projectile is modelled by the sole action of the $\hat{V}_T(\mathbf{r}, \mathbf{R})$ potential. The potential $\hat{V}_T(\mathbf{r}, \mathbf{R})$ induces mixing between the channels formed by target and projectile. Consequently, Eqs. (9.122) and (9.123) will be expressed as coupled-channel equations, which one will be solved using the Berggren basis.

The projectile structure is taken into account by the intrinsic Hamiltonian $H_P$, which will be detailed in the following. $\hat{H}_P$ reads (see Sect. 4.1))

$$\hat{H}_P = \frac{\mathbf{p}^2}{2\mu_P} + \hat{V}_P(r) , \tag{9.124}$$

where $r$ is the relative radius of the projectile, $\mathbf{p}$ is the intrinsic linear momentum of the projectile, $\mu_P$ is the reduced mass of the projectile, and $\hat{V}_P(r)$ represents the interaction between the two subsystems of the projectile.

One will now solve Eqs. (9.122–9.124) by constructing target–projectile channels that verify coupled-channel equations. In the continuum discretized coupled-channel method [58–60, 62], the scattering states $|\Phi_P\rangle$ are dealt with bin states, obtained by integrating scattering states on finite energy interval, so as to obtain a basis of discrete bound states. However, this method does not allow to calculate complex-energy resonance states. Thus, let us consider the Berggren basis associated to the intrinsic Hamiltonian $\hat{H}_P$ of the projectile:

$$\hat{H}_P |\Phi_P\rangle = E_P |\Phi_P\rangle \ . \tag{9.125}$$

$|\Phi_P\rangle$ is straightforward to calculate as $H_P$ only depends on the relative space coordinate $\mathbf{r}$ (see Eq. (9.124)). We can now expand the solution of Eq. (9.122) in a coupled-channel expansion of intrinsic and center-of-mass projectile states:

$$|\Psi\rangle = \sum_n \left[ |\Phi_P^{(n)}\rangle \, |\Psi^{(n)}\rangle \right]^J , \tag{9.126}$$

where $J$ is the total angular momentum of the system, $n$ represents a given target–projectile channel, $|\Phi_P^{(n)}\rangle$ is a Berggren basis eigenstate of $H_P$, and $|\Psi^{(n)}\rangle$ is a center-of-mass projectile state to be determined.

By using the coupled-channel representation of $\hat{V}_T(\mathbf{r}, \mathbf{R})$ and expanding $|\Psi^{(n)}\rangle$ in partial waves (see Eq. (9.126)), one can show that $|\Psi^{(n)}\rangle$ verifies the equations:

$$\Psi^{(n)}(\mathbf{R}) = \sum_L \chi_n^{(L)}(R) Y_L(\Omega_R)$$

$$\times \left( \frac{\hbar^2}{2M_{CM}} \left( -\frac{d^2}{dR^2} + \frac{L(L+1)}{R^2} \right) + V_{nn}^{(T)}(R) - E_n \right) \chi_n^{(L)}(R)$$

$$+ \sum_{n' \neq n} V_{nn'}^{(T)}(R) \chi_{n'}^{(L')}(R) = 0 , \tag{9.127}$$

where

$$V_{nn'}^{(T)}(R) = \left\langle \left[ \Phi_P^{(n')} Y_{L'} \right]_J |\hat{V}_T| \left[ \Phi_P^{(n)} Y_L \right]_J \right\rangle . \tag{9.128}$$

In this equation, $L$ is the orbital angular momentum of the system, $n$ represents the different quantum numbers associated to channels, and $E_n = E - E_P^{(n)}$, with $E_P^{(n)}$ the energy of state $|\Phi_P^{(n)}\rangle$ (see Eq. (9.125)).

Equation (9.128) has the same form as Eq. (9.149), so that it can be solved with the Green's function method described in Sect. 9.3.3.2. As this method replaces the direct integration of coupled-channel equations by a linear system, it should be efficient, insofar as the number of channels in Faddeev equations is usually very large.

It is possible to generalize Eq. (9.128) to the use of projectiles bearing more than one internal degree of freedom. For this, one replaces $V^{(\mathrm{T})}(\mathbf{r}, \mathbf{R})$ by $V^{(\mathrm{T})}(\mathbf{r}_1, \ldots, \mathbf{r}_a, \mathbf{R})$, where $\mathbf{r}_1, \ldots, \mathbf{r}_a$ are the relative radii associated to the $a + 1$ subsystems of the projectile. Consequently, Eq. (9.124) now depends on $\mathbf{r}_1, \ldots, \mathbf{r}_a$, so that it is now a multidimensional equation, to be solved with, e.g., Gamow shell model. However, the numerical cost of solving Eq. (9.124) and calculating the matrix elements of Eq. (9.128) increases quickly with $a$, so due to computational limitations one can expect this method to be practical only if $a = 2, 3$.

## 9.2.9   Elastic and Inelastic Scattering Cross Sections

The cross section expression for direct reaction is very close to that presented in Sect. 9.1.3, where $r$ must be substituted by $R_{\mathrm{CM}}$ and where the internal angular momentum of the projectile, equal to 1/2 for one-nucleon projectiles, is now equal to $J_{\mathrm{int}}$. Therefore, one will restate cross sections formulas in the case of many-nucleon projectiles for clarity, insisting only on the differences with the one-nucleon case.

The scattering amplitude in terms of the $T$-matrix elements (see Eq. (9.36)) writes

$$f_{\tilde{e} M_{\mathrm{P}}^{\tilde{e}} M_{\mathrm{T}}^{\tilde{e}} \to \tilde{c} M_{\mathrm{P}}^{\tilde{c}} M_{\mathrm{T}}^{\tilde{c}}}(\theta, \phi) = \delta_{\tilde{c}\tilde{e}}\, \delta_{M_{\mathrm{P}}^{\tilde{c}} M_{\mathrm{P}}^{\tilde{e}}} \delta_{M_{\mathrm{T}}^{\tilde{c}} M_{\mathrm{T}}^{\tilde{e}}}\, f_C(\theta)$$

$$+ \sum_{L_{\mathrm{CM}}^e J_{\mathrm{P}}^e L_{\mathrm{CM}}^c J_{\mathrm{P}}^c J_A} \frac{C^{L_{\mathrm{CM}}^e J_{\mathrm{P}}^e J_{\mathrm{T}}^{\tilde{e}} L_{\mathrm{CM}}^c J_{\mathrm{P}}^c J_{\mathrm{T}}^{\tilde{c}} J_A}_{M_{\mathrm{P}}^{\tilde{e}} M_{\mathrm{T}}^{\tilde{e}} M_{\mathrm{P}}^{\tilde{c}} M_{\mathrm{T}}^{\tilde{c}}}}{K_{\mathrm{CM}}^{\tilde{e}}}$$

$$\times T^{J_A}_{\tilde{e} L_{\mathrm{CM}}^e J_{\mathrm{P}}^e, \tilde{c} L_{\mathrm{CM}}^c J_{\mathrm{P}}^c} Y^{L_{\mathrm{CM}}^c}_{M_{\mathrm{P}}^{\tilde{e}} + M_{\mathrm{T}}^{\tilde{e}} - M_{\mathrm{P}}^{\tilde{c}} - M_{\mathrm{T}}^{\tilde{c}}}(\theta, \phi), \qquad (9.129)$$

where

$$C^{L_{\mathrm{CM}}^e J_{\mathrm{P}}^e J_{\mathrm{T}}^{\tilde{e}} L_{\mathrm{CM}}^c J_{\mathrm{P}}^c J_{\mathrm{T}}^{\tilde{c}} J_A}_{M_{\mathrm{P}}^{\tilde{e}} M_{\mathrm{T}}^{\tilde{e}} M_{\mathrm{P}}^{\tilde{c}} M_{\mathrm{T}}^{\tilde{c}}}$$

$$= \langle J_{\mathrm{P}}^e M_{\mathrm{P}}^{\tilde{e}} | L_{\mathrm{CM}}^e 0\ J_{\mathrm{int}} M_{\mathrm{P}}^{\tilde{e}} \rangle\ \langle J_{\mathrm{P}}^c\ M_{\mathrm{P}}^{\tilde{e}} + M_{\mathrm{T}}^{\tilde{e}} - M_{\mathrm{T}}^{\tilde{c}} | L_{\mathrm{CM}}^c\ M_{\mathrm{P}}^{\tilde{e}} + M_{\mathrm{T}}^{\tilde{e}} - M_{\mathrm{P}}^{\tilde{c}} - M_{\mathrm{T}}^{\tilde{c}}\ J_{\mathrm{int}} M_{\mathrm{P}}^{\tilde{c}} \rangle$$

$$\times \langle J_A\ M_{\mathrm{P}}^{\tilde{e}} + M_{\mathrm{T}}^{\tilde{e}} | J_{\mathrm{P}}^e M_{\mathrm{P}}^{\tilde{e}}\ J_{\mathrm{T}}^{\tilde{e}} M_{\mathrm{T}}^{\tilde{e}} \rangle\ \langle J_A\ M_{\mathrm{P}}^{\tilde{e}} + M_{\mathrm{T}}^{\tilde{e}} | J_{\mathrm{P}}^c M_{\mathrm{P}}^{\tilde{e}} + M_{\mathrm{T}}^{\tilde{e}} - M_{\mathrm{T}}^{\tilde{c}}\ J_{\mathrm{T}}^{\tilde{c}} M_{\mathrm{T}}^{\tilde{c}} \rangle$$

$$\times i^{(L_{\mathrm{CM}}^e - L_{\mathrm{CM}}^c)} \sqrt{4\pi (2L_{\mathrm{CM}}^c + 1)}\, e^{i(\sigma_{L_{\mathrm{CM}}^e} + \sigma_{L_{\mathrm{CM}}^c})}. \qquad (9.130)$$

Index $e$ in these expressions pertains to the entrance channel.

From the expression (9.39) of the physical scattering state, the differential cross section for the scattering process to the channel $(\tilde{c}, M_{\mathrm{P}}^{\tilde{c}}, M_{\mathrm{T}}^{\tilde{c}})$ at a given angle $\theta$ reads

$$\frac{d\sigma_{\tilde{e} M_{\mathrm{P}}^{\tilde{e}} M_{\mathrm{T}}^{\tilde{e}} \to \tilde{c} M_{\mathrm{P}}^{\tilde{c}} M_{\mathrm{T}}^{\tilde{c}}}}{d\Omega}(\theta) = \frac{K_{\mathrm{CM}}^{\tilde{c}}}{K_{\mathrm{CM}}^{\tilde{e}}} \left| f_{\tilde{e} M_{\mathrm{P}}^{\tilde{e}} M_{\mathrm{T}}^{\tilde{e}} \to \tilde{c} M_{\mathrm{P}}^{\tilde{c}} M_{\mathrm{T}}^{\tilde{c}}}(\theta, \phi) \right|^2. \qquad (9.131)$$

The differential cross sections of the scattering process to a given target state $\tilde{c}$ thus reads

$$\frac{d\sigma_{\tilde{e}\to\tilde{c}}}{d\Omega}(\theta) = \frac{1}{(2J_{\text{int}}+1)(2J_{\text{T}}^{\tilde{e}}+1)} \sum_{M_{\text{P}}^{\tilde{e}}M_{\text{T}}^{\tilde{e}}M_{\text{P}}^{\tilde{c}}M_{\text{T}}^{\tilde{c}}} \frac{K_{\text{CM}}^{\tilde{c}}}{K_{\text{CM}}^{\tilde{e}}} \left| f_{\tilde{e}M_{\text{P}}^{\tilde{e}}M_{\text{T}}^{\tilde{e}}\to\tilde{c}M_{\text{P}}^{\tilde{c}}M_{\text{T}}^{\tilde{c}}}(\theta,\phi) \right|^2 .$$

$$(9.132)$$

## 9.2.10 Radiative Capture Cross Sections

The consideration of many-nucleon projectiles in radiative capture processes induces additional problems that are absent when projectiles consist of only one nucleon. Indeed, the electromagnetic Hamiltonian is defined with the nucleon coordinates, whereas reaction channels make use of the many-nucleon projectile center-of-mass and relative coordinates in the asymptotic region. Consequently, transformations relating nucleon coordinates to the relative and center-of-mass coordinates of the projectile must be utilized to derive radiative capture cross sections. Moreover, Bessel functions enter electromagnetic matrix elements, with which the latter transformations are not as straightforward as with plane waves or harmonic oscillator states. Therefore, they will be detailed in the following sections.

### 9.2.10.1 Electromagnetic Hamiltonian with Many-Nucleon Clusters
The fundamental difference with respect to the one-nucleon case consists of the calculation of many-body electromagnetic transition operator matrix elements. Indeed, when the radiative capture of a cluster is considered, Eq. (9.27) is replaced by

$$\hat{O}_{ML}^{L} = \sum_{i\in\text{T}} \hat{O}_{ML}^{L}(i) + \sum_{i\in\text{P}} \hat{O}_{ML}^{L}(i) .$$

$$(9.133)$$

---

**Exercise V ★ ★ ★**
The Bessel functions entering the electric and magnetic operators of Eqs. (9.53) and (9.54) are defined with nucleon coordinates. One will then show how to expand these functions in relative and center-of-mass coordinates. These expansions will enter the many-body matrix elements in radiative capture cross section formulas.

**A.** Using the standard expression of the Bessel expansion of plane waves, show that

$$e^{i\mathbf{q}\cdot\mathbf{r}} = 4\pi \sum_{LM_L} i^L Y_{LM_L}^*(\Omega_q) Y_{LM_L}(\Omega) j_L(qr) ,$$

where $\Omega$ and $\Omega_q$ represent the angular coordinates related to $\mathbf{r}$ and $\mathbf{q}$, respectively. Deduce the following integral expression obeyed by Bessel functions and spherical harmonics:

$$\mathscr{J}_{LM_L}(q,\mathbf{r}) \equiv j_L(qr)\,Y_{LM_L}(\Omega_r) = (4\pi\,i^L)^{-1}\int e^{i\mathbf{q}\cdot\mathbf{r}}\,Y_{LM_L}(\Omega_q)\,d\Omega_q\,,$$

where one has introduced the concise notation $\mathscr{J}_{LM_L}(q,\mathbf{r})$ for convenience.

**B.** Using the obtained expression with relative and center-of-mass coordinates, demonstrate that

$$\mathscr{J}_{LM_L}(q,\mathbf{r}) = 4\pi\,i^{-L}\sum_{\substack{\ell_{\text{int}}\,m_{\ell_{\text{int}}}\\L_{\text{CM}}\,M_{L_{\text{CM}}}}} i^{\ell_{\text{int}}+L_{\text{CM}}}\,\mathscr{J}_{\ell_{\text{int}}m_{\ell_{\text{int}}}}(q,\mathbf{r}_{\text{int}})\,\mathscr{J}_{L_{\text{CM}}M_{L_{\text{CM}}}}(q,\mathbf{R}_{\text{CM}})$$

$$\times \int Y_{LM_L}(\Omega_q)\,Y^*_{\ell_{\text{int}}m_{\ell_{\text{int}}}}(\Omega_q)\,Y^*_{L_{\text{CM}}M_{L_{\text{CM}}}}(\Omega_q)\,d\Omega_q\,.$$

**C.** Using the Gaunt coefficient, i.e. the analytical value of the integral involving three spherical harmonics [63], deduce that

$$\mathscr{J}_{LM_L}(q,\mathbf{r}) = \sum_{\ell_{\text{int}},L_{\text{CM}}} A^{L_{\text{CM}}}_{\ell_{\text{int}}}\Big[\,\mathscr{J}_{\ell_{\text{int}}}(q,\mathbf{r}_{\text{int}})\otimes\mathscr{J}_{L_{\text{CM}}}(q,\mathbf{R}_{\text{CM}})\Big]^L_{M_L}\,,$$

where the angular factor $A^{L_{\text{CM}}}_{\ell_{\text{int}}}$ reads

$$A^{L_{\text{CM}}}_{\ell_{\text{int}}} = \sqrt{4\pi}\,(-1)^{(L_{\text{CM}}+\ell_{\text{int}}+L)/2}\,\hat{L}_{\text{CM}}\hat{\ell}_{\text{int}}\begin{pmatrix}\ell_{\text{int}} & L_{\text{CM}} & L\\0 & 0 & 0\end{pmatrix}\,.$$

---

**Exercise VI ★ ★ ★**

In this exercise, one will focus on the angular momentum independent parts of many-body electromagnetic transition operator matrix elements in relative and center-of-mass coordinates. They enter radiative capture cross sections with many-nucleon clusters (see Exercise V).

Show that the following equations hold and explain their role in the electromagnetic Hamiltonian:

$$\nabla(\mathscr{J}_{LM_L}(q,\mathbf{r})) = \sum_{\ell_{\text{int}},L_{\text{CM}}} A^{L_{\text{CM}}}_{\ell_{\text{int}}}\Big[\,\nabla_{\text{int}}(\mathscr{J}_{\ell_{\text{int}}}(q,\mathbf{r}_{\text{int}}))\otimes\mathscr{J}_{L_{\text{CM}}}(q,\mathbf{R}_{\text{CM}})\Big]^L_{M_L}$$

$$\hat{\mathscr{J}}'_{LM_L}(q,\mathbf{r}) = \sum_{\ell_{\text{int}},L_{\text{CM}}} A^{L_{\text{CM}}}_{\ell_{\text{int}}} \left[ \hat{\mathscr{J}}'_{\ell_{\text{int}}}(q,\mathbf{r}_{\text{int}}) \otimes \mathscr{J}_{L_{\text{CM}}}(q,\mathbf{R}_{\text{CM}}) \right]^{L}_{M_L}$$

$$+ \sum_{\ell_{\text{int}},L_{\text{CM}}} A^{L_{\text{CM}}}_{\ell_{\text{int}}} \left[ \mathscr{J}_{\ell_{\text{int}}}(q,\mathbf{r}_{\text{int}}) \otimes \hat{\mathscr{J}}'_{L_{\text{CM}}}(q,\mathbf{R}_{\text{CM}}) \right]^{L}_{M_L}$$

$$- \sum_{\ell_{\text{int}},L_{\text{CM}}} A^{L_{\text{CM}}}_{\ell_{\text{int}}} \left[ \mathscr{J}_{\ell_{\text{int}}}(q,\mathbf{r}_{\text{int}}) \otimes \mathscr{J}_{L_{\text{CM}}}(q,\mathbf{R}_{\text{CM}}) \right]^{L}_{M_L}$$

$$\nabla(\hat{\mathscr{J}}'_{LM_L}(q,\mathbf{r})) = \sum_{\ell_{\text{int}},L_{\text{CM}}} A^{L_{\text{CM}}}_{\ell_{\text{int}}} \left[ \nabla_{\text{int}}(\hat{\mathscr{J}}'_{\ell_{\text{int}}}(q,\mathbf{r}_{\text{int}})) \otimes \mathscr{J}_{L_{\text{CM}}}(q,\mathbf{R}_{\text{CM}}) \right]^{L}_{M_L}$$

$$+ \sum_{\ell_{\text{int}},L_{\text{CM}}} A^{L_{\text{CM}}}_{\ell_{\text{int}}} \left[ \nabla_{\text{int}}(\mathscr{J}_{\ell_{\text{int}}}(q,\mathbf{r}_{\text{int}})) \otimes \hat{\mathscr{J}}'_{L_{\text{CM}}}(q,\mathbf{R}_{\text{CM}}) \right]^{L}_{M_L}$$

$$- \sum_{\ell_{\text{int}},L_{\text{CM}}} A^{L_{\text{CM}}}_{\ell_{\text{int}}} \left[ \nabla_{\text{int}}(\mathscr{J}_{\ell_{\text{int}}}(q,\mathbf{r}_{\text{int}})) \otimes \mathscr{J}_{L_{\text{CM}}}(q,\mathbf{R}_{\text{CM}}) \right]^{L}_{M_L}$$

$$\hat{\mathscr{J}}'_{LM_L}(q,\mathbf{r}) + \left(\frac{qr}{2}\right) \hat{\mathscr{J}}_{LM_L}(q,\mathbf{r})$$

$$= \sum_{\ell_{\text{int}},L_{\text{CM}}} A^{L_{\text{CM}}}_{\ell_{\text{int}}} \left[ \left(\hat{\mathscr{J}}'_{\ell_{\text{int}}}(q,\mathbf{r}_{\text{int}}) + \left(\frac{qr_{\text{int}}}{2}\right) \hat{\mathscr{J}}_{\ell_{\text{int}}}(q,\mathbf{r}_{\text{int}})\right) \otimes \mathscr{J}_{L_{\text{CM}}}(q,\mathbf{R}_{\text{CM}}) \right]^{L}_{M_L}$$

$$+ \sum_{\ell_{\text{int}},L_{\text{CM}}} A^{L_{\text{CM}}}_{\ell_{\text{int}}} \left[ \mathscr{J}_{\ell_{\text{int}}}(q,\mathbf{r}_{\text{int}}) \otimes \hat{\mathscr{J}}'_{L_{\text{CM}}}(q,\mathbf{R}_{\text{CM}}) \right]^{L}_{M_L}$$

$$- \sum_{\ell_{\text{int}},L_{\text{CM}}} A^{L_{\text{CM}}}_{\ell_{\text{int}}} \left[ \mathscr{J}_{\ell_{\text{int}}}(q,\mathbf{r}_{\text{int}}) \otimes \mathscr{J}_{L_{\text{CM}}}(q,\mathbf{R}_{\text{CM}}) \right]^{L}_{M_L}$$

$$+ \frac{1}{2} \sum_{\ell_{\text{int}},L_{\text{CM}}} A^{L_{\text{CM}}}_{\ell_{\text{int}}} \left[ \left(\frac{\hat{\mathscr{J}}_{\ell_{\text{int}}}(q,\mathbf{r}_{\text{int}})}{r_{\text{int}}}\right) \otimes R_{\text{CM}} \hat{\mathscr{J}}_{L_{\text{CM}}}(q,\mathbf{R}_{\text{CM}}) \right]^{L}_{M_L}$$

$$+ \sum_{\ell_{\text{int}},L_{\text{CM}}} A^{L_{\text{CM}}}_{\ell_{\text{int}}} \left[ \hat{\mathscr{J}}_{\ell_{\text{int}}}(q,\mathbf{r}_{\text{int}}) \otimes \hat{\mathscr{J}}_{L_{\text{CM}}}(q,\mathbf{R}_{\text{CM}}) \right]^{L}_{M_L} (\mathbf{e}_{\text{int}} \cdot \mathbf{E}_{\text{CM}}) ,$$

where

$$\hat{\mathscr{J}}_{LM_L}(q,\mathbf{r}) = \hat{j}_L(qr) \, Y_{LM_L}(\Omega_r)$$

$$\hat{\mathscr{J}}'_{LM_L}(q,\mathbf{r}) = \hat{j}'_L(qr) \, Y_{LM_L}(\Omega_r)$$

$$\mathbf{e}_{\text{int}} = \mathbf{r}_{\text{int}}/r_{\text{int}}$$

$$\mathbf{E}_{\text{CM}} = \mathbf{R}_{\text{CM}}/R_{\text{CM}} .$$

**Exercise VII ⋆ ⋆ ⋆**
One will consider hereby the angular momentum dependent parts of the electric and magnetic operators for the calculation of radiative capture cross sections of clusters (see Exercise VI).

**A.** Let us firstly consider the parts of the electromagnetic Hamiltonian that depend on spin **s** and **r**, but not on orbital angular momentum **l**. Explain why the use of **s** in Eqs. (9.53) and (9.55) poses no problem, contrary to **l**. Using the latter result, derive the spin-dependent operators of Eq. (9.55) in relative and center-of-mass coordinates:

$$
\nabla \left( \hat{\mathscr{I}}'_{LM_L}(q, \mathbf{r}) \right) \cdot \mathbf{s}
$$

$$
= \sum_{\ell_{\mathrm{int}}, L_{\mathrm{CM}}} A^{L_{\mathrm{CM}}}_{\ell_{\mathrm{int}}} \left[ \left( \nabla_{\mathrm{int}} \left( \hat{\mathscr{I}}'_{\ell_{\mathrm{int}}}(q, \mathbf{r}_{\mathrm{int}}) \right) \cdot \mathbf{s} \right) \otimes \mathscr{I}_{L_{\mathrm{CM}}}(q, \mathbf{R}_{\mathrm{CM}}) \right]^{L}_{M_L}
$$

$$
+ \sum_{\ell_{\mathrm{int}}, L_{\mathrm{CM}}} A^{L_{\mathrm{CM}}}_{\ell_{\mathrm{int}}} \left[ \left( \nabla_{\mathrm{int}} \left( \mathscr{I}_{\ell_{\mathrm{int}}}(q, \mathbf{r}_{\mathrm{int}}) \right) \cdot \mathbf{s} \right) \otimes \hat{\mathscr{I}}'_{L_{\mathrm{CM}}}(q, \mathbf{R}_{\mathrm{CM}}) \right]^{L}_{M_L}
$$

$$
- \sum_{\ell_{\mathrm{int}}, L_{\mathrm{CM}}} A^{L_{\mathrm{CM}}}_{\ell_{\mathrm{int}}} \left[ \left( \nabla_{\mathrm{int}} \left( \mathscr{I}_{\ell_{\mathrm{int}}}(q, \mathbf{r}_{\mathrm{int}}) \right) \cdot \mathbf{s} \right) \otimes \mathscr{I}_{L_{\mathrm{CM}}}(q, \mathbf{R}_{\mathrm{CM}}) \right]^{L}_{M_L}
$$

$$
q \, \hat{\mathscr{I}}_{LM_L}(q, \mathbf{r}) \, (\mathbf{s} \cdot \mathbf{e})
$$

$$
= q \sum_{\ell_{\mathrm{int}}, L_{\mathrm{CM}}} A^{L_{\mathrm{CM}}}_{\ell_{\mathrm{int}}} \left[ \hat{\mathscr{I}}_{\ell_{\mathrm{int}}}(q, \mathbf{r}_{\mathrm{int}}) \, (\mathbf{s} \cdot \mathbf{e}_{\mathrm{int}}) \otimes \mathscr{I}_{L_{\mathrm{CM}}}(q, \mathbf{R}_{\mathrm{CM}}) \right]^{L}_{M_L}
$$

$$
+ q \sum_{\ell_{\mathrm{int}}, L_{\mathrm{CM}}} A^{L_{\mathrm{CM}}}_{\ell_{\mathrm{int}}} \left[ \mathscr{I}_{\ell_{\mathrm{int}}}(q, \mathbf{r}_{\mathrm{int}}) \, (\mathbf{s} \cdot \mathbf{E}_{\mathrm{CM}}) \otimes \hat{\mathscr{I}}_{L_{\mathrm{CM}}}(q, \mathbf{R}_{\mathrm{CM}}) \right]^{L}_{M_L} ,
$$

where $\mathbf{e} = \mathbf{r}/r$.

**B.** One now concentrates on the **l**-dependent parts of the Hamiltonian of Eq. (9.55).

Using Eqs. (9.135) and (9.136), show that the orbital angular momentum writes

$$
\mathbf{l} = \mathbf{l}_{\mathrm{int}} + \left( \frac{m}{M_{\mathrm{P}}} \right) \mathbf{L}_{\mathrm{CM}} + \left( \frac{m}{M_{\mathrm{P}}} \right) (\mathbf{r}_{\mathrm{int}} \times \mathbf{P}_{\mathrm{CM}}) + \mathbf{R}_{\mathrm{CM}} \times \mathbf{p}_{\mathrm{int}} .
$$

Explain why this formulation is convenient to calculate **l**-dependent matrix elements.

**C.** Express the operators

$$
\hat{\mathscr{I}}'_{LM_L}(q, \mathbf{r})(\mathbf{l} \cdot \mathbf{s})
$$

and

$$\nabla \left( \frac{\hat{\mathscr{J}}_{LM_L}(q, \mathbf{r})}{qr} \right) \cdot \mathbf{1}$$

as a function of $\mathbf{r}_{\mathrm{int}}$, $\mathbf{p}_{\mathrm{int}}$, $\mathbf{l}_{\mathrm{int}}$, $\mathbf{s}$ and $\mathbf{R}_{\mathrm{CM}}$, $\mathbf{P}_{\mathrm{CM}}$, and $\mathbf{L}_{\mathrm{CM}}$ similarly to the spin-dependent operators of $\mathbf{A}$ from the previous results. Explain the physical content of these operators.

---

**Exercise VIII ★ ★ ★**

In this exercise, one will show how to formulate electromagnetic operators involving clusters for the calculation of radiative capture cross sections (see Exercises VI–VII).

**A.** Show that the expressions involving $\nabla_{\mathrm{int}}$ in Eqs. (9.54) and (9.55) read

$$\nabla_{\mathrm{int}} \left( \hat{\mathscr{J}}_{\ell_{\mathrm{int}}}(q, \mathbf{r}_{\mathrm{int}}) \right) \cdot \mathbf{l}'$$

$$= -\sqrt{\frac{\ell_{\mathrm{int}}+1}{2\ell_{\mathrm{int}}+1}} \left( \hat{j}'_{\ell_{\mathrm{int}}}(qr_{\mathrm{int}}) - (\ell_{\mathrm{int}}+1) \left( \frac{\hat{j}_{\ell_{\mathrm{int}}}(qr_{\mathrm{int}})}{qr_{\mathrm{int}}} \right) \right) \left( \frac{1}{r_{\mathrm{int}}} \right)$$

$$\times [Y_{\ell_{\mathrm{int}}+1}(\Omega_{r_{\mathrm{int}}}) \otimes \mathbf{l}']^{\ell_{\mathrm{int}}}$$

$$+ \sqrt{\frac{\ell_{\mathrm{int}}}{2\ell_{\mathrm{int}}+1}} \left( \hat{j}'_{\ell_{\mathrm{int}}}(qr_{\mathrm{int}}) + \ell_{\mathrm{int}} \left( \frac{\hat{j}_{\ell_{\mathrm{int}}}(qr_{\mathrm{int}})}{qr_{\mathrm{int}}} \right) \right) \left( \frac{1}{r_{\mathrm{int}}} \right)$$

$$\times [Y_{\ell_{\mathrm{int}}-1}(\Omega_{r_{\mathrm{int}}}) \otimes \mathbf{l}']^{\ell_{\mathrm{int}}}$$

$$\nabla_{\mathrm{int}} \left( \hat{\mathscr{J}}'_{\ell_{\mathrm{int}}}(q, \mathbf{r}_{\mathrm{int}}) \right) \cdot \mathbf{s}$$

$$= -\sqrt{\frac{\ell_{\mathrm{int}}+1}{2\ell_{\mathrm{int}}+1}} \left[ \left( \frac{\ell_{\mathrm{int}}(\ell_{\mathrm{int}}+1)}{qr_{\mathrm{int}}} - qr_{\mathrm{int}} \right) \hat{j}_{\ell_{\mathrm{int}}}(qr_{\mathrm{int}}) - \ell_{\mathrm{int}} \hat{j}'_{\ell_{\mathrm{int}}}(qr_{\mathrm{int}}) \right]$$

$$\times \left( \frac{1}{r_{\mathrm{int}}} \right) [Y_{\ell_{\mathrm{int}}+1}(\Omega_{r_{\mathrm{int}}}) \otimes \mathbf{s}]^{\ell_{\mathrm{int}}}$$

$$+ \sqrt{\frac{\ell_{\mathrm{int}}}{2\ell_{\mathrm{int}}+1}} \left[ \left( \frac{\ell_{\mathrm{int}}(\ell_{\mathrm{int}}+1)}{qr_{\mathrm{int}}} - qr_{\mathrm{int}} \right) \hat{j}_{\ell_{\mathrm{int}}}(qr_{\mathrm{int}}) + (\ell_{\mathrm{int}}+1) \, \hat{j}'_{\ell_{\mathrm{int}}}(qr_{\mathrm{int}}) \right]$$

$$\times \left( \frac{1}{r_{\mathrm{int}}} \right) [Y_{\ell_{\mathrm{int}}-1}(\Omega_{r_{\mathrm{int}}}) \otimes \mathbf{s}]^{\ell_{\mathrm{int}}} ,$$

where $\mathbf{l}'$ is $\mathbf{L}_{\mathrm{CM}}$, $\mathbf{l}_{\mathrm{int}}$, $\mathbf{r}_{\mathrm{int}} \times \mathbf{P}_{\mathrm{CM}}$, or $\mathbf{R}_{\mathrm{CM}} \times \mathbf{p}_{\mathrm{int}}$.

**B.** One now assumes long wavelength approximation. Show that the electric operator reads

$$r^L Y_{LM_L}(\Omega_r) = \sum_{\ell_{\text{int}}+L_{\text{CM}}=L} B_{\ell_{\text{int}}}^{L_{\text{CM}}} \left[ r_{\text{int}}^{\ell_{\text{int}}} Y_{\ell_{\text{int}}}(\Omega_{\text{int}}) \otimes R_{\text{CM}}^{L_{\text{CM}}} Y_{L_{\text{CM}}}(\Omega_{\text{CM}}) \right]_{M_L}^L ,$$

and that the magnetic operators read

$$\nabla \left( r^L Y_{LM_L}(\Omega_r) \right) \cdot \mathbf{l}$$

$$= \sum_{\ell_{\text{int}}+L_{\text{CM}}=L} B_{\ell_{\text{int}}}^{L_{\text{CM}}} \left[ \left( \nabla_{\text{int}} \left( r_{\text{int}}^{\ell_{\text{int}}} Y_{\ell_{\text{int}}}(\Omega_{\text{int}}) \right) \cdot \mathbf{l}_{\text{int}} \right) \otimes R_{\text{CM}}^{L_{\text{CM}}} Y_{L_{\text{CM}}}(\Omega_{\text{CM}}) \right]_{M_L}^L$$

$$+ \left( \frac{m}{M_P} \right) \sum_{\ell_{\text{int}}+L_{\text{CM}}=L} B_{\ell_{\text{int}}}^{L_{\text{CM}}} \left[ \left( \nabla_{\text{int}} \left( r_{\text{int}}^{\ell_{\text{int}}} Y_{\ell_{\text{int}}}(\Omega_{\text{int}}) \right) \cdot \mathbf{L}_{\text{CM}} \right) \right.$$

$$\left. \otimes R_{\text{CM}}^{L_{\text{CM}}} Y_{L_{\text{CM}}}(\Omega_{\text{CM}}) \right]_{M_L}^L$$

$$+ \left( \frac{m}{M_P} \right) \sum_{\ell_{\text{int}}+L_{\text{CM}}=L} B_{\ell_{\text{int}}}^{L_{\text{CM}}} \left[ \left( \nabla_{\text{int}} \left( r_{\text{int}}^{\ell_{\text{int}}} Y_{\ell_{\text{int}}}(\Omega_{\text{int}}) \right) \cdot (\mathbf{r}_{\text{int}} \times \mathbf{P}_{\text{CM}}) \right) \right.$$

$$\left. \otimes R_{\text{CM}}^{L_{\text{CM}}} Y_{L_{\text{CM}}}(\Omega_{\text{CM}}) \right]_{M_L}^L$$

$$+ \sum_{\ell_{\text{int}}+L_{\text{CM}}=L} B_{\ell_{\text{int}}}^{L_{\text{CM}}} \left[ \left( \nabla_{\text{int}} \left( r_{\text{int}}^{\ell_{\text{int}}} Y_{\ell_{\text{int}}}(\Omega_{\text{int}}) \right) \cdot (\mathbf{R}_{\text{CM}} \times \mathbf{p}_{\text{int}}) \right) \right.$$

$$\left. \otimes R_{\text{CM}}^{L_{\text{CM}}} Y_{L_{\text{CM}}}(\Omega_{\text{CM}}) \right]_{M_L}^L$$

$$\nabla \left( r^L Y_L^{M_L}(\Omega_r) \right) \cdot \mathbf{s}$$

$$= \sum_{\ell_{\text{int}}+L_{\text{CM}}=L} B_{\ell_{\text{int}}}^{L_{\text{CM}}} \left[ \left( \nabla_{\text{int}} \left( r_{\text{int}}^{\ell_{\text{int}}} Y_{\ell_{\text{int}}}(\Omega_{\text{int}}) \right) \cdot \mathbf{s} \right) \otimes R_{\text{CM}}^{L_{\text{CM}}} Y_{L_{\text{CM}}}(\Omega_{\text{CM}}) \right]_{M_L}^L ,$$

where $(2\ell_{\text{int}} + 1)!! \, (2L_{\text{CM}} + 1)!! \, B_{\ell_{\text{int}}}^{L_{\text{CM}}} = (2L + 1)!! \, A_{\ell_{\text{int}}}^{L_{\text{CM}}}$.

Show that the equations involving $\nabla_{\text{int}}$ in **A** with long wavelength approximation now read

$$\nabla_{\text{int}} \left( r_{\text{int}}^{\ell_{\text{int}}} Y_{\ell_{\text{int}}}(\Omega_{\text{int}}) \right) \cdot \mathbf{v} = \sqrt{\ell_{\text{int}}(2\ell_{\text{int}} + 1)} r_{\text{int}}^{\ell_{\text{int}}-1} [Y_{\ell_{\text{int}}-1}(\Omega_{r_{\text{int}}}) \otimes \mathbf{v}]^{\ell_{\text{int}}},$$

$$(9.134)$$

where **v** is **s**, $\mathbf{L}_{\text{CM}}$, $\mathbf{l}_{\text{int}}$, $\mathbf{r}_{\text{int}} \times \mathbf{P}_{\text{CM}}$, or $\mathbf{R}_{\text{CM}} \times \mathbf{p}_{\text{int}}$.

The first operator on the right-hand side of Eq. (9.133) poses no problem as it involves target states only, similarly to the one-nucleon projectile. However, one cannot directly calculate the matrix elements of its second operator with a basis of cluster channels, as it is written in terms of nucleon coordinates, whereas cluster channels are expressed in terms of the $\mathbf{R}_{\mathrm{CM}}$, $\mathbf{P}_{\mathrm{CM}}$ center-of-mass coordinates of the cluster, and $\mathbf{r}_{i,\mathrm{int}}$, $\mathbf{p}_{i,\mathrm{int}}$ relative coordinates of the nucleon $i$ in the cluster with respect to the cluster center-of-mass

$$\mathbf{R}_{\mathrm{CM}} = \frac{1}{M_{\mathrm{P}}} \sum_{i \in \mathrm{P}} m_i \, \mathbf{r}_i \quad \text{and} \quad \mathbf{r}_{i,\mathrm{int}} = \mathbf{r}_i - \mathbf{R}_{\mathrm{CM}} \tag{9.135}$$

$$\mathbf{P}_{\mathrm{CM}} = \sum_{i \in \mathrm{P}} \mathbf{p}_i \quad \text{and} \quad \mathbf{p}_{i,\mathrm{int}} = \mathbf{p}_i - \left(\frac{m_i}{M_{\mathrm{P}}}\right) \mathbf{P}_{\mathrm{CM}} . \tag{9.136}$$

Consequently, the $\hat{O}^L_{M_L}(i)$ operator must be written as a function of the $\mathbf{R}_{\mathrm{CM}}$, $\mathbf{P}_{\mathrm{CM}}$ and $\mathbf{r}_{i,\mathrm{int}}$, $\mathbf{p}_{i,\mathrm{int}}$ coordinates. The determination of the electromagnetic Hamiltonian in relative and center-of-mass coordinates and the practical calculation of its matrix elements are the subject of the next exercises. Exercises V–IX are technical, so that readers can admit their results in a first reading.

---

**Exercise IX** ⋆⋆

In Exercises VI–VIII, one has concentrated on the derivation of electromagnetic operators involving clusters. In this exercise, one will show how to calculate electromagnetic matrix elements from these operators.

**A.** Explain why a direct calculation of the relative matrix elements is impossible in practice when using a Slater determinant expansion of cluster wave functions.
**B.** Show that the use of harmonic oscillator shell model is sufficient for considered projectiles. Find a method to retrieve relative matrix elements with that approach using $N\hbar\omega$ spaces.
**C.** Explain how the matrix elements of the electromagnetic Hamiltonian can then be calculated.

---

### 9.2.10.2 Radiative Capture Cross Section
The radiative capture differential cross section writes similarly to Eq. (9.52):

$$\frac{d\sigma}{d\Omega_\gamma} = \left(\frac{q}{8\pi k}\right)\left(\frac{e^2}{\hbar c}\right)\left(\frac{\mu_u c^2}{\hbar c}\right)\left(\frac{1}{2J_{\mathrm{int}}+1}\right)\left(\frac{1}{2J_{\mathrm{T}}+1}\right)$$
$$\times \sum_{J_i,c_e,L} \langle J_f || \mathcal{M}_L || (J_i)_{c_e} \rangle \sum_{J_i',c_e',L'} \langle J_f || \mathcal{M}_{L'} || (J_i')_{c_e'} \rangle$$

$$\times \sum_{M_L, P_\gamma} [g^L_{M_L, P_\gamma}(k, q, \varphi_\gamma, \theta_\gamma)]^* g^{L'}_{M_L, P_\gamma}(k, q, \varphi_\gamma, \theta_\gamma)$$

$$\times \sum_{M_i, M_f} \begin{pmatrix} J_f & L & J_i \\ -M_f & M_L & M_i \end{pmatrix} \begin{pmatrix} J_f & L' & J'_i \\ -M_f & M_L & M_i \end{pmatrix}$$

$$\times \sum_{M_{int}} \langle \ell_{c_e} 0 \, J_{int} M_{int} | j_{c_e} M_{int} \rangle \langle \ell_{c'_e} 0 \, J_{int} M_{int} | j_{c'_e} M_{int} \rangle$$

$$\times \sum_{M_T} \langle j_{c_e} M_{int} J_T M_T | J_i M_i \rangle \langle j_{c'_e} M_{int} J_T M_T | J'_i M_i \rangle \quad , \quad (9.137)$$

where the notations introduced in Eqs. (9.48–9.55) (see Sect. 9.1.3.3) have been used and $g^L_{M_L, P_\gamma}(k, q, \varphi_\gamma, \theta_\gamma)$ is defined in Eq. (9.49).

One can see that the intrinsic spin of the nucleon has been replaced by the total spin $J_{int}$ of the projectile. This arises from cluster approximation and was already seen in Eq. (9.129). The use of clusters at this level does not lead to complications in the calculation of radiative capture cross section, contrary to the application of the electromagnetic Hamiltonian on clusters. Therefore, the only problem occurring in the calculation of cross sections of cluster radiative capture is related to Gamow shell model dimensions. Indeed, they are approximately the product of the Gamow shell model dimensions related to target and projectile space, so that dealing with Gamow shell model vectors involving cluster projectiles is more expensive numerically than in the one-nucleon case.

## 9.3    Computational Methods for Solving the Coupled-Channel Gamow Shell Model

In this section, one will present different numerical techniques used to calculate the $A$-body scattering states $|\Psi^{J_A}_{M_A}\rangle$. For this, one has to evaluate the radial amplitudes $u_c(r)$ and $u_c(R_{CM})$ for each channel (see Eqs. (9.19), (9.112)). As the method to calculate these radial wave functions is the same for one- or multi-nucleon projectiles, one uses the notation $r$ for the radial coordinate associated to both one- and multi-nucleon projectiles. Consequently, one has $r = R_{CM}$ when dealing with cluster projectiles (see Sect. 9.2). One shall now present how to calculate the radial wave functions $u_c(r)$ numerically.

### 9.3.1    Orthogonalization of Target–Projectile Channels

The coupled-channel formalism leads to a generalized eigenvalue problem because different channel basis states are non-orthogonal. The non-orthogonality of channel states comes from the antisymmetry between the projectile and target states. To formulate coupled-channel Gamow shell model equations as the generalized eigenvalue problem, one will express Eq. (9.20) in the orthogonal channel basis

$\{|(c, r)\rangle_o\}$, as presented in Ref.[64]) (see also Ref.[65]):

$$_o\langle(c', r')|(c, r)\rangle_o = \frac{\delta(r - r')}{rr'}\delta_{cc'} .$$  (9.138)

The transformation from the non-orthogonal channel basis $\{|(c, r)\rangle\}$ to the orthogonal one $\{|(c, r)\rangle_o\}$ is given by the overlap operator $\hat{O}$ such that: $|(c, r)\rangle = \hat{O}^{\frac{1}{2}}|(c, r)\rangle_o$. The coupled-channel equations (9.20) written in the orthogonal basis are

$$\sum_{c'} \int_0^{+\infty} \left( _o\langle(c', r')|\hat{H}_o|(c, r)\rangle_o - E \,_o\langle(c', r')|\hat{O}|(c, r)\rangle_o \right) _o\langle(c', r')|\Psi\rangle_o r'^2 dr' = 0 ,$$  (9.139)

where

$$_o\langle(c', r')|\hat{H}_o|(c, r)\rangle_o = \langle(c', r')|\hat{H}|(c, r)\rangle$$  (9.140)

$$_o\langle(c', r')|\hat{O}|(c, r)\rangle_o = \langle(c', r')|(c, r)\rangle$$  (9.141)

$$_o\langle(c, r)|\Psi\rangle_o = \frac{u_c(r)}{r} .$$  (9.142)

Let us introduce the wave function $|\Phi\rangle_o = \hat{O}^{\frac{1}{2}}|\Psi\rangle_o$ in order to transform Eq. (9.139) to a standard eigenvalue problem. One obtains

$$\sum_{c'} \int_0^{+\infty} \left( _o\langle(c', r')|\hat{H}|(c, r)\rangle_o - E \,_o\langle(c', r')|(c, r)\rangle_o \right) _o\langle(c', r')|\Phi\rangle_o r'^2 dr' = 0 .$$  (9.143)

In the non-orthogonal channel basis, these coupled-channel equations become

$$\sum_{c'} \int_0^{+\infty} rr' \langle(c', r')|\hat{H}_{\text{mod}}|(c, r)\rangle w_{c'}(r')dr' = E w_c(r) ,$$  (9.144)

with

$$_o\langle(c', r')|\hat{H}|(c, r)\rangle_o \equiv \langle(c', r')|\hat{H}_{\text{mod}}|(c, r)\rangle ,$$  (9.145)

$$_o\langle(c, r)|\Phi\rangle_o \equiv \frac{w_c(r)}{r} ,$$  (9.146)

and where $\hat{H}_{\text{mod}} = \hat{O}^{-\frac{1}{2}}\hat{H}\hat{O}^{-\frac{1}{2}}$ is the modified Hamiltonian.

In order to have a more precise treatment of the antisymmetry in the calculation of matrix elements of $\hat{H}_{\text{mod}}$, one introduces a new operator $\hat{A}$:

$$\hat{O}^{-\frac{1}{2}} = \hat{A} + \hat{1} , \tag{9.147}$$

which is associated with the part of $\hat{O}^{-\frac{1}{2}}$ acting on the low-energy channel states. Then, instead of calculating the matrix elements of $\hat{H}_{\text{mod}}$ directly, one separates $\hat{H}_{\text{mod}}$ into $\hat{H}$ plus short-range parts involving $\hat{A}$:

$$\hat{H}_{\text{mod}} = (\hat{A} + \hat{1})\hat{H}(\hat{A} + \hat{1}) = \hat{H} + \hat{H}\hat{A} + \hat{A}\hat{H} + \hat{A}\hat{H}\hat{A} . \tag{9.148}$$

In this formulation, the non-antisymmetrized terms are taken into account exactly with the identity operator. Inserting Eq. (9.148) in coupled-channel equations (9.144), one obtains the coupled-channel equations for the reduced radial wave functions $w_c(r)/r$:

$$\left( \frac{\hbar^2}{2\mu} \left( -\frac{d^2}{dr^2} + \frac{\ell_c(\ell_c + 1)}{r^2} \right) + V_c^{(\text{loc})}(r) \right) w_c(r)$$

$$+ \sum_{c'} \int_0^{+\infty} V_{cc'}^{(\text{non-loc})}(r, r')\, w_{c'}(r')\, dr'$$

$$= (E - E_{\text{T}_c} - E_{\text{int}_c}) w_c(r) , \tag{9.149}$$

where $E_{\text{int}_c}$ is the intrinsic energy of the projectile ($E_{\text{int}_c}$ is zero in the nucleon case) and the nonlocal potential $V_{cc'}^{(\text{non-loc})}(r, r')$ reads

$$\frac{V_{cc'}^{(\text{non-loc})}(r, r')}{rr'} = \frac{\tilde{V}_{cc'}(r, r')}{rr'} + \langle (c', r')|\hat{H}\hat{A}|(c, r)\rangle$$

$$+ \langle (c', r')|\hat{A}\hat{H}|(c, r)\rangle + \langle (c', r')|\hat{A}\hat{H}\hat{A}|(c, r)\rangle. \tag{9.150}$$

The radial channel wave functions $u_c(r)$ can be obtained from the solutions of Eq. (9.149) using the fact that $|\Psi\rangle_o = \hat{O}^{-\frac{1}{2}}|\Phi\rangle_o$. One obtains

$$u_c(r) = w_c(r) + \sum_{c'} \int_0^{+\infty} rr'\,{}_o\langle (c', r')|A|(c, r)\rangle_o w_{c'}(r')\, dr' . \tag{9.151}$$

## 9.3.2  Solution of Coupled Equations in Coordinate Space

### 9.3.2.1 Coupled-Channel Equations Using Local Coupling Potentials

One will now present the numerical method used to calculate the wave functions solutions of coupled-channel equations by direct integration in the coordinate space.

The radial wave functions solutions of coupled-channel equations verify $w_c(r = 0) = 0$ for all channels. When $r \to +\infty$, channel wave functions have an outgoing asymptotic behavior, i.e. $w_c(r) = w_c^{(+)}(r)$, except for the incoming channel "$e$," where one has $w_e(r) = w_e^{(+)}(r) + w_e^{(-)}(r)$. Here, the incoming part $w_e^{(-)}$ is fixed, and the outgoing parts $w_c^{(+)}(r)$ and $w_e^{(+)}(r)$ have to be determined. Due to the channel–channel coupling, the radial wave functions $w_c(r)$ are not always $w_c \sim r^{\ell+1}$ for $r \sim 0$ as in the one-dimensional case.

To solve this problem, the radial wave functions $w_c(r)$ are expanded in the forward basis corresponding to the internal region: $0 \leq r \leq R$, where the nuclear part of the potential is not negligible, and in the backward basis corresponding to the asymptotic region: $R \leq r \leq R_{max}$, where the nuclear part of the potential is negligible. The expansion of the coupled-channel equations in the forward basis is integrated from $r = 0$ to $r = R$, and the expansion in the backward basis is integrated from $r = R_{max}$ to $r = R$. Contrary to the radial wave functions $w_c(r)$, these new basis states have the correct boundary conditions.

Thus the coupled-channel equations can be integrated numerically in each region, with the matching condition at $r = R$. The expansion in the forward basis, i.e. the region $0 \leq r \leq R$, is written as

$$w_c(r) = \sum_b C_b^{(0)} w_{c,b}^{(0)}(r) ,$$ (9.152)

and in the backward basis, i.e. the region $R \leq r \leq R_{max}$, is

$$w_c(r) = \sum_b C_b^{(+)} w_{c,b}^{(+)}(r) + w_e^{(-)}(r) .$$ (9.153)

Equations (9.152) and (9.153) are general for scattering states and resonant states, but for resonant states one has $w_e^{(-)}(r) = 0$ in Eq. (9.153).

The forward basis is defined by $w_{c,b}^{(0)}(r) \sim r^{\ell_b+1}$ at $r \sim 0$ and for $c = b$. The other channels $c \neq b$ are ruled by $w_{c,b}^{(0)}(r) = o(r^{\ell_b+1})$. Then, the backward basis verifies $w_{c,b}^{(+)}(r) \sim C_b^{(+)} H_{\ell_b}^+(\eta_b, k_b r)$ for $c = b$ and $w_{c,b}^{(0)}(r) = 0$ for the other channels $c \neq b$. The latter solution is exact as only the Coulomb+centrifugal part remains for $r > R$, so that it can be applied directly. However, as the centrifugal potential is singular at $r = 0$, therefore one cannot demand that the channel wave functions $w_{c,b}^{(0)}(r)$, which verify $w_{c,b}^{(0)}(r) = o(r^{\ell_b+1})$ for $c \neq b$, are simply put to zero, i.e. $w_{c,b}^{(0)}(r) = 0$ for $r \sim 0$, because this would not be numerically precise. It is therefore necessary to devise the behavior of $w_{c,b}^{(0)}(r)$ for $r \sim 0$.

For that, one writes the coupled-channel differential equations for a channel $c \neq b$ at $r \sim 0$:

$$[w_{c,b}^{(0)}]''(r) = \left( \frac{\ell_c(\ell_c + 1)}{r^2} + a_c \right) w_{c,b}^{(0)}(r) + \sum_{c' \neq c} a_{c'} w_{c'}(r) + o(w_{c,b}^{(0)}(r)) ,$$ (9.154)

where $a_c = (2m/\hbar^2)V_{cc}^{(eq)}(0) - k^2$ and $a_{c'} = (2m/\hbar^2)V_{cc'}^{(eq)}(0)$, with $V_{cc'}^{(eq)}(r)$ the equivalent potential associated to $w_c(r)$ (see Sect. 9.3.2.2).

In Eq. (9.154), all terms in the sum are $o(r^{\ell_b+1})$, except the one for which $c' = b$. One also has $w_{c,b}^{(0)}(r) = o(r^{\ell_b+1})$. Thus, Eq. (9.154) becomes

$$[w_{c,b}^{(0)}]''(r) = \left(\frac{\ell_c(\ell_c+1)}{r^2}\right) w_{c,b}^{(0)}(r) + a_b r^{\ell_b+1} + o(r^{\ell_b+1}). \qquad (9.155)$$

It is then immediate to verify that for $c \neq b$

$$w_{c,b}^{(0)}(r) \left(\frac{a_b}{(\ell_b+2)(\ell_b+3) - \ell_c(\ell_c+1)}\right) r^{\ell_b+3}, \qquad \ell_c \neq \ell_b + 2 \qquad (9.156)$$

$$w_{c,b}^{(0)}(r) \left(\frac{a_b}{2\ell_b+5}\right) r^{\ell_b+3} \ln r, \qquad \ell_c = \ell_b + 2. \qquad (9.157)$$

Matching the linear combinations of the two sets of basis wave functions defined previously at a given radius will provide the full solution of the coupled-channel equations. As all channels $w_c(r)$ must be continuous and have their derivative continuous, one obtains the following equations at a matching radius $r_m$ using Eqs. (9.152) and (9.153):

$$\sum_b \left[C_b^{(0)} w_{e,b}^{(0)}(r_m) - C_b^{(+)} w_{e,b}^{(+)}(r_m)\right] = w_e^{(-)}(r_m) \quad \text{(scattering)} \qquad (9.158)$$

$$\sum_b \left[C_b^{(0)} \frac{dw_{e,b}^{(0)}}{dr}(r_m) - C_b^{(+)} \frac{dw_{e,b}^{(+)}}{dr}(r_m)\right] = \frac{dw_e^{(-)}}{dr}(r_m) \quad \text{(scattering)}$$
$$\qquad (9.159)$$

$$\sum_b \left[C_b^{(0)} w_{c,b}^{(0)}(r_m) - C_b^{(+)} w_{c,b}^{(+)}(r_m)\right] = 0 \quad \text{(resonant)} \qquad (9.160)$$

$$\sum_b \left[C_b^{(0)} \frac{dw_{c,b}^{(0)}}{dr}(r_m) - C_b^{(+)} \frac{dw_{c,b}^{(+)}}{dr}(r_m)\right] = 0 \quad \text{(resonant)}. \qquad (9.161)$$

One has $c = c_0$ in Eqs. (9.158) and (9.159) and $c \neq c_0$ in Eqs. (9.160) and (9.161) when considering scattering states. In the latter case, $w_{e,b}^{(-)}(r)$ is an incoming wave function and reduces to $H_{\ell_{c0}}^-(\eta_{c0}, k_{c0}r)$ when $r > R$. Eqs. (9.160) and (9.161) impose the usual outgoing boundary conditions when calculating resonant states.

For scattering states, Eqs. (9.158–9.161) form a linear system $AX = B$, which is simple to solve. For bound and resonance states, one has $AX = 0$, as there is no incoming channel. Hence, in this case, one has to have $\text{Det}[A] = 0$, which occurs only if there is a $w_c(r)$, which has an outgoing wave function asymptote. $\text{Det}[A]$ is thus the generalization of the Jost function for coupled-channel equations (see Sect. 2.6.4). Once the eigenenergy for which $\text{Det}[A] = 0$ has been found, the constants $C_b^{(0)}$, $C_b^{(+)}$ are given by the eigenvector $X$ of $A$ of zero eigenvalue.

### 9.3.2.2 Use of Nonlocal Coupling Potentials

Integro-differential equations are solved using the equivalent potential method [66, 67]. For this, the one-dimensional equation, $w''(r) = f(r)w(r) + s(r)$, has to be replaced by the matrix differential equation:

$$u''(r) = M^{(eq)}(r)u(r) + S^{(eq)}(r) ,$$

where $M_{cc'}^{(eq)}(r)$ is equal to:

$$M_{cc'}^{(eq)}(r) = (2m/\hbar^2)V_{cc'}^{(eq)}(r) + \left( \frac{\ell_c(\ell_c + 1)}{r^2} - k^2 \right) \delta_{cc'} .$$

$u(r)$ is the coupled-channel wave function, vector of channel wave functions is $u_c(r)$, and $S_c^{(eq)}(r)$ is the residual source coming from the equivalent potential method. One has then for Eq. (9.149)

$$V_{cc'}^{(eq)}(r) = V_{cc'}^{(loc)}(r) + \left( \frac{1 - F_{c'}(r)}{w_{c'}(r)} \right) \int_0^{+\infty} V_{cc'}^{(non\text{-}loc)}(r, r') \, w_{c'}(r') \, dr'$$

$$S_c^{(eq)}(r) = \sum_{c'} F_{c'}(r) \int_0^{+\infty} V_{cc'}^{(non\text{-}loc)}(r, r') \, w_{c'}(r') \, dr' , \qquad (9.162)$$

where $F_{c'}(r)$ is a function removing the singularities due to the zeros of $w_{c'}(r)$.

The fundamental difference with the one-dimensional equation lies within the treatment of boundary conditions, which change according to the type of the wave function (bound state, resonance, or scattering wave function) and demand the introduction of basis functions, from which the full solution of the system of coupled-channel equations can be determined (see Sect. 9.3.2.1). The $V_{cc'}^{(eq)}(r)$ potential must be symmetrized if one considers bound and resonance states. Indeed, even though the nonlocal problem is symmetric, the local equivalent equations are not. The local equivalent equations are indeed symmetric only at convergence, so that symmetrization is necessary for bound states to remain real during the iterative process. Symmetry, however, does not have to be imposed for scattering states, as their energy is fixed. Moreover, for the system to remain homogeneous, $S_c^{(eq)}(r)$ must be divided by the number of channels for the case of bound and resonance states. If one calculates scattering states, the source is suppressed for outgoing parts and is taken into account only for the incoming part (see Sect. 9.3.2.1).

The use of the source term $S_c^{(eq)}(r)$ in Eq. (9.162) makes these equations inhomogenous. Consequently, the fact that $w(r)$ is the solution of the local equivalent coupled-channel Schrödinger equation does not imply that const $\times\, w(r)$ is such a solution as well. Thus, the channel wave functions used in Sect. 9.3.2.1) must be renormalized using the $C_b^{(0)}$ and $C_b^{(+)}$ integration constants of the previous iteration (see Sect. 9.3.2.1), so that $w(r)$ is both solution of the local equivalent coupled-channel Schrödinger equation and of the nonlocal initial Schrödinger equation at convergence.

### 9.3.3   Representation of Coupled-Channel Equations in the Berggren Basis

The methods presented in the previous sections are iterative and hence can present instabilities in cases when channel coupling is too strong. It is then more convenient to solve the coupled-channel equations using the Berggren basis. For this, one starts from the $A$-body scattering state of energy $E$, which is a solution of the Schrödinger equation of Eq. (9.1).

In the following, we aim at calculating the $u_c(r)$ channel functions of the $A$-body state that are decomposed in the basis of reaction channel states using the Berggren basis representation (see Eqs. (9.19) and (9.112)). Due to the antisymmetry of the target–projectile composite, the channel functions $u_c(r)$ are not orthogonal. In order to consider a matrix representation of channels, one uses the method described in Sect. 9.3.1. The $u_c(r)$ channel functions are indeed recovered from $w_c(r)$ channel functions at the end of the calculation using $\hat{N}^{-\frac{1}{2}}$ as well (see Sect. 9.3.1). In order to simplify the notation, one will implicitly consider that the $c$ channels are the orthogonalized channels associated to $w_c(r)$ functions, even though they are in principle linear combinations of the initial channels defining $u_c(r)$ channel functions.

One will now present how to calculate bound, resonance, and scattering states in a Berggren basis representation. Due to their different asymptotic behaviors, different numerical methods are used for resonant and scattering states, namely diagonalization and linear system solution, respectively. They will be detailed in the next two sections.

#### 9.3.3.1   Bound and Resonance Solutions of Coupled-Channel Equations
One will firstly deal with the diagonalization of $\hat{H}$ (see Eq. (9.1)) in the basis of channel Berggren states. This diagonalization provides the bound and resonance states of $\hat{H}$. In the case of an infinite model space, diagonalizing $\hat{H}$ in a basis of Slater determinants (see Chap. 5) or with coupled channels are equivalent approaches. As both bases are complete at the theoretical level, diagonalizing $\hat{H}$ in a basis of Slater determinants (see Sect. 5) or with coupled-channel basis should provide with the same bound and resonance eigenstates of $\hat{H}$. However, model spaces have to be truncated, and model space truncation done at the level of channels is not equivalent to that done at the level of configurations (see Chap. 5). Consequently, the eigenstates of $\hat{H}$ obtained in a basis of Slater determinants or from the solution of coupled-channel equations will be different in practice.

Nevertheless, the considered coupled channels are expected to well reproduce physical states if all important decay channels are included. Consequently, as this is the case in practice, one can expect that the eigenergies calculated either from the diagonalization of $\hat{H}$ with a basis of Slater determinants or with coupled channels will be close to each another.

$u_C(r)$ can be expanded in a coupled-channel Berggren basis as

$$u_C(r) = \sum_{n_c} a_{n_c} u_c^{(n_c)}(r) , \tag{9.163}$$

where the $u_c^{(n_c)}(r)$ functions are eigenstates of $\hat{h} = \hat{T} + \hat{U}_{\text{basis}}$ (see Eq. (9.11)) in the one-nucleon case and of $\hat{H}_{\text{CM}}$ (see Eq. (9.90)) in the multi-nucleon (nuclear) case. The $a_{n_c}$ coefficients arise from the diagonalization of $\hat{H}$ with Berggren channel basis states $u_c^{(n_c)}(r)$. The potentials entering $\hat{H}$ are either finite-ranged or of one-body Coulomb type (see Eqs. (9.23) and (9.121)). Therefore, the matrix elements associated to the potentials present in $\hat{H}$ can be conveniently calculated with direct integration using complex scaling (see Sects. 3.3 and 3.4).

Similarly to the matrix problem generated by the Gamow shell model using Slater determinants, one obtains a complex-symmetric matrix to be diagonalized, whose eigenstates consist of bound, resonance, and scattering eigenstates. The resonance eigenstates are embedded among the scattering eigenstates, so one has to use the overlap method (see Sect. 5.4) to identify them in the coupled-channel Gamow shell model spectrum of $\hat{H}$. In order to diagonalize the obtained Hamiltonian matrix, one uses the Jacobi–Davidson method, which is very successful within the Gamow shell model (see Sect. 5.7) The diagonalization of the coupled-channel Hamiltonian matrix is then as convenient as that done in a basis of Slater determinants. The unification of structure and reaction frameworks is also clear, as Eq. (9.163) can be used alternatively to expand resonant and scattering wave functions. Energy spectra calculated with the coupled-channel Gamow shell model will be presented in Sect. 9.4, with the examples of $^7$Li, $^7$Be, and $^{18}$Ne.

### 9.3.3.2 Scattering Solutions of Coupled-Channel Equations

It is convenient to use the Green's function method to obtain the scattering solution of Eq. (9.149). Let us begin by introducing a zeroth-order Hamiltonian $\hat{H}^{(0)}$ and its eigenvector $|\Psi^{(0)}\rangle$:

$$\hat{H}^{(0)} = \hat{t} + \hat{U}_{\text{basis}} \qquad \text{(nucleon)}$$

$$= \hat{T}_{\text{CM}} + \hat{U}_{\text{CM}} \qquad \text{(cluster)} \tag{9.164}$$

$$\hat{H}^{(0)} |\Psi^{(0)}\rangle = E |\Psi^{(0)}\rangle . \tag{9.165}$$

More specifically $\hat{H}^{(0)}$ is a matrix with off-diagonal matrix elements equal to zero (no coupling with other channels), and only one diagonal element is nonzero (only the entrance channel $c_0$ is activated). Eq. (9.165) is straightforward to solve as $\hat{H}^{(0)}$ leads to a one-dimensional differential equation.

Let us then separate $\hat{H}$ and $|\Psi_{M_A}^{J_A}\rangle$ into two parts, involving $\hat{H}^{(0)}$, $|\Psi^{(0)}\rangle$, and a rest part:

$$\hat{H} = \hat{H}^{(0)} + \hat{H}_{\text{rest}} \tag{9.166}$$

$$|\Psi_{M_A}^{J_A}\rangle = |\Psi^{(0)}\rangle + |\Psi_{\text{rest}}\rangle \ . \tag{9.167}$$

Using Eqs. (9.1) and (9.165–9.167), one obtains

$$(\hat{H} - E)\,|\Psi_{\text{rest}}\rangle = |\Psi_S\rangle \tag{9.168}$$

$$|\Psi_S\rangle = -\hat{H}_{\text{rest}}\,|\Psi^{(0)}\rangle \ , \tag{9.169}$$

where the source term $|\Psi_S\rangle$ has been introduced. Eq. (9.169) implies that as $\hat{H}_{\text{rest}}$ is of finite range, then $\Psi_S(r) \to 0$ when $r \to +\infty$. Hence, $|\Psi_S\rangle$ can be expanded in the Berggren basis generated by $\hat{H}^{(0)}$, so that Eq. (9.168) becomes a linear system in this representation:

$$(\Psi_{\text{rest}})_{n,c} = \langle n, c|\Psi_{\text{rest}}\rangle \tag{9.170}$$

$$(M_E)_{n,c\ n',c'} = \langle n', c'|\hat{H} - E|n, c\rangle \tag{9.171}$$

$$(\Psi_S)_{n,c} = \langle n, c|\Psi_S\rangle \tag{9.172}$$

$$M_E\,\Psi_{\text{rest}} = \Psi_S \ , \tag{9.173}$$

where $|n, c\rangle$ is a Berggren basis state of index $n$ of the channel $c$. The matrix elements $\hat{H}$ in Eq. (9.171) are calculated with complex scaling, as explained in Sect. 9.3.3.1.

Similarly to the Lippmann–Schwinger equation, the fundamental problem of Eq. (9.173) is that $M_E$ is not invertible on the real axis, as $H$ possesses a scattering eigenstate of energy $E$ therein. The standard solution is to replace $E$ by $E + i\epsilon$, with $\epsilon \to 0^+$, so as to make the considered linear system invertible, on the one hand, and to impose an outgoing wave function condition to $u_c(r)$ when $c \neq c_0$, on the other hand.

However, this method becomes unstable for small $\epsilon$ and demands to carefully monitor the limiting process. In order to avoid this problem, the contour defining $|n, c\rangle$ Berggren basis states is chosen in such a way that the energy of basis states always has a nonzero imaginary part. Consequently, $M_E$ is invertible along this contour, so that Eq. (9.173) is numerically solvable without an introduction of the regularization parameter. The outgoing wave function character of $\Psi_{\text{rest}}$ is also guaranteed by the finite norm of $\Psi_{\text{rest}}$ in a Berggren basis representation. Indeed, as $||\Psi_{\text{rest}}||$ is finite, one can deduce from the Parseval equality extended to Berggren bases that $|\Psi_{\text{rest}}\rangle$ is a localized state when complex rotation is applied, i.e. $\Psi_{rest}(z) \rightarrow 0$ if $z \rightarrow +\infty$, where $z = r + (R - r)e^{i\theta}$, $R$ is a radius outside the nuclear zone, and $0 \leq \theta \leq \pi/2$ is being properly chosen. This implies that all channel wave functions in $|\Psi_{rest}\rangle$ have an outgoing wave function asymptote.

Having calculated $|\Psi_{rest}\rangle$ in a Berggren basis, its calculation in coordinate space is straightforward. In cases when the equivalent potential method is numerically stable, it has been checked numerically that both Green function and direct integration methods provide the same solution for $|\Psi_{M_A}^{J_A}\rangle$. Equation (9.173) also leads to an additional numerical advantage when many energies have to be considered. Indeed, the spectrum of $\hat{H} - E$ is the same for all energies. Consequently, it is sufficient to calculate a convenient representation of $\hat{H}$ only once, which can then be reused to solve the linear system of Eq. (9.173) at different energies. Indeed, contrary to the Gamow shell model, where the $\hat{H}$ matrix is sparse, the $\hat{H}$ matrix in the Gamow coupled-channel formalism is dense, so that solving many linear systems with the initial $\hat{H}$ matrix is time-consuming. In fact, the typical numerical cost of solving a dense linear system is about $N^3/3$ per energy, where $N$ is the dimension of the matrix, with, e.g. the standard lower–upper decomposition of the matrix of $\hat{H}$ [68]. Conversely, if the $\hat{H}$ matrix has a diagonal or tridiagonal form, one has to solve diagonal or tridiagonal linear systems. The numerical cost associated to this method is about $2N^2$ per energy, which is much faster. Consequently, before solving the linear systems of Eq. (9.173), the matrix $\hat{H}$ will be firstly transformed into diagonal or tridiagonal form.

## 9.4   Calculation of Spectra and Reaction Cross Sections in the Coupled-Channel Gamow Shell Model

One will present in this section few illustrative applications of the coupled-channel Gamow shell model in physically relevant problems. The chosen examples mainly include proton scattering and the proton or neutron radiative capture reactions of interest for understanding the stellar nucleosynthesis. Deuteron capture will also be presented as an example of scattering reaction involving many-nucleon projectile.

As seen in Sect. 9.3.3.1, the coupled-channel Gamow shell model can be used to calculate the bound and resonance eigenstates of the Hamiltonian. Therefore, one will also present spectra resulting from the diagonalization of the Hamiltonian using either a basis of coupled channels or basis of Slater determinants (see Chap. 5). Both approaches will be shown to be numerically equivalent. Indeed, as one sees in Sect. 9.3.3.1, the bases of Slater determinants and coupled channels span the same many-body space of bound, resonance, and scattering states of complex energies.

### 9.4.1    Spectra of Unbound Nuclei and Excitation Functions of Proton–Proton Elastic Scattering

#### 9.4.1.1 $^{18}$Ne(p,p) Elastic Scattering Cross Section and $^{19}$Na Spectrum

The calculation of $^{18}$Ne(p,p) elastic scattering cross sections is convenient as an introduction to the coupled-channel Gamow shell model. Indeed, it can be modelled by three valence protons above an $^{16}$O core, so that Hamiltonian matrix dimensions are small. Moreover, experimental data are available for several values of cross section angles and projectile energies. Added to that, $^{19}$Na is unbound, so that continuum coupling is necessary therein to describe its spectrum.

In the calculation of $^{18}$Ne(p,p) elastic scattering cross sections, the reaction channels are defined by the ground state $J^{\pi} = 0_1^+$ and the first excited state $J^{\pi} = 2_1^+$ of $^{18}$Ne target coupled to the proton in different partial waves with $\ell_c \leq 2$. The single-proton configurations correspond to $0d_{5/2}$ and $1s_{1/2}$ resonances and 28 states of a discretized contour for each resonance. Moreover, $(d_{3/2}, p_{3/2}, p_{1/2})$ partial waves are included, which are decomposed using a real-energy contour of 21 points because in this case the resonance poles are broad. The Hamiltonian consists of the Woods–Saxon potential describing the $^{16}$O core and the modified surface Gaussian effective two-body interaction among valence nucleons [69]

$$V_{JT}^{(MSG)}(\mathbf{r}_1, \mathbf{r}_2)$$

$$= V_0(J, T) \exp\left(-\left(\frac{r_1 - R_0}{\mu_I}\right)^2 - \left(\frac{r_2 - R_0}{\mu_I}\right)^2\right) \sum_{\ell m} Y_{\ell m}^*(\Omega_1) Y_{\ell m}(\Omega_2)$$

$$\times \left(\frac{1}{1 + \exp\left[(r_1 - 2R_0 + r_F)/d_F\right]}\right) \left(\frac{1}{1 + \exp\left[(r_2 - 2R_0 + r_F)/d_F\right]}\right),$$
$$(9.174)$$

where $r_F = 1$ fm and the diffuseness parameter is $d_F = 0.05$ fm $\ll R_0$, so that the modified surface Gaussian interaction is centered at the nuclear surface and becomes negligible for $r > 2R_0$. Strength of the spin–orbit part of the one-body potential together with parameters of the modified surface Gaussian interaction has been fitted to reproduce excitation energies of low-lying states of $^{18}$Ne and $^{19}$Na with respect to the $^{18}$Ne ground state calculated in the Gamow shell model [69]. The single-particle basis in $^{18}$Ne and $^{19}$Na is generated by the same Gamow–Hartree–Fock potential. This potential is obtained from the Woods–Saxon potential of the core and the effective two-body interaction between valence nucleons. The coupled-channel Gamow shell model equations are solved using the method described in Sect. 9.3.2.

The low-energy excitation function for the reaction p+$^{18}$Ne at different center-of-mass angles is plotted in Fig. 9.1. Coupled-channel Gamow shell model calculation reproduces well the experimental results at all angles. The peak in the excitation function at $\sim$1.1 MeV corresponds to the $1/2_1^+$ resonance (see Table 9.1). The agreement with experiment is a direct consequence of the good reproduction of

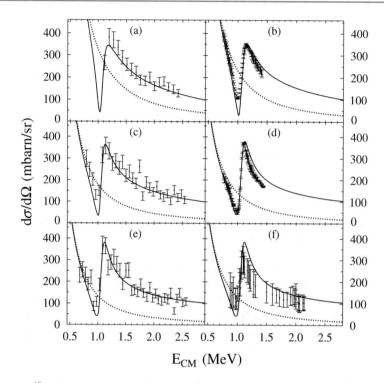

**Fig. 9.1** p+$^{18}$Ne excitation function at different center-of-mass angles $\Theta_{CM}$ in the range from 105 to 180 deg. The solid line shows the coupled-channel Gamow shell model results obtained using the Hamiltonian that includes both nuclear and Coulomb interactions. The dotted line corresponds to the calculation with the Coulomb interaction only. Panels (**a–f**) correspond to the excitation functions at $\Theta_{CM} = 105, 120.2, 135, 156.6, 165$, and 180 deg, respectively. The experimental data at $\Theta_{CM} = 105, 135$, and 165 deg are taken from Ref. [70]. The data at $\Theta_{CM} = 120.2, 156.6$ deg are taken from Ref. [71] and at $\Theta_{CM} = 180$ deg from Ref. [72] (from Ref.[69])

**Table 9.1** Gamow shell model and coupled-channel Gamow shell model results are compared with the experimental data for the low-energy states of $^{19}$Na. The calculations are performed using the modified surface Gaussian two-body interaction (9.174)

| $^{19}$Na | EXP | GSM | GSM-CC |
|---|---|---|---|
| $E(5/2_1^+)$ [MeV] | 0.321 | 0.338 | 0.347 |
| $E(3/2_1^+)$ [MeV] | 0.441 | 0.448 | 0.478 |
| $E(1/2_1^+)$ [MeV] | $1.067 - i0.05 \pm 0.01$ | $1.065 - i0.067$ | $1.115 - i0.079$ |

spectra by the coupled-channel Gamow shell model, especially the $1/2_1^+$ excited state of $^{19}$Na that is responsible for the rapid change of excitation functions near $E_{CM} = 1$ MeV.

The comparison of the Gamow shell model excitation energies of low-lying states in $^{19}$Na with their experimental values is shown in Table 9.1. For the chosen effective two-body interaction, the two-proton separation energy $S_{2p}$ in $^{18}$Ne differs from the experimental value by about 300 keV. All states in $^{19}$Na are narrow resonances that decay by proton emission. The last column in Table 9.1 contains excitation energies of $^{19}$Na calculated in the coupled-channel Gamow shell model. A small difference between results of Gamow shell model and coupled-channel Gamow shell model excitation energies of $^{19}$Na resonances is due to neglecting higher-lying discrete and scattering states of $^{18}$Ne, above the first excited state $J^\pi = 2_1^+$, in constructing the reaction channels.

### 9.4.1.2 $^{14}$O(p,p) Elastic Scattering Cross Section and $^{15}$F Spectrum

The $^{15}$F isotope is unbound, and a few resonance states are known in its spectrum. In particular, it possesses a narrow resonance state above the Coulomb barrier in the vicinity of the two-proton emission threshold. In fact, the unbound spectrum of $^{15}$F and the excitation function of the $^{14}$O(p,p) elastic scattering reaction have been studied experimentally in Ref. [73]. The coupled-channel Gamow shell model approach has been used therein to analyze the experimental data [73].

In the theoretical analysis, we use the Hamiltonian that consists of the Woods–Saxon potential with the spin–orbit term that describes the field of the $^{12}$C core acting on valence nucleons in $^{13}$N, $^{14}$O, and $^{15}$F, the Furutani–Horiuchi–Tamagaki finite-range two-body interaction [74] between valence nucleons, and the recoil term. Parameters of the Hamiltonian are adjusted to reproduce the binding energies of low-lying states, and the one- and two-proton separation energies in $^{15}$F. Parameters of the Hamiltonian are adjusted to reproduce the binding energies of low-lying states and the one- and two-proton separation energies in $^{15}$F. GSM-CC calculations are performed in three resonant shells: $0p_{1/2}$, $0d_{5/2}$, and $1s_{1/2}$, and several shells in the nonresonant continuum along the discretized contours: $L_{d_{5/2}}^+$ and $L_{s_{1/2}}^+$ in the complex momentum $k$ plane.

The eigenstates of the coupled-channel Gamow shell model are expanded in the basis of channel states that are built by coupling the Gamow shell model wave functions for ground state $0_1^+$ and excited states $1_1^-$, $0_2^+$, $3_1^-$, $2_1^+$, $0_1^-$, $2_2^+$, $2_1^-$ of $^{14}$O with the proton wave functions in partial waves: $s_{1/2}$, $p_{1/2}$, $p_{3/2}$, $d_{3/2}$, and $d_{5/2}$. Calculated binding energies of $1/2_1^+$, $5/2_1^+$, and $1/2_1^-$ states in $^{15}$F, relative to the energy of $^{12}$C, are, respectively, $(E, \Gamma)_{\text{GSM-CC}} = (-8.29 \text{ MeV}, 0.437 \text{ MeV})$, $(-6.66 \text{ MeV}, 0.211 \text{ MeV})$, and $(-4.48 \text{ MeV}, 0.031 \text{MeV})$ and well reproduce experimental energies of these resonances: $(E, \Gamma)_{\text{exp}} = (-8.21 \text{ MeV}, 0.376 \text{ MeV})$, $(-6.57 \text{ MeV}, 0.305 \text{ MeV})$, and $(-4.48 \text{ MeV}, 0.036 \text{ MeV})$. The narrow resonance $1/2_1^-$ in the vicinity of $^{13}$N + 2p decay channel can decay either by 1p- or 2p-emission. Its structure corresponds to an almost pure configuration: $[^{13}\text{N}(1/2_1^-) \otimes \pi(p_{1/2})\pi(s_{1/2})^2]$ of one proton in $p_{1/2}$ and two protons in $s_{1/2}$ shells. This is the reason why the width of this state is so narrow. The 2p-decay is hindered by the smallness of the available phase space for this decay. The 1p-decay in the channel $^{14}$O($0_1^+$) + p is suppressed due to the nature of the $1/2_1^-$ resonance, which is dominated by the diproton configuration due to the proximity of the $^{13}$N + 2p-decay channel. This is a splendid example of the continuum-coupling collectivization of a near-threshold shell model state by which the shell model state aligns with a nearby

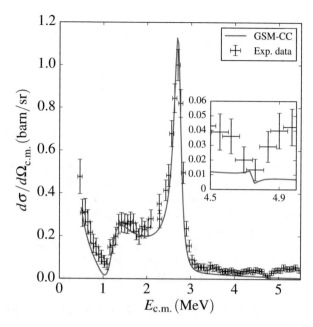

**Fig. 9.2** (Color online) The excitation function of the reaction $^{14}O(p,p)^{14}O$ at $180°$ in the center-of-mass as calculated in the coupled-channel Gamow shell model approach. The excitation function around the $1/2_1^-$ resonance is shown in the inset. The data and the figure are taken from Ref. [73]

decay channel, i.e. carries an imprint of this decay channel [75, 76]. The calculation of the resonance spectrum of $^{15}F$ is effected in Exercise X.

The excitation function for the reaction $^{14}O(p,p)^{14}O$ at $180°$ in the center-of-mass is shown in Fig. 9.2. The overall agreement of the coupled-channel Gamow shell model results with the data is remarkable. This example illustrates the unifying aspects of the coupled-channel Gamow shell model that with the same Hamiltonian and the theoretical formulation describes unbound states and related reaction cross section.

---

**Exercise X ★**

One will calculate the first three resonance states of $^{15}F$, of quantum number $1/2^+$, $5/2^+$, and $1/2^-$, respectively. The cross section of the elastic reaction $^{14}O(p,p)$ will also be calculated using the same Hamiltonian as for the resonance spectrum. While it is not strictly necessary, it is preferable to run codes using a parallel machine with at least 5–10 cores, as calculations can take a few hours otherwise.

**A.** The model space used consists of valence protons interacting in the presence of an inert $^{12}C$ core, where the proton partial waves with $\ell \leq 2$ are included.

The $^{12}$C core is mimicked by a Woods–Saxon potential, while the residual nuclear interaction between valence nucleons is effective and of Furutani–Horiuchi–Tamagaki type.

Explain why this model space is appropriate to calculate the resonance spectrum of $^{15}$F as well as the excitation function of the reaction $^{14}$O(p,p). For the latter point, use the fact that proton projectile energies are smaller than 6 MeV (see Ref.[73]).

**B.** Using physical arguments, show that the $1/2^+$ and $5/2^+$ resonance states have a single-particle character.

The experimental width of the $1/2^-$ resonance is small, of about 40 keV. Its experimental energy, of about 4.8 MeV with respect to proton emission threshold, is well above the Coulomb barrier, situated at 1.2 MeV above proton emission threshold. Consequently, explain why the $1/2^-$ resonance state must have a complicated structure.

**C.** Run the Gamow shell model code to obtain the resonance spectrum of $^{15}$F.

From the configuration mixing associated to the $1/2^+$ and $5/2^+$, and $1/2^-$ many-body states, show that the $1/2^+$ and $5/2^+$ resonance states are single-particle states.

Explain the relative stability of the $1/2^-$ resonance state from its configuration mixing.

**D.** Run the Gamow shell model coupled-channel code to obtain the resonance spectrum of $^{15}$F.

Explain why the latter spectrum and that calculated in **C** slightly differ.

**E.** Run the Gamow shell model coupled-channel code to generate the excitation function of the reaction $^{14}$O(p,p).

From the plot of the $^{14}$O(p,p) excitation function, indicate the energy zones where the influence of the resonance states of $^{15}$F on the cross section is the most important.

Explain the form of the cross section therein.

### 9.4.2 Radiative Capture Cross Sections in the Mass Region $A = 6$–8

Nuclei in the mass region $A = 6$–8 of nuclear chart are of great importance in nuclear astrophysics. The $^7$Be(p, $\gamma$) reaction is responsible of the most energetic solar neutrinos [77]. It occurs at the end of the proton–proton chain and is apparently responsible for most of solar energy release [78]. It is also directly linked to the so-called solar neutrino problem, where the neutrino flux arising from $^8$B in the sun, measured experimentally, is twice smaller than the prediction of the standard model [79]. This problem was solved recently with the discovery of nonzero masses of neutrinos, responsible for neutrino flavor oscillations [80]. In the standard model, the neutrino had always been considered as massless [79].

The mirror radiative capture reaction $^7$Li(n, $\gamma$) at high energies constrains those of the $^7$Be(p, $\gamma$) at solar energies, which are not accessible experimentally [81,82]. The $^7$Li(n, $\gamma$) radiative capture reaction is not found in usual thermonuclear cycles [78, 83] but is, however, a part of the inhomogeneous models of Big Bang nucleosynthesis [84]. Added to that, it is expected that $^{6,7}$Li isotopes, similarly to hydrogen and helium, have been created almost entirely during the Big Bang nucleosynthesis [85]. Consequently, the determination of the universal abundance of $^{6,7}$Li isotopes might allow to better understand the early universe.

This problem is challenging theoretically as at present the calculated abundances of $^{6,7}$Li do not reproduce experimental data [85]. The experimentally inferred primordial abundance of $^6$Li from the baryon-to-photon ratio issued from the Wilkinson Microwave Anisotropy Probe [86] is about 1000 times larger than that provided by theoretical estimates [85]. That same value for $^7$Li is about 3 times smaller than the theoretical predictions [85]. Consequently, a detailed coupled-channel Gamow shell model studies of the reactions involving $^{6,7}$Li isotopes might help understanding this state of affairs.

### 9.4.2.1 $^7$Be(p, $\gamma$) and $^7$Li(n, $\gamma$) Mirror Radiative Capture Reactions

The model space in $^7$Be and $^8$B is limited by the core of $^4$He. The core is described by a Woods–Saxon potential with $\ell$-dependent ($\ell \leq 2$) depth for protons and neutrons, supplemented by the Coulomb potential of radius $r_c = 2.8$ fm [64]. The two-body interaction between valence nucleons is described using the Furutani–Horiuchi–Tamagaki effective interaction (Eq. (6.18)). Parameters of the Hamiltonian have been fitted to reproduce binding energies of low-lying states in $^7$Be and $^8$B [64].

The valence nucleons can occupy the $0p_{3/2}$ and $0p_{1/2}$ discrete single-particle states and several nonresonant single-particle continuum states on discretized contours: $L^+_{s_{1/2}}$, $L^+_{p_{1/2}}$, $L^+_{p_{3/2}}$, $L^+_{d_{3/2}}$, and $L^+_{d_{5/2}}$. Each contour consists of three segments and is discretized with 30 points [64].

The Gamow shell model basis is truncated so as to reduce the size of the Gamow shell model Hamiltonian matrix. For this, the occupation of $p_{3/2}$ and $p_{1/2}$ scattering states in basis Slater determinants is limited to two particles, while the occupation of $s_{1/2}$, $d_{5/2}$, and $d_{3/2}$ scattering states is limited to one particle only. The latter truncation is justified by the fact that the $p_{3/2}$ and $p_{1/2}$ partial waves play a dominant role in the Gamow shell model target states, $s_{1/2}$, $d_{5/2}$, and $d_{3/2}$ states occurring only in the partial wave decomposition of the proton or neutron projectile. In this approximation, the ground state of $^7$Be in Gamow shell model calculations is bound with respect to $^4$He by 9.378 MeV, close to the experimental value $E_{\exp} = 9.304$ MeV. Reaction channels in the coupled-channel Gamow shell model calculations are obtained by the coupling of the ground state $3/2^-$ and the first excited state $1/2^-$ of $^7$Be with the proton partial waves: $s_{1/2}$, $p_{1/2}$, $p_{3/2}$, $d_{3/2}$, and $d_{5/2}$.

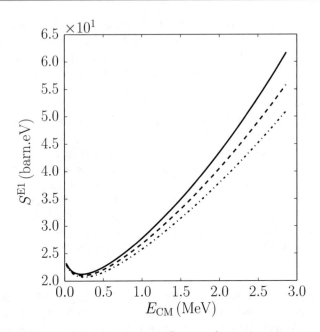

**Fig. 9.3** Plot of the E1 astrophysical factor for the $^7$Be$(p, \gamma)^8$B reaction. The solid line represents the exact, fully antisymmetrized coupled-channel Gamow shell model calculation. The calculations in the long wavelength approximation are represented by the dashed and dotted lines in the fully antisymmetrized and non-antisymmetrized cases, respectively (from Ref. [64])

Results of coupled-channel Gamow shell model cross section calculations for the reaction $^7$Be(p, $\gamma$) are shown in Figs. 9.3 and 9.4. In these calculations, there are no effective charges for M1 transitions, and standard values of effective charges have been used for E1 and E2 transitions. One should mention that the experimentally extracted effective charges show often significant deviations from the standard values [87].

Figure 9.3 shows the E1 astrophysical factor $S^{(E1)}$ (see Eq. (9.62) for its definition) calculated both exactly and approximately using the long wavelength approximation in the calculation of matrix elements of the electromagnetic transitions. For E1, both the long wavelength approximation and the absence of antisymmetry in the channel state $|r, c\rangle$ decrease the E1 contribution to the total astrophysical $S$-factor (see Fig. 9.3). However, whereas the long wavelength approximation reduces the $S$-factor by $\sim$5%, the absence of the antisymmetry in the calculation of the matrix elements of the electromagnetic operators reduces it by almost $\sim$50%.

The calculated total $S$-factor that includes $S^{E1}$, $S^{E2}$, and $S^{M1}$ contributions is compared with the experimental data [88, 89] in Fig. 9.4. Below $E_{CM} = 1$ MeV, the agreement with the data is good if both the ground state of $^7$Be and its first excited state are included. The value of the $S$-factor at zero energy, $S^{GSM-CC}(0)$, is 23.214 b·eV and the slope, $\partial S / \partial E_{CM}|_{E_{CM}} = 0$, is 37.921 b. The accepted experimental

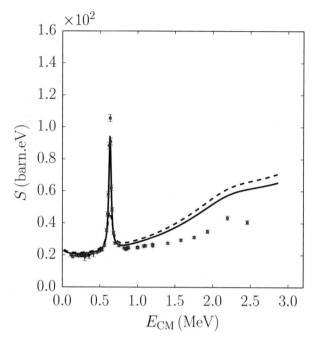

**Fig. 9.4** Plot of the total astrophysical factor for the $^7$Be$(p, \gamma)^8$B reaction. The solid line represents antisymmetrized coupled-channel Gamow shell model calculation that is including both the ground state $J^\pi = 3/2_1^-$ and the first excited state $J^\pi = 1/2_1^-$ of $^7$Be. Calculations neglecting the first excited state of $^7$Be are shown with the dashed line. Data are taken from Refs. [88, 89] (from Ref. [64])

value of the $S$-factor is 20.9±0.6 b·eV, slightly below the coupled-channel Gamow shell model results.

The $S$-factor is almost exclusively built from the E1 and M1 transitions. Indeed, even though included, the E2 component is negligible for all projectile energies. The E1 transition is responsible for the overall shape of the $S$-factor and crucially depends on the low binding energy of $^8$B. The peak seen between 0.5 and 1 MeV is due to the presence of the $1_1^+$ unbound excited state of $^8$B. The second peak, between 2 and 2.5 MeV, arises from the $3_1^+$ unbound excited state of $^8$B. The $3_1^+$ resonance state is much broader, as its width is about 350 keV, so that the $S$-factor varies slowly in this region. At higher energies, coupled-channel Gamow shell model results overshoot the experimental data. This feature could be due to the absence of higher-lying discrete and continuum states of $^7$Be target in the channel basis. Indeed, in the present case, Gamow shell model and coupled-channel Gamow shell model calculations with uncorrected channel–channel coupling potentials $V_{c,c'}$ do not give the same spectra and binding energies of $^7$Be and $^8$B, and the small multiplicative correction factors are necessary.

$^7\text{Li}(n, \gamma)^8\text{Li}$ reaction is described in the analogous model space as its mirror reaction: $^7\text{Be}(p, \gamma)^8\text{B}$. Similarly, the $^4\text{He}$ core is described by a $\ell$-dependent ($\ell \leq 2$) Woods–Saxon potential supplemented by the Coulomb potential of radius $r_c = 2.8\,\text{fm}$. The Furutani–Horiuchi–Tamagaki interaction (see [74, 90] and Eq. (6.18)) describes effective two-body interaction between valence nucleons. Parameters of the Hamiltonian have been fitted to reproduce binding energies of low-lying states in $^7\text{Li}$ and $^8\text{Li}$ [64]. The ground state of $^7\text{Li}$ calculated in the Gamow shell model is bound by 11.228 MeV with respect to $^4\text{He}$, i.e. close to the experimental value $E_{\text{exp}} = 10.948\,\text{MeV}$. Reaction channels in the coupled-channel Gamow shell model calculations are obtained by the coupling of the ground state $3/2^-$ and the first excited state $1/2^-$ of $^7\text{Li}$ with the neutron partial waves: $s_{1/2}$, $p_{1/2}$, $p_{3/2}$, $d_{3/2}$, and $d_{5/2}$.

Discrete states of a composite system $^8\text{Li}$ are $2_1^+$, $1_1^+$ bound states, and $3_1^+$ resonance. To correct for missing reaction channels in coupled-channel Gamow shell model calculations, the channel–channel coupling potentials $V_{c,c'}$ have been modified, and new potentials are: $\tilde{V}_{c,c'} = c(J^\pi)V_{c,c'}$, with $c(2_1^+) = 1.03705$, $c(1_1^+) = 1.04805$ and $c(3_1^+) = 1.03205$.

The total neutron radiative capture cross section is compared with the experimental data [91] in Fig. 9.5. The final nucleus $^8\text{Li}$ has two bound states $J^\pi = 2_1^+$

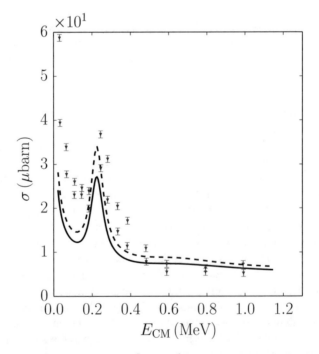

**Fig. 9.5** The total cross section for the $^7\text{Li}(n, \gamma)^8\text{Li}$ reaction. The solid line represents the antisymmetrized coupled-channel Gamow shell model calculation. Calculations neglecting the first excited state of $^7\text{Li}$ are shown with the dashed line. Data are taken from Ref. [91] (from Ref. [64])

and $1_1^+$ below the neutron emission threshold. Neutron separation energies from the ground state and excited states are $S_n = 2.032\,\text{MeV}$ and $S_n = 1.052\,\text{MeV}$, respectively, in good agreement with the experimental data. The $3_1^+$ resonance peak can be seen in M1 and E2 transitions. As in the $^7\text{Be}(p, \gamma)^8\text{B}$ reaction (see Fig. 9.4), E1 and M1 transitions are almost the only contributors to the total neutron radiative capture cross section, the E2 transition cross section being three orders of magnitude smaller. The peak around 200 keV comes from the $3_1^+$ resonance. Its width is about 30 keV, so that the cross section varies quickly therein.

Coupled-channel Gamow shell model calculation underestimates the data of Imhof et al. [91]. The extrapolation of the calculated neutron radiative capture cross section at low $E_{\text{CM}}$ is done using the expansion:

$$\sigma(E_{\text{CM}}) = \frac{4.541}{\sqrt{E_{\text{CM}}}} - 2.360 + 3.387\sqrt{E_{\text{CM}}}\,, \tag{9.175}$$

which yields: $\sigma^{\text{(GSM-CC)}} = 25.41\,\mu\text{b}$ at $E_{\text{CM}} = 25\,\text{keV}$. A simplified calculation of the radiative capture cross sections of $^7\text{Be}(p, \gamma)$ and $^7\text{Li}(n, \gamma)$, i.e. using a small model space, is effected in Exercise XI.

---

**Exercise  XI ⋆**

One will calculate in this exercise the low-lying spectrum of $^8\text{Li}$ and $^8\text{B}$ and the cross sections of the radiative capture reactions $^7\text{Be}(p, \gamma)$ and $^7\text{Li}(n, \gamma)$. While it is not strictly necessary, nevertheless it is preferable to run codes using a parallel machine with at least 5–10 cores, as calculations can take a few hours otherwise.

**A.** To avoid too large model space dimensions, $sd$ partial waves are represented by four harmonic oscillator states. Briefly state the advantages and disadvantages of the use of a few harmonic oscillator states to represent $sd$ partial waves. Explain why the basic features of considered radiative capture cross sections can still be qualitatively reproduced in this approximate picture.

**B.** Run the Gamow shell model code to obtain the bound or resonance states of $^8\text{Li}$ and $^8\text{B}$ with 2p-2h truncations in the nonresonant continuum and without truncation.

   Explain why the nucleon emission widths of the first resonance states of $^8\text{Li}$ and $^8\text{B}$ are not correct when using 2p-2h truncations, whereas they are close to experimental data in the absence of model space truncation.

**C.** Run the Gamow shell model coupled-channel code to obtain the resonant spectrum of $^8\text{Li}$ and $^8\text{B}$.

   State why it is close to that calculated in **B** without model space truncations.

**D.** Run the Gamow shell model coupled-channel code to generate the radiative capture cross sections of $^7$Be(p, $\gamma$) and $^7$Li(n, $\gamma$).

Used proton and neutron effective charges in the calculation for $^7$Be(p, $\gamma$) and $^7$Li(n, $\gamma$) are larger than their standard theoretical values by factors 2.5 and 2, respectively. Explain the physical significance of this modification.

### 9.4.3  $^4He(d, d)$ Cross Section and Deuteron Emission Width

The $^4$He(d,d) elastic scattering reaction is interesting from many points of views. Firstly, it is the simplest reaction involving a many-nucleon projectile. Indeed, it can be modelled by two valence nucleons, one proton and one neutron, above a well-bound $^4$He core, so that dimensions of model space are minimal. Moreover, experimental projectile energies are smaller than 10 MeV, so that the number of reaction channels to consider remains tractable.

$^6$Li isotope also has a very low deuteron emission threshold, and its proton and neutron emission thresholds are 3–4 MeV higher. Consequently, all $T = 0$ states of $^6$Li below 10 MeV are deuteron emitters so deuteron continuum degrees of freedom must be included to properly describe the spectrum of $^6$Li. The cross sections of the $^4$He(d,d) elastic scattering reaction are then dominated by the deuteron channels, and one can neglect other emission channels. $T = 1$ excited states are not important hereby as isospin is almost entirely conserved in the $^4$He(d,d) reaction.

In this study using the unified framework of the coupled-channel Gamow shell model, one will describe with the same Hamiltonian the energy spectrum of $^6$Li, the asymptotic normalization coefficient of the $^6$Li ground state, and the $^4$He(d,d) elastic scattering cross sections. To generate intrinsic states of deuteron, the realistic N$^3$LO interaction [92], fitted on phase shifts properties of proton–neutron elastic scattering reactions [93], is firstly diagonalized in a two-body Berggren basis written in relative coordinates and generated by a Woods–Saxon potential. Diffuseness, radius, central, and spin–orbit strengths of the potential are 0.65 fm, 1.5 fm, 40 MeV, and 7.5 MeV, respectively. The internal structure of deuteron is taken into account using the N$^3$LO interaction [92], fitted on phase shifts properties of proton–neutron elastic scattering reactions (see Ref.[93] for details about basis space truncations).

The basis of relative channel two-body states is generated by Woods–Saxon potential, whose radius is 1.5 fm and whose central and spin–orbit potential depths are equal to 40 MeV and 7.5 MeV, respectively. There is only one bound basis state, which is the ground state of the $^1S_0$ partial wave, of energy –0.457 MeV. Consequently, the basis of relative channel two-body states can be expected to be well adapted to expand intrinsic deuteron wave functions because deuteron has only one bound state, which is dominated by the $^1S_0$ channel. Contours along the real $k$-axis in Berggren basis are discretized with 10 points. It has been checked in Ref.[93] that a finer discretization of contours does not change results significantly. The deuteron ground state energy obtained in this way is $-2.061$ MeV, close to the experimental value $-2.224$ MeV. Deuteron eigenstates are constructed using the method described in Sect. 9.2.3. Thus, the Berggren eigenstates obtained after diagonalization are projected on a basis of harmonic oscillator states, written in cluster orbital shell model coordinates, with $n \leq 5$ and $\ell \leq 4$.

The combined system $[^4\text{He} + \text{d}]$ is described as $^4\text{He}$ core with two valence nucleons, using partial waves up to $\ell = 4$. The $^4\text{He}$ core is mimicked by a Wood–Saxon potential, fitted on phases shifts of elastic scattering reactions involving a neutron on a $^4\text{He}$ target [94]). Interaction among valence nucleons is given by the Furutani–Horiuchi–Tamagaki effective force (see Sect. 6.3.1 and Refs. [74,90]). The shell model space is that of all proton and neutron harmonic oscillator states having $\ell \leq 4$ and $n \leq 5$. The use of Berggren basis at this level is not necessary as the Slater determinants used therein generate only the finite-range Hamiltonian of the coupled-channel Gamow shell model. The Berggren basis of center-of-mass cluster states, used to calculate bound, resonance, and scattering states in $^6\text{Li}$, is generated by the proton and neutron Woods–Saxon potentials [93] using Eqs. (9.84), (9.85). The included partial waves, consisting of $^3S_1$, $^3P_{0,1,2}$, $^3D_{1,2,3}$, $sF_{2,3}$, and $^3G_3$, bear 3, 2, 2, 1, and 0 bound pole states, respectively. All resonance states lie below the Berggren basis contours, so that they do not belong to the considered Berggren bases. The contours consist of three segments, defined by the origin of the $K$-complex plane, with $K$ the linear momentum of the two-body cluster, and the $K$-complex points $0.2 - i0.05 \, \text{fm}^{-1}$, $1.0 - i0.05 \, \text{fm}^{-1}$, and $2 \, \text{fm}^{-1}$. Each segment is discretized with 15 points.

The statistical properties of the Furutani–Horiuchi–Tamagaki interaction parameters for $p$-shell nuclei have been analyzed in Sect. 6.3.1. Consequently, one can modify parameters of this interaction within the bounds of calculated statistical errors. The $T = 1$ part of interaction is negligible both in the $^4\text{He}(\text{d},\text{d})$ reaction and in the $T = 0$ spectrum of $^6\text{Li}$. Moreover, the dependence of energies on $V_C^{00}$ is very weak. Hence, only $V_C^{10}$ and $V_T^{10}$ have been considered when fitting parameters of the Furutani–Horiuchi–Tamagaki interaction. Other parameters of this interaction remain the same as given in Ref. [94].

It has been found that the $^3S_1$ asymptotic normalization coefficient of the $^6\text{Li}$ ground state is too small if only the energy spectrum of $^6\text{Li}$ is used to fit the interaction. The $^3D_1$ asymptotic normalization coefficient in this case is very small, similarly as the value reported in Refs. [95–97]. In order to account for the value of asymptotic normalization coefficients, two different strategies of fitting the Furutani–Horiuchi–Tamagaki interaction parameters are employed. The first strategy corresponds to fitting the $T = 0$ spectrum of $^6\text{Li}$, as already stated, to which the $^3D_1$ asymptotic normalization coefficient of the $^6\text{Li}$ ground state could be added without changing the spectrum and scattering observables significantly. The second strategy corresponds to fitting the $T = 0$ excited states of $^6\text{Li}$, along with the $^3S_1$ and $^3D_1$ asymptotic normalization coefficients of the $^6\text{Li}$ ground state, leaving the ground state energy of $^6\text{Li}$ out of the fit (see Table 9.2). The second strategy yields the ratio of $^3D_1/^3S_1 = -0.0265$, which agrees well with the experimental value -0.025(6)(10) [95–97]. The sets of interaction parameters obtained in these

**Table 9.2** $^3S_1$ and $^3D_1$ asymptotic normalization constants and their ratio for the $1_1^+$ ground state of $^6$Li calculated in coupled-channel Gamow shell model are compared to experiment [95–97]. Both $^3S_1$ and $^3D_1$ asymptotic normalization coefficients have been fitted in FHT(ANC), whereas only the $^3D_1$ asymptotic normalization coefficient has been fitted in FHT(E)

| ANC | FHT(E) | FHT(ANC) | Exp [95–97] |
|---|---|---|---|
| $^3S_1$ (fm$^{-1/2}$) | 1.707 | 2.950 | 2.91 (9) |
| $^3D_1$ (fm$^{-1/2}$) | −0.0788 | −0.077 | −0.077 (18) |
| $^3D_1/^3S_1$ | −0.0462 | −0.0261 | −0.025 (6) (10) |

two fitting strategies are called FHT(E) and FHT(ANC), respectively [93]. With FHT(ANC) fit, ground state energy of $^6$Li is lower by about 1 MeV than its experimental energy. However, as the asymptotic properties of the ground state are closer to experiment, one expects a better reproduction of the cross sections.

To deal with the deuteron projectile in the coupled-channel Gamow shell model, it is necessary to use two different interactions because we have two different physical situations: the deuteron far from the $^4$He target before and after the reaction, where deuteron properties are prominent, and the $^6$Li composite system of $^4$He and deuteron during the reaction. As the Furutani–Horiuchi–Tamagaki interaction is defined from $^6$Li properties, it cannot grasp the cluster structure of the deuteron at large distances. Conversely, the N$^3$LO interaction cannot be used with a core of $^4$He. Moreover, as the N$^3$LO interaction enters only construction of the deuteron basis, it is not explicitly present in the Hamiltonian but just insures that the deuteron projectile has correct asymptote and binding energy. This also implies that the use of both laboratory coordinates and cluster orbital shell model coordinates for the valence nucleons is consistent therein, as both sets of coordinates coincide asymptotically. It has been indeed checked that the use of AV8 [100] and CD-Bonn [101] interactions to generate deuteron projectiles leads to a very small change in energies, asymptotic normalization coefficients, and cross sections. As a consequence, the use of both realistic interaction for projectiles and effective Hamiltonian for composites does not present a problem.

The energies and widths of the $T = 0$ spectrum of $^6$Li calculated using FHT(E) and FHT(ANC) sets of parameters are compared in Fig. 9.6 with centroids and widths determined in Ref. [95]. The calculated spectra are obtained from coupled-channel Gamow shell model calculations, where the Hamiltonian is firstly represented by coupled-channel equations and then diagonalized. As seen in Fig. 9.6, the ground state $1_1^+$ calculated in the coupled-channel Gamow shell model with FHT(ANC) interaction is bound more than found experimentally [95], in order for the $^3S_1$ and $^3D_1$ asymptotic normalization coefficients might have fit the data [95–97] (see Table 9.2). The phase shifts of the $^4$He(d,d) elastic scattering reaction are shown in Fig. 9.7. One can see that the calculated phase shifts reproduce qualitatively the phase shifts extracted from the $R$-matrix analyses [98, 99], except for the $^3D_1$ phase shifts, as the $1_2^+$ state lies too high in energy. The FHT(ANC) interaction provides the best reproduction of phase shifts, especially for the $^3S_1$

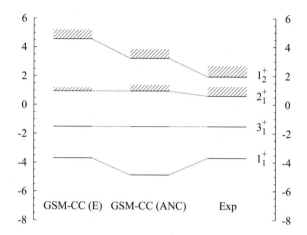

**Fig. 9.6** Energies and widths (in MeV) of the $T = 0$ spectrum of $^6$Li calculated in coupled-channel Gamow shell model approach using the FHT(E) (denoted as GSM-CC(E)) and FHT(ANC) (denoted as GSM-CC(ANC)) interactions are compared to evaluated centroids and width [95]. The energy of $1_1^+$ ground state of $^6$Li is deliberately overbound in the FHT(ANC) fit (from Ref. [93])

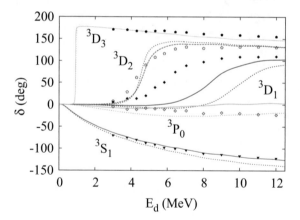

**Fig. 9.7** (Color online) Phase shifts of the $^4$He($^2$H,$^2$H) elastic scattering reaction calculated with the FHT(E) fit (dotted lines) and the FHT(ANC) fit (solid lines) are compared to $R$-matrix analyses of experimental data [98, 99]. $E_d$ is the kinetic energy of the incoming deuteron and is expressed in the laboratory frame (from Ref. [93])

and $^3P_0$ channels. Also the $^3D_1$ phase shifts are closer to the data for FHT(ANC) calculation. This is a direct consequence of a deliberate increasing energy of the $1_1^+$ ground state, which adds binding energy to the first excited $1_2^+$ state, so that its energy is closer to experiment. One may also notice a larger width of the $2_1^+$ state in FHT(ANC) calculation. However, in order to have good $^3D_2$ phase shifts, it was necessary to increase the energy of the $2_1^+$ state by about 300 keV in both GSM-CC(E) and GSM-CC(ANC) calculations, by adding a corrective factor to the

$2^+$ coupled-channel Hamiltonian (see Sect. 9.4.2.1). This is directly related to the large width of the $2_1^+$ state, implying that the many-body $S$-matrix poles provided by Gamow shell model within the resonating group method calculation and the resonance structures seen in reaction observables are not equivalent. Consequently, reproducing resonant states energies in structure calculation does not necessarily guarantee a good reproduction of experimental cross sections.

The center-of-mass differential cross sections for the $^4$He(d,d) elastic scattering reaction have been calculated at two deuteron kinetic energies in the laboratory frame: 2.935 MeV and 7.479 MeV. Calculated cross sections are compared with the data in Fig. 9.8. Experimental data are described much better using the FHT(ANC) interaction. Indeed, even the low-energy differential cross sections are too high in the calculation using the FHT(E).

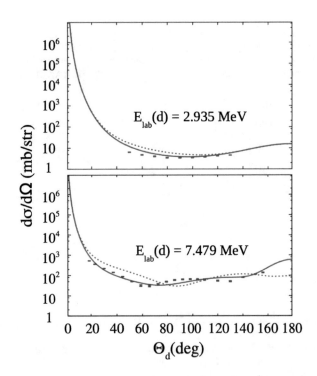

**Fig. 9.8** (Color online) Center-of-mass angular distributions of the $^4$He(d,d) elastic scattering reaction calculated at two different projectile energies, namely 2.935 MeV (upper) and 7.479 MeV (lower) using both the FHT(E) (dashed line) and FHT(ANC) (solid line) interactions, and compared with the experimental data (symbols) [102, 103]. Energy is given in the laboratory frame

This feature of the FHT(E) calculation follows directly from the phase shifts obtained for this interaction (see the dashed line in Fig. 9.7). In particular, the $^3P_0$ phase shifts below 8 MeV are too large in absolute value, what implies the too large values of differential cross section therein. While the $^3S_1$ phase shifts are correct close to 3 MeV, they rapidly become too large in absolute value at higher energies. Also the $^3D_2$ phase shifts increase too quickly around 4 MeV, to remain too high afterwards due to the too small width of the $2_1^+$ state. Both these effects seem to induce wrong positions of minima and maxima in differential cross sections above 4 MeV.

On the contrary, differential cross sections calculated with the FHT(ANC) interaction are always close to experimental data. The $^3D_2$ phase shifts are also very close to experimental data, as the width of the $2_1^+$ state is larger, of 840 keV instead of 1.3 MeV. Hence, the noticed discrepancies with experimental data in elastic deuteron cross sections (see Fig. 9.8) should arise due to the $^3P_0$ and $^3D_1$ phase shifts, which are further away from the experimental data.

### 9.4.4   Inclusion of Nonresonant Channels: The Resonant Spectrum of $^{42}$Sc

In previous applications, reaction channels have been built within a scheme similar to the pole approximation. Indeed, while the projectile can be of scattering character, the target was always a bound or resonance many-body state. This approximation is justified because in practical applications the combined target–projectile systems have low energy, so that the probability to excite high lying nonresonant many-body target states is small. However, in theory, all target states, whether resonant or scattering, enter the coupled-channel representation of Eq. (9.19). The absence of nonresonant target states in reaction channels has already been noticed, as the many-body eigenstate energies are slightly different in Gamow shell model in Slater determinant representation and coupled-channel Gamow shell model (see Sect. 9.4.3 and Exercise X). Thus, in practice, the absence of nonresonant target states is compensated by the corrective factors, i.e. one multiplies all two-body matrix elements by a constant so that the calculated eigenvalues are almost the same in the Gamow shell model and in the coupled-channel Gamow shell model (see Sect. 9.4.3 and Exercise X). However, this phenomenological procedure does not provide with information about the physical importance of the reaction channels built from nonresonant target states. Consequently, it is interesting to calculate the energy spectrum using a basis of reaction channels either possessing both resonant and scattering target states or defined only from resonant target states.

For this, one can only consider two valence nucleons, as otherwise calculations cannot be effected due to the large number of occurring nonresonant target states. Hence, one will consider $^{42}$Sc for that matter, as it can modelled by one valence proton and one valence neutron outside of the $^{40}$Ca core. The used Hamiltonian has the same structure as in Sects. 9.4.1.2 and 9.4.2.1, as it consists of Woods–Saxon potential mimicking the $^{40}$Ca core, to which a residual nucleon–nucleon interaction of Furutani–Horiuchi–Tamagaki type is added [74]. All partial waves bearing an orbital angular momentum $\ell \leq 4$ are taken into account. As the single-particle states of $^{41}$Ca and $^{41}$Sc are built from $fp$ shells, therefore only the $\ell = 1, 3$ parts of the Woods–Saxon potential can be fitted from the spectrum of $A = 41$ nuclei. The $\ell = 0, 2, 4$ parts of the Woods–Saxon potential are then fitted from the $^{40}$Ca(p,p) and $^{40}$Ca(n,n) scattering reactions, where all partial waves are involved. The parameters of the Furutani–Horiuchi–Tamagaki interaction are fitted to the energies of low-lying states in the spectrum of $^{42}$Sc.

Used reaction channels are built from the resonant target states of $^{41}$Ca and $^{41}$Sc, whose angular momentum is equal to $1/2^-, \ldots, 7/2^-$, and from the first scattering target states of $^{41}$Ca and $^{41}$Sc of angular momentum equal to $1/2^\pm, \ldots, 7/2^\pm, 9/2^+$. The obtained energies of the nuclear eigenstates of $^{42}$Sc are shown in Fig. 9.9. In order to ponder out the importance of nonresonant targets in the eigenstates of $^{42}$Sc, no corrective factor has been used in the coupled-channel Hamiltonian. It is clear from Fig. 9.9 that the inclusion of nonresonant target states in the basis of coupled channels is necessary in general. Discrepancies between the energies calculated with Gamow shell model and Gamow shell model with coupled channels, where nonresonant targets are neglected, can be as large as 600 keV, as is the case for the $0^+$ ground state of $^{42}$Sc. The effect of nonresonant target states is relatively smaller for the excited states of $^{42}$Sc. This is especially the case for the $5^+$ and $7^+$ eigenstates, where energy difference is about 100 keV in all used calculation schemes. In other excited states, the energy difference remains close to 200–300 keV. Clearly, these differences arise from the different nucleon–nucleon

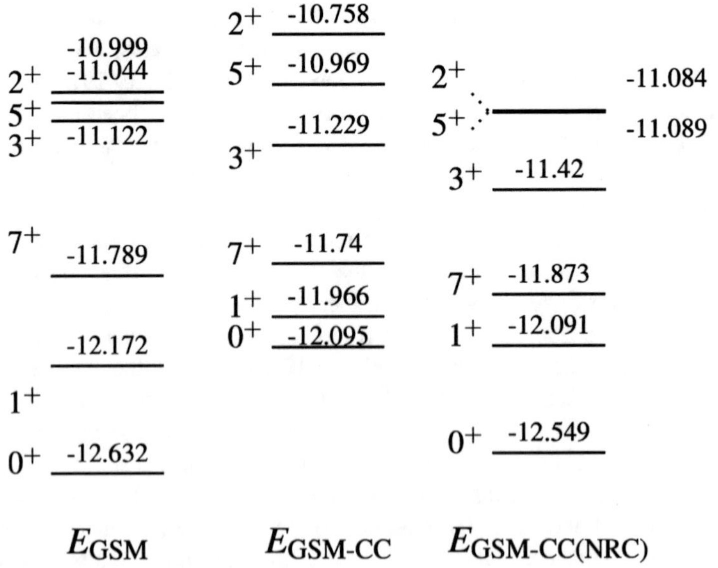

**Fig. 9.9** Low-lying spectrum of $^{42}$Sc calculated in Gamow shell model ($E_{\text{GSM}}$) and coupled-channel Gamow shell model with channels built from resonant target states only ($E_{\text{GSM-CC}}$) or from both resonant and nonresonant target states ($E_{\text{GSM-CC(NRC)}}$). Calculated energies are given with respect to the $^{40}$Ca core (from Ref. [104])

correlations present in the eigenstates of $^{42}$Sc. As the nucleon–nucleon interaction is typically the strongest when nucleons are coupled to $0^+$, it is expected to obtain the largest discrepancy in the ground state of $^{42}$Sc. Conversely, the eigenstates of highest spins do not differ significantly in all used calculation schemes. Indeed, one expects the nucleon–nucleon interaction to be the least binding for these nuclear states. However, it is difficult to predict quantitatively the energy differences between exact and approximate calculations.

As a consequence, one has shown that nonresonant target states are necessary to quantitatively reproduce Gamow shell model eigenvalues when using Gamow shell model in coupled-channel representation. The importance of nonresonant target states differs from one eigenstate to another, so that it is difficult to predict the amount of missing correlations in each eigenstate a priori. One has seen that it is possible to renormalize the effect of nonresonant targets on energies phenomenologically via corrective factors in Hamiltonian matrix elements. However, coupled-channel wave functions might change significantly when considering nonresonant target states, so that other observables as energies could differ.

## Solutions to Exercises[1]

### Exercise I.

**A.** The Berggren basis expansion of $H_{\text{if}}$ (see Eq. (9.10)) is readily obtained from Eq. (9.6):

$$H_{\text{if}} = \sum_{n_i n_f} u_{n_i}(r_i)\, u_{n_f}(r_f)\ \langle \hat{\mathscr{A}} \{ \langle \Psi_{T_f}^{J_{T_f}} | \otimes \langle u_{n_f} \ell_f j_f \tau_f | \}_{M_A}^{J_A} | \hat{H} | \hat{\mathscr{A}} \{ | \Psi_{T_i}^{J_{T_i}} \rangle \otimes | u_{n_i} \ell_i j_i \tau_i \rangle \}_{M_A}^{J_A} \rangle .$$

(9.176)

In order to separate $H_{\text{if}}$ in several parts according to the values of $n_i$, $n_f$, one uses the fundamental assumption imposed on $H$ (see Sect. 9.1.1), i.e. that $V_{\text{res}}$ vanishes when $n > n_{\max}$ in Eq. (9.12). Therefore, $\hat{H} = \hat{H}_T + \hat{H}_P$ when acted on Berggren basis channel states verifying $n_i > n_{\max}$ or $n_f > n_{\max}$. One has also seen that antisymmetry requirements disappear when $n_i > n_{\max}$ or $n_f > n_{\max}$, so that antisymmetrizers $\mathscr{A}$ can been removed from Eq. (9.176) in this situation. The demanded expression of $H_{\text{if}}$ comes forward by suppressing the $\hat{V}_{\text{res}}$ residual interaction and antisymmetrizers $\mathscr{A}$ when $n_i > n_{\max}$ or $n_f > n_{\max}$

$$H_{\text{if}} = \sum_{n \leq n_{\max}} u_n(r_i)\, u_n(r_f)\ \langle \hat{\mathscr{A}} \{ \langle \Psi_{T_f}^{J_{T_f}} | \otimes \langle u_n \ell_f j_f \tau_f | \}_{M_A}^{J_A} | \hat{H} | \hat{\mathscr{A}} \{ | \Psi_{T_i}^{J_{T_i}} \rangle \otimes | u_n \ell_i j_i \tau_i \rangle \}_{M_A}^{J_A} \rangle$$

$$+ \sum_{\substack{n_i \neq n_f \\ n_i, n_f \leq n_{\max}}} u_{n_i}(r_i)\, u_{n_f}(r_f)\ \langle \hat{\mathscr{A}} \{ \langle \Psi_{T_f}^{J_{T_f}} | \otimes \langle u_{n_f} \ell_f j_f \tau_f | \}_{M_A}^{J_A} | \hat{H} | \hat{\mathscr{A}} \{ | \Psi_{T_i}^{J_{T_i}} \rangle \otimes | u_{n_i} \ell_i j_i \tau_i \rangle \}_{M_A}^{J_A} \rangle$$

---

[1]The input files, codes and code user manual associated to computer-based exercises can be found at https://github.com/GSMUTNSR.

$$+ \sum_{n>n_{\max}} u_n(r_i)\, u_n(r_f) \; \langle [\langle \Psi_{T_f}^{J_{T_f}}| \otimes \langle u_n \ell_f j_f \tau_f|]_{M_A}^{J_A} |\hat{H}_T + \hat{H}_P|[|\Psi_{T_i}^{J_{T_i}}\rangle \otimes |u_n \ell_i j_i \tau_i\rangle]_{M_A}^{J_A}\rangle$$

$$+ \sum_{\substack{n_i \neq n_f \\ n_i > n_{\max} \\ n_f \leq n_{\max}}} u_{n_i}(r_i)\, u_{n_f}(r_f) \; \langle \hat{\mathscr{A}}\{\langle \Psi_{T_f}^{J_{T_f}}| \otimes \langle u_{n_f} \ell_f j_f \tau_f|\}_{M_A}^{J_A} |\hat{H}_T + \hat{H}_P|[\Psi_{T_i}^{J_{T_i}}\rangle \otimes |u_{n_i} \ell_i j_i \tau_i\rangle]_{M_A}^{J_A}\rangle$$

$$+ \sum_{\substack{n_i \neq n_f \\ n_i \leq n_{\max} \\ n_f > n_{\max}}} u_{n_i}(r_i)\, u_{n_f}(r_f) \; \langle [|\Psi_{T_f}^{J_{T_f}} \otimes \langle u_{n_f} \ell_f j_f \tau_f|]_{M_A}^{J_A} |\hat{H}_T + \hat{H}_P|\hat{\mathscr{A}}\{|\Psi_{T_i}^{J_{T_i}}\rangle \otimes |u_{n_i} \ell_i j_i \tau_i\rangle\}_{M_A}^{J_A}\rangle \,.$$

$$(9.177)$$

**B.** The sum in which $n_i \neq n_f$ and $n_i > n_{\max}$ will be shown to vanish. Indeed, if $n_i > n_{\max}$ and $n_f \geq 0$, the state $|u_{n_i} \ell_i j_i \tau_i\rangle$ is orthogonal both to all states occupied in $|\Psi_{T_f}^{J_{T_f}, M_{T_f}}\rangle$ and to $|u_{n_f} \ell_f j_f \tau_f\rangle$. Moreover, $(\hat{V}_{res} - \hat{U}_0) = (\hat{V}_{res} - \hat{U}_0)^{A-1}$ in this case, as all the matrix elements of $(\hat{V}_{res} - \hat{U}_0)$ involving $|u_{n_i} \ell_i j_i \tau_i\rangle$ are negligible. Hence, one has

$$\langle \hat{\mathscr{A}}\{\langle \Psi_{T_f}^{J_{T_f}}| \otimes \langle u_{n_f} \ell_f j_f \tau_f|\}_{M_A}^{J_A} |\hat{H}|[|\Psi_{T_i}^{J_{T_i}}\rangle \otimes |u_{n_i} \ell_i j_i \tau_i\rangle]_{M_A}^{J_A}\rangle$$

$$= \langle \hat{\mathscr{A}}\{\langle \Psi_{T_f}^{J_{T_f}}| \otimes \langle u_{n_f} \ell_f j_f \tau_f|\}_{M_A}^{J_A} |\hat{T} + \hat{U}_{basis} + (\hat{V}_{res} - \hat{U}_0)|\{|\Psi_{T_i}^{J_{T_i}}\rangle \otimes |u_{n_i} \ell_i j_i \tau_i\rangle\}_{M_A}^{J_A}\rangle$$

$$= \langle \hat{\mathscr{A}}\{\langle \Psi_{T_f}^{J_{T_f}}| \otimes \langle u_{n_f} \ell_f j_f \tau_f|\}_{M_A}^{J_A} |(\hat{V}_{res} - \hat{U}_0)|\{|\Psi_{T_i}^{J_{T_i}}\rangle \otimes |u_{n_i} \ell_i j_i \tau_i\rangle\}_{M_A}^{J_A}\rangle$$

$$= \langle \hat{\mathscr{A}}\{\langle \Psi_{T_f}^{J_{T_f}}| \otimes \langle u_{n_f} \ell_f j_f \tau_f|\}_{M_A}^{J_A} |(\hat{V}_{res} - \hat{U}_0)^{A-1}|\{|\Psi_{T_i}^{J_{T_i}}\rangle \otimes |u_{n_i} \ell_i j_i \tau_i\rangle\}_{M_A}^{J_A}\rangle$$

$$= 0 \,, \tag{9.178}$$

where one has used the fact that $[|\Psi_{T_i}^{J_{T_i}}\rangle \otimes |u_{n_i} \ell_f j_f \tau_f\rangle]_{M_A}^{J_A}$ and $\hat{\mathscr{A}}\{|\Psi_{T_f}^{J_{T_f}}\rangle \otimes |u_{n_f} \ell_i j_i \tau_i\rangle\}_{M_A}^{J_A}$ do not have a common Slater determinant in their expansions, implying that the $\hat{T} + \hat{U}_{basis}$ matrix element vanishes, and that $(V_{res} - U_0)^{A-1}$ does not act on $|u_{n_i} \ell_i j_i \tau_i\rangle$. Evidently, this argument also applies if $n_i$ and $n_f$ are exchanged, so that the sum in which $n_i \neq n_f$ and $n_f > n_{\max}$ vanish as well. Consequently, only the sum involving $n_i \neq n_f$ and $n_i, n_f \leq n_{\max}$ remains.

If $|\ell_i\, j_i\, \tau_i\rangle \neq |\ell_f\, j_f\, \tau_f\rangle$, the terms of the sum involving $n > n_{\max}$ in Eq. (9.177) all vanish. Indeed,

$$\langle [\langle \Psi_{T_f}^{J_{T_f}}| \otimes \langle u_n \ell_f j_f \tau_f|]_{M_A}^{J_A} |\hat{H}_T + \hat{H}_P|[|\Psi_{T_i}^{J_{T_i}}\rangle \otimes |u_n \ell_i j_i \tau_i\rangle]_{M_A}^{J_A}\rangle$$

$$= \langle \Psi_{T_f}^{J_{T_f}}|\hat{H}_T|\Psi_{T_i}^{J_{T_i}}\rangle \langle u_n \ell_f j_f \tau_f|u_n \ell_i j_i \tau_i\rangle + \langle \Psi_{T_f}^{J_{T_f}}|\Psi_{T_i}^{J_{T_i}}\rangle \langle u_n \ell_f j_f \tau_f|\hat{H}_P|u_n \ell_i j_i \tau_i\rangle$$

$$= 0 \,, \tag{9.179}$$

as $|u_n \ell_f j_f \tau_f\rangle$ and $|u_n \ell_i j_i \tau_i\rangle$ are mutually orthogonal eigenvectors of $\hat{H}_P$.

**C.** One has demonstrated that $\hat{H}$ is diagonal if $n_i > n_{max}$ or $n_f > n_{max}$, so that $n_i = n_f = n$, therein. As one assumes that $|\ell_i \, j_i \, \tau_i\rangle = |\ell_f \, j_f \, \tau_f\rangle = |\ell j \tau\rangle$ as well, $H_{if}$ reads

$$
H_{if} = \sum_{n_i \neq n_f} u_{n_i}(r_i) \, u_{n_f}(r_f) \, \langle \hat{\mathscr{A}}\{\langle \Psi_{T_f}^{J_{T_f}}| \otimes \langle u_{n_f}\ell j\tau|\}_{M_A}^{J_A} |\hat{H}| \hat{\mathscr{A}}\{|\Psi_{T_i}^{J_{T_i}}\rangle \otimes |u_{n_i}\ell j\tau\rangle\}_{M_A}^{J_A}\rangle
$$
$$
+ \sum_{n \leq n_{max}} u_n(r_i) \, u_n(r_f) \, \langle \hat{\mathscr{A}}\{\langle \Psi_{T_f}^{J_{T_f}}| \otimes \langle u_n\ell j\tau|\}_{M_A}^{J_A} |\hat{H}| \hat{\mathscr{A}}\{|\Psi_{T_i}^{J_{T_i}}\rangle \otimes |u_n\ell j\tau\rangle\}_{M_A}^{J_A}\rangle
$$
$$
+ \sum_{n \geq 0} u_n(r_i) \, u_n(r_f) \, \langle [\langle \Psi_{T_f}^{J_{T_f}}| \otimes \langle u_n\ell j\tau|]_{M_A}^{J_A} |\hat{H}_T + \hat{H}_P| [|\Psi_{T_i}^{J_{T_i}}\rangle \otimes |u_n\ell j\tau\rangle]_{M_A}^{J_A}\rangle
$$
$$
- \sum_{n \leq n_{max}} u_n(r_i) \, u_n(r_f) \, \langle [\langle \Psi_{T_f}^{J_{T_f}}| \otimes \langle u_n\ell j\tau|]_{M_A}^{J_A} |\hat{H}_T + \hat{H}_P| [|\Psi_{T_i}^{J_{T_i}}\rangle \otimes |u_n\ell j\tau\rangle]_{M_A}^{J_A}\rangle \ .
$$
$$(9.180)$$

The infinite sum in the $H_{if}$ matrix element has been separated in two parts: one where $n \geq 0$ and the other one where $n \leq n_{max}$. It is indeed more convenient to deal with a sum starting at $n = 0$ to have completeness relations appear. Let us then calculate the infinite sum of Eq. (9.180):

$$
\sum_{n \geq 0} u_n(r_i) u_n(r_f) \, \langle \{\langle \Psi_{T_f}^{J_{T_f}}| \otimes \langle u_n\ell j\tau|\}_{M_A}^{J_A} |\hat{H}_T + \hat{H}_P| \{|\Psi_{T_i}^{J_{T_i}}\rangle \otimes |u_n\ell j\tau\rangle\}_{M_A}^{J_A}\rangle
$$
$$
= \sum_{n \geq 0} u_n(r_i) u_n(r_f) \Big[ \langle \Psi_{T_f}^{J_{T_f}}| \hat{T} + \hat{U}_{core} + \hat{V}_{res} |\Psi_{T_i}^{J_{T_i}}\rangle \, \langle u_n\ell j\tau|u_n\ell j\tau\rangle
$$
$$
+ \langle \Psi_{T_f}^{J_{T_f}}|\Psi_{T_i}^{J_{T_i}}\rangle \, \langle u_n\ell j\tau|\hat{t} + \hat{U}_{basis}|u_n\ell j\tau\rangle \Big]
$$
$$
= \langle \Psi_{T_f}^{J_{T_f}}|\hat{H}|\Psi_{T_i}^{J_{T_i}}\rangle \sum_{n \geq 0} u_n(r_i) u_n(r_f) + \langle \Psi_{T_f}^{J_{T_f}}|\Psi_{T_i}^{J_{T_i}}\rangle \sum_{n \geq 0} e_n \, u_n(r_i) u_n(r_f) \ ,
$$
$$(9.181)$$

where $e_n$ is the single-particle energy of $|u_n\ell j\tau\rangle$, and where Eq. (9.13) has been used.

As $|\Psi_{T_i}^{J_{T_i}}\rangle$ and $|\Psi_{T_f}^{J_{T_f}}\rangle$ are eigenvectors of $\hat{H}$, the matrix element of Eq. (9.181) is nonzero only if $|\Psi_{T_i}^{J_{T_i}}\rangle = |\Psi_{T_f}^{J_{T_f}}\rangle = |\Psi_T^{J_T}\rangle$. Hence, in this case, the matrix element of Eq. (9.181) is equal to

$$
\left( \frac{\hbar^2}{2m} \left( -\frac{d^2}{dr_i^2} + \frac{\ell(\ell+1)}{r_i^2} \right) + E_T \right) \delta(r_i - r_f) + U_{basis}(r_i, r_f) \ ,
$$
$$(9.182)$$

where $E_T$ is the energy of $|\Psi_T^{J_T}\rangle$, and where the following one-body completeness relations have been used

$$\sum_n u_n(r_i)u_n(r_f) = \delta(r_i - r_f)$$

$$\sum_n e_n u_n(r_i)u_n(r_f) = \frac{\hbar^2}{2m}\left(-\frac{d^2}{dr_i^2} + \frac{\ell(\ell+1)}{r_i^2}\right)\delta(r_i - r_f) + U_{basis}(r_i, r_f) .$$

$$(9.183)$$

**D.** One can now calculate $H_{if}$ from the Berggren basis expansion involving arbitrary targets $|\Psi_{T_i}^{J_{T_i}}\rangle$ and $|\Psi_{T_f}^{J_{T_f}}\rangle$ and projectile channels $|r_i \ell_i j_i \tau_i\rangle$ and $|r_f \ell_f j_f \tau_f\rangle$. The incoming and outgoing channels will be denoted by $c_i$ and $c_f$, respectively. The expansion coefficients of many-body composite states in a basis of Slater determinants will be denoted as $c_{\alpha_i}^{(i)}$ and $c_{\alpha_f}^{(f)}$ for initial and final $A$-body wave functions, respectively (see Eq. (9.7)):

$$[|\Psi_{T_i}^{J_{T_i}}\rangle \otimes |u_{n_i}\ell j\tau\rangle]_{M_A}^{J_A} = \sum_{\alpha_i} c_{\alpha_i}^{(i)}|SD_{\alpha_i}\rangle$$

$$[|\Psi_{T_f}^{J_{T_f}}\rangle \otimes |u_{n_f}\ell j\tau\rangle]_{M_A}^{J_A} = \sum_{\alpha_f} c_{\alpha_f}^{(f)}|SD_{\alpha_f}\rangle .$$

Let us evaluate Eq. (9.177) using the results of **A**, **B**, **C**:

$$H_{if} = \left(\frac{\hbar^2}{2m}\left(-\frac{d^2}{dr_i^2} + \frac{\ell(\ell+1)}{r_i^2}\right) + E_T\right)\delta(r_i - r_f)\,\delta_{c_i c_f} + U_{basis}(r_i, r_f)\,\delta_{c_i c_f}$$

$$+ \sum_{n_i n_f}^{n_{max}} u_n(r_i)\,u_n(r_f)\,\langle\hat{\mathscr{A}}\{\langle\Psi_{T_f}^{J_{T_f}}| \otimes \langle u_{n_f}\ell j\tau|\}_{M_A}^{J_A}|\hat{H}|\hat{\mathscr{A}}\{|\Psi_{T_f}^{J_{T_f}}\rangle \otimes |u_{n_i}\ell j\tau\rangle\}_{M_A}^{J_A}\rangle$$

$$- \delta_{c_i c_f}\sum_{n\leq n_{max}} u_n(r_i)\,u_n(r_f)\,\langle[\langle\Psi_{T_f}^{J_{T_f}}| \otimes \langle u_n\ell j\tau|]_{M_A}^{J_A}|\hat{H}_T + \hat{H}_P|[|\Psi_{T_f}^{J_{T_f}}\rangle \otimes |u_n\ell j\tau\rangle]_{M_A}^{J_A}\rangle$$

$$= \left(\frac{\hbar^2}{2m}\left(-\frac{d^2}{dr_i^2} + \frac{\ell(\ell+1)}{r_i^2}\right) + E_T\right)\delta(r_i - r_f)\,\delta_{c_i c_f} + U_{basis}(r_i, r_f)\,\delta_{c_i c_f}$$

$$+ \sum_{n_i n_f \alpha_i \alpha_f}^{n_{max}} u_{n_i}(r_i)u_{n_f}(r_f)c_{\alpha_i}^{(i)}c_{\alpha_f}^{(f)}\langle SD_{\alpha_f}|\hat{H}|SD_{\alpha_i}\rangle - \delta_{c_i c_f}\sum_{n\leq n_{max}}(E_T + e_n)\,u_n(r_i)u_n(r_f)$$

$$\equiv \left(\frac{\hbar^2}{2m}\left(-\frac{d^2}{dr_i^2} + \frac{\ell(\ell+1)}{r_i^2}\right) + E_T\right)\delta(r_i - r_f)\,\delta_{c_i c_f} + U_{basis}(r_i, r_f)\,\delta_{c_i c_f}$$

$$+ V_{c_i c_f}(r_i, r_f) ,$$

$$(9.184)$$

where

$$V_{c_i c_f}(r_i, r_f) = \sum_{n_i n_f \alpha_i \alpha_f}^{n_{max}} u_{n_i}(r_i) u_{n_f}(r_f) c_{\alpha_i}^{(i)} c_{\alpha_f}^{(f)}$$

$$\times \langle SD_{\alpha_f} | \hat{H} | SD_{\alpha_i} \rangle - \delta_{c_i c_f} \sum_{n \leq n_{max}} (E_T + e_n) u_n(r_i) u_n(r_f)$$

is the finite-range coupling potential induced by the nucleon–nucleon interaction between $c_i$ and $c_f$.

## Exercise II.

**A.** $\hat{Q}$ is a Hermitian operator because all the one-body states entering $\hat{Q}$ are bound states. As these one-body states are orthogonal, $\hat{Q}^2$ reads

$$\hat{Q}^2 = \sum_{s \in \text{core}} |s\rangle \langle s| \sum_{s' \in \text{core}} |s'\rangle \langle s'| = \sum_{s,s' \in \text{core}} |s\rangle \langle s|s'\rangle \langle s'|$$

$$= \sum_{s,s' \in \text{core}} \delta_{ss'} |s\rangle \langle s'| = \sum_{s \in \text{core}} |s\rangle \langle s| = \hat{Q} .$$

$\hat{Q}$ is a projector because $\hat{Q}$ is Hermitian and $\hat{Q}^2 = \hat{Q}$. As $\hat{P} = \hat{1} - \hat{Q}$, therefore $\hat{P}$ is Hermitian and commutes with $\hat{Q}$. One directly obtains

$$\hat{P}^2 = (\hat{1} - \hat{Q})^2 = \hat{1} + \hat{Q}^2 - 2\hat{Q} = \hat{1} - \hat{Q} = \hat{P}$$

$$\hat{Q}\hat{P} = \hat{P}\hat{Q} = \hat{Q} - \hat{Q}^2 = 0 .$$

Consequently, $\hat{P}$ is a projector orthogonal to $\hat{Q}$.

If $|s\rangle \in$ core, then $\hat{Q}|s\rangle = |s\rangle$, and one has $\hat{Q}|s\rangle = 0$ if $|s\rangle \notin$ core. Conversely, $\hat{P}|s\rangle = |s\rangle$ if $|s\rangle \notin$ core, so that $\hat{P}|s\rangle = 0$ if $|s\rangle \in$ core. The one-body states inside and outside of the core form a complete basis, so that one cannot generate other eigenvectors as those stated. Thus, the eigenvectors of $\hat{Q}$ and $\hat{P}$ are the states inside and outside the core. Their eigenvalues are 0 or 1, as is the case for all projectors.

**B.** One considers an arbitrary operator $\hat{A}$ acting on the Hilbert space. If one wants $\hat{A}$ to act on the $P$ space, one multiplies $\hat{A}$ on the right by $\hat{P}$ to eliminate all the core components of input states.

Output states should not have any component in the $Q$ space either. Hence, one must also multiply $\hat{A}$ by $\hat{P}$ from the left. Therefore, the $\hat{A}$ operator restricted to the $P$ space is $\hat{P}\hat{A}\hat{P}$.

If $\hat{A}$ is Hermitian, one has

$$(\hat{P}\hat{A}\hat{P})^\dagger = \hat{P}^\dagger(\hat{P}\hat{A})^\dagger = \hat{P}^\dagger \hat{A}^\dagger \hat{P}^\dagger = \hat{P}\hat{A}\hat{P} .$$

Consequently, if $\hat{A}$ is Hermitian, $\hat{P}\hat{A}\hat{P}$ is also Hermitian.

**C.** $\hat{Q}$ projects a given state on the subspace of core states, by suppressing all components of the $P$ space. $\hat{P}$ projects a given state out of the $Q$ space. Thus, $\hat{P}$ projects a given state in the physical valence space of the cluster orbital shell model, which has no core components.

The Hamiltonian matrix in the cluster orbital shell model representation is complex symmetric due to the use of the Berggren basis. However, it is necessary for a Hamiltonian to be real symmetric when represented by real-energy basis states, so that many-body bound states have real energies and resonance states have positive widths. Thus, $\hat{Q}$ and $\hat{P}$ have to be Hermitian, and not complex symmetric, as otherwise projecting Hamiltonians on the $P$ space would lead to complex energies for bound eigenstates and negative widths for resonance eigenstates.

### Exercise III.

**A.** Let us assume that the two conditions $k'_i \ll k_{max}$ and $k'_j \gtrsim k_{max}$ are not verified $\forall (i,j) \in \{a, b, c, d\}$. Consequently, using the definitions of Sect. 9.2.6, one has either $k'_i \geq k_{max}^{(H)}$ or $k'_i \leq k_{max} - k_{max}^{(H)}$ $\forall i \in \{a, b, c, d\}$.

Let us firstly consider that $k'_{i'} \geq k_{max}^{(H)}$ $\forall i' \in \{a, b, c, d\}$. It follows from Eq. (9.96) that there is a state $|i\rangle$, with $i \in \{a, b, c, d\}$, with linear momentum $k_i \ll k_{max}$, i.e. so that $k_i < k_{max}^{(H)}$ (see Sect. 9.2.6). Therefore, $\langle k_i \ell_i m_i | k'_{i'} \ell_i m_i \rangle_{\mathscr{B}} = 0$ and $\langle i^+ | k'_{i'} \ell_i m_i \rangle_{\mathscr{B}} = 0$ (see Sect. 9.2.6.2).

Symmetrically, let us consider that $k'_{i'} \leq k_{max} - k_{max}^{(H)}$ $\forall i' \in \{a, b, c, d\}$. Similarly to the previous situation, it follows from Eq. (9.96) that one can find a state $|i\rangle$, with $i \in \{a, b, c, d\}$, with linear momentum $k_i \gtrsim k_{max}$, i.e. so that $k_i > k_{max} - k_{max}^{(H)}$ (see Sect. 9.2.6). Hence, using the same arguments as before, $|i\rangle$ verifies $\langle k_i \ell_i m_i | k'_{i'} \ell_i m_i \rangle_{\mathscr{B}} = 0$ and $\langle i^+ | k'_{i'} \ell_i m_i \rangle_{\mathscr{B}} = 0$.

As a result, $\langle f_i | k'_{i'} \ell_i m_i \rangle_{\mathscr{B}} = 0$ for a given state $i \in \{a, b, c, d\}$ for both considered cases. The integrand of Eq. (9.106) is then negligible when: $\nexists (i,j) \in \{a, b, c, d\}$ so that $k'_i \ll k_{max}$ and $k'_j \gtrsim k_{max}$.

**B.** Let us expand $\langle k'_a \ell_a m_a \, k'_b \ell_b m_b | \hat{V}_{res} - \hat{U}_0 | k'_c \ell_c m_c \, k'_d \ell_d m_d \rangle_{\mathscr{B}}$ in plane waves:

$$\langle k'_a \ell_a m_a \, k'_b \ell_b m_b | \hat{V}_{res} - \hat{U}_0 | k'_c \ell_c m_c \, k'_d \ell_d m_d \rangle_{\mathscr{B}}$$

$$= \int_{\mathscr{B}} \langle k'_a \ell_a m_a \, k'_b \ell_b m_b | \mathbf{k}''_a \, \mathbf{k}''_b \rangle \, \langle \mathbf{k}''_c \, \mathbf{k}''_d | k'_c \ell_c m_c \, k'_d \ell_d m_d \rangle_{\mathscr{B}}$$

$$\times \langle \mathbf{k}''_a \, \mathbf{k}''_b | \hat{V}_{res} - \hat{U}_0 | \mathbf{k}''_c \, \mathbf{k}''_d \rangle \, d\mathbf{k}''_a \, d\mathbf{k}''_b \, d\mathbf{k}''_c \, d\mathbf{k}''_d$$

$$= \int \langle \ell_a m_a \, \ell_b m_b | \Omega_a \Omega_b \rangle \, \langle \Omega_c \Omega_d | \ell_c m_c \, \ell_d m_d \rangle$$

$$\times \langle \mathbf{k}'_a \, \mathbf{k}'_b | \hat{V}_{res} - \hat{U}_0 | \mathbf{k}'_c \, \mathbf{k}'_d \rangle \, d\Omega_a \, d\Omega_b \, d\Omega_c \, d\Omega_d \, , \quad (9.185)$$

where, $\forall i \in \{a, b, c, d\}$, $\mathbf{k}'_i$ is a linear momentum of modulus $k'_i$ and of angle $\Omega_i$, and where one has used the equality $k''_i = k'_i$. As the $k'_i$ linear momenta verify Eq. (9.96), one has $\langle \mathbf{k}'_a \, \mathbf{k}'_b | \hat{V}_{\text{res}} - \hat{U}_0 | \mathbf{k}'_c \, \mathbf{k}'_d \rangle \simeq 0$ from the results of Sect. 9.2.6.1.

Consequently, $\langle k'_a \ell_a m_a \, k'_b \ell_b m_b | \hat{V}_{\text{res}} - \hat{U}_0 | k'_c \ell_c m_c \, k'_d \ell_d m_d \rangle_{\mathscr{B}} \simeq 0$. Therefore, the integrand of Eq. (9.106) is negligible when the $k'_i$ linear momenta verify Eq. (9.96).

C. The linear momenta $k'_a, k'_b, k'_c, k'_d$ must verify either $\exists (i, j) \in \{a, b, c, d\}$ so that $k'_i \ll k_{\text{max}}$ and $k'_j \gtrsim k_{\text{max}}$ or $\not\exists (i, j) \in \{a, b, c, d\}$ so that $k'_i \ll k_{\text{max}}$ and $k'_j \gtrsim k_{\text{max}}$. These two cases have been treated above. Consequently, $\langle f_a \, f_b | \hat{V}_{\text{res}} - \hat{U}_0 | f_c \, f_d \rangle \simeq 0$ when Eq. (9.96) is fulfilled by the $|f_i\rangle$ states. Thus, using Eqs. (9.104) and (9.105), one obtains that $\langle ab | \hat{V}_{\text{res}} - \hat{U}_0 | cd \rangle \simeq 0$ for the real-energy neutron Berggren basis states that verify Eq. (9.96).

## Exercise IV.

A. Let us expand $H_{cc'}^{J_A M_A}(R_{\text{CM}}, R'_{\text{CM}})$ with the eigenbasis of $\hat{H}_{\text{CM}}$:

$$H_{cc'}(R_{\text{CM}}, R'_{\text{CM}}) = \sum_{N_{\text{CM}}, N'_{\text{CM}}} H_{cc'}^{J_A M_A}(N_{\text{CM}}, N'_{\text{CM}}) \, U_{N_{\text{CM}}}(R_{\text{CM}}) \, U_{N'_{\text{CM}}}(R'_{\text{CM}}) \,,$$

(9.186)

where

$$H_{cc'}^{J_A M_A}(N_{\text{CM}}, N'_{\text{CM}})$$
$$= \langle \hat{\mathscr{A}} \{ \langle \Psi_T^{J_T} | \otimes \langle N_{\text{CM}} \, L_{\text{CM}} \, J_{\text{int}} \, J_P \, M_P | \}_{M_A}^{J_A} | \hat{H} | \hat{\mathscr{A}}$$
$$\{ | \Psi_{T'}^{J_{T'}} \rangle \otimes | N'_{\text{CM}} \, L'_{\text{CM}} \, J'_{\text{int}} \, J'_P \, M'_P \rangle \}_{M_A}^{J_A} \rangle \,.$$

(9.187)

Equation (9.186) can be decomposed in four sums:

$$H_{cc'}(R_{\text{CM}}, R'_{\text{CM}})$$
$$= \sum_{\substack{N_{\text{CM}} \leq N_{\text{CM max}} \\ N'_{\text{CM}} \leq N_{\text{CM max}}}} H_{cc'}^{J_A M_A}(N_{\text{CM}}, N'_{\text{CM}}) \, U_{N_{\text{CM}}}(R_{\text{CM}}) \, U_{N'_{\text{CM}}}(R'_{\text{CM}})$$

$$+ \sum_{\substack{N_{\text{CM}} \leq N_{\text{CM max}} \\ N'_{\text{CM}} > N_{\text{CM max}}}} H_{cc'}^{J_A M_A}(N_{\text{CM}}, N'_{\text{CM}}) \, U_{N_{\text{CM}}}(R_{\text{CM}}) \, U_{N'_{\text{CM}}}(R'_{\text{CM}})$$

$$+ \sum_{\substack{N_{\text{CM}} > N_{\text{CM max}} \\ N'_{\text{CM}} \leq N_{\text{CM max}}}} H_{cc'}^{J_A M_A}(N_{\text{CM}}, N'_{\text{CM}}) \, U_{N_{\text{CM}}}(R_{\text{CM}}) \, U_{N'_{\text{CM}}}(R'_{\text{CM}})$$

$$+ \sum_{\substack{N_{\text{CM}} > N_{\text{CM max}} \\ N'_{\text{CM}} > N_{\text{CM max}}}} H_{cc'}^{J_A M_A}(N_{\text{CM}}, N'_{\text{CM}}) \, U_{N_{\text{CM}}}(R_{\text{CM}}) \, U_{N'_{\text{CM}}}(R'_{\text{CM}}) \,.$$

(9.188)

**B.** The first term in Eq. (9.188) is a finite sum and can be calculated numerically using Gamow shell model formulas as the many-body states therein can be expanded using Slater determinants. The second sum with $N_{CM} \leq N_{CM\,max}$ and $N'_{CM} > N_{CM\,max}$ will be shown to be negligible:

$$H_{cc'}^{J_A M_A}\left(N_{CM}, N'_{CM}\right)$$

$$= \left\langle \left[ \langle \Psi_T^{J_T} | \otimes \langle N_{CM} \, L_{CM} \, J_{int} \, J_P \, M_P | \right]_{M_A}^{J_A} \left| \hat{\mathscr{A}} \hat{H} \hat{\mathscr{A}} \right| \left[ |\Psi_{T'}^{J_{T'}}\rangle \right. \right.$$

$$\left. \otimes |N'_{CM} \, L'_{CM} \, J'_{int} \, J'_P \, M'_P\rangle \right]_{M_A}^{J_A} \right\rangle$$

$$= \left\langle \left[ \langle \Psi_T^{J_T} | \otimes \langle N_{CM} \, L_{CM} \, J_{int} \, J_P \, M_P | \right]_{M_A}^{J_A} \left| \hat{H}_T + \hat{H}_P + \hat{\mathscr{A}} \hat{H}_{TP} \hat{\mathscr{A}} \right| \left[ |\Psi_{T'}^{J_{T'}}\rangle \right. \right.$$

$$\left. \otimes |N'_{CM} \, L'_{CM} \, J'_{int} \, J'_P \, M'_P\rangle \right]_{M_A}^{J_A} \right\rangle$$

$$= \left\langle \left[ \langle \Psi_T^{J_T} | \otimes \langle N_{CM} \, L_{CM} \, J_{int} \, J_P \, M_P | \right]_{M_A}^{J_A} \left| E_{CM} + E_T + E_{int} \right| \left[ |\Psi_{T'}^{J_{T'}}\rangle \right. \right.$$

$$\left. \otimes |N'_{CM} \, L'_{CM} \, J'_{int} \, J'_P \, M'_P\rangle \right]_{M_A}^{J_A} \right\rangle$$

$$+ \sum_{n,n'} c_n c'_{n'} \langle SD_n | \hat{H}_{TP} | SD_{n'} \rangle$$

$$= (E_{CM} + E_T + E_{int}) \, \delta_{cc'} \delta_{N_{CM} N'_{CM}} + \sum_{n,n'} c_n c'_{n'} \langle SD_n | \hat{H}_{TP} | SD_{n'} \rangle$$

$$= \sum_{n,n'} c_n c'_{n'} \langle SD_n | \hat{H}_{TP} | SD_{n'} \rangle$$

$$= \sum_{\substack{n,n' \\ \alpha,\beta,\gamma,\delta \in T,p}} c_n c'_{n'} \langle \alpha\beta | \hat{V}_{res} - \hat{U}_0 | \gamma\delta \rangle \langle SD_n | a_\alpha^+ a_\beta^+ a_\delta a_\gamma | SD_{n'} \rangle,$$

$$(9.189)$$

where $|\alpha\rangle, |\beta\rangle, |\gamma\rangle, \text{ and } |\delta\rangle$ stand for single-particle states that are occupied in target and projectile and where $c_n$ and $c'_{n'}$ are the coefficients of the Slater determinants expansion of many-body composites (see Eq. (9.111)):

$$\left[ |\Psi_T^{J_T}\rangle \otimes |N_{CM} \, L_{CM} \, J_{int} \, J_P \, M_P\rangle \right]_{M_A}^{J_A} = \sum_n c_n \, |SD_n\rangle$$

$$\left[ |\Psi_{T'}^{J_{T'}}\rangle \otimes |N'_{CM} \, L'_{CM} \, J'_{int} \, J'_P \, M'_P\rangle \right]_{M_A}^{J_A} = \sum_{n'} c'_{n'} \, |SD_{n'}\rangle.$$

Antisymmetrizers have been suppressed in Eq. (9.187) except for $\hat{H}_{TP}$ due to Eqs. (9.117) and (9.118).

Due to the finite range of $\hat{H}_{TP}$, the coefficients $c_n$ and $c'_{n'}$ decrease exponentially with the energy of basis scattering Slater determinants, so that they can be considered to be in finite number up to an error that can be made arbitrarily small. As $\hat{H}_{TP}$ couples target and projectile states, using the fact that targets are low-energy many-body states and that $N'_{CM} > N_{CMmax}$, the product $c_n c'_{n'}$ in the last sum of Eq. (9.189) is non-negligible only if $k_\alpha \ll k_{max}$, $k_\beta \ll k_{max}$, $k_\gamma \gtrsim k_{max}$, and $k_\delta \gtrsim k_{max}$. In this case, $\langle \alpha\beta | \hat{V}_{res} - \hat{U}_0 | \gamma\delta \rangle \simeq 0$ in the last sum of Eq. (9.189) because the conditions of Eq. (9.96) are then verified.

As a consequence, the second sum of Eq. (9.188) is equal to zero. The third sum of Eq. (9.188) can be treated identically for symmetry reasons and then vanishes as well.

C. Let us assume identical partitions for the incoming and outgoing channels and consider the sum of Eq. (9.189), where $N_{CM}$, $N'_{CM} > N_{CMmax}$. Similarly to the previous studied case, antisymmetrizers disappear in Eq. (9.187), and the matrix elements involving $\hat{H}_{TP}$ are negligible. Therefore, one has

$$H_{cc'}^{J_A M_A}(N_{CM}, N'_{CM})$$

$$= \left\langle \left[ \langle \Psi_T^{J_T} | \otimes \langle N_{CM} \, L_{CM} \, J_{int} \, J_P \, M_P | \right]_{M_A}^{J_A} \left| \hat{H} \right| \left[ | \Psi_{T'}^{J_{T'}} \rangle \right. \right.$$

$$\left. \otimes | N'_{CM} \, L'_{CM} \, J'_{int} \, J'_P \, M'_P \rangle \right]_{M_A}^{J_A} \right\rangle$$

$$= \left\langle \left[ \langle \Psi_T^{J_T} | \otimes \langle N_{CM} \, L_{CM} \, J_{int} \, J_P \, M_P | \right]_{M_A}^{J_A} \left| \hat{H}_T + \hat{H}_P \right| \left[ | \Psi_{T'}^{J_{T'}} \rangle \right. \right.$$

$$\left. \otimes | N'_{CM} \, L'_{CM} \, J'_{int} \, J'_P \, M'_P \rangle \right]_{M_A}^{J_A} \right\rangle$$

$$= \left\langle \left[ \langle \Psi_T^{J_T} | \otimes \langle N_{CM} \, L_{CM} \, J_{int} \, J_P \, M_P | \right]_{M_A}^{J_A} \left| \hat{H}_T + \hat{H}_{int} + \hat{H}_{CM} \right| \left[ | \Psi_{T'}^{J_{T'}} \rangle \right. \right.$$

$$\left. \otimes | N'_{CM} \, L'_{CM} \, J'_{int} \, J'_P \, M'_P \rangle \right]_{M_A}^{J_A} \right\rangle$$

$$= (E_T + E_{int} + E_{CM}) \, \delta_{cc'} \delta_{N_{CM} N'_{CM}} . \tag{9.190}$$

Let us now consider different partitions in these two channels with $a' \neq a$ and $N_{CM}$, $N'_{CM} > N_{CMmax}$. As $| \Psi_T^{J_T} \rangle$ is an eigenvector of $\hat{H}_T$ and $| N_{CM} \, L_{CM} \, J_{int} \, J_P \, M_P \rangle$ is an eigenvector of $\hat{H}_P$, one has

$$\left\langle \left[ \langle \Psi_T^{J_T} | \otimes \langle N_{CM} \, L_{CM} \, J_{int} \, J_P \, M_P | \right]_{M_A}^{J_A} \left| \hat{H}_T + \hat{H}_P \right| \left[ | \Psi_{T'}^{J'_T} \rangle \right. \right.$$

$$\left. \otimes | N'_{CM} \, L'_{CM} \, J'_{int} \, J'_P \, M'_P \rangle \right]_{M_A}^{J_A} \right\rangle$$

$$= \left\langle \left[ \langle \Psi_T^{J_T} | \otimes \langle N_{CM} \ L_{CM} \ J_{int} \ J_P \ M_P | \right]_{M_A}^{J_A} \left| \hat{H}_T + \hat{H}_{int} + \hat{H}_{CM} \right| \left[ |\Psi_{T'}^{J_T'} \rangle \right. \right.$$

$$\left. \left. \otimes \ |N'_{CM} \ L'_{CM} \ J'_{int} \ J'_P \ M'_P \rangle \right]_{M_A}^{J_A} \right\rangle$$

$$= (E_T + E_{int} + E_{CM}) \ \langle \Psi_T^{J_T} | \Psi_{T'}^{J_T'} \rangle$$

$$\times \langle N_{CM} \ L_{CM} \ J_{int} \ J_P \ M_P | N'_{CM} \ L'_{CM} \ J'_{int} \ J'_P \ M'_P \rangle = 0 \ , \qquad (9.191)$$

where the overlap between projectile states vanishes because $a' \neq a$. As a consequence, $H_{cc'}^{J_A M_A}(N_{CM}, N'_{CM}) = 0$ when $a' \neq a$ and $N_{CM}, N'_{CM} > N_{CM\,max}$.

**D.** Using previous results, one can write the matrix element $H_{cc'}^{J_A M_A}(R_{CM}, R'_{CM})$ as a sum of two terms:

$$H_{cc'}^{J_A M_A}(R_{CM}, R'_{CM})$$

$$= \sum_{\substack{N_{CM} \leq N_{CM\,max} \\ N'_{CM} \leq N_{CM\,max}}} H_{cc'}^{J_A M_A}(N_{CM}, N'_{CM}) U_{N_{CM}}(R_{CM}) U_{N'_{CM}}(R'_{CM})$$

$$+ \delta_{cc'} \sum_{N_{CM} > N_{CM\,max}} (E_T + E_{int} + E_{CM}) \ U_{N_{CM}}(R_{CM}) U_{N'_{CM}}(R'_{CM}) \ . \quad (9.192)$$

The sums in Eq. (9.192) involving $N_{CM} > N_{CM\,max}$ and $N'_{CM} > N_{CM\,max}$ can be written as

$$\sum_{N_{CM} > N_{CM\,max}} (E_T + E_{int} + E_{CM}) \ U_{N_{CM}}(R_{CM}) U_{N'_{CM}}(R'_{CM})$$

$$= \sum_{N_{CM}} (E_T + E_{int} + E_{CM}) \ U_{N_{CM}}(R_{CM}) U_{N'_{CM}}(R'_{CM})$$

$$- \sum_{N_{CM} \leq N_{CM\,max}} (E_T + E_{int} + E_{CM}) \ U_{N_{CM}}(R_{CM}) U_{N'_{CM}}(R'_{CM}) \ , \quad (9.193)$$

where the first sum in Eq. (9.193) can be expressed with Dirac delta's due to completeness properties of $\hat{H}_{CM}$ eigenstates:

$$\sum_{N_{CM}} (E_T + E_{int} + E_{CM}) \ U_{N_{CM}}(R_{CM}) U_{N_{CM}}(R'_{CM})$$

$$= \left( \frac{\hbar^2}{2M_P} \left( -\frac{d^2}{d R_{CM}^2} + \frac{L_{CM}(L_{CM} + 1)}{R_{CM}^2} \right) + E_T + E_{int} \right)$$

$$\times \delta(R_{CM} - R'_{CM}) + U_{CM}^{L_{CM}}(R_{CM}, R'_{CM}) \ , \qquad (9.194)$$

where $U_{CM}^{L_{CM}}(R_{CM}, R'_{CM})$ stands for the potential part of Eq. (9.90).

As a consequence, one can write $H_{cc'}^{J_A M_A}(R_{CM}, R'_{CM})$ as a sum of kinetic and coupling potential operators from the previous results:

$$H_{cc'}^{J_A M_A}(R_{CM}, R'_{CM})$$

$$= \left( \frac{\hbar^2}{2M_P} \left( -\frac{d^2}{dR_{CM}^2} + \frac{L_{CM}(L_{CM}+1)}{R_{CM}^2} \right) + E_T + E_{int} \right) \delta(R_{CM} - R'_{CM}) \delta_{cc'}$$

$$+ U_{CM}^{L_{CM}}(R_{CM}, R'_{CM}) \delta_{cc'} + \sum_{n,n'} c_n c'_{n'} \langle SD_n| \hat{H} |SD_{n'} \rangle$$

$$- \sum_{N_{CM} \leq N_{CM max}} (E_T + E_{int} + E_{CM}) U_{N_{CM}}(R_{CM}) U_{N'_{CM}}(R'_{CM}) \delta_{cc'}$$

$$\equiv \left( \frac{\hbar^2}{2M_P} \left( -\frac{d^2}{dR_{CM}^2} + \frac{L_{CM}(L_{CM}+1)}{R_{CM}^2} \right) + E_T + E_{int} \right) \delta(R_{CM} - R'_{CM}) \delta_{cc'}$$

$$+ U_{CM}^{L_{CM}}(R_{CM}, R'_{CM}) \delta_{cc'} + V_{cc'}^{J_A M_A}(R_{CM}, R'_{CM}), \tag{9.195}$$

where

$$V_{cc'}^{J_A M_A}(R_{CM}, R'_{CM}) = \sum_{n,n'} c_n c'_{n'} \langle SD_n| \hat{H} |SD_{n'} \rangle$$

$$- \delta_{cc'} \sum_{N_{CM} \leq N_{CM max}} (E_T + E_{int} + E_{CM}) U_{N_{CM}}(R_{CM}) U_{N'_{CM}}(R'_{CM})$$

is the finite-range coupling potential arising from channel coupling between $c$ and $c'$.

## Exercise V.

**A.** Plane waves whose linear momentum $q$ is along the $z$-axis can be expanded with Bessel functions:

$$e^{i\mathbf{q} \cdot \mathbf{r}} = \sum_{L=0}^{+\infty} i^L (2L+1) \, P_L(\cos(\theta)) \, j_L(qr) \,,$$

where $\theta$ is the angle between $\mathbf{q}$ and $\mathbf{r}$. Note that this series converges very quickly, similarly to $(c/L)^L$, with $c > 0$, so that series and integrals over angles can always be inverted without problems.

The general plane wave expansion, in which $\mathbf{q}$ has an arbitrary direction, is recovered by using the addition theorem of spherical harmonics:

$$P_L(\cos(\theta)) = \left(\frac{4\pi}{2L+1}\right) \sum_{M_L=-L}^{L} Y^*_{LM_L}(\Omega_q)\, Y_{LM_L}(\Omega) .$$

Multiplying the general plane wave expansion by $Y_{LM_L}(\Omega_q)$ on left- and right-hand sides and integrating on $\Omega_q$ provide with the integral expression of $\mathscr{J}_{LM_L}(q, \mathbf{r})$.

**B.** One will express $\mathscr{J}_{LM_L}(q, \mathbf{r})$ as a finite linear combination of functions depending on either relative or center-of-mass coordinates. Let us write $\mathscr{J}_{LM_L}(q, \mathbf{r})$ using its Bessel function expansion:

$$\mathscr{J}_{LM_L}(q, \mathbf{r}) = (4\pi\, i^L)^{-1} \int e^{i\mathbf{q}\cdot\mathbf{r}_{\text{int}}}\, e^{i\mathbf{q}\cdot\mathbf{R}_{\text{CM}}}\, Y_{LM_L}(\Omega_q)\, d\Omega_q ,$$

where the equality $\mathbf{r}_{\text{int}} + \mathbf{R}_{\text{CM}} = \mathbf{r}$ has been used. The exponential functions in this equation, depending only on $\mathbf{r}_{\text{int}}$ or $\mathbf{R}_{\text{CM}}$, can be expanded with relative and center-of-mass Bessel functions, respectively. One then obtains the demanded result:

$$\mathscr{J}_{LM_L}(q, \mathbf{r}) = (4\pi\, i^L)^{-1} \int Y_{LM_L}(\Omega_q)\, d\Omega_q$$

$$\times\, 4\pi \sum_{\ell_{\text{int}} m_{\ell_{\text{int}}}} i^{\ell}_{\text{int}}\, Y^*_{\ell_{\text{int}} m_{\ell_{\text{int}}}}(\Omega_q)\, Y_{\ell_{\text{int}} m_{\ell_{\text{int}}}}(\Omega_{\text{int}})\, j_{\ell_{\text{int}}}(q r_{\text{int}})$$

$$\times\, 4\pi \sum_{L_{\text{CM}} M_{L_{\text{CM}}}} i^{L_{\text{CM}}}\, Y^*_{L_{\text{CM}} M_{L_{\text{CM}}}}(\Omega_q)\, Y_{L_{\text{CM}} M_{L_{\text{CM}}}}(\Omega_{\text{CM}})\, j_{L_{\text{CM}}}(q R_{\text{CM}})$$

$$=\, 4\pi \sum_{\substack{\ell_{\text{int}}\, m_{\ell_{\text{int}}} \\ L_{\text{CM}}\, M_{L_{\text{CM}}}}} i^{\ell_{\text{int}} + L_{\text{CM}} - L}\, \mathscr{J}_{\ell_{\text{int}} m_{\ell_{\text{int}}}}(q, \mathbf{r}_{\text{int}})\, \mathscr{J}_{L_{\text{CM}} M_{L_{\text{CM}}}}(q, \mathbf{R}_{\text{CM}})$$

$$\times \int Y_{LM_L}(\Omega_q)\, Y^*_{\ell_{\text{int}} m_{\ell_{\text{int}}}}(\Omega_q)\, Y^*_{L_{\text{CM}} M_{L_{\text{CM}}}}(\Omega_q)\, d\Omega_q . . \tag{9.196}$$

**C.** The integral of three spherical harmonics is analytical, and its value is called the Gaunt coefficient [63]:

$$\int Y_{LM_L}(\Omega_q)\, Y^*_{\ell_{\text{int}} m_{\ell_{\text{int}}}}(\Omega_q)\, Y^*_{L_{\text{CM}} M_{L_{\text{CM}}}}(\Omega_q)\, d\Omega_q$$

$$= (-1)^{\ell_{\text{int}} + L_{\text{CM}}} \sqrt{\frac{(2\ell_{\text{int}} + 1)(2L_{\text{CM}} + 1)}{4\pi}}$$

$$\times \begin{pmatrix} \ell_{\text{int}} & L_{\text{CM}} & L \\ 0 & 0 & 0 \end{pmatrix} \langle \ell_{\text{int}}\, m_{\ell_{\text{int}}}\, L_{\text{CM}}\, M_{L_{\text{CM}}} | L\, M_L \rangle .$$

One obtains the result by using the Clebsch–Gordan coefficient present in the Gaunt coefficient to couple spherical harmonics to $L$ and $M_L$.

**Exercise VI.** The equations are provided from basic differentiations of $e^{i\mathbf{q}\cdot\mathbf{r}}$, followed by the use of the plane wave expansion of $e^{i\mathbf{q}\cdot\mathbf{r}}$ (see **A** in Exercise V) and integral representation of Bessel function $\mathscr{J}_{LM_L}(q,\mathbf{r})$ as a function of $\mathscr{J}_{\ell_{int}m_{\ell_{int}}}(q,\mathbf{r}_{int})$ and $\mathscr{J}_{L_{CM}M_{L_{CM}}}(q,\mathbf{R}_{CM})$ (see **B** in Exercise V). One will only present how to apply differentiation operators on $e^{i\mathbf{q}\cdot\mathbf{r}}$, as the use of expansions is then identical to those presented in Exercise V.

The equality providing with $\nabla(\mathscr{J}_{LM_L}(q,\mathbf{r}))$ is derived immediately by noticing that

$$\nabla e^{i\mathbf{q}\cdot\mathbf{r}} = i\mathbf{q}\, e^{i\mathbf{q}\cdot\mathbf{r}} = i\mathbf{q}\, e^{i\mathbf{q}\cdot\mathbf{r}_{int}}\, e^{i\mathbf{q}\cdot\mathbf{R}_{CM}} = \left(\nabla_{int}\, e^{i\mathbf{q}\cdot\mathbf{r}_{int}}\right) e^{i\mathbf{q}\cdot\mathbf{R}_{CM}}\ .$$

The differentiation associated to $\hat{\mathscr{J}}'_{LM_L}(q,\mathbf{r})$ is also straightforward, as it arises from

$$\frac{\partial}{\partial r'}\left(r'\, e^{i\mathbf{q}\cdot\mathbf{r}'}\right)_{r=r'} = (i(\mathbf{q}\cdot\mathbf{r})+1)\, e^{i\mathbf{q}\cdot\mathbf{r}}$$

$$= (i(\mathbf{q}\cdot\mathbf{r}_{int})+1)\, e^{i\mathbf{q}\cdot\mathbf{r}_{int}}\, e^{i\mathbf{q}\cdot\mathbf{R}_{CM}} + i(\mathbf{q}\cdot\mathbf{R}_{CM})\, e^{i\mathbf{q}\cdot\mathbf{r}_{int}}\, e^{i\mathbf{q}\cdot\mathbf{R}_{CM}}$$

$$= \frac{\partial}{\partial r'_{int}}\left(r'_{int}\, e^{i\mathbf{q}\cdot\mathbf{r}'_{int}}\right)_{r_{int}=r'_{int}}\, e^{i\mathbf{q}\cdot\mathbf{R}_{CM}}$$

$$+ e^{i\mathbf{q}\cdot\mathbf{r}_{int}}\left(\frac{\partial}{\partial R'_{CM}}\left(R'_{CM}\, e^{i\mathbf{q}\cdot\mathbf{R}'_{CM}}\right)_{R_{CM}=R'_{CM}} - e^{i\mathbf{q}\cdot\mathbf{R}_{CM}}\right)\ .$$

The expression allowing to calculate $\nabla(\hat{\mathscr{J}}'_{LM_L}(q,\mathbf{r}))$ arises from the two last equations:

$$\nabla\left(\frac{\partial}{\partial r'}\left(r'\, e^{i\mathbf{q}\cdot\mathbf{r}'}\right)_{r=r'}\right) = \nabla\left(i(\mathbf{q}\cdot\mathbf{r})e^{i\mathbf{q}\cdot\mathbf{r}} + e^{i\mathbf{q}\cdot\mathbf{r}}\right)$$

$$= i\mathbf{q}\left(2e^{i\mathbf{q}\cdot\mathbf{r}} + i(\mathbf{q}\cdot\mathbf{r})e^{i\mathbf{q}\cdot\mathbf{r}}\right)$$

$$= i\mathbf{q}\left(2e^{i\mathbf{q}\cdot\mathbf{r}_{int}} + i(\mathbf{q}\cdot\mathbf{r}_{int})e^{i\mathbf{q}\cdot\mathbf{r}_{int}}\right) e^{i\mathbf{q}\cdot\mathbf{R}_{CM}} + (i\mathbf{q}\, e^{i\mathbf{q}\cdot\mathbf{r}_{int}})$$

$$\times\left(i(\mathbf{q}\cdot\mathbf{R}_{CM})\, e^{i\mathbf{q}\cdot\mathbf{R}_{CM}}\right)$$

$$= \nabla_{int}\left(\frac{\partial}{\partial r'_{int}}\left(r'_{int}\, e^{i\mathbf{q}\cdot\mathbf{r}'_{int}}\right)_{r_{int}=r'_{int}}\right) e^{i\mathbf{q}\cdot\mathbf{R}_{CM}}$$

$$+ \left(\nabla_{int}\, e^{i\mathbf{q}\cdot\mathbf{r}_{int}}\right)\left(\frac{\partial}{\partial R'_{CM}}\left(R'_{CM}\, e^{i\mathbf{q}\cdot\mathbf{R}'_{CM}}\right)_{R_{CM}=R'_{CM}} - e^{i\mathbf{q}\cdot\mathbf{R}_{CM}}\right)\ .$$

The last equation, provided with $\hat{\mathscr{J}}'_{LM_L}(q, \mathbf{r}) + (qr/2)\,\hat{\mathscr{J}}_{LM_L}(q, \mathbf{r})$, is derived from the equality related to $\hat{\mathscr{J}}'_{LM_L}(q, \mathbf{r})$ and the following equation:

$$qr\,\hat{\mathscr{J}}_{LM_L}(q, \mathbf{r}) = (qr)^2\,\mathscr{J}_{LM_L}(q, \mathbf{r})$$
$$= (q^2 r_{\text{int}}^2 + q^2 R_{\text{CM}}^2 + 2q^2\,(\mathbf{r}_{\text{int}} \cdot \mathbf{R}_{\text{CM}}))\,\mathscr{J}_{LM_L}(q, \mathbf{r}).$$

The first three equalities enter the magnetic operator $\mathscr{M}_{L,M_L}^{M}$ (see Eq. (9.54)). The first and third equalities, involving a gradient, are used in scalar products involving the spin $\mathbf{s}$ and orbital angular momentum $\mathbf{l}$. The second equality arises after the evaluation of the action of gradients on Bessel functions and spherical harmonics (see Eq. (9.55)). The fourth equation is the electric charge operator (see Eq. (9.53)).

**Exercise VII.**

**A.** Spin is independent of center-of-mass coordinates, so that spin-dependent sub-operators act only on the intrinsic part of the wave function. Consequently, one can apply the Wigner–Eckart theorem on the intrinsic and center-of-mass parts of the wave function to calculate matrix elements. However, as orbital angular momentum possesses both relative and center-of-mass parts, its associated matrix elements cannot be calculated using this method. Equations provided with spin-dependent sub-operators are then straightforward to obtain from the previous results, contrary to that depending on $\mathbf{l}$, which will be considered in the following.

**B.** As $\mathbf{l} = \mathbf{r} \times \mathbf{p}$, one obtains the demanded equation by replacing $\mathbf{r}$ and $\mathbf{p}$ by their expression in terms of $\mathbf{r}_{\text{int}}$, $\mathbf{R}_{\text{CM}}$, $\mathbf{p}_{\text{int}}$, and $\mathbf{P}_{\text{CM}}$ (see Eqs. (9.135), (9.136)):

$$\mathbf{l} = (\mathbf{r}_{\text{int}} + \mathbf{R}_{\text{CM}}) \times \left( \mathbf{p}_{\text{int}} + \left( \frac{m}{M_{\text{P}}} \right) \mathbf{P}_{\text{CM}} \right) = \mathbf{l}_{\text{int}} + \left( \frac{m}{M_{\text{P}}} \right) \mathbf{L}_{\text{CM}}$$
$$+ \left( \frac{m}{M_{\text{P}}} \right) \mathbf{r}_{\text{int}} \times \mathbf{P}_{\text{CM}} + \mathbf{R}_{\text{CM}} \times \mathbf{p}_{\text{int}}.$$

The operators entering that expression of $\mathbf{l}$ depend on either relative or center-of-mass coordinates. Therefore, the matrix elements of these operators can be separated in relative and center-of-mass parts and thus can be conveniently calculated.

**C.** Expressions are straightforward to derive from the results of Exercise VI and the decomposition of $\mathbf{l}$ in intrinsic and center-of-mass operators provided in **B**. For this, one writes

$$\hat{\mathscr{J}}'_{LM_L}(q,\mathbf{r})(\mathbf{l}\cdot\mathbf{s}) = \sum_{\ell_{\text{int}},L_{\text{CM}}} A^{L_{\text{CM}}}_{\ell_{\text{int}}} \left[ \hat{\mathscr{J}}'_{\ell_{\text{int}}}(q,\mathbf{r}_{\text{int}}) \otimes \mathscr{J}_{L_{\text{CM}}}(q,\mathbf{R}_{\text{CM}})(\mathbf{l}\cdot\mathbf{s}) \right]^L_{M_L}$$

$$+ \sum_{\ell_{\text{int}},L_{\text{CM}}} A^{L_{\text{CM}}}_{\ell_{\text{int}}} \left[ \mathscr{J}_{\ell_{\text{int}}}(q,\mathbf{r}_{\text{int}}) \otimes \hat{\mathscr{J}}'_{L_{\text{CM}}}(q,\mathbf{R}_{\text{CM}})(\mathbf{l}\cdot\mathbf{s}) \right]^L_{M_L}$$

$$- \sum_{\ell_{\text{int}},L_{\text{CM}}} A^{L_{\text{CM}}}_{\ell_{\text{int}}} \left[ \mathscr{J}_{\ell_{\text{int}}}(q,\mathbf{r}_{\text{int}}) \otimes \mathscr{J}_{L_{\text{CM}}}(q,\mathbf{R}_{\text{CM}})(\mathbf{l}\cdot\mathbf{s}) \right]^L_{M_L}$$

and

$$\nabla\left(\frac{\hat{\mathscr{J}}_{LM_L}(q,\mathbf{r})}{qr}\right)\cdot\mathbf{1} = \sum_{\ell_{\text{int}},L_{\text{CM}}} A^{L_{\text{CM}}}_{\ell_{\text{int}}} \left[ \left(\nabla_{\text{int}}(\mathscr{J}_{\ell_{\text{int}}}(q,\mathbf{r}_{\text{int}}))\cdot\mathbf{1}\right) \otimes \mathscr{J}_{L_{\text{CM}}}(q,\mathbf{R}_{\text{CM}}) \right]^L_{M_L},$$

and one replaces $\mathbf{l}$ by its expression from $\mathbf{B}$, which is a function of $\mathbf{r}_{\text{int}}$, $\mathbf{p}_{\text{int}}$, $\mathbf{l}_{\text{int}}$, $\mathbf{s}$ and $\mathbf{R}_{\text{CM}}$, $\mathbf{P}_{\text{CM}}$, $\mathbf{L}_{\text{CM}}$ quantities. One then obtains expressions where sub-operators depend only on intrinsic and center-of-mass degrees of freedom. Using standard recoupling algebra, it is then possible to calculate matrix elements by applying the Wigner–Eckart theorem on the intrinsic and center-of-mass parts of the wave function.

The first operator is the electric current operator (see Eq. (9.53)), while the second operator is the magnetic orbital operator (see Eq. (9.55)). The electric current operator is typically small, as it vanishes in long wavelength approximation, whereas the magnetic orbital operator remains nonzero after applying the long wavelength approximation.

## Exercise VIII.

**A.** The formulas therein are directly obtained by replacing $\mathbf{r}$ and $L$ by $\mathbf{r}_{\text{int}}$ and $\ell_{\text{int}}$, respectively. The transformation formula involving gradients can indeed be applied directly with relative coordinates.

**B.** The long wavelength approximation generates a $q^L$ factor on the left-hand side of Eq. (9.196) and a $q^{\ell_{\text{int}}+L_{\text{CM}}}$ factor on the right-hand side of this equation. As the coupling between $\ell_{\text{int}}$ and $L_{\text{CM}}$ to $L$ implies that $\ell_{\text{int}} + L_{\text{CM}} \geq L$, the $q \to 0$ limit borne by the long wavelength approximation implies that only the terms for which $\ell_{\text{int}} + L_{\text{CM}} = L$ remain. Equation (9.134) is obtained by replacing laboratory by intrinsic coordinates in Eq. (9.57).

## Exercise IX.

**A.** The determination of relative electromagnetic operator matrix elements between projectile wave functions is apparently problematic. Indeed, the relative and center-of-mass parts of projectile wave functions are intertwined in their Slater determinant expansion. Moreover, $r_{int}$ is given in Eq. (9.135) as a function of $r$ and $R_{CM}$, so that the intrinsic electromagnetic operator becomes an $A$-body operator in terms of nucleon coordinates. Therefore, $A$-dimensional integrals would have to be calculated, which is too expensive numerically.

**B.** The projectiles are well bound, so that their intrinsic wave function can be expanded in a basis of harmonic oscillator states. Consequently, one can calculate cluster states using harmonic shell model with the Lawson method, so that they have the form:

$$|\Psi_P\rangle = |\Psi_{CM}\rangle \otimes |\Psi_{P,int}\rangle ,$$

where $|\Psi_P\rangle$ is the cluster state, $|\Psi_{CM}\rangle$ is the harmonic oscillator center-of-mass ground state, and $|\Psi_{P,int}\rangle$ is the intrinsic state of the cluster. As $|\Psi_{CM}\rangle$ has an angular momentum equal to zero, the only electric and magnetic sub-operators (see Exercises V–VIII) leading to non-vanishing matrix elements when applied to $|\Psi_P\rangle$ are those whose center-of-mass part is a scalar. Hence, all relative matrix elements can be retrieved from matrix elements calculated with laboratory coordinates.

**C.** Relative matrix elements are recovered from shell model matrix elements by rewriting associated sub-operators using the results of Exercises V–VIII and Eqs. (9.135), (9.136). Thus, from the knowledge of sub-operators matrix elements issued from shell model and of one-body integrals in $R_{CM}$, whose calculation is straightforward, one can deduce the relative electromagnetic operator matrix elements entering $\langle \Psi_{P,int}^{(f)} | \sum_{i \in P} \hat{O}_{M_L}^L (i) | \Psi_{P,int}^{(i)} \rangle$ in Eqs. (9.54) and (9.55).

All many-body matrix elements can be calculated using recoupling algebra once relative and center-of-mass matrix elements are known. Consequently, the calculation of electromagnetic matrix elements in the many-nucleon cluster case is as conveniently done as in the one-nucleon case, albeit more tedious.

## Exercise X.

**A.** The main features concerning the use of a $^{12}$C have already been considered in Exercise IV of Chap. 7. Using only proton partial waves with $\ell \leq 2$ is justified for the calculation of the low-energy spectrum of $^{15}$F because single-particle states belong only to the $p$ and $sd$ shells. Additionally, the $0p_{1/2}$ proton shell is occupied in $^{14}$O and $^{15}$F at Hartree–Fock level. Consequently, one can assume that the excitations from the $^{12}$C core to the valence space are small. Hence, they can be neglected in this model.

Clearly, all neutrons are spectators in the $^{14}$O(p,p) reaction when using an inert core of $^{12}$C. Moreover, as proton projectile energies are smaller than 6 MeV, the partial wave expansion of projectile wave functions has small components with $\ell \geq 3$. Consequently, the used model is also appropriate to describe the excitation function of the elastic scattering reaction $^{14}$O(p,p).

**B.** At the Hartree–Fock level, the $1/2^+$ and $5/2^+$ resonance states consist of a fully occupied $0p_{1/2}$ proton shell, with a proton occupying a $1s_{1/2}$ or $0d_{5/2}$ one-body resonance state, respectively. As there is a gap of several MeVs between the $0p_{1/2}$ shell and the $sd$ shell, the Hartree–Fock configurations can be expected to form a good approximation of the exact $1/2^+$ and $5/2^+$ resonance states.

The $1/2^-$ has an excitation energy well above the Coulomb barrier. Consequently, one could expect its width to be of several MeVs if its structure is mainly that of a single proton above an $^{14}$O core. As the width of the $1/2^-$ resonance is much smaller, of a few tens of keVs, its many-body wave function must bear a sizable configuration mixing in the continuum, so that its width is much reduced compared to its one-particle estimate.

**C.** About 85% of the many-body wave functions of the $1/2^+$ and $5/2^+$ resonance states consist of configurations of the form $[0p_{1/2}^2\,s_{1/2}]$ or $[0p_{1/2}^2\,d_{5/2}]$, respectively. This implies that the $1/2^+$ and $5/2^+$ resonance states are of single-particle character.

On the contrary, the $1/2^-$ resonance state has a fairly large configuration mixing. Indeed, it consists mainly of $[0p_{1/2}\,s_{1/2}^2]$ and $[0p_{1/2}\,d_{5/2}^2]$ configurations, whose weights are about 75% and 23%, respectively. Therefore, the $1/2^-$ resonance state consists mainly of two correlated protons occupying the $s_{1/2}$ and $d_{5/2}$ resonant and nonresonant shells. Moreover, the pole configurations $[0p_{1/2}\,1s_{1/2}^2]$ and $[0p_{1/2}\,0d_{5/2}^2]$ have the largest weights, of about 69% and 22%, respectively. Consequently, one can assume that the two correlated protons are very likely to be coupled to $J^\pi = 0^+$ in the $1/2^-$ resonance wave function. Thus, the $1/2^-$ resonance state wave function is principally made of a proton occupying the $0p_{1/2}$ shell and of two protons occupying the $sd$ shells. The relative stability of the $1/2^-$ resonance is then explained by the fact that the $0p_{1/2}$ shell is well bound, on the one hand, and because of the smallness of the phase space available for diproton decay, on the other hand (see Sect. 9.4.1.2).

**D.** Even though the many-body spaces spanned by Slater determinants and many-body channels are identical, they differ in practice due to the necessary truncations applied to both model spaces. Consequently, the energies and widths obtained with the Gamow shell model code and Gamow shell model coupled-channel code are slightly different.

**E.** From the cross section plot obtained after running the Gamow shell model coupled-channel code, the energy zones where the effect of $1/2^+$, $5/2^+$, and $1/2^-$ resonances are clearly visible.

For projectile energies around 1.5 MeV, the effect of the broad $1/2^+$ many-body state is dominant. Indeed, the cross section exhibits a sudden increase around 1.25 MeV. The cross section is significantly spread around 1.5 MeV due to the large width of the $1/2^+$ many-body state, of about 400 keV.

Conversely, the cross section takes the form of a narrow peak around 2.8 MeV. This is entirely due to the presence of the $5/2^+$ resonance at that energy. The narrowness of the cross section therein comes from the small width borne by the $5/2^+$ resonance, of about 100 keV.

The effect of the very narrow $1/2^-$ resonance is seen at the projectile energy of 4.8 MeV. A similar drop of the cross section at that energy is also seen experimentally [73].

**Exercise XI.**

**A.** The use of four harmonic oscillator states to represent a partial wave heavily reduces model space dimension. Indeed, a Berggren basis representation would demand at least 15 discretized scattering states for a given partial wave. However, one can no longer have correct asymptotes in the $sd$ partial waves present in target nuclear wave functions, which can be of importance in the cross sections of radiative capture, even though they are subdominant compared to $p$ partial waves.

**B.** If one truncates model space with 2p-2h truncations in the nonresonant continuum, the Berggren basis is no longer complete at the many-body level. Consequently, the part of configuration mixing that is necessary to have a precise many-body wave function in the asymptotic region is missing. As a consequence, the particle-emission width of the first resonance states of $^8$Li and $^8$B, which is about 30 keV, is equal to zero when using 2p-2h truncations in the nonresonant continuum. When one has no model space truncation, the many-body completeness is recovered and the particle-emission widths are correctly reproduced.

**C.** When one calculates resonances with the Gamow shell model coupled-channel code, one solves a system of coupled-channel equations involving one-body nucleon projectiles. As only one-nucleon emission is present, coupled-channel wave functions can almost exactly represent many-body resonance wave functions in the asymptotic region. Thus, widths are correctly calculated and are close to those obtained with non-truncated Gamow shell model.

**D.** The calculation of E2 transitions shows that they are orders of magnitude smaller than E1 and M1 transitions. All other electromagnetic transitions are negligible (see Eq. (9.48)). Therefore, as E1 and M1 transitions are dominant in $^7$Be$(p, \gamma)$ and $^7$Li$(n, \gamma)$ reactions, electromagnetic transitions of orbital angular momentum larger than two can be safely neglected.

Obtained cross sections compare well with experimental data. However, an important effect comes to the fore when considering E1 radiative capture cross sections. Even though protons have a unit charge and neutrons have no charge, the effective charge entering Eq. (9.53) is not equal to one for proton or zero for neutron. Indeed, the effective charge is modified by target recoil [2], so that one has in fact $e_p = 1 - Z/A$ and $e_n = -Z/A$, where $e_p$ and $e_n$ are the effective charges of proton and neutron, respectively, and $Z$ and $A$ are the number of protons and nucleons of the nuclear composite, respectively [2]. Added to that, the effect of missing configurations in truncated model spaces, such as that induced by core polarization, is taken into account by a modification of effective charges [2].

The proton and neutron effective charge for E1 transitions arising from target recoil only is 0.375 in absolute value, as $e_p$ and $e_n$ play no role in the used model for $^7$Li(n, $\gamma$) and $^7$Be(p, $\gamma$) reactions, respectively. The proton effective charge allowing to reproduce experimental data is 2.5 times larger than the theoretical proton effective charge, which is acceptable as effective charge typically varies by about one unit compared to its bare value [2]. Effective charges $e_p$ and $e_n$ can be as large as 1.5 and 0.95, respectively, for example, [105]. However, the neutron effective charge needed to have the $^7$Li(n, $\gamma$) E1 cross section reproduced at small energies would be 3–4 times the value of the theoretical neutron effective charge. It would then exceed the previous value of 0.95. Consequently, one preferred to use a neutron effective charge of twice its theoretical value, as it allows to obtain a good reproduction of the E1 cross section at energies around 1 MeV. It is then the optimal choice between having not too large a neutron effective charge, as it should not depart significantly from its theoretical value, and reproducing the overall features of the E1 cross section.

# References

1. P.J. Brussard, P.W.M. Glaudemans, *Shell Model Applications in Nuclear Spectroscopy* (North Holland, Amsterdam, 1977)
2. K.L.G. Heyde, *The Nuclear Shell Model* (Springer, Berlin, 1994)
3. G.R. Satchler, *Direct Nuclear Reactions* (Clarendon Press, Oxford, 1983)
4. N.K. Glendenning, *Direct Nuclear Reactions* (Academic Press, 1983)
5. B.A. Brown, B.H. Wildenthal, Ann. Rev. Nucl. Part. Sci. **38**, 29 (1988)
6. P. Fröbrich, R. Lipperheide, *Theory of Nuclear Reactions* (Oxford Science Publications, Clarendon Press, Oxford, 1996)
7. E. Caurier, G. Martínez-Pinedo, F. Nowacki, A. Poves, A.P. Zuker, Rev. Mod. Phys. **77**, 427 (2005)
8. A. Korff, P. Haefner, C. Bäumer, A.M. van den Berg, N. Blasi, B. Davids, D. De Frenne, R. de Leo, D. Frekers, E.W. Grewe, M.N. Harakeh, F. Hofmann, M. Hunyadi, E. Jacobs, B.C. Junk, A. Negret, P. von Neumann-Cosel, L. Popescu, S. Rakers, A. Richter, H.J. Wörtche, Phys. Rev. C **70**, 067601 (2004)
9. C.W. Towsley, P.K. Bindal, K.I. Kubo, K.G. Nair, K. Nagatani, Phys. Rev. C **15**, 281 (1977)
10. F.D. Becchetti, G.W. Greenlees, Phys. Rev. **182**, 1190 (1969)
11. A.J. Koning, J.P. Delaroche, Nucl. Phys. **A713**, 231 (2003)
12. C. Mahaux, H. Weidenmüller, *Shell-Model Approach to Nuclear Reactions* (North-Holland, Amsterdam, 1969)
13. H.W. Barz, I. Rotter, J. Höhn, Nucl. Phys. A **275**, 111 (1977)
14. I. Rotter, H.W. Barz, J. Höhn, Nucl. Phys. A **297**, 237 (1978)
15. J. Okołowicz, M. Płoszajczak, I. Rotter, Phys. Rep. **374**, 271 (2003)
16. K. Bennaceur, F. Nowacki, J. Okołowicz, M. Płoszajczak, Nucl. Phys. A **651**, 289 (1999)
17. K. Bennaceur, F. Nowacki, J. Okołowicz, M. Płoszajczak, Nucl. Phys. A **671**, 203 (2000)
18. R. Shyam, K. Bennaceur, J. Okołowicz, M. Płoszajczak, Nucl. Phys. A **669**, 65 (2000)
19. K. Bennaceur, N. Michel, F. Nowacki, J. Okołowicz, M. Płoszajczak, Phys. Lett. B **488**, 75 (2000)
20. N. Michel, J. Okołowicz, F. Nowacki, M. Płoszajczak, Nucl. Phys. A **703**, 202 (2002)
21. J. Rotureau, J. Okołowicz, M. Płoszajczak, Nucl. Phys. A **767**, 13 (2006)
22. J. Okołowicz, M. Płoszajczak, R.J. Charity, L.G. Sobotka, Phys. Rev. C **97**, 044303 (2018)

23. R.J. Charity, K.W. Brown, J. Okołowicz, M. Płoszajczak, J.M. Elson, W. Reviol, L.G. Sobotka, W.W. Buhro, Z. Chajecki, W.G. Lynch, J. Manfredi, R. Shane, R.H. Showalter, M.B. Tsang, D. Weisshaar, J.R. Winkelbauer, S. Bedoor, A.H. Wuosmaa, Phys. Rev. C **97**, 054318 (2018)
24. S. Quaglioni, P. Navrátil, Phys. Rev. C **79**, 044606 (2009)
25. S. Baroni, P. Navrátil, S. Quaglioni, Phys. Rev. Lett. **110**, 022505 (2013)
26. G. Hupin, S. Quaglioni, P. Navrátil, Phys. Rev. Lett. **114**, 212502 (2015)
27. C. Romero-Redondo, S. Quaglioni, P. Navrátil, G. Hupin, Phys. Rev. Lett. **117**, 222501 (2016)
28. P. Navrátil, S. Quaglioni, G. Hupin, C. Romero-Redondo, A. Calci, Phys. Scr. **91**, 053002 (2016)
29. F. Raimondi, G. Hupin, P. Navrátil, S. Quaglioni, Phys. Rev. C **93**, 054606 (2016)
30. G. Hupin, J. Langhammer, P. Navrátil, S. Quaglioni, A. Calci, R. Roth, Phys. Rev. C **88**, 054622 (2013)
31. P. Navrátil, R. Roth, S. Quaglioni, Phys. Rev. C **82**, 034609 (2010)
32. S. Baroni, P. Navrátil, S. Quaglioni, Phys. Rev. C **87**, 034326 (2013)
33. S. Quaglioni, P. Navrátil, Phys. Rev. Lett. **101**, 092501 (2008)
34. A. Tichai, J. Müller, K. Vobig, R. Roth, Phys. Rev. C **99**, 034321 (2019)
35. R. Lazauskas, Phys. Rev. C **97**, 044002 (2018)
36. S. Bacca, N. Barnea, A. Schwenk, Phys. Rev. C **86**, 034321 (2012)
37. G. Hagen, N. Michel, Phys. Rev. C **86**, 021602 (2012)
38. J. Rotureau, P. Danielewicz, G. Hagen, F.M. Nunes, T. Papenbrock, Phys. Rev. C **95**, 024315 (2017)
39. J. Rotureau, P. Danielewicz, G. Hagen, G.R. Jansen, F.M. Nunes, Phys. Rev. C **98**, 044625 (2018)
40. A. Idini, C. Barbieri, P. Navrátil, Phys. Rev. Lett. **123**, 092501 (2019)
41. M. Hjorth-Jensen, T.T.S. Kuo, E. Osnes, Phys. Rep. **261**, 125 (1995)
42. N. Tsunoda, T. Otsuka, N. Shimizu, M. Hjorth-Jensen, K. Takayanagi, T. Suzuki, Phys. Rev. C **95**, 021304 (2017)
43. H. Hergert, S. Binder, A. Calci, J. Langhammer, R. Roth, Phys. Rev. Lett. **110**, 242501 (2013)
44. H. Hergert, S.K. Bogner, T.D. Morris, S. Binder, A. Calci, J. Langhammer, R. Roth, Phys. Rev. C **90**, 041302 (2014)
45. S.K. Bogner, H. Hergert, J.D. Holt, A. Schwenk, S. Binder, A. Calci, J. Langhammer, R. Roth, Phys. Rev. Lett. **113**, 142501 (2014)
46. H. Hergert, S.K. Bogner, T.D. Morris, A. Schwenk, K. Tsukiyama, Phys. Rep. **621**, 165 (2016). Memorial Volume in Honor of Gerald E. Brown
47. H. Hergert, S.K. Bogner, S. Binder, A. Calci, J. Langhammer, R. Roth, A. Schwenk, Phys. Rev. C **87**, 034307 (2013)
48. B. Hu, Q. Wu, J. Li, Y. Ma, Z. Sun, N. Michel, F. Xu, Phys. Lett. B **802**, 135206 (2020)
49. T.A. Tombrello, P.D. Parker, Phys. Rev. **131**, 2582 (1963)
50. S.B. Dubovichenko, A.V. Dzhazairov-Kakhramanov, Fiz. Elem. Chastits. At. Yadra **28**, 1529 (1997)
51. P. Descouvemont, D. Baye, Rep. Prog. Phys. **73**, 036301 (2010)
52. A. De Shalit, I. Talmi, *Nuclear Shell Theory* (Dover Publications, New York, 2004)
53. I. Talmi, Helv. Phys. Acta **25**, 185 (1952)
54. M. Moshinsky, Nucl. Phys. **13**, 104 (1959)
55. J. Raynal, J. Revai, Nuovo Cim. A **68**, 612 (1970)
56. A. Messiah, *Quantum Mechanics, Vol. 1 and 2* (North Holland, Amsterdam, 1961)
57. G. Hagen, M. Hjorth-Jensen, N. Michel, Phys. Rev. C **73**, 064307 (2006)
58. R.A.D. Piyadasa, M. Kawai, M. Kamimura, M. Yahiro, Phys. Rev. C **60**, 044611 (1999)
59. N. Austern, M. Yahiro, M. Kawai, Phys. Rev. Lett. **63**, 2649 (1989)
60. A.M. Moro, J.M. Arias, J. Gómez-Camacho, I. Martel, F. Pérez-Bernal, R. Crespo, F. Nunes, Phys. Rev. C **65**, 011602 (2001)
61. D. Baye, P. Descouvemont, N.K. Timofeyuk, Nucl. Phys. A **577**, 624 (1994)
62. Shubhchintak, P. Descouvemont, Phys. Rev. C **100**, 034611 (2019)

63. J.A. Gaunt, Philos. Trans. Roy. Soc. (Lond.) Ser. A **228**, 151 (1929)
64. K. Fossez, N. Michel, M. Płoszajczak, Y. Jaganathen, R.M. Id Betan, Phys. Rev. C **91**, 034609 (2015)
65. E.W. Schmid, Z. Phys. A. **311**, 67 (1983)
66. D. Vautherin, M. Veneroni, Phys. Lett. B **25**, 175 (1967)
67. N. Michel, Eur. Phys. J. A **73**, 523 (2009)
68. W.H. Press, S.A. Teukolsky, W.T. Vetterling, B.P. Flannery, *Numerical Recipes in C* (Cambridge University Press, Cambridge, 1988,1992)
69. Y. Jaganathen, N. Michel, M. Płoszajczak, Phys. Rev. C **89**, 034624 (2014)
70. B.B. Skorodumov, G.V. Rogachev, P. Boutachkov, A. Aprahamian, J.J. Kolata, L.O. Lamm, M. Quinn, A. Woehr, Phys. Atom. Nucl. **12**, 69 (2006)
71. C. Angulo, G. Tabacaru, M. Couder, M. Gaelens, P. Leleux, A. Ninane, F. Vanderbist, T. Davinson, P.J. Woods, J.S. Schweitzer, N.L. Achouri, J.C. Angélique, E. Berthoumieux, F. de Oliveira Santos, P. Himpe, P. Descouvemont, Phys. Rev. C **67**, 014308 (2003)
72. F. de Oliveira Santo, H. P, M. Lewitowicz, I. Stefan, N. Smirnova, N.L. Achouri, A. J.C., C. Angulo, L. Axelsson, D. Baiborodin, F. Becker, M. Bellegui, E. Berthoumieux, B. Blank, C. Borcea, A. Cassimi, J.M. Daugas, G. de France, F. Dembinski, C.E. Demonchy, A. Dlouhy, P. Dolégiéviez, C. Donzaud, G. Georgiev, L. Giot, S. Grévy, D. Guillemaud Mueller, V. Lapoux, E. Liénard, M.J. Lopez Jimenez, K. Markenroth, I. Matea, W. Mittig, F. Negoita, G. Neyens, N. Orr, F. Pougheon, P. Roussel Chomaz, M.G. Saint Laurent, F. Sarazin, H. Savajols, M. Sawicka, O. Sorlin, M. Stanoiu, C. Stodel, G. Thiamova, D. Verney, A.C.C.C. Villari, Eur. Phys. J. A - Hadr. Nucl. **24**, 237 (2005)
73. F. de Grancey, A. Mercenne, F. de Oliveira Santos, T. Davinson, O. Sorlin, J. Angélique, M. Assié, E. Berthoumieux, R. Borcea, A. Buta, I. Celikovic, V. Chudoba, J. Daugas, G. Dumitru, M. Fadil, S. Grévy, J. Kiener, A. Lefebvre-Schuhl, N. Michel, J. Mrazek, F. Negoita, J. Okołowicz, D. Pantelica, M. Pellegriti, L. Perrot, M. Płoszajczak, G. Randisi, I. Ray, O. Roig, F. Rotaru, M. Saint Laurent, N. Smirnova, M. Stanoiu, I. Stefan, C. Stodel, K. Subotic, V. Tatischeff, J. Thomas, P. Ujić, R. Wolski, Phys. Lett. B **758**, 26 (2016)
74. H. Furutani, H. Horiuchi, R. Tamagaki, Prog. Theor. Phys. **62**, 981 (1979)
75. J. Okołowicz, M. Płoszajczak, W. Nazarewicz, Prog. Theor. Phys. Supp. **196**, 230 (2012)
76. J. Okołowicz, W. Nazarewicz, M. Płoszajczak, Fortschr. Phys. **61**, 66 (2013)
77. S. Turck-Chièze, S. Couvidat, Rep. Prog. Phys. **74**(8), 086901 (2011)
78. C.A. Barnes, D.D. Clayton, D.N. Schramm, *Essays in Nuclear Astrophysics. Presented to William A. Fowler* (Cambridge University Press, Cambridge, 1982)
79. J. Bahcall, *Neutrino Astrophysics* (Cambridge University Press, New York, 1989)
80. Y. Fukuda, T. Hayakawa, E. Ichihara, K. Inoue, K. Ishihara, H. Ishino, Y. Itow, T. Kajita, J. Kameda, S. Kasuga, K. Kobayashi, Y. Kobayashi, Y. Koshio, M. Miura, M. Nakahata, S. Nakayama, A. Okada, K. Okumura, N. Sakurai, M. Shiozawa, Y. Suzuki, Y. Takeuchi, Y. Totsuka, S. Yamada, M. Earl, A. Habig, E. Kearns, M.D. Messier, K. Scholberg, J.L. Stone, L.R. Sulak, C.W. Walter, M. Goldhaber, T. Barszczxak, D. Casper, W. Gajewski, P.G. Halverson, J. Hsu, W.R. Kropp, L.R. Price, F. Reines, M. Smy, H.W. Sobel, M.R. Vagins, K.S. Ganezer, W.E. Keig, R.W. Ellsworth, S. Tasaka, J.W. Flanagan, A. Kibayashi, J.G. Learned, S. Matsuno, V.J. Stenger, D. Takemori, T. Ishii, J. Kanzaki, T. Kobayashi, S. Mine, K. Nakamura, K. Nishikawa, Y. Oyama, A. Sakai, M. Sakuda, O. Sasaki, S. Echigo, M. Kohama, A.T. Suzuki, T.J. Haines, E. Blaufuss, B.K. Kim, R. Sanford, R. Svoboda, M.L. Chen, Z. Conner, J.A. Goodman, G.W. Sullivan, J. Hill, C.K. Jung, K. Martens, C. Mauger, C. McGrew, E. Sharkey, B. Viren, C. Yanagisawa, W. Doki, K. Miyano, H. Okazawa, C. Saji, M. Takahata, Y. Nagashima, M. Takita, T. Yamaguchi, M. Yoshida, S.B. Kim, M. Etoh, K. Fujita, A. Hasegawa, T. Hasegawa, S. Hatakeyama, T. Iwamoto, M. Koga, T. Maruyama, H. Ogawa, J. Shirai, A. Suzuki, F. Tsushima, M. Koshiba, M. Nemoto, K. Nishijima, T. Futagami, Y. Hayato, Y. Kanaya, K. Kaneyuki, Y. Watanabe, D. Kielczewska, R.A. Doyle, J.S. George, A.L. Stachyra, L.L. Wai, R.J. Wilkes, K.K. Young, Phys. Rev. Lett. **81**, 1562 (1998)
81. P. Descouvemont, D. Baye, Nucl. Phys. A **567**, 341 (1994)

82. F. Barker, Nucl. Phys. A **588**, 693 (1995)
83. E.G. Adelberger, A. García, R.G.H. Robertson, K.A. Snover, A.B. Balantekin, K. Heeger, M.J. Ramsey-Musolf, D. Bemmerer, A. Junghans, C.A. Bertulani, J.W. Chen, H. Costantini, P. Prati, M. Couder, E. Uberseder, M. Wiescher, R. Cyburt, B. Davids, S.J. Freedman, M. Gai, D. Gazit, L. Gialanella, G. Imbriani, U. Greife, M. Hass, W.C. Haxton, T. Itahashi, K. Kubodera, K. Langanke, D. Leitner, M. Leitner, P. Vetter, L. Winslow, L.E. Marcucci, T. Motobayashi, A. Mukhamedzhanov, R.E. Tribble, K.M. Nollett, F.M. Nunes, T.S. Park, P.D. Parker, R. Schiavilla, E.C. Simpson, C. Spitaleri, F. Strieder, H.P. Trautvetter, K. Suemmerer, S. Typel, Rev. Mod. Phys. **83**, 195 (2011)
84. M. Heil, F. Kaeppeler, M. Wiescher, A. Mengoni, Astrophys. J. **507**, 997 (2009)
85. M. Kusakabe, T. Kajino, R. Boyd, T. Yoshida, Astrophys. J. **680**, 846 (2008)
86. C.L. Bennett, M. Halpern, G. Hinshaw, N. Jarosik, A. Kogut, M. Limon, S.S. Meyer, L. Page, D.N. Spergel, G.S. Tucker, E. Wollack, E.L. Wright, C. Barnes, M.R. Greason, R.S. Hill, E. Komatsu, M.R. Nolta, N. Odegard, H.V. Peiris, L. Verde, J.L. Weiland, Astrophys. J. : Suppl. **148**, 1 (2003)
87. A. Bohr, B.R. Mottelson, *Nuclear Structure, Vol. 2: Nuclear Deformations* (World Scientific, Singapore, 1998)
88. L.T. Baby, C. Bordeanu, G. Goldring, M. Hass, L. Weissman, V.N. Fedoseyev, U. Köster, Y. Nir-El, G. Haquin, H.W. Gäggeler, R. Weinreich, Phys. Rev. C **67**, 065805 (2003)
89. A.R. Junghans, K.A. Snover, E.C. Mohrmann, E.G. Adelberger, L. Buchmann, Phys. Rev. C **81**, 012801 (2010)
90. H. Furutani, H. Kanada, T. Kaneko, S. Nagata, H. Nishioka, S. Okabe, S. Saito, T. Sakuda, M. Seya, Prog. Theor. Phys. Supp. **68**, 193 (1980)
91. W.L. Imhof, R.G. Johnson, F.J. Vaughn, M. Walt, Phys. Rev. **114**, 1037 (1959)
92. D.R. Entem, R. Machleidt, Phys. Rev. C **68**, 041001(R) (2003)
93. A. Mercenne, N. Michel, M. Płoszajczak, Phys. Rev. C **99**, 044606 (2019)
94. Y. Jaganathen, R.M. Id Betan, N. Michel, W. Nazarewicz, M. Płoszajczak, Phys. Rev. C **96**, 054316 (2017)
95. D. Tilley, C. Cheves, J. Godwin, G. Hale, H. Hofmann, J. Kelley, C. Sheu, H. Weller, Nucl. Phys. A **708**, 3 (2002)
96. L.D. Blokhintsev, V.I. Kukulin, A.A. Sakharuk, D.A. Savin, E.V. Kuznetsova, Phys. Rev. C **48**, 2390 (1993)
97. K.D. Veal, C.R. Brune, W.H. Geist, H.J. Karwowski, E.J. Ludwig, A.J. Mendez, E.E. Bartosz, P.D. Cathers, T.L. Drummer, K.W. Kemper, A.M. Eiró, F.D. Santos, B. Kozlowska, H.J. Maier, I.J. Thompson, Phys. Rev. Lett. **81**, 1187 (1998)
98. W. Grüebler, Schmelzbach, V. König, R. Risler, D. Boerma, Nucl. Phys. A **242**, 265 (1975)
99. B. Jenny, W. Grüebler, V. König, Schmelzbach, C. Schweizer, Nucl. Phys. A **397**, 61 (1983)
100. R.B. Wiringa, S.C. Pieper, Phys. Rev. Lett. **89**, 182501 (2002)
101. R. Machleidt, Phys. Rev. C **63**, 024001 (2001)
102. J.H. Jett, J.L. Detch, N. Jarmie, Phys. Rev. C **3**, 1769 (1971)
103. L.S. Senhouse, T.A. Tombrello, Nucl. Phys. **57**, 624 (1964)
104. A. Mercenne, *Nuclear reactions in the Gamow shell model and solutions of the pairing Hamiltonian based on the rational Gaudin model*. Ph.D. thesis, Université de Caen Normandie, 2017
105. G. Neyens, Rep. Prog. Phys. **66**(4), 633 (2003)

# Index

© Springer International Publishing AG 2021
N. Michel, M. Płoszajczak, *Gamow Shell Model*, Lecture Notes in Physics 983,
https://doi.org/10.1007/978-3-030-69356-5

Printed in the United States
by Baker & Taylor Publisher Services